Epigenetic Technological Applications

Epigenetic Technological Applications

Edited by

Y. George Zheng

ELSEVIER

AMSTERDAM • BOSTON • HEIDELBERG • LONDON
NEW YORK • OXFORD • PARIS • SAN DIEGO
SAN FRANCISCO • SINGAPORE • SYDNEY • TOKYO

Academic Press is an imprint of Elsevier

Academic Press is an imprint of Elsevier
125, London Wall, EC2Y 5AS
525 B Street, Suite 1800, San Diego, CA 92101-4495, USA
225 Wyman Street, Waltham, MA 02451, USA
The Boulevard, Langford Lane, Kidlington, Oxford OX5 1GB, UK

Notices

Knowledge and best practice in this field are constantly changing. As new research and experience broaden our understanding, changes in research methods, professional practices, or medical treatment may become necessary.

Practitioners and researchers may always rely on their own experience and knowledge in evaluating and using any information, methods, compounds, or experiments described herein. In using such information or methods they should be mindful of their own safety and the safety of others, including parties for whom they have a professional responsibility.

To the fullest extent of the law, neither the Publisher nor the authors, contributors, or editors, assume any liability for any injury and/or damage to persons or property as a matter of products liability, negligence or otherwise, or from any use or operation of any methods, products, instructions, or ideas contained in the material herein.

Library of Congress Cataloging-in-Publication Data
A catalog record for this book is available from the Library of Congress

British Library Cataloguing-in-Publication Data
A catalogue record for this book is available from the British Library

ISBN: 978-0-12-801080-8

For information on all publications visit
our website at http://store.elsevier.com

Typeset by MPS Limited, Chennai, India
www.adi-mps.com

Working together
to grow libraries in
developing countries

www.elsevier.com • www.bookaid.org

Publisher: Mica Haley
Acquisition Editor: Catherine Van Der Laan
Editorial Project Manager: Lisa Eppich
Production Project Manager: Melissa Read
Designer: Alan Studholme

Contents

v

CHAPTER 7 Crystallography-Based Mechanistic Insights into Epigenetic Regulation 125

Shuai Zhao and Haitao Li

CHAPTER 8 Chemical and Genetic Approaches to Study Histone Modifications................................. 149

Abhinav Dhall and Champak Chatterjee

List of Contributors

Alison I. Bernstein
Department of Human Genetics, Emory University, Atlanta, GA, USA

Kelly M. Biette
Massachusetts General Hospital Cancer Center and Department of Medicine, Harvard Medical School, Charlestown, MA, USA

Joshua C. Black
Massachusetts General Hospital Cancer Center and Department of Medicine, Harvard Medical School, Charlestown, MA, USA

Champak Chatterjee
Department of Chemistry, University of Washington, Seattle, WA, USA

Aili Chen
Divisions of Experimental Hematology and Cancer Biology, Cincinnati Children's Hospital Medical Center, Cincinnati, Ohio, USA; Division of Pathology, Cincinnati Children's Hospital Medical Center, Cincinnati, Ohio, USA; Key Laboratory of Genomic and Precision Medicine, Beijing Institute of Genomics, Chinese Academy of Sciences, Beijing, China

Xiangyun Amy Chen
Center for Eukaryotic Gene Regulation, Department of Biochemistry and Molecular Biology, Pennsylvania State University, PA, USA

Abhinav Dhall
Department of Chemistry, University of Washington, Seattle, WA, USA

Alfonso Dueñas-González
Instituto de Investigaciones Biomédicas, Universidad Nacional Autónoma de México and Unidad de Investigación Biomédica en Cáncer, Instituto Nacional de Cancerología, México

Yuhong Fan
School of Biology, Georgia Institute of Technology, Atlanta, GA, USA; The Petit Institute for Bioengineering and Bioscience, Georgia Institute of Technology, Atlanta, GA, USA

Benjamin A. Garcia
Department of Biochemistry and Biophysics, Perelman School of Medicine, University of Pennsylvania, Philadelphia, PA, USA

Zhen Han
Department of Pharmaceutical and Biochemical Sciences, The University of Georgia, Athens, Georgia, USA

Alexander-Thomas Hauser
Institute of Pharmaceutical Sciences, Albert-Ludwigs-University of Freiburg, Germany

Gang Huang
Divisions of Experimental Hematology and Cancer Biology, Cincinnati Children's Hospital Medical Center, Cincinnati, Ohio, USA; Division of Pathology, Cincinnati Children's Hospital Medical Center, Cincinnati, Ohio, USA

Kenneth Huang
Department of Chemistry, Georgia State University, Atlanta, GA, USA

Peng Jin
Department of Human Genetics, Emory University, Atlanta, GA, USA

Manfred Jung
Institute of Pharmaceutical Sciences, Albert-Ludwigs-University of Freiburg, Germany

Shukkoor Kondengaden
Department of Chemistry, Georgia State University, Atlanta, GA, USA

James R. Kornacki
Departments of Chemistry and Biomedical Engineering, Northwestern University, Evanston, IL, USA

Haitao Li
Department of Basic Medical Sciences, Center for Structural Biology, School of Medicine, Tsinghua University, Beijing, P.R. China

Keqin Kathy Li
Department of Chemistry, Georgia State University, Atlanta, GA, USA; State Key Lab of Medical Genomic, Ruijin Hospital Affiliated to Medical School of Shanghai Jiaotong University, Shanghai, P.R. China

Junyan Lu
Drug Discovery and Design Center, State Key Laboratory of Drug Research, Shanghai Institute of Materia Medica, Chinese Academy of Sciences, Shanghai, China

Yepeng Luan
Department of Pharmaceutical and Biochemical Sciences, The University of Georgia, Athens, Georgia, USA

Cheng Luo
Drug Discovery and Design Center, State Key Laboratory of Drug Research, Shanghai Institute of Materia Medica, Chinese Academy of Sciences, Shanghai, China

Minkui Luo
Molecular Pharmacology and Chemistry Program, Memorial Sloan Kettering Cancer Center, New York, NY, USA

Antonia Masch
Department of Enzymology, Institute of Biochemistry & Biotechnology, Martin-Luther-University Halle-Wittenberg, Halle, Germany

José L. Medina-Franco
Facultad de Química, Departamento de Farmacia, Universidad Nacional Autónoma de México, Mexico City, Mexico

Milan Mrksich
Departments of Chemistry and Biomedical Engineering, Northwestern University, Evanston, IL, USA

Liza Ngo
Department of Pharmaceutical and Biochemical Sciences, The University of Georgia, Athens, Georgia, USA

Adegboyega K. Oyelere
Parker H. Petit Institute for Bioengineering and Bioscience, Georgia Institute of Technology, Atlanta, GA, USA; School of Chemistry and Biochemistry, Georgia Institute of Technology, Atlanta, GA, USA

Chenyi Pan
School of Biology, Georgia Institute of Technology, Atlanta, GA, USA; The Petit Institute for Bioengineering and Bioscience, Georgia Institute of Technology, Atlanta, GA, USA

Sriharsa Pradhan
Genome Biology Division, New England Biolabs, Ipswich, Massachusetts, USA

Zhang Qing
Department of Chemistry, Georgia State University, Atlanta, GA, USA

Meihua Qu
College of Pharmaceutical and Biological Sciences, Weifang Medical University, Weifang, Shandong, P.R. China

Ulf Reimer
JPT Peptide Technologies GmbH, Berlin, Germany

Hamed Reyhanfard
Department of Chemistry, Georgia State University, Atlanta, GA, USA

Martin Roatsch
Institute of Pharmaceutical Sciences, Albert-Ludwigs-University of Freiburg, Germany

Dina Robaa
Department of Pharmaceutical Chemistry, Martin Luther University of Halle-Wittenberg, Germany

Johannes Schulz-Fincke
German Cancer Consortium, Heidelberg, Germany; German Cancer Research Center, Heidelberg, Germany; Institute of Pharmaceutical Sciences, Albert-Ludwigs-University of Freiburg, Germany

Mike Schutkowski
Department of Enzymology, Institute of Biochemistry & Biotechnology, Martin-Luther-University Halle-Wittenberg, Halle, Germany

Simone Sidoli
Department of Biochemistry and Biophysics, Perelman School of Medicine, University of Pennsylvania, Philadelphia, PA, USA

Wolfgang Sippl
Department of Pharmaceutical Chemistry, Martin Luther University of Halle-Wittenberg, Germany

Quaovi H. Sodji
School of Chemistry and Biochemistry, Georgia Institute of Technology, Atlanta, GA, USA

Jinquan Sun
Center for Eukaryotic Gene Regulation, Department of Biochemistry and Molecular Biology, Pennsylvania State University, PA, USA

Jolyon Terragni
Genome Biology Division, New England Biolabs, Ipswich, Massachusetts, USA

Xiao-Jun Tian
Department of Computational and Systems Biology, School of Medicine, University of Pittsburgh, Pittsburgh, PA, USA

Peng George Wang
Department of Chemistry, Georgia State University, Atlanta, GA, USA

Xiaoshi Wang
Department of Biochemistry and Biophysics, Perelman School of Medicine, University of Pennsylvania, Philadelphia, PA, USA

Xuejian Wang
College of Pharmaceutical and Biological Sciences, Weifang Medical University, Weifang, Shandong, P. R. China

Yanming Wang
Center for Eukaryotic Gene Regulation, Department of Biochemistry and Molecular Biology, Pennsylvania State University, PA, USA

Johnathan R. Whetstine
Massachusetts General Hospital Cancer Center and Department of Medicine, Harvard Medical School, Charlestown, MA, USA

Jonathan Wooten
Department of Chemistry, Georgia State University, Atlanta, GA, USA

Jianhua Xing
Beijing Computational Science Research Center, Beijing, China; Department of Computational and Systems Biology, School of Medicine, University of Pittsburgh, Pittsburgh, PA, USA

Wei Xu
McArdle Laboratory for Cancer Research, University of Wisconsin-Madison, Madison, WI, USA

Jakyung Yoo
Life Science Research Institute, Daewoong Pharmaceutical Co., Ltd, Gyeonggi-do, Republic of Korea

Jin Yu
Beijing Computational Science Research Center, Beijing, China

Hao Zeng
McArdle Laboratory for Cancer Research, University of Wisconsin-Madison, Madison, WI, USA

Johannes Zerweck
JPT Peptide Technologies GmbH, Berlin, Germany

Bingxue Chris Zhai
Department of Biology, Georgia State University, Atlanta, GA, USA

Hang Zhang
Beijing Computational Science Research Center, Beijing, China; Department of Biological Sciences, Virginia Polytechnic Institute and State University, Blacksburg, VA, USA

Hao Zhang
Drug Discovery and Design Center, State Key Laboratory of Drug Research, Shanghai Institute of Materia Medica, Chinese Academy of Sciences, Shanghai, China

Liyi Zhang
Drug Discovery and Design Center, State Key Laboratory of Drug Research, Shanghai Institute of Materia Medica, Chinese Academy of Sciences, Shanghai, China

Yunzhe Zhang
School of Biology, Georgia Institute of Technology, Atlanta, GA, USA; The Petit Institute for Bioengineering and Bioscience, Georgia Institute of Technology, Atlanta, GA, USA

Shuai Zhao
Department of Basic Medical Sciences, Center for Structural Biology, School of Medicine, Tsinghua University, Beijing, P.R. China

Y. George Zheng
Department of Pharmaceutical and Biochemical Sciences, The University of Georgia, Athens, Georgia, USA

Foreword

Epigenetics refers to heritable differences in phenotype that cannot be attributed to differences in DNA sequence. Epigenetics is emerging as a cornerstone of modern biology with important and promising practical implications for current and future therapeutics. Over the past 15 years we have come to recognize that patterns of chemical modifications on chromatin — on both proteins and DNA — establish heritable and dynamic gene expression programs responsible for cell type identity through development, adaptation to environmental conditions, and evolution and maintenance of disease states. Our ability to translate this growing body of knowledge into useful strategies to understand, diagnose and treat human disease depends critically on the development and application of technologies to characterize and manipulate the "epigenetic state" of normal and diseased cells and to correlate those states with phenotype. This book summarizes many of these key technologies with emphasis on epigenetic regulation mediated by methylation and acetylation — the two most abundant and diverse epigenetic "marks."

Much progress has been made in recent years in identifying and characterizing the protein factors that "write," "read" and "erase" methyl and acetyl marks using methodologies ranging from cell and molecular biology, genomics and bioinformatics, structural biology, protein biochemistry and enzymology. These studies have described the major protein families and complexes that interact with and modify chromatin and have revealed many of them to be "druggable." Indeed, the discovery of novel inhibitors of histone deacetylases and DNA hypomethylation agents with clinical benefits in oncology has spurred the continued development of new drugs and chemical tools for epigenetics research and therapy.

However, there remains a significant gap in our understanding of the patterns of methylation and acetylation that define both normal and diseased epigenetic states. Recent large-scale projects (such as the Encode Project) to catalogue the localization of epigenetic marks along the genome in different common cell lines have provided a good starting point, but the community now needs to better understand how these patterns differ in human disease and how they can be selectively manipulated for safe therapies. The use and further development of genome- and proteome-wide technologies that can map the locations of epigenetic marks along the genome are particularly powerful in this regard. Further optimization and more widespread application of these "omics" technologies to disease tissue will be essential to the full realization of new epigenetic therapies.

The following survey of epigenetic technologies will serve as an important benchmark for our current capabilities and will inspire further technological advances to reduce the knowledge gap and facilitate more rapid and successful translation of epigenetics research into new diagnoses and therapies.

Cheryl H. Arrowsmith
University of Toronto

THE STATE OF THE ART OF EPIGENETIC TECHNOLOGIES

Y. George Zheng

Department of Pharmaceutical and Biochemical Sciences,
The University of Georgia, Athens, Georgia, USA

CHAPTER OUTLINE

1.1 EPIGENETICS AND CHROMATIN FUNCTION

The field of epigenetics is rapidly booming, evolving, and expanding. The term epigenetics (the prefix "epi" comes from Greek, meaning over, upon, above, in addition to) was created by Waddington in the context of connecting developmental biology and genetics [1]. To date, epigenetics is most frequently referred to as the study of meiotically and mitotically heritable changes in gene activities without alterations in the genetic DNA sequence [1−4]. However, in recent years, it has been debated whether the heritability requirement of the term is necessary or too restrictive given that many (perhaps most) changes in gene activity modulated by chromatin modifications can occur in terminally differentiated and nondividing cells [3,5,6]. Examples include short-lived alterations in histone acetylation and methylation caused by DNA repair, cell-cycle phase, or transcription factor binding. Thus, there exist arguments in which epigenetic changes could be more broadly defined to encompass structural and biochemical alterations of the chromatin at any point in time under the condition that the genetic sequence is kept invariable [5,6]. Under this scheme, epigenetics does not merely refer to heritable chromatin states, but also embraces those that are transient or occur in nondividing cells. A new term, memigenetics, has recently been suggested to

Y.G. Zheng (Ed): Epigenetic Technological Applications. DOI: http://dx.doi.org/10.1016/B978-0-12-801080-8.00001-6

particularly denote transmissible epigenetics, which pertains to the propagation of a chromatin activity state across cell generations [5].

Chromatin lies at the very core of epigenetic biology. In multicelled organisms, DNA, the macromolecule that contains genetic information, is packed into highly ordered chromatin complexes, accommodative to the narrow space of the nucleus. As the structural framework in which DNA embeds, chromatin executes various nuclear functions such as transcription, replication, and differentiation [7,8]. The basal structural building unit of the chromatin is the nucleosome, which is composed of a core histone octamer (containing two copies of each of H2A, H2B, H3, and H4 molecules) wrapped by 146 bp of DNA in approximately 1.7 turns [9,10]. The linker DNAs between neighboring nucleosomes connect nucleosome core particles, forming the 10-nm "beads-on-a-string" chromatin thread that can be observed under electron microscope [11,12]. A fifth histone, linker histone H1, binds to linker DNA and the entry/exit point of nucleosomal DNA, facilitating the folding of chromatin into higher order chromatin structures, such as the 30 nm filaments [13,14]. All five histones, H1, H2A, H2B, H3, and H4, are highly basic (the calculated pKa values are $10.3-11.4$) and rich in positively charged amino acids, most of which are lysines and arginines. The strong basic physicochemical nature of the histones promotes their tight association with the DNA phosphate backbones to form stable chromatin complexes. The nucleosome complex structure represents a physical barrier that hampers the binding of gene regulatory factors, such as the RNA polymerases to gene promoters, thereby suppressing gene expression [15,16].

1.2 MECHANISMS OF EPIGENETIC REGULATION

Though stable, the chromatin structure is a dynamic entity rather than static. Its structural state is directly associated with the activity status of the underlying DNA sequences. Regulation of chromatin structure and DNA activity is highly complex and involves several types of epigenetic control mechanisms, including DNA methylation, various histone posttranslational modifications (PTMs), exchange of canonical histones with histone variants, ATP-dependent chromatin remodeling, and recruitment of long and short noncoding RNAs (Figure 1.1). DNA 5-cytosine methylation is the most studied epigenetic mechanism in the field. Methylation of DNA occurs at the carbon-5 position of the cytosine ring, typically in a CpG dinucleotide context which is a characteristic feature of many eukaryotic genomes. Approximately 70% of CpG residues in the mammalian genomes are methylated, leaving a small portion of the genome methylation-free [17]. CpG methylation is an important mechanism for repressing the transcription of repeat elements and transposons, and plays a crucial role in genomic imprinting and X-chromosome inactivation [18].

Compared to DNA methylation, histone modification encompasses much more diverse chemical groups, including methylation, acetylation, phosphorylation, citrullination, ubiquitination, SUMOylation, ADP-ribosylation, etc. [19]. So far, the majority of histone PTMs are found to be dynamically regulated. A huge number of histone modifiers that can enzymatically introduce or remove different histone marks in a site-specific manner (i.e. mark writers and erasers) have been discovered. Furthermore, epigenetic effectors (i.e. mark readers) can specifically recognize histone modifications, bringing about downstream outcomes. Acetylation is one of the most thoroughly studied histone modifications. The modification is dynamically controlled by the opposing enzyme activities of histone

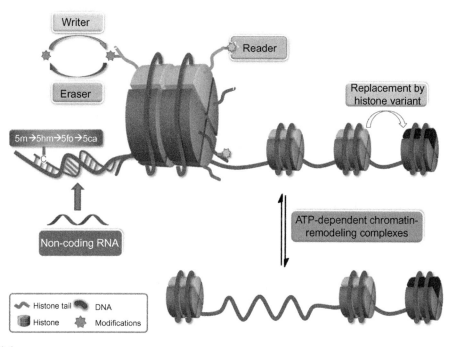

FIGURE 1.1

Chromatin structure and activity are epigenetically regulated by multiple molecular mechanisms, which include DNA modification on the cytosine residue with methyl and related one-carbon groups, assorted histone modifications, ATP-driven chromatin remodeling, exchange with histone variants, and interaction with noncoding RNAs.

acetyltransferases (HATs) and histone deacetylases (HDACs). Histone acetylation tags can be recognized by bromodomain-containing proteins, many of which are chromatin-associated regulators [20]. In general, histone acetylation is associated with open chromatin and transcriptional activation whereas histone deacetylation concurs to heterochromatin and gene silencing [21,22]. However, caution should be taken that this model may be overly simple because acetylated, open chromatin could also allow the access of repressor factors to the DNA template [23,24]. Further, HATs may repress gene activity via other mechanisms such as targeting transcriptional regulators [25,26].

Histone methylation has more complicated patterns and has varied effects on gene activity which has no definite predictive index. Methylation can occur on the ω-amino group of arginine residues or the ε-amino group of lysine residues. Methylations of histone H3K4 and H3K36 are frequently associated with increasing transcriptional activity [27,28], whereas methylations of H3K9 and K3K27 are associated with gene silencing [29]. So far, multiple histone methyltransferases and demethylases were discovered which dynamically regulate the cellular methylation level. Reader proteins recognizing specific methylations have also been broadly studied. For example, chromodomain-containing proteins recognize methylated lysines and the tudor domain binds to methylated arginines [28,30].

Although histone modifications were originally concentrated on the N-terminal tail region, current mass spectrometry (MS)-based detection reveals that they are widely spread over all the regions of the core histones, including the globular fold domain (Table 1.1) [31]. Many modifications are novel either because the modification sites are previously uncharted or the type of chemical groups is new. Based on a recent survey conducted by Huang et al. [31], 26% of amino acid residues in the four core histones were modified, and the majority of the modified sites are on arginine and lysine residues. It appears that virtually every lysine residue in the core histones is chemically modified. This raises many mechanistic questions regarding the timing, mechanism, and function of the modifications. For example, many residues in the globular domain are buried inside the nucleosome and cannot be accessed by enzymes. Posttranslational modifications of these residues must occur either prior to nucleosome assembly or following nucleosome disassembly. It is mysterious whether all histone modifications are enzymatically driven or whether some are due to accidental nonenzymatic chemical reactions. Further, it remains unknown whether all modifications have meaningful biological readouts or some of them are trivial and redundant. Illumination of biological consequences of individual histone modification tags will be a daunting challenge to epigeneticists.

Table 1.1 The Various Chromatin Modification Marks

See a recent summary in [Ref. 31]

Type of Modification	Structure of Chemical Group	Mass Shift	Modification Sites in DNA or Histones
DNA methylation		+14	CpG islands
DNA hydroxymethylation		+30	CpG islands
DNA formylation		+28	CpG islands
DNA carboxylation		+43	CpG islands

Table 1.1 The Various Chromatin Modification Marks *Continued*

Type of Modification	Structure of Chemical Group	Mass Shift	Modification Sites in DNA or Histones
Lysine acetylation		+42	**H3**-K4, K9, K14, K18, K23, K27, K36, K37, K56, K64, K79, K115, K122
			H4-K5, K8, K12, K16, K20, K31, K77, K79, K91
			H2A-K5, K9, K13, K15, K36, K74, K95, K118, K127, K129
			H2B-K5, K11, K12, K15, K16, K20, K23, K24, K43, K46, K57, K85, K108, K116, K120, K125
			H1-K16, K21, K25, K33, K45, K51, K62, K63, K74, K84, K89, K96, K105, K167, K168, K190; K18, K34, K90, K154, K191 [Ref. 32]
Lysine formylation		+28	**H3**-K18, K23, K56, K64, K79, K122
			H4-K12, K31, K59, K77, K79, K91
			H2A-K36, K95, K118
			H2B-K5, K34, K43, K46, K108, K116
			H1-K16, K33, K45, K62, K63, K74, K80, K84, K89, K96, K139, K159; K34, K55, K90 [Ref. 32]
Lysine propionylation		+56	**H3**-K23, K27, K56
			H4-K5, K8, K12, K16, K31, K77, K79, K91
			H2A-K95, K125
Lysine butyrylation		+70	**H3**-K9, K14, K18, K23, K27, K79, K115, K122
			H4-K5, K8, K12, K16, K31, K77, K79, K91
			H2A-K5, K95
			H2B-K5, K11, K12, K15, K16, K20, K23, K46
Lysine crotonylation		+68	**H3**-K4, K9, K18, K23, K27, K56, K79, K122
			H4-K5, K8, K12, K16, K59, K77, K91
			H2A-K36, K95, K118, K125
			H2B-K5, K11, K12, K15, K16, K20, K23, K108, K116
			H1-K33, K63, K84, K89, K96, K158, K167

(Continued)

Table 1.1 The Various Chromatin Modification Marks *Continued*

Type of Modification	Structure of Chemical Group	Mass Shift	Modification Sites in DNA or Histones
Lysine 2-hydroxy isobutyrylation		+86	**H3**-K4, K9, K14, K18, K23, K27, K36, K56, K64, K79, K122
			H4-K5, K8, K12, K16, K31, K59, K77, K79, K91
			H2A-K5, K9, K36, K74, K75, K95, K118
			H2B-K5, K12, K20, K23, K24, K34, K43, K46, K57, K85, K108, K116, K120
			H1-K22, K25, K26, K33, K45, K51, K62, K63, K74, K80, K84, K89, K96, K120, K128, K135, K147, K158, K167, K212
Lysine long fatty acylation		+210	
Cysteine long fatty acylation		+238	**H3**-C110 [Ref. 33]
Lysine malonylation		+86	**H3**-K56
			H2B-K116
Lysine succinylation		+100	**H3**-K14, K23, K27, K56, K79, K122
			H4-K12, K31, K77, K79, K91
			H2A-K9, K36, K95
			H2B-K5, K34, K43, K46, K85, K108, K116, K120
			H1-K33, K45, K62, K63, K84, K89, K96, K105, K120
Lysine 5-hydroxylation		+16	H3, H4, H2A, H2B [Ref. 34]
Lysine hydroxymethylation		+30	Involved in LSD1-mediated demethylation process.
Lysine ubiquinylation		+8.5 kDa (76aa)	**H3**-K14, K18, K23, K27, K36, K56, K79, K122
			H4-K31, K59, K91
			H2A-K13, K15, K36, K95, K118, K119, K125
			H2B- K20, K34, K46, K57, K108, K116, K120
			H1- K33, K45, K51, K63, K74, K89, K96, K105, K109, K126, K139, K159, K167

Table 1.1 The Various Chromatin Modification Marks *Continued*

Type of Modification	Structure of Chemical Group	Mass Shift	Modification Sites in DNA or Histones
Lysine sumoylation	Lys—N(H)—C(O)—SUMO1	+11 kDa (101aa)	H2A, H2B, H3, H4 (sequence not consistent with snapshot) [Ref. 35]
Lysine methylation	Lys—$^+$NH$_2$; Lys—$^+$NH; Lys—$^+$N	+14, 28, 42	**H3**-K4, K9, K14, K18, K23, K27, K36, K37, K56, K64, K79, K122
			H4-K5, K12, K16, K20, K31, K59, K77, K79
			H2A-K9, K74, K95, K99, K118, K125
			H2B-K5, K12, K15, K20, K23, K34, K43, K46, K57, K85, K108, K116
			H1-K16, K21, K25, K26, K33, K51, K62, K63, K89, K96, K105, K109, K118, K128, K147, K167, K168, K186
Arginine methylation	H$_2$N$^+$=C(NH—)(NH); H$_2$N$^+$=C(N—)(NH); (—N)(H)—C$^+$=(N—)(H)(NH)	+14, 28	**H3**-R2, R8, R17, R26, R40, R42, R63, R83, R128
			H4-R3, R17, R19, R23, R35, R55, R67, R78, R92
			H2A-R3, R11, R29, R42, R71, R88
			H2B-R79, R86, R92, R99
Arginine citruillination	O=C(NH$_2$)(NH)	**0**	**H3**-R2, R8, R17, R26
			H4-R3, R17, R19
			H2A-R3
			H1-R53
Serine/threonine acetylation	O—C(O)—CH$_3$	+ 42	**H3**-S10, S28; T22, T80
			H2B-T19
			H1-S35, S50, S112
Serine/threonine/ tyrosine phosphorylation	O—P(O)(O$^-$)(O$^-$)	+78	**H3**-S10, S28, S57, S86; T3, T6, T11, T45, T80, T107; Y41
			H4-S1, S47; Y51, Y72, Y88
			H2A-S1, S122; T59, T101, T120; Y50
			H2B-S6, S14, S32, S36, S56, S64, S75, S87, S91; T19, T52, T88, T115, T119
			H1-S1, S35, S39, S40, S54, S103, S172; T3, T145, T164, T179; Y70, S156 [32]

(Continued)

Table 1.1 The Various Chromatin Modification Marks *Continued*

Type of Modification	Structure of Chemical Group	Mass Shift	Modification Sites in DNA or Histones
Histidine phosphorylation		+78	H4-H18, H75
Serine/threonine O-GlcNAcylation		+201	**H3**-S10; T32 **H4**-S48 **H2A**-T101 **H2B**-S36, T19,
Lysine/glutamate ADP-ribosylation		+539	**H3**-K27, K37, **H4**-K16, **H2A**-K13, **H2B**-K30; E1 **H1**-E2
Biotinylation of lysine		+226	**H3**-K4, K99, K18, K23 [Ref. 36] **H4**-K8, K12, K16 **H2A**-K9,K13,K125,K127, K129 [Ref. 37]
Tyrosine hydroxylation		+16	**H4**-Y51, Y88 **H2A**-Y39 **H2B**-Y37, Y83 **H1**-Y70
Asparagine/ glutamine deamidation		+1	**H1**-N76 [Ref. 32]
Proline isomerization	 *trans* *cis*	0	**H3**-P16, P30, P38. [Ref. 38,39] **H1.1**-P153, P184 [Ref. 40,41]

1.3 TECHNOLOGIES ARE CRITICAL FOR THE ADVANCEMENT OF EPIGENETIC DISCOVERY

The progress in epigenetics is not possible without matching technologies. All the important break-through discoveries in epigenetic findings are made with the support of finely designed tools and strategies to detect, quantify, and image the chromatin state at multiple dimensions. A number of technologies were developed as the epigenetic field moves forward. These epigenetic technologies (Epi-Tech) include locus specific and genome-wide sequencing, creation of high-quality antibodies, chromatin functional assays, imaging tools, bioinformatics, and chemistry development.

The field of DNA methylation has progressed considerably over the past decades. We now know that the 5-methyl group of cytosine can be enzymatically oxidized to hydroxymethyl, formyl, and even carboxyl groups [42]. Detecting and mapping technologies for studying DNA methylation and the newer cytosine modifications has gained rapid development [43]. In Chapters 2 and 3, colleagues from New England Biolabs and Emory University review the developing history of DNA methylation and discuss in detail various formats of high-throughput and high-resolution sequencing technologies for detecting cytosine modifications; the former emphasizes DNA methylation and hydroxymethylation, and the latter extends to cytosine 5-formylation and 5-carboxylation.

Histone modifications are epigenetic tags as important as DNA methylation. The number and extent of histone modifications are far more complex than DNA methylation. MS has served as a powerful technology in epigenetics and has yielded significant impacts in providing a comprehensive picture of histone modifications [44] and understanding the consequence of the combinatorial histone modification patterns in the regulation of gene activities. Chapter 4 summarizes recent accomplishments of MS-based proteomics in the field of epigenetics, including histone sample preparation, enrichments of low abundant PTMs, different MS-based analytical strategies and bioinformatics tools for identification and quantification of histone PTMs, and histone variants. MS analysis is also able to determine chromatin modifications and interactome in a locus-specific manner with the combination of the chromatin immunoprecipitation (ChIP) technique.

A number of technologies have been developed to study functions and dynamics of histone modification. Most of these technologies rely on antibodies that specifically recognize a particular histone modification mark. ChIP is a particularly useful common technique in studying epigenetic marks and has revolutionized chromatin research. ChIP in conjunction with DNA sequencing allows for assigning locus-specific functions to histone marks [45]. For example, ChIP analysis revealed that acetylation at H4 Lys16 is typically found in the promoters of transcriptionally active genes, whereas trimethylation at H3 Lys9 is observed predominantly in silenced heterochromatin [46]. Chapter 5 reviews the technique of ChIP and its application in elucidating histone modifications and chromatin regulators at specific loci. This technique is important in revealing that the different sequential recruitment of HAT and SWI/SNF-remodeling complex is promoter-type dependent: SWI/SNF-remodeling complex comes in first, followed by HATs on the HO gene promoter, while recruitment of HAT occurs prior to SWI/SNF on the *IFN-β* gene promoter [47,48].

In addition to the four core histones, the linker histone H1 participates in the formation of higher order chromatin structures and adds another layer of complexity for the epigenetic regulation of gene expression. There are 11 mammalian H1 variants that are differentially regulated during development and cell cycle [49]. Chapter 6 discusses a suite of analytical tools and technologies used to study the functions of linker histone H1 variants, including qRT-PCR, HPLC analysis, gene targeting for H1 depletion, *in vivo* H1 tagging, ChIP of tagged H1 variants, and mutation analysis.

Structural biology plays a critical role in elucidating the molecular basis of chromatin function and epigenetic regulation. As a powerful tool capable of providing atomic images of molecular recognition, macromolecular X-ray crystallography has greatly facilitated the mechanistic understanding of epigenetics phenomena in three-dimensional and atomic details. For instance, the 2.8 angstrom crystal structure of reconstituted nucleosomes resolved by Luger et al. in 1997 represents a real *tour de force* of discovery in the epigenetics field [10]. It provides atomic information regarding the nucleosome assembly and organization and has stimulated numerous efforts later to investigate the biochemistry and function of histone modifications in epigenetic regulation. Also, the rational drug design that is based on X-ray crystallographic and NMR structures of epigenetic proteins is an indispensable approach toward the development of novel selective epigenetic inhibitors. In Chapter 7, Zhao and Li review crystallographic studies of important epigenetic targets including writers, erasers, and readers of chromatin marks. As the authors point out, the application of macromolecular X-ray crystallography in epigenetic research will continue to provide novel mechanistic insights into epigenetic regulation and prompt the translation of mechanistic discoveries in epigenetics into therapeutic strategies impacting on medicine and its clinical ramifications.

Histone proteins typically contain multiple PTM marks. The number is huge and even more complex is the possible vast combinations of histone marks. It remains challenging to elucidate the mechanistic role of individual marks as well as to identify the functional cross-talk between different PTMs. Molecular tools that enable dissecting function of individual histone marks and determining the biological consequence of cross-talks between different histone marks are highly demanded. In Chapter 8, Chatterjee and Dhall highlight chemical and genetic tools that allow access to homogeneously site-specifically modified histones for mechanistic investigations. Herein, there are two main strategies generally useful for preparing homogeneous histones containing desired PTM tags. The first strategy is direct biosynthesis via genetic codon expansion [50]. This method relies on an evolved orthogonal aminoacyl-tRNA synthetase/tRNA pair that specifically recognizes a specific unnatural amino acid (e.g. acetylated lysine) and directs its incorporation into recombinant proteins such as histones. The insertion site is precisely controlled by the position of an amber codon in the mRNA sequence of the protein target. An alternative approach is to use native chemical ligation and expressed protein ligation technology. Both methods have been extensively used in site-specific installation of histone PTM marks and yielded useful insights on the function of histone modifications [51]. In Chapter 9, Schutkowski and coworkers introduce the histone peptide microarray platform as a tool to profile epigenetic targets efficiently and comprehensively. In this technology, a histone microarray is made presenting thousands of histone peptides derived from human histone isoforms and also incorporating popular histone PTMs, such as lysine acylation, lysine/arginine methylation, serine/threonine/tyrosine phosphorylation, and various combinations of these modifications. This array approach enables rapid and efficient profiling of binding specificities of histone code readers and the determination of substrate specificity of the writers and erasers of histone marks. Importantly, addition of chemically introduced PTMs including recently discovered succinyl and malonyl lysines greatly facilitates to address the complex synergistic or antagonistic cross-talking relationship between different PTMs.

Although many key PTMs, especially methylation and acetylation, were first studied on nucleosomal histones, of great interest is that they were later found to be present widely in nonhistone proteins. Hundreds of methylated proteins and thousands of acetylated proteins have been discovered from MS-based proteomic analysis. However, we do not understand well the dynamics of these modifications; especially the enzymatic modification mechanism is poorly understood [52]. In this regard, bioorthogonal chemical probes have been designed and applied to investigate protein substrates of HATs [53] and methyltransferases [54]. In Chapter 10, Luo reviews the techniques used for protein methylation profiling at the proteomic level. In particular, the author's group developed a chemical biology strategy of bioorthogonal profiling of protein methylation, which is a powerful tool to dissect subsets of methylome for individual protein methyltransferases.

High-throughput sequencing technologies have led to an explosion of biological data and provide the opportunity of mining disease markers and drug targets at the systematic level. Computational methods for epigenetic analysis have developed rapidly in answer to the call for efficient, accurate, fast management and processing of large volumes of data that are derived from epigenomic sequencing [43]. Bioinformatic and biostatistic tools are critical for dealing with large data sets and fishing out useful information in this "omic" era. In Chapter 11, Cheng et al. discuss the power of bioinformatic and biostatistic tools for the discovery of epigenetic biomarkers and drug targets in human diseases. Epigenetics is indispensable for the maintenance of cell phenotypes. The molecular mechanism for the inheritance of modification patterns over cell generations remains to be sorted out. In Chapter 12, Xing and coworkers illustrate a mathematical-modeling approach to explore the dynamics of histone modification, providing a theoretical understanding of the mechanisms of epigenetic inheritance.

1.4 EPIGENETIC INHIBITORS AS NOVEL THERAPEUTICS

Epigenetic aberrations have been recognized in a diverse assortment of diseases such as cancer [55−58], obesity [59,60], diabetes [61,62], cardiovascular disorders [63,64], and neurological diseases [2,55,65,66]. Targeting epigenetic targets and processes represents an innovative strategy in disease treatment as it is aimed to restore the normal chromatin state of important genes, by either silencing or activating specific chromatin regulators. The success of DNMT and HDAC inhibitors as approved clinical drugs for treating hematological malignancies demonstrates that targeting epigenetic targets is a viable therapeutic approach [67]. Great efforts have been seen in recent years to develop chemical regulators of various epigenetic modifiers and readers [68−71].

DNA methyltransferases (DNMTs) are promising epigenetic targets for the development of novel, single, or combined therapies in disease treatment [58,72]. 5-Azacytidine and decitabine are DNMT inhibitors approved by the Food and Drug Administration of the United States for the treatment of myelodysplastic syndromes (MDS). In Chapter 13, Medina-Franco et al. discuss the therapeutic significance of DNMTs in cancer and the current status of DNMT inhibitor development. The authors also present advances in experimental assays to screen compound collections, and elaborate on data sources and applications of chemoinformatics and molecular modeling that exemplify the significance of ChemEpinformatics in translational epigenetics.

Dynamic regulation of cellular acetylation states is achieved by the functionally opposing enzymes — HATs and HDACs. Both HATs and HDACs are associated with various human diseases, particularly cancer. In Chapter 14, Luan et al. give a review on the current study of HAT enzymes, covering the biochemical assays for HAT activity measurement and the development of HAT chemical modulators. Development of small-molecule HDAC inhibitors was among the most advanced in epigenetic drug discovery [73]. Two HDAC inhibitors, romidepsin and SAHA, have already been used by cancer patients. In Chapter 15, Oyelere and coworkers review various assay technologies used for HDAC inhibitor screening and characterization, and highlight the strengths and weaknesses of each assay technology. It is worthwhile to stress that regulating HATs or HDACs is not the sole way of targeting the acetylation pathway. In recent years, success has also been witnessed in developing acetylation inhibitors by targeting the acetyl-lysine reader proteins, such as JQ1 and I-BET762 for the BET family of bromodomain proteins [74,75].

Methylation is widespread in the nucleosomal histones and nonhistone proteins. In Chapter 16, Zeng and Xu comprehensively review various formats of histone methyltransferase assays including radiometric assay, fluorescence polarization assay, coupled enzyme assay, and antibody-based assay, with detailed discussion of their mechanisms, pros and cons, and potential applications. In Chapter 17, Li et al. give an account of inhibitor development for the family of histone methyltransferase enzymes, including lysine methyltransferases (PKMTs) and arginine methyltransferases (PRMTs).

Lysine methylation is antagonized by lysine demethylases (KDMs). Jung and coworkers in Chapter 18 review the different types of histone demethylating enzymes and highlight key findings of their roles in the development of diseases. The authors present various formats of screening technologies for the discovery of histone demethylase inhibitors and demonstrated some key structural features of inhibitor binding. In Chapter 19, Whetstine et al. discuss several approaches for the purification and analysis of enzymatic activity of histone demethylases, and share technical protocols that allow for cellular characterization of these enzymes.

Animal models are obligatory tools, along with cell culture–based systems, in the preclinical evaluation of epigenetic inhibitors. Promising drugs from biochemical or cellular screening usually proceed to be tested in animal models which provide data on dosing, toxicity, and other important pharmacological attributes that prelude to safe and efficient clinical trials. In Chapter 20, Huang et al. describe the use of animal models (both GEMMs and xenotransplant) for the evaluation and development of epigenetic inhibitors. The advantages, challenges, and limitations associated with animal models are discussed.

1.5 PERSPECTIVE

Epigenetic research is rapidly flourishing and many directions are emerging. With many new sites and new types of modifications discovered in histones, precise yet systematic studies are warranted to assess the functions of those modifications. Also, proteomics studies have shown that most modifications found on histones occur across a myriad of proteins in and out of the nucleus, suggesting regulatory roles of these PTMs in cellular homeostasis. Thus, chromatin modification is not the sole domain of histone modifiers and readers. Dissecting the dynamics and functions of the PTM marks that were originally uncovered in chromosomal histones in a much larger context, across multiple cellular organelles and in different animal models, will provide a better, more precise understanding of their biological functions. Elucidating the broad substrate specificity and non-chromatin function of epigenetic proteins is expected. These efforts will in turn benefit the understanding of their regulatory function in chromatin restructuring and epigenetic inheritance. It could be speculated that histone modification is the result of a sophisticated selection of the bulk pool of posttranslational modification biochemistry during the history of evolution, which eventually manifests itself with specialized function for stabilization and regulation of the genetic material.

The connection of epigenetics with metabolism is attracting intensive attention. Methyl donor nutrients (e.g. folate, glycine, and betaine) can increase DNA methylation, thereby altering gene expression [76]. Friis et al. [77] showed that inducing glucose catabolism and production of acetyl-CoA in quiescent cells activates global histone acetylation. Illumination of the functional linkage of metabolism to epigenetics is important, not only for inventing new disease therapy, but also for providing cognitive guidelines for disease prevention and long-term health maintenance. We know that there are many environmental mutagens that can cause cancer-related genetic mutations. Analogously, a number of efforts have been made to investigate the mechanism of environmental chemicals in acting as "epimutagens" to affect the epigenome of somatic cells and reproductive cells, which may lead to pathophenotypes at later stages [78,79]. The influence of nutrition on epigenetic states belongs to the subfield of *nutri-epigenomics* [80]. Possibly, many natural products, dietary supplements, and food components are essential to maintain and propagate the healthy state of the epigenotype. Alteration of nutritional compositions may lead to chronic accumulation of unstable epigenetic marks, which cumulatively promote the predisposition to tumorigenesis or other pathological transformations. Certainly, these new epigenetic findings will provide invaluable information for recommending healthy diet styles for early disease prevention.

Given the great significance of epigenetic inhibitors in the pharmaceutical industry, development of drug-like inhibitors for various epigenetic targets will continue to be an important research focus in this field. Integration of *in silico* design and high-throughput screening of chemical libraries is an effective approach to discovering new drug leads. Natural compounds represent a rich source of chemical structures that could be used for the design of epigenetic drugs [81]. In pursuit of epi-drugs, discrimination of the selected target as a disease marker vs. a therapeutic target will be a priority step. One would wish to target only important driver genes (those that are epigenetically indispensable for disease to occur) instead of passenger genes (those that are epigenetically altered but are dispensable for disease to occur) [58]. Investigating mechanisms of drug action *in vivo* is critical. This is important for at least two reasons: first, many histone-modifying enzymes have numerous non-chromatin targets. For instance, thousands of cellular proteins were found to contain acetylated lysines, highlighting that histones are just the tip of the iceberg of the entire protein acetylome. Under this scenario, pharmacologic impacts of those inhibitors that target HATs or HDACs could be caused by blockage of acetylation-regulated pathways that are not associated with epigenetics. Another reason is regarding the polypharmacologic properties of small molecule drugs. Most drugs have multiple targets in the cell, so it is critical to examine whether the pharmacologic effect caused by the drug is authentically due to its action on epigenetic targets.

In keeping up the pace with the advancement of epigenetic science, Epi-Tech will continue to play indispensable roles in the epigenetic field. The invention of powerful techniques to interrogate function and mechanism of the epigenome and epigenetic regulators has driven the remarkable advances in epigenetics. Numerous methods have been developed that provide means to accurately assess epigenetic processes, most notably DNA methylation and chromatin modifications as the most prevalent epigenetic mechanisms. The technical methods described in this book represent some of the exciting developments in this area. Advances in epigenetics are occurring at a remarkable pace and it is no doubt that this area will continue to herald critical breakthroughs in years to come. We anticipate that the advances in Epi-Tech will continually illuminate unresolved problems and further advance the field of epigenetics. To determine the changes of the epigenome under various contexts and environmental conditions, sequencing DNA methylome with improved technologies, such as SeqCapEpi, will likely become more financially inexpensive and accessible to investigators [82]. Deep sequencing coupled with Hi-C technology is a powerful approach for mapping genetic and epigenetic states with three-dimensional chromatin conformational structures [83,84]. The combination of microfluidic technology, modification-specific antibodies, and single-molecule imaging techniques enables measuring the degree of co-localization of histone modifications and DNA methylation on single nucleosomes [85,86]. The single molecule approach likely will play a more important role in chromatin analysis [87]. One advantage of the single-molecule method is its overcoming of cell source limitation. This is important for many cases; for example, progenitor cells from animals are often limited, which does not allow for standard ChIP analysis. Along with the rapid advance in epigenomic mapping using sophisticated technologies have come increasing challenges in managing massive amounts of information. Further developments in computational epigenetics is compulsory in deciphering the colossal amounts of data derived from experimental epigenomics [43]. With new progress, many key questions in this area, such as epigenetic etiology of disease, epigenetic inheritance, impact by nature and nurture, connection with mental health, and predictable pharmaco-epigenomics, will become illuminated in greater width and depth.

ACKNOWLEDGMENTS

I acknowledge the National Institutes of Health and the American Heart Association for financial support. I thank H. Hu and K. Qian for providing the figure and the table in this chapter.

REFERENCES

[1] Holliday R. Epigenetics: a historical overview. Epigenetics 2006;1(2):76−80.
[2] Egger G, Liang G, Aparicio A, Jones PA. Epigenetics in human disease and prospects for epigenetic therapy. Nature 2004;429(6990):457−63.
[3] Felsenfeld G. The evolution of epigenetics. Perspect Biol Med 2014;57(1):132−48.
[4] Egger G, Liang G, Aparicio A, Jones PA. Epigenetics in human disease and prospects for epigenetic therapy. Nature 2004;429:457−63.
[5] Mann JR. Epigenetics and memigenetics. Cell Mol Life Sci 2014;71(7):1117−22.
[6] Bird A. Perceptions of epigenetics. Nature 2007;447(7143):396−8.
[7] Ravindra KC, Narayan V, Lushington GH, Peterson BR, Prabhu KS. Targeting of histone acetyltransferase p300 by cyclopentenone prostaglandin Delta(12)-PGJ(2) through covalent binding to Cys(1438). Chem Res Toxicol 2012;25(2):337−47.
[8] Campos EI, Reinberg D. Histones: annotating chromatin. Annu Rev Genet 2009;43:559−99.
[9] Davis PKB, Brackman RK. Chromatic remodeling and cancer. Cancer Biol Ther 2003;2(1):22−9.
[10] Luger K, Mader AW, Richmond RK, Sargent DF, Richmond TJ. Crystal structure of the nucleosome core particle at 2.8 A resolution. Nature 1997;389(6648):251−60.
[11] Olins AL, Olins DE. Spheroid chromatin units (v bodies). Science 1974;183(4122):330−2.
[12] Woodcock CL, Safer JP, Stanchfield JE. Structural repeating units in chromatin. I. Evidence for their general occurrence. Exp Cell Res 1976;97:101−10.
[13] van Holde KE. Chromatin (Springer Series in Molecular and Cell Biology). 1988 ed. Springer; 1988. p. 497.

[14] Wolffe A. Chromatin: Structure and function. San Diego, CA: Elsevier Science & Technology Books; 1999.

[15] Lenfant F, Mann RK, Thomsen B, Ling X, Grunstein M. All four core histone N-termini contain sequences required for the repression of basal transcription in yeast. Embo J 1996;15(15):3974−85.

[16] Gregory PD, Wagner K, Horz W. Histone acetylation and chromatin remodeling. Exp Cell Res 2001;265(2):195−202.

[17] Zheng YG, Wu J, Chen Z, Goodman M. Chemical regulation of epigenetic modifications: opportunities for new cancer therapy. Med Res Rev 2008;28(5):645−87.

[18] Reik W. Stability and flexibility of epigenetic gene regulation in mammalian development. Nature 2007;447(7143):425−32.

[19] Buczek-Thomas JA, Hsia E, Rich CB, Foster JA, Nugent MA. Inhibition of histone acetyltransferase by glycosaminoglycans. J Cell Biochem 2008;105(1):108−20.

[20] Marmorstein R, Zhou MM. Writers and readers of histone acetylation: structure, mechanism, and inhibition. Cold Spring Harb Perspect Biol 2014;6(7):a018762.

[21] Kouzarides T. Chromatin modifications and their function. Cell 2007;128:693−705.

[22] Grunstein M. Histone acetylation in chromatin structure and transcription. Nature 1997;389 (6649):349−52.

[23] Ooi L, Belyaev ND, Miyake K, Wood IC, Buckley NJ. BRG1 chromatin remodeling activity is required for efficient chromatin binding by repressor element 1-silencing transcription factor (REST) and facilitates REST-mediated repression. J Biol Chem 2006;281(51):38974−80.

[24] Wu SY, Chiang CM. The double bromodomain-containing chromatin adaptor Brd4 and transcriptional regulation. J Biol Chem 2007;282(18):13141−5.

[25] Eckey M, Kraft F, Kob R, Escher N, Asim M, Fischer H, et al. The corepressor activity of Alien is controlled by CREB-binding protein/p300. Febs J 2013;280(8):1861−8.

[26] Perez-Campo FM, Costa G, Lie ALM, Stifani S, Kouskoff V, Lacaud G. MOZ-mediated repression of p16(INK) (4) (a) is critical for the self-renewal of neural and hematopoietic stem cells. Stem Cells 2014;32(6):1591−601.

[27] Wagner EJ, Carpenter PB. Understanding the language of Lys36 methylation at histone H3. Nat Rev Mol Cell Biol 2012;13(2):115−26.

[28] Ruthenburg AJ, Li H, Patel DJ, Allis CD. Multivalent engagement of chromatin modifications by linked binding modules. Nat Rev 2007;8(12):983−94.

[29] Barski A, Cuddapah S, Cui K, Roh TY, Schones DE, Wang Z, et al. High-resolution profiling of histone methylations in the human genome. Cell 2007;129(4):823−37.

[30] Taverna SD, Li H, Ruthenburg AJ, Allis CD, Patel DJ. How chromatin-binding modules interpret histone modifications: lessons from professional pocket pickers. Nat Struct Mol Biol 2007;14(11): 1025−40.

[31] Huang H, Sabari BR, Garcia BA, Allis CD, Zhao Y. SnapShot: histone modifications. Cell 2014;159(2): 458−458.e1.

[32] Sarg B, Lopez R, Lindner H, Ponte I, Suau P, Roque A. Identification of novel post-translational modifications in linker histones from chicken erythrocytes. J Proteomics 2015;113:162−77.

[33] Wilson JP, Raghavan AS, Yang YY, Charron G, Hang HC. Proteomic analysis of fatty-acylated proteins in mammalian cells with chemical reporters reveals S-acylation of histone H3 variants. Mol Cell Proteomics 2011;10(3): M110 001198.

[34] Unoki M, Masuda A, Dohmae N, Arita K, Yoshimatsu M, Iwai Y, et al. Lysyl 5-hydroxylation, a novel histone modification, by Jumonji domain containing 6 (JMJD6). J Biol Chem 2013;288 (9):6053−62.

[35] Nathan D, Ingvarsdottir K, Sterner DE, Bylebyl GR, Dokmanovic M, Dorsey JA, et al. Histone sumoylation is a negative regulator in *Saccharomyces cerevisiae* and shows dynamic interplay with positive-acting histone modifications. Genes Dev 2006;20(8):966−76.

[36] Pestinger V, Wijeratne SS, Rodriguez-Melendez R, Zempleni J. Novel histone biotinylation marks are enriched in repeat regions and participate in repression of transcriptionally competent genes. J Nutr Biochem 2011;22(4):328−33.

[37] Hassan YI, Zempleni J. A novel, enigmatic histone modification: biotinylation of histones by holocarboxylase synthetase. Nutr Rev 2008;66(12):721−5.

[38] Nelson CJ, Santos-Rosa H, Kouzarides T. Proline isomerization of histone H3 regulates lysine methylation and gene expression. Cell 2006;126(5):905−16.

[39] Howe FS, Boubriak I, Sale MJ, Nair A, Clynes D, Grijzenhout A, et al. Lysine acetylation controls local protein conformation by influencing proline isomerization. Mol Cell 2014;55(5):733−44.

[40] Raghuram N, Strickfaden H, McDonald D, Williams K, Fang H, Mizzen C, et al. Pin1 promotes histone H1 dephosphorylation and stabilizes its binding to chromatin. J Cell Biol 2013;203(1):57−71.

[41] Hanes SD. Prolyl isomerases in gene transcription. Biochim Biophys Acta 2014.

[42] Xu Z, Li H, Jin P. Epigenetics-based therapeutics for neurodegenerative disorders. Curr Transl Geriatr Exp Gerontol Rep 2012;1(4):229−36.

[43] Tollefsbol TO. Advances in epigenetic technology. Methods Mol Biol 2011;791:1−10.

[44] Garcia BA, Pesavento JJ, Mizzen CA, Kelleher NL. Pervasive combinatorial modification of histone H3 in human cells. Nat Methods 2007;4(6):487−9.

[45] Schmidt D, Wilson MD, Spyrou C, Brown GD, Hadfield J, Odom DT. ChIP-seq: using high-throughput sequencing to discover protein-DNA interactions. Methods 2009;48(3):240−8.

[46] Wang Z, Zang C, Rosenfeld JA, Schones DE, Barski A, Cuddapah S, et al. Combinatorial patterns of histone acetylations and methylations in the human genome. Nat Genet 2008;40(7):897−903.

[47] Krebs JE, Kuo MH, Allis CD, Peterson CL. Cell cycle-regulated histone acetylation required for expression of the yeast HO gene. Genes Dev 1999;13(11):1412−21.

[48] Agalioti T, Chen G, Thanos D. Deciphering the transcriptional histone acetylation code for a human gene. Cell 2002;111(3):381−92.

[49] Happel N, Doenecke D. Histone H1 and its isoforms: contribution to chromatin structure and function. Gene 2009;431(1−2):1−12.

[50] Liu CC, Schultz PG. Adding new chemistries to the genetic code. Annu Rev Biochem 2010; 79:413−44.

[51] Li KK, Luo C, Wang D, Jiang H, Zheng YG. Chemical and biochemical approaches in the study of histone methylation and demethylation. Med Res Rev 2012;32(4):815−67.

[52] Choudhary C, Weinert BT, Nishida Y, Verdin E, Mann M. The growing landscape of lysine acetylation links metabolism and cell signalling. Nat Rev Mol Cell Biol 2014;15(8):536−50.

[53] Yang C, Mi J, Feng Y, Ngo L, Gao T, Yan L, et al. Labeling lysine acetyltransferase substrates with engineered enzymes and functionalized cofactor surrogates. J Am Chem Soc 2013;135(21): 7791−4.

[54] Islam K, Bothwell I, Chen Y, Sengelaub C, Wang R, Deng H, et al. Bioorthogonal profiling of protein methylation using azido derivative of S-adenosyl-L-methionine. J Am Chem Soc 2012;134 (13):5909−15.

[55] Fouse SD, Costello JF. Epigenetics of neurological cancers. Future Oncol 2009;5(10):1615−29.

[56] Gnyszka A, Jastrzebski Z, Flis S. DNA methyltransferase inhibitors and their emerging role in epigenetic therapy of cancer. Anticancer Res 2013;33(8):2989−96.

[57] Dawson MA, Kouzarides T. Cancer epigenetics: from mechanism to therapy. Cell 2012;150(1):12−27.

[58] Kelly TK, De Carvalho DD, Jones PA. Epigenetic modifications as therapeutic targets. Nat Biotechnol 2010;28(10):1069−78.

[59] Iyer A, Fairlie DP, Brown L. Lysine acetylation in obesity, diabetes and metabolic disease. Immunol Cell Biol 2012;90(1):39−46.

[60] Kendrick AA, Choudhury M, Rahman SM, McCurdy CE, Friederich M, Van Hove JL, et al. Fatty liver is associated with reduced SIRT3 activity and mitochondrial protein hyperacetylation. Biochem J 2011;433(3):505−14.

[61] Villeneuve LM, Natarajan R. The role of epigenetics in the pathology of diabetic complications. Am J Physiol Renal Physiol 2010;299(1):F14−25.

[62] Cooper ME, El-Osta A. Epigenetics: mechanisms and implications for diabetic complications. Circ Res 2010;107(12):1403−13.

[63] McKinsey TA, Olson EN. Cardiac histone acetylation − therapeutic opportunities abound. Trends Genet 2004;20(4):206−13.

[64] Abi Khalil C. The emerging role of epigenetics in cardiovascular disease. Ther Adv Chronic Dis 2014;5 (4):178−87.

[65] Feng J, Fan G. The role of DNA methylation in the central nervous system and neuropsychiatric disorders. Int Rev Neurobiol 2009;89:67−84.

[66] Valor LM, Viosca J, Lopez-Atalaya JP, Barco A. Lysine acetyltransferases CBP and p300 as therapeutic targets in cognitive and neurodegenerative disorders. Curr Pharm Des 2013;19(28):5051−64.

[67] Minucci S, Pelicci PG. Histone deacetylase inhibitors and the promise of epigenetic (and more) treatments for cancer. Nat Rev Cancer 2006;6(1):38−51.

[68] Miyamoto K, Ushijima T. Diagnostic and therapeutic applications of epigenetics. Jpn J Clin Oncol 2005;35(6):293−301.

[69] Zelent A, Waxman S, Carducci M, Wright J, Zweibel J, Gore SD. State of the translational science: summary of Baltimore workshop on gene re-expression as a therapeutic target in cancer January 2003. Clin Cancer Res 2004;10(14):4622−9.

[70] Gilbert J, Gore SD, Herman JG, Carducci MA. The clinical application of targeting cancer through histone acetylation and hypomethylation. Clin Cancer Res 2004;10(14):4589−96.

[71] Mai A, Altucci L. Epi-drugs to fight cancer: from chemistry to cancer treatment, the road ahead. Int J Biochem Cell Biol 2009;41(1):199−213.

[72] Gros C, Fahy J, Halby L, Dufau I, Erdmann A, Gregoire J-M, et al. DNA methylation inhibitors in cancer: recent and future approaches. Biochimie 2012;94(11):2280−96.

[73] Gryder BE, Sodji QH, Oyelere AK. Targeted cancer therapy: giving histone deacetylase inhibitors all they need to succeed. Future Med Chem 2012;4(4):505−24.

[74] Filippakopoulos P, Qi J, Picaud S, Shen Y, Smith WB, Fedorov O, et al. Selective inhibition of BET bromodomains. Nature 2010;468(7327):1067−73.

[75] Mirguet O, Gosmini R, Toum J, Clement CA, Barnathan M, Brusq JM, et al. Discovery of epigenetic regulator I-BET762: lead optimization to afford a clinical candidate inhibitor of the BET bromodomains. J Med Chem 2013;56(19):7501−15.

[76] Handy DE, Castro R, Loscalzo J. Epigenetic modifications: basic mechanisms and role in cardiovascular disease. Circulation 2011;123(19):2145−56.

[77] Friis RM, Wu BP, Reinke SN, Hockman DJ, Sykes BD, Schultz MC. A glycolytic burst drives glucose induction of global histone acetylation by picNuA4 and SAGA. Nucleic Acids Res 2009;37 (12):3969−80.

[78] Collotta M, Bertazzi PA, Bollati V. Epigenetics and pesticides. Toxicology 2013;307:35−41.

[79] Sutherland JE, Costa M. Epigenetics and the environment. Ann N Y Acad Sci 2003;983:151−60.

[80] Gallou-Kabani C, Vige A, Gross MS, Junien C. Nutri-epigenomics: lifelong remodelling of our epigenomes by nutritional and metabolic factors and beyond. Clin Chem Lab Med 2007;45(3):321−7.

[81] Miceli M, Bontempo P, Nebbioso A, Altucci L. Natural compounds in epigenetics: a current view. Food Chem Toxicol 2014;73C:71−83.

[82] Duhaime-Ross A. Revved-up epigenetic sequencing may foster new diagnostics. Nat Med 2014;20(1):2.

[83] van Berkum NL, Lieberman-Aiden E, Williams L, Imakaev M, Gnirke A, Mirny LA, et al. Hi-C: a method to study the three-dimensional architecture of genomes. J Vis Exp 2010;(39):e1869.

[84] Khrameeva EE, Mironov AA, Fedonin GG, Khaitovich P, Gelfand MS. Spatial proximity and similarity of the epigenetic state of genome domains. PloS One 2012;7(4):e33947.

[85] Hagarman JA, Motley MP, Kristjansdottir K, Soloway PD. Coordinate regulation of DNA methylation and H3K27me3 in mouse embryonic stem cells. PLoS One 2013;8(1):e53880.

[86] Murphy PJ, Cipriany BR, Wallin CB, Ju CY, Szeto K, Hagarman JA, et al. Single-molecule analysis of combinatorial epigenomic states in normal and tumor cells. Proc Natl Acad Sci USA 2013;110(19): 7772−7.

[87] Hyun BR, McElwee JL, Soloway PD. Single molecule and single cell epigenomics. Methods 2015;72: 41−50.

TECHNOLOGIES FOR THE MEASUREMENT AND MAPPING OF GENOMIC 5-METHYLCYTOSINE AND 5-HYDROXYMETHYLCYTOSINE

2

Jolyon Terragni and Sriharsa Pradhan

Genome Biology Division, New England Biolabs, Ipswich, Massachusetts, USA

CHAPTER OUTLINE

2.1 INTRODUCTION

One of the most common and studied DNA modifications in the mammalian genome arises from the enzymatic methylation of cytosine to 5-methylcytosine (5mC) [1]. This cytosine modification, which is created *de novo* and is maintained by the DNA methyltransferase (DNMTs) family of enzymes, is critical in the regulation of gene expression [2,3]. Further illustrating the importance of this DNA modification, irregular patterning of cytosine methylation has been linked to many different diseases, such as acute myeloid and mixed lineage leukemias, AML and MLL respectively [4,5]. Until recently 5mC was thought to be the only mammalian DNA modification that exists, but new evidence acquired from human brain cells has revealed the presence of an additional DNA modification, 5-hydroxymethylcytosine (5hmC) [6]. 5hmC was initially identified as an oxidative DNA damage residue [7,8]. However, recent reports have demonstrated that the 5hmC modification

Y.G. Zheng (Ed): Epigenetic Technological Applications. DOI: http://dx.doi.org/10.1016/B978-0-12-801080-8.00002-8

arises by enzymatic oxidation of 5mC by the Ten-Eleven translocation (TET) family of enzymes [9]. Since the existence of 5hmC in the mammalian genome was a fairly recent discovery, researchers are still in the process of functionally characterizing this DNA modification. However, the current consensus among the epigenetic research community is that 5hmC is an essential intermediate in the active DNA demethylation pathway [9–13]. The epigenetic tools that have been created for the study of 5mC and 5hmC have focused solely on the measurement and mapping of these two epigenetic marks in the vertebrate genomes. In this chapter, we will discuss the creation and implementation of these different epigenetic technologies for the study of 5mC or 5hmC. Furthermore, each of these technologies will be discussed within the context of both cytosine modifications, since it is imperative to understand their ability to discriminate between 5mC and 5hmC.

2.2 MEASURING GENOMIC LEVELS OF 5-METHYLCYTOSINE AND 5-HYDROXYMETHYLCYTOSINE

Methylated cytosine in mammalian genomes was originally discovered over half a century ago, in a study that utilized a paper chromatographic technique. In this foundational study, researchers first hydrolyzed genomic DNA isolated from the thymus of a calf and then separated the single nucleosides by paper chromatographic techniques, which revealed a fifth chromatography spot, identified as methylated cytosine [14]. As is often the case with a newly developed technology, there were several shortcomings with this method, such as the guanine not being completely hydrolyzed or the phenylalanine appearing on the paper in the region occupied by cytosine. It was not until 20 years later, when this assay was performed with ion pair–reversed phase high-pressure liquid chromatography (RP-HPLC), that it finally became the accepted standard for the measurement of genomic 5mC levels [15,16]. In the RP-HPLC method, an enzyme cocktail of DNase I, nuclease P1, and alkaline phosphatase is first utilized to digest genomic DNA down to single nucleosides. The individual pools of DNA bases are then separated through RP-HPLC and the individual bases measured for their UV absorbances (254 and 280 nm). Further improvements to this method have subsequently been made by incorporating mass spectrometry, downstream of standard HPLC nucleotide separation, to measure nucleotides with higher sensitivity and accuracy [17]. Time-of-flight and triple quadrupole mass spectrometers are two common instruments for the qualitative or quantitative analysis of modified cytosine nucleosides. The multiple reactions monitoring (MRM) method can detect 5mC and 5hmC at fmol level, thus making the discovery of low abundance cytosine modification possible [18].

Over 50 years later after the initial discovery of 5mC, chromatographic and mass spectrometry techniques were also instrumental in the discovery of 5hmC in the mammalian genomes. In 2009, Kriaucionis et al. were initially trying to compare the abundance of genomic 5mC between Purkinje and granule cell nuclei, yet to their surprise they also discovered a sixth DNA base corresponding to hydroxymethylated cytosine [6]. In another study, published alongside this chromatographic study, it was also shown that the MLL Partner TET1 enzyme is responsible for the oxidation/conversion of 5mC into 5hmC [9]. Even though numerous improvements have been made through the years to the HPLC-mass spectrometry method for the measurement of modified nucleotides, there are some technical hurdles to this method that need to be considered when incorporating it into an epigenetic study. The most apparent limitation is the need for HPLC-mass spectrometry instrumentation and expertise, which can be problematic at research facilities with constrained resources. Additionally the HPLC-mass spectrometry method can take several days to complete, due to the requirement of genomic DNA having to be completely digested to single nucleosides before HPLC-mass spectrometry analysis. Lastly, this method is only capable of providing relative measurements of modified cytosine.

An immunochemical assay was developed for the quantitative measurement of genomic 5mC, and sometime later, for genomic 5hmC. In the early 1980s, antibodies were raised that could specifically target and bind 5mC on genomic DNA. Soon afterward a study detailed a new method utilizing a 5mC antibody to quantitatively measure genomic 5mC. In the proposed method, genomic DNA is extracted and immobilized to nitrocellulose paper. Total levels of genomic 5mC were then revealed by probing the membrane with an antibody recognizing 5mC and then visualizing with a chemiluminescent secondary antibody [19]. Even in its earliest stages, the sensitivity of this method was very high, capable of measuring nanogram (10^{-9}) quantities of 5mC in intact genomic DNA. This method was further adapted to work in an ELISA-based approach (enzyme-linked immunosorbent assay), thus increasing its high-throughput capabilities [20]. There were also other variations of this immunological method, with one alternative method utilizing radiolabeling of DNA, followed by 5mC pull-down with rabbit polyclonal antibodies and subsequently measuring radioactivity of the enriched DNA to determine overall levels of 5mC [21]. After the discovery of 5hmC in mammalian genomes, researchers quickly created antibodies for the specific recognition of genomic 5hmC. Not surprisingly, after the development of these 5hmC antibodies, an ELISA-based method was proposed for the quantitative measurement of total genomic 5hmC. Since the 5mC and 5hmC modifications are structurally very similar there was some concern about the 5mC and 5hmC antibodies cross reactivity. Fortunately, a research group has recently reexamined the specificity of these antibodies and determined that they were indeed highly specific with little to no cross-reactivity [22].

Enzymatic techniques were also developed for the measurement of either genomic 5mC or 5hmC [23,24]. A study by Wu et al. was one of the first that detailed a method of utilizing DNMT and S-adenosyl-L-[methyl-^{3}H]methionine (SAM[^{3}H]) to measure total levels of unmethylated CpGs in the genome [23]. In this study they were examining the tumorigenic effects of prolonged exposure to overexpressed mouse Dnmt1 in nontransformed mouse fibroblast cells. They found that overexpression of Dnmt1 increased both endogenous DNA methylase activity and hypermethylation of the genome, which subsequently led to tumorigenic transformation of the mouse fibroblasts. To measure the increased genomic methylation levels, they first extracted genomic DNA from both control and Dnmt1-transfected mouse fibroblasts. Then they used the CpG methyltransferase SssI, with (SAM[^{3}H]) cosubstrate, to *in vitro* methylate the genomic DNA extracted from the control and Dnmt1-overexpressed cells. Wu et al. proposed that the total amount of endogenous methylation in these cells is inversely proportional to the total amount of incorporated and measured methyl-^{3}H (Figure 2.1A). As a result, increasing levels of DNA methylation in the Dnmt1-transfected cells will directly correspond to a decrease of SssI-incorporated methyl-^{3}H in the genome, when compared to the untransfected control. As a result, this proposed comparative method can be used to quickly measure changes in genomic DNA methylation that result from a biological phenomenon or treatment of a tissue or cell. Yet even though this method provides a relative measurement of DNA methylation, it is not the appropriate method to utilize if absolute measurements of DNA methylation are desired. Several decades later, an enzymatic technique was developed for the

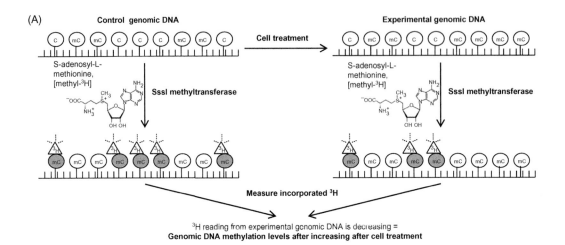

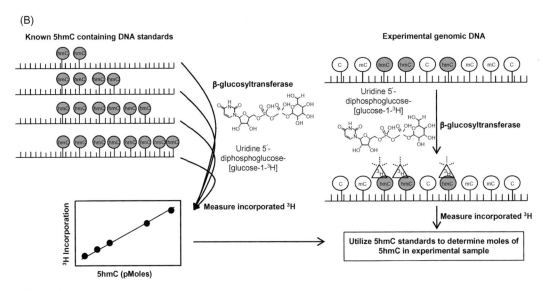

FIGURE 2.1

Measuring genomic levels of 5mC or 5hmC. (A) Utilizing SssI DNA methyltransferases and S-adenosyl-L-methionine, [methyl-³H] (SAM[³H]) substrate to measure genomic levels of 5mC. Control and experimental genomic DNA are incubated with SssI and SAM[³H]. Incorporated ³H is measured by scintillation counting. Changes in genomic DNA methylation levels inversely correlate to the change in the ³H counts between the control and experimental sample. An example is provided for this method where genomic DNA methylation increases in the experimental sample, thus decreasing SssI ³H labeling. (B). Utilizing β-glucosyltransferase (β-GT) and uridine 5′-diphosphoglucose-[glucose-1-³H] (UDP-Glc[³H]) substrate to measure genomic levels of 5hmC. 5hmC containing DNA standards and an experimental genomic DNA sample are incubated with β-GT and UDP-Glc[³H]. Incorporated ³H is measured by scintillation counting. Standard curve is then utilized to determine the moles of 5hmC in the experimental genomic DNA sample.

measurement of genomic 5hmC. This method, unlike the indirect 5mC labeling method, provides a direct measurement of genomic 5hmC, because the 5hmC itself is radiolabeled and measured. This is accomplished by utilizing T4 phage β-GT enzyme to transfer glucose-^{3}H to genomic 5hmC. Then the total levels of radiolabeled glucose incorporated into the genome provide a near-linear measurement of the absolute levels of genomic 5hmC (Figure 2.1B) [25,26]. Illustrating the strength and simplicity of this enzymatic method, it was shown that the sensitivity of this method is similar to the HPLC-mass spectrometry measurement of 5hmC. Furthermore, the β-GT method can be completed in under an hour and at a fraction of the cost required for HPLC-mass spectrometry analysis.

2.3 LOCUS-SPECIFIC ANALYSIS OF 5-METHYLCYTOSINE AND 5-HYDROXYMETHYLCYOTSINE

Prior to the advent of current genome-wide technologies, epigenetic methods were developed to reveal locus-specific methylation or hydroxymethylation at very high resolutions. These methods measure the differential sensitivity of either a single, or multiple, cytosine (C, 5mC, or 5hmC) to either restriction enzyme digestion or chemical conversion. Even though these methods are limited in their genomic scope, only being able to visualize cytosine modification within a very small region of genomic DNA, they have still been instrumental to the study of both 5mC and 5hmC. The original implementation of these technologies will be discussed along with improvements made as a result of technological advances.

The discovery in 1978 of MspI, an isoschizomer of the HpaII restriction enzyme, allowed for the first enzymatic epigenetic tool for the detection and measurement of the 5mC DNA modification [27]. The HpaII and MspI restriction enzymes have identical DNA cut sites, C$_\Delta$CGG, but HpaII can only digest this site when the internal cytosine is not methylated, while MspI completely digests it regardless of the methylation state of the cytosine. In a study by Cedar et al., they proposed a technique that utilized these two restrictions enzymes to directly detect methylation at the internal cytosine of CCGG [28]. In their method genomic DNA was first digested with either MspI or HpaII, then end labeled with the ^{32}P and subsequently digested down to single nucleosides. The nucleosides are then TLC separated to visualize the pool of radiolabeled methylated cytosines. Since MspI completely digests CCGG regardless of the methylation state of the cytosine, this pool of digested DNA provided the baseline measurement for the experiment, while HpaII cannot digest methylated CCGG. So, the drop in total radiolabeled cytosine in the HpaII pool, when compared to the MspI pool, directly reveals the total amount of methylation at interrogated genomic CCGG sites. A limitation to this proposed method was that it does not provide an absolute measurement of methylated cytosine. This was later addressed through the incorporation of the polymerase chain reaction (PCR) downstream of the HpaII enzymatic digestion [29]. This PCR-based technique is done by comparing the efficiency of a PCR reaction, using primers flanking the CCGG site, before and after HpaII digestion. If the PCR reaction is not saturated and is in the linear range of amplification, then the total amount of methylation at the interrogated CCGG site directly correlates to the decrease of PCR efficiency following HpaII digestion of the genomic DNA. For example, a PCR efficiency drop of twofold after HpaII digestion is indicative that 50% of the interrogated internal cytosine, within the

CCGG digestion site, is methylated. Surprisingly, even though it was known as early as 1997 that HpaII cannot cleave CCGG when the internal cytosine is hydroxymethylated [30], scientists were still readily utilizing HpaII-PCR in their studies to measure 5mC levels because they believed that 5hmC was not present in the mammalian genome. Not surprisingly, when 5hmC was later shown to be in the mammalian genome, the HpaII-PCR assay had to be modified in order to distinguish between 5mC and 5hmC in its measurements. This was addressed in a study by Kinney et al. who discovered and took advantage of the fact that MspI digestion was sensitive to glucosylated 5hmC (Figure 2.2A) [31]. The method they proposed entailed treating DNA with, or without, T4 β-GT and then digesting these DNA pools with either HpaII or MspI followed by PCR with primers flanking the CCGG of interest. Levels of 5mC and 5hmC could then be determined by comparing the sensitivity of HpaII digestion to the sensitivity of MspI digestion, both before and after β-GT treatment of the genomic DNA. For example, if an interrogated cytosine contains no modification then it will be completely digested by both HpaII and MspI, regardless of β-GT treatment. If the cytosine is 100% methylated then HpaII digestion will be completely blocked, while MspI digestion still occurs, regardless of the β-GT treatment. Lastly, if the cytosine is 100% hydroxymethylated then both the HpaII and MspI digestion will be blocked before β-GT treatment, but after β-GT treatment MspI will no longer be able to digest the glucosylated 5hmC (Figure 2.2B). Provided that the PCR is not saturated, is in the linear range of amplification and DNA standards have been amplified alongside the experimental samples, the β-GT/HpaII/MspI method will provide accurate and reproducible measurements for both 5mC and 5hmC at an interrogated CCGG site. Simplifying this method even further, recent studies have incorporated quantitative PCR downstream of the β-GT, HpaII and MspI digestion assay, thus removing the difficulties of quantitatively measuring PCR products on a gel [32,33].

The first epigenetic method that was capable of detecting every methylated cytosine on both strands of any target DNA sequence was detailed in studies by Clark et al. and Frommer et al. [34,35]. In these studies, which are often referenced as the beginning of the "bisulfite revolution," sodium bisulfite was utilized to convert cytosine to uracil residues in denatured single-stranded DNA, while methylated cytosine residues, which are not reactive to sodium bisulfite, did not convert. The bisulfite conversion reaction is carried out through three steps with the first being sulphonation of cytosine, followed by hydrolytic deamination and subsequent alkali desulphonation, which yields the final product of uracil. The chemically converted DNA is then PCR amplified with primers flanking a genomic region of interest (primers designed to recognize the bisulfite-converted DNA and devoid of CpG sites) and with a DNA polymerase capable of reading uracil, such as Taq. The PCR step accomplishes two things: it allows for Taq polymerase to replace uracil with thymine and it also allows for the amplification of the bisulfite-converted DNA. After PCR amplification, the PCR product is then cloned, transformed, and a large number of the colonies sequenced. The methylation levels at a CpG site among a large population of cells is determined by correlating the number of cytosine or thymine bases identified at the respective cytosine position in the interrogated CpG dinucleotides. As an example, if the sequencing reads for a CpG dinucleotide come back as ten cytosines and ten thymines, then the total cellular methylation at that CpG is 50%. The bisulfite chemical conversion method has some limitations, given the fact that a large number of clones have to be sequenced when examining either low levels of cytosine methylation or when trying to measure small changes in cytosine methylation. This problem has been addressed in some studies that have suggested using pyrosequencing to directly sequence the bisulfite PCR product. Pryosequencing can provide accurate measurements of the cytosine and thymine content at a single

(A)

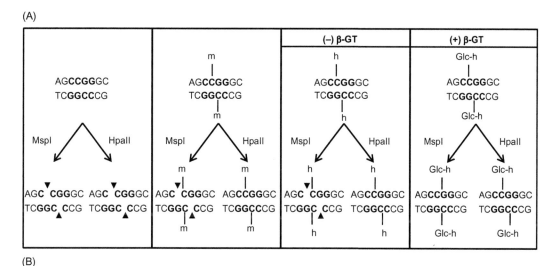

(B)

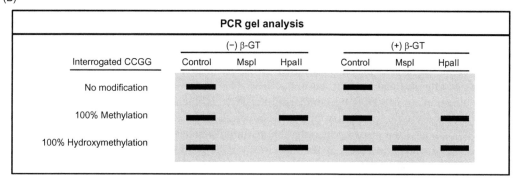

FIGURE 2.2

Locus-specific analysis of genomic 5mC and 5hmC. (A) Sensitivity of MspI and HpaII restriction enzymes to unmodified cytosine, methylated cytosine (m), hydroxymethylated cytosine (h), and glucosylated hydroxymethylated cytosine (Glc-h) within their CCGG restriction enzyme cut site (indicated in bold). (B) PCR gel analysis of genomic DNA samples digested with MspI or HpaII to determine 5mC or 5hmC content. Genomic DNA samples, which are identified to the left of the gel, are divided into two different pools with one untreated (−) and the other treated with β-GT (+), shown at the top of gel. These two DNA pools are then subdivided into three individual pools that are either untreated (control) or treated with MspI or HpaII restriction enzymes, identified at the top of the gel. PCR is carried out on these six different samples with PCR primers flanking the CCGG of interest. The PCR products are shown as dark bands on the gel.

CpG after bisulfite conversion with a detection limit of 1% differential methylation [36,37]. As a result, pyrosequencing can easily determine the complete distribution of CpG methylation in a PCR product of bisulfite-converted DNA, without the need for cloning.

The bisulfite conversion of DNA method has been incorporated into countless epigenetic studies over the last decade and has illustrated both the location and importance of cytosine methylation in the genome. Furthermore, this method can reveal 5mC in preserved DNA samples such as DNA obtained from formalin fixed paraffin embedded (FFPE) tissue samples [38]. The discovery of 5hmC in the mammalian genome complicated our epigenomic interpretation considerably, because the bisulfite conversion method could not distinguish between 5mC and 5hmC [39]. Considering how heavily the epigenetic community had leaned on the bisulfite conversion method for their genomic methylation studies, there was a lot of concern about the past interpretation of data. As a result, this method has recently been modified in a couple of different ways to address this issue, which will be discussed in the next section, when genome-wide techniques for the analysis of 5mC and 5hmC are examined. Regardless, the scientific community already has evidence revealing that both 5mC and 5hmC are highly overlapping in the genome [40]. Therefore identifying the unique biological function of these two DNA modifications, which is independent of the other modification, will always be difficult.

The "bisulfite revolution" has created additional innovative assays for the study of locus-specific methylation that do not require downstream sequencing. One of these methods is methylation-specific PCR (MS-PCR) [41]. For this particular method two different PCR primer pairs are designed, with one only capable of amplifying bisulfite-converted DNA with no cytosine methylation, while the other is only capable of amplifying this same DNA with cytosine methylation. These two primer pairs share the same exact sequence, except for the last base of their forward primer, which is either a cytosine for the primer measuring methylation and thymine for the primer measuring unmodified-cytosine. Since PCR is highly sensitive to a mismatch in the last base of its primer, designing these forward primers with this difference in their last base will make each highly specific for either unmethylated- or methylated-bisulfite-converted DNA. As with all quantitative PCR methods, the PCR cannot be saturated and needs to be in the linear phase of amplification. Cytosine methylation levels are measured in MS-PCR by comparing the PCR efficiency from the reaction containing the primer pair only able to amplify bisulfite-converted DNA lacking cytosine modification to the reaction containing the primer pair only able to amplify bisulfite-converted DNA containing a methylated cytosine. For instance, if the two linear PCR products are equal in their intensity, after gel visualization, then the interrogated cytosine contains 50% methylation. Yet even though this method appears to be simple and easy to interpret, the actual implementation of this assay can be quite difficult. Determining the exact PCR conditions, so that a quantitative difference can be visualized between the two corresponding PCR products, is extremely difficult. As a result, every unique CpG examined will require a considerable time investment to determine the optimal PCR conditions. Technological advances have led to subsequent improvements of this method, such as methylation-sensitive single nucleotide primer extension (MS-SNuPE) method [42], methylation-sensitive single-strand conformation analysis (MS-SSCA) [43], and high-resolution melting (HRM) analysis [44].

2.4 TECHNOLOGIES FOR GENOME-WIDE ANALYSIS OF 5-METHYLCYTOSINE AND 5-HYDROXYMETHYLCYTOSINE

2.4.1 DNA MICROARRAY APPROACHES

The first implementation of DNA microarrays in an epigenetic study occurred in 1999, when researchers utilized microarrays to quickly screen for hypermethylated-CpG islands in human breast cancer cells [45]. They accomplished this by utilizing a methylation-insensitive restriction enzyme, such as MseI, to digest the DNA into small fragments. Linkers were then ligated to the ends of the digested fragments and the DNA was then split into two separate pools. One of these pools was digested further with a methylation-sensitive restriction enzyme, such as BstUI, while the other pool was left untouched. As a result, only the DNA fragments containing methylation remained intact after digestion with the methylation-sensitive restriction enzyme. DNA methylation of the CpG islands was then easily identified by hybridizing the two pools of DNA on their custom CpG island microarray and by measuring the differential hybridization to the microarray. The microarrays that they originally created were very small, containing only 300 CpG islands on each array, which is a far cry from the current capacity of modern printed DNA microarrays. This has been addressed with recent technological advances in DNA microarray printing, which allows for high-density microarrays containing thousands of unique DNA probes on a single glass slide. Another limitation of this method, which is more difficult to address, arises from the constraints imposed by the restriction enzyme recognition sequence. If any of the DNA fragments treated with BstUI contain methylated cytosine that is not in the context of a BstUI restriction cut site, then these fragments will not be digested, and more importantly they will not be identified as containing methylation. In this original study they were only examining methylation in long stretches of genomic DNA containing repeats of CGs, which perfectly mirrors the BstUI recognition sequence, $CG_\Delta CG$.

DNA microarrays were later adapted so that they could be utilized downstream of chromatin-immunoprecipitation (ChIP) for the genome-wide identification of a specific chromatin-associated factor, an assay commonly referred to as ChIP on chip [46]. In this method, an antibody is utilized to immunoprecipitate a specific chromatin-binding protein, along with its bound and cross-linked genomic DNA fragments. The bound DNA fragments are then removed, collected, and extracted from the immunoprecipitated protein. These DNA fragments are visualized through their hybridization to a DNA microarray containing intergenic DNA probes of interest. Since antibodies specific for 5mC were developed in advance of the ChIP on chip assay, scientists realized they could quickly adapt this method for the identification of DNA fragments containing cytosine methylation. This method, which is commonly referred to as methylated DNA immunoprecipitation on a microarray (MeDIP-chip), utilizes antibodies specific for 5mC to immunoprecipitate DNA fragments containing 5mC, which are then hybridized to a microarray containing DNA probes of interest, such as gene promoters and CpG islands. It is important to note that, unlike the ChIP on chip method which utilized chromatin, the MeDIP method uses extracted DNA that is free of bound proteins. This clean DNA extraction will ensure that the 5mC modification is exposed on the genomic DNA and is available to bind the antibody utilized in the immunoprecipitation. The first application of the MeDIP-chip method was used to identify the methylation patterning at a 80-kb resolution for all of the human chromosomes and for a large set of CpG islands [47]. Naturally, after the discovery of 5hmC in the mammalian genome, researchers quickly adapted this affinity-enrichment microarray technology to work with 5hmC antibodies. This technique is commonly referred to as hydroxymethylated DNA immunoprecipitation on a microarray (hMeDIP-chip). The hMeDIP-chip technology has been used to successfully map 5hmC in the genomes of human brains [48] and carcinogen-treated mice [49,50]. The two major limitations to the affinity microarray assay involve the specificity and reactivity of the antibody or the design of the actual microarray. The specificity and reactivity of the antibody has been addressed in other studies, which have proposed using antibody alternatives for either 5mC or 5hmC affinity enrichment. For example, affinity enrichment of

5mC-containing DNA fragments, for subsequent microarray analyses, has also been performed with methyl-binding domain (MBD) proteins. Interestingly, a study revealed that 5mC antibodies and MBD proteins actually provide different specificities for 5mC enrichment, with the antibody enriching for methylated regions containing low CpG density, while the MBD protein enriched for regions containing higher CpG densities, which resulted in the MBD protein providing the greatest proportion of CpG islands; [51] these observations should merit additional studies. There are also different affinity-enrichment approaches for 5hmC that do not utilize antibodies. One popular method utilizes the T4 β-GT to transfer a modified glucose to 5hmC on genomic DNA. A biotin conjugate is then chemically attached to the modified glucose allowing for the subsequent capture and enrichment of 5hmC DNA fragments with streptavidin beads [52].

The restriction enzyme and affinity-enrichment microarray approach have similar limitations that need to be considered. The first, which is an inherent to DNA microarrays, is that genome representation is dictated by the number and base pair (bp) size of the DNA probes printed on the microarray. In very early microarray-based epigenetic studies, when most microarrays contained less than 1,000 DNA probes, only a small portion of the genome was represented. Fortunately, technological advances have led to a drastic increase in the density of DNA microarrays, with commercially available tiling microarrays now containing all CpG islands and full-length promoters in the human genome. The other limitation to both the restriction enzyme and affinity-enrichment microarray approach concerns their poor genomic resolution of detection. Both of these methods rely on the production of large genomic fragments, around 500 bp or more in size, which are subsequently hybridized to DNA microarrays for the detection of 5mC or 5hmC. Only being able to identify either 5mC or 5hmC within a rather large genomic fragment severely limits the appeal of these microarray approaches in future epigenetic studies. This is especially true when considering recent evidence illustrating that the methylation of a single CpG is capable of dictating gene expression [53,54], thus illustrating the need for base resolution detection of 5mC and 5hmC. This was later addressed when a microarray method was proposed that combined bisulfite conversion of genomic DNA with synthetic oligonucleotide microarrays [55,56]. In this method, genomic DNA is first bisulfite converted and fragmented. The bisulfite-converted DNA fragment is then hybridized to an oligonucleotide microarray containing two identical probes, except for a single base difference, for each CpG that is being interrogated. The single base difference is in the base complementary to the genomic CpG that is being interrogated. So, the microarray probe containing guanine will only recognize the modified cytosine after bisulfite conversion, while the probe containing adenine will only recognize the unmodified cytosine after bisulfite conversion. The strength of this method is that the relative amount of modified and unmodified cytosine can be determined at a base resolution by comparing the hybridization between the two corresponding microarray probes. However, the design of the microarray itself can be very difficult due to the loss of genomic complexity resulting from the bisulfite conversion of genomic DNA. Essentially, after all unmodified cytosine are converted to thymine it is difficult to design microarray probes specific for the region of DNA containing the CpG of interest.

2.4.2 NEXT-GENERATION SEQUENCING APPROACHES

The inception of next-generation sequencing technologies, such as Ion Torrent, SOLiD, and Illumina sequencing, has significantly increased the capabilities and efficiencies of epigenetic genome-wide studies. The benefits of utilizing next-generation sequencing in epigenetic studies are numerous, with the most obvious being that the entire genome can be examined with none of the genomic preconditions or constraints that are applicable to other technologies, such as microarrays. Furthermore, next-generation sequencing provides single nucleotide identification, which allows for highly similar DNA sequences to be easily distinguished. However, it is important to consider the high cost of both the sequencing equipment and reagents and also the necessity for computational experts for the downstream analysis of sequencing data. Also, although sequencing costs have significantly reduced in recent years, bioinformatic analysis still remains a bottleneck. Regardless, both the commercial and academic adoption of next-generation sequencing will continue to rise over the coming years, especially when considering current commercial endeavors that are trying to make sequencing more affordable and easy to interpret. Some of the more popular techniques for mapping 5mC and 5hmC by next-generation sequencing will now be discussed.

One of the first next-generation sequencing methods developed for the study of epigenetic DNA modifications utilized an affinity-enrichment approach for the capture of DNA fragments containing 5mC or 5hmC commonly referred to as either MeDIP-Seq or hMeDIP-Seq, respectively (which stand for *methylated DNA immunoprecipitation sequencing* and *hydroxymethylated DNA immunoprecipitation sequencing*, respectively). The DNA immunoprecipitation part of this DIP-Seq is almost identical to the DNA immunoprecipitation done with microarrays (DIP-chip) detailed in earlier sections. In the DIP-Seq method, genomic DNA is first extracted, sheared (mechanical or enzymatic), and then immunoprecipitated with antibodies against either 5mC or 5hmC. End repair is then performed on the antibody-enriched DNA followed by ligation of sequencing adapters, which are specific to the next-generation sequencing system. Interestingly, the first detailed MeDIP-Seq technique was done in conjunction with a proposed cross-platform Bayesian deconvolution strategy that allowed for the first absolute measurement of DNA methylation levels [39]. This Bayesian strategy could be applied to either MeDIP-chip or -Seq studies and addressed a major limitation of these methods' inability to provide absolute quantification of genomic methylation levels. The first implementation of an hMeDIP-Seq occurred shortly after the discovery of 5hmC in mammalian genomes and its application has been instrumental in illustrating the role of 5hmC during embryonic and germ cell development [13,57−59].

The use of bisulfite sequencing (BS-Seq) for the analysis of 5mC and 5hmC is complex, primarily due to the fact that BS-Seq has taken on many different forms since the discovery of 5hmC in the mammalian genome. For simplicity, the "standard" BS-Seq method will be discussed first. In the BS-Seq method, genomic DNA is first sheared, end repaired, adapters ligated, then the DNA is bisulfite converted, PCR amplified with Taq like polymerases, and sequenced with next-generation sequencing systems. Interpretation of the data is performed by computationally comparing the next-generation sequencing reads to a reference genome. In a single sequencing read, if a reference cytosine in a CpG context reads TpG, then the cytosine is unmethylated and if it reads CpG then it is methylated (Figure 2.3A). When this analysis is expanded across all the sequencing reads for a single interrogated CpG dinucleotide, this will yield the percent methylation of that cytosine across the entire cell population or tissue sample being examined. One of the biggest hurdles with this method, and for that matter all bisulfite-related methods, is that it requires a substantial amount of sequencing depth to acquire both a sensitive and accurate measurement of methylated cytosine. For example, if the percent methylation of a CpG dinucleotide is less than 10%, then probability dictates that out of a total of ten sequencing reads, nine of them will contain an unmodified cytosine, while

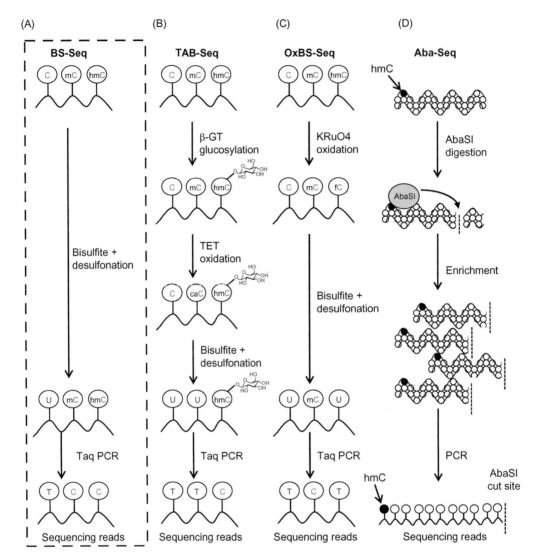

FIGURE 2.3

Genome-wide mapping technologies for 5mC and 5hmC. (A). In "standard" BS-Seq, cytosine modification is revealed genome-wide, but this method cannot specify 5mC and 5hmC. (B). TAB-Seq method utilizes β-GT to first glucosylate 5hmC. Glucosylated 5hmC is then protected from subsequent oxidation by the TET enzyme, unlike 5mC which is oxidized to 5caC. During bisulfite treatment, the glucosylated 5hmC is left unchanged, while the 5caC is chemically converted to uracil. Comparative sequence analysis between standard BS-Seq and TAB-Seq will specify genomic 5mC and 5hmC. (C). OxBS-Seq method utilizes KRuO4 to first selectively oxidize 5hmC to 5fC on genomic DNA. The 5fC will then convert to uracil during subsequent bisulfite conversion of DNA, while 5hmC is left unchanged. Comparative sequence analysis between standard BS-Seq and OxBS will specify genomic 5mC and 5hmC. (D). Aba-Seq method utilizes AbaSI restriction enzyme to digest and enrich for small DNA fragments containing 5hmC. Genomic 5hmC is determined by sequencing AbaSI-digested ends (5hmC is always 9−11 bp from AbaSI cut site). Comparative sequence analysis between standard BS-Seq and Aba-Seq will specify genomic 5mC and 5hmC.

only one will contain a methylated cytosine. As a result, the statistical confidence of this 10% measurement would be very poor, when considering that it was derived from a single positive sequencing read. When considering this and also the fact that the bisulfite conversion reaction is never 100% complete, it becomes apparent that a lot of sequencing reads will need to be obtained. In the scientific community, it has become openly understood that bisulfite sequencing requires a sequencing depth of around 15−20 times coverage for statistically relevant mapping of cytosine methylation. Furthermore, a large portion of mammalian genome is AT rich and does not contain CpG methylation information. Fortunately, there have been variations of the standard BS-Seq method proposed that will greatly increase its sequencing depth, without increasing the actual amount of next-generation sequencing required. One such approach, which has become increasingly adopted by many different research groups, is termed reduced representation bisulfite sequencing (RRBS). The method is based on the size selection of restriction fragments to generate a reduced representation of the genome. In this method a restriction enzyme that is not sensitive to cytosine methylation is used to first digest genomic DNA and then small digested DNA fragments are extracted by size selection, which will enrich for both methylated and unmethylated fractions of the genome. The small DNA fragments are then purified, sequencing adapters ligated, bisulfite converted, PCR amplified (Taq), and sequenced with next-generation sequencing systems. Since this RRBS library only represents a fraction of the entire genome, the sequencing depth will be greatly improved over full genome-wide bisulfite sequencing. The first study to implement the RRBS technique utilized the BglII restriction enzyme, which cuts at $A_\triangle GATCT$ in the genome [60]. The BglII digestion generated around 21,939 small genomic DNA fragments (0.5% of the genome) that were then extracted by a standard size selection method. Four years after this foundational RRBS study, a greatly improved RRBS method was proposed that replaced the BglII restriction enzyme with MspI and also reduced the size selection to 40−220 bp [61]. Since the core sequence of the MspI restriction cut site, $C_\triangle CGG$, is a CpG dinucleotide, utilizing this enzyme in the RRBS method would ensure that every sequence read will contain at least one unmethylated or methylated CpG. Furthermore, by reducing the size selection, the enrichment of DNA fragments containing multiple CpGs would increase. When this modified MspI version of RRBS was applied to the genome of *M. musculus*, ~90% of all CpG islands were covered in the sequencing reads, indicating the strength of this assay in the study of DNA methylation [61].

Soon after the discovery of 5hmC in mammalian genomes, there was a need to modify the standard BS-Seq method, so that it could readily differentiate between 5mC and 5hmC. To date, only two modifications of the bisulfite-sequencing method have been proposed that have addressed this issue and they are commonly referred to as Tet enzyme Assisted Bisulfite sequencing (TAB-Seq) and oxidative Bisulfite Sequencing (OxBS-Seq). Both of these competing bisulfite methods, which are also commercially available, incorporate a selective oxidation step before the bisulfite conversion of genomic DNA. Interestingly, even though these two methods addressed the shortcoming of bisulfite DNA conversion in a similar manner, each have taken completely different approaches to how they selectively oxidize genomic DNA. TAB-Seq utilizes an enzymatic approach, while OxBS-Seq incorporates a chemical approach to selectively oxidize genomic DNA. These methods, which both provide base resolution mapping of genomic 5mC and 5hmC, will be discussed.

The TAB-Seq method was originally revealed in a study by Yu et al. [62,63]. They detailed a comparative method that combined their new technique, TAB-Seq, with standard BS-Seq, which would allow for the mapping of both genomic 5mC and 5hmC at base resolution. This was performed by first dividing their single genomic DNA sample into two separate pools. One of the DNA pools is then subjected to traditional BS-Seq, where the genomic DNA is bisulfite converted, PCR amplified, and then sequenced with next-generation sequencing systems. The second DNA pool is then subjected to their new TAB-Seq method. The TAB-Seq method is carried out by first glucosylating genomic DNA with T4-βGT and then oxidizing it with the TET enzyme. Afterward the glucosylated-oxidized genomic DNA is bisulfite converted, amplified, and sequenced with next-generation sequencing systems. The glucosylation of 5hmC will prevent the oxidation of 5hmC to 5caC by the TET oxidizing enzyme, while 5mC is left unprotected, allowing TET to completely oxidize 5mC to 5caC. Subsequent bisulfite conversion of the oxidized DNA will then convert the 5caC to 5caU, which is replaced with thymine after Taq amplification. After both genomic DNA pools are sequenced and a comparative analysis is carried out with the BS-Seq utilized to map all the modified cytosine, while the Tab-Seq is subsequently utilized to specify the type of modification, 5mC or 5hmC (Figure 2.3B). Even though this comparative sequencing technique provides base resolution identification of both 5mC and 5hmC, there are still concerns about this method. The first is that the TET enzymatic conversion of 5mC to 5Cac has not been completely biochemically characterized. Any 5mC that is not completely converted to 5caC will be recorded as 5hmC in the TAB-Seq assay, so it is imperative that the TET enzyme incorporated into this step is capable of converting 100% of 5mC to 5caC. Furthermore, the TAB-Seq method utilized a truncated version of the TET enzyme that only contains the catalytic domain of this protein, so it could behave very differently from the full-length protein. Also, in the study by Yu et al., they utilized a large amount of truncated TET enzyme ($\sim$13.5 μg enzyme in 50 μL buffer) in their oxidation step, which is indicative of this recombinant enzyme not being catalytically efficient [63]. Another concern about utilizing the TAB-Seq method in a genome-wide study is the extremely high next-generation sequencing load requirement. TAB-Seq only works when performed in conjunction with BS-Seq. Considering that BS-Seq already has a requirement of at least 15 $\times$ sequencing depth, the addition of TAB-Seq will essentially add an additional 15 $\times$ sequencing depth, thus doubling the amount of sequencing required. This sequencing load will require multiple lanes of the most sensitive next-generation sequencing instrument and would likely be cost prohibitive for smaller research labs. One possible solution to reducing the sequencing requirement would be to perform RRBS downstream of the oxidation step. Also, another solution would be to incorporate a traditional locus cloning and sequencing approach downstream of the oxidation-bisulfite conversion step. Yet, even though these proposals would greatly reduce the amount of next-generation sequencing required, and hence decrease costs substantially, the commercially available TAB-Seq assay itself is also fairly expensive, thus making even single locus analysis costly. Regardless of these sequencing hurdles, the TAB-Seq method is quickly becoming adopted in the scientific community, thus illustrating its usefulness in epigenetic research.

The OxBS-Seq method was first detailed in a study by Booth et al. around the same time the competing TAB-Seq method was first proposed [64,65]. While the TAB-Seq technique differentiates 5mC from 5hmC by selectively oxidizing the 5mC before bisulfite conversion, the OxBS-Seq technique makes this differentiation by selectively oxidizing the 5hmC. Like TAB-Seq the OxBS-Seq method is done in conjunction with standard BS-Seq to allow for both the identification of 5mC and 5hmC throughout the genome. This comparative method is carried out by first splitting a single genomic DNA sample into two separate samples, with the first sample undergoing traditional bisulfite conversion followed by sequencing. This first bisulfite converted sample will identify all the modified cytosine in the genome, but it will be unable to specify whether the modification is 5mC from 5hmC. To make this distinction the second DNA sample is chemically oxidized with potassium perruthenate (KRuO4), which will selectively oxidize 5hmC to 5-formylcytosine (5fC). When this chemically oxidized DNA sample is subsequently bisulfite converted, the 5fC will be chemically converted to uracil, which will be replaced by thymine after PCR amplification with Taq. So, comparative analysis of the sequencing from both of the DNA pools, BS-Seq and OxBS-Seq, will map all the 5mC and 5hmC at base resolution in the genome (Figure 2.3C). Limitations and concerns of the xBS-Seq method are very similar to the TAB-Seq method. As has already been discussed with the TAB-Seq method, genome-wide analyses with the OxBS-Seq method require a lot of sequencing depth, which can be cost-prohibitive to many small research groups. A simple solution to this problem is to incorporate either RRBS or locus-specific sequencing into the OxBS-Seq method, thus increase sequencing depth without requiring additional sequencing. Interestingly, the adoption of TAB-Seq in epigenetic genome-wide studies has occurred much quicker than OxBS-Seq. One possibility for this difference could be related to issues arising from the chemical oxidation of genomic DNA. Hopefully, the recent release of a commercially available OxBS-Seq kit, thus standardizing and simplifying this technique, will promote its adoption in the epigenetic research community.

It has been emphasized in this section that Tab-Seq and OxBS-Seq both require much higher sequencing depth when performing genome-wide 5mC and 5hmC analyses. Recently, an additional method, Aba-Seq (AbaSI enzyme—assisted sequencing), has been proposed that requires substantially less sequencing depth than either of these methods, when combined with BS-Seq for the mapping of 5mC and 5hmC. The Aba-Seq method, which was first detailed in a study by Sun et al., utilizes the AbaSI restriction enzyme, which is a member of the PvuRTS1I family of restriction enzymes that specifically cleaves glucosylated 5hmC [40]. In the proposed method, genomic 5hmC is first glucosylated with β-GT, DNA then digested with the AbaSI restriction enzyme, ligated to biotin-tagged sequencing adapters, sheared to 300−500 bp, extracted with streptavidin beads, end repaired/ligated to another sequencing adapter, and lastly the DNA is amplified and sequenced. Next-generation sequencing of this DNA library will reveal all the AbaSI restriction cut sites, thus providing 5hmC mapping at a very high resolution in the genome (Figure 2.3D). Furthermore, it was shown that the sequencing read number for each strand of DNA, created by AbaSI digestion, strongly correlated to the amount of 5hmC on each strand. Also, unlike the TAB-Seq and OxBS-Seq methods, the Aba-Seq enzymatic method inherently enriches for genomic DNA containing 5hmC. This enrichment combined with the fact that all sequenced DNA contains 5hmC allows for a much lower sequencing depth requirement than either TAB-Seq or OxBS-Seq. As a proof of principle, the Aba-Seq method was utilized to map 5hmC in mouse embryonic stem cell DNA. This study showed that Aba-Seq method could completely map genomic 5hmC at high resolution with less genomic DNA input and with far less next-generation sequencing than any other method that has been proposed for 5hmC mapping. The results of this study indicate that the Aba-Seq method is a viable alternative that can be used in conjunction with BS-Seq to completely map 5mC and 5hmC in genomic DNA.

Another emerging technology for epigenetic modification detection is single molecule real-time sequencing (SMRT), which is based on parallelized single molecule DNA sequencing by synthesis technology [66]. Here DNA sequencing is performed on a chip containing many zero-mode waveguides (ZMWs). Each ZMW contains a single active DNA polymerase with a single immobilized molecule of single-strand DNA through which light can penetrate and create a visualization chamber, thus allowing for the monitoring of DNA polymerization at a single molecule level. Signal from a phospho-linked nucleotide incorporated by the DNA polymerase is detected as DNA synthesis occurs, resulting in DNA sequencing being recorded in real time. Depending on the epigenetic modification different signals will be obtained that can be subsequently analyzed to determine sequence identity and nucleotide modification. This SMRT technology has shown promise in detection of N6-methyladenine in bacterial genome. However, in SMRT 5mC signals are weak and as a result hard to differentiate from unmodified cytosine, making the specificity of C and 5mC difficult [67]. However, further development of SMRT technology and software may solve this problem in the future.

2.5 CONCLUSIONS

There have been many different methods created over the last half century for the study of either, or both, genomic 5mC or 5hmC (Figure 2.4). Illustrating the strength and versatility of these

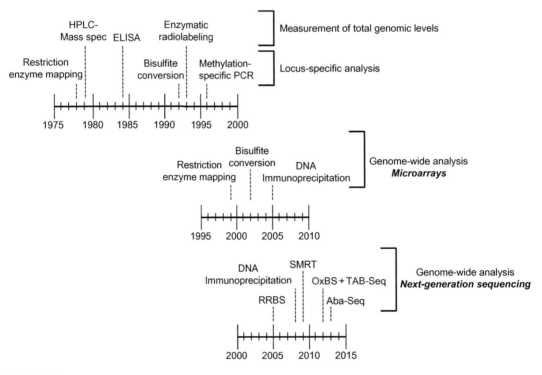

FIGURE 2.4

Timeline of epigenetic technologies used in the study of 5mC or 5hmC. The original implementation date for each of the discussed epigenetic technologies has been detailed on a timeline. Also, these technologies have been subdivided into the following application categories: Measurement of total genomic levels, Locus-specific analysis, Genome-wide analysis (Microarrays), and Genome-wide analysis (Next-generation sequencing).

methods, most if not all are still being incorporated into modern epigenetic studies. This has been instrumental to the progress of epigenetic research by granting research labs many different choices of methods for the study of 5mC or 5hmC, which has allowed for the most efficient and cost effective studies to be carried out. Some of these methods, such as ELISAs or HPLC-mass spectrometry techniques, were utilized to provide highly sensitive and reproducible measurements of genomic 5mC or 5hmC. Other methods, such as methyl- and hydroxymethyl sensitive restriction enzymes or bisulfite conversion followed by locus-specific sequencing, have been utilized to identify and quantitate locus-specific 5mC or 5hmC. Lastly, methods such as TAB-Seq, OxBS-Seq, and Aba-Seq have been developed that can take advantage of next-generation sequencing to provide mapping of genome-wide 5mC or 5hmC at very high resolutions. It is the combination of all these technological advances, both past and present, which will provide the foundation and fuel the epigenetic revolution as we move forward. Furthermore, it is highly probable that new and unimaginable epigenetic tools will be proposed in the future that might completely change our perspective on mammalian epigenetic DNA modification.

ACKNOWLEDGMENTS

We would like to thank Drs. Pierre-Olivier Estève and Guoqiang Zhang for constructive comments and editing. We are grateful to Drs. Donald Comb and Rich Roberts at New England Biolabs Inc. for research support and encouragement.

REFERENCES

[1] Ziller MJ, Gu H, Muller F, Donaghey J, Tsai LT, Kohlbacher O, et al. Charting a dynamic DNA methylation landscape of the human genome. Nature 2013;500(7463):477−81.

[2] Holliday R, Pugh JE. DNA modification mechanisms and gene activity during development. Science 1975;187(4173):226−32.

[3] Riggs AD. X inactivation, differentiation, and DNA methylation. Cytogenet Cell Genet 1975;14(1):9−25.

[4] Schoofs T, Berdel WE, Muller-Tidow C. Origins of aberrant DNA methylation in acute myeloid leukemia. Leukemia 2014;28(1):1−14.

[5] Bernt KM, Armstrong SA. Targeting epigenetic programs in MLL-rearranged leukemias. Hematology Am Soc Hematol Educ Program 2011;2011:354−60.

[6] Kriaucionis S, Heintz N. The nuclear DNA base 5-hydroxymethylcytosine is present in Purkinje neurons and the brain. Science 2009;324(5929):929−30.

[7] Burdzy A, Noyes KT, Valinluck V, Sowers LC. Synthesis of stable-isotope enriched 5-methylpyrimidines and their use as probes of base reactivity in DNA. Nucleic Acids Res 2002;30(18):4068−74.

[8] Valinluck V, Tsai HH, Rogstad DK, Burdzy A, Bird A, Sowers LC. Oxidative damage to methyl-CpG sequences inhibits the binding of the methyl-CpG binding domain (MBD) of methyl-CpG binding protein 2 (MeCP2). Nucleic Acids Res 2004;32(14):4100−8.

[9] Tahiliani M, Koh KP, Shen Y, Pastor WA, Bandukwala H, Brudno Y, et al. Conversion of 5-methylcytosine to 5-hydroxymethylcytosine in mammalian DNA by MLL partner TET1. Science 2009;324(5929):930−5.

[10] He YF, Li BZ, Li Z, Liu P, Wang Y, Tang Q, et al. Tet-mediated formation of 5-carboxylcytosine and its excision by TDG in mammalian DNA. Science 2011;333(6047):1303−7.

[11] Ito S, Shen L, Dai Q, Wu SC, Collins LB, Swenberg JA, et al. Tet proteins can convert 5-methylcytosine to 5-formylcytosine and 5-carboxylcytosine. Science 2011;333(6047):1300−3.

[12] Williams K, Christensen J, Pedersen MT, Johansen JV, Cloos PA, Rappsilber J, et al. TET1 and hydroxymethylcytosine in transcription and DNA methylation fidelity. Nature 2011;473(7347):343−8.

[13] Wu H, D'Alessio AC, Ito S, Xia K, Wang Z, Cui K, et al. Dual functions of Tet1 in transcriptional regulation in mouse embryonic stem cells. Nature 2011;473(7347):389−93.

[14] Hotchkiss RD. The quantitative separation of purines, pyrimidines, and nucleosides by paper chromatography. J Biol Chem 1948;175(1):315−32.

[15] Ehrlich M, Ehrlich K. Separation of six DNA bases by ion pair − reversed phase high pressure liquid chromatography. J Chromatogr Sci 1979;17(9):531−4.

[16] Kuo KC, McCune RA, Gehrke CW, Midgett R, Ehrlich M. Quantitative reversed-phase high performance liquid chromatographic determination of major and modified deoxyribonucleosides in DNA. Nucleic Acids Res 1980;8(20):4763−76.

[17] Singer J, Schnutc Jr. WC, Shively JE, Todd CW, Riggs AD. Sensitive detection of 5-methylcytosine and quantitation of the 5-methylcytosine/cytosine ratio in DNA by gas chromatography − mass spectrometry using multiple specific ion monitoring. Anal Biochem 1979;94(2):297−301.

[18] Le T, Kim KP, Fan G, Faull KF. A sensitive mass spectrometry method for simultaneous quantification of DNA methylation and hydroxymethylation levels in biological samples. Anal Biochem 2011;412(2):203−9.

[19] Achwal CW, Chandra HS. A sensitive immunochemical method for detecting 5mC in DNA fragments. FEBS Lett 1982;150(2):469−72.

[20] Achwal CW, Ganguly P, Chandra HS. Estimation of the amount of 5-methylcytosine in *Drosophila melanogaster* DNA by amplified ELISA and photoacoustic spectroscopy. EMBO J 1984;3(2):263−6.

[21] Adouard V, Dante R, Niveleau A, Delain E, Revet B, Ehrlich M. The accessibility of 5-methylcytosine to specific antibodies in double-stranded DNA of *Xanthomonas* phage XP12. Eur J Biochem 1985;152(1):115−21.

[22] Jin SG, Kadam S, Pfeifer GP. Examination of the specificity of DNA methylation profiling techniques towards 5-methylcytosine and 5-hydroxymethylcytosine. Nucleic Acids Res 2010;38(11):e125.

[23] Wu J, Issa JP, Herman J, Bassett Jr. DE, Nelkin BD, Baylin SB. Expression of an exogenous eukaryotic DNA methyltransferase gene induces transformation of NIH 3T3 cells. Proc Natl Acad Sci USA 1993;90(19):8891−5.

[24] Balaghi M, Wagner C. DNA methylation in folate deficiency: use of CpG methylase. Biochem Biophys Res Commun 1993;193(3):1184−90.

[25] Terragni J, Bitinaite J, Zheng Y, Pradhan S. Biochemical characterization of recombinant beta-glucosyltransferase and analysis of global 5-hydroxymethylcytosine in unique genomes. Biochemistry 2012;51(5):1009−19.

[26] Szwagierczak A, Bultmann S, Schmidt CS, Spada F, Leonhardt H. Sensitive enzymatic quantification of 5-hydroxymethylcytosine in genomic DNA. Nucleic Acids Res 2010;38(19):e181.

[27] Waalwijk C, Flavell RA. MspI, an isoschizomer of hpaII which cleaves both unmethylated and methylated hpaII sites. Nucleic Acids Res 1978;5(9):3231−6.

[28] Cedar H, Solage A, Glaser G, Razin A. Direct detection of methylated cytosine in DNA by use of the restriction enzyme MspI. Nucleic Acids Res 1979;6(6):2125−32.

[29] Singer-Sam J, Grant M, LeBon JM, Okuyama K, Chapman V, Monk M, et al. Use of a HpaII-polymerase chain reaction assay to study DNA methylation in the Pgk-1 CpG island of mouse embryos at the time of X-chromosome inactivation. Mol Cell Biol 1990;10(9):4987−9.

[30] Tardy-Planechaud S, Fujimoto J, Lin SS, Sowers LC. Solid phase synthesis and restriction endonuclease cleavage of oligodeoxynucleotides containing 5-(hydroxymethyl)-cytosine. Nucleic Acids Res 1997;25 (3):553−9.

[31] Kinney SM, Chin HG, Vaisvila R, Bitinaite J, Zheng Y, Esteve PO, et al. Tissue-specific distribution and dynamic changes of 5-hydroxymethylcytosine in mammalian genomes. J Biol Chem 2011;286 (28):24685−93.

[32] Tsumagari K, Baribault C, Terragni J, Varley KE, Gertz J, Pradhan S, et al. Early de novo DNA methylation and prolonged demethylation in the muscle lineage. Epigenetics 2013;8(3):317−32.

[33] Tsumagari K, Baribault C, Terragni J, Chandra S, Renshaw C, Sun Z, et al. DNA methylation and differentiation: HOX genes in muscle cells. Epigenetics Chromatin 2013;6(1):25.

[34] Clark SJ, Harrison J, Paul CL, Frommer M. High sensitivity mapping of methylated cytosines. Nucleic Acids Res 1994;22(15):2990−7.

[35] Frommer M, McDonald LE, Millar DS, Collis CM, Watt F, Grigg GW, et al. A genomic sequencing protocol that yields a positive display of 5-methylcytosine residues in individual DNA strands. Proc Natl Acad Sci U S A 1992;89(5):1827−31.

[36] Uhlmann K, Brinckmann A, Toliat MR, Ritter H, Nurnberg P. Evaluation of a potential epigenetic biomarker by quantitative methyl-single nucleotide polymorphism analysis. Electrophoresis 2002;23 (24):4072−9.

[37] Dupont JM, Tost J, Jammes H, Gut IG. De novo quantitative bisulfite sequencing using the pyrosequencing technology. Anal Biochem 2004;333(1):119−27.

[38] Anderson AE, Haines TR, Robinson DP, Butcher DT, Sadikovic B, Rodenhiser DI. Sodium bisulfite analysis of the methylation status of DNA from small portions of paraffin slides. Biotechniques 2001;31 (5): 1004, 1006, 1008.

[39] Down TA, Rakyan VK, Turner DJ, Flicek P, Li H, Kulesha E, et al. A Bayesian deconvolution strategy for immunoprecipitation-based DNA methylome analysis. Nat Biotechnol 2008;26(7):779−85.

[40] Sun Z, Terragni J, Borgaro JG, Liu Y, Yu L, Guan S, et al. High-resolution enzymatic mapping of genomic 5-hydroxymethylcytosine in mouse embryonic stem cells. Cell Rep 2013;3(2):567−76.

[41] Herman JG, Graff JR, Myohanen S, Nelkin BD, Baylin SB. Methylation-specific PCR: a novel PCR assay for methylation status of CpG islands. Proc Natl Acad Sci U S A 1996;93(18):9821−6.

[42] Gonzalgo ML, Jones PA. Rapid quantitation of methylation differences at specific sites using methylation-sensitive single nucleotide primer extension (Ms-SNuPE). Nucleic Acids Res 1997;25 (12):2529−31.

[43] Bianco T, Hussey D, Dobrovic A. Methylation-sensitive, single-strand conformation analysis (MS-SSCA): a rapid method to screen for and analyze methylation. Hum Mutat 1999;14(4):289−93.

[44] Wojdacz TK, Dobrovic A. Methylation-sensitive high resolution melting (MS-HRM): a new approach for sensitive and high-throughput assessment of methylation. Nucleic Acids Res 2007;35(6):e41.

[45] Huang TH, Perry MR, Laux DE. Methylation profiling of CpG islands in human breast cancer cells. Hum Mol Genet 1999;8(3):459−70.

[46] Ren B, Robert F, Wyrick JJ, Aparicio O, Jennings EG, Simon I, et al. Genome-wide location and function of DNA binding proteins. Science 2000;290(5500):2306−9.

[47] Weber M, Davies JJ, Wittig D, Oakeley EJ, Haase M, Lam WL, et al. Chromosome-wide and promoter-specific analyses identify sites of differential DNA methylation in normal and transformed human cells. Nat Genet 2005;37(8):853−62.

[48] Jin SG, Wu X, Li AX, Pfeifer GP. Genomic mapping of 5-hydroxymethylcytosine in the human brain. Nucleic Acids Res 2011;39(12):5015−24.

[49] Thomson JP, Lempiainen H, Hackett JA, Nestor CE, Muller A, Bolognani F, et al. Non-genotoxic carcinogen exposure induces defined changes in the 5-hydroxymethylome. Genome Biol 2012;13(10):R93.

[50] Thomson JP, Hunter JM, Lempiainen H, Muller A, Terranova R, Moggs JG, et al. Dynamic changes in 5-hydroxymethylation signatures underpin early and late events in drug exposed liver. Nucleic Acids Res 2013;41(11):5639−54.

[51] Nair SS, Coolen MW, Stirzaker C, Song JZ, Statham AL, Strbenac D, et al. Comparison of methyl-DNA immunoprecipitation (MeDIP) and methyl-CpG binding domain (MBD) protein capture for genome-wide DNA methylation analysis reveal CpG sequence coverage bias. Epigenetics 2011;6(1):34−44.

[52] Song CX, Szulwach KE, Fu Y, Dai Q, Yi C, Li X, et al. Selective chemical labeling reveals the genome-wide distribution of 5-hydroxymethylcytosine. Nat Biotechnol 2011;29(1):68−72.

[53] Murayama A, Sakura K, Nakama M, Yasuzawa-Tanaka K, Fujita E, Tateishi Y, et al. A specific CpG site demethylation in the human interleukin 2 gene promoter is an epigenetic memory. EMBO J 2006;25 (5):1081−92.

[54] Nile CJ, Read RC, Akil M, Duff GW, Wilson AG. Methylation status of a single CpG site in the IL6 promoter is related to IL6 messenger RNA levels and rheumatoid arthritis. Arthritis Rheum 2008;58 (9):2686−93.

[55] Adorjan P, Distler J, Lipscher E, Model F, Muller J, Pelet C, et al. Tumour class prediction and discovery by microarray-based DNA methylation analysis. Nucleic Acids Res 2002;30(5):e21.

[56] Gitan RS, Shi H, Chen CM, Yan PS, Huang TH. Methylation-specific oligonucleotide microarray: a new potential for high-throughput methylation analysis. Genome Res 2002;12(1):158−64.

[57] Hackett JA, Sengupta R, Zylicz JJ, Murakami K, Lee C, Down TA, et al. Germline DNA demethylation dynamics and imprint erasure through 5-hydroxymethylcytosine. Science 2013;339(6118):448−52.

[58] Ficz G, Branco MR, Seisenberger S, Santos F, Krueger F, Hore TA, et al. Dynamic regulation of 5-hydroxymethylcytosine in mouse ES cells and during differentiation. Nature 2011;473(7347):398−402.

[59] Pastor WA, Pape UJ, Huang Y, Henderson HR, Lister R, Ko M, et al. Genome-wide mapping of 5-hydroxymethylcytosine in embryonic stem cells. Nature 2011;473(7347):394−7.

[60] Meissner A, Gnirke A, Bell GW, Ramsahoye B, Lander ES, Jaenisch R. Reduced representation bisulfite sequencing for comparative high-resolution DNA methylation analysis. Nucleic Acids Res 2005;33 (18):5868−77.

[61] Smith ZD, Gu H, Bock C, Gnirke A, Meissner A. High-throughput bisulfite sequencing in mammalian genomes. Methods 2009;48(3):226−32.

[62] Yu M, Hon GC, Szulwach KE, Song CX, Jin P, Ren B, et al. Tet-assisted bisulfite sequencing of 5-hydroxymethylcytosine. Nat Protoc 2012;7(12):2159−70.

[63] Yu M, Hon GC, Szulwach KE, Song CX, Zhang L, Kim A, et al. Base-resolution analysis of 5-hydroxymethylcytosine in the mammalian genome. Cell 2012;149(6):1368−80.

[64] Booth MJ, Ost TW, Beraldi D, Bell NM, Branco MR, Reik W, et al. Oxidative bisulfite sequencing of 5-methylcytosine and 5-hydroxymethylcytosine. Nat Protoc 2013;8(10):1841−51.

[65] Booth MJ, Branco MR, Ficz G, Oxley D, Krueger F, Reik W, et al. Quantitative sequencing of 5-methylcytosine and 5-hydroxymethylcytosine at single-base resolution. Science 2012;336(6083):934−7.

[66] Eid J, Fehr A, Gray J, Luong K, Lyle J, Otto G, et al. Real-time DNA sequencing from single polymerase molecules. Science 2009;323(5910):133−8.

[67] Flusberg BA, Webster DR, Lee JH, Travers KJ, Olivares EC, Clark TA, et al. Direct detection of DNA methylation during single-molecule, real-time sequencing. Nat Methods 2010;7(6):461−5.

HIGH-THROUGHPUT SEQUENCING-BASED MAPPING OF CYTOSINE MODIFICATIONS

3

Alison I. Bernstein and Peng Jin

Department of Human Genetics, Emory University, Atlanta, GA, USA

CHAPTER OUTLINE

3.1 INTRODUCTION

The epigenetic landscape was originally defined by Conrad Waddington in 1939 as events that led to changes in developmental genetics programs [1]. The modern definition of epigenetics includes

Y.G. Zheng (Ed): Epigenetic Technological Applications. DOI: http://dx.doi.org/10.1016/B978-0-12-801080-8.00003-X

all meiotically and mitotically heritable changes in gene expression that are not coded in the DNA sequence itself and alter phenotype without changing genotype [2]. Initiation and maintenance of epigenetic regulation are modulated by three mechanisms: cytosine modifications, histone modifications, and noncoding RNA (ncRNA)-mediated processes, which are critical to mammalian development, specifically in stem cell self-renewal and differentiation and neurodevelopment [3−7]. While the concept of epigenetics was articulated 75 years ago, recent technological advances have led to a renewed interest and new discoveries in this field.

The role of epigenetic mechanisms has long been recognized in cancer. However, disturbance of epigenetic systems results in an array of multisystem disorders, including immunodeficiency, centromeric region instability, facial anomalies syndrome, Rett syndrome, and autism [8−15]. Also, several inherited syndromes are due to faulty genomic imprinting (parent-specific, monoallelic expression of a gene) [16,17]. It is also thought that epigenetic mechanisms, and in particular cytosine modifications, are especially important for development and function of the central nervous system (CNS) [18−23]. Consequently, dysregulation of appropriate patterns of cytosine modifications may contribute to neurodevelopmental and neurodegenerative disorders. In the following sections, we briefly summarize the known cytosine modifications and discuss the high-throughput sequencing-based techniques currently available for detecting and mapping 5mC, 5hmC, 5fC, and 5caC within the genome.

3.2 CYTOSINE MODIFICATIONS

3.2.1 5-METHYLCYTOSINE

5-methylcytosine (5mC), methylation of the fifth position of cytosine, is the most-studied cytosine modification. The methylation of cytosine is catalyzed by DNA methyltransferases (DNMTs), which fall into two groups: (1) maintenance DNMTs, which maintain cytosine methylation during DNA replications; and (2) *de novo* DNMTs, which methylate DNA during development with help of accessory proteins such as UHRF1 and DNMT3L [24−33]. Proteins such as methyl-CpG binding protein 2 (MeCP2) and methyl-CpG-binding domain proteins 1−4 (MBD1-4) recognize 5mC and bind to methylated DNA [14,34−36].

5mC exists symmetrically at CpG dinucleotides and regulates gene transcription [6]. An exception to this is CpG islands, which tend to be located near promoters and are usually unmethylated. Methylation is generally associated with transcriptional repression and has been demonstrated to play an important role in transcriptional regulation of gene expression, chromatin structure, gene imprinting, X-chromosome inactivation, and genomic stability [4,37,38]. Dysregulation of proper DNA methylation patterns has been associated with many diseases, including neurological diseases and cancers [39−44].

3.2.2 5-HYDROXYMETHYLCYTOSINE

5mC has historically been viewed as a stable covalent modification to DNA; however, 5mC can be enzymatically modified to 5-hydroxymethylcytosine (5hmC) by Tet family proteins through Fe(II) α-KG-dependent hydroxylation and is recognized by its own set of binding proteins [45−47].

Although 5hmC was identified many decades ago [48,49], its importance was not recognized until the past 5 years. In 2009, two groups described the presence of 5hmC in mouse Purkinje neurons and embryonic stem cells [46,50]. This sparked extensive study of 5hmC in the CNS, where the dynamic regulation of 5hmC is critical [51].

5hmC is important in the process of demethylation as it is generated from 5mC and subsequently converted to 5fC and 5caC by Tet enzymes. 5hmC mediates demethylation by interfering with the binding of DNA and proteins responsible for DNA methylation maintenance [52−55]. 5hmC is also important for DNA-repair-based DNA methylation. Tet enzymes further oxidize 5hmC to 5fC and 5caC, which can be excised by thymine-DNA glycosylase (TDG), triggering DNA base excision repair to generate an unmethylated cytosine [56−59]. Alternately, deamination and removal of the uracil can also trigger base excision repair [60]. Finally, DNMT3A and DNMT3B may also have dehydroxymethylation function, in addition to its well-studied methylation function [61].

In addition to its roles in demethylation, it is thought that 5hmC, like 5mC, acts as a distinct and stable epigenetic mark with its own set of 5hmC-binding protein [47,54,62,63]. As we have recently reviewed, this seems to be particularly important in the CNS [43]. 5hmC is highly enriched in the CNS, with levels approximately ten-fold higher than in embryonic stem cells [19]. We have also reported that 5hmC is acquired by mouse neuronal cells throughout development, from the postnatal stage through adulthood [20]. In addition, we have identified conserved patterns of 5hmC in the mouse cerebellum and hippocampus and found that 5hmC is depleted along the X chromosome, but accumulated at intragenic and exon-enriched loci [20]. We have also identified fetal- and adult-specific differentially hydroxymethylated loci (DhMLs) in human cerebellum [22]. There is also an age-dependent enrichment of intragenic 5hmC in genes linked to neurodegenerative disease [20,21]. A recent study also found that the distribution of 5mC and 5hmC and interactions with their respective binding proteins changes throughout development parallel to the processes of neuronal differentiation [23]. Based on this data, it is thought that the patterns of cytosine modifications acquired throughout neuronal development are critical for proper neurodevelopment and neurological function in the adult brain. In fact, recent studies have implicated epigenetic regulation in both Alzheimer's and Huntington's diseases [7,44,64]. Together, these observations strongly suggest that 5hmC plays a critical role in brain development and related diseases [43].

3.2.3 5-FORMYLCYTOSINE AND 5-CARBOXYLCYTOSINE

Very recently, two additional cytosine modifications were discovered, 5-formylcytosine (5fC) and 5-carboxylcytosine (5caC) [56,57,59]. These are generated by further oxidation of 5hmC by the Tet enzymes and are removed from the genome by TDG, initiating base excision repair and resulting in demethylation. It is thought that 5fC and 5caC are dynamic regulators of distinct categories of functional regulatory elements in the mammalian genome [65]. Specifically, 5fC appears to be important at low methylated regions and poised enhancers, where it functions in the epigenetic priming of regulatory elements.

Very little is known about the roles of these two modifications because their abundance is extremely low and techniques to enrich and map them are very new [57]. Our group and others have recently developed techniques to map 5fC and 5caC [65,66]. These initial studies revealed that they have specific patterns in the genome, suggesting that active DNA methylation may play a larger role in mammalian cells than previously thought by dynamically regulating the epigenetic state at function regulatory elements. Specifically, in mouse embryonic stem cells, 5fC is preferentially located at enhancers, with a preference for poised enhancers compared to active enhancers, suggesting a role in epigenetic priming [65]. Furthermore, active oxidation of 5mC/5hmC to 5fC and 5caC, along with TDG-coupled base excision repair, dynamically regulates the epigenetic state at regulatory elements [65].

3.3 HIGH-THROUGHPUT SEQUENCING METHODS FOR DETECTING CYTOSINE MODIFICATIONS

Techniques for mapping these modifications can be roughly categorized as genome-wide or single-base resolution. In genome-wide techniques, regions of the genome enriched for the modification of interest are identified. In contrast, single-base resolution methods define the specific modified cytosines in the genome. Single-base resolution techniques can be carried out on the entire genome or on targeted regions of the genome when combined with capture techniques or oligo-based enrichment of DNA regions of interest. Below we describe genome-wide methods, followed by single-base resolution methods, for detecting and mapping 5mC, 5hmC, 5fC, and 5caC.

3.3.1 GENOME-WIDE MAPPING OF CYTOSINE MODIFICATIONS

All of the genome-wide mapping techniques involve capturing DNA containing the modification of interest to create a library of DNA enriched for that modification. The enriched DNA is then used for high-throughput sequencing to sequence all DNA containing that modification. Bioinformatics packages are then used to align these reads to the reference genome and determine the regions that are enriched for the modification of interest by comparing to an unenriched library. Capture methods include immuno-affinity-based techniques, protein-affinity-based capture, and enzymatic modification or chemical labeling, and can be used in isolation or combination. Here, we describe the enrichment techniques that have been developed to create enriched DNA libraries for downstream analysis by high-throughput sequencing. These are summarized in Table 3.1 with the preferred analysis method for each modification in bold.

Table 3.1 Techniques for Genome-Wide Mapping of Cytosine Modifications

Protocols for detecting cytosine modifications are listed with the preferred methods indicated in bold

	DIP	**Protein Affinity**	**Chemical Labeling**
5mC	**MeDIP**	MBD	n/a
5hmC	hMeDIP	JBP1	**hMe-Seal**
5fC	fDIP	n/a	**fC-Seal**
			oxyamine
5caC	caDIP	n/a	n/a

3.3.1.1 DNA immunoprecipitation

In DNA immunoprecipitation (DIP), antibodies specific to 5mC, 5hmC, 5fC, or 5caC are utilized to enrich a single-stranded genomic DNA library for the cytosine modification of interest via immunoprecipitation [45,66,67]. This enriched library is then sequenced by high-throughput sequencing. Methylated DNA IP (MeDIP) has been extremely useful for genome-wide mapping of 5mC and has been reviewed previously by Laird et al. [67–69]. MeDIP is sensitive for the detection of differentially methylated regions and has been shown to enrich for regions with a low CpG density [70].

While this pure MeDIP technique has been useful for mapping 5mC, similar techniques for the 5hmC enrichment (hMeDIP) have been less successful as the antibodies lack the specificity to adequately discriminate DNA strands containing the modification from those without the modification in an immunoprecipitation protocol [45,71]. Later attempts were more successful; however, there is a bias in the enrichment of CpG-containing regions [72].

Attempts at using DIP for enrichment of 5fC- and 5caC-containing DNA have been slightly more successful [66]. Under the conditions utilized by Shen et al., the antibodies against both 5fC and 5caC were able to specifically enrich for DNA containing these modifications. However, as will all antibody-based immunoprecipitation techniques, this performs well in areas of dense modifications but poorly in areas with sparse modifications [66,73]. This is especially a concern for 5fC and 5caC, which occur at a very low frequency throughout the genome. Thus, these immuno-affinity-based techniques have proven more useful for 5mC than the other modifications and other techniques have been developed that provide better enrichment of 5hmC-, 5fC-, and 5caC-containing DNA.

3.3.1.2 Protein-based capture

An alternative to immuno-affinity-based capture of DNA is the use of beads coated with proteins that bind to modified DNA, such as methyl-CpG binding domain (MBD) proteins for 5mC enrichment [70,74–79] or J-binding protein 1 (JBP1) for 5hmC enrichment [72,80]. In these protocols, double-stranded DNA fragments are captured with beads coated with the DNA binding protein of interest to create an enriched library for downstream sequencing.

3.3.1.2.1 MBD capture

Enrichment of 5mC-containing DNA can also be carried out by affinity-based capture of MDB proteins [70,74–79]. In this strategy, double-stranded methylated DNA fragments are captured with MBD-coated beads. This enriched library can then be sequenced or separated by salt fractionation prior to sequencing. Salt fractionation separates the DNA into fractions based on the level of methylation: higher salt fractions contain more highly methylated DNA and lower salt fractions contain less methylated DNA [74]. A comparison of MBD capture with MeDIP found that both techniques are sensitive for the detection of methylated regions of DNA [70]. In contrast to MeDIP, which is biased toward areas of low CpG density, MBD capture is biased toward regions of higher CpG density and identifies a larger proportion of CpG islands.

3.3.1.2.2 JBP1 capture

A similar protein-based capture method has been developed for the enrichment of 5hmC-containing sequences; it combines an enzymatic modification with protein-based capture [80]. The first step of this technique is the glucosylation of 5hmC by β-glucosyltransferase (β-GT) to produce β-glucosyl-5hmC. JBP1 can specifically recognize this glucosylated 5hmC and can thus be used to specifically enrich a double-stranded DNA library for 5hmC-containing DNA. This technique has two levels of specificity in that β-GT only modifies on 5hmC and JBP1 only interacts with modified 5hmC. However, a comparison of JBP1-based capture, hMeDIP, and hMe-Seal found that JBP-1-based capture performs poorly in the generation of genome-wide 5hmC profiles compared to hMeDIP and hMe-Seal (described next) [72].

3.3.1.3 Chemical-labeling-based capture

A third class of techniques for genome-wide mapping of cytosine modifications are the chemical labeling approaches. These techniques have a higher specificity than the antibody and protein-based enrichment protocols and can overcome some of the enrichment biases of the other methods [72].

3.3.1.3.1 hMe-seal

Our group developed a technique for the selective chemical labeling of 5hmC [21]. The first step of this technique, like the JBP1-based capture approach, involves glucosylation of 5hmC by β-GT to form β-glucosyl-5hmC using chemically modified UDP-glucose as a co-factor to add an azide group. This azide group is then labeled with a biotin moiety using click chemistry, allowing for affinity purification and downstream sequencing. The combination of covalent chemical labeling and biotin-based affinity purification provides many advantages over noncovalent, antibody or protein-based enrichment since it ensures the accurate and comprehensive capture of 5hmC-containing DNA fragments and provides high selectivity. In the study discussed previously that compared hMe-Seal, hMeDIP, and JBP1-based capture, hMeSeal was shown to be the preferred method for genome-wide mapping of 5hmC due to the increased specificity and reduced bias for repetitive genomic regions [72].

3.3.1.3.2 Oxyamine-based 5fC enrichment

This approach involves an oxyamine-mediated reaction to covalently attach biotin to 5fC in a genomic DNA library [71]. This chemical labeling is carried out using a hydroxylamine aldehyde reactive probe to attach biotin to 5fC. Specificity for 5fC arises from the fact that formyl groups react selectively with hydroxylamines to produce stable oxime derivatives [81]. Pull down of biotinylated DNA is then carried out to collect 5fC-containing DNA [73]. A drawback of this labeling technique is that it captures a high percentage of nonspecific DNA, possibly through the capture of abasic sites, leading to high background noise [65]. This greatly limits its utility as a pull-down based approach.

3.3.1.3.3 fC-seal

Our group developed the 5fC-selective chemical labeling (fC-Seal) technique [65]. This technique takes advantage of the fact that 5fC can be selectively reduced to 5hmC by sodium borohydride (NaBH$_4$) [82]. The first step of this technique is to block 5hmC in a genomic DNA library with unmodified glucose using β-GT. 5fC is then reduced with NaBH4 to 5hmC. The hMe-Seal chemical labeling approach can then be utilized to capture only the newly generated 5hmC [21]. This method is specific for 5fC since the glucose-protected 5hmC, as well as 5mC and 5caC, is not reduced by $NaBH_4$ under the conditions used. Additionally, in comparison to the hydroxylamine-based method, this technique captures significantly less nonspecific DNA; the hydroxylamine-method pulls down 16% of all genomic DNA, while fC-Seal pulls down only 0.08%, an amount much more in line with estimates of 5fC content of the genome [65].

3.3.2 SINGLE-BASE RESOLUTION METHODS OF DETECTING CYTOSINE MODIFICATIONS

A major limitation of all the genome-wide techniques is that the resolution (several hundred bases to larger than 1 kb) of the maps is limited by the size of the enriched DNA fragments. In contrast, base-resolution methods identify the specific cytosines across the whole genomes that are modified. In addition, these base-resolution methods are quantitative, whereas genome-wide methods are not. The methods described below are all variations on bisulfite sequencing (BS-Seq) and must all be compared to the results of BS-Seq to properly discriminate the various modified cytosines. The steps of each of these protocols are illustrated in Figure 3.1 and the results of sequencing for each technique are summarized in Table 3.2.

3.3.2.1 BS-Seq for 5mC

BS-Seq involves bisulfite conversion of DNA prior to high-throughput sequencing [83–85]. Briefly, bisulfite conversion utilizes sodium bisulfite to convert all unmodified cytosines to uracil, by deamination, but does not deaminate 5mC unaltered. The converted cytosines are read as thymines during sequencing, while the unconverted cytosines are read as cytosine. The unconverted 5mC is read as a cytosine. Thus, the percent of methylation at each cytosine is then determined from the proportion of cytosines and thymines detected at each position.

A limitation of BS-Seq is that it cannot distinguish 5mC from 5hmC or 5fC and 5aC from unmodified cytosine [86–89] (Figure 3.1). Therefore, a proportion of bases identified at 5mC by BS-Seq may actually be 5hmC and bases identified as unmodified make actually be 5fC or 5caC. To differentiate these modifications from each other, chemical labeling and enzymatic modification techniques have been developed to differentiate 5hmC from 5mC and cytosine from 5fC and 5caC (Figure 3.1). By comparing the results of these techniques with BS-Seq results, a base resolution map of each of these modifications can be generated (Table 3.1).

3.3.2.2 TAB-Seq for 5hmC

Our group developed Tet-assisted BS-Seq (TAB-Seq) to discriminate 5hmC from 5mC by taking advantage of Tet-mediated oxidation of 5mC and 5hmC to 5caC and the fact that 5caC acts like unmodified cytosine after bisulfite treatment [56,57]. In this method, β-GT is used to add a glucose onto 5hmC to generate 5gmC, protecting 5hmC from further Tet oxidation [89]. After 5hmC is blocked, all 5mC is converted to 5caC by an excess of recombinant Tet1 protein so that all 5mC is 5caC and all 5hmC is 5gmC. Normal bisulfite conversion of this modified DNA then converts all the cytosine and 5caC (derived from 5mC) to uracil or 5caU, respectively, while the original 5hmC remain

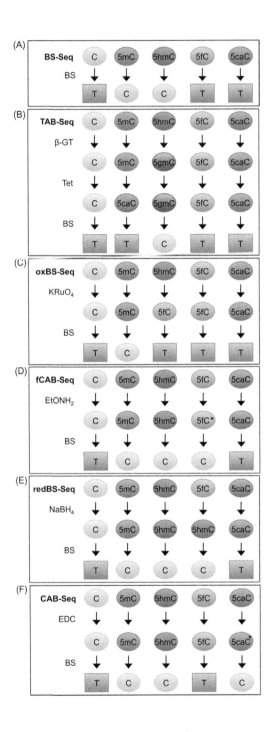

Table 3.2 Sequencing Results of Single-Base Resolution Methods

	C	5mC	5hmC	5fC	5caC
BS-Seq	T	C	C	T	T
TAB-Seq	T	T	C	T	T
oxBS-Seq	T	C	T	T	T
redBS-Seq	T	C	C	C	T
fCAB-Seq	T	C	C	C	T
CAB-Seq	T	C	C	T	C

as 5gmC (Figure 3.1). Sequencing of the converted library reports 5hmC as C and comparison to traditional BS-Seq yields an accurate map of the abundance of 5hmC and 5mC at each cytosine (Table 3.2). A limitation of this technique is that it requires highly active Tet enzymes to achieve a high conversion rate of 5mC to 5caC; recombinant Tet1 must be expressed and purified from insect cells [90].

3.3.2.3 oxBS-Seq for 5hmC

Oxidative BS-Seq (oxBS-Seq) is another method for the discrimination of 5mC and 5hmC that uses chemical, rather than enzyme-mediated, oxidation of 5hmC [88]. Specific oxidation of 5hmC to 5fC is carried out with potassium perruthenate ($KRuO_4$). Since 5fC behaves like unmodified cytosine in bisulfite conversion, treatment of the chemically modified DNA with bisulfite converts any base that was originally 5hmC to uracil and is read as thymine in subsequent sequencing (Figure 3.1). 5mC remains unconverted and is read as cytosine (Table 3.2). As with TAB-Seq, this requires comparison to traditional BS-Seq to yield a map of 5hmC and 5mC abundance at each cytosine. A limitation of this method is that it must be carefully optimized to avoid damage and degradation of DNA by the chemical oxidation conditions and by the repeated bisulfite treatments required to fully deaminate 5fC.

◀ **FIGURE 3.1**

Single-base resolution mapping of cytosine modifications. The steps of each single-base resolution mapping method are illustrated. (A) In BS-Seq, BS treatment converts unmodified cytosines, 5fC and 5caC, so they are read as thymine in sequencing, while 5mC and 5hmC are unconverted and read as cytosines. (B) In TAB-Seq, 5hmC is protected from Tet-mediated oxidation by glucosylation by β-GT. Recombinant Tet is then used to oxidize 5mC to 5caC. After BS treatment, unmodified cytosines, 5mC, 5fC, and 5caC, are read as thymines in sequencing while 5hmC is read as cytosine. (C) In oxBS-Seq, 5hmC is selectively chemically oxidized by $KRuO_4$ to 5fC. Treatment with BS then converts everything except 5mC so that 5mC is read as cytosine and all other cytosine modifications are read as thymine in sequencing. (D) In fCAB-Seq, 5fC is protected from BS-mediated deamination by hydroxylamine treatment. This modified 5fC is not modified by BS and is thus read as a cytosine by sequencing. (E) In redBS-Seq, selective chemical reduction of 5fC by $NaBH_4$ converts it to 5hmC. BS treatment converts only unmodified cytosines and 5caC so that these are read as thymines in sequencing and 5mC, 5hmC, and 5fC are read as cytosines. (F) In CAB-Seq, 5caC is selectively labeled using EDC. The resulting modified 5caC is protected by BS-mediated deamination so that it is read as a C during sequencing.

3.3.2.4 Aba-Seq for 5hmC

A third method for single-base resolution description of 5hmC has been developed and takes advantage of the 5hmC-specific endonuclease, AbaS1 [91]. AbaSI specifically recognizes 5-glucosylatedmethylcytosine (5gmC) compared to 5mC and cytosine and cleaves at a specific distance from the recognized base. As in TAB-Seq, the first step of this technique is DNA glucosylation to convert 5hmC to 5gmC. This is followed by digestion of glucosylated DNA with AbaS1, creating a library of DNA fragments cleaved only at 5hmC sites. Upon sequencing, the AbaS1 cleavage sites can be mapped by mapping the sequencing reads to the reference genome. The authors state that this technique carries distinct advantages over TAB-Seq and oxBS-Seq including: (1) reduced sequencing depth required to detect low-level 5hmCs and the increased ability to detect sites of low 5hmC, (2) simplified library preparation compatible with low amounts of input DNA, and (3) simpler data processing. A possible limitation may arise from the varying efficiency of AbaS1-mediated cleavage at different sites [92]. This results in ambiguity in the assignment of 5hmC at a small subset of sites.

3.3.2.5 fCAB-Seq for 5fC

Our group recently developed a method for base-resolution detection of 5fC using chemically assisted bisulfite sequencing (fCAB-Seq) [65]. This technique takes advantage of the specific reaction of hydroxylamine with 5fC described above. While this reaction is limited in its utility for the pull-down method previously described, it is useful for base-resolution mapping [71]. In fCAB-Seq, 5fC is protected from bisulfite-mediated deamination by treatment with O-ethylhydroxylamine (EtONH$_2$) (Figure 3.1). While abasic sites are a problem for the pull-down approach described above, they are not a problem for base-resolution mapping as abasic sites read as thymine [93]. We also tested whether reduction by NaBH$_4$ (see redBS-Seq below) protected 5fC from bisulfite-mediated deamination, but found that EtONH$_2$ achieves better protection for 5fC after bisulfite treatment. The hydroxylamine-protected 5fC is then read as a cytosine instead of a thymine as it is in BS-Seq (Table 3.2). Comparison of the results of BS-Seq and fCAB-Seq can differentiate the locations of 5fC from unmodified cytosine and 5caC.

3.3.2.6 redBS-Seq for 5fC

Reduced bisulfite sequencing (redBS-Seq) is based on the selective chemical reduction of 5fC and subsequent bisulfite conversion [94]. Upon treatment with bisulfite, 5fC is deformylated and then deaminated to a uracil so that it is read as a thymine and identified as an unmodified cytosine in BS-Seq [88]. However, if 5fC is first reduced to 5hmC with NaBH$_4$; then it reacts like 5hmC when treated with bisulfite (Figure 3.1). By comparing the results of redBS-Seq with standard BS-Seq, 5fC can be identified as sites identified as cytosine in BS-Seq but thymine in redBS-Seq. All other forms of cytosine (C, 5mC, 5hmC, and 5caC) have the same readout by BS-Seq and redBS-Seq, allowing the specific mapping of 5fC (Table 3.2). When combined with oxBS-Seq, the identity of 5mC, 5hmC, and 5fC can be differentiated.

3.3.2.7 CAB-Seq for 5caC

We have also developed a method of chemical modification-assisted bisulfite sequencing for single-base resolution detection of 5caC [95]. In this technique, 5caC is selectively labeled by

1-ethyl-3-[3—dimethylaminopropyl]-carbodiimide hydrochloride (EDC) using a xylene-based primary amine (Figure 3.1). This specifically labels 5caC, but not other modified cytosines. This modified 5caC is not deaminated by bisulfite as unmodified 5caC is and thus sequenced as a cytosine instead of a thymine (Table 3.2). As with the other base resolution methods, comparison to BS-Seq data allows the mapping of 5caC specifically.

3.4 CONCLUSIONS

The discovery of novel cytosine modifications in recent years has necessitated technical advances in the detection and mapping of these modifications. Genome-wide techniques can be used to identify regions that contain cytosine modifications, while single-base resolution techniques are able to map the specific bases in the genome that are modified. Techniques for the mapping of 5mC and 5hmC, both genome-wide and single-base resolution, have been incorporated in the epigenetic field and have been utilized in many systems to yield new and exciting information about the role of epigenetics in biological functions. The very new techniques described here for detection and mapping of 5fC and 5caC will enable researchers to study the role of these modifications as well. Furthermore, as the cost of high-throughput sequencing continues to drop, it will become more feasible to incorporate these techniques into the standard arsenal of epigenetic methods.

REFERENCES

[1] Waddington CH. An Introduction to Modern Genetics. New York: The Macmillan Company; 1939.
[2] Egger G, Liang G, Aparicio A, Jones PA. Epigenetics in human disease and prospects for epigenetic therapy. Nature 2004;429:457—63.
[3] Davis BN, Hilyard AC, Lagna G, Hata A. SMAD proteins control DROSHA-mediated microRNA maturation. Nature 2008;454(7200):56—61.
[4] Li E. Chromatin modification and epigenetic reprogramming in mammalian development. Nat Rev Genet 2002;3(9):662—73.
[5] Li X, Zhao X. Epigenetic regulation of mammalian stem cells. Stem Cells Dev 2008;17(6):1043—52.
[6] Smith ZD, Meissner A. DNA methylation: roles in mammalian development. Nat Rev Genet 2013;14(3): 204—20.
[7] Yao B, Jin P. Cytosine modifications in neurodevelopment and diseases. Cell Mol Life Sci 2014;71: 405—18.
[8] Amir RE, Van den Veyver IB, Wan M, Tran CQ, Francke U, Zoghbi HY. Rett syndrome is caused by mutations in X-linked MECP2, encoding methyl-CpG-binding protein 2. Nat Genet 1999;23(2):185—8.
[9] Chen WG, Chang Q, Lin Y, Meissner A, West AE, Griffith EC, et al. Derepression of BDNF transcription involves calcium-dependent phosphorylation of MeCP2. Science 2003;302(5646):885—9.
[10] Jones PA, Takai D. The role of DNA methylation in mammalian epigenetics. Science 2001;293(5532): 1068—70.
[11] Klose R, Bird A. Molecular biology. MeCP2 repression goes nonglobal. Science 2003;302(5646): 793—5.
[12] Li H, Yamagata T, Mori M, Yasuhara A, Momoi MY. Mutation analysis of methyl-CpG binding protein family genes in autistic patients. Brain Dev 2005;27(5):321—5.

[13] Martinowich K, Hattori D, Wu H, Fouse S, He F, Hu Y, et al. DNA methylation-related chromatin remodeling in activity-dependent BDNF gene regulation. Science 2003;302:890−3.

[14] Okano M, Bell DW, Haber DA, Li E. DNA methyltransferases Dnmt3a and Dnmt3b are essential for de novo methylation and mammalian development. Cell 1999;99(3):247−57.

[15] Xu GL, Bestor TH, Bourc'his D, Hsieh CL, Tommerup N, Bugge M, et al. Chromosome instability and immunodeficiency syndrome caused by mutations in a DNA methyltransferase gene. Nature 1999;402 (6758):187−91.

[16] Maher ER, Reik W. Beckwith-Wiedemann syndrome: imprinting in clusters revisited. J Clin Invest 2000; 105:247−52.

[17] Nicholls RD, Saitoh S, Horsthemke B. Imprinting in Prader-Willi and Angelman syndromes. Trends Genet 1998;14:194−200.

[18] Fan G, Beard C, Chen RZ, Csankovszki G, Sun Y, Siniaia M, et al. DNA hypomethylation perturbs the function and survival of CNS neurons in postnatal animals. J Neurosci 2001;21:788−97.

[19] Globisch D, Münzel M, Müller M, Michalakis S, Wagner M, Koch S, et al. Tissue distribution of 5-hydroxymethylcytosine and search for active demethylation intermediates. PloS One 2010;5(12):e15367.

[20] Szulwach KE, Li X, Li Y, Song C, Wu H, Dai Q, et al. 5-hmC−mediated epigenetic dynamics during postnatal neurodevelopment and aging. Nat Neurosci 2011;14:1607−16.

[21] Song CX, Szulwach KE, Fu Y, Dai Q, Yi C, Li X, et al. Selective chemical labeling reveals the genome-wide distribution of 5-hydroxymethylcytosine. Nat Biotechnol 2011;29:68−72.

[22] Wang T, Pan Q, Lin L, Szulwach KE, Song CX, He C, et al. Genome-wide DNA hydroxymethylation changes are associated with neurodevelopmental genes in the developing human cerebellum. Hum Mol Genet 2012;21(26):5500−10.

[23] Chen Y, Damayanti NP, Irudayaraj J, Dunn K, Zhou FC. Diversity of two forms of DNA methylation in the brain. Front Genet 2014;5:46.

[24] Unoki M, Nishidate T, Nakamura Y. ICBP90, an E2F-1 target, recruits HDAC1 and binds to methyl-CpG through its SRA domain. Oncogene 2004;23:7601−10.

[25] Jeanblanc M, Mousli M, Hopfner R, Bathami K, Martinet N, Abbady AQ, et al. The retinoblastoma gene and its product are targeted by ICBP90: a key mechanism in the G1/S transition during the cell cycle. Oncogene 2005;24:7337−45.

[26] Bostick M, Kim JK, Estève P-O, Clark A, Pradhan S, Jacobsen SE. UHRF1 plays a role in maintaining DNA methylation in mammalian cells. Science 2007;317:1760−4.

[27] Neri F, Krepelova A, Incarnato D, Maldotti M, Parlato C, Galvagni F, et al. Dnmt3L antagonizes DNA methylation at bivalent promoters and favors DNA methylation at gene bodies in ESCs. Cell 2013; 155:121−34.

[28] Jia D, Jurkowska RZ, Zhang X, Jeltsch A, Cheng X. Structure of Dnmt3a bound to Dnmt3L suggests a model for de novo DNA methylation. Nature 2007;449:248−51.

[29] Jurkowska RZ, Rajavelu A, Anspach N, Urbanke C, Jankevicius G, Ragozin S, et al. Oligomerization and binding of the Dnmt3a DNA methyltransferase to parallel DNA molecules: heterochromatic localization and role of Dnmt3L. J Biol Chem 2011;286:24200−7.

[30] Chedin F, Lieber MR, Hsieh C-L. The DNA methyltransferase-like protein DNMT3L stimulates de novo methylation by Dnmt3a. Proc Natl Acad Sci USA 2002;99:16916−21.

[31] Chen Z-X, Mann JR, Hsieh C-L, Riggs AD, Chédin F. Physical and functional interactions between the human DNMT3L protein and members of the de novo methyltransferase family. J Cell Biochem 2005; 95:902−17.

[32] Gowher H, Liebert K, Hermann A, Xu G, Jeltsch A. Mechanism of stimulation of catalytic activity of Dnmt3A and Dnmt3B DNA-(cytosine-C5)-methyltransferases by Dnmt3L. J Biol Chem 2005;280: 13341−8.

[33] Kareta MS, Botello ZM, Ennis JJ, Chou C, Chédin F. Reconstitution and mechanism of the stimulation of de novo methylation by human DNMT3L. J Biol Chem 2006;281:25893−902.

[34] Bogdanovic O, Veenstra GJ. DNA methylation and methyl-CpG binding proteins: developmental requirements and function. Chromosoma 2009;118(5):549−65.

[35] Hendrich B, Bird A. Identification and characterization of a family of mammalian methyl-CpG binding proteins. Mol Cell Biol 1998;18(11):6538−47.

[36] Jones PA, Liang G. Rethinking how DNA methylation patterns are maintained. Nat Rev Genet 2009;10 (11):805−11.

[37] Suzuki MM, Bird A. DNA methylation landscapes: provocative insights from epigenomics. Nat Rev Genet 2008;9(6):465−76.

[38] Gopalakrishnan S, Van Emburgh BO, Robertson KD. DNA methylation in development and human disease. Mutat Res 2008;647(1−2):30−8.

[39] Jones PA, Baylin SB. The fundamental role of epigenetic events in cancer. Nat Rev Genet 2002;3(6): 415−28.

[40] Jakovcevski M, Akbarian S. Epigenetic mechanisms in neurological disease. Nat Med 2012;18(8): 1194−204.

[41] Ma DK, Marchetto MC, Guo JU, Ming GL, Gage FH, Song H. Epigenetic choreographers of neurogenesis in the adult mammalian brain. Nat Neurosci 2010;13:1338−44.

[42] Baylin SB, Jones PA. A decade of exploring the cancer epigenome − biological and translational implications. Nat Rev Cancer 2011;11(10):726−34.

[43] Cheng Y, Bernstein A, Chen D, Jin P. 5-Hydroxymethylcytosine: a new player in brain disorders? Exp Neurol 2014.

[44] Irier HA, Jin P. Dynamics of DNA methylation in aging and Alzheimer's disease. DNA Cell Biol 2012; 31(Suppl. 1):S42−8.

[45] Ito S, D'Alessio AC, Taranova OV, Hong K, Sowers LC, Zhang Y. Role of Tet proteins in 5mC to 5hmC conversion, ES-cell self-renewal and inner cell mass specification. Nature 2010;466:1129−33.

[46] Tahiliani M, Koh KP, Shen Y, Pastor WA, Bandukwala H, Brudno Y, et al. Conversion of 5-methylcytosine to 5-hydroxymethylcytosine in mammalian DNA by MLL partner TET1. Science 2009;324:930−5.

[47] Spruijt CG, Gnerlich F, Smits AH, Pfaffeneder T, Jansen PW, Bauer C, et al. Dynamic readers for 5-(hydroxy)methylcytosine and its oxidized derivatives. Cell 2013;152(5):1146−59.

[48] Wyatt GR, Cohen SS. The bases of the nucleic acids of some bacterial and animal viruses: the occurrence of 5-hydroxymethylcytosine. Biochem J 1953;55(5):774−82.

[49] Penn NW, Suwalski R, O'Riley C, Bojanowski K, Yura R. The presence of 5-hydroxymethylcytosine in animal deoxyribonucleic acid. Biochem J 1972;126(4):781−90.

[50] Kriaucionis S, Heintz N. The nuclear DNA base 5-hydroxymethylcytosine is present in Purkinje neurons and the brain. Science 2009;324(5929):929−30.

[51] Riccio A. Dynamic epigenetic regulation in neurons: enzymes, stimuli and signaling pathways. Nat Neurosci 2010;13:1330−7.

[52] Bostick M, Kim JK, Esteve PO, Clark A, Pradhan S, Jacobsen SE. UHRF1 plays a role in maintaining DNA methylation in mammalian cells. Science 2007;317(5845):1760−4.

[53] Sharif J, Muto M, Takebayashi S, Suetaki I, Iwamatsu A, Endo TA, et al. The SRA protein Np95 mediates epigenetic inheritance by recruiting Dnmt1 to methylated DNA. Nature 2007;450(7171):908−12.

[54] Frauer C, Rottach A, Meilinger D, Bultmann S, Fellinger K, Hasenöder S, et al. Different binding properties and function of CXXC zinc finger domains in Dnmt1 and Tet1. PloS One 2011;6(2):e16627.

[55] Hashimoto H, Liu Y, Upadhyay AK, Chang Y, Howerton SB, Vertino PM, et al. Recognition and potential mechanisms for replication and erasure of cytosine hydroxymethylation. Nucleic Acids Res 2012;40(11): 4841−9.

[56] He YF, Li BZ, Li Z, Liu P, Wang Y, Tang Q, et al. Tet-mediated formation of 5-carboxylcytosine and its excision by TDG in mammalian DNA. Science 2011;333:1303−7.

[57] Ito S, Shen L, Dai Q, Wu SC, Collins LB, Swenberg JA, et al. Tet proteins can convert 5-methylcytosine to 5-formylcytosine and 5-carboxylcytosine. Science 2011;333:1300−3.

[58] Guo JU, Su Y, Zhong C, Ming GL, Song H. Hydroxylation of 5-methylcytosine by TET1 promotes active DNA demethylation in the adult brain. Cell 2011;145(3):423−34.

[59] Maiti A, Drohat AC. Thymine DNA glycosylase can rapidly excise 5-formylcytosine and 5-carboxyl-cytosine: potential implications for active demethylation of CpG sites. J Biol Chem 2011;286:35334−8.

[60] Cortellino S, Xu J, Sannai M, Moore R, Caretti E, Cigliano A, et al. Thymine DNA glycosylase is essential for active DNA demethylation by linked deamination-base excision repair. Cell 2011;146:67−79.

[61] Chen CC, Wang KY, Shen CK. The mammalian de novo DNA methyltransferases DNMT3A and DNMT3B are also DNA 5-hydroxymethylcytosine dehydroxymethylases. J Biol Chem 2012;287(40): 33116−21.

[62] Mellen M, Ayata P, Dewell S, Kriaucionis S, Heintz N. MeCP2 binds to 5hmC enriched within active genes and accessible chromatin in the nervous system. Cell 2012;151(7):1417−30.

[63] Yildirim O, Li R, Hung JH, Chen PB, Dong X, Ee LS, et al. Mbd3/NURD complex regulates expression of 5-hydroxymethylcytosine marked genes in embryonic stem cells. Cell 2011;147(7):1498−510.

[64] Wang F, Yang Y, Lin X, Wang J-Q, Wu Y-S, Xie W, et al. Genome-wide loss of 5-hmC is a novel epigenetic feature of Huntington's disease. Hum Mol Genet 2013;22(18):3641−53.

[65] Song C-X, Szulwach KE, Dai Q, Fu Y, Mao SQ, Lin L, et al. Genome-wide profiling of 5-formylcytosine reveals its roles in epigenetic priming. Cell 2013;153:678−91.

[66] Shen L, Wu H, Diep D, Yamaguchi S, D'Alessio AC, Fung HL, et al. Genome-wide analysis reveals TET- and TDG-dependent 5-methylcytosine oxidation dynamics. Cell 2013;153:692−706.

[67] Weber M, Davies JJ, Wittig D, Oakeley EJ, Haase M, Lam WL, et al. Chromosome-wide and promoter-specific analyses identify sites of differential DNA methylation in normal and transformed human cells. Nat Genet 2005;37:853−62.

[68] Szulwach KE, Li X, Smrt RD, Li Y, Luo Y, Lin L, et al. Cross talk between microRNA and epigenetic regulation in adult neurogenesis. J Cell Biol 2010;189:127−41.

[69] Laird PW. Principles and challenges of genomewide DNA methylation analysis. Nat Rev Genet 2010; 11:191−203.

[70] Nair SS, Coolen MW, Stirzaker C, Song JZ, Statham AL, Strbenac D, et al. Comparison of methyl-DNA immunoprecipitation (MeDIP) and methyl-CpG binding domain (MBD) protein capture for genome-wide DNA methylation analysis reveal CpG sequence coverage bias. Epigenetics 2011;6:34−44.

[71] Raiber E-A, Beraldi D, Ficz G, Burgess HE, Branco MR, Murat P, et al. Genome-wide distribution of 5-formylcytosine in embryonic stem cells is associated with transcription and depends on thymine DNA glycosylase. Genome Biol 2012;13:R69.

[72] Thomson JP, Hunter JM, Nestor CE, Dunican DS, Terranova R, Moggs JG, et al. Comparative analysis of affinity-based 5-hydroxymethylation enrichment techniques. Nucleic Acids Res 2013;41:e206.

[73] Pastor WA, Pape UJ, Huang Y, Henderson HR, Lister R, Ko M, et al. Genome-wide mapping of 5-hydroxymethylcytosine in embryonic stem cells. Nature 2011;473:394−7.

[74] Serre D, Lee BH, Ting AH. MBD-isolated genome sequencing provides a high-throughput and comprehensive survey of DNA methylation in the human genome. Nucleic Acids Res 2010;38:391−9.

[75] Klose RJ, Sarraf SA, Schmiedeberg L, McDermott SM, Stancheva I, Bird AP. DNA binding selectivity of MeCP2 due to a requirement for A/T sequences adjacent to methyl-CpG. Mol Cell 2005;19:667−78.

[76] Brinkman AB, Simmer F, Ma K, Kaan A, Zhu J, Stunnenberg HG. Whole-genome DNA methylation profiling using MethylCap-seq. Methods 2010;52:232−6.

[77] Rauch T, Pfeifer GP. Methylated-CpG island recovery assay: a new technique for the rapid detection of methylated-CpG islands in cancer. Lab Invest 2005;85:1172−80.

[78] Rauch TA, Pfeifer GP. The MIRA method for DNA methylation analysis. Methods Mol Biol 2009; 507:65−75.

[79] Gebhard C, Schwarzfischer L, Pham T-H, Schilling E, Klug M, Andreesen R, et al. Genome-wide profiling of CpG methylation identifies novel targets of aberrant hypermethylation in myeloid leukemia. Cancer Res 2006;66:6118−28.

[80] Robertson AB, Dahl JA, Ougland R, Klungland A. Pull-down of 5-hydroxymethylcytosine DNA using JBP1-coated magnetic beads. Nat Protoc 2012;7:340−50.

[81] Pfaffeneder T, Hackner B, Truss M, Münzel M, Müller M, Deiml CA, et al. The discovery of 5-formylcytosine in embryonic stem cell DNA. Angew Chem Int Ed 2011;50:7008−12.

[82] Dai Q, He C. Syntheses of 5-formyl- and 5-carboxyl-dC containing DNA oligos as potential oxidation products of 5-hydroxymethylcytosine in DNA. Org Lett 2011;13(13):3446−9.

[83] Rein T, DePamphilis ML, Zorbas H. Identifying 5-methylcytosine and related modifications in DNA genomes. Nucleic Acids Res 1998;26:2255−64.

[84] Clark SJ, Statham A, Stirzaker C, Molloy PL, Frommer M. DNA methylation: bisulphite modification and analysis. Nat Protoc 2006;1:2353−64.

[85] Beck S, Rakyan VK. The methylome: approaches for global DNA methylation profiling. Trends Genet 2008;24:231−7.

[86] Huang Y, Pastor Wa, Shen Y, Tahiliani M, Liu DR, Rao A. The behaviour of 5-hydroxymethylcytosine in bisulfite sequencing. PloS One 2010;5:e8888.

[87] Jin SG, Kadam S, Pfeifer GP. Examination of the specificity of DNA methylation profiling techniques towards 5-methylcytosine and 5-hydroxymethylcytosine. Nucleic Acids Res 2010;38(11):e125.

[88] Booth MJ, Branco MR, Ficz G, Oxley D, Krueger F, Reik W, et al. Quantitative sequencing of 5-methylcytosine and 5-hydroxymethylcytosine at single-base resolution. Science 2012;336:934−7.

[89] Yu M, Hon GC, Szulwach KE, Song CX, Zhang L, Kim A, et al. Base-resolution analysis of 5-hydroxymethylcytosine in the mammalian genome. Cell 2012;149(6):1368−80.

[90] Song C-X, Yi C, He C. Mapping recently identified nucleotide variants in the genome and transcriptome. Nat Biotechnol 2012;30:1107−16.

[91] Sun Z, Terragni J, Borgaro JG, Liu Y, Yu L, Guan S, et al. High-resolution enzymatic mapping of genomic 5-hydroxymethylcytosine in mouse embryonic stem cells. Cell Rep 2013;3(2):567−76.

[92] Wang H, Guan S, Quimby A, Cohen-Karni D, Pradhan S, Wilson G, et al. Comparative characterization of the PvuRts1I family of restriction enzymes and their application in mapping genomic 5-hydroxymethylcytosine. Nucleic Acids Res 2011;39(21):9294−305.

[93] Sikorsky JA, Primerano DA, Fenger TW, Denvir J. DNA damage reduces Taq DNA polymerase fidelity and PCR amplification efficiency. Biochem Biophys Res Commun 2007;355(2):431−7.

[94] Booth MJ, Marsico G, Bachman M, Beraldi D, Balasubramanian S. Quantitative sequencing of 5-formylcytosine in DNA at single-base resolution. Nat Chem 2014;6:435−40.

[95] Lu X, Song C, Szulwach K. Chemical modification-assisted bisulfite sequencing (CAB-SEQ) for 5-carboxylcytosine detection in DNA. J Am Chem Soc 2013;135:9315−17.

APPLICATION OF MASS SPECTROMETRY IN TRANSLATIONAL EPIGENETICS

4

Xiaoshi Wang, Simone Sidoli, and Benjamin A. Garcia

Department of Biochemistry and Biophysics, Perelman School of Medicine,
University of Pennsylvania, Philadelphia, PA, USA

CHAPTER OUTLINE

4.1 INTRODUCTION

Epigenetics describes a second layer of information that causes the changes of gene expression [1,2]. This regulation is above genome, resulting from environmental factors and through a number of processes such as DNA methylation, posttranslational modification of chromatin-associated proteins, nucleosome location, and noncoding RNA [3]. It involves a variety of biomolecules and different cellular pathways. Chromatin is a highly organized but dynamic structure. The basic repeat components of chromatin are nucleosomes consisting of 147 bp DNA wrapped around an octamer core of histones including H2A, H2B, H3, and H4. Histone H1 is a linker between nucleosomes, consolidating the nucleosome binding with DNA and stabilizing the zig-zagged

Y.G. Zheng (Ed): Epigenetic Technological Applications. DOI: http://dx.doi.org/10.1016/B978-0-12-801080-8.00004-1

chromatin fiber. DNA and histones are largely modified with covalent modifications. DNA is methylated at CpG sites, which is linked to transcriptional silencing [4]. Histones are heavily decorated with a variety of PTMs, such as methylation, acetylation, phosphorylation, ubiquitination, citrullination, and ADP-ribosylation on distinct amino acids. Such modifications frequently coexist in complex combinations [5,6]. Figure 4.1 shows a summary of common and several novel modifications on histones identified in the literature so far [7–9]. In addition, canonical histones are replaced by histone variants in specific regions of the chromatin under certain conditions [10]. For instance, in mammals, there are at least six isotypes of linker histone H1 in somatic cells, namely H1.1–1.5 and H1°, with additional one oocyte-specific and two testis-specific histones (H1t and HILS1).

The dynamic alternation of the chromatin architecture enables the transcriptional machinery to access packaged DNA, regulating gene expression, and determining cellular phenotypes. The complexity of both the number and types of PTMs and histone variants leads to potentially millions of different histone forms. It has been hypothesized that combinations of histone PTMs may form a "histone code" that would link specific histone PTM combinations with gene activation or silencing, as these marks would recruit other "reader" proteins that act to alter chromatin structures or to promote transcription [11,12]. We currently know the functions of only a few combinatorial histone modifications and histone interacting proteins with multiple domains that recognize multiple PTMs [13].

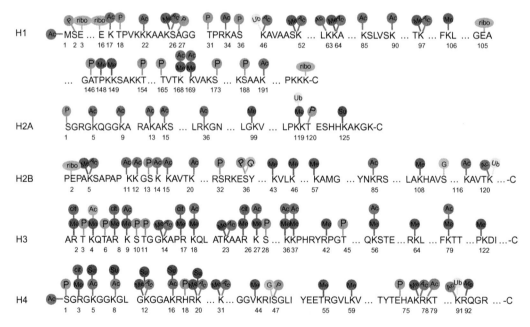

FIGURE 4.1

Histones are heavily decorated by PTMs. Acetylation is shown in blue, methylation is shown in red (mono, di, tri), phosphorylation in green, ubiquitination in yellow, citrullination in magenta, ADP-ribosylation in cyan, sumoylation in indigo, and GlcNAc in peach. (*This figure is reproduced in color in the color plate section.*)

The intricacy of the epigenetic marks, in particular the potential histone codes, has not been fully understood yet. As discussed by Brian Turner, the genetic code has defined "signs" (64 combinations of nucleotide triplets) and each of these signs has a meaning (the 20 amino acids) [14]. In epigenetics we are still not aware of how many signs are possible, due to the possibilities of DNA methylations/histone PTMs/noncoding RNA combinations, and how many meanings they can generate, which are the various regulation activities.

Epigenetics research has thus become one of the fastest-growing areas of science and a central topic in biological studies toward disease development and progression [9]. Aberrations in histone modifications have been found to be associated with a number of diseases such as cancer, diabetes, and neurological disorders. For example, the reduction of H4K20me3 and H4K16ac was shown to be related to an early event of tumorigenesis in both cancer tissues and animal studies [15]. Low levels of H3K27me3 have been reported in breast, ovarian, and pancreatic cancers [16]. In several studies, identifying global histone PTM patterns were also found to be essential in improving clinical diagnostics and predictive capacity. For instance, Seligson and coworkers identified that H3K18ac coupled with H3K4me2 are significant predictors associating with a lower recurrence of prostate cancer [17]. Due to the fact that epigenetic changes are reversible, epigenetic therapy might be less toxic than conventional chemotherapy. Therefore, epigenetic therapy is a promising strategy for disease treatment [18]. Currently, a number of epigenetic drugs (epi-drugs) have already been approved by the FDA. Vorinostat (a pan-histone deacetylases inhibitor) and romidepsin (a class I-specific histone deacetylases inhibitor) were approved for cancer treatment. Ruxolitinib (a Janus kinase 2 inhibitor) was approved for the treatment of intermediate or high-risk myelofibrosis. Many more epi-drugs are currently under clinical trials. Preclinical investigation is also undergoing toward discovering novel epigenetic targets and more effective inhibitors.

Appropriate technologies are required for a deeper understanding of the role of histone PTMs and histone PTM coexistence. As histone tails can be decorated with up to 8−10 PTMs concurrently, it is necessary to develop techniques that can characterize all marks simultaneously in a given peptide. Traditional histone PTM analysis has commonly been performed by using antibody-based techniques. Chandra et al. showed independent chromatin regions for H3K9me3, H3K27me3, and H3K36me3 by using antibodies with fluorescent tags [19]. In another study, Yuan et al. proved that H3K27me3 and H3K36me3 are mutually exclusive in HeLa cells by using immunoprecipitation followed by western blot [20]. But antibody-based approaches are very limited for the following reasons: (1) only very few antibodies recognize multiple PTMs; (2) only known modifications can be studied with this approach; (3) antibody-based techniques rarely can be adopted in a large-scale manner, necessary to investigate combinatorial codes; and (4) the specificity of antibodies for histone PTMs, even if commercial, is sometimes not sufficient.

Mass spectrometry based proteomics has evolved as a sensitive and accurate strategy for protein characterization, in particular to identify and quantify protein PTMs [7,21−24]. Therefore, it has rapidly become a favored technique in epigenetic research, as it greatly facilitates such research. Acetylation was the first PTM discovered in histones. One of the several early MS-based studies on histone marks focused on acetylation of histone H4, and they led to the discovery of the "zip" model, which showed that the acetylation of H4 proceeds from K16 to K5 and deacetylation proceeds in the opposite direction [25]. Methylation and phosphorylation are also extensively studied by using MS. For instance, phosphorylation sites on histone H1 isoforms were identified, also thanks to peptide enrichment with immobilized metal affinity chromatography prior to MS analysis. Such experiments extend the "methyl/phos" switch hypothesis to linker histones [26,27]. For the last several decades, mass spectrometry has been considered as one of the most important analytic technologies for studying molecular mechanisms in epigenetics, in particular histone PTMs and chromatin-binding proteomes. In this chapter, we describe the principles and highlight current strategies that make MS applicable for epigenetic research.

4.2 APPLICATIONS OF MASS SPECTROMETRY IN EPIGENETIC RESEARCH

4.2.1 PRINCIPLES OF MASS SPECTROMETRY IN PROTEOMICS

In principle, mass spectrometers measure the mass-to-charge ratio (m/z) of freely moving gas-phase ions in electric and/or magnetic fields. Ions are generated by an ionization source, while the m/z ratio is calculated by the mass analyzer. The invention of "soft-ionization" technologies, including matrix-assisted laser desorption ionization (MALDI) and electrospray ionization (ESI), made possible the ionization of large biomolecules such as peptides and proteins, revolutionizing the field of proteomics. In the last decade, advances in the development of mass analyzers, such as the Fourier transform ion cyclotron resonance (FTICR) mass spectrometer and the Orbitrap mass spectrometer, have dramatically improved the resolution (>100,000), the accuracy (<5 ppm), the sensitivity (femtomole), and the speed (up to 15 Hz) of proteomic analysis [28]. "Hybrid" instruments have also been designed. The combination of multiple mass analyzers and a fragmentation cell allowed tandem mass spectrometry (MS/MS).

MS/MS consists of a scan event where precursor masses are isolated and fragmented. The determination of both precursor and product masses increases the confidence of the identification and it allows peptide/protein sequencing. The sequence, and therefore the mass, of a peptide can be predicted by knowing the sequence of the intact protein plus knowledge about the specificity of cleavage of the enzyme used for the digestion. The fragmentation pattern can be predicted based on the type of fragmentation applied. The most commonly used fragmentation methods are collision-induced dissociation (CID), higher-energy C-trap dissociation (HCD), and electron capture dissociation (ECD) or electron transfer dissociation (ETD), which generate distinct and specific fragment ions. Figure 4.2 shows the generation of type a, x, b, y, c, z fragment ions with the use of different fragmentation methods. The match between predicted and observed masses is usually performed by bioinformatics tools, and this leads to peptide/protein identification from MS/MS spectra. In addition, fragmentation allows mapping of PTMs on the protein sequence, if the mass of the PTM is known. The fragmentation pattern of a modified sequence is distinguished from an unmodified one, due to the mass shift of the fragment ions that carry the PTM. The fragments with a shifted mass indicate the localization of the PTM in the sequence.

High-resolution mass spectrometers, such as the Orbitrap, provide sufficiently high accuracy to discriminate the small difference in the mass shifts between acetylation and trimethylation. The acetylation on a lysine residue generates a mass shift of 42.011 Da, whereas the lysine trimethylation results in a mass shift of 42.047 Da. In addition, CID or HCD produces characteristic neutral loss signals that are also useful for PTM identification. For instance, a tri-methylated lysine produces a neutral loss ion of 59.073 Da. Table 4.1 is a summary of the mass shifts, diagnostic

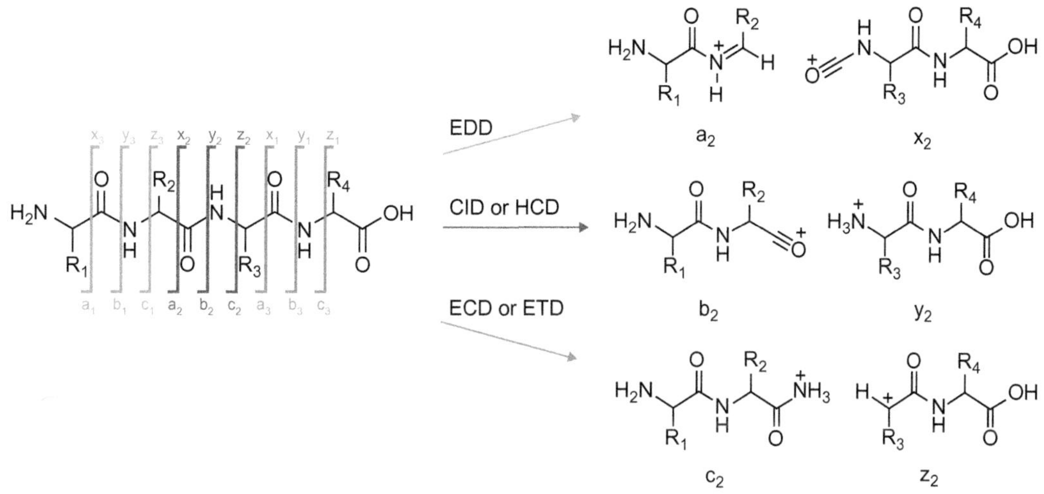

FIGURE 4.2

Generation of a, x, b, y, c, z type ions with the use of different fragmentation methods. (*This figure is reproduced in color in the color plate section.*)

Table 4.1 List of Common and Several Novel Histone Modifications with Mass Shifts, Diagnostic Ions, Neutral Losses, and Enrichment Methods

PTM	Residues	Mass Shift (Da)	Diagnostic Ions (m/z) [29] Neutral Loss (Da)	Enrichment Methods
Acetylation	K	42.011	84.081, 143.118, 126.091	Immunoaffinity [30]
Mono-methylation		14.016	[84.081, 98.096, 143.118] K [32.049, 74.071, 57.055] R	
Di-methylation	K/R	28.031	[84.081] K [46.065 (sym), 32.049 (unsym), 71.060, 88.087] R	Immunoaffinity [31,32]/ 3 × MBT [33]
Tri-methylation	K	42.047	84.081, 143.154 −59.073	
Phosphorylation	S/T/Y/H	79.966	−97.982 [34]	IMAC [35,36]/TiO$_2$ [37,38] Immunoaffinity [35]
ADP-ribosylation	K/E	541.061	−347.10 [39]	TiO$_2$ [40]
Ubiquitination	K	114.043	−	immunoaffinity [41] epitope tag [42]
Sumoylation	K	242.102	−	epitope tag [43]
Crotonylation	K	68.026	−	immunoaffinity [44]
O-GlcNAcylation	S/T	203.079	−203.079 [45]	LWAC [45]/chemical derivatization [46,47]

ions, neutral losses and enrichment methods of common and several novel histone PTMs. Recently, LTQ-Orbitrap has become a powerful instrument in protein identification, high mass accuracy PTM characterization, quantification, and top-down proteomics.

4.2.2 MAJOR APPROACHES FOR MASS-SPECTROMETRY-BASED HISTONE PTM ANALYSIS

4.2.2.1 Sample preparation for histone analysis

One of the major challenges in histone analysis lies in their extensive heterogeneity and dynamic ranges. Therefore, analysis of histones requires accurate sample preparation as histones are divided into numerous variants and degrees of modifications. In this section, we will summarize the preparation of histones from cells and tissues, separation of histone isotypes, and enrichment of low abundant PTMs.

4.2.2.1.1 Histone isolation

Histones are highly alkaline proteins bearing a positively charged N-terminus caused by a large number of lysine and arginine residues. Taking advantage of this chemical property of histones, two common methods are used to extract a large amount of histones from cells or tissues [48]. The acid extraction method utilizes 0.4 N of H_2SO_4 to dissolve histones into acidic solution, followed by a step of precipitation using trichloroacetic acid (TCA). The pellet is rinsed with acetone and resuspended in water. Alternatively, the high-salt extraction is a gentle method that prevents the loss of some acid-labile PTMs, such as histidine phosphorylation. Nuclei are lysed with no salt buffer (3 mM EDTA, 0.2 mM EGTA). After centrifugation, the supernatant containing nucleoplasm is discarded and the chromatin pellet is resuspended in high salt buffer (50 mM Tris-HCl 2.5 M NaCl and 0.05% NP40). Extracted histones are then dialyzed to lower the salt concentration. The high-salt extraction method may also reduce the amount of insoluble material that is frequently produced with acid extraction.

4.2.2.1.2 Histone isotype separation

After isolation of crude histones, different techniques can be employed to separate various histone isotypes. While gel electrophoresis is for histone visualization by staining techniques, reversed-phase liquid chromatography (RPLC) is the method of choice to purify histone isotypes. The eluted histone is dissolved in H_2O/acetonitrile and is ready for downstream sample preparation or MS analysis. An alternative approach to separate histone isotypes is to utilize hydrophilic interaction liquid chromatography (HILIC) [49]. HILIC was first introduced to meet the need for the analysis of polar compounds that reversed phase resins do not bind with. HILIC separation with a combination of middle-down and top-down proteomics of intact histones or histone tails can preserve the interdependence of modification. In addition, HILIC separation is a promising additional enrichment step for glycosylation, N-acetylation, and phosphorylation [50]. Lastly, taking advantage of the orthogonal nature of RP-HPLC and HILIC, multidimensional separation proves to be a gold standard for the separation of complex mixtures in proteomics applications [51].

4.2.2.1.3 PTM enrichment

Acetylation and methylation are the two most abundance histone PTMs. H3K27me2 is present on approximately 70% of total histone H3 [52]. But many other types of PTMs are in low abundance so that proper enrichment methods are necessary to prevent signal suppression and characterize them. Tremendous progress has been made on the enrichment methods for specific types of PTMs over the past several decades. Such enrichments have been proven to dramatically improve the sensitivity of analysis and assist the understanding of the pivotal roles of PTMs in cellular physiology and diseases [53].

Enrichment is mostly performed at the peptide level. A widely used strategy for PTM enrichment is antibody-based immunoaffinity purification. It utilizes a pan-specific PTM antibody to enrich the peptides bearing the PTM of interest. This method has been successfully used for the global analysis of lysine acetylation, arginine methylation, postdigested lysine-GG tag of ubiquitination, and so on, as shown in Table 4.1. This approach is likely to be applicable to most PTMs but the selectivity and efficiency of the immunoaffinity enrichment relies on a high-quality antibody, use of which is costly and not always feasible. Therefore, non-antibody enrichment methods become attractive. A number of strategies utilize the chemical nature of a specific PTM and its interaction with a stationary phase. For instance, phosphorylation can be enriched with immobilized metal ion affinity chromatography (IMAC) or titanium dioxide (TiO_2), which have both proven to be highly specific and efficient [37,38]. For example, Gygi and coworkers detected 13,720 total phosphorylation sites on proteins extracted from *Drosophila* embryos, with a false positive rate lower than 1%. This method has also been successfully utilized in histone phosphorylation analysis [54]. Hunt and coworkers localized 19 novel phosphorylation sites on human histone H1 isotypes [27]. In another study, potential binary switches were observed on histone H3 between T3ph/K4me1, K9me1−3/S10ph, and K27me1−3/S28ph. Recently, Garcia and coworkers developed a technique that adopts stable isotopic labeling (γ-$^{18}O_4$) to determine rates of phosphorylation of over 500 sites with a wide range of rate constants from 0.34 $\min^{-1}$ to 0.001 $\min^{-1}$ in Hela cells by using TiO_2 enrichment [55]. The ion-interaction-based method has been employed for the enrichment of other kinds of PTMs. For example, TiO_2 has been utilized to enrich sialic-acid-containing glycopeptides [50].

Recently, other types of PTM enrichment methods were also developed. Gozani and coworkers designed engineered three malignant brain tumor domain repeats of L3MBTL1 ($3 \times$ MBT) as a universal affinity reagent for the enrichment of proteome-wide lysine methylation [33,56]. This method utilized specific PTM−domain interaction and was able to recognize a wider range of methylated proteins than any available method using antibodies. Another example is the enrichment of glycosylated peptides by lectin weak affinity chromatography. With this method, over 60 nuclear proteins were identified to be O-GlcNAcylated in mouse embryonic stem cells [45]. Molecular biology approaches were also used for PTM enrichment. Typically, an epitope tag, for example a six histidines tag, an HA tag, or a flag tag, was fused with PTM moiety, such as ubiquitin and SUMO proteins [41]. Metabolic incorporation of a tag, such as an azide functional group, was also used and it successfully identified proteins that are modified with O-GlcNAc [46], farnesylation [57], palmitoylation [58], and so on. Lastly, chemical or enzymatic derivatization of a specific PTM with a tag was performed for enrichment. For example, β-elimination of O-GlcNAc Ser/Thr residues followed by Michael addition of a free thiol tag permit the identification of four O-GlcNAc histone peptides [47].

4.2.2.2 Mass spectrometry approaches for histone PTM analysis

Histones are most likely the protein family with the most heterogeneous variety of modifications, which is even more impressive given their small size (10−15 kDa). Cross-talk for histone PTM analysis describes two marks that maintain their coexistence frequency with each other independently from their relative abundance [59]. Several examples describe PTM cross-talk in histone proteins. For instance, the combinatorial mark H3S10K14ac on the gene p21 activates its transcription, which would not occur with only one of the two PTMs [60]. The protein HP1 recognizes H3K9me2/me3 and it spreads this mark along the chromatin to compact it, but it releases the binding in case S10 is phosphorylated [61]. Acetylation of H3K4 was found to inhibit the binding of the protein spChp1 to H3K9me2/me3 in *Schizosaccharomyces pombe* [62]. As well, the histone lysine demethylase PHF8 has its highest binding efficiency to the nucleosome when the three marks H3K4me3K9acK14ac are present [63]. Finally, combinatorial PTMs can be categorized as orthosteric or allosteric [64]. The former defines two modifications that occur nearby or at least in the same active region, and act via direct recognition or by blocking active sites; the latter describes two PTMs or sites distant in the protein that cross-talk with each other through conformational changes or via indirect influence of PTM-reading proteins. This is important to consider while choosing the analytical technique for their characterization, as marks far from each other in the protein sequence are generally more challenging to be detected simultaneously.

Three different MS-based proteomics strategies are used to map the complicated histone codes, identify coexisting PTMs, and study PTM cross-talk. They are the bottom-up, middle-down, and top-down methods, which are different in the size of their analytes [7,65]. Bottom-up is the most popular method used in proteomics for protein identification and characterization of protein PTMs. Proteins of interest are digested with a trypsin endoprotease into smaller peptides of 7−25 amino acids in length. As trypsin is highly specific in cleaving at the C-terminus of lysines and arginines, leaving each peptide with at least one basic amino acid residue, it enhances positive ionization efficiency. Such small peptides ionize more efficiently than intact proteins, separate better with RPLC and are identified with higher mass accuracy due to their small molecular weight. In addition, they are suitable for CID. However, a caveat of the bottom-up approach is that the information on PTM combinations is mostly lost, unless PTMs are localized within the short peptide. The top-down approach allows the investigation of coexisting distant PTMs as the protein of interest is analyzed as intact into the mass spectrometer. Intact proteins generally fragment more efficiently with ECD or ETD. But the limitation of the top-down method is that it is still not amenable to high-throughput analysis, as this technique is technically more challenging, mostly due to limitations in efficient separation of proteoforms and efficient sequencing, caused by incomplete fragmentation. The limitations of both bottom-up and top-down can be ameliorated by the middle-down strategy, a compromise between the other two strategies. This method involves limited enzymatic digestion and therefore long polypeptides (40−50 aa residues) are generated. In particular, histones can be digested by using enzymes such as AspN or GluC, which cleave the N-terminal of aspartic acid and the C-terminal of glutamic acid, respectively. Such amino acids occur more than 40 residues downstream of the sequence in almost every histone, allowing the cleaving of the entire N-terminal tail from the nucleosome core. This is able to provide a semiglobal overview of the PTM combinations on histone tails. Middle-down is gaining interest also due to improvements in peptide separation and bioinformatics [21]. At the moment, this strategy achieves hundreds of combinatorial

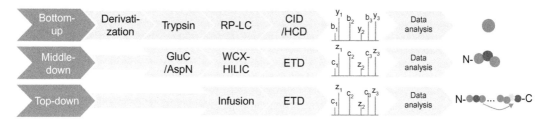

FIGURE 4.3

Typical experimental workflow for bottom-up, middle-down, and top-down MS analysis of histone PTMs.

PTMs identified on histone tails, which is where most of the PTMs reside. The use of this strategy requires MS capable of high mass resolution and ETD fragmentation, due to the high mass and the high charge state of histone tails. In addition, the novelty of the middle-down and the top-down strategies as large-scale proteomics platforms implies that only a few software applications are currently available to process such data. Figure 4.3 shows the workflow of each approach including sample digestion, separation, fragmentation methods, and typical MS results. In the following section, examples and practical considerations of three of the approaches in histone PTM analysis will be discussed. Figure 4.4 shows the base peak chromatogram and MS results of typical bottom-up and middle-down experiments.

4.2.2.2.1 Bottom-up

Most of the bottom-up proteomics experiments employ trypsin for digestion because of its efficiency, specificity, and robustness. However, histone tails are rich in lysine and arginine residues, so they are cleaved into excessively short peptides which do not retain on traditional RP chromatographic columns and thus cannot be efficiently detected by MS. In addition, the PTMs localized on lysine and arginine residues of the targeted histone reduce proteolytic efficiency, generating a mixture of peptides with and without missed cleavages, which reduces the reproducibility of the analysis. Alternatively, it is possible to digest histones with other approaches that cleave only the arginine residues and generate peptides of proper sizes. For instance, ArgC is an endoprotease specific for the C-terminal of arginines. This method has been used to determine the abundance of lysine modifications in canonical histone H3 and H3.3 in *Drosophila* cells. H3.3 was found to be enriched in modifications associated with transcriptional activity but deficient in H3K9me2 [67]. Alternatively, chemical derivatization of lysine residues followed by trypsin digestion can be used to generate an ArgC-like digestion. This method is more commonly adopted than ArgC digestion, as it does not require the purchase of additional enzymes and the derivatization process is commonly performed with inexpensive chemicals. In addition, the reactive group can be used with heavy isotopes, so that label-based relative quantification can be performed. For example, Smith et al. chemically acetylated the endogenously unacetylated lysines in the N-terminal tail of histone H4 with deuterated acetic anhydride [68]. The 3 Da mass difference between protiated and deuterated acetyl groups was used to determine the endogenous level of acetylation using MS. Garcia et al. developed a chemical derivatization method based on propionic anhydride derivatization; free amines of N-termini and unmodified and mono-methylated lysines react with propionic

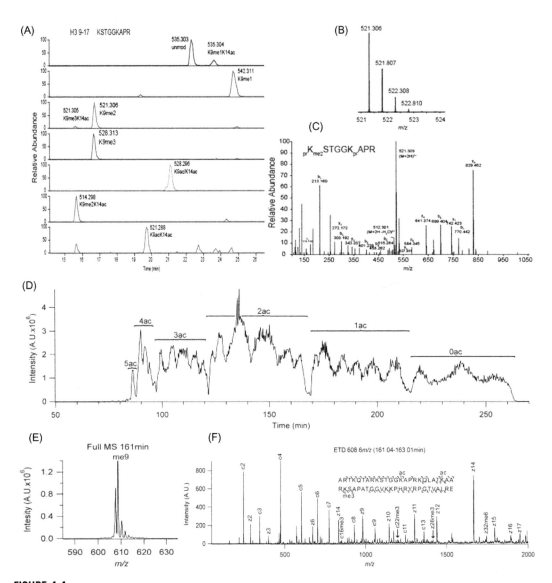

FIGURE 4.4

Examples of bottom-up and middle-down [66] analyses of the combinatorial histone codes of histone H3.
(A) Selective chromatogram of the $[M + 2H]^{2+}$ ions of histone H3 9−17 peptides. The peptides are propionylated
and digested by trypsin. Modifications and m/z values are indicated. (B) Full mass spectrum of the precursor ion
of 521.306 m/z. (C) MS/MS spectrum of the 521.306 m/z precursor ion at 17 min. (D) Base peak chromatogram
of the WCX-HILIC separation. (E) Full MS at 161 min showing relatively few co-eluting forms. The peaks
labeled with the number of methyl equivalences necessary to account for the mass shift. (F) ETD MS/MS
spectrum of the 608.6 m/z precursor ion at 162 min.

anhydride to form propionyl amides [69]. This method provides an additional advantage as compared to ArgC digestion or deuterated acetyl group modification because propionylated peptides have higher hydrophobicity. Therefore, the retention on the RP columns is enhanced, resulting in a more efficient chromatographic separation. This method has been successfully applied on histones for both in-solution and in-gel digestion. For instance, patterns of histone PTMs associating with bromodomain-containing proteins (Brd) and chromodomain-containing heterochromatin proteins (HP1) were enriched using chromatin immunoprecipitation (ChIP) and quantified by bottom-up MS [70]. Histones in both ChIP and input samples were derivatized with either d0- or isotopically stable d5-propionyl groups, so that two samples could be compared in a single MS experiment [71]. Such an approach could be expanded to map histone PTMs of local chromatin environment with other associating proteins.

Bottom-up proteomics, also named shotgun proteomics, was widely used to identify novel histone PTMs. For example, the Zhao group identified that 28 lysine residues of core histones and histone H1 were modified with crotonyl groups with a characteristic 68 Da mass shift [44]. This novel histone mark is concentrated on sex chromosome regions in postmeiotic male germinal cells and could be a signal of the male germ cell differentiation. Several other novel histone marks were also identified, including lysine formylation [72], succinylation [73], malonylation [73], propionylation [74], butyrylation [74], tyrosine hydroxylation [44], and so on [8]. In 2010, serine and threonine O-GlcNAcylation was also discovered as a part of the histone code [47]. Fujiki et al. suggests that H2B K120 GlcNAcylation promotes H2B K120 monoubiquitination for transcriptional activation, providing new evidence on histone PTM cross-talk [75]. Similarly, novel ubiquitination marks on N-terminal tails of histone H2A and H2A.X were identified at H2AK13 and H2AK15 by LC-MS/MS [76]. Functional studies indicated these sites are DNA-damage dependent and are targets by the E3 ubiquitin ligase RNF8 and RNF168. Collectively, the bottom-up strategy has proven to be the most sensitive, robust, and accurate technique to map PTMs. It is suitable for high-throughput and large-scale screenings.

4.2.2.2.2 Top-down

The major limitation of the bottom-up strategy is that trypsin digestion causes the loss of information about which PTMs coexist on the same histone protein. In addition, due to the high similarity in the sequence of histone variants, the bottom-up strategy does not always allow determination of which histone a certain PTM belongs to. For instance, the peptide containing K9 on histone H3 (KSTGGKAPR) is identical for histone H3.1, H3.2, H3.3, H3.1t and H3.3C. To address this problem, top-down proteomics was developed to investigate the entire protein sequence including all combinatorial PTMs. However, top-down MS analysis is still not as developed as bottom-up, as both LC and MS analysis are limited in resolution and efficient sequence analysis. Therefore, the common approach is to perform fractionation of histone isotypes offline previous MS direct infusion. Intact histones can be separated on the basis of their sizes using gel-filtration or of their isoelectric point (pI) using ion-exchange chromatography. However, these two methods achieve poor resolution; in particular, they cannot discriminate two histones with the same sequence but different PTMs. A successful top-down MS analysis relies on many key steps, including intact protein separation strategies, fragmentation methods, high-resolution mass spectrometers, mapping of the mass spectral data with bioinformatic software, and so on. Progress has been made toward the top-down MS sample preparation. Isoelectric focusing by free-flow electrophoresis (FFE), HILIC and multidimensional chromatography all start to play an important role in histone fractionating for top-down MS analysis. For example, with the combination of RP-HPLC, HPCE (High-Performance Capillary Electrophoresis) and HILIC, Lindner and co-workers separated and isolated histone H2A variants, acetylated H2A variants and acetylated H4 histones of butyrate-treated Friend erythroleukaemic cells (line B8) at both analytical and semi-preparative scales [77]. Comprehensive top-down MS analysis of intact histone H4 characterized and quantified 42 different proteoforms with combinations of methylation and acetylation, some of which were cell-cycle specific [78]. In another study, using an on-line two-dimensional RP-WCX-HILIC LC-MS/MS platform, 708 histone isoforms were identified with only 7.5 µg of histone sample in a single MS run [79]. In 2011, four-dimensional protein separation methods based on protein isoelectric point, size, and hydrophobicity were used to identify histone PTMs in a large-scale manner. More than 3,000 proteoforms from human cells were identified, among which 400 histones were included [80]. This demonstrates that top-down MS is emerging as a promising strategy for histone analysis in proteomics. However, the LC setup and the sample preparation are still complex and time consuming.

Intact histones are usually fragmented by using ECD or ETD because the high charge state of an electrospray-ionized intact protein is suitable for this type of MS/MS. ECD/ETD fragmentation consists of the addition of an electron to the analyte, donated from an electron donor (usually a molecule with an aromatic ring). This additional electron to the analyte rearranges the covalent bonds of the peptide backbone and it breaks at the N-Cα bond, generating mostly "c" and "z" ions (Figure 4.2). ECD is older than ETD, but it did not become widely adopted, since it requires the molecule to be immersed in a dense population of near-thermal electrons [81]. This specific condition is possible in cases when the fragmentation cell can generate a static magnetic or electric field, typical of expensive instruments such as FTICR MS. This is not applicable to instruments that isolate ions with radio-frequency electrostatic fields, which are the more widespread mass spectrometers. Afterward, ETD was developed to be suitable for more cost-effective mass analyzers such as linear trap quandrupoles. Linear traps are currently modified to apply secondary radio-frequency fields that trap both cations and anions, making the reaction between protein/peptide and the electron donor possible [82]. ECD and ETD of intact proteins usually require averaging of multiple MS/MS scans to increase the quality of the final spectrum. Therefore, it is recommended that intact proteins are not injected into the mass spectrometer for a very short amount of time (<1−2 s). By using direct infusion and an LTQ-Orbitrap equipped with ETD, 74 unique forms of histone H4 in human embryonic stem cells were identified and quantified during cell differentiation [83]. Similar analysis on histone H2A, H2B, H3 all revealed the presence of multiple variants and modifications [10].

4.2.2.2.3 Middle-down

To overcome the limitations of top-down but still gain information on histone variants and coexisting PTMs, the middle-down strategy was developed as a compromise between bottom-up and top-down. As mentioned, proteins are digested at low frequency occurring amino acid residues, such as glutamic (E) or aspartic acid (D), which produce peptides of 5−6 kDa, often corresponding to the intact histone N-terminal tails. This method has been successfully applied for the analysis of histone H3 tails, both cleaved at the first glutamic acid (50 aa) by GluC or at the first aspartic acid (77 aa) by AspN [84]. Garcia and coworkers developed a salt-less pH gradient based on weak cation exchange (WCX)-HILIC for online middle-down analysis of histone H3 and H4 tails [66]. Over 200 isoforms of histone H3.2 and over 70 isoforms of histone H4 in one MS experiment were reported. This WCX-HILIC method was further optimized and the histone H3 tails from mouse embryonic stem cells were analyzed, revealing major site-specific coexisting H3 marks, for example K9me1K23acK27me2K36me2 and K9me3K23acK27me2K36me2 quadruplet marks [85]. The middle-down strategy has also been used to study different phosphorylation states of linker histone H1t during spermatogenesis [86]. Phosphorylation sites in H1t were found at three nonconsensus motifs and two CDK consensus motifs.

4.2.3 BIOINFORMATICS FOR HISTONE CODE ANALYSIS

The match between theoretical and observed masses to unambiguously identify histone PTMs requires bioinformatics tools, in particular for large-scale studies where manual analysis of MS/Ms spectra would become unsustainable. Several software applications are currently available for database searching, such as MASCOT (Matrix Science) [87], SEQUEST (ThermoFisher Scientific) [88], X!Tandem [89], MaxQuant [90], and pFind [91]. Such tools are suitable for bottom-up proteomics, due to the relatively simple MS/MS spectra generated. For top-down and middle-down, limited software and tools are available for PTM characterization. Currently, only BIG Mascot (Matrix Science) and ProSightPC (ThermoFisher Scientific) are commercially available; the first one is an extension of Mascot that can process polypeptides or proteins up to 110 kDa; the second one is a commercially optimized version of ProSight PTM [92]. Such limited availability is due both to the lower number of requests for tools to process top-down or middle-down spectra and to the complexity of developing software that handles MS/MS spectra of large polypeptides. These spectra contain a large number of fragments, including non-backbone ones, and mixed species, due to the difficulties in discriminating nearly isobaric or isobaric polypeptides at the precursor mass selection. High-resolution mass analyzers resolve multiply-charged ions in MS/MS spectra in order to define the charge state of the identified signals. Software for middle-down and top-down analysis is required to deconvolute the mass of precursor and fragment ions from the observed m/z. The most common algorithms for spectra deconvolution are THRASH [93], Xtract, MS-Deconv [94], Y.A.D. A [95]. and Mascot Distiller (Matrix Science). In addition, software for histone modification analysis must allow for multiple different, coexisting PTMs in database searching. By including multiple variable modifications we dramatically raise the number of molecular candidates to be considered during the search. This demands high computational power and it increases the probability of false-positive identifications. Because of this, most software has a built-in upper limit for the number of variable PTMs allowed. For instance, VEMS (Virtual Expert Mass Spectrometrist) [96] is free software available on the web which can, among other things, discriminate between acetylation (42.011 Da) and trimethylation (42.047 Da) even from relatively low-resolution spectra by using diagnostic ions and neutral loss signals present in MS/MS spectra. Specifically for middle-down LC-MS/MS analysis, DiMaggio et al. proposed an algorithm based on a mixed integer linear optimization [97], which was implemented in the publicly available PILOT_PTM [98]. In summary, while for bottom-up proteomics several software packages are already available and widely accepted as reliable, middle-down and top-down studies are still based on few and little-exploited bioinformatics tools.

4.2.4 QUANTIFICATION AND DYNAMICS OF HISTONE PTMS BY MASS SPECTROMETRY

Chromatin is a highly dynamic architecture. Therefore, relative and absolute quantification of histone variants and PTMs is necessary when comparing different sets of experiments. Traditionally, the quantitative analysis of histone PTMs has been performed by immunochemical methods, such as western blots. As MS is a quantitative technology, strategies to analyze the dynamics of histone variants and PTMs have been extensively developed. Several chemicals and strategies have been published or commercialized to reach the best compromises between costs, efficacy, and multiplexing (analysis of multiple conditions simultaneously) of quantification. These strategies can be broadly divided into two major groups, which are label-free and stable isotopic labeling methods, as shown in Figure 4.5.

4.2.4.1 Label-Free Method

Compared with stable isotopic labeling methods, label-free quantification is cost-effective and it only requires minimal sample preparation. It also allows comparison across multiple experimental conditions. In principle, the label-free approach is based on peptide intensity profiling. The relative abundance of a specific modified peptide is obtained by integrating the area under its peak and

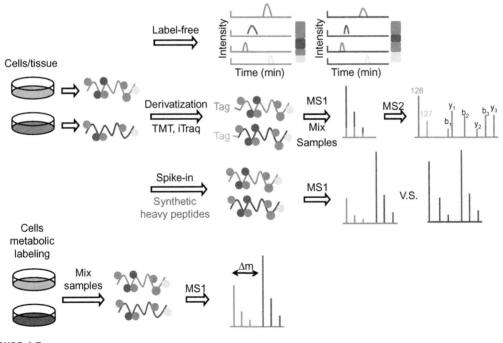

FIGURE 4.5

Histone quantification methods (label-free and stable isotopic labeling methods).

then dividing it by the sum area of that peptide in unmodified and all modified forms. In recent years, several groups have shown that histone modifications can be reliably quantified with the label-free approach. For example, Jenuwein and coworkers studied the histone lysine methylation states in partitioning chromosomal subdomains. Histone H3K9me3 and H3K27me3 were found to be enriched in pericentric heterochromatin and dependent on Suv39h histone methyltransferases [99]. In another study, Reinberg and coworkers studied the symmetricity of histone PTMs in nucleosomes [100]. Relative abundances of unmodified and three methylation forms of histone H3 K27 in nucleosomes containing at least one H3K27me2/3 in E14 ES cells were quantified. Results showed that 79% of all histone H3 tail in H3K27Me2/3 containing mononucleosomes have the H3K27me2/3 mark, whereas 21% are either unmethylated or monomethylated, which means H3K27me2/3 are modified with both symmetrically (58%) and asymmetrically (42%). The asymmetrical modification increases the range of attainable histone mark combinations.

Histone peptides without modification can be utilized as internal standards, similar to the loading control in western blot analysis. In each data set, the intensity of the peptide of interest is normalized based on the internal standards peptides and then compared across different experimental conditions. With peptides HLQLAIR from H2A, STELLIR from H3.3, and VFLENVIR and ISGLIYEETR from H4 used as internal peptides, Jensen and coworkers unbiasedly quantified the dose-response effect of the histone deacetylase inhibitor (HDACi) PDX101 on histone acetylation in human cell cultures [101]. With the higher concentration of PDX101 treatment, a dose-dependent increase in acetylated peptides from histone H2A, H2B, and H4 was detected with a decrease of H2B K57me2.

4.2.4.2 Stable isotopic labeling method

Even though label-free quantification is the most straightforward method for PTM analysis, its quality and reproducibility can be affected by many factors, such as peptide ionization efficiency. Stable isotopic labeling usually reduces the variability of the analysis. The common methods to incorporate an isotopic labeling for quantification are chemical derivatization with a tag, metabolic labeling or spike-in of a standard.

4.2.4.2.1 Metabolic labeling

Stable isotope labeling by amino acids in cell culture (SILAC) is a multiplexing quantitative proteomic method that uses labeled isotopically heavy amino acids, for example $^{13}C_6, ^{15}N_2$-lysine and $^{13}C_6, ^{15}N_4$-arginine, incorporated metabolically into the whole proteome [102,103]. Upon trypsin digestion at the C-terminal of lysine and arginine, peptides cultured in heavy media contain at least one heavily labeled amino acid residues, producing a mass difference from the same peptides cultured in traditional light media. The same population of cells from heavy or light media are mixed, as they can be distinguishable based on the peptide masses. The ratios between heavy and light peptides represent the relative abundance of the given peptides in the two conditions. The advantage of SILAC in quantification lies in its sample preparation. Because samples are mixed prior to protein extraction, the downstream treatments are all the same for both heavy and light samples, minimizing experimental errors. Therefore, SILAC produces relatively low variability and it is particularly useful to detect relatively small but significant changes in protein or PTM abundance between experimental conditions.

SILAC was originally designed for the analysis of the whole proteome. Rates of protein synthesis, degradation, proteome abundance, and dynamics have been investigated. SILAC can be used to quantify any modification because it labels the peptides. For example, Burlingame and coworkers studied the global O-GlcNAc modification influenced by Polycomb repressive complex 2 in mouse embryonic stem cells [45]. In their study, PRC2-null ESCs was cultured in light media and wild-type ESCs in heavy media; relative abundance of O-GlcNAcylated peptides between WT and eed$^{-/-}$ nuclear proteins were measured by ETD MS/MS after lectin enrichment. SILAC can also be performed by using nonnatural amino acid labeling; for example, azide-bearing methionine, L-azidohomoalanine (AHA), was used to label newly synthesized proteins [104]. After cell lysis, AHA carrying new proteins were enriched with an alkyne-bearing biotinylated tag using a click reaction, and protein expression was quantified from H/L ratios. SILAC has also been used to analyze histone PTM profiles as well. In the combination of mass spectrometry with SILAC, histone modification signatures in breast cancer cells were compared with normal epithelial breast cells [105]. In this study, breast cancer cells were labeled with heavy arginine (Arg10), as endoproteinase Arg-C was employed for digestion of histones. The so-called "breast cancer-specific epigenetic signatures" were identified with significant changes, such as loss of H3K9me3 and reduction of H4K16ac. Freitas and coworkers reported a method to study the cross-talk network between PTMs by using SILAC with MS [106]; 44 yeast histone mutants were generated to mimic acetylation or phosphorylation on the most commonly modified amino acid residues on all core histones, and they compared the effects by measuring the relative abundance of H3 K79 methylation and H3 K56 acetylation to those of wild type. This study recapitulated several known cross-talk and identified novel ones.

Cells can also be labeled in a pulsed manner, so that a fraction of proteins are labeled. By doing the pulse-chase-style labeling, kinetics of histone synthesis and degradation can be obtained by measuring the appearance or disappearance of the isotopic labels at different time points. Imhof and coworkers studied the kinetics of acetylation and methylation on newly synthesized histones in the S phase of the cell cycle by using $^{15}N_4$-arginine (R4) pulse-labeling method [107]. Different modification patterns on parental vs. newly deposited histones were determined. Kinetics of histones modifications were also studied after S phase by chase-labeling cells with $^{13}C_6,^{15}N_4$-arginine (R10) until achieving histone modification patterns indistinguishable from the parental ones.

Quantification of PTMs can also be achieved by introducing the isotopic labels into modifications. It is an extension of SILAC and has been applied for the quantification of methylation, phosphorylation, and acetylation. Replacing the normal methionine with methyl group isotopically labeled methionine in cell culture results in methyl-group-labeled S-adenosyl methionine, which is the sole methyl group donor; therefore all cellular methyl groups can be labeled [31]. In this way, Garcia and coworkers demonstrated histone methylation dynamics and stability in Hela cells [108]. Methylations associated with gene activation were found to have faster turnover rates than those associated with silent genes. Similarly, using [γ-$^{18}O4$]ATP as the phosphate source, global site-specific phosphorylation rates were determined, with a wide range of phosphorylation rate constants from 0.34 min^{-1} to 0.001 min^{-1} [55]. Dynamics of several histone phosphorylation sites were also measured across the cell cycle. Additionally, histone acetylation is regulated by metabolic intermediates of glucose and one major enzyme, ATP-citrate lyase, which converts citrate produced by the mitochondria into acetyl-CoA. Taking advantage of this metabolic pathway, acetyl groups of histone lysines have been shown to be isotopically labeled by using ^{13}C-glucose [109]. Turnovers of histone acetylation are generally slower than phosphorylation, but faster relative to methylation and the rates may vary depending on the modification sites as well as the presence of neighboring modifications.

Quantitative proteomics plays an important role in studying protein—protein interaction. In the context of epigenetics, the SILAC-based histone peptide pull-down experiment is a way to achieve unbiased identification of the PTM—readers interaction [63]. In the forward pull-down experiment, modified histone peptides at a specific site (baits) are immobilized on beads and incubated with nuclear extract which are labeled with heavy amino acids. The unmodified control peptides are incubated with light nuclear extract. In the reverse pull-down experiment, bait peptides are incubated with light nuclear extract while controls are incubated with heavy nuclear extract. After incubation and enrichment, the eluates in the forward or reverse experiments are mixed, and then SILAC H/L ratios are determined. Reader candidates give large H/L values in the forward experiment and small H/L ratios in the reverse experiment. On the other hand, background and unspecific binders are discriminated, giving H/L ratios of about 1. With this method, binders of H3K4me3 were identified, such as TFIID complex member TAF1. This approach can be extended to study all histone PTM readers. It holds great promise in defining the histone PTM interactome.

4.2.4.2.2 Chemical tags

The main limitation of metabolic labeling is that isotopic labels have to be incorporated during a stable cell culture; usually 6—8 passages are necessary to achieve >97% isotopic incorporation. This is not amenable in case of analyzing clinical samples (e.g., biological fluids, tissue samples) and in cell lines which could not pass a certain number of passages. Chemical derivatization post-sample collection is a popular method to overcome these issues. For instance, heavy isotope-labeled d_{10}-propionic anhydride can be used to label the N-terminus of histone peptides after the first round of propionylation and trypsin digestion. Equal amount of duplex isotopically labeled samples can then be mixed and analyzed in a single LC-MS run. Chemically identical peptides elute at the same time but have a mass difference of 5 Da.

Peptide derivatization by using isobaric tags is another method for relative quantification. A unique isobaric tag for each sample is labeled at the N-terminus amines of the peptides after trypsin digestion. All tagged samples are then equally mixed for MS analysis; iTRAQ and TMT are the most common isobaric tags [110]. Each tag is composed of three parts: a reporter, a balancer, and a reactive group. The reactive group creates a covalent bond with the peptides, generally with free amines; the reporter is released upon MS/MS fragmentation, and it has a different mass for each condition labeled; the balancer has a mass that corresponds to the total mass of the tag (identical for all conditions) minus reactive group and reporter. In this way, the same peptides from various samples have the same precursor masses but upon CID or ETD fragmentation, the reporters are released. Relative quantification information can be read out from the intensity of the reporters at the MS2 level. iTRAQ has been used to quantify differences in the expression of histone variants between samples, such as highly tumorigenic cancer lines and less tumorigenic parental lines. Isobaric tags can also be read out at the MS3 level in order to minimize the interference on the reporters' intensity caused by co-isolated multiple precursor ions [111]. To increase the multiplexing in bottom-up MS analysis, Gygi and coworkers combined 6-plex TMT with triple SILAC labeling to quantify 18 different samples in a single run. With this hyperplexing method, they quantified the protein abundance changes of rapamycin treated yeast at different time points.

4.2.4.2.3 Internal standards spike-in

A drawback of labeling methods is that the chemical derivatization of the analyte could be incomplete or generate side products not considered during data analysis. Each additional step during sample preparation might lead to variability or sample loss. Additionally, instrumental bias, such as that caused by different ionization efficiencies between unmodified and modified peptides, cannot be eliminated with standard spike-in quantification. Briefly, synthetic and isotopically labeled peptides are spiked in as internal standards for relative or absolute quantification. Intensities of peptides of interest are normalized to their corresponding isotope-encoded synthetic peptides. With this method, the Garcia lab recently developed a new method in histone PTM analysis, in which a library of 93 synthetic histone peptides were used as spiked-in internal standards [112]. This method improved the accuracy of histone PTM bottom-up quantification. By spiking-in known concentrations of isotopically labeled synthetic peptides, it is also possible to measure the absolute quantification for the analyzed targeted. Even though most experiments analyze the changes of a model system between two or more conditions, knowledge of the absolute amount of a protein or PTM might be important in analysis where the amount of a biomolecule is relevant for a diagnosis. Darwanto and coworkers identified a cross-talk between H3K79 methylation with H2BK120 ubiquitination in U927 lymphoma cells with the spike-in method [113]. Heavy label cells can also be utilized as internal standards. For example, the Neuro2A cell line has been metabolically labeled with isotopic leucine and used as internal standards to quantify the protein abundance in mouse brains [114]. In another study, the hepatoma cell line Hepa1−6 was labeled to study mouse primary hepatocytes [115]. Later, the Super-SILAC method was developed in order to overcome the problem of high ratio differences between internal standards of cell lines and samples from primary tissues or lack of presence of a specific peptide in cell lines [116].

4.3 CONCLUSION

In summary, the high-throughput, large-scale, and quantitative characters of mass spectrometry enable it to serve as a versatile and powerful analytical tool in epigenetics, complementary to other valuable technologies. Over the past decade, MS-based proteomics has already made dramatic impacts on comprehensive understanding of epigenetic mechanisms in health and disease. High-throughput bottom-up mass spectrometry plays a very important role in deciphering the histone codes and elucidating the complex chromatin interactome. MS analysis may be able to provide not only a global picture of histone PTMs by studying bulk chromatin, but also may be able to determine chromatin modifications and interactome in a locus-specific manner with the combination of ChIP. Additionally, in conjunction with next-generation sequencing technologies, such as ChIP-seq, gene-specific function analysis of individual histone PTM would be known. Sample preparation prior to mass spectrometry analysis, such as methodologies of PTM enrichment, also significantly improves the selectivity and accuracy of MS analysis. Bioinformatics tools post-mass spectrometry support efficient and reliable data analysis. Software packages are becoming more accessible and new algorithms are developing. Top-down and middle-down methodologies have also experienced a tremendous growth and have proven their significant roles in the field of epigenetics in resolving comprehensive multivalent and coexisting marks. Although applications of high-throughput top-down and middle-down proteomics are still limited, they might yet become achievable in the near future, thanks to the accelerating speed of the ongoing development of mass spectrometry, computational methods, and chromatography technologies.

REFERENCES

[1] Russo VEA, Martienssen RA, Riggs AD. Epigenetic Mechanisms of Gene Regulation. Plainview, NY: Cold Spring Harbor Laboratory Press; 1996.

[2] Bernstein BE, Meissner A, Lander ES. The mammalian epigenome. Cell 2007;128(4):669−81.

[3] Allis CD, Jenuwein T, Reinberg D. Epigenetics. Cold Spring Harbor, New York: Cold Spring Harbor Laboratory Press; 2007.

[4] Bird A. DNA methylation patterns and epigenetic memory. Genes Dev 2002;16(1):6−21.

[5] Bannister AJ, Kouzarides T. Regulation of chromatin by histone modifications. Cell Res 2011;21(3): 381−95.

[6] Kouzarides T. Chromatin modifications and their function. Cell 2007;128(4):693−705.

[7] Young NL, Dimaggio PA, Garcia BA. The significance, development and progress of high-throughput combinatorial histone code analysis. Cell Mol Life Sci 2010;67(23):3983−4000.

[8] Arnaudo AM, Garcia BA. Proteomic characterization of novel histone post-translational modifications. Epigenetics Chromatin 2013;6.

[9] Portela A, Esteller M. Epigenetic modifications and human disease. Nat Biotechnol 2010;28(10): 1057−68.

[10] Arnaudo AM, Molden RC, Garcia BA. Revealing histone variant induced changes via quantitative proteomics. Crit Rev Biochem Mol Biol 2011;46(4):284−94.

[11] Allis CD. Translating the histone code: a tale of tails. Speech. Available at: < http://www.nigms.nih.gov/News/meetings/pages/Stetten_2001.aspx>.

[12] Jenuwein T, Allis CD. Translating the histone code. Science 2001;293(5532):1074−80.

[13] Wang ZX, Patel DJ. Combinatorial readout of dual histone modifications by paired chromatin-associated modules. J Biol Chem 2011;286(21):18363−8.

[14] Turner BM. Defining an epigenetic code. Nat Cell Biol 2007;9(1):2−6.

[15] Fraga MF, Ballestar E, Villar-Garea A, Boix-Chornet M, Espada J, Schotta G, et al. Loss of acetylation at Lys16 and trimethylation at Lys20 of histone H4 is a common hallmark of human cancer. Nat Genet 2005;37(4):391−400.

[16] Wei Y, Xia W, Zhang Z, Liu J, Wang H, Adsay NV, et al. Loss of trimethylation at lysine 27 of histone H3 is a predictor of poor outcome in breast, ovarian, and pancreatic cancers. Mol Carcinog 2008;47(9):701−6.

[17] Seligson DB, Horvath S, McBrian MA, Mah V, Yu H, Tze S, et al. Global levels of histone modifications predict prognosis in different cancers. Am J Pathol 2009;174(5):1619−28.

[18] Yoo CB, Jones PA. Epigenetic therapy of cancer: past, present and future. Nat Rev Drug Discov 2006;5(1): 37−50.

[19] Chandra T, Kirschner K, Thuret JY, Pope BD, Ryba T, Newman S, et al. Independence of repressive histone marks and chromatin compaction during senescent heterochromatic layer formation. Mol Cell 2012;47(2):203−14.

[20] Yuan W, Xu M, Huang C, Liu N, Chen S, Zhu B. H3K36 methylation antagonizes PRC2-mediated H3K27 methylation. J Biol Chem 2011;286(10):7983−9.

[21] Sidoli S, Cheng L, Jensen ON. Proteomics in chromatin biology and epigenetics: elucidation of post-translational modifications of histone proteins by mass spectrometry. J Proteomics 2012;75(12): 3419−33.

[22] Eberl HC, Mann M, Vermeulen M. Quantitative proteomics for epigenetics. Chembiochem 2011;12(2): 224−34.

[23] Zee BM, Young NL, Garcia BA. Quantitative proteomic approaches to studying histone modifications. Curr Chem Genomics 2011;5(Suppl. 1):106−14.

[24] Karch KR, Denizio JE, Black BE, Garcia BA. Identification and interrogation of combinatorial histone modifications. Front Genet 2013;4:264.

[25] Zhang KL, Williams KE, Huang L, Yau P, Siino JS, Bradbury EM, et al. Histone acetylation and deacetylation − Identification of acetylation and methylation sites of HeLa histone H4 by mass spectrometry. Mol Cell Proteomics 2002;1(7):500−8.

[26] Fischle W, Tseng BS, Dormann HL, et al. Regulation of HP1-chromatin binding by histone H3 methylation and phosphorylation. Nature 2005;438(7071):1116−22.

[27] Garcia BA, Busby SA, Barber CM, Shabanowitz J, Allis CD, Hunt DF. Characterization of phosphorylation sites on histone H1 isoforms by tandem mass spectrometry. J Proteome Res 2004;3(6):1219−27.

[28] Han X, Aslanian A, Yates III JR. Mass spectrometry for proteomics. Curr Opin Chem Biol 2008;12(5): 483−90.

[29] Trelle MB, Salcedo-Amaya AM, Cohen AM, Stunnenberg HG, Jensen ON. Global histone analysis by mass spectrometry reveals a high content of acetylated lysine residues in the malaria parasite *Plasmodium falciparum*. J Proteome Res 2009;8(7):3439−50.

[30] Kim SC, Sprung R, Chen Y, Xu Y, Ball H, Pei J, et al. Substrate and functional diversity of lysine acetylation revealed by a proteomics survey. Mol Cell 2006;23(4):607−18.

[31] Ong SE, Mittler G, Mann M. Identifying and quantifying in vivo methylation sites by heavy methyl SILAC. Nat Methods 2004;1(2):119−26.

[32] Cao XJ, Arnaudo AM, Garcia BA. Large-scale global identification of protein lysine methylation in vivo. Epigenetics 2013;8(5):477−85.

[33] Carlson SM, Moore KE, Green EM, Martin GM, Gozani O. Proteome-wide enrichment of proteins modified by lysine methylation. Nat Protoc 2014;9(1):37−50.

[34] Schlosser A, Pipkorn R, Bossemeyer D, Lehmann WD. Analysis of protein phosphorylation by a combination of elastase digestion and neutral loss tandem mass spectrometry. Anal Chem 2001;73(2):170−6.

[35] Kee JM, Oslund RC, Perlman DH, Muir TW. A pan-specific antibody for direct detection of protein histidine phosphorylation. Nat Chem Biol 2013;9(7):416−21.

[36] Ficarro SB, McCleland ML, Stukenberg PT, Burke DJ, Ross MM, Shabanowitz J, et al. Phosphoproteome analysis by mass spectrometry and its application to *Saccharomyces cerevisiae*. Nat Biotechnol 2002;20(3):301−5.

[37] Thingholm TE, Jorgensen TJD, Jensen ON, Larsen MR. Highly selective enrichment of phosphorylated peptides using titanium dioxide. Nat Protoc 2006;1(4):1929−35.

[38] Larsen MR, Thingholm TE, Jensen ON, Roepstorff P, Jorgensen TJD. Highly selective enrichment of phosphorylated peptides from peptide mixtures using titanium dioxide microcolumns. Mol Cell Proteomics 2005;4(7):873−86.

[39] Fedorova M, Frolov A, Hoffmann R. Fragmentation behavior of Amadori-peptides obtained by non-enzymatic glycosylation of lysine residues with ADP-ribose in tandem mass spectrometry. J Mass Spectrom 2010;45(6):664−9.

[40] Laing S, Koch-Nolte F, Haag F, Buck F. Strategies for the identification of arginine ADP-ribosylation sites. J Proteomics 2011;75(1):169−76.

[41] Kirkpatrick DS, Denison C, Gygi SP. Weighing in on ubiquitin: the expanding role of mass-spectrometry-based proteomics. Nat Cell Biol 2005;7(8):750−7.

[42] Peng J, Schwartz D, Elias JE, Thoreen CC, Cheng D, Marsischky G, et al. A proteomics approach to understanding protein ubiquitination. Nat Biotechnol 2003;21(8):921−6.

[43] Wohlschlegel JA, Johnson ES, Reed SI, Yates III JR. Global analysis of protein sumoylation in *Saccharomyces cerevisiae*. J Biol Chem 2004;279(44):45662−8.

[44] Tan MJ, Luo H, Lee S, Jin F, Yang JS, Montellier E, et al. Identification of 67 histone marks and histone lysine crotonylation as a new type of histone modification. Cell 2011;146(6):1015−27.

[45] Myers SA, Panning B, Burlingame AL. Polycomb repressive complex 2 is necessary for the normal site-specific O-GlcNAc distribution in mouse embryonic stem cells. Proc Natl Acad Sci USA 2011;108 (23):9490−5.

[46] Vocadlo DJ, Hang HC, Kim EJ, Hanover JA, Bertozzi CR. A chemical approach for identifying O-GlcNAc-modified proteins in cells. Proc Natl Acad Sci USA 2003;100(16):9116−21.

[47] Sakabe K, Wang ZH, Hart GW. Beta-N-acetylglucosamine (O-GlcNAc) is part of the histone code. Proc Natl Acad Sci USA 2010;107(46):19915−20.

[48] Shechter D, Dormann HL, Allis CD, Hake SB. Extraction, purification and analysis of histones. Nat Protoc 2007;2(6):1445−57.

[49] Boersema PJ, Mohammed S, Heck AJ. Hydrophilic interaction liquid chromatography (HILIC) in proteomics. Anal Bioanal Chem 2008;391(1):151−9.

[50] Palmisano G, Lendal SE, Engholm-Keller K, Leth-Larsen R, Parker BL, Larsen MR. Selective enrichment of sialic acid-containing glycopeptides using titanium dioxide chromatography with analysis by HILIC and mass spectrometry. Nat Protoc 2010;5(12):1974−82.

[51] Di Palma S, Boersema PJ, Heck AJ, Mohammed S. Zwitterionic hydrophilic interaction liquid chromatography (ZIC-HILIC and ZIC-cHILIC) provide high resolution separation and increase sensitivity in proteome analysis. Anal Chem 2011;83(9):3440−7.

[52] Ferrari KJ, Scelfo A, Jammula S, Cuomo A, Barrozi I, Stützer A, et al. Polycomb-dependent H3K27me1 and H3K27me2 regulate active transcription and enhancer fidelity. Mol Cell 2014;53(1):49−62.

[53] Zhao Y, Jensen ON. Modification-specific proteomics: strategies for characterization of post-translational modifications using enrichment techniques. Proteomics 2009;9(20):4632−41.

[54] Zhai B, Villen J, Beausoleil SA, Mintseris J, Gygi SP. Phosphoproteome analysis of *Drosophila melanogaster* embryos. J Proteome Res 2008;7(4):1675−82.

[55] Molden RC, Goya J, Khan Z, Garcia BA. Stable isotope labeling of phosphoproteins for large-scale phosphorylation rate determination. Mol Cell Proteomics 2014;13(4):1106−18.

[56] Moore KE, Carlson SM, Camp ND, Cheung P, James RG, Chua KF, et al. A general molecular affinity strategy for global detection and proteomic analysis of lysine methylation. Mol Cell 2013;50 (3):444−56.

[57] Kho Y, Kim SC, Jiang C, Barma D, Kwon SW, Cheng J, et al. A tagging-via-substrate technology for detection and proteomics of farnesylated proteins. Proc Natl Acad Sci USA 2004;101(34):12479−84.

[58] Kostiuk MA, Corvi MM, Keller BO, Plummer G, Prescher JA, Hangauer MJ, et al. Identification of palmitoylated mitochondrial proteins using a bio-orthogonal azido-palmitate analogue. FASEB J 2008;22(3): 721−32.

[59] Schwammle V, Aspalter CM, Sidoli S, Jensen ON. Large-scale analysis of co-existing post-translational modifications on histone tails reveals global fine-structure of crosstalk. Mol Cell Proteomics 2014.

[60] Simboeck E, Sawicka A, Zupkovitz G, Senese S, Winter S, Dequiedt F, et al. A phosphorylation switch regulates the transcriptional activation of cell cycle regulator p21 by histone deacetylase inhibitors. J Biol Chem 2010;285(52):41062−73.

[61] Hirota T, Lipp JJ, Toh BH, Peters JM. Histone H3 serine 10 phosphorylation by Aurora B causes HP1 dissociation from heterochromatin. Nature 2005;438(7071):1176−80.

[62] Xhemalce B, Kouzarides T. A chromodomain switch mediated by histone H3 Lys 4 acetylation regulates heterochromatin assembly. Genes Dev 2010;24(7):647−52.

[63] Vermeulen M, Eberl HC, Matarese F, Marks H, Denissov S, Butter F, et al. Quantitative interaction proteomics and genome-wide profiling of epigenetic histone marks and their readers. Cell 2010;142(6): 967–80.

[64] Nussinov R, Tsai CJ, Xin F, Radivojac P. Allosteric post-translational modification codes. Trends Biochem Sci 2012;37(10):447–55.

[65] Garcia BA. What does the future hold for top down mass spectrometry? J Am Soc Mass Spectrom 2010;21(2):193–202.

[66] Young NL, DiMaggio PA, Plazas-Mayorca MD, Baliban RC, Floudas CA, Garcia BA. High throughput characterization of combinatorial histone codes. Mol Cell Proteomics 2009;8(10):2266–84.

[67] McKittrick E, Gaften PR, Ahmad K, Henikoff S. Histone H3.3 is enriched in covalent modifications associated with active chromatin. Proc Natl Acad Sci USA 2004;101(6):1525–30.

[68] Smith CM, Gafken PR, Zhang ZL, Gottschling DE, Smith JB, Smith DL. Mass spectrometric quantification of acetylation at specific lysines within the amino-terminal tail of histone H4. Anal Biochem 2003;316(1):23–33.

[69] Garcia BA, Mollah S, Ueberheide BM, Busby SA, Muratore TL, Shabanowitz J, et al. Chemical derivatization of histones for facilitated analysis by mass spectrometry. Nat Protoc 2007;2(4):933–8.

[70] Leroy G, Chepelev I, DiMaggio PA, Blanco MA, Zee BM, Zhao K, et al. Proteogenomic characterization and mapping of nucleosomes decoded by Brd and HP1 proteins. Genome Biol 2012;13:8.

[71] Plazas-Mayorca MD, Bloom JS, Zeissler U, Leroy G, Young NL, DiMaggio PA, et al. Quantitative proteomics reveals direct and indirect alterations in the histone code following methyltransferase knockdown. Mol Biosyst 2010;6(9):1719–29.

[72] Jiang T, Zhou XF, Taghizadeh K, Dong M, Dedon PC. N-formylation of lysine in histone proteins as a secondary modification arising from oxidative DNA damage. Proc Natl Acad Sci USA 2007;104(1): 60–5.

[73] Xie ZY, Dai JBA, Dai LZ, Tan M, Cheng Z, Wu Y, et al. Lysine succinylation and lysine malonylation in histones. Mol Cell Proteomics 2012;11(5):100–7.

[74] Chen Y, Sprung R, Tang Y, Ball H, Sangras B, Kim SC, et al. Lysine propionylation and butyrylation are novel post-translational modifications in histones. Mol Cell Proteomics 2007;6(5):812–19.

[75] Fujiki R, Hashiba W, Sekine H, Yokoyama A, Chikanishi T, Ito S, et al. GlcNAcylation of histone H2B facilitates its monoubiquitination. Nature 2011;480(7378):557–60.

[76] Gatti M, Pinato S, Maspero E, Soffientini P, Polo S, Penengo L. A novel ubiquitin mark at the N-terminal tail of histone H2As targeted by RNF168 ubiquitin ligase. Cell Cycle 2012;11(13):2538–44.

[77] Lindner H, Sarg B, Meraner C, Helliger W. Separation of acetylated core histones by hydrophilic-interaction liquid chromatography. J Chromatogr A 1996;743(1):137–44.

[78] Pesavento JJ, Bullock CR, Leduc RD, Mizzen CA, Kelleher NL. Combinatorial modification of human histone H4 quantitated by two-dimensional liquid chromatography coupled with top down mass spectrometry. J Biol Chem 2008;283(22):14927–37.

[79] Tian ZX, Tolic N, Zhao R, Moore RJ, Hengel SM, Robinson EW, et al. Enhanced top-down characterization of histone post-translational modifications. Genome Biol 2012;13:10.

[80] Tran JC, Zamdborg L, Ahlf DR, Lee JE, Catherman AD, Durbin KR, et al. Mapping intact protein isoforms in discovery mode using top-down proteomics. Nature 2011;480(7376):254–8.

[81] Zubarev RA, Horn DM, Fridriksson EK, Kelleher NL, Kruger NA, Lewis MA, et al. Electron capture dissociation for structural characterization of multiply charged protein cations. Anal Chem 2000;72(3): 563–73.

[82] Syka JE, Coon JJ, Schroeder MJ, Shabanowitz J, Hunt DF. Peptide and protein sequence analysis by electron transfer dissociation mass spectrometry. Proc Natl Acad Sci USA 2004;101(26):9528–33.

[83] Phanstiel D, Brumbaugh J, Berggren WT, Conard K, Feng X, Levenstein ME, et al. Mass spectrometry identifies and quantifies 74 unique histone H4 isoforms in differentiating human embryonic stem cells. Proc Natl Acad Sci USA 2008;105(11):4093−8.

[84] Garcia BA, Pesavento JJ, Mizzen CA, Kelleher NL. Pervasive combinatorial modification of histone H3 in human cells. Nat Methods 2007;4(6):487−9.

[85] Jung HR, Sidoli S, Haldbo S, Sprenger RR, Schwämmle V, Pasini D, et al. Precision mapping of coexisting modifications in histone h3 tails from embryonic stem cells by ETD-MS/MS. Anal Chem 2013;85(17):8232−9.

[86] Sarg B, Chwatal S, Talasz H, Lindner HH. Testis-specific linker histone H1t is multiply phosphorylated during spermatogenesis. Identification of phosphorylation sites. J Biol Chem 2009;284(6):3610−18.

[87] Perkins DN, Pappin DJ, Creasy DM, Cottrell JS. Probability-based protein identification by searching sequence databases using mass spectrometry data. Electrophoresis 1999;20(18):3551−67.

[88] Eng JK, McCormack AL, Yates JR. An approach to correlate tandem mass spectral data of peptides with amino acid sequences in a protein database. J Am Soc Mass Spectrom 1994;5(11):976−89.

[89] Craig R, Beavis RC. TANDEM: matching proteins with tandem mass spectra. Bioinformatics 2004;20(9): 1466−7.

[90] Cox J, Mann M. MaxQuant enables high peptide identification rates, individualized p.p.b.-range mass accuracies and proteome-wide protein quantification. Nat Biotechnol 2008;26(12):1367−72.

[91] Li D, Fu Y, Sun R, Ling CX, Wei Y, Zhou H, et al. pFind: a novel database-searching software system for automated peptide and protein identification via tandem mass spectrometry. Bioinformatics 2005;21 (13):3049−50.

[92] LeDuc RD, Taylor GK, Kim YB, Januszyk TE, Bynum LH, Sola JV, et al. ProSight PTM: an integrated environment for protein identification and characterization by top-down mass spectrometry. Nucleic Acids Res 2004;32(Web Server issue):W340−5.

[93] Horn DM, Zubarev RA, McLafferty FW. Automated reduction and interpretation of high resolution electrospray mass spectra of large molecules. J Am Soc Mass Spectrom 2000;11(4):320−32.

[94] Liu X, Inbar Y, Dorrestein PC, Wynne C, Edwards N, Souda P, et al. Deconvolution and database search of complex tandem mass spectra of intact proteins: a combinatorial approach. Mol Cell Proteomics 2010;9(12):2772−82.

[95] Carvalho PC, Xu T, Han X, Cociorva D, Barbosa VC, Yates III JR. YADA: a tool for taking the most out of high-resolution spectra. Bioinformatics 2009;25(20):2734−6.

[96] Matthiesen R, Trelle MB, Hojrup P, Bunkenborg J, Jensen ON. VEMS 3.0: algorithms and computational tools for tandem mass spectrometry based identification of post-translational modifications in proteins. J Proteome Res 2005;4(6):2338−47.

[97] DiMaggio Jr. PA, Young NL, Baliban RC, Garcia BA, Floudas CA. A mixed integer linear optimization framework for the identification and quantification of targeted post-translational modifications of highly modified proteins using multiplexed electron transfer dissociation tandem mass spectrometry. Mol Cell Proteomics 2009;8(11):2527−43.

[98] Baliban RC, DiMaggio PA, Plazas-Mayorca MD, Young NL, Garcia BA, Floudas CA. A novel approach for untargeted post-translational modification identification using integer linear optimization and tandem mass spectrometry. Mol Cell Proteomics 2010;9(5):764−79.

[99] Peters AH, Kubicek S, Mechtler K, O'Sullivan RJ, Derijck AA, Perez-Burgos L, et al. Partitioning and plasticity of repressive histone methylation states in mammalian chromatin. Mol Cell 2003;12 (6):1577−89.

[100] Voigt P, LeRoy G, Drury WJ, Zee BM, Son J, Beck DB, et al. Asymmetrically modified nucleosomes. Cell 2012;151(1):181−93.

[101] Beck HC, Nielsen EC, Matthiesen R, Jensen LH, Sehested M, Finn P, et al. Quantitative proteomic analysis of post-translational modifications of human histones. Mol Cell Proteomics 2006;5(7): 1314−25.

[102] Ong SE, Blagoev B, Kratchmarova I, Kristensen DB, Steen H, Pandey A, et al. Stable isotope labeling by amino acids in cell culture, SILAC, as a simple and accurate approach to expression proteomics. Mol Cell Proteomics 2002;1(5):376−86.

[103] Mann M. Functional and quantitative proteomics using SILAC. Nat Rev Mol Cell Biol 2006;7(12): 952−8.

[104] Howden AJM, Geoghegan V, Katsch K, Efstathiou G, Bhushan B, Boutureira O, et al. QuaNCAT: quantitating proteome dynamics in primary cells. Nat Methods 2013;10(4):343−6.

[105] Cuomo A, Moretti S, Minucci S, Bonaldi T. SILAC-based proteomic analysis to dissect the "histone modification signature" of human breast cancer cells. Amino Acids 2011;41(2):387−99.

[106] Guan XY, Rastogi N, Parthun MR, Freitas MA. Discovery of histone modification crosstalk networks by stable isotope labeling of amino acids in cell culture mass spectrometry (SILAC MS). Mol Cell Protcomics 2013;12(8):2048−59.

[107] Scharf AND, Barth TK, Imhof A. Establishment of histone modifications after chromatin assembly. Nucleic Acids Res 2009;37(15):5032−40.

[108] Zee BM, Levin RS, Xu B, LeRoy G, Wingreen NS, Garcia BA. In vivo residue-specific histone methylation dynamics. J Biol Chem 2010;285(5):3341−50.

[109] Evertts AG, Zee BM, DiMaggio PA, Gonzales-Cope M, Coller HA, Garcia BA. Quantitative dynamics of the link between cellular metabolism and histone acetylation. J Biol Chem 2013;288(17):12142−51.

[110] Ross PL, Huang YN, Marchese JN, Williamson B, Parker K, Hattan S, et al. Multiplexed protein quantitation in *Saccharomyces cerevisiae* using amine-reactive isobaric tagging reagents. Mol Cell Proteomics 2004;3(12):1154−69.

[111] Ting L, Rad R, Gygi SP, Haas W. MS3 eliminates ratio distortion in isobaric multiplexed quantitative proteomics. Nat Methods 2011;8(11):937−40.

[112] Lin S, Wein S, Gonzales-Cope M, Otte GL, Yuan Z-F, Afjehi-Sadat L, et al. Stable isotope labeled histone peptide library for histone post-translational modification and variant quantification by mass spectrometry. Mol Cell Proteomics 2014;13:2450−66.

[113] Darwanto A, Curtis MP, Schrag M, Kirsch W, Liu P, Xu G, et al. A modified "cross-talk" between histone H2B Lys-120 ubiquitination and H3 Lys-79 methylation. J Biol Chem 2010;285(28):21868−76.

[114] Ishihama Y, Sato T, Tabata T, Miyamoto N, Sagane K, Nagasu T, et al. Quantitative mouse brain proteomics using culture-derived isotope tags as internal standards. Nat Biotechnol 2005;23(5):617−21.

[115] Pan CP, Kumar C, Bohl S, Klingmueller U, Mann M. Comparative proteomic phenotyping of cell lines and primary cells to assess preservation of cell type-specific functions. Mol Cell Proteomics 2009;8 (3):443−50.

[116] Geiger T, Cox J, Ostasiewicz P, Wisniewski JR, Mann M. Super-SILAC mix for quantitative proteomics of human tumor tissue. Nat Methods 2010;7(5):383−5.

TECHNIQUES ANALYZING CHROMATIN MODIFICATIONS AT SPECIFIC SINGLE LOCI

Xiangyun Amy Chen, Jinquan Sun, and Yanming Wang

Center for Eukaryotic Gene Regulation, Department of Biochemistry and Molecular Biology,
Pennsylvania State University, PA, USA

CHAPTER OUTLINE

Y.G. Zheng (Ed): Epigenetic Technological Applications. DOI: http://dx.doi.org/10.1016/B978-0-12-801080-8.00005-3

79

There are over 20,000 genes in the human genome. These genes are precisely regulated so that the right amount of gene product is expressed at the right time and space during early development and adult life cycles in a multicellular organism. Why are some genes expressed while others in the vicinity are not? Why are housekeeping genes always expressed while cell cycle regulatory genes are expressed only at certain stages of the cell cycle? Why are cell death regulatory genes expressed only under certain physiological or pharmacological conditions? Those questions all converge on a central question: How is a single gene uniquely regulated?

5.1 GENE EXPRESSION PATTERN IS RELATED TO THE COVALENT MODIFICATION OF CORE HISTONES

In eukaryotic nuclei, about 200 bp double-stranded DNA wrap around the four core histones (H2A, H2B, H3, and H4) to form a nucleosome, the basic building block of eukaryotic chromatin [1]. Those core histones are subject to many covalent modifications such as methylation, acetylation, phosphorylation, and citrullination [2]. A single modification represents a unique gene expression feature that could differ under different genomic contexts. For example, methylation of histone H3 lysine 9 in the promoter region of a singular gene is related to heterochromatin formation and gene repression [3], while this modification can contribute to gene activation if located in the gene body [1]. In addition, methylation of histone H3 lysine 4 is often the hallmark of euchromatin and gene activation [4]. The N-terminal of histone H3 tail is heavily modified with several modification clusters/cassettes (Figure 5.1A). Of these, Arg8, Lys9, and Ser10 represent an interesting cluster where citrullination of Arg 8 by chromatin modifier PAD4 (peptidylarginine deiminase 4) leads to dissociation of HP1 (heterochromatin protein 1) that recognizes Lys 9 methylation [5,6], while Lys9/Ser10 might also serve as a binary switch and regulate the interaction of HP1 with Lys9 methylation [7]. Therefore, different combinations of histone modifications form numerous codes for the gene expression state. It is thus important to study these histone modifications on a target gene locus to understand the multifaceted role they play in transcription.

5.2 HETEROCHROMATIN IS TIGHTLY PACKED TO REPRESS GENE EXPRESSION VIA INTERACTION OF HP1 AND HISTONE H3 LYS9 METHYLATION

5.2.1 POSITION EFFECT VARIEGATION: A CLASSICAL EPIGENETIC PHENOMENON TO UNDERSTAND CHROMOSOME ORGANIZATION AND GENE EXPRESSION

Heitz [8] distinguished two distinct types of chromatin in 1928 by observing interphase nuclei of eukaryotic cells: the highly condensed chromatin, named heterochromatin, and the less condensed chromatin, named euchromatin (Figure 5.1B). For example, the specialized *Drosophila* Y chromatin contains very few functioning genes, a large fraction of which are in the heterochromatin status [9]. Therefore, heterochromatin was considered as an "inert" region [10]. In 1930, Muller described a striking repressive role of heterochromatin through studying a gene that controls the pigmentation of the *Drosophila* eye [11]. He recovered several surprising *Drosophila* strains carrying the mutated *white +* (w^+) gene with red and white mosaic compound eyes (Figure 5.1C). Further investigations revealed that the mutant *white-mottled* (w^m) alleles were introduced by rare inversions that place the gene near to heterochromatin (Figure 5.1D). Transcription machinery can be prevented by the presence of repressive transcription factors or repressive histone markers. The former can be a repressive transcription factor or a protein complex that occupies the promoter of a gene, while the latter mark forbidden chromatin states such as tightly packed heterochromatin. In normal cells, buffering barriers separate the euchromatin and heterochromatins from each other, thereby

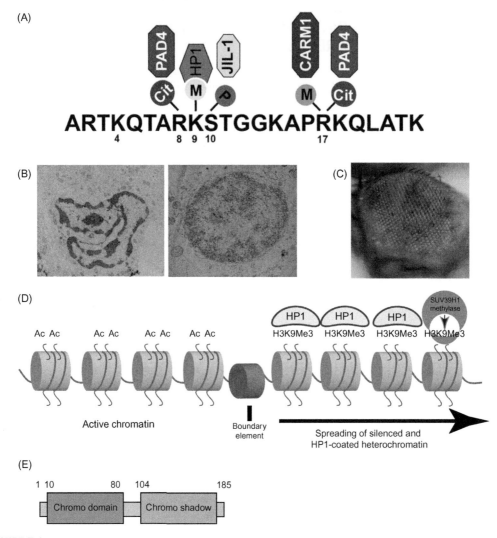

FIGURE 5.1

Position effect variation and the role of histone methylation in heterochromatin formation. (A) Clustered histone modification and its corresponding modifying enzyme on the N-terminal of histone H3. (B) TEM image of heterochromatin (left) and euchromatin (right). (C) Representative image of a mosaic *Drosophila* eye harboring mutant *white-mottled* (*w^m*) alleles. (D) Illustration of histone H3 lysine 9 methylation spreading and heterochromatin formation.

maintaining a desirable chromatin state (Figure 5.1C). In the case of *position effect variation* (PEV), each single ommatidium of the compound eye has the intact *white* gene, and would have normal red pigmentation if the heterochromatin did not spread over the gene. In contrast, if the heterochromatin spreads across the gene, the w^+ would be silenced and leave a white-colored ommatidium. Therefore, although the gene itself is intact, it assumes an inactive state due to chromatin structure changes after gene inversion. The balance between heterochromatin and euchromatin as it affects gene expression is now known as PEV, representing a classical epigenetic phenomenon.

5.2.2 HETEROCHROMATIN FORMATION DEPENDS ON HISTONE H3 LYSINE 9 METHYLATION AND HP1

PEV holds true in many model organisms such as *yeast*, *arabidopsis*, and *mouse* [12,13,14]. This biological phenomenon not only reflects the repressive function of heterochromatin but also offers excellent genetic tools to screen for secondary mutations that suppress or enhance the PEV effects. Such screening using a fly line resulted in around 150 loci that were named as *Su(var)* (suppressor of variegation) or *E(var)* (enhancer of variegations) [15]. *Su(var)3–9* locus was found to encode the HP1a protein that recognizes histone H3 lysine 9 trimethylation (H3K9Me3) and localizes to heterochromatin. Further crystallography studies showed that HP1a was composed of evolutionarily conserved chromo- and chromo-shadow domains (Figure 5.1E) [16]. The chromodomain of HP1 can recognize and bind H3K9Me3 at the histone H3 N-terminus [16]; the chromo-shadow domain interacts with *Su(var)3–9*, a histone lysine methyltransferase that generates the H3K9Me3 modification [17,18]. The HP1 protein then interacts with *Su(var)3–9* to spread H3K9Me3 modification mark and further packs the chromatin into heterochromatin [3,19,20]. However, the presence of HP1a does not necessarily lead to gene repression; occasionally, HP1a can associate with an active gene during transcription [21]. In addition to this genetic approach, the amount of HP1 and H3K9Me3 can be monitored via *chromatin immunoprecipitation* (ChIP) to analyze gene regulation mechanisms.

5.2.3 PEV SCREENING: USING WHITE GENE AS A REPORTER AND THE P ELEMENT INSERTION AS A MUTAGEN

The genetic screen of PEV modifiers has been very powerful in searching for the potential chromatin remodelers that affect PEV. The screening in *Drosophila* has been done systematically, and this method requires specific phenotypes, that is, the mosaic eyes with the inverted w^+ gene, to facilitate the genetic screen. As each ommatidium expresses the w^+ pigmentation gene independently, the compound eye offers a unique tool to observe the PEV effects [22].

5.3 IN SITU OBSERVATION OF CHROMATIN REORGANIZATION DURING TRANSCRIPTION ACTIVATION

5.3.1 CHROMATIN IS REORGANIZED DURING TRANSCRIPTION ACTIVATION

In eukaryotes, DNA is wrapped around histones to form chromatin, which results in nearly 10^4-fold condensation of the genome. During transcription activation, to expose the promoter region at a specific gene locus is required. Factors regulating this process were largely unknown in the 1990s.

At that point, electron microscopy could discern 10 nm chromatin fibers. However, the lack of an imaging method to study chromatin structures above 30 nm chromatin fibers made it challenging to study higher order chromatin structures. Thus, scientists were eager in searching out methods to inspect unfolded chromosomes under live conditions without perturbing chromatin organization. In 1996, Dr. Andrew S. Belmont's group came up with a novel imaging tool to observe large-scale chromosome organization in live cells by adapting the lac operator and repressor system into eukaryotes [23].

5.3.2 LAC REPRESSOR SYSTEM AS A POWERFUL TOOL FOR LIVE CELL IMAGING TO INVESTIGATE FACTORS THAT REGULATE CELL CYCLE PROGRESSION

The lac repressor (LacI) was first isolated by Walter Gilbert and Benno Muller-Hill in 1966 [24]. The protein is a homo-tetramer, which contains two DNA binding surfaces that recognize a DNA sequence named "operator." Each dimer binds to the operator sequence via a helix-turn-helix motif. As such, two operator sequences are recognized to increase specificity. Lac repressor inhibits the expression of genes regulating lactose metabolism in bacteria so that bacteria express lactose metabolic genes only when lactose is present, ensuring efficient energy investment.

This specific operator-binding repressor system is used as an imaging tool for *in situ* study of specific gene locus by inserting the operator sequence into the genome. The LacI system is initially used for visualizing homogeneous staining regions (HSRs), which are produced by gene amplification [25,26]. HSRs typically contain tens to hundreds of copies of a certain DNA sequence (in the length of hundreds to thousands of kb). Amplification of the lac operator sequence in HSRs ensures increased signal-to-noise ratio for imaging.

5.3.2.1 *LacI in studying sister chromatin cohesion in budding yeast*

Andrew S. Belmont and colleagues first adapted the lac operator/repressor system to visualize sister chromatic cohesion in budding yeast [27] (Figure 5.2). Using budding yeast as the model system, they inserted 256 tandem repeats of lac operators into the genome. This insertion can be targeted at a specific site via homologous recombination. A plasmid containing the lac repressor gene with an in-frame nuclear localization signal is then transfected into the cells. Expression of the lac repressor gene is driven by the HIS3 gene promoter (imidazoleglycerol-phosphate dehydratase: catalyzes the sixth step in histidine biosynthesis) and induced with 3-aminotriazole. Additionally, fusing eGFP (enhanced GFP) to the lac repressor gene in the same cassette allows direct observation of this specific locus on chromatin.

In this specific work, after induction with 3-aminotriazole for half an hour, cells are returned to a normal medium for observations. Alpha-factor is used to synchronize cells at the G1 stage by activating the mating pheromone signaling pathway. After release, they were able to see separation and transportation of the eGFP-labeled chromosome III from a single parental cell to its daughter cells. With this powerful imaging system, they were able to observe the effect of multiple mutations on sister chromatid separation [27]. Since wild-type LacI forms a tetramer that binds two operator sequences together, they also proved LacI binding to two sister chromatids is enough to inhibit their separation.

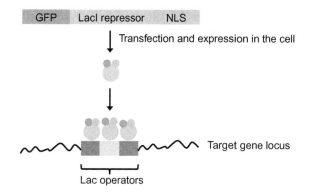

FIGURE 5.2

Lac repressor adapted into a powerful imaging technique. Lac repressor is engineered into a plasmid with GFP and nuclear localization signal at the same cassette. Then it is transfected into cells that are engineered with lac operator sequence repeats in the genome. (*This figure is reproduced in color in the color plate section.*)

5.3.2.2 The LacI system facilitates the discovery of COBRA1 in assisting large-scale chromatin unfolding during the cell cycle

Dr. Belmont's group later extended lac repressor imaging technique from budding yeast to mammalian Chinese hamster ovary (CHO) cells [28], which is dihydrofolate reductase (DHFR) negative; 256 copies of lac operator repeat is engineered into a DHFR-containing vector. Then a progressive increase of methotrexate (MTX) concentration was used for selection of successfully transfected cells. Using this protocol, they were able to obtain one clone named A03_1, which contains a stable single HSR with multiple copies of the lac operator.

Using the A03_1 cell line, Rong Li and colleagues were able to show that *BRCA1* (Breast Cancer 1) induces large-scale chromatin unfolding [29]. By fusing *BRCA1* gene with lac repressor, they observed loose and irregular shaped subnuclear structure in 14% of transfected cells. They were able to discover the unfolding activity is independently conferred by subdomains in the BRCA1 transactivation domain, which eventually led to the discovery of a novel cofactor of BRCA1 (COBRA1).

5.3.2.3 LacI in investigating post-mitotic transcription in vivo

In a recent paper by David Spector's group, the authors constructed a real-time expression system with LacI to investigate factors that induce postmitotic transcription [30]. They used a modified cell line U2OS-2-6-3 (U2OS = homo sapiens bone osteosarcoma cell) which stably expresses pTet-on, mCherry-PolII, and MS2-YFP. Lac repressors in the same cassette allow visualization of the gene locus. Expression of the nascent transcript MS2 is driven by tetracycline using Tet-ON (tetracycline-inducible expression system) downstream of the LacI-binding sites. MS2 transcripts contain loop structures, which are recognized by an MS2-binding fluorescent protein, allowing observation of newly transcribed mRNA. Fixing cells at different stages of the cell cycle, they observed significantly faster kinetics of MS2 transcription in postmitotic cells than in interphase cells. This rapid kinetics was proposed as a result of higher levels of H4K5 acetylation on the MS2 locus, which recruits BRD4 (bromodomain containing 4) to facilitate reactivation in postmitotic cells.

5.3.3 POLYTENE CHROMOSOMES IN *DROSOPHILA* PROVIDE CONVENIENT WAY TO OBSERVE CHROMATIN REMODELING

5.3.3.1 Squash method for polytene chromosome extraction

Since Thomas Hunt Morgan made it famous in the 1910s, *Drosophila* has served as a valuable animal model for genetic study for more than 100 years [31,32]. Besides the advantages of a simple chromosome and powerful genetic tools, *Drosophila* polytene chromosomes have been especially helpful in studying chromatin organization [33,34]. Polytene chromosomes form in the *Drosophila* giant salivary gland during development by multiple rounds of genome replication without cell division to generate polytene chromosomes containing up to 1000 copies of DNA per chromosome. Despite the high copy number, polytene chromosomes still retain similar features to those of the interphase chromatin, such as interbands and bands with relatively loose and condensed chromatin, respectively, as well as chromatin decondensation to form puffs during gene activation. Polytene banding patterns are highly reproducible. Moreover, proteins (e.g. polymcrase and transcription factors) and histone modifications are specifically localized to certain genomic regions. These features allow researchers to study molecules required for transcriptional regulation using polytene chromosomes as a model.

In a conventional squash extraction, first, salivary glands were pulled by forceps from PBS (phosphate-buffered saline)-rinsed larvae. Then the glands were fixed sequentially on a slide for 1−2 min with 4% paraformaldehyde, making sure to remove excess fat tissue from the salivary glands [35]. The glands were transferred into a lactoacetic acid solution on a coverslip, gently moving the coverslip back and forth on a slide to spread the chromosomes while avoiding any vertical pressure. This was checked under a 20 × phase contrast microscope and this process repeated until the chromosome was spread well on the coverslip. Then a Kim-wipe sandwich was made with the sample in the middle and coverslip-side down, pressing vertically with the thumb to avoid horizontal movement, which could shear the chromosome. This was examined under microscope again and, if suitable for the following procedure, fixed with liquid nitrogen. Then the coverslip was flipped off, while making sure glands are not on the coverslip but the slide. For the last step, immunostaining with the interested antibody will show localization of targeted proteins.

To reach higher resolution, a precise device can be adapted for higher vertical pressure and less horizontal force that could shear the chromosome [35]. At the same time, GFP-tagged proteins imaged in live tissue show better dynamics of transcription and avoid resolution sacrifice resulted from a bad antibody [35].

5.3.3.2 Polytene squash extraction has been adapted to study of Hsp70 and JIL-1 expression pattern at native locus

Using live cell imaging techniques and taking advantage of the high DNA copy numbers in polytene chromosomes, single-cell imaging was performed to observe a native locus [36,37]. John Lis and colleagues conducted live cell imaging of heat shock 70 (*Hsp70*) gene activation after heat shock (HS) [38]. Using a spinning disk confocal microscopy, they were able to visualize temporal recruitment of factors, such as Pol II, p-TEFb, and topoisomerase I, onto the Hsp70 gene in a single salivary gland nucleus to better understand the mechanisms of gene activation.

Using the squash method for larval polytene extraction, it was found that JIL-1, the kinase responsible for H3S10 phosphorylation, affects higher order chromatin structure [39].

Kristen M. Johansen and colleagues constructed hypomorphic or null mutants of JIL-1 and extracted larval polytene from these mutants. Immunostaining with MSL2 antibody, which labels X chromosome, Hoechst, and/or JIL-1 antibody, it was observed that there was a decrease in the euchromatic interbands after the reduction of the JIL-1 protein, providing physiological evidence for a role of JIL-1 in maintaining higher order chromatin structure.

5.4 CHROMATIN IMMUNOPRECIPITATION
5.4.1 ChIP METHOD DEVELOPMENT AND APPLICATION

In the last several decades, numerous efforts have been put into identification of regulatory factors of gene expression. Classically, artificial methods, for example the reporter assay, have been widely used to identify regulatory genomic regions [40]. However, these approaches are most successful in rapid inducible genes and produce artificial results under certain conditions. ChIP was developed in 1988 by Solomon to map protein−DNA interactions after formaldehyde fixation [41]. Since then, ChIP has revolutionized epigenetics and chromatin research. To measure the modifications at a specific locus, we need to first prepare chromatin using an appropriate method of choice. We can either isolate chromatin under native cellular conditions − native ChIP (N-ChIP) − or isolate the chromatin fraction under fixed conditions − cross-linking ChIP (X-ChIP) − depending on the starting materials and experimental aims. The advantage of X-ChIP is that it requires small cell numbers to start with and small amounts of antibodies. In addition, cross-link between the DNA and protein allows X-ChIP to study transcription factors that are loosely attached to DNA. However, X-ChIP has its own limitations: precipitation is very inefficient, so the specificity of the results is compromised by the insufficient protein recovery. In contrast, N-ChIP can only be used to study strong DNA-binding partners such as histones. The precipitation is very efficient in N-ChIP so that researchers can get sufficient protein materials. The DNA recovered can be quantified directly or with many fewer PCR (polymerase chain reaction) cycles to amplify. Additionally, antibodies are generally raised against nature peptide rather than fixed peptide so the specificity of the antibody binding to the natural chromatin proteins is more reliable. In summation, for the study of core histone modifications, N-ChIP is preferable as histones are tightly associated with chromatin, thus allowing easier precipitation and enhanced signal-to-noise ratio [42].

5.4.2 N-ChIP IS POPULAR METHOD TO OBSERVE HISTONE MODIFICATIONS AT SPECIFIC LOCI
5.4.2.1 Experimental design and controls

Figure 5.3B illustrates the experimental flow for analyzing chromatin modifications at specific single loci. The whole procedure includes five steps and takes around 3 days to complete. These steps are: (1) purification of nuclei; (2) chromatin fractionation using MNase digestion; (3) immunoprecipitation using antibody targeting specific factors or histone modifications; (4) DNA recovery; and (5) DNA quantification and analysis.

To permit statistical analyses, one should design the ChIP experiment with positive controls, negative controls, and include at least three biological replicates. It is also necessary to check

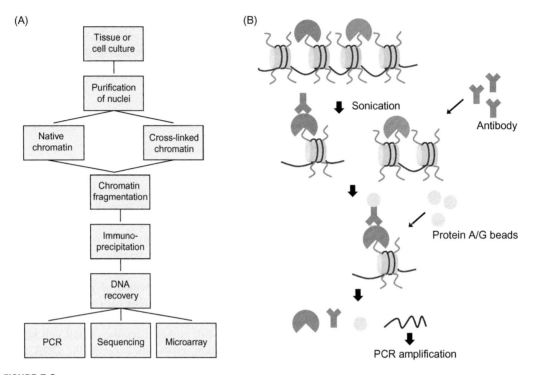

FIGURE 5.3

Chromatin immunoprecipitation. (A) Flow chart to describe the procedures used to investigate histone modifications at a specific locus using X-ChIP and N-ChIP. (B) Chromatin-binding proteins are firstly cross linked to chromatin for X-ChIP. Sonication breaks chromatin into single or oligo nucleosomes. Antibody binds its target on the fragmented chromatin. Following purification with protein A/G beads, amplification with PCR allows examination of the DNA sequence bound to the protein.

antibody specificity and titer via Western blot or immunostaining before proceeding to ChIP. Readers are directed to popular protocols for Western blot and immunostaining characterization of antibodies [42−44].

5.4.2.2 Purification of nuclei

If the number of cells is not limited, roughly 10^7 cells are good starting materials for ChIP experiments. For adherent cells, wash with PBS once and treat the cells with 0.05% trypsin solution for 1 min or until the cells are detached, and neutralize trypsin by FBS-containing medium. This step is not applicable to cells in suspension. Centrifuge at 1000 rpm for 5 min at 4°C to obtain the cell pellet. Wash the cell pellet with PBS supplemented with protease inhibitors. Aspire the remaining PBS and wash cells with buffer I (0.3 M sucrose, 60 mM KCl, 15 mM NaCl, 5 mM MgCl2, 0.1 mM EGTA, 15 mM Tris.HCl pH 7.5, 0.5 mM DTT, 0.1 mM PMSF, and 3.6 ng/mL aprotinin). Resuspend cells in 1 mL ice cold buffer I to obtain single-cell suspension, then vortex at low speed while adding 1 mL ice cold 2× lysis buffer (buffer I plus 10% vol/vol NP-40) drop wise. Sit on ice for 10 min before centrifuging and save the white nuclear pellet. Wash the nuclear pellet with buffer I again and resuspend in 100 μL buffer I. Adjust the concentration of nucleic acid using nanodrop UV-Vis spectrophotometer to final concentration about 1 μg/μL.

5.4.2.3 MNase digestion and quality check

Mix gently after adding appropriate amount of MNase into 100 μL nuclei suspension at a ratio of 50 mU/40 μg of nucleic acid. Incubate in 37°C water bath for 10 min, and then stop the digestion by adding 10% of total volume of stop solution (20 mM EDTA, pH8.0). Check the nucleosome size by running 1.2% agarose gel to make sure the digestion is mainly distributed within mono-nucleosome to tri-nucleosomes.

If X-ChIP is used, fragment the nucleosome via sonication for 3−5 min in ice cold water at 30 s pulse, 30 s pause cycles at high-energy output. See further X-ChIP discussion in Section 5.5.2.

5.4.2.4 Pre-clearing and immunoprecipitation

After chromatin fragmentation, centrifuge the sample at 12,000 rpm for 15 min at 4°C and save the supernatant. Incubate and rotate about 100 μg chromatin at final volume of 300 μL with 50 μL protein A/G beads for 2 h at 4°C. Pair the chromatin and antibody at a ratio of 5 μg antibody per 100 μg chromatin per tube in 1.5 mL tubes to rotate at 4°C for overnight.

5.4.2.5 IP washes and DNA recovery

After the targeted DNA is bound to the antibody, add 50 μL of protein A/G beads to the chromatin sample and rotate for another 2 h at 4°C. Centrifuge briefly to collect drops on the wall and let the beads settle down naturally on ice for around 10 min. Save 5% of total volume of supernatant as total input DNA from the beads alone control tubes. Transfer the rest of the supernatant into a new tube carefully without disturbing the beads. Wash the beads by adding 1 mL buffer, then rotate for 5 min at 4°C. The sequential wash includes three times ChIP buffer I (25 mM Tris.HCl pH 8.0, 5 mM $MgCl_2$, 100 mM KCl, 10% vol/vol glycerol, and 0.1% NP-40), two times ChIP buffer II (25 mM Tris.HCl pH8.1, 5 mM $MgCl_2$, 300 mM KCl, 10% vol/vol glycerol, and 0.1% NP40), and two times TE buffer. Resuspend beads as well as the input DNA sample in 300 μL of TE buffer containing 10 μg proteinase K to release the DNA and protein complex. Incubate at 65°C for 45 min to solubilize the DNA. The DNA will stay in the supernatant after centrifuge at 12,000 rpm for 10 min. Transfer the supernatant to a new tube and purify the DNA using the method of choice; we prefer to use spin-columns from QIAGEN according to the manufacturer's instructions. The purified DNA will be ready for real-time PCR with primers designed for specific genomic locus.

5.4.3 COMPARISON OF ChIP-PCR, ChIP-ChIP, AND ChIP-SEQ

After ChIP DNA is recovered, PCR is a suitable method to quickly study the epigenetic modifications at a specific locus owning to its time and cost efficiency. One can get the results for several specific gene loci within one day with a cost of less than $100. If the purpose of the experiment is to analyze thousands of gene promoters, exons, and introns, PCR is obviously not the best choice due to its low throughput. ChIP-Chip (ChIP-coupled microarray) would be a reliable choice for studying the histone modifications at each individual genomic feature across the genome or through custom designed arrays. All the methods discussed in this subsection — ChIP-PCR, ChIP-Chip, and ChIP-Seq (ChIP-coupled high-throughput sequencing) — require researchers to know their specific interest loci. To explore novel regulation mechanisms without specific chromosome knowledge, the next generation sequencing has been applied more and more widely. However, due to the enormous data generated, digging out the dynamic regulation and expression of a single gene locus remains challenging. This becomes even more obscure when we look into the epigenetic landscape across a specific gene locus where signals spread widely while the sequencing resolution is hundred base pairs. Rhee and Pugh recently developed the ChIP-exo method to improve the resolution of mapping results to up to base pair resolution at a higher but still affordable cost [45]. The choice of ChIP DNA analyses should be made according to the experimental goals.

5.4.4 ChIP TROUBLESHOOTING

There are two common problems for newcomers to the ChIP experiments: one is high background signal and the other is low DNA recovery. If high background was detected using the negative control antibodies, check the following steps: (1) make sure enough beads were used in the preclearing step; (2) make sure residue liquid was removed completely during the washing step; and (3) make sure buffers were not contaminated by genomic DNA. If a robust specific signal was not detected, first check the antibody to make sure it is specific and effectively precipitates the protein/ DNA complex from fragmented chromatin. Make sure the size of the fragmented chromatin is in an appropriated range. If the cross-link ChIP method is used, the cross-link time should be controlled precisely according to the experimental design, as too long or too short cross-link time affects the efficacy.

5.4.5 LIMITATIONS OF THE ChIP METHOD

Given the popularity and wide application of ChIP methods, there are limitations using ChIP to detect epigenetic modifications at specific loci. The most urgent issue is the lack of precise and specific antibody. The antibodies for specific histone modifications, say H3K4Me3, may cross-react with the original histone H3, thus giving high background noise. To avoid cross-reaction of the antibody, before spending a lot of time undertaking ChIP, parallel controls should be used to do ChIP analysis in cells lacking the specific modifications, if available. For example, the SETD2 (SET domain-containing protein 2) depleted cell line is deprived of H3K36me3 and therefore can be used as a negative control for H3K36me3 antibody specificity [46]. The rapid development of antibody pipelines will enable us to perform ChIP analyses of our favorite genes with the proper histone modification antibodies to study molecular and cellular functions. However, we need to keep in mind that ChIP alone cannot demonstrate the function of a specific factor or histone modification. ChIP helps reveal the localization of epigenetic marks at a specific genomic locus but needs to be coupled with other experiments such as gene expression assay, reporter assay, and genetic analyses to dissect the biological functions of the modification.

5.5 CROSS-LINKING CHROMOSOME IMMUNOPRECIPITATION AS A POWERFUL TOOL TO INVESTIGATE SEQUENTIAL RECRUITMENT OF HISTONE ACETYLTRANSFERASE AND SWI/SNF CHROMATIN-REMODELING COMPLEX TO THE PROMOTERS

5.5.1 HISTONE ACETYLTRANSFERASES AND CHROMATIN-REMODELING COMPLEX (SUCH AS SWI/SNF) REQUIRED FOR GENE ACTIVATION

Histone code hypothesis states that gene regulation is partly dependent on histone modifications that primarily occur on histone tails [17,47]. It is clear that the recruitment of transcription factors to their target sites requires access to histones and DNA thereby unwrapping chromatin structures [48,49]. During this process, at least two functions are necessary. One is acetylation by histone acetyltransferases (HATs), which assists recruiting factors to facilitate chromatin unfolding; the other is chromosome remodeling by SWI/SNF (SWItch/Sucrose NonFermentable, a nucleosome-remodeling complex) or other remodeling complexes, which open the chromatin by altering the DNA binding with core histones [50,51]. However, little was known about the sequential order of different machineries to cooperatively initiate transcription. Studies on the activation of the HO gene and the IFN-β gene (interferon type I gene) have shed some light onto this fundamental question. Biochemistry experiments, such as ChIP and *in vitro* pull down assays, are important methodologies to unveil this process.

5.5.2 X-ChIP REVEALS SEQUENTIAL RECRUITMENT OF HATs AND SWI/SNF-REMODELING COMPLEX AT DIFFERENT PROMOTERS

To perform X-ChIP to analyze protein−DNA interaction (Figure 5.3B), chromatin is first fixed by formaldehyde to cross-link the binding proteins with its cognate-binding sites. Following immunoprecipitation with specific antibodies and PCR with primers targeting a specific locus, protein−DNA interaction is quantified. In the following section, we will discuss successful adaption of biochemical techniques to monitor protein interaction with the specific gene, which is a relatively close reflection of the *in vivo* situations.

5.5.2.1 Recruitment of SWI/SNF-remodeling complex and then HATs on the HO gene promoter

In 1999, Craig Peterson and colleagues investigated acetylation on the HO gene promoter region during the cell cycle in yeast [52]. The HO gene product regulates the yeast mating type switch, offering a powerful genetic system to study development of gene expression. The authors performed chromatin immunoprecipitation experiments on yeast cells at different time points of the cell cycle. A burst of acetylation in mid-G1 correlates with HO gene expression, and declines after gene activation. They further performed ChIP with strains harboring mutations on the SWI/SNF ATPase subunits, SWI2, SWI5, or SWI4. After deletion of these genes, acetylation on histone H3 in the HO gene promoter region is largely abolished. Based on these results, the authors proposed the recruitment of SWI/SNF-remodeling complex followed by HATs during HO gene activation (Figure 5.4A).

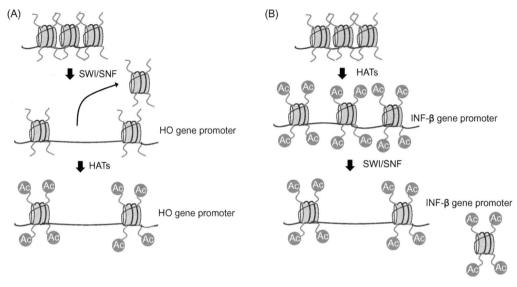

FIGURE 5.4

HAT and chromatin-remodeling complex is recruited sequentially onto HO gene promoter (A) and INF-β gene promoter (B) in different manner.

5.5.2.2 Recruitment of HAT then SWI/SNF on IFN-β gene promoter

Dimitris Thanos and colleagues at the Columbia University tested the histone code hypothesis by elucidating the regulation of the IFN-β gene promoter by histone acetylation [53]. They discovered a sequential recruitment of GCN5 and SWI/SNF (SWItch/Sucrose NonFermentable, a nucleosome-remodeling complex) onto the IFN-β gene promoter, whereby acetylation of histone occurs before chromatin remodeling (Figure 5.4B). First, HeLa cells were infected with the Sendai virus to induce the IFN-β gene expression. At different time points after infection, cells were fixed with formaldehyde to stabilize protein−protein and protein−DNA interactions. Cross-linked chromatin was immunoprecipitated with H3/H4 acetylation antibodies. Finally, Western blot revealed sequential onset of acetylation on histone tails. To identify factors required for SWI/SNF-remodeling complex and TFIID recruitment, they synthesized a series of enhanceosomes containing different parts of the IFN-β gene promoter [54]. They reconstituted *in vitro* transcription with either wild type or mutant H3 with the acetylation site altered, without or with acetyl-coA (the cofactor required for acetylation). The biotinylated IFN-β gene promoter fragment containing different enhanceosome is attached to Dyna beads. After incubation with nuclear extract from HeLa cells, they are able to purify the enhanceosome and probe with specific acetylation antibodies for Western blot analysis [53].

This work confirmed a different sequential recruitment of HAT and SWI/SNF-remodeling complex onto the IFN-β gene promoter region compared to that at the HO gene promoter. It elucidated promoter-to-promoter variability in the manner of chromosome reorganization during gene activation.

5.6 CONVERGENCE OF BIOINFORMATICS AND BIOPHYSICS TECHNIQUES AT OBSERVATION IN LIVE SINGLE CELL

After decades of epigenetic research, bioinformatics techniques finally allow us to develop databases that can monitor the changes of multiple loci to the scale of the entire genome [55,56]. Furthermore, advanced imaging tools and genome manipulating techniques converge on chromatin organization studies. Biophysicists and/or bioengineers have been trying to fulfill the goal of *in vivo* manipulation of the genome. We discuss here several new engineering tools for the study of transcription initiation.

5.6.1 SINGLE-CELL LIVE IMAGING REVEALS FUNCTION OF NUCLEOSOME-DEPLETED REGION IN GENE ACTIVATION

In 2010, Drs. Lu Bai and Frederick Cross reported a real-time single-cell observation of the *Cln2* and *HO* gene expression with a goal to study the role of the nucleosome-depleted regions (NDRs) in gene expression [57]. It has long been postulated that binding motifs localized in NDRs are important for the binding of transcription factors and thereby facilitate gene expression [58]. However, some promoters lacking NDR can still be functional [56,59]. Previous studies in this field mainly focused on bulk gene expression analyses based on average signals from a group of cells. Since cell-to-cell variability is increasingly recognized as a vital feature of transcription regulation, the authors concentrated on single-cell imaging in their studies.

5.6.1.1 Unstable GFP and Myo1-mcherry as markers for gene expression

To fulfill the goal of single-cell tracing and imaging under a microscope, Bai et al. cultured yeast cells in between a glass coverslip and a gel pad [57]. Placing an unstable GFP gene driven by a target promoter, such as *Cln2pr* (ceroid-lipofuscinosis, neutonal 2 promoter), allows real-time study of the transcription signal in a dynamic fashion. Since Myo1 (yeast type II myosin heavy chain gene) forms a bud neck ring during bud emergence and disappears sharply at cytokinesis, the authors adapted Myo1-mcherry signal to monitor the cell cycle.

5.6.1.2 Manipulating NDR at the promoter: exploration on Cln2 and HO promoter revealed NDR's function in stabilizing transcription during evolution

To study the biological functions of NDR, Bai et al. explored two cell-cycle-related promoters. The *Cln2* promoter (*Cln2pr*) contains the activator SBF (yeast transcription factor) binding sites in NDR, while the yeast *HO* gene promoter (*HOpr*) contains the same activator-binding sites buried in nucleosomes.

They first mapped SBF-binding sites in the *Cln2* promoter (*Cln2pr*) and found that they are situated in a ~300 bp NDR upstream of the transcription start site (TSS). Mapping information is confirmed by comparison with immunoprecipitation data and sequencing data by others. In the wild-type cells, they monitored expression from *Cln2pr* through several cell cycles and observed a consistent burst of expression once in every cell cycle (Figure 5.5A and 5.5B). Depletion of the binding sites from *Cln2pr* totally abolished cell-cycle-related burst expression (Figure 5.5B). To investigate if NDR positioning of SBF-binding sites affects transcription, they artificially inserted

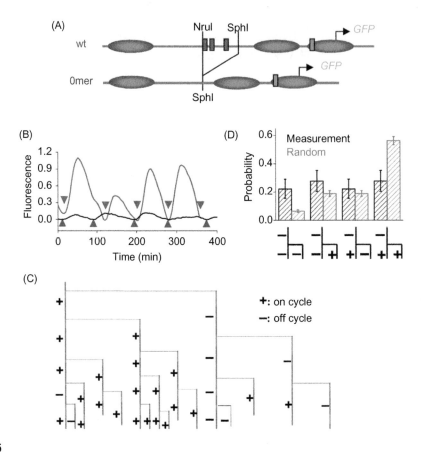

FIGURE 5.5

Single-cell real-time imaging and statistics converge to elucidate NDR function in gene activation. (A) Wild-type *Cln2* promoter and engineered promoter with SBF-binding sites deleted from NDR region. (B) Wild-type *Cln2* promoter drives burst expression of GFP signal in every cell cycle and the burst pattern is abolished by deleting SBF-binding sites. (C) Real-time recorded "on" and "off" expression patterns from a single-cell division event from engineered promoter with nucleosome-buried SBF-binding sites. (D) Statistics prove "on" and "off" pattern is a "memory" mechanism between parental and daughter cells.

Figures are from [Ref. 57].

them into a neighboring nucleosome. This promoter has bimodal, "on" or "off" activity among cell cycles: it can still activate GFP expression, but only fires in 75% of cell cycles (Figure 5.5C). By tracing multiple dividing events, they found this "on/off" switch would be inherited to the daughter cells as a "short-memory" mechanism (Figure 5.5D). Again by generating hybrid promoters from *HOpr* and *Cln2pr*, they observed "on" or "off" activation when SBF-binding sites are located in nucleosome, and robust, once-per-cycle activation when the sites are located in NDR. So they concluded that the biological function of NDR is to reduce transcription variability and is selected during evolution.

This work clearly demonstrated the importance of single-cell assay in studying chromatin and gene regulation. It further elucidated that chromatin organization not only affects the average level of gene expression, but also its cell-to-cell variability.

5.7 FUTURE PERSPECTIVES: EPIGENOME ENGINEERING AND MANIPULATING AT SPECIFIC GENOMIC LOCI IN MAMMALIAN CELLS

5.7.1 PAD4 IS A HISTONE ARG DEIMINASE

Humans have abundant citrulline amino acid in the body, especially in hair, blood, and skin. However, citrulline has no corresponding three-nucleotide codons during the protein synthesis process. Within the cell, one of the important/essential amino acids, arginine, can be converted to citrulline catalyzed by the peptidylarinine deitminase (PAD) enzymes [60]. In addition to traditional histone modifications such as methylation, acetylations, and phosphorylations, citrullination of histone is gaining more and more attention [61]. PAD4 is the only member of the PAD family that contains a nuclear localization signal [62]. Therefore, PAD4 has been the candidate for modifying histone Arg. In fact, PAD4 was one of the first demethylases proven by Yanming Wang at 2004 [62]. PAD4 was proven to play an important role in innate immunity by nuclear extracellular traps (NETs) formation [63]. An elevated expression level of PAD4 has been related to cancer progression and autoimmune disease [64−67]. Recently, citrullination of histone H1R52 has been shown to benefit the formation of pluripotent stem cell [68]. The PAD4 is gaining in interest in the epigenetic field because of its impact on innate immunity, autoimmune disease, cancer, and pluripotency regulation.

5.7.2 CRISPRs ARE PROMISING TOOLS TO EDIT THE EPIGENOME AT SPECIFIC GENOMIC LOCI

Historically, our knowledge of the adaptive immune system was restricted to multicellular creatures. Until 2010 [69], after more than 200 years of using *Escherichia coli* as an object of study, Horvath found that bacteria and archaea can fight with invaders, phages for example, by a series of reactions: acquisition, expression, and interference (Figure 5.6A) [70]. This special mechanism was named *clustered regularly interspaced short palindromic repeats* (CRISPRs) [69]. First, after the phage invades the bacteria, the photospacer within the CRISPRs array would integrate fragments from virus DNA or plasmid DNA to acquire the information on the outsider. If the phage invades again, the expression of Cas (CRISPR-associated proteins) endonuclease protein as well as the complementary tracrRNA (trans-activating RNA) would elevate immediately. The Cas endonuclease would interfere with the exogenous DNA with the guide of crRNA (CRISPR-RNA) that is specific to the invading phage [71]. Several groups adapted the type II CRISPR system from *Streptococcus pyogenes* to induce a sequence-specific double-strand break at targeted genome locations [72,73]. The imperfect natural of nonhomologous repair of double-strand breaks sometimes leads to frame shift or an immature stop codon that will inactivate the target genes. To edit the epigenomic feature at a specific gene locus, we want to preserve the genomic specificity of the CRISPR system while abolishing the endonuclease activity of the Cas protein. Stanley Qi's

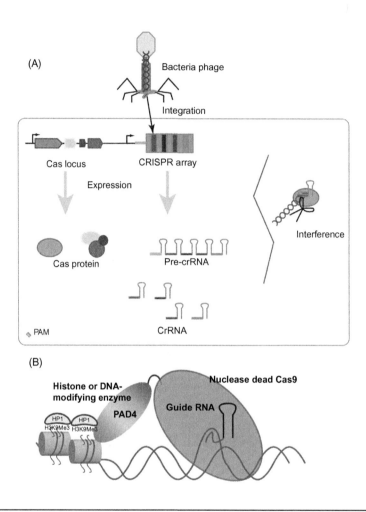

FIGURE 5.6

CRISPR employments in nature used by bacteria to acquire resistance to phage and artificially used by researchers to edit the epigenome. (A) Cartoon of the CRISPR-Cas adaptive immune system for bacteria to defend phage infection. (B) Imaginary illustration of the locus-specific epigenome editing employing the CRISPR/Cas9 system.

group engineered a kinase dead version of dCas9 (with mutation D10A and H840A) suited very much for this purpose [74]. We illustrate a hypothesized working model of dCas9-PAD4 that can in theory specifically interfere with the H3K9Me3 marker and the function of HP1 (Figure 5.6B). One of the challenges for *in vivo* applications is that the Cas9-PAD4 exceeds the upper limit for genome size of 4.8 kb to use AAV (adeno-associated virus) [75,76]. In addition, the side effect of tracrRNA complementing the DNA template should be carefully taken into consideration. However, the ability to artificially introduce histone modifications at specific loci will enable scientists to solve important basic questions such as whether histone modifications are leftover traces of gene regulation or the upstream regulator of gene regulation. Recently, two research groups have employed a more established TALEs (transcription activator-like effectors) method independently to modify the histone state at specific gene loci [77,78]. Given the flexibility and specificity of CRISPRs comparing to TALEs in genome editing, the future of applying CRISPRs in epigenome editing would be very promising [79,80].

REFERENCES

[1] Kornberg RD, Thomas JO. Chromatin structure; oligomers of the histones. Science 1974;184(4139): 865—8.

[2] Li B, Carey M, Workman JL. The role of chromatin during transcription. Cell 2007;128(4):707—19.

[3] Nakayama J, Rice JC, Strahl BD, Allis CD, Grewal SI. Role of histone H3 lysine 9 methylation in epigenetic control of heterochromatin assembly. Science 2001;292(5514):110—13.

[4] Dou Y, Milne TA, Tackett AJ, Smith ER, Fukuda A, Wysocka J, et al. Physical association and coordinate function of the H3 K4 methyltransferase MLL1 and the H4 K16 acetyltransferase MOF. Cell 2005; 121(6):873—85.

[5] Leshner M, Wang S, Lewis C, Zheng H, Chen XA, Santy L, et al. PAD4 mediated histone hypercitrullination induces heterochromatin decondensation and chromatin unfolding to form neutrophil extracellular trap-like structures. Front Immunol 2012;3:307.

[6] Sharma P, Azebi S, England P, Christensen T, Møller-Larsen A, Petersen T, et al. Citrullination of histone H3 interferes with HP1-mediated transcriptional repression. PLoS Genet 2012;8(9):e1002934.

[7] Fischle W, Wang Y, Allis CD. Binary switches and modification cassettes in histone biology and beyond. Nature 2003;425(6957):475—9.

[8] Heitz E. Das heterochromatin der moose. Jahrb Wiss Botanik 1928;69:762—818.

[9] Cooper KW. Cytogenetic analysis of major heterochromatic elements (especially Xh and Y) in *Drosophila melanogaster*, and the theory of "heterochromatin". Chromosoma 1959;10(1—6):535—88.

[10] Hannah A. Localization and function of heterochromatin in *Drosophila melanogaster*. Adv Genet 1951; 4:87—125.

[11] Muller HJ. Further studies on the nature and causes of gene mutations. In: Proceedings of the 6th international congress of genetics; 1932; 1932. p. 3—255.

[12] Allshire RC, Javerzat J-P, Redhead NJ, Cranston G. Position effect variegation at fission yeast centromeres. Cell 1994;76(1):157—69.

[13] Blewitt M, Whitelaw E. The use of mouse models to study epigenetics. Cold Spring Harb Perspect Biol 2013;5(11):a017939.

[14] Rédei G. Biochemical aspects of a genetically determined variegation in *Arabidopsis*. Genetics 1967;56 (3):431.

[15] Elgin SC, Reuter G. Position-effect variegation, heterochromatin formation, and gene silencing in *Drosophila*. Cold Spring Harb Perspect Biol 2013;5(8):a017780.

[16] Jacobs SA, Khorasanizadeh S. Structure of HP1 chromodomain bound to a lysine 9-methylated histone H3 tail. Science 2002;295(5562):2080—3.

[17] Jenuwein T, Allis CD. Translating the histone code. Science 2001;293(5532):1074—80.

[18] Tschiersch B, Hofmann A, Krauss V, Dorn R, Korge G, Reuter G. The protein encoded by the *Drosophila* position-effect variegation suppressor gene Su (var) 3—9 combines domains of antagonistic regulators of homeotic gene complexes. EMBO J 1994;13(16):3822.

[19] Nielsen SJ, Schneider R, Bauer UM, Bannister AJ, Morrison A, O'Carroll D, et al. Rb targets histone H3 methylation and HP1 to promoters. Nature 2001;412(6846):561—5.

[20] Schotta G, Ebert A, Krauss V, Fischer A, Hoffmann J, Rea S, et al. Central role of *Drosophila* SU (VAR) 3−9 in histone H3-K9 methylation and heterochromatic gene silencing. EMBO J 2002;21(5): 1121−31.

[21] Wallrath LL, Elgin SC. Stimulating conversations between HP1a and histone demethylase dKDM4A. Mol Cell 2008;32(5):601−2.

[22] Spradling AC, Stern DM, Kiss I, Roote J, Laverty T, Rubin GM. Gene disruptions using P transposable elements: an integral component of the *Drosophila* genome project. Proc Natl Acad Sci USA 1995;92 (24):10824−30.

[23] Belmont AS. Visualizing chromosome dynamics with GFP. Trends Cell Biol 2001;11(6):250−7.

[24] Gilbert W, Muller-Hill B. Isolation of the lac repressor. Proc Natl Acad Sci USA 1966;56(6):1891−8.

[25] Delidakis C, Swimmer C, Kafatos FC. Gene amplification: an example of genome rearrangement. Curr Opin Cell Biol 1989;1(3):488−96.

[26] Swimmer C, Delidakis C, Kafatos FC. Amplification-control element ACE-3 is important but not essential for autosomal chorion gene amplification. Proc Natl Acad Sci USA 1989;86(22):8823−7.

[27] Straight AF, Belmont AS, Robinett CC, Murray AW. GFP tagging of budding yeast chromosomes reveals that protein-protein interactions can mediate sister chromatid cohesion. Curr Biol 1996;6(12): 1599−608.

[28] Robinett CC, Straight A, Li G, Willhelm C, Sudlow G, Murray A, et al. In vivo localization of DNA sequences and visualization of large-scale chromatin organization using lac operator/repressor recognition. J Cell Biol 1996;135(6 Pt 2):1685−700.

[29] Ye Q, Hu YF, Zhong H, Nye AC, Belmont AS, Li R. BRCA1-induced large-scale chromatin unfolding and allele-specific effects of cancer-predisposing mutations. J Cell Biol 2001;155(6):911−21.

[30] Zhao R, Nakamura T, Fu Y, Lazar Z, Spector DL. Gene bookmarking accelerates the kinetics of post-mitotic transcriptional re-activation. Nat Cell Biol 2011;13(11):1295−304.

[31] Civelekoglu-Scholey G, Sharp DJ, Mogilner A, Scholey JM. Model of chromosome motility in *Drosophila* embryos: adaptation of a general mechanism for rapid mitosis. Biophys J 2006;90(11): 3966−82.

[32] Suzuki T. Current topics on genetic researches using *Drosophila* as a model system. Genes Genet Syst 2014;89(1):1.

[33] Lu X, Wontakal SN, Emelyanov AV, Morcillo P, Konev AY, Fyodorovet DV, et al. Linker histone H1 is essential for *Drosophila* development, the establishment of pericentric heterochromatin, and a normal polytene chromosome structure. Genes Dev 2009;23(4):452−65.

[34] Cave MD. Chromosome replication and synthesis of non-histone proteins in giant polytene chromosomes. Chromosoma 1968;25(4):392−401.

[35] Johansen KM, Cai W, Deng H, Bao X, Zhang W, Girton J, et al. Polytene chromosome squash methods for studying transcription and epigenetic chromatin modification in *Drosophila* using antibodies. Methods 2009;48(4):387−97.

[36] Davies DR, Cullis CA. A simple plant polytene chromosome system, and its use for in situ hybridisation. Plant Mol Biol 1982;1(4):301−4.

[37] Pardue ML. In situ hybridization to polytene chromosomes in *Drosophila* using tritium-labeled probes. Cold Spring Harb Protoc 2011;2011(8):1007−11.

[38] Zobeck KL, Buckley MS, Zipfel WR, Lis JT. Recruitment timing and dynamics of transcription factors at the Hsp70 loci in living cells. Mol Cell 2010;40(6):965−75.

[39] Wang Y, Zhang W, Jin Y, Johansen J, Johansen KM. The JIL-1 tandem kinase mediates histone H3 phosphorylation and is required for maintenance of chromatin structure in *Drosophila*. Cell 2001;105(4): 433−43.

[40] Gorman CM, Moffat LF, Howard BH. Recombinant genomes which express chloramphenicol acetyl-transferase in mammalian cells. Mol Cell Biol 1982;2(9):1044−51.

[41] Solomon MJ, Larsen PL, Varshavsky A. Mapping protein DNA interactions in vivo with formaldehyde: evidence that histone H4 is retained on a highly transcribed gene. Cell 1988;53(6):937−47.

[42] O'Neill LP, Turner BM. Immunoprecipitation of native chromatin: NChIP. Methods 2003;31(1):76−82.

[43] Brand M, Rampalli S, Chaturvedi CP, Dilworth FJ. Analysis of epigenetic modifications of chromatin at specific gene loci by native chromatin immunoprecipitation of nucleosomes isolated using hydroxyapatite chromatography. Nat Protoc 2008;3(3):398−409.

[44] Umlauf D, Goto Y, Feil R. Site-specific analysis of histone methylation and acetylation. Epigenetics Protocols: Springer 2004;99−120.

[45] Rhee HS, Pugh BF. ChIP-exo method for identifying genomic location of DNA-binding proteins with near-single-nucleotide accuracy. In: Frederick M Ausubel [et al.], editor. Current protocols in molecular biology; 2012. 100, 21.24.1−21.24.14.

[46] Li F, Mao G, Tong D, Huang J, Gu L, Yang W, et al. The histone mark h3k36me3 regulates human dna mismatch repair through its interaction with mutsα. Cell 2013;153(3):590−600.

[47] Ajiro K, Allis CD. [Histone code hypothesis]. Tanpakushitsu kakusan koso 2002;47(7):753−60.

[48] Wang GG, Allis CD, Chi P. Chromatin remodeling and cancer, Part II: ATP-dependent chromatin remodeling. Trends Mol Med 2007;13(9):373−80.

[49] Saha A, Wittmeyer J, Cairns BR. Chromatin remodelling: the industrial revolution of DNA around histones. Nat Rev Mol Cell Biol 2006;7(6):437−47.

[50] Imbalzano AN, Imbalzano KM, Nickerson JA. BRG1, a SWI/SNF chromatin remodeling enzyme ATPase, is required for maintenance of nuclear shape and integrity. Comm Integr Biol 2013;6(5): e25153.

[51] Tolstorukov MY, Sansam CG, Lu P, Koellhoffer EC, Helming KC, Alver BH, et al. SWI/SNF chromatin remodeling/tumor suppressor complex establishes nucleosome occupancy at target promoters. Proc Natl Acad Sci USA 2013;110(25):10165−70.

[52] Krebs JE, Kuo MH, Allis CD, Peterson CL. Cell cycle-regulated histone acetylation required for expression of the yeast HO gene. Genes Dev 1999;13(11):1412−21.

[53] Agalioti T, Chen G, Thanos D. Deciphering the transcriptional histone acetylation code for a human gene. Cell 2002;111(3):381−92.

[54] Merika M, Thanos D. Enhanceosomes. Curr Opin Genet Dev 2001;11(2):205−8.

[55] Horak CE, Snyder M. ChIP-chip: a genomic approach for identifying transcription factor binding sites. Methods Enzymol 2002;350:469−83.

[56] Yuan GC, Liu YJ, Dion MF, Slack MD, Wu LF, Altschuler SJ, et al. Genome-scale identification of nucleosome positions in *S. cerevisiae*. Science 2005;309(5734):626−30.

[57] Bai L, Charvin G, Siggia ED, Cross FR. Nucleosome-depleted regions in cell-cycle-regulated promoters ensure reliable gene expression in every cell cycle. Dev Cell 2010;18(4):544−55.

[58] Lee CK, Shibata Y, Rao B, Strahl BD, Lieb JD. Evidence for nucleosome depletion at active regulatory regions genome-wide. Nat Genet 2004;36(8):900−5.

[59] Field Y, Kaplan N, Fondufe-Mittendorf Y, Moore IK, Sharon E, Lubling Y, et al. Distinct modes of regulation by chromatin encoded through nucleosome positioning signals. PLoS Comput Biol 2008;4(11): e1000216.

[60] Arita K, Hashimoto H, Shimizu T, Nakashima K, Yamada M, Sato M. Structural basis for Ca(2+)-induced activation of human PAD4. Nat Struct Mol Biol 2004;11(8):777−83.

[61] Vossenaar ER, Zendman AJ, van Venrooij WJ, Pruijn GJ. PAD, a growing family of citrullinating enzymes: genes, features and involvement in disease. Bioessays 2003;25(11):1106−18.

[62] Wang Y, Wysocka J, Sayegh J, Lee YH, Perlin JR, Leonelli L, et al. Human PAD4 regulates histone arginine methylation levels via demethylimination. Science 2004;306(5694):279−83.

[63] Seri Y, Shoda H, Matsumoto I, Sumida T, Fujio K, Yamamoto K. Peptidylarginine deiminase type4 (PADI4) role in immune system. Nihon Rinsho Meneki Gakkai kaishi 2014;37(3):154−9.

[64] Suzuki A, Yamada R, Chang X, Tokuhiro S, Sawada T, Suzuki M, et al. Functional haplotypes of PADI4, encoding citrullinating enzyme peptidylarginine deiminase 4, are associated with rheumatoid arthritis. Nat Genet 2003;34(4):395−402.

[65] Iwamoto T, Ikari K, Nakamura T, Kuwahara M, Toyama M, Tomatsu T, et al. Association between PADI4 and rheumatoid arthritis: a meta-analysis. Rheumatology 2006;45(7):804−7.

[66] Chang X, Fang K. PADI4 and tumourigenesis. Cancer Cell Int 2010;10:7.

[67] Tommasi C, Petit-Teixeira E, Cournu-Rebeix I, Caponi L, Pierlot C, Fontaine B, et al. PADI4 gene in multiple sclerosis: a family-based association study. J Neuroimmunol 2006;177(1−2):142−5.

[68] Christophorou MA, Castelo-Branco G, Halley-Stott RP, Halley-Stott RP, Oliveira CS, Loos R, et al. Citrullination regulates pluripotency and histone H1 binding to chromatin. Nature 2014;507(7490): 104−8.

[69] Horvath P, Barrangou R. CRISPR/Cas, the immune system of bacteria and archaea. Science 2010;327 (5962):167−70.

[70] Sampson TR, Saroj SD, Llewellyn AC, Tzeng YL, Weiss DS. A CRISPR/Cas system mediates bacterial innate immune evasion and virulence. Nature 2013;497(7448):254−7.

[71] Bondy-Denomy J, Pawluk A, Maxwell KL, Davidson AR. Bacteriophage genes that inactivate the CRISPR/Cas bacterial immune system. Nature 2013;493(7432):429−32.

[72] Shalem O, Sanjana NE, Hartenian E, Shi X, Scott DA, Mikkelsen TS, et al. Genome-scale CRISPR-Cas9 knockout screening in human cells. Science 2014;343(6166):84−7.

[73] Cong L, Ran FA, Cox D, Lin S, Barretto R, Habib N, et al. Multiplex genome engineering using CRISPR/Cas systems. Science 2013;339(6121):819−23.

[74] Qi LS, Larson MH, Gilbert LA, Doudna JA, Weissman JS, Arkin AP, et al. Repurposing CRISPR as an RNA-guided platform for sequence-specific control of gene expression. Cell 2013;152(5):1173−83.

[75] McCown TJ. Adeno-associated virus (AAV) vectors in the CNS. Curr Gene Ther 2011;11(3):181−8.

[76] Huang ZJ, Taniguchi H, He M, Kuhlman S. Cre-dependent adeno-associated virus preparation and delivery for labeling neurons in the mouse brain. Cold Spring Harb Protoc 2014;2014(2):190−4.

[77] Mendenhall EM, Williamson KE, Reyon D, Zou JY, Ram O, Joung JK, et al. Locus-specific editing of histone modifications at endogenous enhancers. Nat Biotechnol 2013.

[78] Konermann S, Brigham MD, Trevino AE, Hsu PD, Heidenreich M, Cong L, et al. Optical control of mammalian endogenous transcription and epigenetic states. Nature 2013;500(7463):472−6.

[79] Voigt P, Reinberg D. Epigenome editing. Nat Biotechnol 2013;31(12):1097−9.

[80] Rusk N. CRISPRs and epigenome editing. Nat Methods 2014;11(1):28.

COMPREHENSIVE ANALYSIS OF MAMMALIAN LINKER-HISTONE VARIANTS AND THEIR MUTANTS

6

Chenyi Pan[1,2], Yunzhe Zhang[1,2], and Yuhong Fan[1,2]

[1]*School of Biology, Georgia Institute of Technology, Atlanta, GA, USA* [2]*The Petit Institute for Bioengineering and Bioscience, Georgia Institute of Technology, Atlanta, GA, USA*

CHAPTER OUTLINE

6.1 INTRODUCTION

In eukaryotic cells, DNA is packaged into chromatin by binding to histone proteins, a family of positively charged proteins, including H1, H2A, H2B, H3, and H4. The fundamental repeating unit of chromatin is the nucleosome [1,2], which consists of about 147 bp of DNA wrapped approximately 1.7 turns around an octameric protein core containing two copies of each core histone: H2A, H2B, H3, and H4. The DNA between nucleosomes is called "linker DNA", which together with nucleosome core particles forms the extended chromatin fiber, the 10 nm "beads-on-a-string" fiber observed

Y.G. Zheng (Ed): Epigenetic Technological Applications. DOI: http://dx.doi.org/10.1016/B978-0-12-801080-8.00006-5

under the electron microscope [3,4]. A fifth histone, the linker histone H1, binds to the entry and exit sites of the nucleosomal DNA and the linker DNA, stabilizing the nucleosomes and facilitating the folding of chromatin into higher order structures, such as the 30 nm filament [1,2].

Modifications on DNA and histones constitute major mechanisms for epigenetics, which refers to the heritable changes in gene expression not associated with changes in DNA sequences. Epigenetic mechanisms are crucial for chromatin structure reprogramming and gene expression during mammalian development and cell differentiation. Dysregulation of epigenetic information is implicated in many human diseases, including cancer, because it disrupts fundamental processes such as imprinting, X chromosome inactivation, and tissue specific and developmental gene regulation. Chromatin and epigenetics have been the subjects of intensive studies across different fields in recent years, and we have seen tremendous advances in our knowledge about chromatin function, especially through the discoveries of nucleosome structure, the chromatin-modifying activities and chromatin-remodeling complexes [5−7]. The technological applications for studying DNA modifications and histone modifications are covered in other chapters in this book. Compared with core histones and their variants, H1 histones have been understudied, and their functions and regulatory mechanisms in the genome and development remain much less well understood. In this chapter, we will focus our discussion on the analysis tools and technologies we have used to study mammalian H1 linker histone and its variants.

6.2 LINKER HISTONE H1 AND ITS VARIANTS

The H1 histone family is the most divergent and heterogeneous group of histones among the highly conserved histone protein families. While core histones are evolved from archaeal histones, linker histones appear to have their evolutionary origin from lysine-rich proteins in eubacteria [8]. All metazoan H1s share the same tripartite domain structure with the central globular domain flanked by a short N- and a long C- terminal-tail region. The structure of H1s from unicellular eukaryotes, yeast, and *Tetrahymena*, however, is rather different from metazoan H1s. *Tetrahymena* H1 lacks the most conserved globular domain found in multicellular organisms [9−11], whereas the yeast H1, Hho1, consists of two such structured globular domains [12−14]. The central globular domain of H1 is highly conserved and necessary for interaction with nucleosomal DNA [15]. Both the globular and the C-terminal domains are involved in high affinity binding of H1 to chromatin [15−18]. *In vivo* studies using H1-GFP (green fluorescent protein) fusion proteins and FRAP (fluorescence recovery after photobleaching) assays show that the binding of H1 to chromatin is dynamic with a rapid exchange rate [19,20].

Early studies with *in vitro* constituted chromatin indicates that H1 can condense chromatin structure and thus has a general repressive role for transcription by all three types of RNA polymerases (reviewed in references [21] and [22]). However, more recent studies show a much more specific role of linker histone H1 on gene regulation *in vivo* [23,24]. Deletion of macronuclear H1 in *Tetrahymena* and the H1 analog *Hho1* in yeast shows subtle phenotypic changes and specific gene expression changes [14,25−27]. Deletion of H1 in *Drosophila* also affects a small number of genes, including active and inactive genes, as well as genes enriched in transposons [28,29]. In *Caenorhabditis elegans*, the H1 variant HIS-24, together with HP-1-like proteins (HPLs), synergistically regulates the transcription of immune-relevant genes and *Hox* genes [30,31]. In *Xenopus*, somatic H1 proteins are involved in regulating differential expression of the oocyte and somatic 5S rRNA genes during *Xenopus* development [32] as well as transcriptional silencing of genes required for mesodermal differentiation pathways [33]. In DT40 chicken B cell line, deletion of H1 variants affects the transcription of many genes, most of which are downregulated [34,35]. In mouse, gene expression analysis of compound H1 knockouts (KOs), such as H1a/H1t double null mouse germ line cells and H1c/H1d/H1e triple knockout (TKO) embryonic stem cells (ESCs), indicates that H1 regulates distinct sets of genes in various cellular systems [36−39].

In higher organisms, the existence of multiple nonallelic linker histone variants provides additional levels of regulation on chromatin structure and function. In mammals, there are 11 H1 variants identified, including seven somatic H1s (H1^0, H1a, H1b, H1c, H1d, H1e, and H1x) and four germ-cell-specific H1s (H1t, HILS1, H1T2, and H1oo), that are differentially regulated during mammalian development and cellular differentiation [24]. H1^0 is expressed mainly in differentiated and nondividing cells [40], whereas H1a−e are ubiquitously expressed somatic H1 histones. H1oo and H1t are germ-cell-specific H1s expressed in oocytes and testes, respectively [41,42]. There are also two H1t-related proteins (H1T2 and HILS1) identified that are specifically expressed in spermatids [43,44]. The expression levels of H1s are tightly regulated in development and cellular differentiation, and their composition differs in different tissues [23,38,45,46]. For example, while H1^0 and H1e are the major H1 variants in adult mouse liver, accounting for approximately 30% and 40% of total H1, respectively, these two H1 variants only constitute a respective ∼2% and ∼10% of total H1 in the mouse thymus [46].

Each H1 variant is encoded by a single copy gene. H1 genes can be categorized into replication-dependent or replication-independent groups based on the dependence of their gene expression on cell cycle [45,47]. In mammals, there are six replication-dependent H1 genes, including five somatic genes, H1a−e, and the testis-specific H1t genes. The mRNA messages of replication-dependent H1 genes lack introns and a poly-A tail, but possess a stem-loop sequence at 3′ end instead, and their proteins are mainly synthesized during the S phase [48]. The replication-dependent H1 genes are clustered together with multiple core histone genes on human chromosome 6 and murine chromosome 13, whereas other H1 genes are scattered in the genome and expressed as polyadenylated mRNA messages throughout the cell cycle independent of DNA replication. Interestingly, the *H1c* gene was found to encode both replication-dependent and replication-independent polyadenylated mRNA transcripts that end at different 3′ regions of the gene. The production of these two types of H1c mRNA messages is regulated independently and may allow for synthesis of H1c protein in both dividing and nondividing cells [49].

Different H1 variants exhibit significant sequence divergence from one another, especially in the C-terminal domain, and yet each somatic H1 variant is highly conserved during evolution in mammals, suggesting distinct functions for these variants in chromatin and mammalian development. Indeed, somatic H1 variants exhibit varied *in vivo*−binding dynamics in oocytes and during ES cell nuclear transfer [42]. Given that H1 variants bind to DNA with distinct affinities *in vitro* and vary in their abilities in chromatin condensation [50], the dramatically different compositions of H1 variants in various tissues could result in fine tuning chromatin condensation in both bulk chromatin and at specific regions in the genome. These properties of H1 variants suggest that they are excellent candidates for mediating chromatin reprogramming during mammalian development.

The functional importance of H1 and its variants in development, particularly during gametogenesis and embryogenesis, has been demonstrated in *C. elegans*, *Xenopus*, and mouse.

In *C. elegans*, one of the major somatic H1 variants is found to be essential for germline transgene silencing and germline development [51,52]. In *Xenopus*, the replacement of oocyte- and embryo-specific H1 variant (named B4 or H1M) by somatic H1 variants correlates with the transcriptional repression of many genes from the mid-blastula transition to neurulation [32,53], and this transition of H1 variants is the rate-limiting step for the loss of mesodermal competence [33]. In mouse, deletion of $H1^0$, H1c, and H1e together causes growth retardation with smaller body size and a shorter life span, and deletion of H1c, H1d, and H1e leads to embryonic lethality [46]. These findings indicate that H1 is essential for mouse development and that the total level of histone H1 is important for embryogenesis and postnatal development in mammals.

Despite the sequence diversity and conservation, and the abundance of individual H1 variants in specific tissues, there appears to be compensation among somatic H1s in cellular function and mammalian development. While TKO of H1c, H1d, and H1e in mice leads to embryonic lethality, deletion of individual somatic H1 variants is tolerated and mice deficient for one or two H1 variants do not show obvious phenotypes [54–58]. These H1 single or double KO mice do not have a reduction in the H1 to nucleosome ratio due to the increase of other H1 variants to compensate for the lost H1s [36,55].

Although the physiological functions of individual H1 variants remain elusive, recent studies show that individual H1 variants are different in gene regulation and genomic localization. Overexpression of $H1^0$ and H1c elicits different effects in gene expression and nucleosome repeat length [59,60]. Using a set of single H1 variant KO mice, Alami et al. find that H1 variants can differentially affect transgene expression as powerful modifiers for position effect variegation of β-globin transgenes *in vivo* [61]. Knockdown of individual H1 variants in human breast cancer cell line T47D identifies specific roles of H1 variants in gene expression and cell growth [62]. H1c has been shown to be a signal transducer in apoptosis induced by double-strand DNA breaks [63]. Recent high-resolution mapping studies have indicated the common and specific occupancy patterns of H1 variants in the genome [64–66]. These studies further demonstrate the specificity of H1 variants.

In the following sections, we compile the methods and models we have utilized and developed to profile H1 expression patterns and to characterize the functions of mammalian histone H1 variants. For studies used to characterize other H1 properties, such as analysis of H1's dynamic binding by FRAP assays and of H1's posttranslational modifications by proteomics approach, we refer the readers to other references and literature [16,19,20,67–71]. Here, we first describe a set of qRT-PCR (quantitative reverse transcription-polymerase chain reaction) assays and HPLC (high-performance liquid chromatography) analysis to quantify the RNA and protein levels of H1 variants. Next we illustrate a series of genetic analyses to examine the functions of H1 variants in mammalian development and stem cell differentiation, to achieve genome-wide mapping of H1 variants in the epigenome, and to assess the chromatin binding of the H1 mutants identified in follicular lymphoma.

6.3 EXPRESSION ANALYSIS OF MAMMALIAN LINKER HISTONE H1 VARIANTS

A remarkable characteristic of the histone H1 family is its heterogeneity, with multiple H1 variants conserved among orthologs but differing in biochemical properties and expression patterns among paralogs [45]. As mentioned above, there are 11 mammalian H1 variants that are differentially regulated during development and the cell cycle [24]. The total level of H1 is a determinant for chromatin compaction and for nucleosome repeat length *in vivo* [23,37,46], and the different levels and compositions of H1 variants modulate and fine-tune higher order chromatin structures. Quantitative assessment of expression levels of H1 variants allows for estimation of the chromatin status and helps in monitoring the spatiotemporal changes in chromatin in various cellular processes and biological contexts. Thus, the quantitative expression analysis of H1 variants we established and described here should have wide applications in chromatin research and epigenetic analysis.

Here, we present a set of assays for comprehensive analysis of mammalian H1 variant expression for both mRNA transcripts and proteins. Properly designed qRT-PCR assays offer highly sensitive and accurate measurements of the expression of histone H1 genes at the RNA level, with the critical step being the synthesis of cDNA (complementary DNA) using random hexamers, whereas HPLC analysis provides a highly reproducible quantitation of H1 proteins, determined as the H1/nucleosome ratio and the percentage of the total H1. For detailed protocols, the readers are referred to our paper in JoVE [72].

6.3.1 ANALYSIS OF H1 GENE TRANSCRIPTS BY QUANTITATIVE REVERSE TRANSCRIPTION PCR

Transcript levels of H1 variants in cells or tissues can be measured by a set of highly sensitive and quantitative reverse transcription PCR (qRT-PCR) assays we have recently developed (Figure 6.1) [72,73]. Like most core histone transcripts, the majority of H1 mRNAs lack a long polyA tail as observed in most cellular mRNA messages, but contain a stem-loop structure at the 3′ untranslated region [45]. Therefore, random hexamers, instead of oligo-dT primers, are used for the preparation of cDNA from total RNA by reverse transcription. Real-time PCR analysis with primers specific to each H1 variant is performed to obtain quantitative measurement of the expression levels of individual H1 variants after normalization with housekeeping genes.

To extract RNA from mouse tissues, the organs of interest are dissected from euthanized mice, and washed in ice cold phosphate buffered saline (PBS, pH 7.4). Sufficient amount of TRIzol® reagent (Invitrogen) is necessary for obtaining high-quality RNA. Roughly, $50-100$ mg tissue samples or $5-10 \times 10^6$ cells require 1 mL TRIzol reagent. The tissue samples in TRIzol reagent are then homogenized with a Polytron® PT2100 homogenizer (or equivalent), followed by RNA extraction according to the manufacturer's manual. If total RNA is to be extracted from cultured cells, the TRIzol reagent can be applied to cells directly following the manufacturer's instructions. The concentration of the isolated RNA is measured using a NanoDrop™ 1000 (Thermo Scientific) and RNA quality is assessed by gel electrophoresis. The typical yields range from 1 to 10 μg RNA per mg of tissue and 5 to 15 μg per 1×10^6 cells. RNA may also be isolated with the RNeasy kit (QIAGEN) or AllPrep DNA/RNA kit (QIAGEN) when both DNA and RNA are desired. To eliminate the potential contamination of genomic DNA, RNA samples are treated with RNase-free DNase (Sigma AMP-D1) according to the manufacturer's instructions; cDNA is synthesized from total RNA using SuperScript® III First-strand Synthesis System (Invitrogen™) with random hexamers. However, if expression analysis of genes with polyadenylated messages at low levels is also

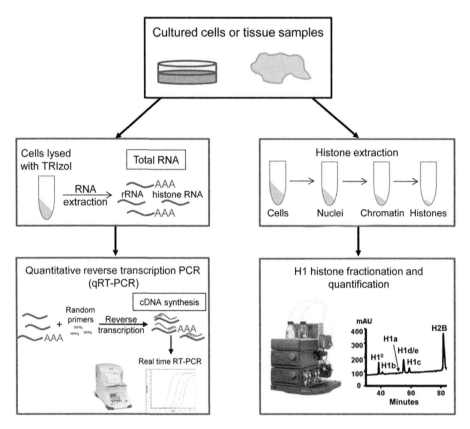

FIGURE 6.1

Overall scheme of expression analysis of mammalian linker histone variants. Flow charts for quantitative RT-PCR (qRT-PCR) analysis of mRNA expression (left) and HPLC analysis (right) of H1 variants are shown.

Reproduced from Medrzycki et al. [72].

desired, a mixture of random hexamers and oligo-dT may be used to improve the reverse transcription efficiency of polyadenylated mRNAs.

We use the Bio-Rad iQ-5™ real-time PCR detection system for quantitative PCR (qPCR) of H1 variant mRNAs. We have designed a set of real-time PCR primers for H1 genes (Table 6.1) [72,73] that are specific to the target H1 variant and do not cross amplify other H1 variants. Since most H1 genes do not contain introns, genomic DNA contamination should be examined by RT (-)-qPCR. Housekeeping genes such as glyceraldehyde-3-phosphate dehydrogenase *(GAPDH)* and beta-actin are chosen as internal reference genes for normalization as described [72,73]. We use the following real-time PCR protocol: 95°C for 3 minutes, followed by 40 cycles of 95°C for 10 seconds, 60°C for 20 seconds, and 72°C for 30 seconds. Triplicate reactions and statistical analysis are performed. After normalization with housekeeping genes, the expression levels of H1 variants are compared among various samples.

Table 6.1 Mouse and Human H1 Subtypes

	Mouse H1 Subtypes		Human H1 Subtypes	
Mammalian Subtypes	**Gene Name**	**Accession No.**	**Gene Name**	**Accession No.**
Histone H1a	Hist1h1a (H1.1)	NM_030609	HIST1H1A (H1.1)	NM_005325
Histone H1b	Hist1h1b (H1.5)	NM_020034	HIST1H1B (H1.5)	NM_005322
Histone H1c	Hist1h1c (H1.2)	NM_015786	HIST1H1C (H1.2)	NM_005319
Histone H1d	Hist1h1d (H1.3)	NM_145713	HIST1H1D (H1.3)	NM_005320
Histone H1e	Hist1h1e (H1.4)	NM_015787	HIST1H1E (H1.4)	NM_005321
Histone $H1^0$	H1f0	NM_008197	H1F0	NM_005318
Histone H1oo	H1foo	NM_183811	H1FOO	NM_153833
Histone H1t	Hist1h1t	NM_010377	HIST1H1T	NM_005323
Histone H1t2	H1fnt	NM_027304	H1FNT	NM_181788
Histone H1x	H1fx	NM_198622	H1FX	NM_006026
Histone Hils1	Hils1	NM_081792	HILS1	AY286318

Adapted from Medrzycki et al [72].

6.3.2 HPLC ANALYSIS OF LINKER HISTONES

The relative abundance of proteins of individual H1 variants can be gauged by reverse-phase high-performance liquid chromatography (RP-HPLC) analysis of total histones extracted from mammalian cells or tissues (Figure 6.1) [11,55,74]. The HPLC method and elution conditions described here give optimal separation of mouse H1 variants. By quantifying the HPLC peaks, the relative proportions of individual H1 variants within the H1 family as well as the H1 to nucleosome ratio in the cells or tissues are derived.

To purify histone proteins, mouse tissues are dissected and rinsed with ice cold PBS, and then minced into small pieces with a razor blade. 1 gram of mince is transferred to a Dounce homogenizer (B pestle) with 10 mL sucrose buffer (0.3 M sucrose, 15 mM NaCl, 10 mM HEPES [pH 7.9], 2 mM EDTA, 0.5 mM PMSF, Complete Mini Protease Inhibitor Cocktail Tablet, added freshly), followed by homogenizing the tissue for 10−15 strokes. The homogenates are subsequently transferred to a 15 mL centrifuge tube and spun at 500 rpm for 30 seconds (Eppendorf 5810R) to precipitate tissue debris. After centrifugation, the supernatant is transferred to a new tube, followed by centrifugation again at 2000 rpm for 5 minutes to pellet cells. If histones are to be extracted from cultured cells, cells are harvested, rinsed with PBS, and pelleted.

The cell pellet isolated from tissues or cultured cells is then resuspended in 10 mL Sucrose buffer supplemented with 0.5% NP-40 (per gram of starting tissue amount or 10^8 cells). The cells are homogenized in a Dounce homogenizer (B pestle) for 10 strokes within 20 minutes, followed by centrifugation at 2000 rpm for 5 minutes to pellet nuclei. The nuclei are then resuspended in 3 mL high salt buffer (0.35 M KCl, 10 mM Tris [pH 7.2], 5 mM $MgCl_2$, 0.5 mM PMSF, added freshly before use) per 1 g of tissue or 10^8 cells, and transferred to a small Dounce homogenizer (B pestle) and homogenized for 5−10 strokes. The suspension is then aliquoted into 3 Eppendorf tubes (1 mL each) and incubated on ice for 20 minutes, followed by centrifugation at 14,000 rpm for 10 minutes to pellet chromatin. To extract histones, the pelleted chromatin is resuspended in 0.8 mL 0.2 N H_2SO_4 and ground with an Eppendorf tube pestle Dounce until complete dissociation. After incubation on a rotating platform at 4°C overnight, the insoluble genomic DNA is removed by centrifugation at 14,000 rpm for 10 minutes, and the supernatant containing histone extracts is obtained. To precipitate the histone proteins, 2.5 volumes of ice-cold ethanol are added to histone extracts, and stored at −20°C overnight. The next day, total histones are precipitated by centrifugation at 14,000 rpm for 10 minutes, washed with 70% ethanol three times and air dried. The histone pellets can be stored at −80°C or dissolved in ddH_2O for immediate HPLC analysis.

The volume of ddH_2O needed to dissolve the histone pellet depends on the capacity of the reverse-phase column and the HPLC instrument. We use the C18 reverse-phase column 250 × 4.6 mm (Vydac) and the ÄKTApurifier UPC 900 instrument (GE Healthcare Life Sciences) for HPLC analysis. Typically, 50−100 µg of total histones are resuspended in 100 µL of ddH_2O. After removal of the insoluble residues by centrifugation at 14,000 rpm for 5 minutes, the proteins are injected into the reverse-phase column on the HPLC system. Linker histones and core histones are fractionated with an increasing acetonitrile gradient as described in detail previously [72]. The effluent signals are monitored at 214 nm, and HPLC profiles are recorded and analyzed with UNICORN 5.11 software (GE Healthcare Life Sciences). A representative HPLC profile is shown in Figure 6.1. The protein fractions can be collected with a fraction auto-collector (Frac-920-GE) for further analysis, such as SDS-PAGE and mass spectrometry. The areas of A_{214} peaks (absorbence at 214 nm) of H1 variants and H2B are normalized by the number of peptide bonds of respective histone proteins, and the normalized values are used for calculation of H1 to nucleosome ratio (H1/nuc) as well as the relative proportions of individual H1 variants among the total H1 histone pools (percent of total H1) as previously described [72,74].

6.4 GENETIC ANALYSIS OF H1 VARIANTS BY GENE INACTIVATION

Through conventional gene targeting by homologous recombination in mouse ESCs, five of the six somatic H1 variants have been individually inactivated in mice. Interestingly, these H1 single KO mice, null for $H1^0$, or H1a, or H1c, or H1d, or H1e, develop normally and display no obvious phenotypes [55,56,58]. Analysis of H1 proteins demonstrates a normal H1/nucleosome ratio in these mice, suggesting a biochemical compensation of the lost H1 proteins by the remaining H1 variants. These studies indicate that the individual H1 variant is not required for mouse development. However, as a group, somatic H1s are essential for mammalian development as shown by embryonic lethality of H1c/H1d/H1e TKO mice [46]. These H1c/H1d/H1e TKO mice, generated by sequential gene targeting, have a 50% reduction in total H1 level in mouse embryos and display embryonic lethality at midgestation. The approach and procedures used to generate single, double, and triple H1 KO mice have been described in detail in a previous chapter [74]. In this section, we will mainly focus on the derivation of triple and single H1 KO ESCs, and the use of H1 KO ESCs as an experimental system for genetic analysis of H1 functions in differentiation. Here, we describe the establishment of ESC lines depleted of one or multiple H1 variants and the usage of these ESCs in characterizing the role of H1 and its variants in stem cell differentiation.

6.4.1 DERIVATION OF TRIPLE AND SINGLE H1 KNOCKOUT EMBRYONIC STEM CELLS

Homozygous H1c/H1d/H1e triple-H1 null embryos survive to E9.5−E11.5 [46] (embryonic day 9.5 to 11.5), so it was possible to derive homozygous triple H1 KO ESCs from H1c/H1d/H1e TKO embryos. H1c/H1d/H1e triple heterozygotes were intercrossed and blastocysts were harvested from pregnant females for ESC derivation from blastocyst outgrowth. Wild-type and H1 TKO ESCs were identified from the derived ESC lines by PCR genotyping as described previously [37,46]. Established H1 TKO ESC lines and wild-type (WT) control ESC lines were cultured as previously described [38]. Similarly, single H1 KO ESCs were derived from $H1c^{-/-}$, or $H1d^{-/-}$, or $H1e^{-/-}$ blastocyst outgrowth, respectively.

Under culture conditions promoting ESC self-renewal, all H1 single- and triple-KO ESCs have normal karyotypes, cell and colony morphology, as well as comparable expression of pluripotency markers, such as Oct3/4 (also known as POU5F1), as WT ESCs [37−39]. HPLC analysis indicates that, compared with WT ESCs, H1 TKO ESCs have a 50% reduction of the total H1 level [37]. Surprisingly, single H1 KO ESCs also show a reduction in total H1 levels, ranging from ∼15% reduction in $H1c^{-/-}$ ESCs to ∼25% reduction in $H1d^{-/-}$ cells [39]. This is in contrast to the normal H1/nucleosome ratio observed in single H1 KO mouse tissues [55], suggesting that the biochemical compensation among H1 family in ESCs is less prominent than that in the differentiated cells and tissues. H1 TKO ESCs have global changes in chromatin structure, including reduced nucleosome repeat length, local reduction in chromatin compaction, and changes in specific histone modifications, but only display limited and specific gene expression changes [37]. H1 TKO and single KO ESCs also show reduction in the expression levels of specific Hox genes [39]. These ESC lines provide useful cell resources for epigenetic analysis of chromatin high-order folding and for studying H1 and its variants in regulating gene expression and epigenetic events during ESC self-renewal and differentiation.

6.4.2 DIFFERENTIATION OF H1 KNOCKOUT EMBRYONIC STEM CELLS

ESCs possess the capacity to differentiate into almost all cell lineages, offering great promise in regenerative medicine and cellular therapy. Compared with lineage committed or differentiated cells, ESCs have a hyperactive global transcriptional activity and possess an open and more relaxed chromatin structure with higher chromatin mobility and hyperdynamic chromatin proteins as hallmarks [74−76,77,78]. In agreement with this view, wild-type ESCs (WT ESCs) have an H1/nucleosome ratio of 0.46 [37], much lower than that of 0.75∼0.83 observed in various differentiated cell types in mouse tissues [23,55]. Since H1 TKO ESCs have an unusually low H1/nucleosome ratio of 0.25, approximately 50% reduction as compared with WT ESCs, and exhibit decondensation in bulk chromatin [37], these ESCs offer an ideal system to test the necessity of chromatin compaction for ESC pluripotency and differentiation. Through a variety of differentiation schemes, we demonstrated, for the first time, that inactivation of multiple H1 variants impairs the differentiation capacity of ESCs, indicating that a sufficient H1 amount is required for proper stem cell differentiation [38].

H1 TKO ESCs appear to be more resistant to spontaneous differentiation in cultures lacking leukemia inhibitory factor (LIF) and mouse embryonic fibroblast feeder layer when compared with

WT ESCs [38]. Furthermore, by utilizing a rotary orbital suspension culture system which enables high efficiency and homogeneity of embryoid body (EB) differentiation, we find that H1 TKO EBs are significantly impaired in spontaneous differentiation and lack cells with morphologies representing three different germ layers [38]. Formation and differentiation of EBs from ESC aggregates mimic some of the early events during embryonic development *in vivo*, and thus serve as a good *in vitro* model for ESC differentiation and early embryogenesis. Upon EB differentiation, the levels of H1c, H1d, H1e, and H1^0 in WT EBs are progressively increased over time, with the H1 to nucleosome ratio increasing to 0.62 in day 10 EBs from 0.46 in WT ESCs (Figure 6.2A and 6.2B). In contrast, H1 TKO ESCs mostly form putative EBs of undifferentiated aggregates of ESCs even under prolonged rotary suspension culture with an H1 to nucleosome ratio merely at 0.36 in day 10 EBs [38] (Figure 6.2B). On the other hand, ESCs with three single H1 KOs (H1c$^{-/-}$, or H1d$^{-/-}$, or H1e$^{-/-}$) exhibit normal differentiation in EBs and teratomas [39]. Thus, the defects of H1 TKO ESCs in these described spontaneous differentiation schemes suggest that the total level of histone H1 plays a critical role in ESC differentiation.

To determine if the impairment of H1 TKO ESCs in spontaneous differentiation is a result of blockade or delay in differentiation, we subject the H1 TKO ESCs to induction of lineage-specific differentiation. For this purpose, we utilize a well-defined and robust neural differentiation scheme (Figure 6.2Bi). EBs were allowed to form using the hanging drop method for 4 days before being transferred to ultra-low attachment plates (Corning) and cultured for additional 2 days in the presence of 1 μM *all-trans* retinoic acid (RA) as described previously [38]. EBs were subsequently transferred to tissue culture dishes coated with poly-L-ornithine and laminin (PLO + L) (5 μg/mL) at 10 EBs/cm^2 and cultured in NeuroCult$^®$ NSC proliferation medium (supplemented with 10 ng/mL b-FGF [basic fibroblast growth factor]; Stemcell Technologies) for continued neural differentiation. By day 6 + 7 of this *in vitro* neural differentiation regimen (Figure 6.2Bi), abundant neurite outgrowth from WT EBs was clearly observed, whereas H1 TKO EBs have dramatically decreased neurite numbers (Figure 6.2Bii). Neurites are enriched in cylindrical bundles of microtubules composed of β-III tubulin (TUBB3) proteins, extending from the body of neurons which eventually differentiate into axons or dendrites. Quantification of neurite outgrowth showed that, under this differentiation protocol, approximately 50% of WT EBs form 18 neurites on average in contrast to only 10% with 8 neurites on average for H1 TKO EBs (Figure 6.2Biii). The defects in neural differentiation of H1 TKO ESCs were also confirmed by immunostaining of neural markers, including the neural stem cell marker, Nestin, and the glial cell marker, GFAP (glial fibrillary acidic protein) [38]. Collectively, these results suggest that neural differentiation is largely blocked by H1 depletion.

6.5 *IN VIVO* TAGGING AND GENOME-WIDE MAPPING OF H1 VARIANTS

Individual H1 variants are conserved in mammals, differentially regulated during development, and have been shown to regulate specific gene expression across different cell types [24,60−62,79]. H1 variants also have distinct biochemical properties and differ significantly in their residence time on chromatin and ability to promote chromatin condensation *in vitro* [50,67,80]. Mapping genome-wide localization of H1 variants would be necessary for understanding the regulatory mechanisms

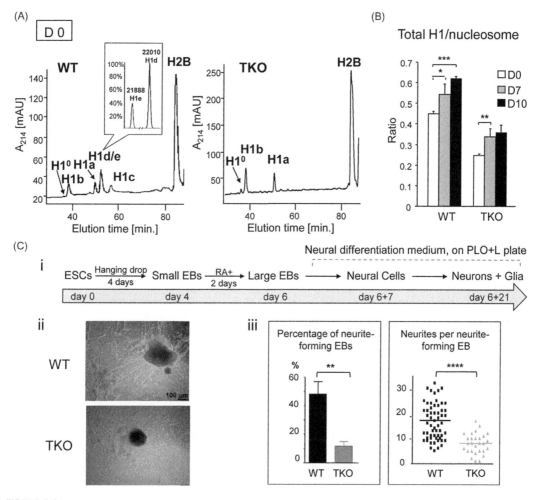

FIGURE 6.2

H1 depletion impairs ESC differentiation. (A) Reverse-phase HPLC and mass spectrometry (inset) analysis of histones from WT and H1 TKO ESCs (D0). X axis, elution time; Y axis, absorbency at A_{214}; mAU, milli-absorbency units. Inset shows the relative signal intensity of H1d and H1e mass spectral peaks in the H1d/H1e fraction collected from indicated HPLC eluates of histone extracts of WT ESCs. (B) H1/nucleosome ratio of the total H1 during EB differentiation. (C) H1 TKO ESCs fail to undergo neural differentiation: (i) Neural differentiation scheme for ESCs; (ii) Phase contrast images show that H1 TKO mutants were unable to adequately form neurites and neural networks; (iii) Left panel: Percentage of neurite-forming EBs. Numbers were averaged from six experiments. 80 EBs were counted per experiment. Right panel: Numbers of neurites per neurite-forming EB. Numbers of neurites were counted from EBs that produced neurites. 58 and 28 neurite-forming EBs from respective WT and TKO cultures were selected and counted for neurite numbers. $**P < 0.01$; $****P < 0.0001$.

Adapted from Zhang et al [38].

of H1 in chromatin structure and gene expression *in vivo*. However, this has been challenging due to the lack of H1 variant antibodies of high specificity and affinity. We achieved the first set of high-resolution maps of H1 variants in ESCs using a tagged-H1 knock-in strategy combined with chromatin immunoprecipitation followed by deep sequencing (ChIP-seq), and others have reported H1 variant mapping in human cells [64−66,81]. We have also demonstrated that the tagged H1 variants are functionally interchangeable with their endogenous counterparts. Thus, the *in vivo* tagging systems can also be utilized for identification of binding partners and cellular localization of specific H1 variants. Here we describe our approach and key findings from mapping H1 variants in ESCs. The knock-in strategy for *in vivo* tagging and high-throughput mapping described here should have wide applications to study highly similar histone variants *in vivo*.

Epigenetic regulation is perceived to play a critical role in stem cell fate determination, and genome-wide profiling of the epigenome has unraveled that ESCs have epigenetic landscapes distinct from differentiated cells [82,83]. By achieving high-resolution mapping of H1 variants, an important yet understudied type of chromatin proteins and epigenetic factors, we integrate their localization into the epigenome landscape of mouse ESCs, and provide new insights into the regulatory roles of H1s in stem cells. We first established a tagged-H1 knock-in system and demonstrated that the N-terminally tagged H1 proteins are functionally interchangeable to their endogenous counterparts *in vivo*. Subsequent mapping of the tagged H1 variants in ESCs by ChIP-seq revealed that H1 is depleted from GC- and gene-rich regions and active promoters, in inverse correlation with the active histone mark H3K4me3 and in positive correlation with the repressive histone mark H3K9me3. Importantly, both H1d and H1c are significantly enriched at major satellites and H1 depletion causes chromocenter clustering and de-repression of major satellites, indicating an important role of histone H1 in pericentromeric integrity. Specific binding differences of H1d, H1c, and $H1^0$ in the ESC epigenome are also noted [64].

6.5.1 EMBRYONIC LETHALITY IN H1c/H1d/H1e TRIPLE NULL MICE IS RESCUED BY FLAG-H1d

To circumvent an obstacle in mapping H1 variants due to the lack of high-quality H1 variant-specific antibodies, we used a knock-in strategy to insert, at the N-terminus of endogenous H1 variants, epitopes, such as FLAG and Myc, for which highly specific antibodies exist. The fact that H1c/H1e double KO mice are normal but H1c/H1d/H1e TKO (H1 TKO) mice are embryonic lethal [46] allows us to stringently test the functional equivalence of tagged H1 variants and the respective endogenous H1 variants by rescue of embryonic lethality in H1c/H1d/H1e triple null mice by FLAG-H1d (FLAG-tagged H1d) (Figure 6.3).

A FLAG-H1d knock-in vector with *H1d* upstream and downstream homology regions flanking the N-terminally FLAG-tagged H1d and the blasticidin-resistance gene was constructed and transfected into the *cis* triply targeted $H1c^{+/-}H1d^{+/-}H1e^{+/-}$ ESCs established previously [46]. ESC colonies resistant to blasticidin (Life Technology) were picked and screened for homologous recombination by Southern blotting as described previously [64]. *Cis* triply targeted $H1c^{+/-}H1d^{+/FLAG}H1e^{+/-}$ ESCs were injected into mouse blastocysts to generate chimeras which produced $H1c^{+/-}H1d^{+/FLAG}H1e^{+/-}$ mice. The $H1c^{-/-}H1d^{FLAG/FLAG}H1e^{-/-}$ mice, obtained from intercrosses of $H1c^{+/-}H1d^{+/FLAG}H1e^{+/-}$ mice, are viable, fertile and phenotypically normal. The rescue of embryonic lethality of H1c/H1d/H1e

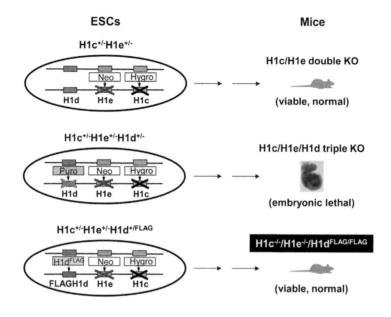

FIGURE 6.3

Rescue of embryonic lethality in H1c/H1d/H1e triple null mice by FLAG-H1d. H1c/H1e knockout mice generated from *cis*-targeted $H1c^{+/-}H1e^{+/-}$ heterozygous ESCs are viable and develop normally [46]. H1c/H1d/H1e triple knockout mice generated from *cis*-targeted $H1c^{+/-}H1d^{+/-}H1e^{+/-}$ heterozygous ESCs through a sequential targeting strategy are embryonic lethal [46]. $H1c^{+/-}H1d^{+/FLAG}H1e^{+/-}$ ESCs, created by knock-in of the $H1d^{FLAG}$ allele in a *cis* configuration in $H1c^{+/-}H1d^{+/-}H1e^{+/-}$ ESCs, were used to produce $H1c^{+/-}H1d^{+/FLAG}H1e^{+/-}$ mice. $H1c^{-/-}H1d^{FLAG/FLAG}H1e^{-/-}$ mice, generated from intercrosses of $H1c^{+/-}H1d^{+/FLAG}H1e^{+/-}$ mice, are viable, fertile, and develop normally.

TKO mice by FLAG-H1d indicates that tagged H1d (FLAG-H1d) is functionally interchangeable with the endogenous H1d.

HPLC, mass spectrometry, and Western blotting analyses of histone extracts from $H1c^{+/-}H1d^{+/FLAG}H1e^{+/-}$ ESCs and $H1c^{-/-}H1d^{FLAG/FLAG}H1e^{-/-}$ mouse tissues demonstrate that FLAG-H1d is associated with chromatin, has the same hydrophobicity and expression as the endogenous H1d [64]. Similar analysis of histone extracts from $H1c^{+/Myc}H1d^{+/-}H1e^{+/-}$ ESCs, established by a tag knock-in at the endogenous *H1c* gene locus, indicated that N-terminally Myc-tagged H1c (Myc-H1c) also displays identical biochemical properties and expression levels as the endogenous H1c. Taken together, these results demonstrate that addition of these small epitopes at the N-terminus of H1 variants do not alter H1 properties and functions and that these tagged H1 variants can functionally substitute their endogenous counterparts *in vivo*. These studies pave the way for analyzing similar histone variants on a genomic scale using an *in vivo* tagging approach.

6.5.2 MAPPING H1 VARIANTS IN ESCs

H1d and H1c are among the most abundant linker histones in mouse ESCs, accounting for 33% and 16% of the total H1 level, respectively [37]. These two variants differ significantly in the

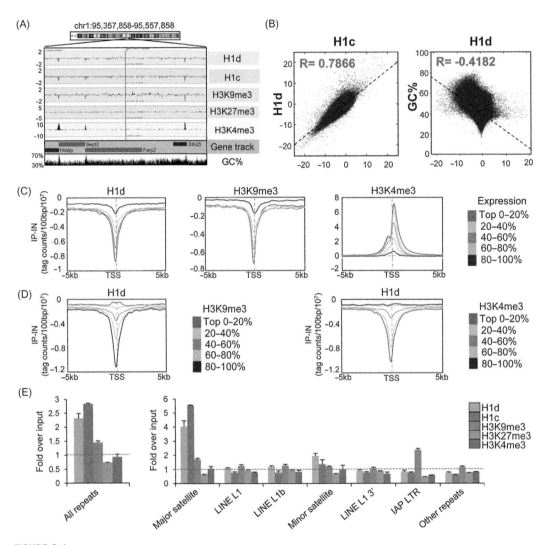

FIGURE 6.4

Genome-wide mapping of H1d and H1c in mouse ESCs. (A) An example of distribution of H1 variants and histone marks at a 200 kb region. The GC density track was obtained from the UCSC genome browser. Genes are color coded according to their transcription directions (Red: sense strand; Blue: anti-sense strand). (B) Genome-wide correlation scatter plots of H1d vs. H1c (left) and GC% vs. H1d (right). The correlation coefficient (R value) and the trend line are generated as described [64]. Pearson's correlation was used to perform the analysis. $P < 10^{-100}$ for all correlation coefficients. (C) Metagene analysis of H1d, H3K9me3, and H3K4me3 in relation to gene expression levels on a 10 kb window centered on TSSs. Genes were partitioned into five groups according to their expression levels. TSS: transcription start sites; Y axis: tag counts per 100 bp window per 10 million mappable reads; IP-IN: normalized signal values of ChIP-seq subtracted by that of input-seq.

(Continued)

binding affinity to chromatin and residence time in chromatin [50,67]. The aforementioned H1d^{FLAG} and H1c^{Myc} knock-in ESCs were used to map the genome-wide distribution profiles of H1d and H1c by ChIP-seq as described previously [64]. Briefly, $\sim 10^8$ ESCs were cross-linked with 1% formaldehyde followed by extraction of nuclei. Crosslinked chromatin was subsequently sheared by sonication and soluble chromatin was incubated overnight at 4°C with anti-FLAG (Sigma-Aldrich F3165) or anti-Myc (Cell Signaling #2272) antibodies pre-incubated with Dynabeads$^®$ Protein G (Life Technologies). The next day, Dynabead−chromatin complexes were washed, immunoprecipitates of DNA−protein complexes were eluted and reverse cross-linked at 65°C, and DNA was purified with a DNA isolation column (QIAGEN). Input control DNA was prepared from reverse-crosslinked soluble chromatin without immunoprecipitation. The libraries of ChIP DNA or input DNA were prepared using ChIP-seq Sample Preparation Kit (Illumina) following the manufacturer's manual and subject to massive parallel sequencing on the Illumina Genome Analyzer II or Illumina HiSeq 2000 systems. Sequence reads were aligned against mouse genome mm9 using the Bowtie aligner program [84]. Approximately 80−90% of reads in each ChIP-seq library were mappable to the mouse genome. IP-IN signals were calculated in 100 bp or 1000 bp sliding windows after normalization of libraries using GenPlay software [85]. ChIP-seq of histone marks, including an active histone mark, H3K4me3, and two repressive histone marks, H3K9me3 (histone H3 lysine 9 tri-methylation) and H3K27me3 (histone H3 lysine 27 tri-methylation), was also performed to facilitate the comparisons of H1 variant binding maps with characteristic epigenetic marks across the ESC epigenome.

Visual examination of sequencing track files reveals that both H1d and H1c are enriched at gene desert, but generally depleted from gene-rich regions of high GC% (guanine-cytosine content) with the deepest dips around transcription start sites (TSSs) of active genes (Figure 6.4A). Both H1d and H1c dips coincide with H3K9me3 dips and H3K4me3 (histone H3 lysine 4 tri-methylation) peaks, suggesting similar binding patterns of H1d and H1c with the repressive H3K9me3 epigenetic mark. Indeed, genome-wide correlation analysis demonstrates that the distributions of H1d and H1c are highly similar (R = 0.7866), and correlate negatively with GC% and H3K4me3, but positively with H3K9me3 (Figure 6.4B and [64]). Metagene analysis of H1 occupancy on genes finely partitioned (into five groups) according to their expression levels, or signals of H3K4me3 and H3K9me3 over a 10 kb region surrounding TSSs, further corroborates its inverse correlation with transcriptional activity (deepest H1 dip in the group of genes with highest expression and the active histone mark H3K4me3) and its similarity with H3K9me3 repressive histone mark

◄ (D) Metagene analysis of H1d in relation to the levels of H3K9me3 (left) and H3K4me3 (right) on regions covering −5 kb to +5 kb of TSSs. Genes were evenly grouped into five categories according to the signals of the respective histone marks. (E) Enrichment of H1d and H1c at the major satellite sequences. Fold enrichment of percent mappable repeats (mapped to RepBase) from H1d, H1c, and histone marks of ChIP-seq libraries over that from the corresponding chromatin input-seq library on all repeats (left), six most abundant repetitive sequences and the remaining other repeats (right). The dashed lines indicate the level of normalized input signal. *P* values calculated with Fisher's exact test comparing ChIP-seq with input-seq libraries are less than 2.5×10^{-5} for all repeat classes shown. Error bars represent the differences between replicates. Data are presented as average $\pm$ SEM (standard error of the mean). (*This figure is reproduced in color in the color plate section.*)

Adapted from Cao et al [64].

(Figure 6.4C and 6.3D) [64]. $H1^0$, the differentiation-associated H1 variant with minimum expression in ESCs, was mapped using FLAG-$H1^0$-overexpressing ESCs. $H1^0$ displays similar features as those of H1d and H1c at active promoters [64], suggesting that the relationships among H1 binding, gene expression, and histone marks are applicable to all somatic H1 variants.

Alignment of H1 ChIP-seq reads to mm9 indicates a higher percentage of multi-match reads than that of input control libraries, 45% for H1c ChIP DNA vs. 22% for Input DNA, suggesting overrepresentation of H1 variants on repetitive sequences. Epi-GRAPH [86] analysis of H1 rich regions identified by GenPlay [85] or SICER [87] (spatial clustering approach for the identification of ChIP-enriched regions) also indicates that H1d/H1c common peaks are enriched at AT-rich sequences, satellite DNA, and chromosome G-bands [64]. Specific differences among H1d-, H1c-, and $H1^0$-enriched regions are also noted. Examination of several top-ranked H1 peak regions reveals that these regions overlap perfectly with major satellite sequences. These data prompted us to perform a comprehensive alignment of sequencing reads against mammalian repeats from RepBase Update [88], a comprehensive database of repetitive elements, using the Bowtie aligner program. The percentage of reads for specific repeat sequences was calculated and the fold enrichment for H1 occupancy and histone marks at each repeat element was derived as described [64].

The enrichment of H1d and H1c in repetitive sequences was predominant at major satellite sequences, with 4.0- and 5.6- fold enrichment for H1d and H1c, respectively (Figure 6.4E). Pericentric major satellites from different chromosomes cluster to form chromocenters, which are important for the structural integrity of chromosomes [89]. The high occupancy of H1 variants at major satellites suggests a role of H1 in establishment and maintenance of pericentromeric heterochromatin. Indeed, H1c/H1d/H1e deletion leads to increased chromocenter clustering as well as elevated transcription from major satellites [64]. Taken together, these results integrate this significant repressive mark, namely H1 occupancy, into the ESC epigenome and provide new insights into the roles of H1 and chromatin folding at various genomic regions.

6.6 MUTATION ANALYSIS OF HISTONE H1

The importance of H1 variants in chromatin, gene regulation, and cell functions has been further highlighted by recent findings recognizing H1 mutations as potential driver mutations in tumorigenesis. Frequent mutations in *H1* genes have been identified in colorectal cancer through genome-wide analysis [90,91]. Recurrent missense mutations in multiple somatic H1 variants have been found in follicular lymphomas [92−95]. Most H1 mutations occur in the C-terminal domain and the globular domain that are directly involved in DNA binding and interaction with core histone particles [15,69,96]. The mechanisms of H1 mutations in promoting the initiation and/or progression of tumorigenesis remain to be determined. Here we describe our recent work in characterizing H1 mutations occurring in follicular lymphomas to illustrate how the methods and cellular systems discussed in previous sections can be utilized to study the oncogenic mutations in H1 genes.

Through genome-wide genomic and exonic sequencing on follicular lymphoma-transformed follicular lymphoma pairs followed by deep sequencing of target genes, Okosun et al. identified a group of genes, including H1 genes, whose recurrent mutations drive the initiation and progression of follicular lymphomas [94]. 55 mutations were identified in four H1 genes, *H1b−H1e*, in

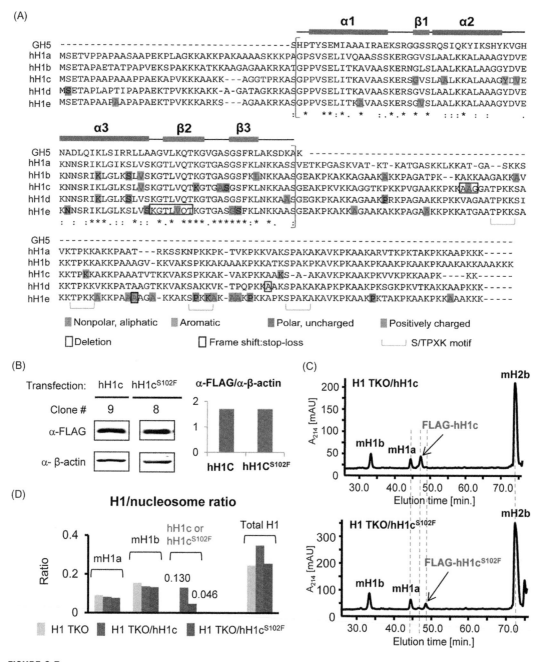

FIGURE 6.5

(Continued)

38 patients, and most of those mutations were missense mutations [94]. Overall, 28% of cases harbor H1 mutations in at least one histone H1 gene, and *H1c* and *H1e* are the most frequently mutated [94]. Sequence alignment of H1 variants and the globular domain of avian H5 [96] indicates that vast majority of the mutations are clustered within the C-terminal domain and the highly conserved globular domain involved in DNA binding and chromatin compaction (Figure 6.5A).

To characterize the mechanism of action of these mutations, we took advantage of H1c/H1d/H1e null ESCs to assess the properties of the recurrent mutations in human H1 proteins. H1 TKO ESCs, lacking the endogenous H1c, H1d, and H1e, provide a clean cellular system for us to compare the differences between the hH1 mutants and the corresponding wild-type hH1 variant in chromatin binding. We started with the S > F (serine to phenylalanine) mutation in hH1c (a.a. 102), a recurrent mutation in both follicular lymphomas [94] and diffuse large B-cell lymphomas (DLBCL) [92].

We first set out to generate H1 TKO ESCs expressing WT hH1c or hH1c(Ser102Phe) mutant (S102F — serine to phenylalanine mutation at amino acid 102). A point mutation (C305T — cytosine to thymidine mutation at nucleotide position 305) in WT hH1c coding region was made using the QuikChange II Site-Directed Mutagenesis Kit (Agilent Technologies), resulting in the Ser102Phe mutation present in follicular lymphomas. To facilitate the screening of ESC colonies expressing WT hH1c or hH1c(S102F) mutant, we inserted the FLAG tag sequence at the N-terminus of the WT or mutant hH1c gene, allowing Western blotting analysis to quantify the expression levels of hH1c or hH1c(S102F) using an antibody against FLAG. As discussed in the previous section, the N-terminal FLAG epitope does not change the biochemical properties or *in vivo* functions of H1 variants [64]. FLAG-hH1c or FLAG-hH1c^{S102F} were subsequently inserted into an expression vector containing mouse H1d regulatory regions and the blasticidin resistance gene established previously [38]. The resulting vectors were stably transfected into H1 TKO ESCs by electroporation, and 24 blasticidin-resistant clones for each construct were picked and cultured. These cells are designated as H1 TKO/

◀ Functional analysis of H1 mutations identified in follicular lymphoma. (A) hH1a—hH1e sequence alignment and the distribution of H1 mutations. GH5, globular domain of chicken histone H5. H1 globular domain is marked with the red bracket. (B) Expression of FLAG tagged WT hH1c or hH1cS102F mutant in H1c/H1d/H1e triple null embryonic stem cells (H1 TKO ESCs). H1 TKO ESCs were transfected with vectors expressing FLAG-hH1c or FLAG-hH1c/S102F mutant. Stable ESC clones were picked for each transfection and screened using an anti-FLAG antibody. Immunoblotting with anti-β-ACTIN antibody were included as loading controls. Two clones with similar expression levels of hH1c and hH1c/S102F were selected for subsequent analysis. FLAG-hH1c and FLAG-hH1cS102F are expressed at the same level in selected clones shown. (C) Reverse-phase (RP)-HPLC profiles of histones extracted from chromatin isolated from histone H1 triple-knockout (H1 TKO) mouse ESCs expressing wild-type or Ser102Phe human histone H1c. The Ser102Phe mutant demonstrated higher hydrophobicity than the wild-type protein. mH1a, mouse histone H1a; mH1b, mouse histone H1b; mH2b, mouse histone H2b; hH1c, human histone H1c. (D) Ratio of individual histone H1 variants (and total histone H1) to the nucleosome of the indicated ESCs. The ratio is calculated from the HPLC analysis in (C) and demonstrates that the total histone H1 levels in histone H1 triple-knockout ESCs expressing human histone H1c Ser102Phe were reduced compared to cells expressing wild-type human histone H1c, as a result of the weaker association of the mutant histone with chromatin (only 35% of wild-type association). (*This figure is reproduced in color in the color plate section.*)

Adapted from Okosun et al. [94]

hH1c and H1 TKO/hH1c^{S102F} ESC lines. ESC lines with equal expression levels of respective FLAG-hH1c and FLAG-hH1c^{S102F} were selected for subsequent analysis (Figure 6.5B).

Histones were extracted from purified chromatin of H1 TKO/hH1c and H1 TKO/hH1c^{S102F} ESC lines and subjected to reserve-phase HPLC (RP-HPLC) analysis as described in Section 6.3. The overexpressed FLAG-hH1c and FLAG-hH1c^{S102F} elute in separate peaks on HPLC profiles with a delay of elusion for the FLAG-hH1c^{S102F} peak, indicating a higher hydrophobicity of FLAG-hH1c^{S102F} than FLAG-hH1c (Figure 6.5C). In addition, quantification of the individual H1/nucleosome ratios shows a much lower level of chromatin-bound FLAG-hH1c^{S102F} (0.046) than that of FLAG-hH1c (0.13) (Figure 6.5D). These results suggest that, despite an equal expression level as hH1c, hH1c^{S102F} has drastically reduced binding affinity and residence in chromatin, most likely due to the change in the biochemical property and the interference of DNA binding caused by this mutation.

Thus, the S102F point mutation in hH1c effectively causes a loss-of-function phenotype by reducing the H1c binding in chromatin which could compromise chromatin compaction and regulation of key genes, contributing to malignant transformation. Given that most of the H1 mutations in follicular lymphoma occur at sites necessary for DNA and chromatin binding, other H1 mutations may cause a similar loss-of-function effect as the hH1c^{S102F} analyzed here. Mutations in H1 variants may also interrupt the interaction of respective H1 variants with their binding partners. It would be interesting to tease out how H1 mutations affect gene regulation in follicular lymphomas and other tumors, which could provide new insights on the oncogenic roles of H1 mutations in tumorigenesis.

6.7 CONCLUSION

As a major chromatin component, linker histone H1 participates in the formation of higher order structure and adds another dimension to the already complicated epigenetic regulation of gene expression and cell differentiation. However, this group of proteins remains largely understudied in mammals due to their heterogeneity and redundancy. Here we described a suite of methodologies to characterize their functions, including random primer-based qRT-PCR, HPLC analysis, gene targeting for H1 depletion, *in vivo* H1 tagging, ChIP of tagged H1 variants, and mutation analysis. These methods serve as powerful tools for studying the functions of linker histone H1 both in terms of variant specificity and as a group.

REFERENCES

[1] van Holde KE. Chromatin. Springer series in molecular and cell biology. New York: Springer; 1988. p. 497.

[2] Wolffe A. Chromatin: structure and function. San Diego, CA: Academic Press; 1999.

[3] Olins AL, Olins DE. Spheroid chromatin units (v bodies). Science 1974;183(4122):330−2.

[4] Woodcock CL, Safer JP, Stanchfield JE. Structural repeating units in chromatin. I. Evidence for their general occurrence. Exp Cell Res 1976;97:101−10.

[5] Chen T, Dent SY. Chromatin modifiers and remodellers: regulators of cellular differentiation. Nat Rev Genet 2014;15(2):93−106.

[6] Li M, Liu GH, Izpisua Belmonte JC. Navigating the epigenetic landscape of pluripotent stem cells. Nat Rev Mol Cell Biol 2012;13(8):524–35.

[7] Zentner GE, Henikoff S. Regulation of nucleosome dynamics by histone modifications. Nat Struct Mol Biol 2013;20(3):259–66.

[8] Kasinsky HE, Lewis JD, Dacks JB, Ausio J. Origin of H1 linker histones. FASEB J 2001;15(1):34–42.

[9] Wu M, Allis CD, Richman R, Cook RG, Gorovsky MA. An intervening sequence in an unusual histone H1 gene of *Tetrahymena thermophila*. Proc Natl Acad Sci USA 1986;83(22):8674–8.

[10] Wolffe AP. Histone H1. Int J Biochem Cell Biol 1997;29(12):1463–6.

[11] Brown DT, Sittman DB. Identification through overexpression and tagging of the variant type of the mouse H1e and H1c genes. J Biol Chem 1993;268(1):713–18.

[12] Ali T, Thomas JO. Distinct properties of the two putative "globular domains" of the yeast linker histone, Hho1p. J Mol Biol 2004;337(5):1123–35.

[13] Freidkin I, Katcoff DJ. Specific distribution of the *Saccharomyces cerevisiae* linker histone homolog HHO1p in the chromatin. Nucleic Acids Res 2001;29(19):4043–51.

[14] Downs JA, Kosmidou E, Morgan A, Jackson SP. Suppression of homologous recombination by the *Saccharomyces cerevisiae* linker histone. Mol Cell 2003;11(6):1685–92.

[15] Brown DT, Izard T, Misteli T. Mapping the interaction surface of linker histone H1(0) with the nucleosome of native chromatin in vivo. Nat Struct Mol Biol 2006;13(3):250–5.

[16] Hendzel MJ, Lever MA, Crawford E, Th'ng JP. The C-terminal domain is the primary determinant of histone H1 binding to chromatin in vivo. J Biol Chem 2004;279(19):20028–34.

[17] Stasevich TJ, Mueller F, Brown DT, McNally JG. Dissecting the binding mechanism of the linker histone in live cells: an integrated FRAP analysis. EMBO J 2010;29(7):1225–34.

[18] Syed SH, Goutte-Gattat D, Becker N, Meyer S, Shukla MS, Hayes JJ, et al. Single-base resolution mapping of H1-nucleosome interactions and 3D organization of the nucleosome. Proc Natl Acad Sci USA 2010;107(21):9620–5.

[19] Lever MA, Th'ng JP, Sun X, Hendzel MJ. Rapid exchange of histone H1.1 on chromatin in living human cells. Nature 2000;408(6814):873–6.

[20] Misteli T, Gunjan A, Hock R, Bustin M, Brown DT. Dynamic binding of histone H1 to chromatin in living cells. Nature 2000;408(6814):877–81.

[21] Wolffe AP, Kurumizaka H. The nucleosome: a powerful regulator of transcription. Prog Nucleic Acid Res Mol Biol 1998;61:379–422.

[22] Vignali M, Workman JL. Location and function of linker histones. Nat Struct Biol 1998;5(12):1025–8.

[23] Woodcock CL, Skoultchi AI, Fan Y. Role of linker histone in chromatin structure and function: H1 stoichiometry and nucleosome repeat length. Chromosome Res 2006;14(1):17–25.

[24] Happel N, Doenecke D. Histone H1 and its isoforms: contribution to chromatin structure and function. Gene 2009;431(1–2):1–12.

[25] Patterton HG, Landel CC, Landsman D, Peterson CL, Simpson RT. The biochemical and phenotypic characterization of Hho1p, the putative linker histone H1 of *Saccharomyces cerevisiae*. J Biol Chem 1998;273(13):7268–76.

[26] Shen X, Yu L, Weir JW, Gorovsky MA. Linker histones are not essential and affect chromatin condensation in vivo. Cell 1995;82(1):47–56.

[27] Shen X, Gorovsky MA. Linker histone H1 regulates specific gene expression but not global transcription in vivo. Cell 1996;86(3):475–83.

[28] Lu X, Wontakal SN, Emelyanov AV, Morcillo P, Konev AY, Fyodorov DV, et al. Linker histone H1 is essential for *Drosophila* development, the establishment of pericentric heterochromatin, and a normal polytene chromosome structure. Genes Dev 2009;23(4):452–65.

[29] Vujatovic O, Zaragoza K, Vaquero A, Reina O, Bernues J, Azorin F. *Drosophila melanogaster* linker histone dH1 is required for transposon silencing and to preserve genome integrity. Nucleic Acids Res 2012;40(12):5402−14.

[30] Studencka M, Konzer A, Moneron G, Wenzel D, Opitz L, Salinas-Riester G, et al. Novel roles of *Caenorhabditis elegans* heterochromatin protein HP1 and linker histone in the regulation of innate immune gene expression. Mol Cell Biol 2012;32(2):251−65.

[31] Studencka M, Wesolowski R, Opitz L, Salinas-Riester G, Wisniewski JR, Jedrusik-Bode M. Transcriptional repression of Hox genes by *C. elegans* HP1/HPL and H1/HIS-24. PLoS Genet 2012;8 (9):e1002940.

[32] Bouvet P, Dimitrov S, Wolffe AP. Specific regulation of *Xenopus* chromosomal 5S rRNA gene transcription in vivo by histone H1. Genes Dev 1994;8(10):1147−59.

[33] Steinbach OC, Wolffe AP, Rupp RA. Somatic linker histones cause loss of mesodermal competence in *Xenopus*. Nature 1997;389(6649):395−9.

[34] Takami Y, Nishi R, Nakayama T. Histone H1 variants play individual roles in transcription regulation in the DT40 chicken B cell line. Biochem Biophys Res Commun 2000;268(2):501−8.

[35] Hashimoto H, Takami Y, Sonoda E, Iwasaki T, Iwano H, Tachibana M, et al. Histone H1 null vertebrate cells exhibit altered nucleosome architecture. Nucleic Acids Res 2010;38(11):3533−45.

[36] Lin Q, Inselman A, Han X, Xu H, Zhang W, Handel MA, et al. Reductions in linker histone levels are tolerated in developing spermatocytes but cause changes in specific gene expression. J Biol Chem 2004;279(22):23525−35.

[37] Fan Y, Nikitina T, Zhao J, Fleury TJ, Bhattacharyya R, Bouhassira EE, et al. Histone H1 depletion in mammals alters global chromatin structure but causes specific changes in gene regulation. Cell 2005;123 (7):1199−212.

[38] Zhang Y, Cooke M, Panjwani S, Cao K, Krauth B, Ho PY, et al. Histone h1 depletion impairs embryonic stem cell differentiation. PLoS Genet 2012;8(5):e1002691.

[39] Zhang Y, Liu Z, Medrzycki M, Cao K, Fan Y. Reduction of Hox gene expression by histone H1 depletion. PLoS One 2012;7(6):e38829.

[40] Zlatanova J, Doenecke D. Histone H1 zero: a major player in cell differentiation? FASEB J 1994;8 (15):1260−8.

[41] Tanaka M, Hennebold JD, MacFarlane J, Adashi EY. A mammalian oocyte-specific linker histone gene H1oo: homology with the genes for the oocyte-specific cleavage stage histone (cs-H1) of sea urchin and the B4/H1M histone of the frog. Development 2001;128(5):655−64.

[42] Becker M, Becker A, Miyara F, Han Z, Kihara M, Brown DT, et al. Differential in vivo binding dynamics of somatic and oocyte-specific linker histones in oocytes and during ES cell nuclear transfer. Mol Biol Cell 2005;16(8):3887−95.

[43] Martianov I, Brancorsini S, Catena R, Gansmuller A, Kotaja N, Parvinen M, et al. Polar nuclear localization of H1T2, a histone H1 variant, required for spermatid elongation and DNA condensation during spermiogenesis. Proc Natl Acad Sci USA 2005;102(8):2808−13.

[44] Yan W, Ma L, Burns KH, Matzuk MM. HILS1 is a spermatid-specific linker histone H1-like protein implicated in chromatin remodeling during mammalian spermiogenesis. Proc Natl Acad Sci USA 2003; 100(18):10546−51.

[45] Wang ZF, Sirotkin AM, Buchold GM, Skoultchi AI, Marzluff WF. The mouse histone H1 genes: gene organization and differential regulation. J Mol Biol 1997;271(1):124−38.

[46] Fan Y, Nikitina T, Morin-Kensicki EM, Zhao J, Magnuson TR, Woodcock CL, et al. H1 linker histones are essential for mouse development and affect nucleosome spacing in vivo. Mol Cell Biol 2003;23 (13):4559−72.

[47] Plumb M, Marashi F, Green L, Zimmerman A, Zimmerman S, Stein J, et al. Cell cycle regulation of human histone H1 mRNA. Proc Natl Acad Sci USA 1984;81(2):434−8.

[48] Dominski Z, Marzluff WF. Formation of the 3′ end of histone mRNA. Gene 1999;239(1):1−14.

[49] Cheng GH, Nandi A, Clerk S, Skoultchi AI. Different 3′-end processing produces two independently regulated mRNAs from a single H1 histone gene. Proc Natl Acad Sci USA 1989;86(18): 7002−6.

[50] Clausell J, Happel N, Hale TK, Doenecke D, Beato M. Histone H1 subtypes differentially modulate chromatin condensation without preventing ATP-dependent remodeling by SWI/SNF or NURF. PLoS One 2009;4(10):e0007243.

[51] Jedrusik MA, Schulze E. A single histone H1 isoform (H1.1) is essential for chromatin silencing and germline development in *Caenorhabditis elegans*. Development 2001;128(7):1069−80.

[52] Jedrusik MA, Schulze E. Linker histone HIS-24 (H1.1) cytoplasmic retention promotes germ line development and influences histone H3 methylation in *Caenorhabditis elegans*. Mol Cell Biol 2007;27 (6):2229−39.

[53] Andrews MT, Loo S, Wilson LR. Coordinate inactivation of class III genes during the Gastrula-Neurula Transition in *Xenopus*. Dev Biol 1991;146(1):250−4.

[54] Drabent B, Saftig P, Bode C, Doenecke D. Spermatogenesis proceeds normally in mice without linker histone H1t. Histochem Cell Biol 2000;113(6):433−42.

[55] Fan Y, Sirotkin A, Russell RG, Ayala J, Skoultchi AI. Individual somatic H1 subtypes are dispensable for mouse development even in mice lacking the H1(0) replacement subtype. Mol Cell Biol 2001;21 (23):7933−43.

[56] Lin Q, Sirotkin A, Skoultchi AI. Normal spermatogenesis in mice lacking the testis-specific linker histone H1t. Mol Cell Biol 2000;20(6):2122−8.

[57] Rabini S, Franke K, Saftig P, Bode C, Doenecke D, Drabent B. Spermatogenesis in mice is not affected by histone H1.1 deficiency. Exp Cell Res 2000;255(1):114−24.

[58] Sirotkin AM, Edelmann W, Cheng G, Klein-Szanto A, Kucherlapati R, Skoultchi AI. Mice develop normally without the H1(0) linker histone. Proc Natl Acad Sci USA 1995;92(14):6434−8.

[59] Brown DT, Gunjan A, Alexander BT, Sittman DB. Differential effect of H1 variant overproduction on gene expression is due to differences in the central globular domain. Nucleic Acids Res 1997;25 (24):5003−9.

[60] Gunjan A, Brown DT. Overproduction of histone H1 variants in vivo increases basal and induced activity of the mouse mammary tumor virus promoter. Nucleic Acids Res 1999;27(16):3355−63.

[61] Alami R, Fan Y, Pack S, Sonbuchner TM, Besse A, Lin Q, et al. Mammalian linker-histone subtypes differentially affect gene expression in vivo. Proc Natl Acad Sci USA 2003;100(10):5920−5.

[62] Sancho M, Diani E, Beato M, Jordan A. Depletion of human histone H1 variants uncovers specific roles in gene expression and cell growth. PLoS Genet 2008;4(10):e1000227.

[63] Konishi A, Shimizu S, Hirota J, Takao T, Fan Y, Matsuoka Y, et al. Involvement of histone H1.2 in apoptosis induced by DNA double-strand breaks. Cell 2003;114(6):673−88.

[64] Cao K, Lailler N, Zhang Y, Kumar A, Uppal K, Liu Z, et al. High-resolution mapping of H1 linker histone variants in embryonic stem cells. PLoS Genet 2013;9(4):e1003417.

[65] Izzo A, Kamieniarz-Gdula K, Ramirez F, Noureen N, Kind J, Manke T, et al. The genomic landscape of the somatic linker histone subtypes H1.1 to H1.5 in human cells. Cell Rep 2013;3(6):2142−54.

[66] Millán-Ariño L, Islam AB, Izquierdo-Bouldstridge A, Mayor R, Terme J-M, Luque N, et al. Mapping of six somatic linker histone H1 variants in human breast cancer cells uncovers specific features of H1.2. Nucleic Acids Res 2014;42(7):4474−93.

[67] Th'ng JP, Sung R, Ye M, Hendzel MJ. H1 family histones in the nucleus. Control of binding and localization by the C-terminal domain. J Biol Chem 2005;280(30):27809−14.

[68] Raghuram N, Carrero G, Th'ng J, Hendzel MJ. Molecular dynamics of histone H1. Biochem Cell Biol 2009;87(1):189−206.

[69] Vyas P, Brown DT. N- and C-terminal domains determine differential nucleosomal binding geometry and affinity of linker histone isotypes H1(0) and H1c. J Biol Chem 2012;287(15):11778−87.

[70] Wisniewski JR, Zougman A, Kruger S, Mann M. Mass spectrometric mapping of linker histone H1 variants reveals multiple acetylations, methylations, and phosphorylation as well as differences between cell culture and tissue. Mol Cell Proteomics 2007;6(1):72−87.

[71] Telu KH, Abbaoui B, Thomas-Ahner JM, Zynger DL, Clinton SK, Freitas MA, et al. Alterations of histone H1 phosphorylation during bladder carcinogenesis. J Proteome Res 2013;12(7):3317−26.

[72] Medrzycki M, Zhang Y, Cao K, Fan Y. Expression analysis of mammalian linker-histone subtypes. J Vis Exp 2012;(61).

[73] Medrzycki M, Zhang Y, McDonald JF, Fan Y. Profiling of linker histone variants in ovarian cancer. Front Biosci 2012;17:396−406.

[74] Fan Y, Skoultchi AI. Genetic analysis of H1 linker histone subtypes and their functions in mice. Methods Enzymol 2004;377:85−107.

[75] Boskovic A, Eid A, Pontabry J, Ishiuchi T, Spiegelhalter C, Ram E, et al. Higher chromatin mobility supports totipotency and precedes pluripotency in vivo. Genes Dev 2014;28(10):1042−7.

[76] Gaspar-Maia A, Alajem A, Meshorer E, Ramalho-Santos M. Open chromatin in pluripotency and reprogramming. Nat Rev Mol Cell Biol 2011;12(1):36−47.

[77] Meshorer E, Yellajoshula D, George E, Scambler PJ, Brown DT, Misteli T. Hyperdynamic plasticity of chromatin proteins in pluripotent embryonic stem cells. Dev Cell 2006;10(1):105−16.

[78] Efroni S, Duttagupta R, Cheng J, Dehghani H, Hoeppner DJ, Dash C, et al. Global transcription in pluripotent embryonic stem cells. Cell Stem Cell 2008;2(5):437−47.

[79] Bhan S, May W, Warren SL, Sittman DB. Global gene expression analysis reveals specific and redundant roles for H1 variants, H1c and H1(0), in gene expression regulation. Gene 2008;414(1−2):10−18.

[80] Orrego M, Ponte I, Roque A, Buschati N, Mora X, Suau P. Differential affinity of mammalian histone H1 somatic subtypes for DNA and chromatin. BMC Biol 2007;5:22.

[81] Li JY, Patterson M, Mikkola HK, Lowry WE, Kurdistani SK. Dynamic distribution of linker histone H1.5 in cellular differentiation. PLoS Genet 2012;8(8):e1002879.

[82] Lu R, Markowetz F, Unwin RD, Leek JT, Airoldi EM, MacArthur BD, et al. Systems-level dynamic analyses of fate change in murine embryonic stem cells. Nature 2009;462(7271):358−62.

[83] Meissner A. Epigenetic modifications in pluripotent and differentiated cells. Nat Biotechnol 2010;28 (10):1079−88.

[84] Johns Hopkins University. Bowtie: an ultrafast memory efficient short read aligner. 2015. Available from: <http://bowtie-bio.sourceforge.net/index.shtml>.

[85] Lajugie J, Bouhassira EE. GenPlay, a multipurpose genome analyzer and browser. Bioinformatics 2011;27(14):1889−93.

[86] Bock C, Halachev K, Buch J, Lengauer T. EpiGRAPH: user-friendly software for statistical analysis and prediction of (epi)genomic data. Genome Biol 2009;10(2):R14.

[87] Zang C, Schones DE, Zeng C, Cui K, Zhao K, Peng W. A clustering approach for identification of enriched domains from histone modification ChIP-Seq data. Bioinformatics 2009;25(15):1952−8.

[88] Jurka J, Kapitonov VV, Pavlicek A, Klonowski P, Kohany O, Walichiewicz J. Repbase Update, a database of eukaryotic repetitive elements. Cytogenet Genome Res 2005;110(1−4):462−7.

[89] Guenatri M, Bailly D, Maison C, Almouzni G. Mouse centric and pericentric satellite repeats form distinct functional heterochromatin. J Cell Biol 2004;166(4):493−505.

[90] Sjoblom T, Jones S, Wood LD, Parsons DW, Lin J, Barber TD, et al. The consensus coding sequences of human breast and colorectal cancers. Science 2006;314(5797):268−74.

[91] Wood LD, Parsons DW, Jones S, Lin J, Sjöblom T, Leary RJ, et al. The genomic landscapes of human breast and colorectal cancers. Science 2007;318(5853):1108−13.

[92] Morin RD, Mendez-Lago M, Mungall AJ, Goya R, Mungall KL, Corbett R, et al. Frequent mutation of histone-modifying genes in non-Hodgkin lymphoma. Nature 2011;476(7360):298−303.

[93] Lohr JG, Stojanov P, Lawrence MS, Auclair D, Chapuy B, Sougnez C, et al. Discovery and prioritization of somatic mutations in diffuse large B-cell lymphoma (DLBCL) by whole-exome sequencing. Proc Natl Acad Sci USA 2012;109(10):3879−84.

[94] Okosun J, Bodor C, Wang J, Araf S, Yang CY, Pan C, et al. Integrated genomic analysis identifies recurrent mutations and evolution patterns driving the initiation and progression of follicular lymphoma. Nat Genet 2014;46(2):176−81.

[95] Li H, Kaminski MS, Li Y, Yildiz M, Ouillette P, Jones S, et al. Mutations in linker histone genes HIST1H1 B, C, D, and E; OCT2 (POU2F2); IRF8; and ARID1A underlying the pathogenesis of follicular lymphoma. Blood 2014;123(10):1487−98.

[96] Ramakrishnan V, Finch JT, Graziano V, Lee PL, Sweet RM. Crystal structure of globular domain of histone H5 and its implications for nucleosome binding. Nature 1993;362(6417):219−23.

CRYSTALLOGRAPHY-BASED MECHANISTIC INSIGHTS INTO EPIGENETIC REGULATION

7

Shuai Zhao and Haitao Li

Department of Basic Medical Sciences, Center for Structural Biology, School of Medicine,
Tsinghua University, Beijing, P.R. China

CHAPTER OUTLINE

7.1 INTRODUCTION

Epigenetics involves heritable phenotypic alternations without changes in DNA sequence. Such inheritance is usually achieved by diverse epigenetic mechanisms including DNA methylation, histone posttranslational modifications, chromatin remodeling, and noncoding RNA, among others. Epigenetic regulation is a highly complex phenomenon and requires the interplay of a plethora of epigenetic factors. These factors can be broadly classified into (a) histones and DNA that constitute the chromatin, with the nucleosome as the fundamental building unit; (b) histone/DNA modifiers

Y.G. Zheng (Ed): Epigenetic Technological Applications. DOI: http://dx.doi.org/10.1016/B978-0-12-801080-8.00007-7

(e.g. "writers," "erasers," and "editors") that can enzymatically alter chromatin states notably by introducing diverse site-specific chemical modifications; (c) epigenetic effectors or "readers" that can specifically recognize histone/DNA modifications, known as epigenetic marks, to bring about downstream outcomes; (d) ATP-dependent chromatin remodelers and histone chaperones that control the dynamics of chromatin and nucleosome assembly; and (e) noncoding RNAs that can regulate chromatin structure through sequence-dependent or -independent mechanisms. An exquisite and faithful establishment of a distinct epigenetic landscape relies on highly regulated and concerted activities of the aforementioned epigenetic regulators over the chromatin template. Thus, elucidating how these epigenetic regulators perform their cellular function in molecular detail is critical for a mechanistic understanding of epigenetic regulation. To this end, structural biology, notably X-ray crystallography, has played a pivotal role in deciphering aspects of the epigenetic code at three-dimensional and atomic levels.

Figure 7.1 provides an overview of the functional discovery and structural elucidation of representative epigenetic regulators in chronological order spanning the past decade. Of note, the year 1996 marked the emergence of epigenetics as a new scientific discipline with two seminal discoveries on the identification of a histone acetyltransferase (HAT) — Gcn5 from the ciliate *Tetrahymena* [1] — and a histone deacetylase (HDAC) — mammalian Rpd3p [2] — as key regulators of gene expression. From this date on, epigenetics became one of the fastest-moving fields in biology and gradually entered nearly all corners of the life sciences. To date, hundreds of nuclear proteins or domains with previously unknown function have been attributed to activities related to epigenetic regulation. These can be exemplified by characterization of bromodomains as histone acetyllysine readers [3], of chromodomains and plant homeodomain (PHD) fingers as histone methyllysine

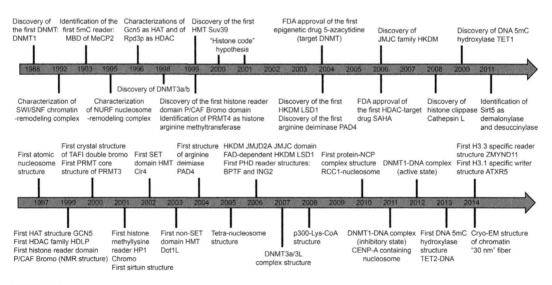

FIGURE 7.1

A chronological view of functional and structural studies of key epigenetic regulators. Upper arrow, Functional discovery and identification of epigenetic regulators; Lower arrow, Structural studies of epigenetic regulators.

readers [4], of SRA domains (SET- and RING-associated domains) as DNA 5-methylcytosine readers [5], of SET-domains as histone lysine methyltransferase writers [6], of LSD1 (lysine-specific demethylase 1) and JMJC (Jumonji domain C domain) family proteins as histone lysine demethylases (HKDMs) [7], and of TET (ten eleven translocation) family members as DNA 5-methylcytosine hydroxylases [8]. In addition, mechanistic studies of these epigenetic regulators have provided deep insights into how these factors execute their function. Structural biology, notably X-ray crystallography, has led to the structure determination of many key epigenetic regulators concerning the generation, elimination, and readout of distinct histone or DNA marks in the contexts of histone peptide, DNA, nucleosome, or even a high-order chromatin structure. Epigenetics and crystallography have mutually boosted each other by blending functional and structural studies to provide mechanistic insights into epigenetic regulation in three-dimensional and at atomic detail since the inception of the epigenetics era in the late 1990s (Figure 7.1).

Many diseases have been shown to have a close relationship to dysregulation of epigenetic factors [9]. For example, aberrant DNA methylation states and histone modification patterns were observed in cancer cells, constituting "cancer epigenomes" [10]. Thus, dysregulation of SIRT1, a member of Class III HDACs, was found to impact on obesity and diabetes. Additionally, neurological diseases and inflammation are also linked to dysfunction of epigenetic factors [11]. Both DNA methylation and histone modifications are reversible; the reversible nature of human epigenomes renders epigenetic factors as ideal targets for clinical therapies [12]. Indeed, research on epigenetic-related diseases has progressed from clinical cases to drug discovery. Target-based drug discovery and drug optimization rely heavily on structural information of target proteins, which can be provided by X-ray crystallography and/or nuclear magnetic resonance (NMR) spectroscopy. Fragment-based drug discovery (FBDD), an increasingly popular strategy toward innovative drug discovery, requires the identification of key fragment—target protein interactions that can be readily obtained by X-ray crystallography. As such, the implementation of macromolecular crystallography in modern drug discovery has started to reinforce the translation of mechanistic discoveries in epigenetics to therapeutic applications for the benefit of our everyday life.

7.2 DEVELOPMENT OF X-RAY CRYSTALLOGRAPHY

Macromolecular X-ray crystallography has already become the fundamental experimental technique of current structural biology, generating about 90% of entries deposited in the Protein Data Bank (PDB). The first protein structure, myoglobin, was determined by the laboratory of John Kendrew in 1957, the fruit of continuous efforts spanning nearly two decades; and this achievement was notably facilitated by the method of multiple isomorphous replacement through heavy atom soaking developed by Max Perutz for phase solution. By contrast, nowadays the determination of macromolecular crystal structures has become much easier, with a turn-over time from crystal generation to structure determination achievable in days based on advances in state-of-the-art hardware and software for X-ray crystallography. Newer technologies impacting on modern macromolecular crystallography include various prokaryotic and eukaryotic systems for heterogeneous protein expression, commercial reagent kits, and automated robotic arms for crystal screening, advanced synchrotron light sources, and user-friendly software packages for structural determination and refinement.

The general procedures of X-ray crystallography involve a flow chart of protein expression and purification, crystal screening and optimization, diffraction data collection and phase determination, as well as model building and refinement (Figure 7.2). The key and rate-limiting step in macromolecular crystallography is the generation of diffraction quality crystals. In a sense, crystallization reflects a combination of art and science. Besides systematic screening and crystallization condition optimization, improvement of solution behavior of the target protein or its complex through biochemical characterization and extensive protein engineering are often key strategies for successful crystallization, especially for the case of difficult proteins. Once diffraction quality crystals are obtained, diffraction data can be collected at a defined wavelength using either home or synchrotron light sources. The diffraction data could be then utilized to calculate electron density maps through Fourier transformation. However, meaningful Fourier transformation requires the phase information of each reflection that is missing during data collection. To solve the phase problem, molecular replacement and experimental phasing strategies are often adopted. The former takes advantage of an available homologous protein structure and extracts the initial phase information

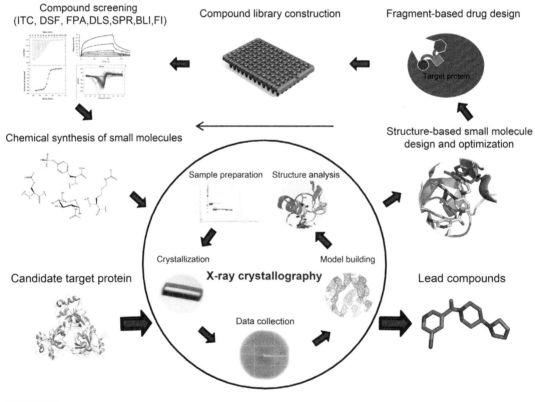

FIGURE 7.2

Flow chart of X-ray crystallography-guided drug discovery. Procedures inside the circle depict key steps in X-ray crystallography. Procedures outside the circle depict key steps in drug discovery guided by X-ray crystallography.

by proper placement of the homologous model in the crystal lattice. The latter comprises methods of isomorphous replacement through heavy atom soaks (e.g. MIR, multiple isomorphous replacement; SIR, single isomorphous replacement), anomalous scattering (e.g. MAD, multi-wavelength anomalous diffraction; SAD, single-wavelength anomalous dispersion), or their hybrids (e.g. SIRAS, single isomorphous replacement with anomalous scattering). Notably, in addition to heavy atom soaking, the ease of recombinant incorporation of heavy atoms (e.g. selenium of selenomethione) into proteins and access to powerful synchrotron radiation facilities have made Se-SAD or Se-MAD the popular method for *de novo* structure determination during the past two decades. After initial phase solution and electron density improvement, model building could be performed based on the sequence of the target protein to fit each amino acid into the electron density map. Finally, refinement and validation of this structural model yields a protein structure of reasonable quality.

Compared to NMR and electron microscopy (EM) methods, X-ray crystallography has the advantages of higher resolution (compared to EM), few size limitations (compared to solution NMR), and rapid turn-over time (provided crystals are available) due to the relative maturity of the methodology. As detailed in the following sections, X-ray crystallography has been widely adopted and has greatly accelerated the pace of epigenetic research at the molecular level, ranging from elucidating basic mechanisms of epigenetic regulation to structure-based epigenetic drug discovery.

7.3 KEY EPIGENETIC PROJECTS SOLVED BY X-RAY CRYSTALLOGRAPHY

Structures of many key epigenetic proteins and complexes have been determined by X-ray crystallography (representative structures are shown in Figure 7.3), revealing detailed molecular mechanisms of epigenetic regulation, and playing an instrumental role for many subsequent studies.

7.3.1 ELUCIDATION OF THE FUNDAMENTAL BUILDING BLOCKS OF CHROMATIN

Structural biology has been an early contributor to the development of epigenetic research. The 2.8 Å *Xenopus* nucleosome core particle (NCP) structure was solved in 1997 [13]. This was the first atomic-resolution structure of an NCP, revealing detailed interactions between different histones and between histone and DNA. Subsequent nucleosomal structural studies extend to NCPs containing major histones of human and other species, various histone variants (e.g. H2A.Z, macroH2A, CENP-A), NCPs in complex with proteins or small molecules, and even the tetranucleosome [14]. The structure of tetranucleosome supports a continuous fiber model of chromatin compaction, explaining how DNA sequences could be highly compacted into the small nucleus [15]. The 2.9 Å crystal structure of the RCC1−NCP complex represented the first atomic protein-nucleosome structure, revealing a mode of histone/DNA co-recognition by RCC1 − a chromatin protein that regulates spindle formation during mitosis [16]. The elucidation of fundamental building blocks of chromatin by X-ray crystallography has provided precise details of the structure of nucleosomes and how they form higher-order chromatin architecture, thereby improving our understanding of chromatin-related biological events that contribute to epigenetic regulation.

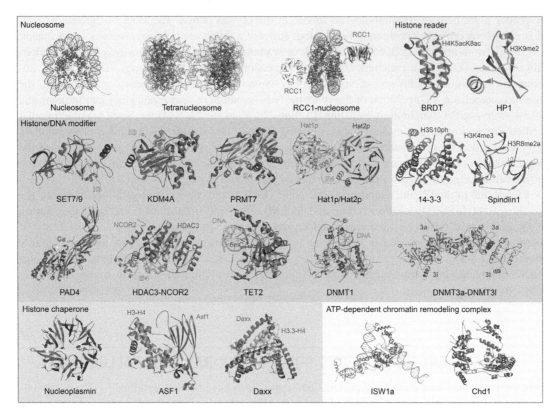

FIGURE 7.3

A structural gallery of nucleosomes and representative epigenetic regulators. Coordinates have PDB codes 1AOI (Nucleosome), 1ZBB (Tetranucleosome), 3MVD (RCC1−nucleosome complex, RCC1 is shown in purple), 1O9S (SET7/9), 2OQ6 (Catalytic domain of KDM4A), 4M38 (Catalytic domain of PRMT7), 4PSX (Hat1p/Hat2p, Hat1p is shown in green, Hat2p is shown in blue, Histone H4 peptide is shown in yellow), 1C3P (HDLP), 4A69 (HDAC3-NCOR2, HDAC3 is shown in blue, Deacetylase activation domain of NCOR2 [Nuclear receptor co-repressor 2] is shown in green), 4NM6 (TET2−DNA complex), 4DA4 (DNMT1−DNA complex), 2QRV (DNMT3a−DNMT3L complex, C-terminal domain of DNMT3L is shown in green, Catalytic domain of DNMT3a is shown in blue), 2WP2 (Bromodomain of Brdt), 1KNA (Chromodomain of HP1), 2C1J (14-3-3), 4MZG (Spindlin1), 1K5J (Nucleoplasmin), 2IO5 (ASF1-H3−H4 complex, ASF1 is shown in blue, H3−H4 is shown in green), 4H9N (DAXX H3−H4 complex, DAXX is shown in blue, H3−H4 is shown in green), 2Y9Z (ISW1a−DNA complex), and 3MWY (chromodomain-ATPase portion of Chd1). (*This figure is reproduced in color in the color plate section.*)

7.3.2 CREATION OF EPIGENETIC MODIFICATION PATTERN BY EPIGENETIC MODIFIERS

Hundreds of posttranslational modifications (PTMs) are known to occur on histones, often clustered in the flexible N- or C-terminal tails that protrude out of the nucleosome disc. These PTMs, referred to as the "histone code" [17], in concert with DNA methylation and its derivatives, are critical in defining distinct chromatin states, and are generated by a diverse array of enzymatic factors. Enzymes creating these modifications are called "writers," while enzymes eliminating these modifications are called "erasers." There also exists a special class of "erasers" — the so-called "editors" that do not simply erase a certain chemical group but rather alter its chemical type. For examples, the PAD4 deiminase enzyme deiminates arginine into citrullin and TET1 hydroxylase oxidizes methylcytosine into hydroxymethylcytosine. Epigenetic writers and erasers are classified into different subfamilies according to their respective modification types. Key writers for epigenetic modifications consist of HKMTs, protein arginine methyltransferases (PRMTs), HATs, histone kinases, histone ubiquitin ligases, histone SUMO ligases, histone poly-ADP ribose polymerases (PARPs), and DNA methyltransferases (DNMTs). Key erasers of epigenetic modifications include HKDM, histone deacetyltransferases (HDAC), protein phosphatases (PPase), deubiquitinases (Dub), SUMOlases, and DNA 5-methyl cytosine hydroxylases.

The enzymatic modifications of histone and DNA are highly regulated. A single site could be modified by different types of modifications; for example, histone H2B lysine 120 can undergo both acetylation or ubiquitination, and histone H4 lysine 20 can undergo mono-, di-, tri-methylation, or acetylation. Meanwhile, the same type of modification can be created in a site-specific way — for example, trimethylation could occur on histone H3 lysine residues at positions 4, 9, 27, 36, and 79 with distinct functional implications. Another layer of regulation is that even different methylation states, that is mono-, di-, or tri-methylation of lysine and mono-, asymmetrical di-, and symmetrical di-methylation of arginine, are precisely controlled by different enzymatic machineries. Hence, how these enzymes attach or subtract a particular epigenetic mark with high chemical type-, site-, and state-specificity represents an active area to be explored by structural biology. In the following paragraphs, we will discuss representative structural studies on enzymatic operation of histone and DNA modifications and highlight emerging common working principles and underlying mechanisms.

The functional discoveries of histone acetyltransferases (HAT) and deacetylases (HDAC) in 1996 marked the birth of modern epigenetics [1,2]. Subsequent structural studies further elucidated the underlying molecular mechanism of histone acetylation and deacetylation. The structure of the *Tetrahymena* GCN5 (tGCN5) in complex with coenzyme A (CoA) and histone H3 peptide has provided insights into the mechanism of GCN5-mediated histone acetylation [18]. The structure reveals the structural basis for recognition of the histone "GKXP" motif and the essential role of CoA for catalysis. The transition-state reaction intermediate is stabilized by a glutamic acid residue and a backbone amide group. The structural studies also call attention to cross-talk among different histone modifications. For example, phosphorylation of Ser10 on H3 could facilitate Lys14 acetylation. The mechanism of this cross-talk was nicely explained by a costructure of tGCN5 with H3S10ph peptide (Figure 7.4A) [19]. The phospho-Ser10 binds within a shallow groove of tGCN5 and mediates additional histone−protein interactions, thereby facilitating substrate binding and H3K14 acetylation. HATs often form large complexes with other protein factors to gain extra specificity or activity for exquisite functional regulation. Recent crystal structural studies of Hat1p/Hat2p complex showcased how additional site specificity is achieved upon complex formation. Hat1p alone can acetylate both H4K5 and H4K12 while Hat1p/Hat2p complex acetylates specifically at H4K12. Structural studies of the complex revealed that histone H4 peptide form extensive contacts at the Hat1p and Hat2p interface. Specific interaction between H4 fragment 16−46 and Hat2p specifically presents H4K12 but not H4K5 to the active site of Hat1p for acetylation (Figure 7.4B) [20].

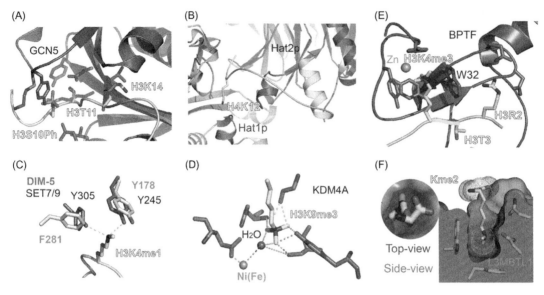

FIGURE 7.4

Site- and state-specific incorporation, elimination and readout of histone marks by representative histone writers, erasers and readers. (A) GCN5 in complex with peptide containing H3S10Ph (1PUA); (B) Hat1p/Hat2p in complex with H4 and H3 peptide (4PSX); (C) SET7/9 (1O9S) and DIM-5 (1PEG) in complex with H3 peptide. SET7/9 is shown in blue and DIM-5 is shown in gray; (D) KDM4A in complex with peptide containing H3K9me3 (2OQ6); (E) BPTF in complex with H3K4me3 peptide (2F6J); (F) L3MBTL1 pocket 2 with inserted dimethyllysine ligand (2RHX). In all panels, histone peptide is colored yellow; PDB entry code of each structure is listed in parentheses. (*This figure is reproduced in color in the color plate section.*)

SAHA (suberoylanilide hydroxamic acid) is an HDAC inhibitor that has been approved by the Food and Drug Administration (FDA). Despite the demonstration that SAHA or its analogs induce transformed cell growth arrest and differentiation in the 1970s, the topology of the target protein and the binding mode of SAHA was not solved until the late 1990s [21]. The crystal structure of an archaebacterial HDAC, HDLP, bound to trichostatin A and SAHA inhibitors revealed that the active site of HDLP contained a zinc-binding site, two aspartic acids, and two histidines. Two inhibitors have been shown to be inserted into the HDLP pocket and coordinated with zinc ions, validating the binding mode of these two inhibitors [22]. This research confirmed that HDAC was the target of SAHA and also explained the mechanism of deacetylation mediated by HDAC. Subsequent studies on the identification of new HDAC inhibitors have been performed based on this HDLP−SAHA/TSA complex crystal structure, including its amide-analogues and phenylbutyrate-derived inhibitors [23,24]. Another important mechanistic discovery on HDAC regulation was driven by crystal structural studies of HDAC3 bound to the co-repressor, SMRT (silencing mediator for retinoid or thyroid-hormone receptors), and inositol tetraphosphate, Ins $(1,4,5,6)P_4$ [25]. During structural solution, a well-ordered small molecule density was identified in the electron density map of the HDAC3−SMRT complex and it was finally assigned to an inositol tetraphosphate that co-purified from a mammalian cell expression system. It was subsequently concluded that Ins(1,4,5,6)P$_4$ served as an intermolecular glue that enhanced interactions between HDAC3 and SMRT, which further enabled histone binding by creating a new substrate-binding surface. This serendipitous discovery explained why recombinant HDAC3 alone is inactive and explained a previously observed role of inositol phosphates and their kinases as transcriptional regulators [26].

Sirtuins, also known as the Class III HDACs, are NAD-dependent deacetylases that catalyze deacetylation of lysine residues [27]. The first structure of Sirtuin family proteins was that of Sir2Af1 in complex with the co-factor NAD [28]. The core domain of Sir2Af1 comprises a large domain containing a Rossmann fold and a small zinc-containing domain, with NAD (nicotinamide adenine dinucleotide) bound in a pocket between the two domains. The structure of this complex supports a two-step catalytic mechanism involving the generation of an oxo-carbenium species and the formation of an active deacetylase. Sirt5, which has weak deacetylase activity, has been recently identified as a protein lysine demalonylase and desuccinylase [29]. The structure of the Sirt5-succinyl peptide−NAD complex reveals that the negatively charged carboxylate group binds Sirt5 better than the acetyl group via interactions with Y102 and R105.

HKMTs usually contain a SET domain (Su(var)3-9, Enhancer-of-zeste and Trithorax domain) and use S-adenosyl methionine (SAM) as a methyl donor for catalysis. To date, only DOT1L [30,31] and the multi-subunit WARD (WDR5, RbBP5, Ash2L, and DPY-30) subcomplex of the MLL1 (mixed lineage leukemia 1) complex [32] are reported to be non-SET HKMTs with activities on histone H3K79 and H3K4, respectively. The histone tail binds to the catalytic groove and often forms a short parallel β-sheet with HKMT. The site-specificity for target lysine methylation is usually determined by co-recognition of the flanking residues. For example, SET7/9, a H3K4 methyltransferase, forms hydrogen bonds with Arg2, Thr3, and Gln5 of the histone H3 tail, thus rendering site-specificity toward Lys4 [33]. Similarly, DIM-5, a H3K9-specific methyltransferase, forms extensive interactions involving H3 residues Arg8, Ser10, and Thr11 in addition to β-sheet formation engaging the Lys9-Ser10 step [34]. In many cases, site-specific methylation by HKMT is also regulated by the nucleosomal context, especially for those sites that are close to or localized at the nucleosomal core. For example, when a histone octamer is used as substrate, the NSD2 (nonspecific dilatation 2) H3K36 methyltransferase can also methylate H4K44 *in vitro*, while in the context of a nucleosome, H3K36 becomes the sole methylation site [35]. A crystal structural study of the NSD1 SET domain revealed autoinhibition of H3K36 peptide binding by a post-SET loop, highlighting an autoregulatory mechanism underlying H3K36 site-selective methylation [36].

Lysine can be mono-, di-, or tri-methylated, and different methylation states even on the same site can code for different functional outcomes. For example, H3K4 monomethylation generated by SET7/9, MLL3, or MLL4 often marks the enhancer region; while H3K4 trimethylation generated by SET1 and MLL1 complexes is usually enriched at transcription start sites (TSSs), thus controlling transcription at different stages. The mechanism of state-specific monomethylation of H3K4 by SET7/9 and trimethylation of H3K9 by DIM-5 is elegantly explained by the Y/F switch mechanism [34]. As shown in Figure 7.4C, residues Y305 of SET7/9 and F281 of DIM-5 affect state-specific lysine methylation via hydrogen bond formation and steric hindrance with the lysine substrate at the active site. In the case of SET7/9, hydrogen bond formation between the phenolic hydroxyl of Y305 (together with Y245) and the H3K4 side chain, as well as a steric clash between Y305 and methylated lysine, restricts free rotation of the histone H3K4 side chain for additional methylation toward higher states. By contrast, the loss of the phenolic hydroxyl group in F281 of DIM-5 enables free rotation of the monomethyllysine for successive di- or tri-methylation. In support, wild type DIM-5 can trimethylate H3K9 while a F281Y mutant DIM-5 primarily mono-methylates the same substrate [34].

In addition to molecular insights revealed for site- and state-specific methylation by HKMT, what has emerged recently is the concept of histone-variant-specific writers, such as histone H3K27 methyltransferases ATXR5 and its paralog, ATXR6 in *Arabidopsis*. ATXR5/6 specifically methylates histone H3K27 in a variant H3.1-specific manner to regulate heterochromatin replication [37]. Histone H3.1 and H3.3 differ at position 31 with alanine for H3.1 and threonine for H3.3 in *Arabidopsis*. Crystal structural studies revealed steric clashes introduced by the slightly oversized Thr31 of H3.3 within the selectivity pocket, thus inhibiting the H3K27 methylation activity of ATXR5/6 [37].

HKDMs comprise the LSD and the JMJC subfamilies [38]. They often function in methylation site- and state-specific manner and help to shape the landscape of histone lysine methylome in concert with HKMTs. Structural studies of two related human JMJC-family demethylases, PHF8 and KIAA1718, revealed a "reading"-guided site-specific demethylation mechanism [39]. Both PHF8 and KIAA1718 harbor an H3K4me3-binding PHD finger positioned prior to the JMJC catalytic domain. Both JMJC domains alone are active on both H3K9me2 and H3K27me2 by recognizing a shared "ARKS" motif. Interestingly, the presence of "PHD-H3K4me3" recognition renders H3K9me2 a preferred site for PHF8 and H3K27me2 a preferred site for KIAA1718 for demethylation. The underlying molecular basis was nicely disclosed by crystallographic studies on the PHD-JMJC cassette of both enzymes, in which PHF8 PHD-JMJC adopts a bent conformation, allowing each of its domains to engage its respective target, whereas KIAA1718 adopts an extended conformation, which prevents its access to H3K9me2 by its JMJC domain when its PHD domain engages H3K4me3.

Molecular basis for state-specific demethylation can be exemplified by JMJD2A and PHF8. JMJD2A is able to demethylate di- and trimethylated lysine, while PHF8 only demethylates mono- or dimethylated lysine. The JMJD2A−H3K9me3 complex structure revealed a deep and snug insertion of K9me3, which is further stabilized by a hydrogen-bonding network at the catalytic center (Figure 7.4D) [40]. Extensive interactions around H3K9me3 positioned one methyl group closer to Fe^{2+} and α-ketoglutarate (α-KG) for further hydroxylation and demethylation. The methyl group of dimethyllysine was slightly away from the catalytic Fe^{2+}, thus being less efficient for catalysis. By contrast, the side chain of monomethyllysine is completely sequestered by the hydrogen-bond network and completely forfeits the ability for catalysis [40]. PHF8 cannot demethylate trimethylated lysine residues because of steric hindrance [39]. It is worth noting that as an amine oxidase, LSD family members can only demethylase mono- or dimethyllysine due to its dependence of a lone electron pair on the lysine ε−nitrogen atom for catalysis. Crystal structure determination of the LSD1−CoREST complex revealed an elongated structure with a long stalk connecting the catalytic domain of LSD1 and the CoREST DNA-binding SANT2 domain, suggesting a multivalent nucleosome binding model for efficient H3K4 demethylation [41].

Methylation on arginine residues is also an important PTM that is involved in various biological events, such as gene transcription, DNA repair, and signal transduction [42]. Histone arginine residue can be monomethylated, symmetrically dimethylated, or asymmetrically dimethylated, resulting in different transcriptional outcomes. Histone arginine methylation is catalyzed by PRMTs. The core structure of the first PRMT was solved in 2000 and this structure revealed a two-domain topology including an AdoMet-binding domain and a barrel-like domain at the C-terminal end [43]. The catalytic pocket resides between the two domains and the catalytic residues are conserved among the PRMT family. The methylation state of histone arginine residues is also elegantly regulated by writers, as exemplified in the crystal structure of *Trypanosoma brucei* PRMT7 (TbPRMT7), a PRMT that specifically catalyzes arginine monomethylation. The complex structure of TbPRMT7 and histone H4 peptide reveals that the catalytic pocket of TbPRMT7 is narrower than that of arginine dimethyltransferases, rendering it unsuitable for further methylation [44].

Enzymes that directly demethylate histone arginine residues are still poorly understood. However, arginine deamination can antagonize arginine methylation and histone editor PAD4 has been shown to catalyze this alteration [45,46]. PAD4 is a Ca^{2+}-dependent enzyme and a structural study revealed that binding of Ca^{2+} can induce conformational changes of PAD4 to generate the catalytic pocket [47]. PAD4 can also convert monomethylated arginine into citrulline and the structural basis for this catalytic activity was explained by the complex structure of PAD4 and histone peptide [48]. PAD4 recognizes the histone peptide mainly through main chain interactions and the catalytic pocket in PAD4 can tolerate the monomethylated arginine residue.

DNA methylation and demethylation has also been a focus in structural epigenetics. DNMT3a/3b has *de novo* DNA methyltransferase activity and DNMT1 harbors maintenance DNA methyltransferase activity. The structure of the Dnmt3L–Dnmt3a complex revealed a Dnmt3L–Dnmt3a-Dnmt3a–Dnmt3L tetramer topology with two active sites, suggesting that CpGs could be simultaneously methylated by Dnmt3a–Dnmt3L [49]. Additionally, both Dnmt3a and Dnmt3L contain an ADD (ATRX-DNMT3-DNMT3L) domain that recognizes unmodified H3K4, thus explaining why H3K4 methylation at differentially methylated regions protects these regions from *de novo* DNA methylation. The structure of the productive DNMT1–DNA complex together with its autoinhibited counterpart revealed the mechanism of maintenance of DNA methylation patterns [50,51]. Autoinhibition protects newly synthesized CpG from *de novo* methylation while hemimethylated CpG is recognized by the TRD subdomain and presented to the active site.

TET family proteins oxidize methyl-cytosine in DNA to hydroxylmethyl-cytosine on the pathway to DNA demethylation. Recently, the structure of the TET2–DNA complex has been determined by X-ray crystallography, explaining how 5mC is recognized and oxidized by the TET2 protein [52]. TET2 specifically recognizes CpG dinucleotide with a preference for 5mC. The methyl group of 5mC orients toward the Fe^{2+} for the oxidization reaction associated with the catalytic cavity. This methyl group does not interact with the TET2, enabling the TET2 catalytic cavity to accommodate oxidized products of 5mC. These structures illustrate molecular mechanisms from *de novo* DNA methylation, maintenance of DNA methylation, as well as DNA demethylation.

7.3.3 DECODING THE EPIGENETIC CODE WITH READER MODULES

Asides from histone "writers" and "erasers" that create or erase histone modifications, there are also histone "readers" that specifically recognize these PTMs. More than 20 distinct histone reader modules have been identified to date and structural studies on these reader modules have significantly advanced our understanding of the molecular details on "histone modification" decoding. Representative reader modules include Bromo, Royal Family members (e.g. Chromo, MBT,

PWWP, Tudor), PHD finger, CW, ADD, Ankyrin repeat, PH, BIR, BAH, WD40 repeats, BRCT, 14-3-3, and most recently YEATS [53] (Figure 7.3 and see Patel and Wang [54] for a comprehensive review).

Like histone writers and erasers, histone readers also display mark-, site-, and state-specificities. Taking the PHD finger as an example, this module can recognize both methylated and acetylated lysine marks, interact with sites R2, K4, K9, K14 of histone H3, and sense methylation states from unmodified to trimethylated H3K4 [55]. For H3K4me3 readout, crystal structural studies revealed an "aromatic cage" on the surface of BPTF PHD finger with recognition of trimethyllysine notably stabilized by cation-π and methyl-π interactions (Figure 7.4E) [56]. The PHD finger is also capable of recognizing lysine acetylation as illustrated in crystal structural studies of MOZ double PHD finger (DPF) bound to H3K14ac peptide [57]. H3K14ac is inserted into a narrow groove formed on the PHD1 β2 surface of MOZ DPF and stabilized by hydrogen bonds and van der Waals interactions. MOZ DPF also induces an α-helical conformation of on the "K4-T11" segment of H3 upon complex formation, revealing a unique mode of H3 recognition. Site-specific readout can also be exemplified by the complex structure of BPTF PHD finger, which specifically recognizes H3K4me3 instead of methylated states of H3K9 or H3K27 [56]. As shown in Figure 7.4E, the H3 peptide staples its long R2 and K4me3 side chains into two surface channels that are separated by a tryptophan residue (W32). Such a recognition mode requires an "R-X-K" sequence motif located at H3K4 (R2-T3-K4me3) instead of an "R-K" motif located at H3K9 (R8-K9me3) and H3K27 (R26-K27me3). Additionally, the free N-terminal amine of H3 also contributes to recognition and further enhancement of H3K4 specificity. Methylation state-specific readout is achieved by fine-tuning the composition and dimension of the aromatic cage. The L3MBTL1 (lethal 3 malignant brain tumor 1) protein utilizes its second MBT domain to recognize mono- or dimethylated lysine, while trimethylation disrupts binding [58]. Crystal structural studies revealed an acidic residue-lined aromatic cage within L3MBTL1 accounting for the lower methylation state readout utilizing a "cavity insertion" mode [58]. As shown in Figure 7.4F, the dimension of the pocket is deep but narrow, allowing dimethyllysine (Kme2) to snugly insert into the pocket and this alignment is further stabilized by charge-stabilized hydrogen bond formation with an acidic aspartate residue. It is conceivable that trimethylation will introduce a steric clash and disrupt hydrogen bonding within the pocket, thus being disfavored by the size-selective filter associated with this exquisitely designed pocket.

Many histone PTMs often coexist as patterns within a histone tail, a single nucleosome or defined chromatin domain. The combinatorial readout of histone marks could be achieved by reader modules paired in a single protein or different proteins of a single complex [59]. The structural basis for tandem reader modules and combinatorial readout has been extensively explored (reviewed in Patel and Wang [54]). One recent example is the structural characterization of Spindlin1 Spin/Ssty repeats as a methyl pattern reader of H3 "K4me3-R8me2a" marks [60]. Co-crystal structure studies revealed concurrent recognitions of K4me3 by Spin/Ssty 2 and of R8me2a by Spin/Ssty 1 with their respective aromatic pockets. A binding K_D of 45 nM was measured for such a dual recognition event, highlighting the potential of combinatorial readout.

Like histone-variant-specific writers, histone-variant-specific readers have also been identified recently. The tandem bromo-PWWP domain of ZMYND11 specifically recognizes H3.3 by forming hydrogen bonds with Ser31 (which is an alanine residue in H3.1) [61]. Additionally, the tandem Bromo-PWWP domain also recognizes H3K36me3. The dual recognition of both histone-variant and histone modification reveals the complexity of decoding histone PTMs.

Methylated cytosine can be recognized by DNA readers, such as the SRA domain and methyl-CpG-binding domain (MBD). The SRA domain of UHRF1 (ubiquitin-like, containing PHD and RING finger domains 1) flips the 5-methylcytosine out of the DNA helix and stabilizes the 5-methylcytosine by planar stacking contacts, hydrogen bonds, and van der Waals interactions [62−64]. The SRA domain has also been identified as a reader for hydroxymethyl-cytosine [65] and the mechanism for this recognition event has been explained by the structure of the complex of UHRF2-SRA with 5hmC-containing DNA [66]. The hydroxyl group of 5hmC forms a hydrogen bond with T508, resulting in higher affinity than that for 5mC. The binding pocket in UHRF1, which prefers 5mC, comprises tyrosine and aspartic acid residues; however, F495 and E498 lining the binding pocket of UHRF2 render it larger thereby accommodating the 5hmC.

7.3.4 GOVERNING CHROMATIN/NUCLEOSOME DYNAMICS BY EPIGENETIC CHAPERONES AND REMODELERS

Histone chaperones and ATP-dependent chromatin-remodeling complexes assemble nucleosome building blocks and govern chromatin dynamics. Histone chaperones interact with histones and participate in histone exchange and storage without the need of ATP. The ATP-dependent chromatin-remodeling complexes consume ATP to mobilize or restructure chromatin or nucleosomes. Major histone chaperones include ASF1, CAF1, DAXX, HIRA, HJURP, Rtt106, Chz1, Nap1, FACT (facilitates chromatin transcription) among others; major ATP-dependent chromatin-remodeling complexes include SWI/SNF (SWItch/Sucrose nonfermentable complex), ISWI (imitation switch complex), Mi-2/CHD, and INO80 [67,68].

Histone chaperones are capable of recognizing different histone type (H2A-H2B, H3-H4, or their variants) and at levels of dimer or tetramer, exhibiting histone type- and oligomerization state-specific recognition activities. For example, CIA/ASF1 binds to an H3/H4 dimer while Rtt106 dimer binds to an (H3-H4)$_2$ tetramer [69,70]. The β-strand of H4 that previously binds to H2A changes its conformation and binds to CIA-I. *In vitro* experiments also verify that CIA-I could disrupt the H3-H4 tetramer. The Spt16 middle domain from FACT also binds to the histone H2A-H2B dimer through a "U-turn" motif [69]. In contrast, Rtt106 forms a dimer and binds to the (H3-H4)$_2$ tetramer. H3K56ac enhances this interaction by increasing the conformational entropy [70].

Histone chaperone could also recognize histone variants. For example, DAXX specifically binds to H3.3-H4 dimer. The structure of the DAXX−histone complex reveals that the Glu225 in DAXX contributes significantly for the H3.3 specificity [71]. Gly90 (in H3.3) forms more stable hydrogen bond networks with the DAXX and H4 than Met90 (in H3.1 and H3.2).

The structural study on Chd1 revealed the inhibitory role of the double chromodomains on the ATPase motor [72]. The complex structure reveals that a helix of the double chromodomains packs against the ATPase motor and such an autoinhibited domain organization prevents the ATPase motor from binding to the duplex DNA. The authors proposed that binding to the nucleosomes could stabilize that double chromodomain and allow the ATPase motor to achieve an active conformation. X-ray crystallography studies also prompted a mechanistic understanding of the ISWI complex [73]. The X-ray structure of the ISW1a when combined with electron microscopy studies suggested a dinucleosome model in which ISW1a interacts with two nucleosomes and sets the length of linker DNA as a "protein ruler." Structural studies of these ATP-dependent chromatin remodeling complexes reveal how nucleosomes are moved and restructured as the building blocks for high-order chromatin structure.

7.4 APPLICATION OF X-RAY CRYSTALLOGRAPHY IN EPIGENETIC DRUG DISCOVERY

Since epigenetic modifications are shown to be highly related to human diseases, such as cancers, inflammation, diabetes, and neurodegenerative diseases, epigenetic regulators have become a promising target class in drug discovery [11]. X-ray crystallography could reveal key structural determinants within the binding pocket or surface, and inform how small molecules interact with the target macromolecules. Such information should prove critical for the rational analysis and optimization of the small molecules. In addition, crystal structures of target proteins also provide an initial model for computer-aided drug discovery, including virtual screening studies. X-ray crystallography has become one of the core technologies for modern drug discovery due to the in-depth and quantitative information it can provide in atomic detail. As such, structure-based drug discovery has already become an important approach toward the development of innovative medicine, notably in the field of epigenetics (Figure 7.5).

7.4.1 TARGETING EPIGENETIC MODIFIERS

Structure-based optimization has the potential to help enhance the binding potency of inhibitors. For example, high-throughput screening (HTS) of over 125,000 compounds identified BIX01294 as an inhibitor of G9a ($IC_{50} = 1.7$ μM) and to a lesser extent of GLP ($IC_{50} = 38$ μM) [74]. BIX01294 contains a quinazoline core and this core scaffold was further modified to enhance binding efficiency. The structure of the GLP−BIX01294 complex revealed that BIX01294 bound to the histone peptide-binding site but did not insert into the lysine-binding channel [75]. Based on this structure of the complex, modification of BIX01294 targeting lysine-binding channel was performed to increase binding potency. The 7-methoxy moiety of the quinazoline was optimized leading to the discovery of UNC0224 [76]. UNC0224 was a more potent G9a inhibitor with $K_D = 23$ nM (measured by ITC) and the structure of the G9a-UNC0224 complex was solved to confirm the binding mode. The structure of the G9a−UNC0224 complex confirmed that the 7-dimethylamino propoxy side chain indeed inserted into the lysine-binding channel but did not fully occupy it. Further optimization of the side chain led to the discovery of UNC0321 that was 40-fold more potent than UNC0224 [77].

Structural information is also crucial to improve specificities of inhibitors. KrennHrubec and colleagues discovered that a subpocket formed near the active site when CRA-A bound to HDAC8 [78]. An exploration for inhibitors targeting both active site and this subpocket identified an HDAC8-specific inhibitor that was more than 100-fold selective relative to HDAC1 and HDAC6. Structural information also assisted in the discovery of HDAC inhibitors specific to HDAC8 from *Schistosoma mansoni*, a pathogen that causes schistomiasis. Since smHDAC8 was shown to be crucial for parasite infectivity, Marek and colleagues solved the crystal structures of smHDAC8 in

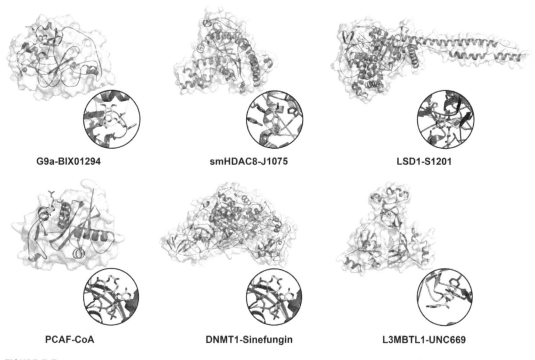

G9a-BIX01294   smHDAC8-J1075   LSD1-S1201

PCAF-CoA   DNMT1-Sinefungin   L3MBTL1-UNC669

FIGURE 7.5

A structural gallery of epigenetic regulators bound to small molecules. Coordinates have PDB codes 3FPD for G9a in complex with BIX-01294, 4BZ9 for smHDAC8 in complex with J1075, 3ABU for LSD1 in complex with S1201, 1CM0 for PCAF in complex with CoA, 3SWR for DNMT1 in complex with Sinefungin, 3P8H for L3MBTL1 in complex with UNC669. Except for smHDAC8 that is from the blood fluke *Schistosoma mansoni*, all other proteins listed here are from human.

complex with SAHA and M344 [79]. The active sites of the two HDAC8 complexes were different, especially at F151. This phenylalanine could adopt a flipped-in or flipped-out conformation in smHDAC8, which adopted only a flipped-in conformation in hHDAC8. Virtual screening was performed to identify inhibitors toward the enlarged active pocket of smHDAC8. These efforts discovered J1075 as a smHDAC8-specific inhibitor and confirmed the binding mode of J1075 to smHDAC8. They discovered that J1075 forced the phenylalanine flip out, which was in contrast to the flipped-in conformation in hHDAC8. This example also provided an additional concept on the development of new drugs: finding specific inhibitors targeting epigenetic regulators of eukaryotic parasites.

Both the potency and specificity of an inhibitor can be significantly improved based on structural information, as demonstrated in the case of the histone LSD1 inhibitor discovery. The first generation of LSD1 inhibitor was 2-PCPA, a common inhibitor of monoamine oxidase (MAO). This inhibitor was not potent or specific enough as a good drug candidate [80]. The crystal structure of the LSD1−PCPA complex was solved, providing structural guidance for further optimization of PCPA as an LSD1 inhibitor [81]. This structure revealed that a large hydrophobic pocket was occupied by a phenyl group of FAD-PCPA. However, this phenyl group did not make extensive interactions with LSD1, suggesting an approach for optimizing this phenyl position to gain more potent binding affinity. This structure also provided the structure of the LSD1 active site for developing LSD1-specific inhibitors. Based on this structure, an optimization of PCPA was performed leading to the discovery of 2-PFPA, S1201 as more potent and specific LSD1 inhibitors [82]. Mimasu and colleagues added halogens in the phenyl ring to gain more hydrogen bonds with LSD1 and added large groups on the phenyl ring to collide with MAO-B. After three rounds of screening and validation, S2101 was discovered and shown to have a K_I value $<1\,\mu M$ and >250-fold selectivity toward LSD1 compared with MAO-B.

Based on structural information, HTS-identified natural products that could also be optimized for higher potency and specificity. Natural products constitute a large family of HAT inhibitors, such as anacardic acid, curcumin, garcinol, and EGCG [83]. Further optimizations of these natural products were performed based on structure docking. For example, anacardic acid identified by HTS was docked into the PCAF (P300/CBP-associated factor) active site by molecular modeling [84]. The salicylated group of anacardic acid was thought to mimic the pyrophosphate group of CoA and formed a hydrogen bond network with PCAF. The salicylate ring of anacardic acid also stacked on Y616 to stabilize the interaction. The study optimized the interaction between anacardic acid and PCAF pantothenic acid pocket and Y616 based on the docking model. A two fold more potent inhibitor toward PCAF was identified through this hypothesis-driven optimization of natural products.

Computational homology model building and virtual screening represents another method for identification of lead compounds. The crystal structure of DNMT1 catalytic domain was not resolved until 2011 [50]. In order to perform virtual screening, a homology model of DNMT1 was generated based on DNMT2 and a bacterial homologue structure [85]. Based on this homology model, Siedlecki and colleagues performed virtual screens to identify lead compounds [86]. NSC303530 and NSC401077 were discovered and confirmed as DNMT1 inhibitors. These two inhibitors were chemical analogs and were conceived to bind to the DNMT1 active site in a similar manner. The docking result revealed that both inhibitors bound deeply into the active site cleft and occupied the cytosine-binding pocket.

7.4.2 TARGETING EPIGENETIC READERS

Drug discovery targeting reader modules has also relied on information from existing crystal structures. Since different reader modules utilize different binding pockets or surfaces, new drug discovery generally focuses on narrow and deep pockets, which are more tractable chemically (such as dimethyllysine-binding pocket). For example, small molecules targeting L3MBTL1 were discovered based on the L3MBTL1 crystal structure [87]. Compared with trimethyllysine reader modules, the MBT domain has a relatively narrow-binding pocket which only recognizes lower methylation states (Figure 7.4F). Herold and colleagues initially performed a high-throughput screen of 100,000 compounds, but no hits were identified by this approach. They next focused on pyrrolidine that mimics methyllysine residue and identified a ligand having five fold higher affinity to L3MBTL1 than the H3K9Me1 peptide through a ligand- and structure-based method. This study solved the first MBT domain-inhibitor structure and found that the inhibitor bound to the second MBT repeat mainly through cation-π and van der Waals interactions. It is worth noting that both UNC669 targeting MBT and UNC638 targeting G9a contain a pyrrolidine moiety, suggesting that the pyrrolidine moiety represents a good candidate that mimics the Kme2 side chain.

7.5 FRAGMENT-BASED DRUG DISCOVERY

Fragment-based drug discovery (FBDD) aims at finding molecules less than 250 Da with low binding affinity to target macromolecules. A small library (containing about 1000 fragments) is utilized to identify desired fragments. The weak binding affinity is generally detected by biophysical technologies such as ITC (isothermal titration calorimetry) and NMR. The binding state and key interactions are determined by solving the structure of fragments and the target protein. Based on these known interactions, fragments are modified into more potent compounds or can be linked together if these fragments bind to proximal sites. Since molecules screened in FBDD are rather simple and their interactions with macromolecules are well known, their optimization tends to be more tractable and efficient. (See Figure 7.6.)

According to Murray's opinion, although binding affinity is weak considering the small size of the fragment, the fragments can form high-quality interactions and can be more easily optimized into high potency inhibitors [88]. The ligand efficiency (LE) considering both binding potency and molecular weight would be a better criterion for judging the optimizability. In fact, screening of small fragments has a higher potential of identifying hits because both desired and undesired moieties often coexist in big compounds. FBDD focuses on more economical fragments without redundant moieties and more rational design based on structural information instead of higher throughput screening. The main challenges in FBDD are detecting the relatively weak interactions and efficiently optimizing fragments. X-ray crystallography is a potent technology in solving both challenges. Asides from DSF- (differential scanning flurimetry), NMR-, and SPR (surface plasmon resonance)-based methods, crystallographic screening is a reliable technology at finding promising fragments. Generally, 10 fragments are mixed as one cocktail, and then a library containing 1000 fragments could be classified into several hundred cocktails. Co-crystallization of cocktail-target protein or soaking cocktail into protein crystals may identify promising fragments [89]. X-ray crystallography is more widely utilized in confirming binding modes of identified fragments. The structure of the complex provides information on key interactions between fragments and target protein, guiding further optimization of fragments.

The bromodomain has become an active field for FBDD. Based on structures of existing bromodomains in complex with small molecules, Chung et al. found a common binding mode between inhibitors and bromodomain: the hydrogen bonds between inhibitors and BRD2 (bromodomain 2) involving Asn156 and Tyr113 are crucial for the recognition and the methyl group of the inhibitors inserts into a small pocket in BRD2 bromodomain [90]. Based on this structural knowledge, the authors focused on fragments mimicking acetyllysine and containing hydrogen bond forming ability to identify new lead drugs. A total of 1376 compounds were tested using fluorescence anisotropy assay and promising compounds were soaked into the BRD2 bromodomain crystals. X-ray crystallography was so powerful a technology that structures of 40 fragment−protein complexes were solved in this study, providing plentiful information on binding modes. The authors reported six representative small fragments and two of them represent new chemotypes binding to

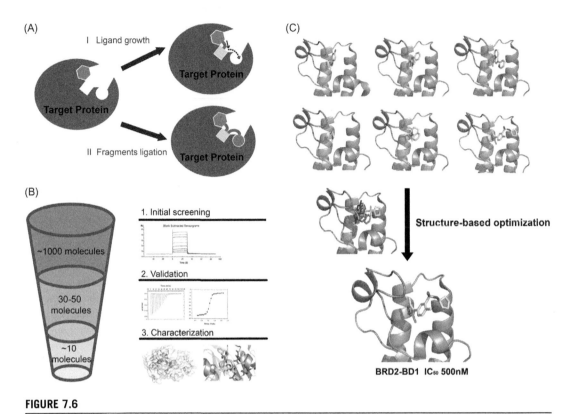

FIGURE 7.6

A schematic model of fragment-based drug discovery. (A) General principle of fragment-based drug discovery (FBDD). Two strategies are adopted in FBDD: ligand growth and fragments ligation. (B) Overall flowchart of FBDD. The procedure of FBDD generally involves initial screening (about 1000 molecules), validation (30–50 molecules) and final characterization (about 10 molecules). (C) A case study of FBDD. The first bromodomain (BD1) of BRD2 is chosen as the target for FBDD. The crystal structures of BRD2-BD1 in complex with fragments are shown in the upper panel. The optimized compound is shown in the lower panel (PDB code, 4A9M).

bromodomains. Since these molecules are generally less than 250 Da and carry high-quality interactions with target protein, the authors optimized a phenyl dimethyl isoxazole chemotype based on the structural knowledge [91]. They exploited BRD2 "WPF shelf" outside of the acetyllysine-binding pocket as the target site to modify initial hits after analyzing the complex structure. Optimized molecules were soaked into target protein and structure-activity relationships (SARs) were explored based on structures of the complex. The optimized molecules exhibiting two orders of magnitude binding potency showed anti-inflammatory activity in cellular assays. Recently, drugable inhibitors of the 2-thiazolidinone family that target the BRD4 (bromodomain 4) were also discovered through FBDD [92]. Binding models calculated by computational docking were checked to identify conserved interactions. Through this *in silico* screening, a library containing 500 fragments was reduced to 41 fragments. Manually chosen fragments were screened by X-ray crystallography.

X-ray crystallography provided validated and detailed information on fragment-target protein interactions. Nine fragments were identified in this study and a 2-thiazolidinone scaffold was optimized to 0.3 μM potency to BRD4 after several rounds of exploring SAR analyses. These studies validated the feasibility of X-ray crystallography screening and revealed that the bromodomain was very suitable for FBDD. With the crystal structure determination of an ever-growing number of epigenetic regulators, crystallography-based FBDD holds great promise for more successful cases in epigenetic-targeted drug discovery.

7.6 CONCLUSION

In the current time frame, epigenetics has become a central player in biomedical research with the potential for significant impact on human health and disease. As a powerful tool capable of providing atomic images of molecular recognition, macromolecular crystallography has greatly facilitated the mechanistic understanding of epigenetics phenomena in three-dimensional and atomic details. In addition, crystal structure-based drug design has already emerged as an important approach toward the development of selective inhibitors against epigenetic targets. Hence, it is anticipated that the application of X-ray crystallography to epigenetic researches shall continue to provide novel mechanistic insights into epigenetic regulation and prompt the translation of mechanistic discoveries in epigenetics into therapeutic strategies impacting on innovative medicine and its clinical ramifications.

ACKNOWLEDGMENTS

We sincerely thank Dr. Dinshaw J. Patel of the Memorial Sloan-Kettering Cancer Center for critical reading and comments on the manuscript. We apologize to all authors whose important contributions could not be acknowledged because of space constraints. This work was supported by The General Program of National Natural Science Foundation of China (31270763) and Program for New Century Excellent Talents in University to H.L.

REFERENCES

[1] Brownell JE, Zhou J, Ranalli T, Kobayashi R, Edmondson DG, Roth SY, et al. Tetrahymena histone acetyltransferase A: a homolog to yeast Gcn5p linking histone acetylation to gene activation. Cell 1996;84 (6):843−51.

[2] Taunton J, Hassig CA, Schreiber SL. A mammalian histone deacetylase related to the yeast transcriptional regulator Rpd3p. Science 1996;272(5260):408−11.

[3] Mujtaba S, Zeng L, Zhou MM. Structure and acetyl-lysine recognition of the bromodomain. Oncogene 2007;26(37):5521−7.

[4] Yap KL, Zhou MM. Keeping it in the family: diverse histone recognition by conserved structural folds. Crit Rev Biochem Mol 2010;45(6):488−505.

[5] Hashimoto H, Horton JR, Zhang X, Bostick M, Jacobsen SE, Cheng XD. The SRA domain of UHRF1 flips 5-methylcytosine out of the DNA helix. Nature 2008;455(7214):826−9.

[6] Dillon SC, Zhang X, Trievel RC, Cheng XD. The SET-domain protein superfamily: protein lysine methyltransferases. Genome Biol 2005;6(8).

[7] Cloos PAC, Christensen J, Agger K, Helin K. Erasing the methyl mark: histone demethylases at the center of cellular differentiation and disease. Genes Dev 2008;22(9):1115−40.

[8] Wu H, Zhang Y. Mechanisms and functions of Tet protein-mediated 5-methylcytosine oxidation. Genes Dev 2011;25(23):2436−52.

[9] Jiang YH, Bressler J, Beaudet AL. Epigenetics and human disease. Annu Rev Genomics Hum Genet 2004;5:479−510.

[10] Baylin SB, Jones PA. A decade of exploring the cancer epigenome − biological and translational implications. Nat Rev Cancer 2011;11(10):726−34.

[11] Arrowsmith CH, Bountra C, Fish PV, Lee K, Schapira M. Epigenetic protein families: a new frontier for drug discovery. Nat Rev Drug Discov 2012;11(5):384−400.

[12] Kelly TK, De Carvalho DD, Jones PA. Epigenetic modifications as therapeutic targets. Nat Biotechnol 2010;28(10):1069−78.

[13] Luger K, Mäder AW, Richmond RK, Sargent DF, Richmond TJ. Crystal structure of the nucleosome core particle at 2.8 A resolution. Nature 1997;389(6648):251−60.

[14] Tan S, Davey CA. Nucleosome structural studies. Curr Opin Struct Biol 2011;21(1):128−36.

[15] Schalch T, Duda S, Sargent DF, Richmond TJ. X-ray structure of a tetranucleosome and its implications for the chromatin fibre. Nature 2005;436(7047):138−41.

[16] Makde RD, England JR, Yennawar HP, Tan S. Structure of RCC1 chromatin factor bound to the nucleosome core particle. Nature 2010;467(7315):562−6.

[17] Strahl BD, Allis CD. The language of covalent histone modifications. Nature 2000;403(6765):41−5.

[18] Rojas JR, Trievel RC, Zhou J, Mo Y, Li X, Berger SL, et al. Structure of *Tetrahymena* GCN5 bound to coenzyme A and a histone H3 peptide. Nature 1999;401(6748):93−8.

[19] Clements A, Poux AN, Lo WS, Pillus L, Berger SL, Marmorstein R. Structural basis for histone and phosphohistone binding by the GCN5 histone acetyltransferase. Mol Cell 2003;12(2):461−73.

[20] Li Y, Zhang L, Liu T, Chai C, Fang Q, Wu H, et al. Hat2p recognizes the histone H3 tail to specify the acetylation of the newly synthesized H3/H4 heterodimer by the Hat1p/Hat2p complex. Genes Dev 2014;28(11):1217−27.

[21] Marks PA, Breslow R. Dimethyl sulfoxide to vorinostat: Development of this histone deacetylase inhibitor as an anticancer drug. Nat Biotechnol 2007;25(1):84−90.

[22] Finnin MS, Donigian JR, Cohen A, Richon VM, Rifkind RA, Marks PA, et al. Structures of a histone deacetylase homologue bound to the TSA and SAHA inhibitors. Nature 1999;401 (6749):188−93.

[23] Van Ommeslaeghe K, Elaut G, Brecx V, Papeleu P, Iterbeke K, Geerlings P, et al. Amide analogues of TSA: synthesis, binding mode analysis and HDAC inhibition. Bioorg Med Chem Lett 2003;13 (11):1861−4.

[24] Lu Q, Wang D-S, Chen C-S, Hu Y-D, Chen C-S. Structure-based optimization of phenylbutyrate-derived histone deacetylase inhibitors. J Med Chem 2005;48(17):5530−5.

[25] Watson PJ, Fairall L, Santos GM, Schwabe JWR. Structure of HDAC3 bound to co-repressor and inositol tetraphosphate. Nature 2012;481(7381):335−40.

[26] Odom AR, Stahlberg A, Wente SR, York JD. A role for nuclear inositol 1,4,5-trisphosphate kinase in transcriptional control. Science 2000;287(5460):2026−9.

[27] Yuan H, Marmorstein R. Structural basis for sirtuin activity and inhibition. J Biol Chem 2012;287 (51):42428−35.

[28] Min J, Landry J, Sternglanz R, Xu RM. Crystal structure of a SIR2 homolog-NAD complex. Cell 2001;105(2):269−79.

[29] Du J, Zhou Y, Su X, Yu JJ, Khan S, Jiang H, et al. Sirt5 is a NAD-dependent protein lysine demalony-lase and desuccinylase. Science 2011;334(6057):806−9.

[30] Feng Q, Wang H, Ng HH, Erdjument-Bromage H, Tempst P, Struhl K, et al. Methylation of H3-lysine 79 is mediated by a new family of HMTases without a SET domain. Curr Biol 2002;12(12):1052−8.

[31] Min J, Feng Q, Li Z, Zhang Y, Xu RM. Structure of the catalytic domain of human DOT1L, a non-SET domain nucleosomal histone methyltransferase. Cell 2003;112(5):711−23.

[32] Patel A, Vought VE, Dharmarajan V, Cosgrove MS. A novel non-SET domain multi-subunit methyl-transferase required for sequential nucleosomal histone H3 methylation by the mixed lineage leukemia protein-1 (MLL1) core complex. J Biol Chem 2011;286(5):3359−69.

[33] Wilson JR, Jing C, Walker PA, Martin SR, Howell SA, Blackburn GM, et al. Crystal structure and func-tional analysis of the histone methyltransferase SET7/9. Cell 2002;111(1):105−15.

[34] Zhang X, Yang Z, Khan SI, Horton JR, Tamaru H, Selker EU, et al. Structural basis for the product specificity of histone lysine methyltransferases. Mol Cell 2003;12(1):177−85.

[35] Li Y, Trojer P, Xu CF, Cheung P, Kuo A, Drury III WJ, et al. The target of the NSD family of histone lysine methyltransferases depends on the nature of the substrate. J Biol Chem 2009;284(49):34283−95.

[36] Qiao Q, Li Y, Chen Z, Wang M, Reinberg D, Xu RM. The structure of NSD1 reveals an autoregulatory mechanism underlying histone H3K36 methylation. J Biol Chem 2011;286(10):8361−8.

[37] Jacob Y, Bergamin E, Donoghue MT, Mongeon V, LeBlanc C, Voigt P, et al. Selective methylation of histone H3 variant H3.1 regulates heterochromatin replication. Science 2014;343(6176):1249−53.

[38] Mosammaparast N, Shi Y. Reversal of histone methylation: biochemical and molecular mechanisms of histone demethylases. Annu Rev Biochem 2010;79:155−79.

[39] Horton JR, Upadhyay AK, Qi HH, Zhang X, Shi Y, Cheng X. Enzymatic and structural insights for sub-strate specificity of a family of Jumonji histone lysine demethylases. Nat Struct Mol Biol 2010;17 (1):38−43.

[40] Ng SS, Kavanagh KL, McDonough MA, Butler D, Pilka ES, Lienard BMR, et al. Crystal structures of histone demethylase JMJD2A reveal basis for substrate specificity. Nature 2007;448(7149):87−91.

[41] Yang M, Gocke CB, Luo X, Borek D, Tomchick DR, Machius M, et al. Structural basis for CoREST-dependent demethylation of nucleosomes by the human LSD1 histone demethylase. Mol Cell 2006;23 (3):377−87.

[42] Yang YZ, Bedford MT. Protein arginine methyltransferases and cancer. Nat Rev Cancer 2013;13 (1):37−50.

[43] Zhang X, Zhou L, Cheng XD. Crystal structure of the conserved core of protein arginine methyltransfer-ase PRMT3. EMBO J 2000;19(14):3509−19.

[44] Wang C, Zhu Y, Caceres TB, Liu L, Peng J, Wang J, et al. Structural determinants for the strict mono-methylation activity by *Trypanosoma brucei* protein arginine methyltransferase 7. Structure 2014;22 (5):756−68.

[45] Cuthbert GL, Daujat S, Snowden AW, Erdjument-Bromage H, Hagiwara T, Yamada M, et al. Histone deimination antagonizes arginine methylation. Cell 2004;118(5):545−53.

[46] Wang Y, Wysocka J, Sayegh J, Lee YH, Perlin JR, Leonelli L, et al. Human PAD4 regulates histone arginine methylation levels via demethylimination. Science 2004;306(5694):279−83.

[47] Arita K, Hashimoto H, Shimizu T, Nakashima K, Yamada M, Sato M. Structural basis for Ca(2+)-induced activation of human PAD4. Nat Struct Mol Biol 2004;11(8):777−83.

[48] Arita K, Shimizu T, Hashimoto H, Hidaka Y, Yamada M, Sato M. Structural basis for histone N-terminal recognition by human peptidylarginine deiminase 4. Proc Natl Acad Sci USA 2006;103 (14):5291−6.

[49] Jia D, Jurkowska RZ, Zhang X, Jeltsch A, Cheng XD. Structure of Dnmt3a bound to Dnmt3L suggests a model for de novo DNA methylation. Nature 2007;449(7159):248−51.

[50] Song JK, Rechkoblit O, Bestor TH, Patel DJ. Structure of DNMT1-DNA complex reveals a role for autoinhibition in maintenance DNA methylation. Science 2011;331(6020):1036−40.

[51] Song JK, Teplova M, Ishibe-Murakami S, Patel DJ. Structure-based mechanistic insights into DNMT1-mediated maintenance DNA methylation. Science 2012;335(6069):709−12.

[52] Hu L, Li Z, Cheng J, Rao Q, Gong W, Liu M, et al. Crystal structure of TET2-DNA complex: insight into TET-mediated 5mC oxidation. Cell 2013;155(7):1545−55.

[53] Li Y, Wen H, Xi Y, Tanaka K, Wang H, Peng D, et al. AF9 YEATS domain links histone acetylation to DOT1L-mediated H3K79 methylation. Cell 2014;159(3):558−71.

[54] Patel DJ, Wang Z. Readout of epigenetic modifications. Annu Rev Biochem 2013;82:81−118.

[55] Li Y, Li H. Many keys to push: diversifying the 'readership' of plant homeodomain fingers. Acta Biochim Biophys Sin 2012;44(1):28−39.

[56] Li H, Ilin S, Wang W, Duncan EM, Wysocka J, Allis CD, et al. Molecular basis for site-specific read-out of histone H3K4me3 by the BPTF PHD finger of NURF. Nature 2006;442(7098):91−5.

[57] Dreveny I, Deeves SE, Fulton J, Yue B, Messmer M, Bhattacharya A, et al. The double PHD finger domain of MOZ/MYST3 induces alpha-helical structure of the histone H3 tail to facilitate acetylation and methylation sampling and modification. Nucleic Acids Res 2014;42(2):822−35.

[58] Li H, Fischle W, Wang W, Duncan EM, Liang L, Murakami-Ishibe S, et al. Structural basis for lower lysine methylation state-specific readout by MBT repeats of L3MBTL1 and an engineered PHD finger. Mol Cell 2007;28(4):677−91.

[59] Ruthenburg AJ, Li H, Patel DJ, Allis CD. Multivalent engagement of chromatin modifications by linked binding modules. Nat Rev Mol Cell Biol 2007;8(12):983−94.

[60] Su X, Zhu G, Ding X, Lee SY, Dou Y, Zhu B, et al. Molecular basis underlying histone H3 lysine-arginine methylation pattern readout by Spin/Ssty repeats of Spindlin1. Genes Dev 2014;28(6):622−36.

[61] Wen H, Li Y, Xi Y, Jiang S, Stratton S, Peng D, et al. ZMYND11 links histone H3.3K36me3 to transcription elongation and tumour suppression. Nature 2014;508(7495):263−8.

[62] Hashimoto H, Horton JR, Zhang X, Bostick M, Jacobsen SE, Cheng X. The SRA domain of UHRF1 flips 5-methylcytosine out of the DNA helix. Nature 2008;455(7214):826−9.

[63] Avvakumov GV, Walker JR, Xue S, Li Y, Duan S, Bronner C, et al. Structural basis for recognition of hemi-methylated DNA by the SRA domain of human UHRF1. Nature 2008;455(7214):822−5.

[64] Arita K, Ariyoshi M, Tochio H, Nakamura Y, Shirakawa M. Recognition of hemi-methylated DNA by the SRA protein UHRF1 by a base-flipping mechanism. Nature 2008;455(7214):818−21.

[65] Spruijt CG, Gnerlich F, Smits AH, Pfaffeneder T, Jansen PW, Bauer C, et al. Dynamic readers for 5-(hydroxy)methylcytosine and its oxidized derivatives. Cell 2013;152(5):1146−59.

[66] Zhou T, Xiong J, Wang M, Yang N, Wong J, Zhu B, et al. Structural basis for hydroxymethylcytosine recognition by the SRA domain of UHRF2. Mol Cell 2014;54(5):879−86.

[67] Eitoku M, Sato L, Senda T, Horikoshi M. Histone chaperones: 30 years from isolation to elucidation of the mechanisms of nucleosome assembly and disassembly. Cell Mol Life Sci 2008;65(3):414−44.

[68] Burgess RJ, Zhang Z. Histone chaperones in nucleosome assembly and human disease. Nat Struct Mol Biol 2013;20(1):14−22.

[69] Natsume R, Eitoku M, Akai Y, Sano N, Horikoshi M, Senda T. Structure and function of the histone chaperone CIA/ASF1 complexed with histones H3 and H4. Nature 2007;446(7133):338−41.

[70] Su D, Hu Q, Li Q, Thompson JR, Cui G, Fazly A, et al. Structural basis for recognition of H3K56-acetylated histone H3-H4 by the chaperone Rtt106. Nature 2012;483(7387):104−7.

[71] Elsasser SJ, Huang H, Lewis PW, Chin JW, Allis CD, Patel DJ. DAXX envelops a histone H3.3-H4 dimer for H3.3-specific recognition. Nature 2012;491(7425):560−5.

[72] Hauk G, McKnight JN, Nodelman IM, Bowman GD. The chromodomains of the Chd1 chromatin remodeler regulate DNA access to the ATPase motor. Mol Cell 2010;39(5):711−23.

[73] Yamada K, Frouws TD, Angst B, Fitzgerald DJ, DeLuca C, Schimmele K, et al. Structure and mechanism of the chromatin remodelling factor ISW1a. Nature 2011;472(7344):448−53.

[74] Kubicek S, O'Sullivan RJ, August EM, Hickey ER, Zhang Q, Teodoro ML, et al. Reversal of H3K9me2 by a small-molecule inhibitor for the G9a histone methyltransferase. Mol Cell 2007;25(3):473−81.

[75] Chang Y, Zhang X, Horton JR, Upadhyay AK, Spannhoff A, Liu J, et al. Structural basis for G9a-like protein lysine methyltransferase inhibition by BIX-01294. Nat Struct Mol Biol 2009;16(3):312−17.

[76] Liu F, Chen X, Allali-Hassani A, Quinn AM, Wasney GA, Dong AP, et al. Discovery of a 2,4-Diamino-7-aminoalkoxyquinazoline as a potent and selective inhibitor of histone lysine methyltransferase G9a. J Med Chem 2009;52(24):7950−3.

[77] Liu F, Chen X, Allali-Hassani A, Quinn AM, Wigle TJ, Wasney GA, et al. Protein Lysine Methyltransferase G9a Inhibitors: design, synthesis, and structure activity relationships of 2,4-Diamino-7-aminoalkoxy-quinazolines. J Med Chem 2010;53(15):5844−57.

[78] KrennHrubec K, Marshall BL, Hedglin M, Verdin E, Ulrich SM. Design and evaluation of 'Linkerless' hydroxamic acids as selective HDAC8 inhibitors. Bioorg Med Chem Lett 2007;17(10):2874−8.

[79] Marek M, Kannan S, Hauser AT, Moraes Mourão M, Caby S, Cura V, et al. Structural basis for the inhibition of histone deacetylase 8 (HDAC8), a key epigenetic player in the blood fluke Schistosoma mansoni. PLoS Pathog 2013;9(9):e1003645.

[80] Hojfeldt JW, Agger K, Helin K. Histone lysine demethylases as targets for anticancer therapy. Nat Rev Drug Discov 2013;12(12):917−30.

[81] Yang MJ, Culhane JC, Szewczuk LM, Jalili P, Ball HL, Machius M, et al. Structural basis for the inhibition of the LSD1 histone demethylase by the antidepressant trans-2-phenylcyclopropylamine. Biochemistry 2007;46(27):8058−65.

[82] Mimasu S, Umezawa N, Sato S, Higuchi T, Umehara T, Yokoyama S. Structurally Designed trans-2-Phenylcyclopropylamine derivatives potently inhibit histone demethylase LSDI/KDM1. Biochemistry 2010;49(30):6494−503.

[83] Furdas SD, Kannan S, Sippl W, Jung M. Small molecule inhibitors of histone acetyltransferases as epigenetic tools and drug candidates. Arch Pharm (Weinheim) 2012;345(1):7−21.

[84] Ghizzoni M, Boltjes A, de Graaf C, Haisma HJ, Dekker FJ. Improved inhibition of the histone acetyltransferase PCAF by an anacardic acid derivative. Bioorg Med Chem 2010;18(16):5826−34.

[85] Siedlecki P, Garcia Boy R, Comagic S, Schirrmacher R, Wiessler M, Zielenkiewicz P, et al. Establishment and functional validation of a structural homology model for human DNA methyltransferase 1. Biochem Biophys Res Commun 2003;306(2):558−63.

[86] Siedlecki P, Boy RG, Musch T, Brueckner B, Suhai S, Lyko F, et al. Discovery of two novel, small-molecule inhibitors of DNA methylation. J Med Chem 2006;49(2):678−83.

[87] James LI, Korboukh VK, Krichevsky L, Baughman BM, Herold JM, Norris JL, et al. Small-molecule ligands of methyl-lysine binding proteins: optimization of selectivity for L3MBTL3. J Med Chem 2013;56(18):7358−71.

[88] Murray CW, Rees DC. The rise of fragment-based drug discovery. Nat Chem 2009;1(3):187−92.

[89] Murray CW, Blundell TL. Structural biology in fragment-based drug design. Curr Opin Struct Biol 2010;20(4):497−507.

[90] Chung CW, Dean AW, Woolven JM, Bamborough P. Fragment-based discovery of bromodomain inhibitors part 1: inhibitor binding modes and implications for lead discovery. J Med Chem 2012;55(2):576−86.

[91] Bamborough P, Diallo H, Goodacre JD, Gordon L, Lewis A, Seal JT, et al. Fragment-based discovery of bromodomain inhibitors part 2: optimization of phenylisoxazole sulfonamides. J Med Chem 2012;55(2):587−96.

[92] Zhao L, Cao D, Chen T, Wang Y, Miao Z, Xu Y, et al. Fragment-based drug discovery of 2-thiazolidinones as inhibitors of the histone reader BRD4 bromodomain. J Med Chem 2013;56(10):3833−51.

CHEMICAL AND GENETIC APPROACHES TO STUDY HISTONE MODIFICATIONS

Abhinav Dhall and Champak Chatterjee

Department of Chemistry, University of Washington, Seattle, WA, USA

CHAPTER OUTLINE

8.1 EUKARYOTIC CHROMATIN AND HISTONES

Chromatin is the massive nucleoprotein complex in which our genomic DNA is stored in the nuclei of our cells [1]. Histones are the major protein component of chromatin and they serve as a scaffold around which DNA is tightly wound to facilitate its storage in the small nuclear space [2,3]. Our understanding of histone function has undergone rapid growth in the last two decades, and histones have been found to play indispensable roles in regulating both chromatin structure and function [4,5]. Indeed, a vast amount of literature has shown that histones play major roles in almost every DNA-templated process including transcription [6−8], replication [9−11], and repair [12−14]. The two histone-centric mechanistic pathways that dictate cell fate are the exchange of core histones with histone variants [15−17], and posttranslational modifications (PTMs) of histone side-chains [18,19]. This chapter will focus on our current knowledge of the latter, touching upon its

Y.G. Zheng (Ed): Epigenetic Technological Applications. DOI: http://dx.doi.org/10.1016/B978-0-12-801080-8.00008-9

dysregulation in human disease states and describing chemical and genetic methods that have been developed to elucidate the mechanistic roles for specific histone modifications.

Two copies of each of the four core histone proteins H2A, H2B, H3, and H4 come together to form a globular octamer around which about 147 bp of double-stranded DNA is wrapped to form the fundamental unit of chromatin — the nucleosome. The bulk of each histone protein is shielded from solution by its association with other histones in the octamer and with double-stranded DNA, leaving largely the N-terminal regions or histone *tails* exposed to the aqueous environment [20,21]. Not surprisingly, these tails are also sites for extensive chemical modification by a large family of histone-modifying enzymes. The reversible modification of histone tails by a range of small chemical groups, as well as by entire proteins, underlies dramatic changes in chromatin structure and function [22]. The dysregulation of these histone modifications, or *marks*, underlies severe developmental disorders and human diseases including leukemias [23], ataxias [24], and fragile X syndrome, to name a few [25]. Indeed, aberrant histone marks have attracted interest for their association with various types of cancers. For example, the absence of trimethylation at histone H3 and acetylation at H4 is found in many cancer cell lines and primary tumors [26] and excessive H3 phosphorylation is associated with colorectal cancer [27]. The success of small-molecule inhibitors of histone deacetylases that are in various phases of clinical trials as anti-cancer agents suggests that targeting specific histone marks and their associated proteins is a viable therapeutic approach [28]. Particularly appealing is the concept of personalized medicines that address patient-specific misregulation of epigenetic marks [29]. However, there is an urgent need to understand the precise mechanistic roles for different histone marks prior to devising such targeted epigenetic therapies.

8.2 THE CHALLENGING DIVERSITY OF HISTONE POSTTRANSLATIONAL MODIFICATIONS

All four core histones, H2A, H2B, H3, and H4 as well as the histone variants and linker histone H1 are subject to extensive chemical modification in our cells [30]. Over 16 chemical PTMs have been identified to date, ranging from well-known modifications such as methylation, acetylation, and phosphorylation to more exotic modifications such as ubiquitylation, SUMOylation (modification by the small ubiquitin (Ub)-like modifier protein), and ADP-ribosylation (Figure 8.1) [5]. Adding combinatorial complexity in addition to the diversity of histone modifications is the fact that multiple amino acid side-chains in a histone can bear the same or different PTMs. In fact, proteomic studies of human histones have revealed over 150 PTM combinations on a single histone H3.2 [31]. The staggering complexity of histone modifications *in vivo* presents a challenge when attempting to decipher their discrete roles in cellular events such as gene transcription and DNA strand-break repair. Another important aspect of histone PTMs that was overlooked in early studies, largely due to the limited set of tools available, is the extent of PTM per nucleosome. Since there are two copies of each histone in a nucleosome, there is the possibility that one or both histones may be modified. This stoichiometry is particularly important when considering processes such as the recruitment of chromatin-binding proteins that bind specific marks on adjacent nucleosomes [32] and the inheritance of histone marks during DNA replication [33].

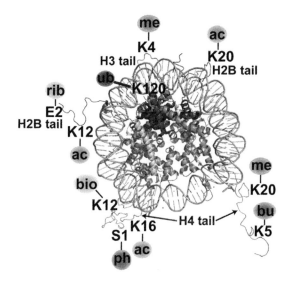

FIGURE 8.1

Histone tail modifications. X-ray crystal structure of a mononucleosome at 1.9 Å resolution (PDB code 1KX5) showing the histone tails and selected sites of posttranslational modification. *Xenopus laevis* core histones are shown in red (H2A), blue (H2B), amber (H3), and green (H4). Ac, acetyl; bio, biotinyl; bu, butyryl; me, methyl; ph, phosphoryl; rib, ADP-ribosyl groups. (*This figure is reproduced in color in the color plate section.*)

Structure rendered with PyMol from DeLano Scientific LLC.

Genetic studies that have taken advantage of methods in molecular biology to mutate key histone residues, and to overexpress or knockout histone-modifying enzymes in order to observe the resultant effects on gene function have provided many clues regarding potential roles for histone marks in normal development and disease states [34]. However, such strategies suffer from several limitations, including the inability to control the degree and diversity of histone modifications in chromatin. Furthermore, since the same histone residue may be modified by different marks that have unique and sometimes functionally opposing outcomes, in many instances phenotypes arising from the mutation of a histone residue cannot be solely ascribed to the absence of a specific mark. Another key technique that has advanced our understanding of the functional consequences of histone marks is the utilization of histone modification-specific antibodies [35]. Chromatin immunoprecipitation (ChIP) experiments [36] with such antibodies in conjunction with DNA microarrays (ChIP-on-chip) and massively parallel DNA sequencing (ChIP-seq) have permitted the assignment of gene and position-specific functions to histone marks [37]. For example, acetylation at H4 Lys16 is typically found in the promoters of transcriptionally active genes, whereas trimethylation at H3 Lys9 is observed predominantly in silenced heterochromatin [38]. Antibodies suffer from certain limitations, such as the occlusion of epitopes by PTMs at adjacent residues that may result in failed detection in ChIP experiments and the inability to distinguish modifications in the same nucleosome from those in adjacent nucleosomes. More recently, the combination of microfluidic technology, modification-specific antibodies, and single-molecule imaging techniques has enabled measuring the degree of co-localization of histone modifications and DNA methylation on single nucleosomes [39,40]. Increasingly powerful mass spectrometric methods have also begun to address the challenge of detecting many simultaneous modifications in one histone variant [31], as well as determining the symmetric or asymmetric nature of marks in a nucleosome [41]. Altogether, these methodologies continue to provide the foundation for mechanistic investigations of the context-dependent roles for various histone marks.

8.3 A CHEMICAL BIOLOGY APPROACH TO INVESTIGATE HISTONE MODIFICATIONS

One approach to understanding the specific mechanistic roles of histone marks is to subject pure site-specifically modified histones to biophysical and biochemical assays under chemically defined conditions. Historically, erythrocyte and thymus cell nuclei have proven to be a good source for quantities of modified histones that can be purified using acid extraction and ion-exchange chromatography, followed by reconstitution into octamers and nucleosomes [42,43]. Histones thus obtained are, however, not homogeneously modified and may contain multiple modifications to varying degrees, which complicates the interpretation of results from biochemical assays. Alternatively, in instances where the specific histone-modifying enzymes are known and may be obtained in sufficient quantities, site-specifically modified histones may in principle be obtained by subjecting bacterially expressed recombinant histones to enzymatic modification. However, it is difficult to drive enzymatic reactions to completion while retaining site-specificity [44].

The fact that histones can be purified under harsh denaturing conditions in organic solvents and reconstituted into mononucleosomes and chromatin arrays with double-stranded DNA facilitates the application of protein chemistry techniques to interrogate the mechanistic roles of histone marks [45]. The following sections will describe chemical strategies that have enabled generating homogeneously site-specifically modified histones and the mechanistic studies undertaken with these tools.

8.4 STRATEGIES OF NATIVE CHEMICAL AND EXPRESSED PROTEIN LIGATION

Although core histones are relatively small proteins with the longest, H3, being ~ 130 amino acids in length, they still pose a significant challenge for linear solid-phase peptide synthesis [46]. Therefore, techniques such as native chemical ligation (NCL) [47] and expressed protein ligation (EPL) [48] that facilitate a fragment-based approach, and therefore require the synthesis of shorter peptides, have greatly expanded access to site-specifically modified histones. NCL was reported by Kent and co-workers as a chemoselective peptide fragment ligation strategy that joins two peptide fragments by means of a native amide bond [47]. The only limitations for applying NCL toward the synthesis of proteins are a C-terminal α-thioester in the N-terminal fragment peptide and an N-terminal Cys residue in the C-terminal fragment. The two peptide fragments may be obtained by solid-phase peptide synthesis using well-established fluorenylmethoxycarbonyl (Fmoc-) or tert-butyloxycarbonyl (Boc-) amine protecting group strategies. Several chemical linkers and derivatization strategies are available for the synthesis of peptide α-thioesters by both chemistries [49,50]. Since most histone marks are found in the N-terminal tails, the C-terminal fragment may be larger and obtained by heterologous expression in *Escherichia coli* (Figure 8.2). However, for producing histones modified at internal sites, for example acetylation at H3 Lys56, the synthesis of both N- and C-terminal peptide fragments in reasonable yields is challenging, due to the inherent limitations of peptide synthesis. One strategy that has been successfully applied to overcome this limitation is the synthesis of full-length histones by a three-fragment approach, which employs two successive NCL reactions with shorter peptide sequences [51]. Among the many advantages of synthetic strategies is the ability to introduce several simultaneous site-specific modifications in histones, the ability to incorporate fluorescent probes, and photoactivatable cross-linkers at key

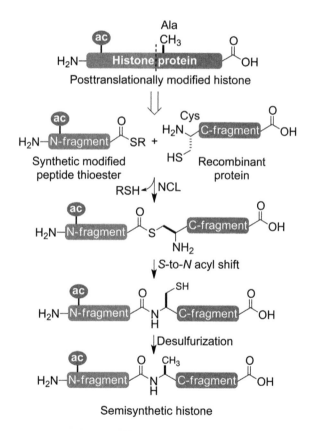

FIGURE 8.2

Native chemical ligation approach for histone semisynthesis. The desired posttranslationally modified histone is split into two fragments. The shorter synthetic peptide α-thioester fragment contains the modified amino acid and the larger recombinant protein fragment has an N-terminal Cys. Native chemical ligation (NCL) of the two fragments is followed by Raney-Ni-mediated desulfurization of Cys to yield the desired semisynthetic histone-bearing site-specific modifications.

sites, and the ability to generate milligram quantities of modified histones for biophysical character-ization by X-ray crystallography [5].

EPL is a technique that greatly extends the scope of NCL by generating longer protein α-thioesters using intein fusions. Inteins are a family of proteins that catalyze their own excision from a larger precursor protein and simultaneously ligate the two flanking fragments (exteins) to generate a smaller protein minus the intein segment (Figure 8.3A) [52]. Thus, inteins may be con-sidered self-splicing protein analogs of introns found in eukaryotic mRNA (messenger RNA). A subset of inteins also exists as naturally fragmented domains that are separately expressed and rapidly associate to catalyze protein splicing *in trans*. When the two halves are genetically fused, this results in a class of ultrafast inteins with highly desirable properties for protein engineering, including greater substrate tolerance and the ability to function at 37°C [53]. The first step in the intein-catalyzed splicing reaction is the generation of a thioester linkage between the N-terminal

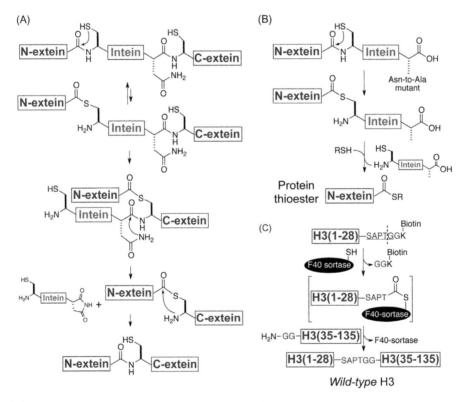

FIGURE 8.3

Intein and sortase-mediated ligation strategies. (A) The chemical mechanism of intein-mediated splicing of N- and C-terminal extein polypeptides. (B) Strategy to generate protein C-terminal α-thioesters using a mutant intein. (C) A sortase-mediated ligation strategy to produce full-length *wild-type* H3. The mutant F40-sortase cleaves the SAPTG recognition sequence as indicated by the dashed line, and the resultant acyl-enzyme intermediate is intercepted with an H3 fragment bearing an N-terminal Gly–Gly motif.

extein and a Cys residue at the intein N-terminus. In a subsequent step, the N-terminal extein fragment is transferred to a Cys or Ser side-chain in the C-terminal extein to form a branched intermediate. Key to the formation of this branched intermediate is the presence of a C-terminal Asn in the intein and the presence of a Cys or Ser at the N-terminus of the C-terminal extein. By mutating the Asn and Cys/Ser residues, all steps succeeding the first step in the splicing reaction may be prevented. Any protein, including histone fragments, may therefore be genetically fused to the N-terminus of such a mutant intein and after recombinant expression in *E. coli* the amide bond between the protein fragment and intein is converted to a thioester bond [54]. This thioester linkage is readily displaced by exogenously added thiols to generate a protein α-thioester in solution that is no longer attached to the intein (Figure 8.3B) [48]. The protein α-thioester may then be employed in a typical NCL reaction with a C-terminal protein fragment to generate the full-length protein. EPL is particularly useful when interrogating modifications occurring at the C-terminus of histones, such as ubiquitylation near the C-termini of histones H2A and H2B. Since any number and type of site-specific chemical modifications may essentially be incorporated during the synthesis of histone fragments, the techniques of NCL and EPL allow access to any physiologically relevant modified histone for mechanistic studies. Another recent development in the area of histone semisynthesis is the application of sortase-mediated ligation to the production of histone H3 (Figure 8.3C) [55]. This eliminates the need for histone-intein fusions or solid-phase peptide synthesis to generate peptide/protein C-terminal α-thioesters. Instead, a short C-terminal sortase-recognition sequence is appended to the histone peptide, which is cleaved by an evolved form of the Cys protease Sortase A to generate a transient histone-enzyme thioester intermediate (Figure 8.3C). The thioester intermediate may be intercepted with peptides/proteins bearing two N-terminal Gly residues to yield the native ligation product. H3 and H4 are ideal targets for sortase-mediated ligation as they each have two Gly-Gly repeats near the N-terminus that can be used as ligation sites.

8.5 **THIALYSINE ANALOGS OF METHYLATED AND ACETYLATED HISTONES**

A chemical strategy that has found wide application in studies pertaining to histone methylation and acetylation is that of Cys alkylation [56]. In comparison with side-chain amine and phenol groups in proteins, the thiol side-chain of Cys is particularly nucleophilic at pH 7.5−8.0 and also forms a thiyl radical in the presence of radical initiators. This has led to the utilization of site-directed mutagenesis in histones to introduce a single Cys at the desired site of modification [57]. Histones are particularly well suited for Cys mutagenesis as there is only a single Cys (Cys110 in H3) among the four histones in higher organisms and this can be mutated to Ala without any changes in the stability or biochemical function of the histone octamer. The single Cys may be directly alkylated with different degrees of *N*-methylated 2-bromoethylamines to generate thialysine analogs bearing one, two, or three methyl groups (Figure 8.4) [57]. Alternatively, a Cys-specific thiyl radical may be generated by treatment of the histone with an appropriate water-soluble radical initiator (Figure 8.4). This reacts with *N*-vinylacetamide to generate the corresponding thia-analog of acetyllysine [58]. Both methods lead to the substitution of the γ-carbon in the Lys side-chain by a sulfur atom; however, this was shown to have minimal detrimental effects on the properties of

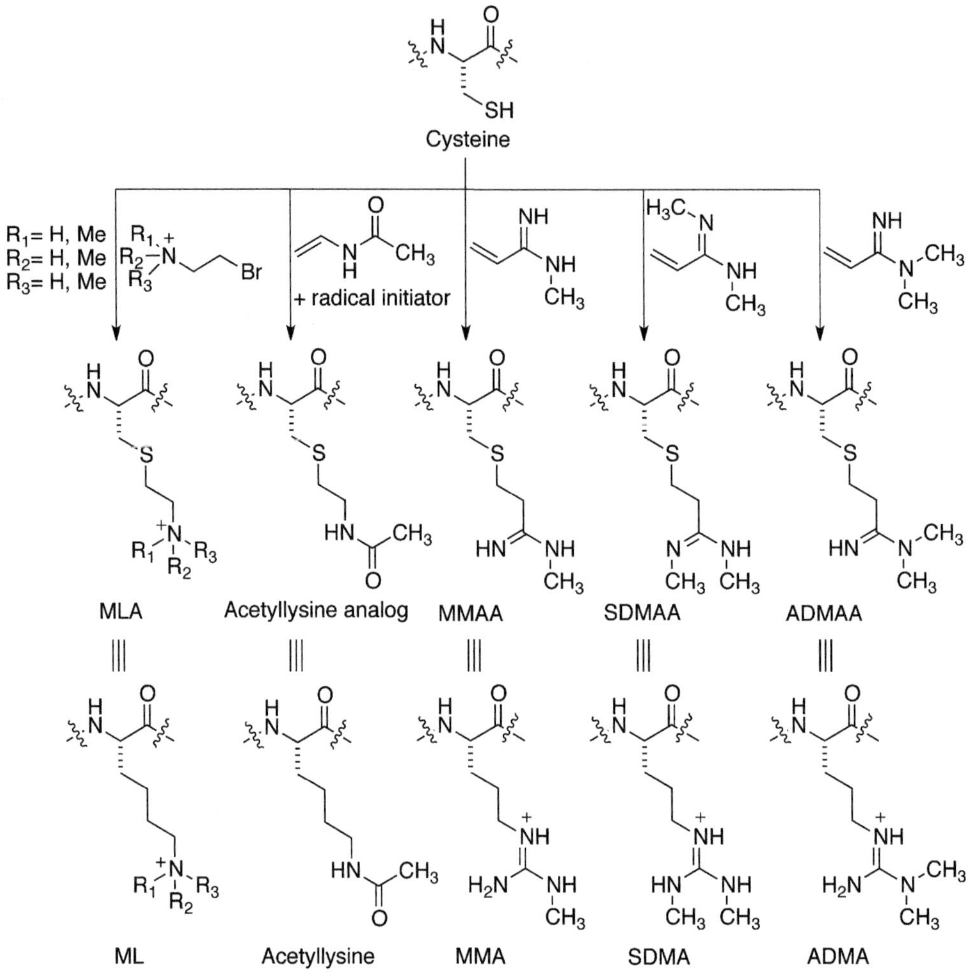

FIGURE 8.4

Cysteine-targeted strategies to generate functional analogs of methylated and acetylated histones. ML, Methyllysine; MLA, Methyllysine analog; MMA, Monomethylarginine; MMAA, Monomethylarginine analog; SDMA, Symmetric dimethylarginine, SDMAA, Symmetric dimethylarginine analog; ADMA, Asymmetric dimethylarginine; ADMAA, Asymmetric dimethylarginine analog.

the methylated and acetylated lysine analogs. Comparable recognition of the *wild-type* and thialysine versions of modifications by specific antibodies suggests their utility in biochemical assays [57]. Furthermore, when incorporated in place of H4 Lys16, the acetyllysine analog displays identical inhibition of compaction in nucleosome arrays as the *wild-type* H4 Lys16ac modification [58]. Although Cys modification strategies have facilitated many studies of methylated and acetylated histones, one obvious limitation of this strategy is that acetylation and methylation cannot be site-specifically incorporated into the same histone protein due to cross-reactivity of the starting materials at all Cys residues in the histone. Hence EPL, NCL, and sortase-mediated ligation are more practical for the introduction of mixed modifications in histones.

8.6 GENETIC INCORPORATION OF MODIFIED AMINO ACIDS IN HISTONES

Amber suppression mutagenesis of protein targets is achieved by incorporating the amber codon UAG in genes with orthogonal pairs of amber suppressor tRNAs (transfer RNAs) and their cognate aminoacyl tRNA synthetases (aaRS). This strategy has been extensively used to incorporate modified amino acids in histone H3 [59−62] and more recently in histone H4 [63]. Histone lysines are important targets for amber suppression mutagenesis as they can be modified by a wide range of functional groups including methyl, acetyl, malonyl, succinyl, butyryl, propionyl, and crotonyl groups [64−66]. These modifications influence histone interactions and thereby gene function by changing net charge, hydrophobicity, and to some degree the flexibility of histone tails. In the last decade, the use of aaRS-tRNA$_{CUA}$ pairs, such as the pyrrolysyl-tRNA synthetase (PylRS) from *Methanosarcina mazei* and *Methanosarcina barkeri* or tyrosyl-tRNA synthetase (TyrRS) from *Methanococcus jannaschii*, which incorporate pyrrolysine (Pyl) and tyrosine, respectively, has facilitated the incorporation of non-natural amino acids and chemically or photochemically modifiable precursors in a range of proteins. Amber suppression has been used to introduce thia-analogs of methylated and acetylated Lys in histones. Schultz and coworkers site-specifically incorporated phenylselenocysteine in place of Lys9 in histone H3 expressed in *E. coli* [59]. Oxidation to phenylselenic acid followed by spontaneous intramolecular elimination yielded the unsaturated amino acid dehydroalanine (Dha), which underwent Michael-type addition with *N*-acetylated and *N*-methylated 2-aminoethanethiols to yield thialysine analogs of acetylated and methylated Lys. An evolved PylRS-tRNA$_{CUA}$ pair from *Methanosarcina barkeri* also permitted the production of human H2B that was site-specifically crotonylated at Lys11 [67]. Surprisingly, Carell and co-workers found that the *wild-type* pyrrolysine tRNA synthetase from *Methanosarcina mazei* efficiently incorporated ε-*N*-propionyl-, ε-*N*-butyryl-, and ε-*N*-crotonyl Lys in place of Lys9 in H3 [68]. Chin and co-workers directly incorporated *N*-ε-acetyllysine in histone H3 in place of Lys56 and used this reagent to demonstrate increased DNA unwrapping from nucleosomes by acetylation [61]. The stability of histones to strong acids also facilitated the incorporation of methylated lysines protected with a Boc- group in histone H3. Surprisingly, the bulky modified amino acid was an efficient substrate for the *M. barkeri* aaRS-tRNA$_{CUA}$ pair. Deprotection of the Boc- group produced H3 monomethylated at Lys9 that was efficiently bound by an anti-H3-Lys9me antibody and the chromodomain from heterochromatin protein 1β (HP1β) [62].

One challenge for amber suppression has been the inability to obtain sufficient quantities of recombinant H4 from heterologous expression in *E. coli*. As recently demonstrated for the production of H4 Lys16ac, a combination of codon optimization and the use of RNAseE (ribonuclease E) mutant *E. coli* may overcome this limitation and greatly expand the scope of amber suppression in studies of histone modifications [63]. Another genetic strategy that has been used to incorporate methylated Lys residues in H3 is UAG codon reassignment [69]. Release Factor-1 (RF-1) recognizes the termination codons UAA and UAG and signals the critical termination of protein synthesis on ribosomes in response to stop codons. Deletion of RF-1 from the *E. coli* genome facilitates reassignment of the UAG stop codon. Therefore, in RF-1-knockout (RFzero) cells, or a cell-free system based on the RFzero cell lysate, the UAG codon may be entirely repurposed as a sense codon for non-natural amino acids. Sakamoto and co-workers demonstrated the incorporation of a Boc-protected *N*-ε-monomethyllysine analog in H3 by a mutant PylRS-tRNA$_{CUA}$ pair from *M. mazei* at multiple positions specified by the UAG codon [70]. The Boc- groups were subsequently removed with strong acid to generate *N*-ε-monomethyllysine residues at positions 4, 9, 27, 36, and/or 79 of human histone H3. In another example, a hyperacetylated form of H4 acetylated at Lys 5, 8, 12, and 16 was successfully produced in a cell-free lysate [71].

8.7 BIOCHEMICAL AND BIOPHYSICAL STUDIES OF HISTONE UBIQUITYLATION

The small protein Ub plays many critical roles in eukaryotic cellular development. Polyubiquitylation (modification by chains of Ub) of eukaryotic proteins typically marks them for degradation by 26*S* proteasomes [72]. On the other hand, monoubiquitylation is associated with multiple cellular signaling roles. Histones are modified by both mono- and polyubiquitin and our understanding of the roles for their ubiquitylation has grown significantly in the last decade [73]. Although all four core histones and the linker histone H1 are modified by Ub, historically monoubiquitylation at Lys120 in H2B (uH2B) and Lys119 in H2A (uH2A) have received the most attention [74,75]. Interestingly, the discovery of Ub itself occurred concomitant with its discovery as a covalent modifier of H2A. uH2B is associated with DNA damage repair, methylation of H3 Lys4, and Lys79 by the methyltransferases hSet1 (human suppressor of variegation, enhancer of zeste and trithorax (SET) domain containing protein 1) and hDot1L (human disruptor of telomeric silencing-like protein 1) in humans, respectively, and with transcription elongation [76,77]. Recently Lys34 in H2B was also found to be ubiquitylated and to play a role in facilitating transcription elongation [78]. Biochemical studies of ubiquitylated histones are challenged by the dynamic nature and heterogeneity of these modifications in cells. However, the semisynthesis of site-specifically ubiquitylated histones has provided important insights on their structural effects on chromatin and the biochemical crosstalk between ubiquitylation and methylation. This crosstalk is particularly important from a clinical point of view as hDot1L interacts with AF10 (acute lymphocytic leukemia 1-fused gene from chromosome 10 protein), an MLL (mixed lineage leukemia) fusion partner involved in acute myeloid leukemia. Genetic fusions of MLL-hDOT1L and MLL-AF10 result in increased transcription of a number of developmentally important genes, such as *hoxa9*, along with hypermethylation of H3 Lys79 and leukemogenesis [79]. Therefore, understanding how uH2B leads to H3 Lys79 methylation may provide new avenues for therapeutic intervention in blood cancers.

Muir and coworkers reported the first synthetic strategy for peptide ubiquitylation by employing a traceless photoremovable ligation auxiliary that permitted attachment of a C-terminally truncated Ub(1–75)-α-thioester to an H2B C-terminal peptide by a *wild-type* isopeptide linkage (Figure 8.5) [80]. Extension of this strategy to full-length H2B was achieved by ligating the ubiquitylated C-terminal H2B fragment with a recombinant H2B-α-thioester N-terminal fragment using a

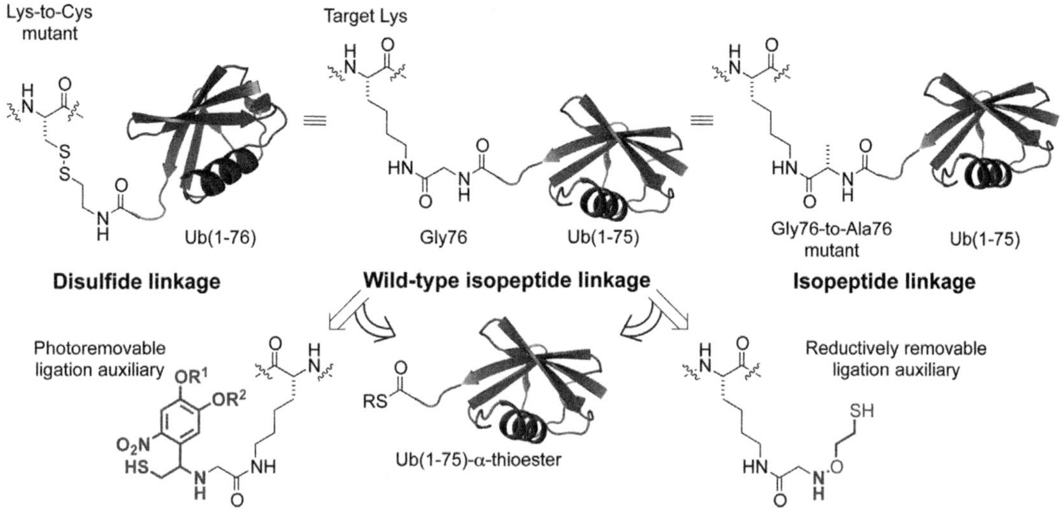

FIGURE 8.5

Chemical strategies to generate *wild-type* ubiquitylated histones and their functional analogs. Two similar strategies for generating *wild-type* ubiquitylated proteins employ temporary native chemical ligation auxiliaries that may be removed by light (bottom left) or by mild reduction with zinc (bottom right). A Gly-to-Cys mutation at the C-terminus of Ub facilitates the synthesis of isopeptide-linked ubiquitylated proteins that differ from the *wild-type* protein by a single Gly-to-Ala mutation at the Ub C-terminus (top right). Disulfide linkage between a site-directed Cys mutant in the target histone and the C-terminus of Ub is a functionally viable alternative to the isopeptide linkage (top left).

standard NCL method [81]. Reconstitution of uH2B in nucleosomes and subsequent biochemical assays with hDot1L revealed that uH2B directly stimulates methylation of H3 Lys79. However, the complex synthesis of the photoremovable auxiliary was a limitation for its broad applicability. This has inspired several alternate approaches to append Ub to peptides and proteins (Figure 8.5) [82–84]. Altogether these approaches have allowed insights on the effects of H2B ubiquitylation on chromatin structure and modification state. For example, structure-activity relationship studies with ubiquitylated nucleosomes revealed that the canonical hydrophobic patch in Ub, which is important for binding Ub-interacting motifs in most proteins [85], was not critical for stimulating hDot1L activity [82]. Surprisingly, studies of the site-specificity of crosstalk between uH2B and hDot1L suggested that multiple sites of ubiquitylation on the nucleosome may stimulate hDot1L activity [83]. The discovery of ubiquitylation at H2B Lys34 and its ability to stimulate hDot1L activity is further proof of the plasticity in crosstalk between Ub and H3 Ly79 methylation. In an extreme example of plasticity in the crosstalk between Ub and histone methylation, Dou and co-workers showed that activation of the methyltransferase activity of MLL at H3 Lys4 is not strictly dependent on uH2B, but may actually be stimulated by N-terminal ubiquitylation of ASH2L (absent, small, or homeotic-like protein 2), a component of the MLL complex [86]. Thus, there are distinct mechanistic differences between the roles for uH2B in methylation by hDot1L and the MLL-complex. For biophysical studies with uH2B, Chatterjee et al. reported a readily reversible disulfide-linked analog (Figure 8.5) [83]. The synthetic accessibility and high yield of this analog permitted a series of biophysical experiments that revealed the inhibitory effect of uH2B on $MgCl_2$-mediated compaction and higher-order fiber formation in nucleosomal arrays [87].

In contrast with uH2B, uH2A is associated with the repression of transcription. It acts to prevent transcription elongation by blocking the release of poised RNA Pol II and facilitates the formation of higher order chromatin structures [88]. Recently, Allis and coworkers demonstrated that semi-synthetic uH2A directly represses H3 K27 methylation in nucleosomes by the polycomb repressing complex 2 (PRC2) while having no effect on hDot1L activity [89]. This revealed important differences in the effects of uH2A and uH2B on methyltransferase enzymes. The semisynthesis of uH2A and uH2B has also resolved some controversy regarding the effect of these modifications on the net stability of nucleosomes. In competition assays with fluorescently labeled histones it was found that uH2A is less effectively incorporated into nucleosomes by the mouse histone chaperone mNap1 (mouse nucleosome assembly protein 1) than unmodified H2A [90]. Further, by employing yeast Nap1 in a chaperone-assisted coupled equilibrium assay, a small nucleosome destabilizing effect was measured for both uH2A and uH2B. This led to the conclusion that uH2A and uH2B have small detrimental effects on nucleosome stability that are only observable in highly sensitive fluorescence-based measurements.

8.8 BIOCHEMICAL AND BIOPHYSICAL STUDIES OF HISTONE METHYLATION

Methylation at Lys in histones was first discovered in 1964 [91] and methylation at protein Arg residues was first reported in 1970 [92]. The side-chains of both residues are capable of undergoing methylation to varying degrees. For example, Lys may be mono-, di-, and trimethylated giving rise to three discrete states (Figure 8.4) [93]. Similarly, methylated Arg may exist in any of three different forms: (1) a single methyl group on a sidechain ω-nitrogen (N^G-methylarginine, MMA), or, the addition of two methyl groups on either (2) the same nitrogen (ω-N^G, N^G-asymmetric dimethylarginine, ADMA), or (3) one each on both ω-nitrogens (ω-N^G, N'^G symmetric dimethylarginine, SDMA) (Figure 8.4). While both mono- and dimethylated arginine are observed *in vitro*, dimethylarginine is the most commonly observed form *in vivo* [94]. The possibility of multiple chemically distinct methylation states at Lys and Arg that can interact with different histone-binding proteins gives rise to combinatorial complexity in signaling. But this also poses a challenge for biochemical studies of methylated histones, since it is difficult to obtain these in any single modification state from natural sources or by enzymatic means. Although Lys and Arg methylation play important roles in cellular signaling, they have been best studied in the context of histone tails, and as modifications of transcription factors that regulate chromatin architecture and transcription state [93].

The first application of a semisynthetic strategy to generate histones with specific methyl marks close to the N-terminus was reported by McCafferty and coworkers [95]. An H3(1−24) peptide C-terminal α-thioester with Lys9me$_3$ was synthesized using Boc-based SPPS. The remainder of H3, namely, H3 (25−135) was obtained by recombinant expression in *E. coli* and an Ala25Cys mutation was introduced at the N-terminus to facilitate NCL. Reaction between the 24-mer peptide α-thioester and H3(25−135)Ala25Cys in the presence of catalytic amounts of thiophenol yielded full-length H3(1−135)Ala25Cys containing Lys9me$_3$. Importantly, the Cys25 mutation was effectively converted to the naturally occurring Ala25 by Raney nickel-mediated desulfurization leading to *wild-type* H3 Lys9me$_3$. This strategy may easily be recapitulated with different methylated peptides to install methylated Lys and Arg residues in histone tails as long as the suitably methylated and Boc- or Fmoc-protected amino acid monomers are accessible. The semisynthesis and simultaneous incorporation of H3 Lys4me$_3$ and H4 Lys16ac in mononucleosomes has also led to the discovery that the two modifications may simultaneously bind the PHD finger and adjacent bromodomain of the NURF (Nucleosome Remodeling Factor) chromatin-remodeling complex subunit BPTF (Bromodomain and PHD finger Transcription Factor), and has demonstrated how combinatorial histone modifications may result in downstream signaling, as originally proposed in the histone code hypothesis [96].

Muir and coworkers highlighted the utility of site-specifically modified histones in interrogating the histone code by generating a chemically defined DNA-barcoded nucleosome library (DNL) [97]. Each library member bore distinct sets of lysine modifications including H3 Lys4me$_3$, Lys9me$_3$, and Lys27me$_3$. The DNL permitted simultaneous biochemical assays with the entire pooled collection of differently modified nucleosomes in solution, and highly sensitive detection of the modified reaction products by *in vitro* ChIP-seq (Figure 8.6). In assays with nuclear extracts from U2OS human osteosarcoma cells, the presence of H3 Lys27me$_3$ or H3 Lys9me$_3$ in nucleosomes was found to inhibit monomethylation at H3 Lys4. Moreover, the presence of H3 Lys4me$_3$ was found to enhance acetylation at H4 Lys14, whereas H3 Lys27me$_3$ had a negative effect on the same acetylation. Thus, the combination of semisynthetic strategies to obtain site-specifically modified histones, in combination with libraries of DNLs, has great potential in future systems-wide investigations of the crosstalk between any number and variety of histone PTMs.

Cys-directed alkylation strategies have also proven useful in generating sufficient quantities of methylated histones for the biophysical characterization of their effects on nucleosome structure. Luger and coworkers used methyllysine analogs to study two functionally opposing histone modifications, H3 Lys79me$_2$ which is enriched at active promoters and H4 Lys20me$_3$ which marks

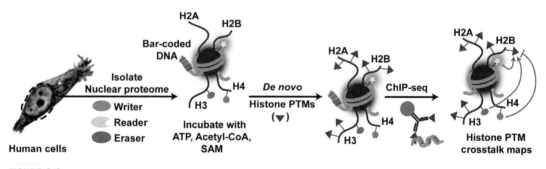

FIGURE 8.6

Systems-level studies of histone PTM crosstalk. Members of a DNA-barcoded library of differently modified nucleosomes (shown as star, ellipse, and hexagon) are incubated with the entire nuclear proteome of a human cell line in the presence of ATP, S-adenosyl methionine (SAM), and acetyl-CoA. The degree of *de novo* chromatin modification (shown as triangles) is analyzed by *in vitro* ChIP-seq in order to obtain histone crosstalk maps. (*This figure is reproduced in color in the color plate section.*)

repetitive elements found in repressed chromatin [98]. Interestingly, while H3 Lys79me$_2$ only induced small changes in local structure and had no effect on chromatin compaction, H4 Lys20me$_3$ led to a more compact chromatin structure in the presence of 1.5 mM MgCl$_2$. This may underlie the high abundance of the latter in pericentric heterochromatin where chromatin is more densely packed. Fujimori and coworkers recently reported the synthesis of thia-analogs of methylated arginines [99]. In their approach, the site-specific installation of all three methylated arginines was accomplished by the chemoselective conjugate addition of a single Cys residue in the histone to α,β-unsaturated amidine scaffolds with varying degrees of N-methylation (Figure 8.4). The analog precursors were reacted with histones bearing H3 Arg2Cys and H4 Arg3Cys mutations to generate a variety of symmetrically and asymmetrically methylated Arg analogs. Satisfyingly, antibodies directed toward the *wild-type* modifications showed good recognition of the methylated analogs, despite the presence of an ε-methylene in place of a nitrogen atom and the resultant loss of planarity at this position in the analogs. Furthermore, by incorporating the asymmetric dimethylated arginine analog in a 15-mer peptide the authors demonstrated binding by the methylarginine-specific Tudor domain-containing effector protein, TDRD3, which is known to bind asymmetrically dimethylated H4 Arg3, albeit about 1.5-fold less effectively. Therefore, while qualitatively similar to the *wild-type* modifications, a degree of caution is necessary when interpreting results with thia-analogs of methylated Arg and Lys in quantitative biochemical assays. For example, in assays with the catalytic domain of the Fe(II)- and α-ketoglutarate-dependent Jumonji histone demethylase, Jmjd2, it was found that the thia-lysine analog of H3 Lys9me$_3$ exhibited $\sim$4-fold slower turnover, but $\sim$5-fold tighter binding when present in the context of an 8-mer H3 tail peptide [100].

8.9 OUTLOOK

DNA-templated processes in eukaryotic cells depend upon the dynamic interplay between histone modifying enzymes (*writers* of the *histone code*), modification-specific binding proteins (*readers*), and enzymes that remove histone modifications (*erasers*). Not surprisingly, the dysregulation of any of these proteins and the processes they control has the potential to result in severe human diseases [101]. For example, mutations in the bromodomain of the nonspecific histone acetyltransferase, p300, are associated with acute myeloid and lymphoblastic leukemias, diffuse large B-cell lymphoma, and cancers of the breast and pancreas. Mutations of the H3 Lys27-specific methyltransferase, EZH2, are associated with myelodysplastic syndromes and myeloproliferative diseases. Mutations of bromodomain and chromodomain containing chromatin readers, which bind acetyl and methyl marks, respectively, are associated with midline, papillary thyroid, and colorectal carcinomas. One particular challenge in ascribing disease phenotypes to a specific epigenetic mechanism, however, is the fact that along with epigenetic mutations, DNA mutations are also widely found in cancer cells. Thus, although recurrent epigenetic mutations that silence tumor suppressor genes or activate *hox* genes implicate epigenetic mechanisms in oncogenesis, by the time a cancer is detected in the clinic it has typically accumulated several genetic and epigenetic defects. Hence it is difficult to establish the mechanistic origins of the disease [102]. Only in a handful of instances such as the H3 Lys27Met mutation, MLL-AF10 fusion, and hypermethylation of the MLH1 (MutL homolog 1) gene promoter region have epigenetic changes been shown to drive human diseases such as pediatric glioblastomas, acute myeloid leukemia, and hereditary non-polyposis colorectal cancer, respectively [103].

A key strength of the chemical and genetic approaches described in this chapter is their ability to facilitate detailed mechanistic studies of epigenetic events. Such investigations have already provided much-needed molecular insights into the direct effects of histone acetylation, methylation, and ubiquitylation on chromatin structure and function [5]. In the future, site-specifically modified nucleosomes could in principle be assembled from any combination of chemically modified core histones and DNA sequence to mimic chromatin at specific promoter sites. When employed in combination with the appropriate biochemical and biophysical experiments, such *designer* nucleosomes and nucleosome arrays may reveal the modification- and site-specific effects of histone PTMs in a density-dependent manner at specific chromosomal locations. In addition to studies of the biochemical crosstalk between various histone modifications with purified enzymes, histone-mediated mechanisms that are implicated in disease progression may also be investigated in complex nuclear proteomes. The ability to reconstitute complex epigenetic signaling events and to employ next-generation DNA-sequencing technologies to measure changes in histone modifications is particularly appealing for screening inhibitors of chromatin-modifying enzymes. Indeed, such strategies could provide an important bridge between purely *in vitro* and cellular phenotype studies, improving the chances of finding small-molecule modulators of histone code *writers*, *binders*, and *erasers*. Finally, as advanced methods for the detection of protein PTMs continue to discover new chemical reactions at histone side-chains, the technologies described in this chapter will become increasingly important for mechanistic studies of the expanding histone code.

ACKNOWLEDGMENTS

We gratefully acknowledge financial support from the Department of Chemistry and the Royalty Research Fund at the University of Washington, Seattle, and the National Institutes of Health (grant 1R01GM110430-01). We apologize to those authors whose excellent work we could not include, and thank our many collaborators for their invaluable contributions to our research on site-specific histone modifications.

REFERENCES

[1] Mirsky AE. The structure of chromatin. Proc Natl Acad Sci U S A 1971;68(12):2945−8.

[2] Kornberg RD. Chromatin structure: a repeating unit of histones and DNA. Science 1974;184 (139):868−71.

[3] Kornberg RD, Thomas JO. Chromatin structure; oligomers of the histones. Science 1974;184 (139):865−8.

[4] Chatterjee C, Muir TW. Chemical approaches for studying histone modifications. J Biol Chem 2010;285 (15):11045−50.

[5] Dhall A, Chatterjee C. Chemical approaches to understand the language of histone modifications. ACS Chem Biol 2011;6(10):987−99.

[6] Kornberg RD, Lorch Y. Interplay between chromatin structure and transcription. Curr Opin Cell Biol 1995;7(3):371−5.

[7] Li B, Carey M, Workman JL. The role of chromatin during transcription. Cell 2007;128(4):707−19.

[8] Smolle M, Workman JL. Transcription-associated histone modifications and cryptic transcription. Biochim Biophys Acta 2013;1829(1):84–97.

[9] Aoki E, Schultz RM. DNA replication in the 1-cell mouse embryo: stimulatory effect of histone acetylation. Zygote 1999;7(2):165–72.

[10] Dorn ES, Cook JG. Nucleosomes in the neighborhood: new roles for chromatin modifications in replication origin control. Epigenetics 2011;6(5):552–9.

[11] Fu H, Maunakea AK, Martin MM, Huang L, Zhang Y, Ryan M, et al. Methylation of histone H3 on lysine 79 associates with a group of replication origins and helps limit DNA replication once per cell cycle. PLoS Genet 2013;9(6):e1003542.

[12] Rogakou EP, Pilch DR, Orr AH, Ivanova VS, Bonner WM. DNA double-stranded breaks induce histone H2AX phosphorylation on serine 139. J Biol Chem 1998;273(10):5858–68.

[13] van Attikum H, Gasser SM. The histone code at DNA breaks: a guide to repair? Nat Rev Mol Cell Biol 2005;6(10):757–65.

[14] Mailand N, Bekker-Jensen S, Faustrup H, Melander F, Bartek J, Lukas C, et al. RNF8 ubiquitylates histones at DNA double-strand breaks and promotes assembly of repair proteins. Cell 2007;131 (5):887–900.

[15] Ahmad K, Henikoff S. Histone H3 variants specify modes of chromatin assembly. Proc Natl Acad Sci USA 2002;99(Suppl. 4):16477–84.

[16] Banaszynski LA, Allis CD, Lewis PW. Histone variants in metazoan development. Dev Cell 2010;19 (5):662–74.

[17] Maze I, Noh KM, Soshnev AA, Allis CD. Every amino acid matters: essential contributions of histone variants to mammalian development and disease. Nat Rev Genet 2014;15(4):259–71.

[18] Strahl BD, Allis CD. The language of covalent histone modifications. Nature 2000;403(6765):41–5.

[19] Jenuwein T, Allis CD. Translating the histone code. Science 2001;293(5532):1074–80.

[20] Herrera JE, Schiltz RL, Bustin M. The accessibility of histone H3 tails in chromatin modulates their acetylation by P300/CBP-associated factor. J Biol Chem 2000;275(17):12994–9.

[21] Davey CA, Sargent DF, Luger K, Maeder AW, Richmond TJ. Solvent mediated interactions in the structure of the nucleosome core particle at 1.9 Å resolution. J Mol Biol 2002;319(5):1097–113.

[22] Kouzarides T. Chromatin modifications and their function. Cell 2007;128(4):693–705.

[23] Linggi BE, Brandt SJ, Sun Z-W, Hiebert SW. Translating the histone code into leukemia. J Cell Biochem 2005;96(5):938–50.

[24] Bamforth SD, Braganca J, Eloranta JJ, Murdoch JN, Marques FI, Kranc KR, et al. Cardiac malformations, adrenal agenesis, neural crest defects and exencephaly in mice lacking Cited2, a new Tfap2 co-activator. Nat Genet 2001;29(4):469–74.

[25] Coffee B, Zhang F, Warren ST, Reines D. Acetylated histones are associated with FMR1 in normal but not fragile X-syndrome cells. Nat Genet 1999;22(1):98–101.

[26] Santos-Rosa H, Caldas C. Chromatin modifier enzymes, the histone code and cancer. Eur J Cancer 2005;41(16):2381–402.

[27] Ota T, Suto S, Katayama H, Han Z-B, Suzuki F, Maeda M, et al. Increased mitotic phosphorylation of histone H3 attributable to AIM-1/Aurora-B overexpression contributes to chromosome number instability. Cancer Res 2002;62(18):5168–77.

[28] Minucci S, Pelicci PG. Histone deacetylase inhibitors and the promise of epigenetic (and more) treatments for cancer. Nat Rev Cancer 2006;6(1):38–51.

[29] Stefansson OA, Esteller M. Epigenetic modifications in breast cancer and their role in personalized medicine. Am J Pathol 2013;183(4):1052–63.

[30] Villar-Garea A, Imhof A. Fine mapping of posttranslational modifications of the linker histone H1 from *Drosophila melanogaster*. PLoS ONE 2008;3(2):e1553.

[31] Garcia BA, Pesavento JJ, Mizzen CA, Kelleher NL. Pervasive combinatorial modification of histone H3 in human cells. Nat Methods 2007;4(6):487−9.

[32] Canzio D, Chang EY, Shankar S, Kuchenbecker KM, Simon MD, Madhani HD, et al. Chromodomain-mediated oligomerization of HP1 suggests a nucleosome-bridging mechanism for heterochromatin assembly. Mol Cell 2011;41(1):67−81.

[33] Abmayr SM, Workman JL. Holding on through DNA replication: histone modification or modifier? Cell 2012;150(5):875−7.

[34] Chi P, Allis CD, Wang GG. Covalent histone modifications − miswritten, misinterpreted and mis-erased in human cancers. Nat Rev Cancer 2010;10(7):457−69.

[35] Suka N, Suka Y, Carmen AA, Wu J, Grunstein M. Highly specific antibodies determine histone acetylation site usage in yeast heterochromatin and euchromatin. Mol Cell 2001;8(2):473−9.

[36] Jackson V. Studies on histone organization in nucleosome using formaldehyde as a reversible cross-linking agent. Cell 1978;15(3):945−54.

[37] Schmidt D, Wilson MD, Spyrou C, Brown GD, Hadfield J, Odom DT. ChIP-seq: using high-throughput sequencing to discover protein−DNA interactions. Methods 2009;48(3):240−8.

[38] Wang Z, Zang C, Rosenfeld JA, Schones DE, Barski A, Cuddapah S, et al. Combinatorial patterns of histone acetylations and methylations in the human genome. Nat Genet 2008;40(7):897−903.

[39] Hagarman JA, Motley MP, Kristjansdottir K, Soloway PD. Coordinate regulation of DNA methylation and H3K27me3 in mouse embryonic stem cells. PLoS One 2013;8(1):e53880.

[40] Murphy PJ, Cipriany BR, Wallin CB, Jud CY, Szetoc K, Hagarman JA, et al. Single-molecule analysis of combinatorial epigenomic states in normal and tumor cells. Proc Natl Acad Sci U S A 2013;110 (19):7772−7.

[41] Voigt P, LeRoy G, Drury III WJ, Zee BM, Son J, Beck DB, et al. Asymmetrically modified nucleosomes. Cell 2012;151(1):181−93.

[42] Smith EL, DeLange RJ, Bonner J. Chemistry and biology of the histones. Physiol Rev 1970;50 (2):159−70.

[43] Allan J, Fey SJ, Cowling GJ, Gould HJ, Maryanka D. Pathway-dependent reconstitution of chromatin structure from separated constituents. J Biol Chem 1979;254(21):11061−5.

[44] Robinson PJ, An W, Routh A, Martino F, Chapman L, Roeder RG, et al. 30 nm chromatin fibre decompaction requires both H4-K16 acetylation and linker histone eviction. J Mol Biol 2008; 381(4):816−25.

[45] Moore SC, Rice P, Iskandar M, Ausió J. Reconstitution of native-like nucleosome core particles from reversed-phase−HPLC-fractionated histones. Biochem J 1997;328(Pt 2):409−14.

[46] Kent SBH. Chemical synthesis of peptides and proteins. Annu Rev Biochem 1988;57:957−89.

[47] Dawson PE, Muir TW, Clark-Lewis I, Kent SB. Synthesis of proteins by native chemical ligation. Science 1994;266(5186):776−9.

[48] Muir TW, Sondhi D, Cole PA. Expressed protein ligation: a general method for protein engineering. Proc Natl Acad Sci U S A 1998;95(12):6705−10.

[49] Camarero JA, Adeva A, Muir TW. 3-thiopropionic acid as a highly versatile multidetachable thioester resin linker. Lett Pept Sci 2000;7(1):17−21.

[50] Zheng JS, Tang S, Huang YC, Liu L. Development of new thioester equivalents for protein chemical synthesis. Acc Chem Res 2013;46(11):2475−84.

[51] Shimko JC, North JA, Bruns AN, Poirier MG, Ottesen JJ. Preparation of fully synthetic histone h3 reveals that acetyl-lysine 56 facilitates protein binding within nucleosomes. J Mol Biol 2011;408 (2):187−204.

[52] Klabunde T, Sharma S, Telenti A, Jacobs W, Sacchettini J. Crystal structure of GyrA intein from *Mycobacterium xenopi* reveals structural basis of protein splicing. Nature Struct Biol 1998;5(1):31−6.

[53] Shah NH, Dann GP, Vila-Perello M, Liu Z, Muir TW. Ultrafast protein splicing is common among cyanobacterial split inteins: implications for protein engineering. J Am Chem Soc 2012;134(28):11338−41.

[54] Xu MQ, Evans Jr. TC. Purification of recombinant proteins from *E. coli* by engineered inteins. Methods Mol Biol 2003;205:43−68.

[55] Piotukh K, Geltinger B, Heinrich N, Gerth F, Beyermann M, Freund C, et al. Directed evolution of sortase A mutants with altered substrate selectivity profiles. J Am Chem Soc 2011;133(44):17536−9.

[56] Bochar DA, Tabernero L, Stauffacher CV, Rodwell VW. Aminoethylcysteine can replace the function of the essential active site lysine of *Pseudomonas mevalonii* 3-hydroxy-3-methylglutaryl coenzyme A reductase. Biochemistry 1999;38(28):8879−83.

[57] Simon M, Chu F, Racki L, de la Cruz CC, Burlingame AL, Panning B, et al. The site-specific installation of methyl-lysine analogs into recombinant histones. Cell 2007;128(5):1003−12.

[58] Li F, Allahverdi A, Yang R, Lua GBJ, Zhang X, Cao Y, et al. A direct method for site-specific protein acetylation. Angew Chem Int Ed Engl 2011;50(41):9611−14.

[59] Guo J, Wang J, Lee JS, Schultz PG. Site-specific incorporation of methyl- and acetyl-lysine analogues into recombinant proteins. Angew Chem Int Ed Engl 2008;47(34):6399−401.

[60] Neumann H, Peak-Chew SY, Chin JW. Genetically encoding N(epsilon)-acetyllysine in recombinant proteins. Nat Chem Biol 2008;4(4):232−4.

[61] Neumann H, Hancock SM, Buning R, Routh A, Chapman L, Somers J, et al. A method for genetically installing site-specific acetylation in recombinant histones defines the effects of H3 K56 acetylation. Mol Cell 2009;36(1):153−63.

[62] Nguyen DP, Alai MMG, Kapadnis PB, Neumann H, Chin JW. Genetically encoding N(epsilon)-methyl-L-lysine in recombinant histones. J Am Chem Soc 2009;131(40):14194−5.

[63] Kallappagoudar S, Dammer EB, Duong DM, Seyfried NT, Lucchesi JC. Expression, purification and proteomic analysis of recombinant histone H4 acetylated at lysine 16. Proteomics 2013;13 (10−11):1687−91.

[64] Chen Y, Sprung R, Tang Y, Ball H, Sangras B, Kim SC, et al. Lysine propionylation and butyrylation are novel post-translational modifications in histones. Mol Cell Proteomics 2007;6(5):812−19.

[65] Tan MJ, Luo H, Lee S, Jin F, Yang JS, Montellier E, et al. Identification of 67 histone marks and histone lysine crotonylation as a new type of histone modification. Cell 2011;146(6):1015−27.

[66] Xie Z, Dai J, Dai L, Tan M, Cheng Z, Wu Y, et al. Lysine succinylation and lysine malonylation in histones. Mol Cell Proteomics 2012;11(5):100−7.

[67] Kim CH, Kang M, Kim HJ, Chatterjee A, Schultz PG. Site-specific incorporation of epsilon-N-crotonyllysine into histones. Angew Chem Int Ed Engl 2012;51(29):7246−9.

[68] Gattner MJ, Vrabel M, Carell T. Synthesis of epsilon-N-propionyl-, epsilon-N-butyryl-, and epsilon-N-crotonyl-lysine containing histone H3 using the pyrrolysine system. Chem Commun 2013;49(4):379−81.

[69] Mukai T, Hayashi A, Iraha F, Sato A, Ohtake K, Yokoyama S, et al. Codon reassignment in the *Escherichia coli* genetic code. Nucleic Acids Res 2010;38(22):8188−95.

[70] Yanagisawa T, Takahashi M, Mukai T, Sato S, Wakamori M, Shirouzu M, et al. Multiple site-specific installations of N(epsilon)-monomethyl-l-lysine into histone proteins by cell-based and cell-free protein synthesis. ChemBioChem 2014;15(12):1830−8.

[71] Mukai T, Yanagisawa T, Ohtake K, Wakamori M, Adachi J, Hino N, et al. Genetic-code evolution for protein synthesis with non-natural amino acids. Biochem Biophys Res Commun 2011;411(4):757−61.

[72] Hershko A, Ciechanover A. The ubiquitin system. Annu Rev Biochem 1998;67:425−79.

[73] Weller CE, Pilkerton ME, Chatterjee C. Chemical strategies to understand the language of ubiquitin signaling. Biopolymers 2014;101:144−55.

[74] Goldknopf IL, Busch H. Isopeptide linkage between nonhistone and histone 2A polypeptides of chromosomal conjugate-protein A24. Proc Natl Acad Sci U S A 1977;74(3):864−8.

[75] West MH, Bonner WM. Histone 2B can be modified by the attachment of ubiquitin. Nucleic Acids Res 1980;8(20):4671−80.

[76] Weake VM, Workman JL. Histone ubiquitination: triggering gene activity. Mol Cell 2008;29 (6):653−63.

[77] Kim J, Guermah M, McGinty RK, Lee JS, Tang Z, Milne TA, et al. RAD6−mediated transcription-coupled H2B ubiquitylation directly stimulates H3K4 methylation in human cells. Cell 2009;137 (3):459−71.

[78] Wu LP, Zee BM, Wang YM, Garcia BA, Dou YL. The RING finger protein MSL2 in the MOF complex is an E3 ubiquitin ligase for H2B K34 and is involved in crosstalk with H3 K4 and K79 methylation. Mol Cell 2011;43(1):132−44.

[79] Okada Y, Feng Q, Lin Y, Jiang Q, Li Y, Coffield VM, et al. hDOT1L links histone methylation to leukemogenesis. Cell 2005;121(2):167−78.

[80] Chatterjee C, McGinty RK, Pellois J-P, Muir TW. Auxiliary-mediated site-specific peptide ubiquitylation. Angew Chem Int Ed Engl 2007;46(16):2814−18.

[81] McGinty R, Kim J, Chatterjee C, Roeder R, Muir T. Chemically ubiquitylated histone H2B stimulates hDot1L-mediated intranucleosomal methylation. Nature 2008.

[82] McGinty RK, Köhn M, Chatterjee C, Chiang KP, Pratt MR, Muir TW. Structure-activity analysis of semisynthetic nucleosomes: mechanistic insights into the stimulation of Dot1L by ubiquitylated histone H2B. ACS Chem Biol 2009;4(11):958−68.

[83] Chatterjee C, McGinty RK, Fierz B, Muir TW. Disulfide-directed histone ubiquitylation reveals plasticity in hDot1L activation. Nat Chem Biol 2010;6(4):267−9.

[84] Weller CE, Huang W, Chatterjee C. Facile synthesis of native and protease-resistant ubiquitylated peptides. ChemBioChem 2014;15(9):1263−7.

[85] Hicke L, Schubert HL, Hill CP. Ubiquitin-binding domains. Nat Rev Mol Cell Biol 2005;6(8):610−21.

[86] Wu L, Lee SY, Zhou B, Nguyen UT, Muir TW, Tan S, et al. ASH2L regulates ubiquitylation signaling to MLL: trans-regulation of H3 K4 methylation in higher eukaryotes. Mol Cell 2013; 49(6):1108−20.

[87] Fierz B, Chatterjee C, McGinty RK, Bar-Dagan M, Raleigh DP, Muir TW. Histone H2B ubiquitylation disrupts local and higher-order chromatin compaction. Nat Chem Biol 2011;7(2):113−19.

[88] Stock JK, Giadrossi S, Casanova M, Brookes E, Vidal M, Koseki H, et al. Ring1-mediated ubiquitination of H2A restrains poised RNA polymerase II at bivalent genes in mouse ES cells. Nat Cell Biol 2007;9 (12):1428−35.

[89] Whitcomb SJ, Fierz B, McGinty RK, Holt M, Ito T, Muir TW, et al. Histone monoubiquitylation position determines specificity and direction of enzymatic cross-talk with histone methyltransferases Dot1L and PRC2. J Biol Chem 2012;287(28):23718−25.

[90] Fierz B, Kilic S, Hieb AR, Luger K, Muir TW. Stability of nucleosomes containing homogenously ubiquitylated H2A and H2B prepared using semisynthesis. J Am Chem Soc 2012;134(48):19548−51.

[91] Allfrey VG, Faulkner R, Mirsky AE. Acetylation and methylation of histones and their possible role in the regulation of RNA synthesis. Proc Natl Acad Sci U S A 1964;51:786−94.

[92] Paik WK, Kim S. Omega-N-methylarginine in protein. J Biol Chem 1970;245(1):88−92.

[93] Slade DJ, Subramanian V, Fuhrmann J, Thompson PR. Chemical and biological methods to detect post-translational modifications of arginine. Biopolymers 2014;101(2):133−43.

[94] Smith BC, Denu JM. Chemical mechanisms of histone lysine and arginine modifications. Biochim Biophys Acta 2009;1789(1):45−57.

[95] He S, Bauman D, Davis JS, Loyola A, Nishioka K, Gronlund JL, et al. Facile synthesis of site-specifically acetylated and methylated histone proteins: reagents for evaluation of the histone code hypothesis. Proc Natl Acad Sci U S A 2003;100(21):12033−8.

[96] Ruthenburg AJ, Li H, Milne TA, Dewell S, McGinty RK, Yuen M, et al. Recognition of a mononucleo-somal histone modification pattern by BPTF via multivalent interactions. Cell 2011;145(5):692−706.

[97] Nguyen UT, Bittova L, Müller MM, Fierz B, David Y, Houck-Loomis B, et al. Accelerated chromatin biochemistry using DNA-barcoded nucleosome libraries. Nat Methods 2014;11(8):834−40.

[98] Lu X, Simon MD, Chodaparambil JV, Hansen JC, Shokat KM, Luger K. The effect of H3K79 dimethy-lation and H4K20 trimethylation on nucleosome and chromatin structure. Nat Struct Mol Biol 2008;15 (10):1122−4.

[99] Le DD, Cortesi AT, Myers SA, Burlingame AL, Fujimori DG. Site-specific and regiospecific installa-tion of methylarginine analogues into recombinant histones and insights into effector protein binding. J Am Chem Soc 2013;135(8):2879−82.

[100] Shiau C, Trnka MJ, Bozicevic A, Ortiz Torres I, Al-Sady B, Burlingame AL, et al. Reconstitution of nucleosome demethylation and catalytic properties of a Jumonji histone demethylase. Chem Biol 2013;20(4):494−9.

[101] Dawson MA, Kouzarides T. Cancer epigenetics: from mechanism to therapy. Cell 2012;150(1):12−27.

[102] Martin DI, Cropley JE, Suter CM. Epigenetics in disease: leader or follower? Epigenetics 2011;6 (7):843−8.

[103] Lewis PW, Müller MM, Koletsky MS, Cordero F, Lin S, Banaszynski LA, et al. Inhibition of PRC2 activity by a gain-of-function H3 mutation found in pediatric glioblastoma. Science 2013;340 (6134):857−61.

PEPTIDE MICROARRAYS FOR PROFILING OF EPIGENETIC TARGETS

9

Antonia Masch[1], Ulf Reimer[2], Johannes Zerweck[2], and Mike Schutkowski[1]

[1]*Department of Enzymology, Institute of Biochemistry & Biotechnology, Martin-Luther-University Halle-Wittenberg, Halle, Germany* [2]*JPT Peptide Technologies GmbH, Berlin, Germany*

CHAPTER OUTLINE

9.1 INTRODUCTION

Epigenetic modulation takes place mainly at the nucleosome. The nucleosome is composed of four pairs of histone molecules (H2A, H2B, H3, and H4) forming an octamer, which is wrapped 2.5 times with DNA. Binding of histone H1 between the nucleosomes stabilizes the chromatin structure. Modification of the histone proteins could either lead to interference with the DNA interaction or to the condensation into the chromosome [1−5]. Particularly interesting are the N-terminal tails of the histone proteins reaching out of the histone octamer. At this N-terminal tail most of the identified posttranslational modifications (PTMs) are located. Enzymes transferring or removing such PTMs are introduced as writers or erasers of epigenetic marks, respectively. The resulting decoration of the histone proteins with PTMs results in a histone code which could be read-out by the binding to protein domains (readers) specifically recognizing different PTMs or combinations of

Y.G. Zheng (Ed): Epigenetic Technological Applications. DOI: http://dx.doi.org/10.1016/B978-0-12-801080-8.00009-0

PTMs. Often, PTMs could influence the substrate recognition of writers and erasers known as crosstalk [6−9]. Some of these modifications are relatively stable like histone H3 methylation at lysine residues 9 or 27 [10], but other modifications like histone acetylation and phosphorylation are transient and more dynamic [11,12].

Besides well-known phosphorylations of serine, threonine, and tyrosine residues, methylations of arginines and lysines, and acetylations of lysines, in the last few years a number of other acylations of lysines have been identified in histones, like propionylation, butyrylation, malonylation, succinylation, glutarylation, crotonoylation, or hydroxyisobutyrylation [13−20].

Systematic analysis of histone modifications is difficult because of the high number of theoretically possible combinations of PTMs. Most of the research is focusing on the N-terminal tails of the histones H3 and H4 (reviewed in [21]). Resin-bound histone H4 tail libraries were used for deciphering the histone code [22]. An unbiased 5000-member combinatorial library representing all possible combinations of modifications in the histone H3 N-terminal 10-meric peptide was used to profile PTM crosstalk for six different histone code readers known to recognize the methylation status of H3K4 [23]. Arrays of 384 microspheres coated with peptide-cellulose conjugates derived from histones were used to characterize a panel of 40 antibodies directed against differently modified histones [24]. Moreover, it is assumed that the histone core region does not play an important role in interaction with other molecules and that nucleosome structure does not allow efficient writing or reading of PTMs. However, recent work has shown that phosphorylation of core histone sites have cell-cycle-dependent half-lives varying dramatically between interphase and mitosis [25]. Additionally, Wee1-mediated phosphorylation of histone H2B at Tyr 37 at the border of the histone core [26], methylation of histone H3 on Lys 79 (H3K79) mediated by lysine methyltransferase Dot1 [27], or acetylation of H3K122 by p300/CBP (binding protein 300/CREB-binding protein) [28] were reported, demonstrating that histone core regions are accessible for enzymatic modifications.

We developed a high-content histone peptide microarray for the systematic profiling of binding specificities of readers and substrate specificities of writers/erasers of the histone code. Recently, we used very similar peptide microarray technology for the analysis of human SH2- and SH3- (Src Homology 2 and 3) reader domains with microarrays displaying 6202 phosphotyrosine-containing peptides [29] and 9192 prolin-rich peptides [30], respectively. The profiling of enzymatic activities on peptide microarrays is possible as shown by us for erasers like all human sirtuin isoforms [31,32] and for writers like protein kinases [33−37] and methyltransferases [38].

The high content histone peptide microarray presented here displays 3874 purified peptides in triplicates resulting in more than 11,000 data points per experiment. The preparation of the peptides by SPOT-synthesis and the purification via chemoselective immobilization of the N-terminal end onto modified glass slides is described in [32]. The knowledge-based part of the histone microarray displays 20-meric peptides of the histones H2A, H2B, H3, and H4 for which multiple sequence variants are observed (i.e. H3.1, H3.2, H3.3). Here the PTMs at all reported sites as well as the known PTM combinations (up to six PTMs per peptide) are presented, too. We focused at acetylated (Kac), propionylated (Kprop), butyrylated (Kbut), methylated (mono-, di-, and tri-methylated, Kme1, Kme2, and Kme3, respectively) lysine residues, mono- and di-methylated arginine residues (Rme1, Rme2a, Rme2s for symmetrically and asymmetrically dimethylated arginines, respectively), arginine-citrulline exchange as well as phosphorylated serine (Sp), threonine (Tp), and tyrosine (Yp) residues. Furthermore, the knowledge-based part of the microarray contains overlapping peptides derived from histone H1 together with all PTMs published by Wisniewski et al. [39]. We added a systematic part to the presented library containing scans through histones H2A, H2B, H3, and H4 and their isoforms (i.e. H3.1, H3.2, H3.3) with single modifications of all PTMs mentioned above at all possible sites and additionally malonylated (Kmal) and succinylated (Ksucc) lysine residues to the histone peptide microarray to enable the identification of novel histone-based enzyme substrates or binding sites. All peptides derived from the N-termini of the histones are immobilized alternatively via their C-termini to address the question whether free N-terminus is needed for effective interaction with the target protein.

This content is clearly different from and more comprehensive than the histone peptide arrays published so far, such as histone peptides immobilized on cellulose membranes [40−43], biotinylated histone peptides immobilized on streptavidin-coated glass slides [44−49], biotinylated peptide−neutravidin complexes immobilized on activated glass slides [50], or peptide-cellulose conjugates immobilized on glass slides [41,51−54] or nitrocellulose-coated glass slides [42].

In this chapter we would like to introduce the histone peptide microarray platform as a tool to profile epigenetic targets efficiently and comprehensively. We will demonstrate that both the profiling of binding specificities of histone code readers and the determination of substrate specificity of histone code writers and erasers is possible using the same technology. Additionally, we will show that the crosstalk between different PTMs could be uncovered for enzymatic activities.

9.2 APPLICATIONS OF HISTONE PEPTIDE MICROARRAYS

9.2.1 PROFILING OF BINDING SPECIFICITY OF HISTONE CODE READERS

Binding of target protein to immobilized peptides could be visualized by three different read-out methods. The first approach uses antibodies that are specific for the target reader protein. Subsequent to treatment of the peptide microarray with the target-protein, the anti-target-protein-antibody followed by a fluorescently labeled secondary antibody is used to generate a fluorescence signal on positions of the microarray where the reader is bound to an immobilized peptide. The epitope of the anti-target-protein-antibody should be known to avoid any interference with the recognition of the immobilized peptide. Alternatively, the anti-target-protein-antibody itself could be fluorescently labeled, like the detection of anti-histone-peptide-antibodies binding using anti-IgG (anti-immunoglobulin) antibodies or profiling of pan-specific antibodies [32,55] on peptide microarrays (Figure 9.1A).

The second approach uses the chemical labeling of the target reader protein either by fluorophores enabling direct read-out by fluorescence scanning [56] or with biotin enabling read-out subsequent to treatment with fluorescently labeled streptavidin or avidin. Very often, chemical labeling using lysine or cysteine side chains is not regio-selective, resulting in a complex mixture of differently labeled reader protein molecules. This could be circumvented by regio-selective labeling using genetic methods like the introduction of GST-fusions [29,30], FLAG- (Figure 9.1B), Myc-, His-, and HA-tags or biotinylation sites. Specific anti-tag-antibodies or fluorescently labeled streptavidin/avidin enable a read-out by fluorescence scanning. The third approach uses a protein kinase to phosphorylate the GST-fusion of the reader domain in the presence of radio-isotopically labeled

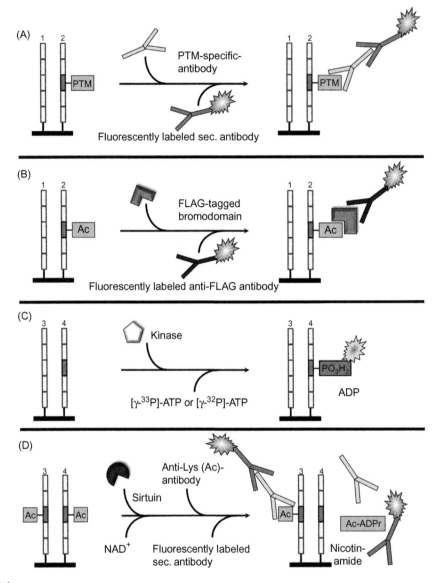

FIGURE 9.1

Assay principles on the histone peptide microarrays described in this chapter. (A) Binding of PTM-specific antibodies, like anti-succinyl-lysine-antibodies, followed by fluorescently labeled secondary antibodies can be used to determine subsite and acyl specificity. (B) Determination of subsite and acyl specificity of histone code reading domains using FLAG-tagged bromodomains in combination with fluorescently labeled anti-FLAG-antibody. (C) Enzymatic activity of kinases can be profiled using radio-isotopically labeled ATP, leading to phosphorylated substrate and the release of ADP. Incorporated $[\gamma\text{-}^{32}P]$- or $[\gamma\text{-}^{33}P]$-phosphate could be visualized by phosphorimaging. (D) Sirtuins are profiled in presence of co-substrate NAD^+ followed by treatment with anti-acetyl-lysine-antibodies and fluorescently labeled secondary antibodies resulting in signal decrease due to transfer of the acetyl residue from the immobilized peptide to the ADP-ribose moiety of the co-substrate forming acetylated ADP-ribose (Ac-ADPr) and nicotinamide. (1 = non-binder, 2 = binder, 3 = no substrate, 4 = substrate).

ATP leading to radioactive spots on the microarray at positions of bound reader which could be detected by phosphorimaging or exposure to X-ray film.

9.2.1.1 *Profiling of subsite and acyl specificity of PTM-specific antibodies*

Acetylation of lysine side chains in proteins is one of the most abundant PTMs in all the domains of life. Recently, additional acylations of lysine side chains were found *in vivo*, such as malonylation [15], succinylation [15], and glutarylation [20]. Identification of several thousand succinylation sites in *Escherichia coli*, *S. cerevisiaem* and human cells [17,57−59] underline that succinylation is a frequently occurring modification of lysine residues in all domains of life and that this acylation overlaps with acetylation [17]. Most of the bottom-up proteomics reports on protein succinylation make use of a polyclonal anti-succinyl-lysine-antibody (anti-Ksucc-Ab) which is claimed to be specific for the succinyl moiety but pan-specific (which means not specific) for the amino acid residues surrounding the succinylated lysine residue (anti-Ksucc-Ab of PTM Biolabs, Chicago, USA). This antibody was produced by immunizing rabbits with chemically succinylated bovine serum albumin and purified by succinyl-lysine agarose affinity chromatography. We profiled the binding specificity of this anti-Ksucc-Ab by treatment with the histone peptide microarray (2.5 μg/mL in 5% bovine serum albumin) at 4°C overnight. Subsequent to the washing steps the histone peptide microarray was incubated with fluorescently labeled secondary antibody followed by fluorescence scanning at 10 μm resolution. Image processing resulted in a list of peptide sequences with respective fluorescence intensities.

We focused here on two different aspects: first, the subsite specificity of the anti-Ksucc-Ab, which means if there is any preference for amino acid residues in different positions relative to the succinylated lysine, and second, specificity for the chemical nature of the acyl residue (acyl specificity). Generally, acyl specificity has two aspects, either the cross-reactivity vs. alternative acyl modifications or the discrimination between acylated and non-acylated peptides.

To simplify the presentation of the data we focused here on histone H3 containing 13 lysine residues (Figure 9.2A), because histone H3 and its modifications are the most studied among the histone proteins [52,54,60]. Nevertheless, the same experiments yielded similar data sets for all other histone isoforms. In Figure 9.2A the corresponding fluorescence signals for all 13 succinylated lysines within the histone H3 are shown. The maximum signal (H3_K56succ) was set to 100% for normalization of the results.

All succinylated lysine residues are recognized successfully by the anti-Ksucc-Ab with at least 25% of the maximum signal for H3_K37succ. Seven out of the 13 succinyl-lysine containing peptides show signals with more than 60% signal intensity as compared to the succinylated H3_K56 site, demonstrating that this pan-specific antibody has nearly no subsite specificity.

Calculating the ratios of signal intensities for the different histone H3 peptides in the succinylated form and nonmodified form, almost half of the presented peptides were recognized more than 30-fold better in the succinylated form up to a ratio of almost 120 for the H3_K79succ peptide. For all histone peptides a more than ten fold preference (as the ratio of signal intensities of succinylated lysine to nonmodified peptide) of the succinylated variants was found using anti-Ksucc-Ab from PTM Biolabs (data not shown).

To profile the acyl specificity of anti-Ksucc-Ab in more detail, we plotted the detected relative signal intensities for the lysine residues (Figure 9.2B) known to be succinylated in human histone H3 protein [61]. Signal intensity for H3_K122succ was set to 100%. Figure 9.2B clearly

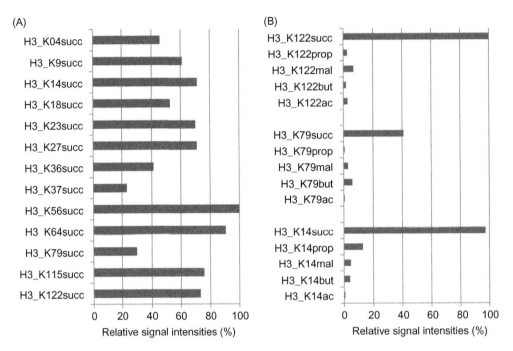

FIGURE 9.2

Subsite and acyl specificity of anti-succinyl-lysine-antibody. (A) Relative signal intensities subsequent to fluorescence scanning are shown for all histone H3-derived peptides containing succinylated lysines. Signal intensity for H3_K56succ was set to 100%. (B) Relative signal intensities subsequent to fluorescence scanning are shown for histone H3-derived peptides containing different acyl residues on the respective lysine. Signal intensity for H3_K122succ was set to 100%.

demonstrates that the recognition of the succinylated lysine residue is acyl specific because alternative acyl residues in histone H3-derived peptides like closely related malonyl, or acetyl, propionyl and butyryl residues yielded about 10−100-fold lower signal intensities.

In summary, the anti-Ksucc-Ab recognized all succinylated lysine residues we presented on the histone peptide microarray with high specificity for succinyl residues as acyl moiety. Using a similar assay principle (Figure 9.1A) we profiled commercially available polyclonal and monoclonal anti-acetyl-lysine-antibodies on peptide microarrays [32] including the histone peptide microarray and found a subsite specificity for all of them. Nevertheless, a combination of three anti-Kac-Abs [32] enables the reliable recognition of all acetylated peptides displayed on the histone microarray. The acyl specificity is high for this antibody mixture because, besides the acetylated lysine residues, only propionylated residues are recognized but with about ten fold lower signal intensity. No binding to butyrylated, malonylated and succinylated peptides could be detected using a concentration of 0.5 µg/mL for each individual antibody within the mixture.

Histone peptide microarrays could be used for quality control of antibodies directed against PTMs in a defined histone protein. We profiled several commercially available antibodies, some of them often used for chromatin immunoprecipitation experiments (Abcam: H3K9me1, H3K9me2; Diagenode: H3K9me3; Millipore: H4K16ac, H3S28ph, H2A/H4S1ph; Santa Cruz: H4K16ac; Active Motif: H3K4me1, H3K4me2, H3K27me3, H3K36me1, H3K36me2, H3K36me3, data not shown). Several of the antibodies showed suboptimal subsite specificity, as for example anti-H3K36me1-Ab. This antibody generated fluorescence signals for H2AK126me1 and H1K37me1 peptides with similar intensities as compared to the target site H3K36me1. Moreover, at least on our peptide microarray, the binding to the off-target H3K64me1 seems to be better than to the H3K36me1 target peptide.

9.2.1.2 Profiling of subsite and acyl specificity of bromodomains

Bromodomains represent an evolutionarily conserved protein fold which is involved in the lysine-acetylation dependent assembly of protein complexes regulating transcription of genes. The 61 different bromodomains found in the human proteome are located in 46 proteins and several of them represent interesting pharmaceutical targets for the treatment of inflammatory diseases, viral infections and cancer [62].

Bromodomain-containing protein 4 (BRD4) acts as a transcriptional regulator [63] and contains two N-terminal bromodomains, BRD4 (1) and BRD4 (2). BRD4 associates with chromatin through the bromodomains which are claimed to recognize acetylated lysine residues in histones. BRD4 has been identified recently as a therapeutic target in many cancers, including acute myeloid leukemia, multiple myeloma, colon cancer and breast cancer [64,65].

The recombinant BRD4 proteins, BRD4 (1) (amino acid residues 44−168) and BRD4 (2) (amino acid residues 333−460), were purchased from Active Motif (Carlsbad, USA). The recombinant proteins contain an N-terminal His-tag and a C-terminal Flag-tag. We profiled the subsite and acyl specificity of these two bromodomains onto the histone peptide microarray using 20 μg of target reader protein per assay (Figure 9.1B).

Figures 9.3A and 9.3B illustrate the subsite specificity of BRD4 (1) and BRD4 (2), respectively, for the amino acid sequence around the 11 lysines found in histone H4. Obviously, there is a preference in binding of BRD (1) to H4K91ac, followed by H4K44ac and H4K59ac. The same top binders and similar signal intensities were observed for BRD4 (2) as depicted in Figure 9.3B. This result is very similar to a recent work using a limited number of histone peptides immobilized on cellulose membranes for profiling the subsite specificity of BRD4 (1) and BRD4 (2) [43]. Fillipakopoulos and coworkers also obtained the H4K91ac and H4K44ac peptides as top binder for both BRD4 bromodomains, but detected a high signal for H4K79ac, too. This difference could be caused by the extremely different loadings of the two different surfaces. SPOT-membranes are loaded in a range of 5 nmol peptide per mm^2 which is in sharp contrast to loadings reported for peptide microarrays (15−250 fmol per mm^2). Based on this difference binding events with binding constants higher than about 100 μM could not be detected using the peptide microarray approach. Protein-peptide interactions with much lower binding affinities (binding constants up to 1 mM) could be investigated on cellulose membranes produced by SPOT-synthesis because virtual peptide concentration within the spots during the assay is about 500 mM.

Figures 9.3C and 9.3D show the acetyl selectivity of the BRD4 (1) and BRD (2) recognition, respectively. We compared the signal intensities of the histone H4 derived, lysine containing peptides with the respective signal intensities for the acetylated counterparts resulting in signal ratios between 0.5 (H4K77) and 2 (H4K44) for BRD4 (1). Thus, no clear acetyl selectivity could be

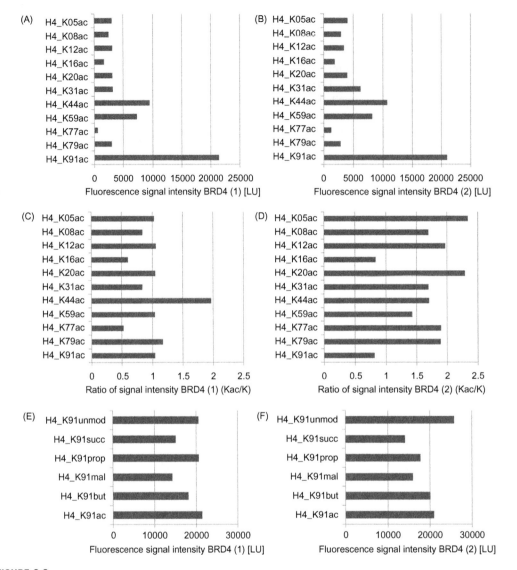

FIGURE 9.3

Subsite and acyl specificity of bromodomains BRD4 (1) and BRD4 (2). (A) Signal intensities subsequent incubation with BRD4 (1) and fluorescence scanning are shown for all histone H4-derived peptides containing acetylated lysines. (B) Signal intensities subsequent incubation with BRD4 (2) and fluorescence scanning are shown for all histone H4-derived peptides containing acetylated lysines. (C) Ratio of signal intensities is shown for histone H4-derived peptide pairs (acetylated lysine/non-modified lysine) after incubation with BRD4 (1). (D) Ratio of signal intensities is shown for histone H4-derived peptide pairs (acetylated lysine/nonmodified lysine) after incubation with BRD4 (2). (E) Signal intensities subsequent incubation with BRD4 (1) and fluorescence scanning are shown for all histone H4K91-derived peptides containing differently acylated lysines and nonmodified lysine. (F) Signal intensities subsequent incubation with BRD4 (2) and fluorescence scanning are shown for all histone H4K91-derived peptides containing differently acylated lysines and nonmodified lysine.

found for BRD4 (1). Similar results were obtained for BRD4 (2) although here in general signal intensities for the acetylated peptides were slightly higher, resulting in signal ratios higher than 1.5 for most of the peptide pairs (Figure 9.3D). For the best binder H4K91ac we collected the corresponding signals for alternative acyl residues (Figure 9.3E and 9.3F). Both bromodomains show no significant difference in signal intensities for the differently acylated lysine residues. Taken together, it seems that BRD4 bromodomains have subsite specificity but no clear acyl specificity. This corresponds to the conclusions drawn by Fillipakopoulos et al. using histone peptide arrays on cellulose that the Kac-mark only weakly contributes to the binding affinity of BRD proteins to their target sites [43].

9.2.2 PROFILING OF SUBSTRATE SPECIFICITY OF HISTONE CODE WRITERS

Histone-code-writing enzymes, like kinases or methyl-, acetyl-, glycosyl-, and ADP-ribosyl-transferases, are transferring moieties from a co-substrate onto the target histone site, resulting in a modified histone and a co-product. Generally, there are two strategies, differing in the nature of the co-substrate, used for the read-out of enzymatic activities on peptide microarrays comprehensively reviewed in [66,67] (Figure 9.4).

For the majority of the used assays on peptide microarrays the enzyme is incubated with a tagged co-substrate, like radio-isotopically labeled ATP, acetyl-CoA (acetyl-coenzyme A), S-adenosylmethionine or NAD^+ (nicotinamide adenine dinucleotide) for profiling of kinases, acetyltransferases, methyltransferases, or (poly)ADP-ribosyltransferases, respectively. Incorporation of radioactivity could be detected by phosphorimaging or exposure to X-ray film. Alternatively, biotinylated ATP [68] or NAD^+ as co-substrate were used for the profiling of kinases or ADP-ribosyltransferases subsequent to treatment with fluorescently labeled streptavidin. In addition, ferrocene-conjugated ATP [69] or γ-thio-ATP [70] in combination with electrochemical detection or dansylated ATP in combination with fluorescence read-out was applied for protein kinase profiling [71]. In the case of ADP-ribosyl-transferases use of etheno-NAD^+ in combination with highly specific anti-etheno-NAD^+-antibodies generates robust signals on peptide microarrays (unpublished results).

If the writer is incubated with the peptide microarray in the presence of nonmodified co-substrate, a reagent specific for the incorporated PTM is needed for facilitating read-out of enzymatic activity (Figure 9.4). This reagent could be either a protein (an anti-PTM-antibody, a lectin or a PTM-specific reader domain), a small molecule, specifically recognizing the PTM introduced into the immobilized peptide (i.e. fluorescently labeled [72] or biotinylated phosphate monoester chelators [73,74]), or a chemical reaction specific for the PTM. One example for the use of a PTM-specific antihistone antibody as a sensor for the enzymatic activity of arginine methyltransferase PRMT5-MEP50 (protein arginine *N*-methyltransferase5-methylosome protein 50) using a histone peptide microarray was described recently [38]. A chemical reaction specific for a PTM is the transformation of phosphoserine/phosphothreonine-containing peptides generated by the action of a serine/threonine protein kinase into dehydroalanine- or 2-amino-dehydrobutyryl derivatives, respectively, by base catalyzed β-elimination [75]. The resulting double bond is unique within the immobilized peptide and could be used for chemoselective labeling with a fluorophore or a biotin moiety. Shults et al. reported carbodiimide-mediated formation of a covalent bond between a fluorescence dye derivative and the phosphate monoester moiety subsequent to kinase reaction [76].

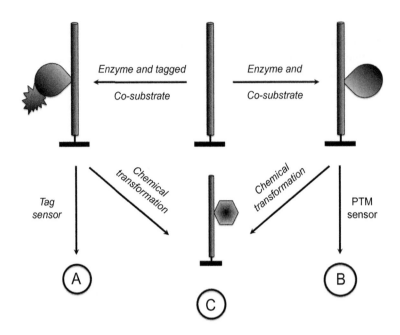

FIGURE 9.4

Different approaches for profiling enzymatic activities on peptide microarrays. (A) Enzyme-mediated transfer of a tagged moiety to the immobilized peptide creates a tagged peptide which could be detected using a reagent or method sensing the tag. Often, radio-isotopically labeled co-substrates like $[\gamma\text{-}^{32}\text{P}]$-ATP or $[\gamma\text{-}^{32}\text{P}]$-NAD$^+$ in combination with phosphorimaging are used. Alternatively, ferrocene-labeled ATP or biotinylated ATP enables electrochemical or surface plasmon resonance/fluorescence read-out, respectively, for detection of kinase activities on peptide microarrays. Use of etheno-NAD$^+$ in combination with anti-etheno-NAD$^+$-antibody allowed fluorescence read-out of ADP-ribosyl-transferase activities. (B) Enzyme in the presence of non-modified co-substrate generates modified peptides on the microarray which could be recognized by PTM-specific reagents (PTM sensor) like phospho-specific antibodies or phospho-specific fluorescence dyes. Bound PTM sensor can be detected directly by fluorescence if the sensor itself is fluorescently labeled, or indirectly by incubation with secondary reagents with interact specifically with the bound PTM sensor. (C) Modified PTMs introduced by the use of tagged co-substrates could be transformed by a chemical reaction into peptide derivatives which react chemoselectively with fluorophores allowing read-out by fluorescence scanning. One example is the kinase-mediated transfer of thio phosphate residues to microarray-bound peptides followed by chemical transformation into biotinylated derivatives via Michael addition. Incorporated biotin could be visualized using fluorescently labeled (strept)avidin. Alternatively, the PTM itself could be transformed into unique structural elements within a peptide. Base catalyzed β-elimination of phosphorylated threonine or serine residues generates 2-amino-dehydro butyric acid residues or dehydroalanines, respectively, which could be used for chemoselective addition of fluorophores or biotin residues. Moreover, carbodiimide-mediated selective formation of a covalent bond between a fluorescence dye derivative and a phosphate moiety in microarray-immobilized peptides was reported.

Phosphorylation of histone proteins represents a widespread modification, often inducing or preventing the activity of other writers. Additionally, the created phosphorylated histone sites could be recognized by phospho-specific readers like SH2-domains or 14-3-3 domains (reviewed in [25]). Growth factor and stress-mediated activation of p38-MAPK pathways lead to activation of MSK1 and MSK2 kinase recognizing serine residues in the histone H3 ARKS-motifs [77]. This motif is found two times in histone H3 resulting in MSK1-mediated phosphorylation of S10 and S28. These modifications are involved in the regulation of the transcription of genes like *Jun* and *Fos* which were shown to be up-regulated in various cancers. The involvement of deregulated MSK1 kinase activity (Life Technologies, Carlsbad, USA) in human cancers makes the kinase itself an attractive target for drug discovery [78].

We profiled MSK1 activity in the presence of $[\gamma\text{-}^{33}\text{P}]$-ATP using histone peptide microarray and detected besides the known sites H2AS1 [79], H3S10, and H3S28 [77] additional sites in histone H2B (S37) and histone H4 (S48). At least histone H2B phosphorylation of the protein could be demonstrated using incorporation of radio-isotopically labeled ATP for detection but no information about the phosphorylation site was given [80]. The signals for the sites in histones H2B and H4 are about five times stronger than for the H3S10 site. All sites contain the arginine residue in the -2 position relative to the phosphate accepting serine residue and modification of this arginine by methylation or replacement by a citrulline dramatically impede the kinase-mediated phosphorylation of the immobilized peptides (data not shown).

Wee1 is a nuclear tyrosine kinase that negatively regulates the activity of the cyclin-dependent kinase Cdc2 preventing mitotic entry before completion of DNA synthesis [26]. Mahajan et al. have found that phosphorylation of H2B on tyrosine residue 37 by Wee1 suppresses the expression of replication-dependent core histone genes. Their work presents the first tyrosine phosphorylation site in H2B. We profiled Wee1 (Invitrogen, Grand Island, USA) using the histone peptide microarray to analyze the crosstalk between different PTMs. We detected robust signals using either $[\gamma\text{-}^{33}\text{P}]$-ATP (Figure 9.1C) in combination with phosphorimaging or anti-phosphotyrosine-antibody in combination with fluorescence imaging.

Exemplarily, Figure 9.5 shows the signal intensities of phosphorimaging for differently modified histone H2B [30−37,39−50] peptides containing the tyrosine residue 37. Similar data could be generated for the other detected phosphorylation sites (data not shown). A crosstalk between the modification at lysine 43 and the ability of the kinase to phosphorylate tyrosine 37 could be uncovered. While acetylation or monomethylation of K43 did not change the signal intensity for the phosphorylation, decoration of K43 with other acyl residues like malonyl-, succinyl-, propionyl-, and butyryl- residues improves the substrate properties. In contrast, introduction of Kme2 or Kme3 at position 43 seems to abrogate Y37 phosphorylation. One could argue that K43 in the +6 position is too far away to cause crosstalk, but it was demonstrated (using peptide microarrays displaying systematic substitutions of a substrate) that kinases are able to recognize amino acid residues up to +8 position specifically [36]. Priming phosphorylations in the surroundings of the Y37 site (Y40, Y42, S36, and S38) resulted in higher signals for the Wee1-mediated phosphorylation. Such priming phosphorylations are systematically identified for other kinases using a (phospho)peptide microarray approach [37].

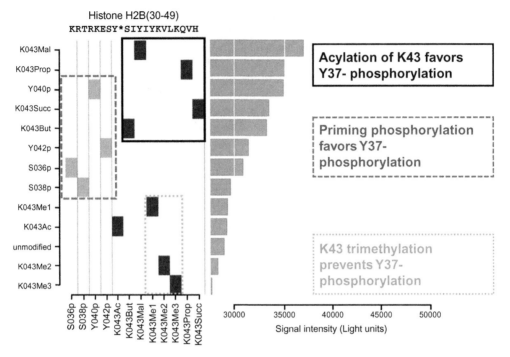

FIGURE 9.5

Wee1 kinase activity is modulated by crosstalk with PTMs. Histone peptide microarray was incubated with Wee1 kinase and [γ-^{33}P]-ATP. Read-out was performed by phosphorimaging. Data for several histone H2B peptides covering sequence [30−37,39−50] are shown. Identified phosphorylation site is indicated by Y*. Left panel shows a selection of PTMs present in the peptides. The boxes indicate the presence of the PTM at a given position. In the histogram on the right panel, Wee1 activity (signal in phosphorimage) is shown for each peptide.

9.2.3 PROFILING OF SUBSTRATE SPECIFICITY OF HISTONE CODE ERASERS

Erasers such as phosphatases, protein arginine deimidases, lysine demethylases, and lysine deac(et)-ylases are the counteracting enzymes to the appropriate writers, removing modifications from histone proteins thereby either generating the starting material or introducing novel modifications (i.e. citrulline residues instead of (methylated) arginine introduced by protein arginine deimidases or aldehyde functions instead of methylated lysines introduced by lysyl oxidases [81]). Very often, assay principles similar to the detection of the histone code writers could be used but in opposite direction. Nevertheless, the content of the microarray has to be changed, displaying the chemically modified peptides as substrates.

Sirtuins are NAD$^+$-dependent lysine deac(et)ylases with additional ADP-ribosylation activity for some isoforms [82,83]. They regulate enzyme activity, metabolic processes and stress response and play a major role in life span, metabolism, and age-related diseases. Some sirtuins are considered to be tumor suppressors [84−86]. The subsite specificity of all seven human sirtuin isoforms was analyzed comprehensively using peptide microarrays displaying more than 6800 peptides derived from human acetylation sites [32]. For each sirtuin, a well-defined subsite specificity could be found, allowing the conclusion that sirtuins should be susceptible to crosstalk. PTMs often dramatically change the chemical nature of an amino acid residue. Especially, phosphorylation of serine or threonine residues represents such a dramatic change converting a residue with a small neutral side chain into a larger double negatively charged one. Sirtuins represent important regulators of the acetylation status of defined sites in histones, including H3K9 [83,87], which contain S10 and T11 as phosphate acceptor sites.

We profiled sirtuin 3 activity on the histone peptide microarray using the optimized anti-Kac-Ab mix [32] and fluorescently labeled secondary antibody for signal read-out (Figure 9.1D). Because we have to measure signal decrease, a negative control has to be made to define the start signal for each peptide. We used similar assay conditions but without the co-substrate NAD$^+$ as a control experiment. The ratio of the signal intensities without and with co-substrate defines substrate quality.

Figure 9.6 displays the activity of sirtuin 3 for N-terminal histone H3 peptides containing the biologically important H3K9ac site. As expected, a dramatic impact on deacetylase activity by adjacent phosphorylations is visible. It does not matter whether single phosphorylated residues

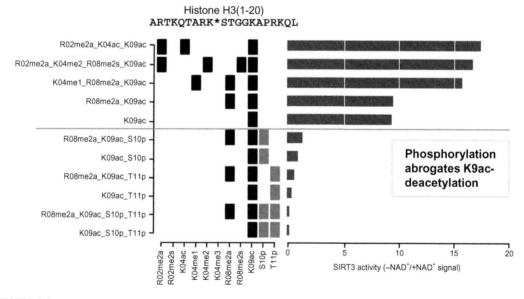

FIGURE 9.6

Sirtuin 3 activity on H3K9ac-site is influenced by crosstalk with PTMs. Histone peptide microarray was incubated with sirtuin 3 in the presence and absence of co-substrate NAD$^+$. Read-out was performed by anti-acetyl-lysine antibodies and fluorescently labeled secondary antibodies. Identified deacetylation site is indicated by K*. Data for several N-terminal histone H3 peptides covering sequence [1−20] are shown. Left panel shows a selection of PTMs present in the peptide. The boxes indicate the presence of the PTM at a given position. In the histogram on the right panel, Sirt3 relative activity (ratio of signals w/o or with NAD$^+$) is shown for each peptide.

Ser10 or Thr11 or both together are phosphorylated, the Lys9-deacetylation is strongly abrogated by the factor of 9, compared to the control peptide (H3K09ac). In contrast, deacetylation activity can be improved by specific combinations of acetylated lysines and methylated lysine/arginine residues in the N-terminal part of the substrate.

9.3 CONCLUSION AND OUTLOOK

The peptide microarray platform presented here enables rapid and efficient profiling of binding specificities of histone code readers and the identification of substrates for enzymes acting on histones including sequences within the core region of the histones. The presentation of all possible histone-derived peptides allows for the first time the discovery of novel substrates or binding sites for histone code writers/erases or readers, respectively. The additional peptides on the microarray with chemically introduced PTMs, including recently discovered succinylated and malonylated lysines, help to understand the complex crosstalk between different modifications. Nevertheless, the number of presented combinations of modifications is one limitation of the histone peptide microarray. The physical limit of the contact-printing technology using split pins is in the range of 100,000 deposited spots per standard industry glass slide, but there are emerging technologies like the light-directed synthesis of peptides directly on the glass surface, resulting in about 2 million peptides per slide (and announced 10 million peptides per slide). If companies developing such technologies are able to extend the number of possible building blocks (from 24 to at least 50 to include all known phosphorylations, methylations and acylations), a much better coverage of PTM combinations could be displayed, dramatically improving our understanding of the histone code and epigenetics.

ACKNOWLEDGMENTS

We acknowledge financial support through BMBF grant ProNet-T3 to A.M. and M.S. and BMBF grant EP111300 to U.R. and J.Z.

REFERENCES

[1] Bannister AJ, Kouzarides T. Regulation of chromatin by histone modifications. Cell Res 2011;21 (3):381–95.
[2] Chi P, Allis CD, Wang GG. Covalent histone modifications — miswritten, misinterpreted and mis-erased in human cancers. Nat Rev Cancer 2010;10(7):457–69.
[3] Rossetto D, Avvakumov N, Cote J. Histone phosphorylation: a chromatin modification involved in diverse nuclear events. Epigenetics 2012;7(10):1098–108.
[4] Cosgrove MS, Boeke JD, Wolberger C. Regulated nucleosome mobility and the histone code. Nat Struct Mol Biol 2004;11(11):1037–43.
[5] Rothbart SB, Strahl BD. Interpreting the language of histone and DNA modifications. Biochim Biophys Acta 2014;1839(8):627–43.

[6] Musselman CA, Lalonde ME, Cote J, Kutateladze TG. Perceiving the epigenetic landscape through histone readers. Nat Struct Mol Biol 2012;19(12):1218−27.

[7] Baek SH. When signaling kinases meet histones and histone modifiers in the nucleus. Mol Cell 2011;42(3):274−84.

[8] Lohse B, Helgstrand C, Kristensen JB, Leurs U, Cloos PA, Kristensen JL, et al. Posttranslational modifications of the histone 3 tail and their impact on the activity of histone lysine demethylases in vitro. PLoS One 2013;8(7):e67653.

[9] Liokatis S, Stutzer A, Elsasser SJ, Theillet FX, Klingberg R, van Rossum B, et al. Phosphorylation of histone H3 Ser10 establishes a hierarchy for subsequent intramolecular modification events. Nat Struct Mol Biol 2012;19(8):819−23.

[10] Smith E, Shilatifard A. The chromatin signaling pathway: diverse mechanisms of recruitment of histone-modifying enzymes and varied biological outcomes. Mol Cell 2010;40(5):689−701.

[11] Jackson V, Shires A, Chalkley R, Granner DK. Studies on highly metabolically active acetylation and phosphorylation of histones. J Biol Chem 1975;250(13):4856−63.

[12] Barth TK, Imhof A. Fast signals and slow marks: the dynamics of histone modifications. Trends Biochem Sci 2010;35(11):618−26.

[13] Chen Y, Sprung R, Tang Y, Ball H, Sangras B, Kim SC, et al. Lysine propionylation and butyrylation are novel post-translational modifications in histones. Mol Cell Proteomics 2007;6(5):812−19.

[14] Peng C, Lu Z, Xie Z, Cheng Z, Chen Y, Tan M, et al. The first identification of lysine malonylation substrates and its regulatory enzyme. Mol Cell Proteomics 2011;10(12): M111 012658.

[15] Xie Z, Dai J, Dai L, Tan M, Cheng Z, Wu Y, et al. Lysine succinylation and lysine malonylation in histones. Mol Cell Proteomics 2012;11(5):100−7.

[16] Zhang Z, Tan M, Xie Z, Dai L, Chen Y, Zhao Y. Identification of lysine succinylation as a new post-translational modification. Nat Chem Biol 2011;7(1):58−63.

[17] Weinert BT, Scholz C, Wagner SA, Iesmantavicius V, Su D, Daniel JA, et al. Lysine succinylation is a frequently occurring modification in prokaryotes and eukaryotes and extensively overlaps with acetylation. Cell Rep 2013;4(4):842−51.

[18] Tan M, Luo H, Lee S, Jin F, Yang JS, Montellier E, et al. Identification of 67 histone marks and histone lysine crotonylation as a new type of histone modification. Cell 2011;146(6):1016−28.

[19] Dai L, Peng C, Montellier E, Lu Z, Chen Y, Ishii H, et al. Lysine 2-hydroxyisobutyrylation is a widely distributed active histone mark. Nat Chem Biol 2014;10(5):365−70.

[20] Tan M, Peng C, Anderson KA, Chhoy P, Xie Z, Dai L, et al. Lysine glutarylation is a protein posttranslational modification regulated by SIRT5. Cell Metab 2014;19(4):605−17.

[21] Voigt P, Reinberg D. Histone tails: ideal motifs for probing epigenetics through chemical biology approaches. Chembiochem 2011;12(2):236−52.

[22] Garske AL, Craciun G, Denu JM. A combinatorial H4 tail library for exploring the histone code. Biochemistry 2008;47(31):8094−102.

[23] Garske AL, Oliver SS, Wagner EK, Musselman CA, LeRoy G, Garcia BA, et al. Combinatorial profiling of chromatin binding modules reveals multisite discrimination. Nat Chem Biol 2010;6(4):283−90.

[24] Heubach Y, Planatscher H, Sommersdorf C, Maisch D, Maier J, Joos TO, et al. From spots to beads-PTM-peptide bead arrays for the characterization of anti-histone antibodies. Proteomics 2013;13(6):1010−15.

[25] Sawicka A, Seiser C. Sensing core histone phosphorylation − A matter of perfect timing. Biochim Biophys Acta 2014;1839(8):711−18.

[26] Mahajan K, Fang B, Koomen JM, Mahajan NP. H2B Tyr37 phosphorylation suppresses expression of replication-dependent core histone genes. Nat Struct Mol Biol 2012;19(9):930−7.

[27] Frederiks F, Stulemeijer IJ, Ovaa H, van Leeuwen F. A modified epigenetics toolbox to study histone modifications on the nucleosome core. Chembiochem 2011;12(2):308−13.

[28] Tropberger P, Pott S, Keller C, Kamieniarz-Gdula K, Caron M, Richter F, et al. Regulation of transcription through acetylation of H3K122 on the lateral surface of the histone octamer. Cell 2013;152 (4):859−72.

[29] Tinti M, Kiemer L, Costa S, Miller ML, Sacco F, Olsen JV, et al. The SH2 domain interaction landscape. Cell Rep 2013;3(4):1293−305.

[30] Carducci M, Perfetto L, Briganti L, Paoluzi S, Costa S, Zerweck J, et al. The protein interaction network mediated by human SH3 domains. Biotechnol Adv 2012;30(1):4−15.

[31] Lakshminarasimhan M, Rauh D, Schutkowski M, Steegborn C. Sirt1 activation by resveratrol is substrate sequence-selective. Aging (Albany NY) 2013;5(3):151−4.

[32] Rauh D, Fischer F, Gertz M, Lakshminarasimhan M, Bergbrede T, Aladini F, et al. An acetylome peptide microarray reveals specificities and deacetylation substrates for all human sirtuin isoforms. Nat Commun 2013;4:2327.

[33] Papadopoulos C, Arato K, Lilienthal E, Zerweck J, Schutkowski M, Chatain N, et al. Splice variants of the dual specificity tyrosine phosphorylation-regulated kinase 4 (DYRK4) differ in their subcellular localization and catalytic activity. J Biol Chem 2011;286(7):5494−505.

[34] Mah AS, Elia AE, Devgan G, Ptacek J, Schutkowski M, Snyder M, et al. Substrate specificity analysis of protein kinase complex Dbf2-Mob1 by peptide library and proteome array screening. BMC Biochem 2005;6:22.

[35] Rychlewski L, Kschischo M, Dong L, Schutkowski M, Reimer U. Target specificity analysis of the Abl kinase using peptide microarray data. J Mol Biol 2004;336(2):307−11.

[36] Lizcano JM, Deak M, Morrice N, Kieloch A, Hastie CJ, Dong L, et al. Molecular basis for the substrate specificity of NIMA-related kinase-6 (NEK6). Evidence that NEK6 does not phosphorylate the hydrophobic motif of ribosomal S6 protein kinase and serum- and glucocorticoid-induced protein kinase in vivo. J Biol Chem 2002;277(31):27839−49.

[37] Schutkowski M, Reimer U, Panse S, Dong L, Lizcano JM, Alessi DR, et al. High-content peptide microarrays for deciphering kinase specificity and biology. Angew Chem Int Ed Engl 2004;43(20):2671−4.

[38] Ho MC, Wilczek C, Bonanno JB, Xing L, Seznec J, Matsui T, et al. Structure of the arginine methyltransferase PRMT5-MEP50 reveals a mechanism for substrate specificity. PLoS One 2013;8(2):e57008.

[39] Wisniewski JR, Zougman A, Kruger S, Mann M. Mass spectrometric mapping of linker histone H1 variants reveals multiple acetylations, methylations, and phosphorylation as well as differences between cell culture and tissue. Mol Cell Proteomics 2007;6(1):72−87.

[40] Nady N, Min J, Kareta MS, Chedin F, Arrowsmith CH. A SPOT on the chromatin landscape? Histone peptide arrays as a tool for epigenetic research. Trends Biochem Sci 2008;33(7):305−13.

[41] Bock I, Kudithipudi S, Tamas R, Kungulovski G, Dhayalan A, Jeltsch A. Application of Celluspots peptide arrays for the analysis of the binding specificity of epigenetic reading domains to modified histone tails. BMC Biochem 2011;12:48.

[42] Su Z, Boersma MD, Lee JH, Oliver SS, Liu S, Garcia BA, et al. ChIP-less analysis of chromatin states. Epigenetics Chromatin 2014;7:7.

[43] Filippakopoulos P, Picaud S, Mangos M, Keates T, Lambert JP, Barsyte-Lovejoy D, et al. Histone recognition and large-scale structural analysis of the human bromodomain family. Cell 2012;149(1):214−31.

[44] Bua DJ, Kuo AJ, Cheung P, Liu CL, Migliori V, Espejo A, et al. Epigenome microarray platform for proteome-wide dissection of chromatin-signaling networks. PLoS One 2009;4(8):e6789.

[45] Fuchs SM, Krajewski K, Baker RW, Miller VL, Strahl BD. Influence of combinatorial histone modifications on antibody and effector protein recognition. Curr Biol 2011;21(1):53−8.

[46] Matthews AG, Kuo AJ, Ramon-Maiques S, Han S, Champagne KS, Ivanov D, et al. RAG2 PHD finger couples histone H3 lysine 4 trimethylation with V(D)J recombination. Nature 2007;450(7172):1106−10.

[47] Levy D, Kuo AJ, Chang Y, Schaefer U, Kitson C, Cheung P, et al. Lysine methylation of the NF-kappaB subunit RelA by SETD6 couples activity of the histone methyltransferase GLP at chromatin to tonic repression of NF-kappaB signaling. Nat Immunol 2011;12(1):29−36.

[48] Hung T, Binda O, Champagne KS, Kuo AJ, Johnson K, Chang HY, et al. ING4 mediates crosstalk between histone H3 K4 trimethylation and H3 acetylation to attenuate cellular transformation. Mol Cell 2009;33(2):248−56.

[49] Kleine-Kohlbrecher D, Christensen J, Vandamme J, Abarrategui I, Bak M, Tommerup N, et al. A functional link between the histone demethylase PHF8 and the transcription factor ZNF711 in X-linked mental retardation. Mol Cell 2010;38(2):165−78.

[50] Liu H, Galka M, Iberg A, Wang Z, Li L, Voss C, et al. Systematic identification of methyllysine-driven interactions for histone and nonhistone targets. J Proteome Res 2010;9(11):5827−36.

[51] Bock I, Dhayalan A, Kudithipudi S, Brandt O, Rathert P, Jeltsch A. Detailed specificity analysis of antibodies binding to modified histone tails with peptide arrays. Epigenetics 2011;6(2):256−63.

[52] Zhang Y, Jurkowska R, Soeroes S, Rajavelu A, Dhayalan A, Bock I, et al. Chromatin methylation activity of Dnmt3a and Dnmt3a/3L is guided by interaction of the ADD domain with the histone H3 tail. Nucleic Acids Res 2010;38(13):4246−53.

[53] Dhayalan A, Tamas R, Bock I, Tattermusch A, Dimitrova E, Kudithipudi S, et al. The ATRX-ADD domain binds to H3 tail peptides and reads the combined methylation state of K4 and K9. Hum Mol Genet 2011;20(11):2195−203.

[54] Rothbart SB, Dickson BM, Ong MS, Krajewski K, Houliston S, Kireev DB, et al. Multivalent histone engagement by the linked tandem Tudor and PHD domains of UHRF1 is required for the epigenetic inheritance of DNA methylation. Genes Dev 2013;27(11):1288−98.

[55] Zerweck J, Masch A, Schutkowski M. Peptide microarrays for profiling of modification state-specific antibodies. Methods Mol Biol 2009;524:169−80.

[56] Funkner A, Parthier C, Schutkowski M, Zerweck J, Lilie H, Gyrych N, et al. Peptide binding by catalytic domains of the protein disulfide isomerase-related protein ERp46. J Mol Biol 2013;425(8):1340−62.

[57] Rardin MJ, He W, Nishida Y, Newman JC, Carrico C, Danielson SR, et al. SIRT5 regulates the mitochondrial lysine succinylome and metabolic networks. Cell Metab 2013;18(6):920−33.

[58] Park J, Chen Y, Tishkoff DX, Peng C, Tan M, Dai L, et al. SIRT5-mediated lysine desuccinylation impacts diverse metabolic pathways. Mol Cell 2013;50(6):919−30.

[59] Colak G, Xie Z, Zhu AY, Dai L, Lu Z, Zhang Y, et al. Identification of lysine succinylation substrates and the succinylation regulatory enzyme CobB in *Escherichia coli*. Mol Cell Proteomics 2013;12(12):3509−20.

[60] Wen H, Li J, Song T, Lu M, Kan PY, Lee MG, et al. Recognition of histone H3K4 trimethylation by the plant homeodomain of PHF2 modulates histone demethylation. J Biol Chem 2010;285(13):9322−6.

[61] Nardelli SC, Che FY, Silmon de Monerri NC, Xiao H, Nieves E, Madrid-Aliste C, et al. The histone code of *Toxoplasma gondii* comprises conserved and unique posttranslational modifications. MBio 2013;4(6): e00922−13.

[62] Filippakopoulos P, Knapp S. Targeting bromodomains: epigenetic readers of lysine acetylation. Nat Rev Drug Discov 2014;13(5):337−56.

[63] Yang Z, He N, Zhou Q. Brd4 recruits P-TEFb to chromosomes at late mitosis to promote G1 gene expression and cell cycle progression. Mol Cell Biol 2008;28(3):967−76.

[64] Wu SY, Lee AY, Lai HT, Zhang H, Chiang CM. Phospho switch triggers Brd4 chromatin binding and activator recruitment for gene-specific targeting. Mol Cell 2013;49(5):843−57.

[65] Liu Y, Wang X, Zhang J, Huang H, Ding B, Wu J, et al. Structural basis and binding properties of the second bromodomain of Brd4 with acetylated histone tails. Biochemistry 2008;47(24):6403−17.

[66] Thiele A, Zerweck J, Schutkowski M. Peptide arrays for enzyme profiling. Methods Mol Biol 2009;570:19−65.

[67] Thiele A, Stangl GI, Schutkowski M. Deciphering enzyme function using peptide arrays. Mol Biotechnol 2011;49(3):283−305.

[68] Green KD, Pflum MK. Kinase-catalyzed biotinylation for phosphoprotein detection. J Am Chem Soc 2007;129(1):10−11.

[69] Song H, Kerman K, Kraatz HB. Electrochemical detection of kinase-catalyzed phosphorylation using ferrocene-conjugated ATP. Chem Commun (Camb) 2008;28(4):502−4.

[70] Kerman K, Kraatz HB. Electrochemical detection of kinase-catalyzed thiophosphorylation using gold nanoparticles. Chem Commun (Camb) 2007;21(47):5019−21.

[71] Green KD, Pflum MK. Exploring kinase cosubstrate promiscuity: monitoring kinase activity through dansylation. Chembiochem 2009;10(2):234−7.

[72] Martin K, Steinberg TH, Cooley LA, Gee KR, Beechem JM, Patton WF. Quantitative analysis of protein phosphorylation status and protein kinase activity on microarrays using a novel fluorescent phosphorylation sensor dye. Proteomics 2003;3(7):1244−55.

[73] Shimomura T, Han X, Hata A, Niidome T, Mori T, Katayama Y. Optimization of peptide density on microarray surface for quantitative phosphoproteomics. Anal Sci 2011;27(1):13−17.

[74] Kinoshita E, Kinoshita-Kikuta E, Sugiyama Y, Fukada Y, Ozeki T, Koike T. Highly sensitive detection of protein phosphorylation by using improved Phos-tag Biotin. Proteomics 2012;12(7):932−7.

[75] Akita S, Umezawa N, Kato N, Higuchi T. Array-based fluorescence assay for serine/threonine kinases using specific chemical reaction. Bioorg Med Chem 2008;16(16):7788−94.

[76] Shults MD, Kozlov IA, Nelson N, Kermani BG, Melnyk PC, Shevchenko V, et al. A multiplexed protein kinase assay. Chembiochem 2007;8(8):933−42.

[77] Soloaga A, Thomson S, Wiggin GR, Rampersaud N, Dyson MH, Hazzalin CA, et al. MSK2 and MSK1 mediate the mitogen- and stress-induced phosphorylation of histone H3 and HMG-14. EMBO J 2003;22 (11):2788−97.

[78] Healy S, Khan P, He S, Davie JR. Histone H3 phosphorylation, immediate-early gene expression, and the nucleosomal response: a historical perspective. Biochem Cell Biol 2012;90(1):39−54.

[79] Zhang Y, Griffin K, Mondal N, Parvin JD. Phosphorylation of histone H2A inhibits transcription on chromatin templates. J Biol Chem 2004;279(21):21866−72.

[80] Shimada M, Nakadai T, Fukuda A, Hisatake K. cAMP-response element-binding protein (CREB) controls MSK1-mediated phosphorylation of histone H3 at the c-fos promoter in vitro. J Biol Chem 2010;285(13):9390−401.

[81] Herranz N, Dave N, Millanes-Romero A, Morey L, Diaz VM, Lorenz-Fonfria V, et al. Lysyl oxidase-like 2 deaminates lysine 4 in histone H3. Mol Cell 2012;46(3):369−76.

[82] Sauve AA, Wolberger C, Schramm VL, Boeke JD. The biochemistry of sirtuins. Annu Rev Biochem 2006;75:435−65.

[83] Morris BJ. Seven sirtuins for seven deadly diseases of aging. Free Radic Biol Med 2013;56:133−71.

[84] Ming M, Qiang L, Zhao B, He YY. Mammalian SIRT2 inhibits keratin 19 expression and is a tumor suppressor in skin. Exp Dermatol 2014;23(3):207−9.

[85] Xiao K, Jiang J, Wang W, Cao S, Zhu L, Zeng H, et al. Sirt3 is a tumor suppressor in lung adenocarcinoma cells. Oncol Rep 2013;30(3):1323−8.

[86] Csibi A, Fendt SM, Li C, Poulogiannis G, Choo AY, Chapski DJ, et al. The mTORC1 pathway stimulates glutamine metabolism and cell proliferation by repressing SIRT4. Cell 2013;153(4):840−54.

[87] Michishita E, McCord RA, Berber E, Kioi M, Padilla-Nash H, Damian M, et al. SIRT6 is a histone H3 lysine 9 deacetylase that modulates telomeric chromatin. Nature 2008;452(7186):492−6.

CURRENT METHODS FOR METHYLOME PROFILING

10

Minkui Luo

Molecular Pharmacology and Chemistry Program, Memorial Sloan
Kettering Cancer Center, New York, NY, USA

CHAPTER OUTLINE

Y.G. Zheng (Ed): Epigenetic Technological Applications. DOI: http://dx.doi.org/10.1016/B978-0-12-801080-8.00010-7

10.1 INTRODUCTION

Posttranslational methylation plays essential epigenetic roles by modulating how proteins interact with their binding partners and thus altering their downstream functions implicated in transcription, RNA processing, signal transduction, and metabolism through diverse mechanisms [1–4]. Although protein methylation has been reported to occur on the side chains of at least eight amino acids as well as on terminal α-amine and carboxylate groups [5], methylation of arginine (Arg) and lysine (Lys) has attracted by far the most attention because of its conversation from yeast to human, cellular omnipresence, and functional relevance to many essential biological processes [6–8]. In humans, methylation of Arg and Lys is catalyzed by around 70 protein methyltransferases (PMTs) with S-adenosyl-L-methionine (SAM) as the universal methyl-donor cofactor (co-substrate) [6–12]. The catalytic activities of individual PMTs on their targets are further tuned by other factors such as their localization and binding partners [6–8,13]. The PMTs further diverge into two subfamilies with nine protein arginine methyltransferases (PRMTs) and around 60 protein lysine methyltrans-ferases (PKMTs) encoded by the human genome (Figure 10.1) [6–13].

More than 90% of human PKMTs (>50) are SET domain-containing PKMTs (Class V methyl-transferases, SET for Suppressor of variegation 3–9, Enhancer of zeste and Trithorax, three genetic

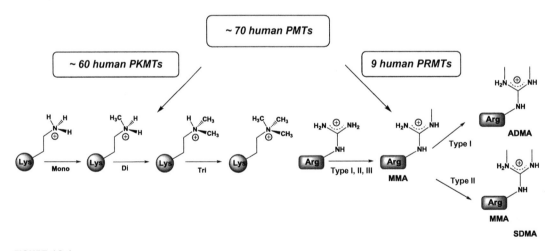

FIGURE 10.1

PMTs encoded by the human genome. The human genome encodes around 70 protein methyltransferases (PMTs): 9 PRMTs and around 60 PKMTs. PKMTs methylate lysine's ε-amino groups of their substrates to Kme1/2/3. PRMTs methylate arginine's ω-guanidino nitrogen of their substrates into three forms: MMA, ADMA, and SDMA.

phenotypes of *Drosophila*), which rely on the characteristic 130-aa SET domain for SAM binding and enzyme catalysis [13]. DOT1L, METTL21A, METTL21C, METTL21D, METTL10, and the orphan catalytic site of human MLL1 are few non-SET domain PKMTs characterized so far [9−16]. Besides sequence and structural features, PKMTs are also distinct on the basis of their ability to methylate lysine's ε-amino groups to different degrees (mono-, di-, and trimethylation, Kme1/2/3) (Figure 10.1) [13]. In comparison, all nine human PRMTs share four conserved motifs (I, post-I, II, and III) and the characteristic THW loop for SAM binding [8,13]. PRMTs modify arginine's ω-guanidino nitrogen of their substrates into three forms: MMA (monomethylarginine), ADMA (asymmetric dimethylarginine), and SDMA (symmetric dimethylarginine) (Figure 10.1) [13]. Such product specificity further classifies PRMTs into Type I PRMTs including the predominant PRMT1 with MMA and ADMA as the products, Type II PRMTs such as the predominant PRMT5 with MMA and ADMA as the products, and Type III PRMTs with MMA as the only product (Figure 10.1) [13].

PMTs function epigenetically through methylating diverse histone and nonhistone substrates [13,17]. For instance, EuHMT1 and EuHMT2 methylate lysine 9 of histone H3 (H3K9), a mark implicated for transcription repression [6,7]. In contrast, MLL1 and SETD2 methylate lysine 4 and lysine 36 of histone H3 (H3K4 and H3K36), respectively [6,7]. These marks are often associated with gene activation. In a similar manner, PRMT1 methylates arginine 3 of histone H4 (H4R3) as a mark of transcriptional activation; whereas PRMT5 and PRMT6 methylate H4R3 and H3R2 as the marks of transcriptional repression [6,7,13]. The physiological and pathogenic roles of PMTs are further complicated via methylation of nonhistone targets [13,17]. Efforts over the past decade have led to the characterization of <100 PKMT nonhistone substrates [13,17]. Among many examples of biologically relevant substrates are the tumor suppressor p53 as the substrates of EuHMT1, EuHMT2, SET7/9, SMYD2, SETD8, PRMT1 and PRMT5 [18−22], and transcription factors STAT1, RUNX1, and FOXO1 as the substrates of PRMT1 [23−25]. PMTs often act in a temporal and cell-type-specific manner. For instance, the methylation of Reptin by EuHMT2 occurs only under hypoxic conditions [26]. CARM1 is required for maintaining pluripotency in embryonic stem (ES) cells through methylating H3R17/26 but for promoting ES cell differentiation in myocyte lineage through methylating the transcription factor Pax7 [27,28].

PMT-mediated histone and nonhistone methylation, together with other posttranslational modifications (e.g. acetylation and phosphorylation), can regulate engagement of the target proteins with their binding partners, and thus render meaningful outcomes [2,8,29]. Besides the roles of PMTs in normal biology, their dysregulation has been found in cancer. Overexpression of EuHMT1, SUV39H2, NSD2, NSD3, SMYD3, and PRDM14 has been reported in primary tumors [2,30]. Oncogenic mechanisms of PMTs can be associated with the downregulation of tumor suppressors or upregulation of oncogenes in a methylation-dependent manner [2,30]. Dissecting context−specific methylome, a collection of proteins that contain methylated lysine or arginine (Kme/Rme), is a key step toward elucidating the diverse biological and pathogenic roles of PMTs.

Although it has become more common to profile proteome-wide phosphorylation and acetylation events and to characterize individual protein methylation events through a candidate-based approach [31,32], it is not trivial to profile the methylome with conventional tools [13,33]. There are at least two reasons that make it difficult to reveal Kme/Rme in comparison with any other posttranslational modifications such as acetylation and phosphorylation: (a) as a neutral reaction, Lys/Arg methylation does not change the formal charge of parent Lys/Arg residues; (b) as the

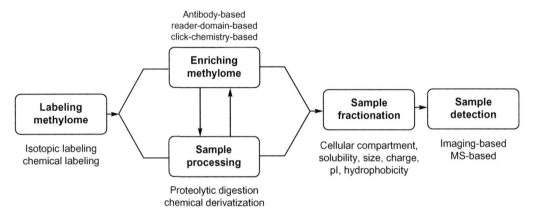

FIGURE 10.2

Dissecting technological modules for methylome profiling. As organized in the experimental order, the workflow of methylome profiling includes labeling methylome, enriching targets, sample processing, sample fractionation, and sample detection. The order of enriching methylome and sample processing can be changed on the basis of the need of enrichment reagents.

smallest posttranslational modification, methylation does not significantly change the size of Lys/Arg residues. There is therefore a critical need to develop reagents and methods to distinguish such subtle differences and enrich or trace the methylome [33].

This chapter summarizes the technological advancements for methylome profiling in the past decade, in particular within the past 2 years. With the focus on technology rather than biology, the experimental components of many successful cases are dissected and reorganized according to technological merits with the goal for methylome profiling in broader contexts (Figure 10.2). As the order to organize the chapter aligned with how a profiling experiment is carried out, the methods of methylome labeling are summarized first. Thereafter discussed are current strategies for methylome enrichment. Successful sample processing and methylome detection methods are then outlined. Also discussed briefly are bioinformatics tools to predict methylation sites and the database to access prior methylomic information. Several representative examples are described, with the emphasis on how a combination of technological modules facilitates methylome profiling. To minimize redundancy of the topics that have been covered by several excellent reviews [5,33], the chapter mainly focuses on technological aspects of recently published work. The author apologizes for the omission of many high-quality works because of space limitations.

10.2 LABELING PROTEIN METHYLATION

PMTs modify Lys and Arg residues of their substrates with SAM as the methyl-donor cofactor (co-substrate) [13]. Profiling methyltransferase activities of PMTs in native contexts can be accelerated by activity-based labeling of PMT substrates with heavy isotopic methyl groups or distinct chemical tags replacing the methyl group (Figure 10.3) [5,33]. Heavy isotope labeling and

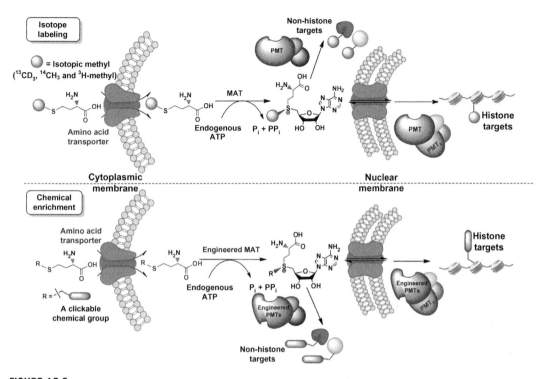

FIGURE 10.3

Dual approaches to label methylome within living cells: heavy isotope labeling and bioorthogonal chemical labeling. To label PMT targets with the isotopic SAM cofactor in living cells, the light methionine is replaced with S-[heavy-methyl] methionines as the substrate of MAT. The resultant synthesized SAM will be used as the cofactor by native PMTs to label their substrates. For BPPM within living cells, engineered human MATs will process membrane-permeable methionine analogues into SAM analogues, which are used as the cofactors by engineered PMTs to label their substrates.

bioorthogonal chemical labeling are two emerging activity-based labeling strategies for tracing proteome-wide methylation events under cellular settings (Figure 10.3) [13,17,34]. The heavy isotope labeling method with stable isotopes further allows characterization of methylation sites and quantitative comparison of multiple sets of methylome data with high confidence, while the chemical labeling method can introduce noncanonical chemical tags for facile target characterization and enrichment [13,35,36]. The bioorthogonal chemical labeling further enables assignment of the distinctly labeled targets to specific PMTs in an unambiguous manner [36].

10.2.1 CHALLENGES FOR METHYLOME PROFILING WITH CONVENTIONAL METHODS

Within living cells, the degrees of the Lys/Arg methylation are regulated by multiple factors including enzymatic activities and local concentrations of specific PMTs, the accessibility of demethylases for removal of methylation marks, as well as the relative lifetime of PMT substrates versus their methylation products. Perturbation of these factors is expected to directly alter the levels of the associated methylation events. However, such perturbation may also serve as an upstream signal that affects not only the methylation levels of immediate downstream targets but also the associated methylome in a cascade manner. Individual methylation events can be probed by αKme/Rme antibodies, mass spectrometry (MS), or autoradiography with native SAM or S-[methyl-^{14}C/^{3}H]-labeled radioactive SAM as methyl-donor cofactors [37]. However, it is challenging to profile the methylome under specific cellular settings solely depending on antibodies for Western blot or MS for quantification of methylated peptides, given the poor quality of many pan-αKme/Rme antibodies and the limitation of MS to resolve a large number of low abundant peptide ions [33,38]. The direct use of radioactive S-[methyl-^{14}C/^{3}H]-SAM as the methyl-donor cofactor is often applied under well-defined *in vitro* settings because of the poor membrane permeability, weak β-emission, high cost, and potential environmental contamination of these radioactive reagents [37]. The conventional methods using antibodies, MS, or autoradiography alone are thus more suitable for probing individual methylation events rather than methylome, unless they can be coupled with advanced sample processing or target-array strategies as detailed later.

10.2.2 METHYLOME LABELING WITH ISOTOPES INSIDE LIVING CELLS

At least two isotopic labeling strategies have been adapted to uncover proteome-wide methylation events (Figure 10.4): heavy methyl labeling of methylation sites and global SILAC (stable isotope labeling by amino acids) of proteome containing methylated PMT targets [5,33,35]. These isotopic labeling methods allow PMT-associated methylome to be characterized with more confidence or be compared in a quantitative manner through exploiting the specific properties of these isotope tags (e.g. characteristic β-emission and mass shifts) [5,33,35,37].

Because of the poor membrane permeability of SAM, the direct use of isotopic SAM as a PMT cofactor is largely limited for *in vitro* settings [37,39]. SAM is synthesized *in vivo* by methionine adenosyltransferases (MATs) with endogenous ATP and methionine as precursors (Figures 10.3 and 10.4) [17,36]. To label PMT targets with the isotopic SAM cofactor in living cells, the MAT-catalyzed biosynthesis of SAM can be exploited by replacing the light methionine with S-[heavy-methyl] methionines. The heavy methionines can be internalized by living cells, likely through methionine transporter(s), and processed into the corresponding SAM cofactors by endogenous MATs for target labeling. A heavy methyl-labeling process can also be perturbed under specific settings to examine the changes of the methylome. It is worth noting that the heavy methyl labeling of PMT targets by the isotopic SAM cofactors through methionine precursors needs to be carried out in the presence of protein translation inhibitor(s) such as cycloheximide to prevent the incorporation of heavy methionine into proteins through RNA translation process [37,40].

Both radioactive S-[methyl-^{14}C/^{3}H]-methionine and stable S-[methyl-^{13}CD$_3$]-methionine have been used for the heavy methyl labeling (Figures 10.3, 10.4) [35,37]. The radioactive methionine is often used as tracers in a mixture containing a large percentage of the light methionine carrier because of potential environmental contamination and high cost using excessive radioactive materials. The resultant labeled methylome is anticipated to contain the trace amount of [methyl-^{14}C/^{3}H] and can be visualized by autoradiography [37]. The detection thresholds and signal-to-noise ratios with the [methyl-^{14}C/^{3}H]-SAM as PMT cofactors can be improved in combination with

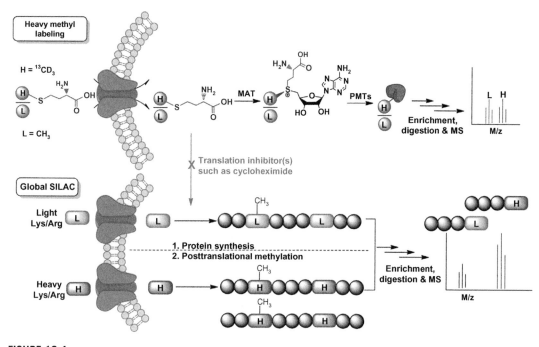

FIGURE 10.4

Isotopic labeling approaches for quantitive comparison of methylome. Heavy methyl labeling is featured by the use of stable S-[methyl-$^{13}CD_3$] methionine with the resultant S-[methyl-$^{13}CD_3$] labeling PMT substrates. This method is often used to map the methylation sites. The global SILAC approach is to label the whole proteome with heavy Lys or Arg. The result light versus heavy methylome can be quantified by MS after enrichment.

advanced sample enrichment and processing strategies (see Section 10.3, "Recognizing and Enriching Methylome" and Section 10.5, "Deconvolution of Methylome Sample"). In general, the radioisotope-based approach cannot be coupled with MS/MS technologies to reveal proteome-wide target identification and methylation sites of PMT substrates, because of low isotope abundance and potential environmental contamination of these radioactive materials. In comparison with the heavy methyl labeling using radioactive S-[methyl-$^{14}C/^3H$]-methionine, alternative approaches have been developed to label methylome of PMTs in a cellular context as detailed below.

In order to formulate a heavy methyl labeling method compatible with MS for subsequent target identification (see Section 10.6, "Detection Methods for Methylome Profiling"), Ong and Mann pioneered the heavy methyl-SILAC labeling approach with S-[$^{13}CD_3$] methionine as a SAM precursor and identified close to 100 PMT methylation sites from the proteome of HeLa cells [33,35]. In contrast with radioactive S-[methyl-$^{14}C/^3H$]-methionine, S-[$^{13}CD_3$] methionine is more environmentally friendly and less expensive. The heavy methyl labeling experiment can thus be carried out by replacing the methionine of natural abundance with S-[$^{13}CD_3$]-enriched methionine (>98% isotopic abundance). PMT-mediated methylation can be characterized by the mass increase of 14 Da and 18 Da per methylation with the light SAM and the heavy S-[$^{13}CD_3$]-SAM as cofactors, respectively.

After digesting the cell lysates with proteases and enriching methylated peptides with antibodies (see Section 10.3, "Recognizing and Enriching Methylome" and Section 10.4, "Choices of Processing Methylome Samples"), the $[CH_3]/[^{13}CD_3]$-modification sites can be mapped with MS/MS (see Section 10.6, "Detection Methods for Methylome Profiling"). The existence of the characteristic pairs of 14 versus 18 Da mass shifts significantly increases the confidence of identifying bona fide methylation sites (Figure 10.4) [33,35]. In contrast, a +14 Da mass shift alone can be an artifact such as the methanol-mediated conversion of acidic residues into their methyl esters during sample workup [5,33,35].

SILAC has been broadly applied for proteomic comparison in a quantitative manner and demonstrates increased use for methylome profiling [5,33,34,41]. In a general SILAC experiment, cells are grown in the media containing light and heavy amino acids, respectively. After genetic or chemical perturbation, the resultant proteomes are subject to MS analysis with the differentiated abundance of target proteins quantified by their characteristic mass shifts matched with the used light and heavy amino acids (Figure 10.4) [33,34,41]. Alternatively, three sets of otherwise identical cells can be labeled with light, middle, and heavy isotopes, respectively, for a three-way comparison [34]. Coupled with the suitable enrichment strategies such as pan-αKme/Rme antibodies (see Section 10.3, "Recognizing and Enriching Methylome"), the global SILAC approach can be readily adapted to interrogate methylome [34]. In comparison with the heavy methyl-SILAC labeling approach, the global SILAC approach is distinguished by its much higher peptide recovery for MS analysis because the whole proteins rather than only a few methylated peptides are subject to MS analysis for target identification. Because Lys and Arg are often used as the stable isotopes in SILAC experiments, the resultant labeling can assure the isotopic amino acids are incorporated into Kme/Rme-containing peptides and thus facilitate the identification of the modification sites through the characteristic isotopic mass shifts.

10.2.3 CHEMICAL LABELING OF PROTEOME-WIDE PMT SUBSTRATES *IN VITRO* AND WITHIN LIVING CELLS

To label PMT substrates with a chemical moiety, a well-appreciated approach is to develop SAM surrogates in which SAM's sulfonium-methyl group is replaced with functionalized chemical groups [42–44]. These chemical moieties can then be transferred to PMT substrates in an enzymatically dependent manner [17,36,45]. To facilitate the subsequent target enrichment and characterization, the main focus in the past decade has been to develop the SAM analogue cofactors containing clickable sulfonium-alkyl moieties such as terminal alkyne or azide groups because of their amenable conjugation with other probes (e.g. dye and biotin) via well-established Huisgen cycloaddition (the click chemistry) (Figure 10.5) [17]. While the utility of the sulfonium-alkyl SAM analogues as cofactors of DNA/RNA/small-molecule methyltransferases has been reported since 2006 [17], their application for PMTs had not been documented until 2010 [17,46–50]. The Weinhold group first showed that EnYn-SAM (*S*-(*E*)-pent-2-en-4-ynyl analogue of SAM) can be utilized by a fungal PKMT Dim-5 and two human PKMTs (MLL and MLL4) to a certain degree (Figure 10.5) [17]. To develop more reactive SAM analogue cofactors containing clickable sulfonium-alkyl moieties, many prior efforts focused on the smallest candidate, propargyl-SAM (*S*-propargylic analogue of SAM, Figure 10.5) [51]. Unfortunately, propargyl-SAM is not stable under

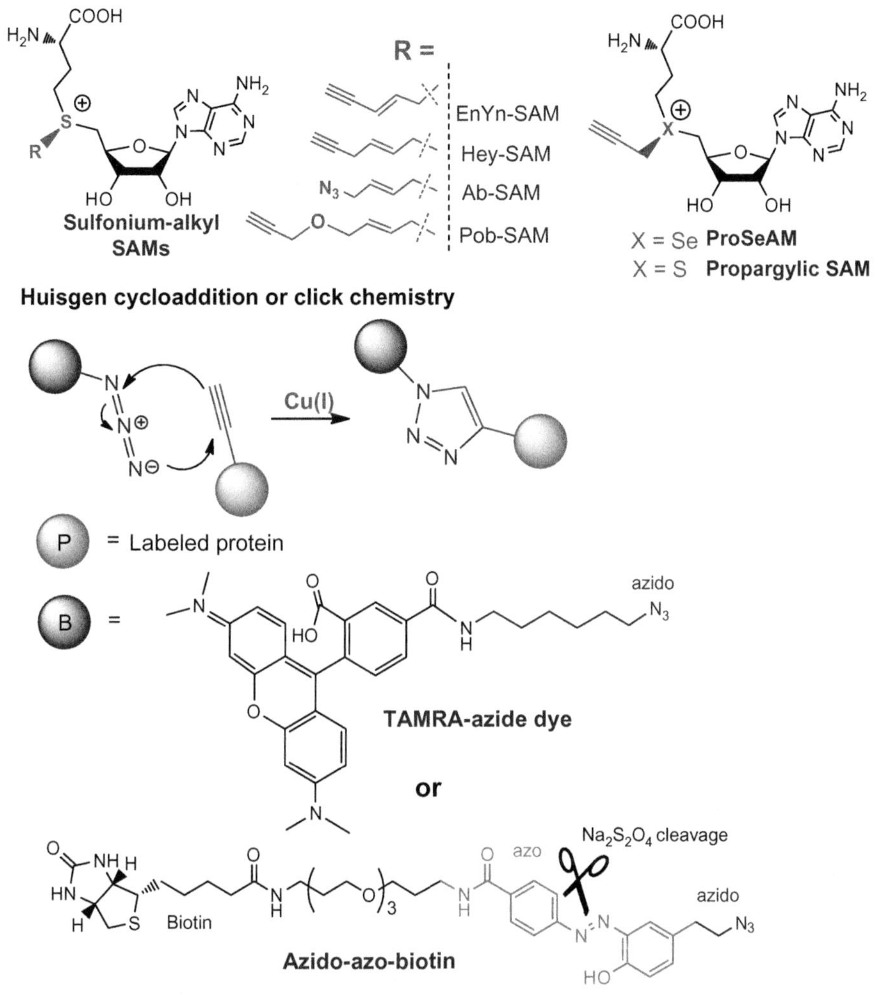

FIGURE 10.5

Structures of SAM analogues for BPPM together with BPPM reagents for the click reaction.

physiological pH and therefore its use is limited [51]. Our laboratory as well as the Weinhold group solved this instability issue by developing ProSeAM (propargylic *Se*-adenosyl-*L*-selenomethionine) [51,52]. Our laboratory showed that, among eight native human PMTs (PRMT1, PRMT3, CARM1, SUV39H2, SET7/9, SET8, G9a, and GLP1), only SUV39H2, G9a, and GLP1 demonstrated decent activities with ProSeAM as a cofactor but not bulky SAM analogues, such as EnYn-SAM, Hey-SAM (*S*-(*E*)-hex-2-en-5-ynyl analogue of SAM), and Pob-SAM (*S*-4-propargyloxy-but-2-enyl analogue of SAM) (Figure 10.5) [43]. The general lack of the cofactor activity of bulky sulfonium-alkyl SAM analogues for native PMTs is also consistent with the structures of PMTs in which

SAM-binding pockets are not spacious enough to accommodate sterically hindered SAM analogues [53,54]. This limitation thus inspired us to engineer native PMTs to act on these bulky SAM analogues for more efficient chemical labeling of PMT targets [13,17].

Conventional methods to correlate the methylome with specific methyltransferase activities are genetic knockout, shRNA/siRNA-mediated knockdown and overexpression of the corresponding PMTs. However, it is challenging to establish the direct cause-consequence connection between the perturbed PMT and the resultant methylome, given the potential cascade effect of such perturbation as well as likely PMT redundancy. To address this challenge as well as low activities of native PMTs with sulfonium-alkyl SAM analogue cofactors in general, we have formulated a technology, called bioorthogonal profiling of protein methylation (BPPM), to profile the methyltransferase activities of designated (engineered) PMTs (Figure 10.3) [13]. Herein the SAM-binding pockets of PMTs are engineered to accommodate S-alkyl SAM analogues, which are otherwise too sterically bulky to serve as the cofactors of native PMTs. The engineered PMTs can then transfer the distinct alkyl moieties to the substrates of the PMTs. The BPPM approach allows the distinctly labeled targets to be assigned to individual (engineered) PMTs in an unambiguous manner.

To identify suitable PMT variants and SAM analogues for BPPM, our laboratory carried out systematic mutagenesis and in parallel generated a collection of SAM analogues with SAM's methyl group replaced by structurally diverse S-alkyl substituents [13,36,42,44,45,55−58]. Upon screening the mutants of representative human PMTs against the collection of SAM analogues, multiple mutant-cofactor pairs have been identified as suitable BPPM reagents. For instance, Y1211A/Y1154A mutants of EuHMT1/2, Y39FM48G mutant of PRMT1, and M233G mutant of PRMT3 are promiscuous toward several S-alkyl SAM analogues including Hey-SAM, Pob-SAM, or Ab-SAM (S-4-azidobut-2-enyl analogue of SAM) [17]. For BPPM in vitro, PMT variants are expressed in a cellular context or added exogenously into cell lysates [42,44,55,58], followed by incubation of lysates with S-alkyl SAM analogues as the BPPM cofactors. Because the S-alkyl substituents of these SAM analogues contain a terminal-alkyne or azide group, the resultant modified PMT substrates can be coupled, via the click reaction, with terminal-azide/alkyne-containing dyes or biotin for subsequent visualization or enrichment (see Section 10.3, "Recognizing and Enriching Methylome" and Section 10.6, "Detection Methods for Methylome Profiling") (Figure 10.5) [17]. The enriched proteins can also be analyzed by MS/MS to elucidate target identification and modification sites (see Section 10.6, "Detection Methods for Methylome Profiling") [55].

Similar to SAM, the bulky SAM analogues show poor membrane permeability and thus cannot directly label targets inside cells [37,39]. To overcome the issue of membrane permeability, Wang et al. engineered human MATs to process membrane-permeable methionine analogues into SAM analogues for BPPM within living cells (Figure 10.3) [36]. The three-step BPPM within living cells involves the biosynthesis of SAM analogues from cell-permeable methionine analogues by MAT2A I117A mutant, in situ substrate modification by engineered PMTs, and the subsequent enrichment of the uniquely modified targets. With this technology in combination with next-generation DNA-sequencing technology, Wang et al. dissected the chromatin-modifying activities of closely related EuHMT1 and EuHMT2, and showed that the two PMTs have complementary methyltransferase activities for chromatin modification with no preference between promoter loci and gene bodies [36]. The BPPM within living cells is expected to be applicable to profile nonhistone targets and map the methylation sites of designated PMTs if MS analysis instead of DNA sequencing analysis is applied to characterize the enriched targets. Besides the unambiguous assignment of the revealed substrates to designated (engineered) PMTs, the merit of the BPPM technology further lies in its ability to couple with the efficient alkyne-azide click chemistry for target visualization and enrichment (see Section 10.3, "Recognizing and Enriching Methylome" and Section 10.6, "Detection Methods for Methylome Profiling") [17,36,45].

10.3 **RECOGNIZING AND ENRICHING METHYLOME**

Because the methylome generally accounts for a small portion of proteome under a cellular setting, distinguishing methylome from other irrelevant proteins is essential to increase the signal-to-noise ratios for subsequent target characterization. However, it remains challenging for conventional antibodies to efficiently recognize or enrich methylome because protein Lys/Arg methylation, unlike other posttranslational modifications (e.g. acetylation or phosphorylation), does not alter the overall charges and only slightly increases the sizes of the targeted residues. Many high-quality αKme/Rme antibodies were developed to recognize Kme/Rme in the context of the surrounding sequences [59]. Although several broadly specific αKme/Rme antibodies have also been documented, multiple reports argue a great margin to further improve the qualities of pan-antibodies for recognizing Kme/Rme epitopes with high selectivity and affinity but without depending upon their neighboring sequences [33,60]. In the course of searching high-quality pan-αKme/Rme antibodies, complementary approaches have also been developed to recognize and enrich methylome with naturally occurring Kme reader domains [34,41]. The bioorthogonal chemical labeling of methylome by terminal-alkyne/azide-containing moieties further facilitates the visualization and enrichment of the PMT targets using chemical dye or biotin probes via the well-established click reaction (Figure 10.5) [56,57]. These methods, together with advanced MS technologies, have accelerated the finding of thousands of PMT targets [33].

10.3.1 **ANTIBODY-BASED RECOGNITION OF METHYLOME**

An early work to enrich methylome with a broadly specific antibody was conducted by the Richard group [61]. Over 200 proteins from HeLaS3 cells were immunoenriched with the two antibodies that were developed at that time to recognize a previously characterized Arg-Gly motif containing ADMA and SDMA, respectively. Later, the Mann group relied on an αRme antibody, together with the heavy-methyl—labeling technology, to immunoenrich Rme-containing proteins [35]. The subsequent LC-MS/MS (liquid chromatography—mass spectrometry) characterization uncovered 31 Rme-containing proteins and 57 Rme sites from HeLaS3 cells. However, similar work using an αKme antibody only led to the identification of H3K27 and H4K20 as Kme sites [35], likely reflecting the low quality of the αKme antibody available at that time. Since 2013, development of diverse αRme/Kme antibodies, together with advanced MS technology, helps uncover thousands of PMT targets and methylation sites. For instance, the Bonaldi group combined a broad collection of antibodies as enrichment reagents against 11 Rme/Kme-containing epitopes [62]. With the additional aid of heavy methyl labeling, multistep sample fractionation strategies and high-resolution MS, 139 proteins and 397 methylation sites were identified from HeLaS3 cells [62]. The Read group also carried out similar work with αRme2 antibodies and identified 676 Rme-containing proteins and 1,332 Rme sites from *Trypanosoma brucei*, including 167 Rme proteins and 253 Rme sites from mitochondria of the pathogen. Using three sets of in-house developed pan-αKme antibodies, coupled with the heavy methyl labeling, fractionation, sample processing, and MS analysis, the Garcia group identified 413 Lys-methylated proteins and 552 Lys methylation sites from HeLa cells [63,64]. In a similar manner, Guo et al. developed a collection of broadly specific antibodies: four for Rme and three for Kme [65]. With the pan-αRme/Kme antibodies, >800 Rme-containing proteins, >1,000 Rme sites, 130 Kme-containing proteins, and 165 Kme sites are uncovered by MS analysis from human cell line HCT116 [65]. With a pan-antibody against the Rme1 epitope, the Nielsen group carried out a single-step immunoenrichment, coupled with high-resolution MS, to uncover 1,027 MMA sites and a total of 494 human proteins from human U20S cells [66]. These antibodies can also be used as general blotting reagents to profile proteome-wide Kme/Rme-containing proteins. It shall be noted that there is no significant overlap among the methylation sites revealed in the different sets of experiments. Such unexpected difference can be due to the difference of the used cell lines, sample processing protocols and MS. It is also likely that these immunoenrichment reagents lack the broadly context-independent recognition against Rme/Kme epitopes. It remains to be examined whether more PMT targets can be identified by combining multiple pools of the immunoenrichment reagents.

10.3.2 RECOGNIZING METHYLOME WITH KME READER DOMAINS

Although Arg/Lys methylation only slightly alters the physical properties (charge and size) of the targeted residues, such difference can be distinguished to a certain degree by natural Kme/Rme reader domains [67–69]. More importantly, some reader domains solely recognize these epitopes with no significant need to interact with their surrounding amino acid sequences. These reader domains can thus be used in a similar manner as pan−αKme antibodies. The structure of L3MBTL1 indicates that its three Malignant Brain Tumor domain (3 × MBT) preferentially recognizes Kme1/2 over Kme3 and free Lys epitopes through the cation-π interaction and the hydrogen bonding of Kme1/2's methylammonium with a size-matched aromatic pocket and D355 residue of 3 × MBT [70,71]. Here no other significant contact is engaged between the 3 × MBT domain and the neighboring residues of the Kme1/2 epitopes. Such structural arrangement is consistent with the comparable K_d values of 3 × MBT in complex with H3K4me1/2, H3K9me1/2, H4K20me1/2, and p53 K382me1/2 peptides [70–72]. The Gozani group first advanced the 3 × MBT domain as an enrichment reagent for Kme1/2-containing proteins [34,41]. With the aid of the GST-tagged 3 × MBT domain, combined with the global SILAC and MS analysis, Moore et al. identified hundreds of potential Kme1/2-containing proteins from the nuclear extract of HEK293T cells [34,41]. In contrast with pan-αKme antibodies for which little is known about the mechanism of their recognition on Kme-containing peptides, the well-defined structure of the 3 × MBT domain further allowed Moore et al. to pair 3 × MBT and its D355N variant (negative control) to distinguish Kme-containing proteins from nonspecifically bound proteins. One limitation of 3 × MBT is that it cannot enrich Kme1/2-containing peptides [34,41]. It is thus less efficient to use the 3 × MBT domain as an enrichment reagent to map Kme sites.

Another example using Kme reader domains as enrichment reagents was reported by the Li group [60]. Although many Kme reader domains are originally described as the binder to methylated histones, recent work shows that many Kme reader domains also interact with diverse methylated nonhistone targets [33,67,69]. Li and colleagues hypothesized that the chromodomain of HP1β can recognize other Kme3-containing proteins besides the well-characterized H3K9me2/3. With the immobilized HP1β chromodomain as an enrichment reagent, they identified 109 interacting proteins [60]. Since only HP1β chromodomain but no inactive HP1β chromodomain mutant (negative control) was used in the work, further analysis is required to determine whether these proteins interact with the HP1β chromodomain in a Kme-dependent manner. The decent target overlap between the 3 × MBT domain and the HP1β chromodomain argues that the two reader domains act on a shared pool of Kme-containing targets. The 3 × MBT domain and the HP1β chromodomain therefore provide alternative options to enrich Kme-containing proteins.

10.3.3 RECOGNITION OF METHYLOME LABELED WITH CLICKABLE CHEMICAL TAGS

In comparison with weak noncovalent interactions between Kme/Rme-containing proteins and many enrichment reagents such as pan-αKme/Rme antibodies and Kme reader domains, enrichment via a covalently chemical interaction can be a robust alternative. The BPPM technology, as described above, provides a platform for chemical enrichment of proteome-wide PMT targets (Figure 10.6) [17]. A subset of native PMTs can also transfer the propargyl moiety of ProSeAM to their respective substrate pools [51,57]. With the vision of coupling the chemical-labeling approach with the well-established click chemistry, the SAM analogues developed for the BPPM as well as ProSeAM all contain the clickable S-alkyl moieties such as terminal alkyne or azide groups to be transferred to PMT targets [17]. For the methylome labeled with a terminal alkyne moiety, a terminal azide-containing probe such as TAMRA-azide fluorescent dye or azido-azo-biotin can be used for the subsequent chemical conjugation (Figure 10.6) [56,57]. The former can be used to visualize the labeled methylome via fluorescence imaging; the latter allows the ready enrichment of the labeled methylome with streptavidin beads (Figures 10.5 and 10.6) [56,57]. Another merit of the azido-azo-biotin probe lies in its distinct azo-linker, which can be cleaved under a mild condition with sodium dithionite ($Na_2S_2O_4$) (Figures 10.5 and 10.6) [56,57]. The chemically selective cleavage is expected to further improve the signal-to-noise ratios upon characterizing the enriched methylome because only the azo-linker-containing proteins rather than nonspecific binders of streptavidin beads can be eluted out after the treatment of $Na_2S_2O_4$ [56,57]. In comparison with the use of αKme/Rme antibodies and Kme reader domains to profile the methylome, the click-reaction-based chemical strategy is distinguished by its robustness and specificity for a designated PMT. With the advancement of BPPM technology beyond EuHMT1/2 and PRMT1, and PRMT3 [42,45,55], the approach is expected to demonstrate more broad application.

10.4 CHOICES OF PROCESSING METHYLOME SAMPLE

After isotopic or chemical labeling of the methylome, the resultant cell lysate needs to be properly processed for target enrichment, fractionation, and characterization. Most of the methods for

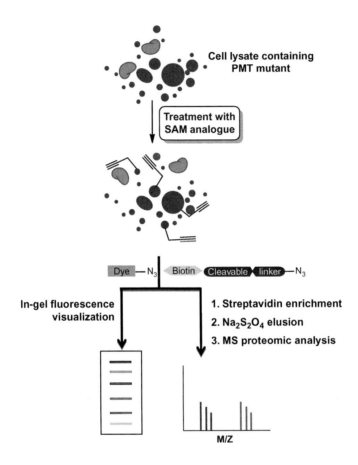

FIGURE 10.6

Descriptive workflow of methylome profiling by BPPM. Cells were transfected with a PMT mutant plasmid and then lysed, followed by treatment with SAM analogue cofactors. The labeled targets were then conjugated with the fluorescent dye for in-gel fluorescence or with cleavable azido-azo-biotin probe for target enrichment. (*This figure is reproduced in color in the color plate section.*)

methylome profiling have been significantly advanced in the past except for the reagents for methylome recognition or enrichment. For instance, the isotopic labeling of proteome-wide PMT targets has become routine for many laboratories (see Section 10.2, "Labeling Protein Methylation"). Many prior proteomic fractionation methods have been adapted to analyze methylome (see Section 10.5, "Deconvolution of Methylome Sample"). The rapid advancement of MS instruments further allows protein/peptide samples to be analyzed with unprecedented sensitivity and accuracy (see Subsection 10.6.2, "MS-Based Detection Modules"). The potential technical bottleneck for methylome enrichment largely determines how the samples can be processed for maximal target recovery. After the step of sample enrichment, additional isotopic labeling and sample processing steps can also be included to further improve the overall quality of the sample for subsequent characterization.

10.4.1 **DISSECTING METHYLOME AS PEPTIDES OR FULL-LENGTH PROTEINS**

The available enrichment reagents such as αKme/Rme antibodies, Kme reader domains and azide/alkyne-containing chemical probes have different compatibility toward peptide samples versus intact full-length proteins. For instance, so-far-characterized Kme reader domains generally favor enriching methylome from a mixture of intact full-length proteins [33,60]. Pan-Kme/Rme antibodies have been used as methylome-enrichment reagents to immunoprecipitate peptides and sometimes full-length proteins [5,33]. Such scenarios are largely determined by how the antibodies were developed and characterized. In contrast, azide/alkyne-containing chemical probes are expected to be tolerant of both full-length proteins and peptides. Besides the consideration of methylome enrichment reagents, how proteomic samples are processed also depends on the methods by which the enriched methylome will be characterized later. For instance, intact full-length proteins are suitable for examining methylation crosstalk via top-down MS analysis or profiling methylome via in-gel fluorescence or protein array (Figure 10.7) (see Section 10.6, "Detection Methods for Methylome Profiling"). For target identification, Kme/Rme-containing full-length proteins are first enriched and then digested into short peptides for MS analysis. In comparison, this shotgun approach allows all the peptide sequences of target proteins to be subject to MS analysis. To map Kme/Rme sites, the full-length proteins are often digested into short peptides with only Kme/Rme-containing peptides enriched for MS analysis. This approach is expected to improve signal-to-noise

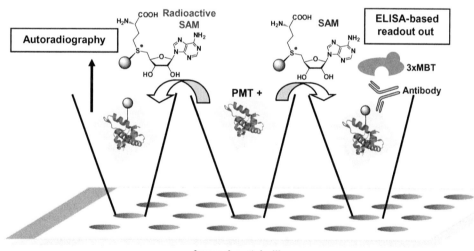

FIGURE 10.7

Application of the protein array for methylome profiling with single-target resolution. The arrayed substrate candidates are incubated with either radioactive SAM or nonradioactive SAM in the presence of a PMT. Autoradiography will be the readout of the former and the modified ELISA with antibody or reader domain will be the readout of the latter.

ratios by separating Kme/Rme-containing peptides from other irrelevant peptides. The methylome samples can also be subject to additional chemical derivatization with isotopic tags for MS quantification (see Section 10.4.3, "Chemical Derivatization of Methylome for MS Analysis").

10.4.2 CHOICES OF PROTEOLYTIC REAGENTS

Although trypsin is most frequently used to digest proteomic samples because of its high efficiency to cleave the carboxyl-side amide bond of the Lys or Arg, it may not always be the best choice of proteases for methylome profiling, in particular for mapping methylation sites. Among the Kme/Rme-harboring sequences, some may contain K/R-rich motifs, whose trypsin-digested products can be too short to be analyzed by LC-MS/MS. As alternative proteases for proteomic analysis, Arg-C, Glu-C, and Asp-N recognize the amide bonds at the carboxylic side of Arg and Glu or the amino side of Asp, respectively, and thus generate the different sets of digested products in comparison with those of trypsin-processed samples. Some Kmc/Rme-containing peptides can therefore be maintained to enough length for MS analysis with the alternative proteases but be fully digested with trypsin. One limitation using Arg-C is its low protease efficiency in comparison with trypsin [5]. To overcome the low digestion efficiency of Arg-C, Garcia et al. developed a chemical derivatization method in combination with trypsin to mimic Arg-C-like digestion patterns [5,73,74]. Here a protein sample is treated with propionic anhydride to protect free amine groups on the N-termini and ε-amino group of free Lys and Kme1/2 (see Section 10.4.3, "Chemical Derivatization of Methylome for MS Analysis"). The resultant samples were then treated with trypsin for which propionyl Lys and Kme1/2 sites will be spared from proteolytic cleavage. Although trypsin is expected to recognize free Lys and Arg as its cleavage sites better than Kme/Rme, it shall be noted that no systematic work has been done to explore the selectivity in a quantitive manner. We often note that trypsin-treated peptide products contain C-termini Kme/Rme (unpublished results), indicating that trypsin has residual activity to cleave the amide bonds after certain Kme/Rme sites. For K/R-deficient peptides, trypsin-digested products can be too long to be ionized efficiently for MS analysis. For this situation, a combination of proteases can be used to prepare protein samples for LC-MS/MS.

10.4.3 CHEMICAL DERIVATIZATION OF METHYLOME FOR MS ANALYSIS

Although the global SILAC is a powerful approach to differentially label PMT targets for MS quantification, this method can be costly and time consuming. In addition, it is also difficult for a common group to perform global SILAC experiments beyond cell lines for animal studies. Multiple chemical derivatization methods have been developed to differentially label digested peptides with isotopic chemical bar codes and thus applicable for animal and patient samples (Figure 10.8) [5,73–75]. The characteristic light-to-heavy isotopic mass shifts can then be used to quantify the relative ratios of multiple sets of samples. Garcia et al. first introduced a two-step chemical derivatization method to quantify histone methylation as well as other posttranslational modifications [73,74]. The first step is to label the N-termini and ε-amino group of free Lys and Kme1/2 with propionic anhydride, followed by trypsin digestion. This treatment essentially spares the Lys sites and leaves all Arg sites for trypsin cleavage to generate an ArgC-like proteolytic digestion pattern regardless of the levels of the prior Lys methylation. After the treatment with

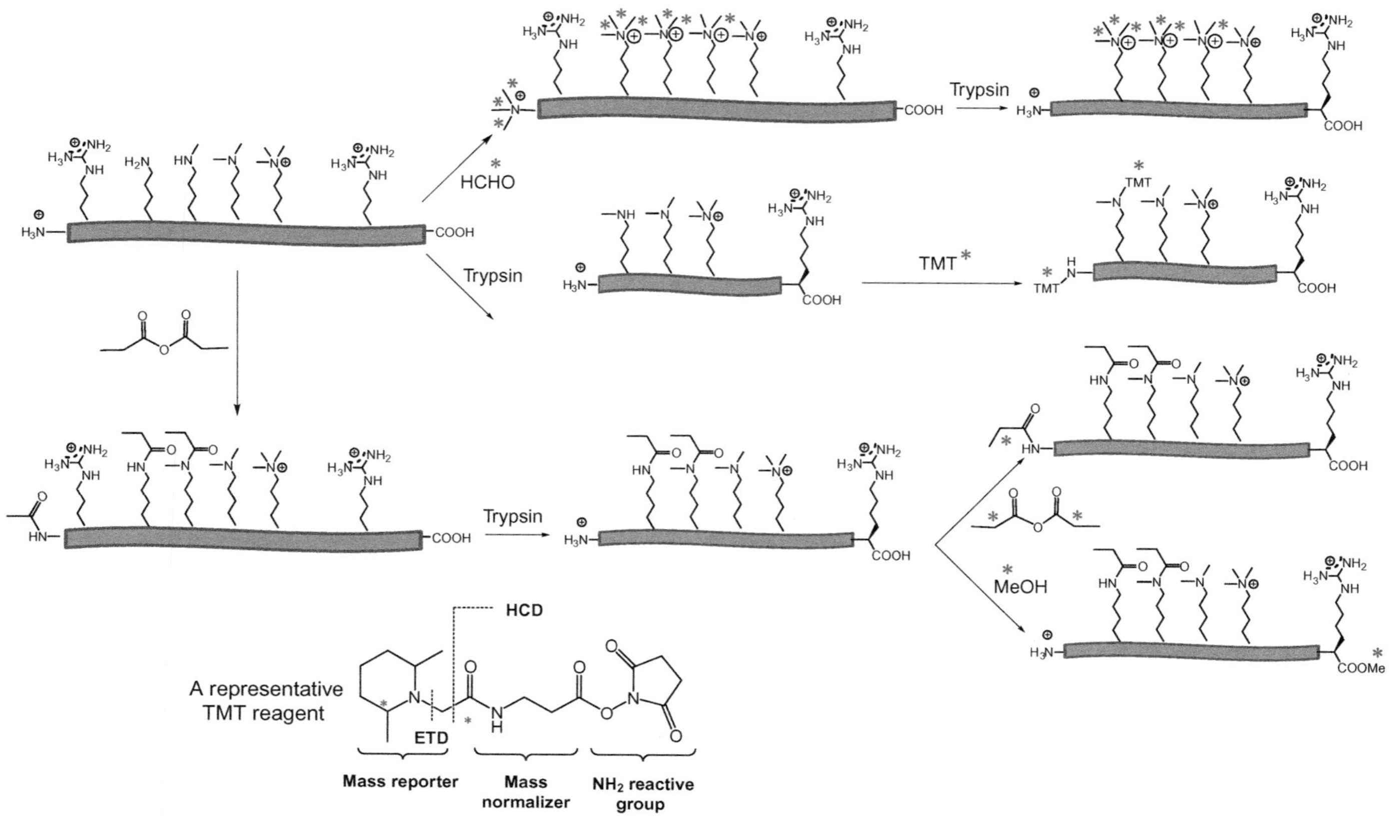

FIGURE 10.8

Derivatization of methylome with isotopic chemical bar codes. The characteristic light-to-heavy isotopic mass shifts can be introduced at various stages of sample processing. Their relative MS ratios will be used for quantification by MS. Free amine group can be labeled with isotopic propionic anhydride, formaldehyde or TMT; free carboxylic group can be labeled with isotopic methanol. A representative structure of TMT reagents is shown with the ETD/HCD cleavage sites highlighted. *The position for potential isotopic labeling.

trypsin, the second-step derivation involves the esterification of the carboxylic acid groups of the two sets of peptide samples with methanol and stable-isotope-labeled d4-methanol, respectively. Alternatively, after the ArgC-like digestion with trypsin, the two sets of peptide samples can be derivatized with d0- and d10-propionic anhydride, respectively, to label the newly released free N-terminal amino group. These methylated peptides can be directly quantified by comparing the relative ratios of the peak pairs of a 4 Da and 5 Da mass shift, respectively.

Another chemical derivatization method for MS quantification of PMT targets involves isotopic reductive methylation (MS quantification using isotopic reductive methylation or MassSQUIRM) (Figure 10.8) [75]. For MassSQUIRM, the peptides regardless of the levels of Lys methylation (free Lys and Kme1/2) are processed into K-methylation-saturated products via reductive methylation with isotopic formaldehyde. Each formaldehyde-derivatized chemical methylation is distinguished by an additional 2 Da mass increase in comparison with pre-existing native methylation. Besides the merit of quantification by the characteristic isotopic mass shift, all the peptides are converted into identical chemical entities and thus have the same ionization efficiency for MS quantification.

Fisher Thermo recently commercialized tandem mass tags (TMT) as alternative isotopic derivatization reagents for MS quantification (Figure 10.8). Amine-reactive TMT is an isobaric mass tag containing several regions (Figure 10.8): a mass reporter (cleaved off from the peptide), normalization fragment (linked to peptide after cleavage), MS-based cleavable linker (the residue for cleavage), and protein reactive group (amine-reactive functionality). After covalently labeling multiple sets of digested peptides at the N-terminus and side-chain amines with the protein reactive group of paired TMT tags, the samples are pooled and analyzed by LC-MS/MS. The ratios of isotopically characteristic MS/MS-fragment reporter ions will reflect the relative amounts of the differentially labeled peptides. The newly commercialized TMT kits allow ten sets of samples to be analyzed and compared in a single experiment.

10.5 DECONVOLUTION OF METHYLOME SAMPLE

Despite various labeling or enrichment strategies to resolve proteome-wide methylation targets from irrelevant background proteins, additional sample fractionation steps are often required to further improve the resolution of the methylome, which often contains tens of thousands of proteins or peptides with ion signals that differ by many orders of magnitude. Protein array technology allows the methylome to be revealed with single-target resolution (Figure 10.7), whereas the stepwise fractionation is more often used to characterize the methylome enriched from living cells. Prototypes of fractionation strategies can be based upon localization (e.g. nucleus, cytoplasm, or mitochondria), simple physical properties (e.g. acid extraction for basic proteins such as histones), and molecular weights (e.g. 1-D/2-D SDS-PAGE) to dissect methylome in a rough manner. More sophisticated strategies involve various types of chromatography to resolve methylome samples according to their slight differences in charges, isoelectric points, and hydrophilicity. Different chromatography methods can also be coupled in a continuous mode, if compatible eluents can be used, to improve their overall throughput or in a parallel mode to generate complementary data sets [33].

10.5.1 PROTEIN ARRAY TO REVEAL METHYLOME WITH SINGLE-TARGET RESOLUTION

Although several peptide array approaches have been used to profile the substrates of designed PMTs [33,76−80], arrayed proteins have merit as substrate candidates for methylome profiling given that they better mimic native substrates. Protein array allows thousands of substrate candidates to be screened against PMTs at single-target resolution (Figure 10.7) [34,38]. The modified targets in an array sample can be detected by autoradiography with radioactive SAM as a cofactor or by pan-αKme/Rme antibodies or Kme reader domains [34,38]. The Bedford group first used a protein array to screen the substrates of PRMT1 and CARM1. They identified five targets for PRMT1 and 2 for CARM1 [81]. The Gozani group recently demonstrated the use of a protein array to profile proteome-wide substrates of PMTs [38]. With the ProtoArray® glass slide coated with 9,500 proteins (Life Technologies), they screened SETD6 and identified 206 proteome-wide targets of the PMT [38]. Besides the ProtoArray as the protein array platform, at least two academic laboratories developed the technology to generate protein array chips of comparable quality [82−86]. The Zhu group developed the protocols to purify >17 K proteins and deposit them onto different slide surfaces, whereas the LaBaer group developed an *in vitro* transcription-translation protocol to immobilize newly synthesized proteins *in situ* [82−86]. The former is featured by high concentrations of proteins that can be deposited on chip spots. The latter approach is robust and convenient because only DNA templates are required to generate a protein array chip. The two protein array platforms have been used to explore various types of protein chemistry including several posttranslational modifications [82−86]. They are expected to be readily adapted for methylome profiling. One concern of protein array is the quality of proteins. It is thus necessary to validate the targets revealed by protein array in native contacts. In addition, the methyltransferase activities of PMTs in a cellular context may not be recapitulated with recombinant PMTs as enzyme sources. Although cell lysates could be used as enzyme sources to profile methylome in a protein array format, it would be challenging to assign the revealed methylation events to a specific PMT given the existence of multiple, even redundant, PMTs in each cellular context. However, this issue is expected to be solvable by combining a protein array with the bioorthogonal chemical-labeling technology (BPPM on protein chip) (Figure 10.3) [84].

10.5.2 FRACTIONATION WITH CHROMATOGRAPHY

After trypsin digestion and methylome enrichment (see Section 10.3, "Recognizing and Enriching Methylome" and Section 10.4, "Choices of Methylome Processing"), the resultant peptides often contain extra positively charged internal Kme/Rme sites in comparison with unmethylated peptides, which typically contain the N-terminal amino and C-terminal K/R residues (at least +3 charge for methylated peptides versus +2 for most unmethylated peptides at pH < 3) [87]. The Actuo group leveraged this difference and used strong cation-exchange (SCX) for further sample fractionation, coupled with MS, to reveal 39 Rme sites [87]. Isoelectric focusing (IEF) is a technique to resolve molecular entities by the difference of their isoelectric points (*p*I). The Actuo group noticed that many methylated peptides are basic with their isoelectric points close to 11 [87]. With aid of the conventional IEF technique, coupled with MS, they can identify 66 methylation sites [87]. The Actuo group further observed that many methylated peptides are hydrophilic [87]. These peptides can thus be resolved by hydrophilic interaction liquid chromatography (HILIC). After fractionating a mixture of peptide samples with a HILIC, they identified 215 methylation sites. Interestingly, despite being from the same sample, individual fractionation methods (SCX, IEF, and HILIC) revealed few overlaid targets [87]. Such observation argues that the methylome can be too complicated to be fully resolved by a single fractionation method. Combining several approaches can be more effective toward revealing a comprehensive landscape of methylome.

One limitation of SCX is that the samples are eluted with ionic solution, which is not compatible with HILIC and MS. An offline desalting step is needed after SCX fractionation, a process that makes it difficult to implement SCX in a continuous mode. To improve the throughput of cation exchange chromatography, the Garcia group developed a protocol to use a novel salt-free pH gradient for weak-cation exchange-hydrophilic interaction chromatography (WCX-HILIC) [88]. In contrast with the previous methods, which required >100 µg of sample and 50−100 h of analysis time to fully characterize a single histone extract, the WCX-HILIC method allows comparable work to be done with <1 µg of sample and within a few hours [88]. Since reversed phase HPLC (high-performance liquid chromatography) elutes out samples with MS-compatible solvents, this chromatography is often used in the last step of fractionation, also as a desalting step, prior to MS analysis.

10.6 DETECTION METHODS FOR METHYLOME PROFILING

After stepwise labeling, enrichment, processing, and deconvolution, the resultant methylome samples can be subject to various detection methods for target identification (Figure 10.2). As the most popular application of MS, protease-digested peptide samples are subject to shotgun analysis with MS to map sequences and methylation sites. In contrast, full-length protein samples can be examined by top-down MS as well as by imaging-based methods including autoradiography for radiolabeled samples, in-gel/on-chip fluorescence for dye-conjugated samples, and ELISA (enzyme-linked immunosorbent assay) [17,33,34,38,41,56,57]. Except for the protein array sample, which reaches the single-target resolution, the imaging-based methods are often used to roughly outline methylome rather than for target identification. The MS-based methods are classified on the basis of the difference of sample sizes (bottom-up, middle-down, and top-down) and instrumental modules for sample ionization, fragmentation, and detection [5].

10.6.1 IMAGING-BASED DETECTION METHODS

Prior sections have outlined several strategies to label methylome with radioactive materials. For instance, S-[methyl-^{14}C/^{3}H]-labeled radioactive SAM can be used as a cofactor *in vitro* to label PMT substrates arrayed on a protein chip (Figure 10.7) [33,34,38]. Alternatively, cell-membrane permeable S-[methyl-^{14}C/^{3}H]-labeled radioactive methionine can be processed by MATs into SAM within living cells, which can then be used *in situ* by endogenous PMTs. The resultant [methyl-^{14}C/^{3}H]-labeled products in both cases can be probed with autoradiography [37]. However, the broad application of this approach is limited because of the weak β-emission of [methyl-^{14}C/^{3}H] and thus low signal-to-noise ratios. In addition, S-[methyl-^{14}C/^{3}H]-labeled methylome can be subject to SDS-PAGE separation, although the resolution of this approach is too low to dissect individual protein bands for target identification.

Chemical labeling of methylome is distinguished by its ability to deliver a terminal-alkyne/azide group to target proteins [17]. Through the well-established click chemistry, the terminal-alkyne/ azide moieties can selectively react with a terminal-azide/alkyne-containing dye [56,57]. Many azide/alkyne-containing fluorescent dyes are commercially available with their emission of $570 \sim 670$ nm wavelength for red fluorescence (e.g. Cy3, tetramethylrhodamine, and Cy5). The resultant labeled methylome can be resolved by SDS-PAGE and visualized by in-gel fluorescence (Figure 10.7). Similar to other PAGE-resolving methods, the low resolution only allows the method to the rough profile of a methylome. Given the increased use of IRDye and VRDye secondary antibodies, their combination with multiple high-quality primary pan-αKme/Rme antibodies can be promising for multicolor fluorescent imaging of methylome resolved by SDS-PAGE or arrayed on a protein chip [33]. Alternatively, pan-αKme/Rme antibodies can also be used as detection reagents in conventional contexts such as ELISA [89] for methylome profiling [34,38,41].

10.6.2 MS-BASED DETECTION MODULES

Current MS instruments can handle three types of methylome samples classified by their approximate sizes: full-length methylated proteins (top-down analysis, >20 kDa), medium-length polypeptides (middle-down, $2 \sim 20$ kDa), and short peptides (bottom-up, <2 kDa) [5]. Analyzing intact methylated proteins containing multiple methylation sites requires instruments with high mass accuracy like FT-ICR-MS and occasionally hybrid linear quadrupole ion trap-orbitrap MS [5]. In contrast, short peptides are more suitable for tandem-based MS instruments with high capacity for sample quantification. Middle-down has a comparable capacity to bottom-up for sample quantification and to top-down for characterization of a combination of multiple methylations as well as other posttranslational modifications. There are several ways of ionization to prepare methylome samples for MS detection. In the matrix-assisted laser desorption/ionization (MALDI), an analyte (a methylome sample) is premixed with a matrix such as α-cyano-4-hydroxycinnamic acid, usually in a liquid phase. This mixture is deposited and then crystallizes on a metal plate. The analyte is ionized by a laser with the assistance of the matrix. The ions are then mobilized by a vacuum into an MS analyzer. In electrospray ionization (ESI), an analyte (a methylome sample) is ionized in a solvent (e.g. through protonation). The mixture is sprayed into a MS instrument via a force of a small spring, accompanied with the evaporation of the solvent. Electron capture dissociation (ECD) and electron transfer dissociation (ETD) can be used to ionize medium-length polypeptides $(2 \sim 20$ kDa) [5].

After sample ionization, ion fragmentation is often included to generate tandem mass spectra in modern MS instruments [5]. Among popular fragmentation techniques for tandem mass analysis are collision-induced dissociation [90], ETD and higher energy collisional dissociation (HCD). Here the patterns of the fragmentation of individual ions, together with their accurate mass, can provide valuable information such as peptide sequences and methylation sites through their b-type/y-type daughter ions. Given the existence of two isoforms of di-methylated arginine (ADMA and SDMA), the ETD-mediated fragmentation can be further implemented to analyze methyl-arginine-associated mass losses [5]. Here the characteristic fragment ions of dimethylammonium ion (m/z 46.06) and dimethylcarbodiimidium ion (mz 71.06), respectively, can be used to distinguish ADMA sites from

SDMA sites [91,92]. By adding the ammonium ion scanning module, the Wilkins group has improved the discovery of methylated peptides by four fold [93].

For methylome quantification, both shotgun and targeted MS methods have been developed [5]. The former is distinguished by its large-scale capacity by intensity-based product ion scanning without predefining analyzed ions. In contrast, the targeted MS quantification methods such as selected reaction monitoring (SRM) decouple the selection and analysis modes by first identifying a specific ion and then analyzing its intensity [5]. Given that only a specified set of ions is subject to MS analysis, SRM gives better sensitivity and broader detection range in comparison with the shotgun method. Parallel reaction monitoring (PRM) with a Q-Extractive instrument further improved the dynamic range, specificity, and detection efficiency of SRM by preselecting a specified m/z range and detecting multiple sets of product ions in a simultaneous manner. Here the methylome samples can be quantified as label-free peptides or isotopically labeled ions on the basis of the intensity of product ions or by comparing their isotopic ratios, respectively (see Section 10.2, "Labeling Protein Methylation").

10.7 BIOINFORMATICS OF METHYLOME

Given the current challenges to experimental methylome profiling, many efforts have also been made during the past decade to develop complementary bioinformatic approaches to predict methylation sites. In 2005, Daily et al. first proposed that disordered peptide regions are more susceptible for methylation and developed the corresponding algorithm to predict methylation sites [94]. MeMo is the first methylation modification prediction server, launched in 2006 [95]. MeMo predicts the potential methylation sites on the basis of the primary sequence features of the prior known methylation targets. BPB-PPMS (bi-profile Bayes-prediction of protein methylation server) and MASA (methylation accessible surface area) were introduced in 2009 with the capability to predict methylation sites on the basis of multiple parameters including primary sequences, accessible surface areas, and secondary structures of target peptides [96,97]. Two years later, Hu et al. developed a computational program to predict methylation sites with more parameters including amino acid frequencies, aromatic content, flexibility scalar, net charge, hydrophobic moment, beta entropy, disorder information as well as the PSI-BLAST (position-specific iterative basic local alignment search tool) profile of target peptides [98]. Many of these parameters were reflected in the later PLMLA (prediction of lysine methylation and lysine acetylation) method to predict lysine methylation in the context of full-length proteins [99]. Among the recently launched computational programs are PMeS, Methcrf, and iMethyl-PseAAC, which are featured by their further enhanced encoding parameters such as the sparse property coding, normalized van der Waals volume, position weight, and sequence evolution with the latest version containing a 346-dimensional vector [100–102]. Although the accuracy of these methylation prediction programs has gradually improved, there is still a significant margin to further improve the accuracy of these computational approaches to predict proteome-wide methylation sites.

Besides the intention to predict methylation sites, significant attention has been paid to establishing methylome databases such as PhosphoSitePlus[®]. PhosphoSitePlus is an online resource containing comprehensive information and tools to study methylation as well as other protein posttranslational modifications. As of November 2014, PhosphoSitePlus has collected information on 5,000 sites of mono-methylation, 2,559 sites of di-methylation, and 321 sites of tri-methylation, the majority of which are on Lys or Arg. UniProtKB/Swiss-Prot is a protein sequence database including methylation and other protein posttranslational modifications extracted from the scientific literature and biocurator-evaluated computational analysis. Many sequence analysis tools are associated with UniProtKB/Swiss-Prot entries through the ExPASy server. Interestingly, no tool on the present ExPASy server is available to predict methylation sites at this moment.

10.8 EXAMPLES OF INTEGRATIVE MODULES FOR METHYLOME PROFILING

Significant progress has been made in the past 2 years in revealing proteome-wide methylation events. These changes include novel methylome-specific labeling technology, better methylome enrichment reagents, and high-end methylome-analyzing instruments in particular MS. The combination of these modules allows for the discovery of thousands of novel PMT targets as well as methylation sites from the human proteome. Here a few representative examples were chosen to highlight the current capacity for methylome profiling.

10.8.1 GLOBAL IDENTIFICATION OF PROTEIN LYSINE METHYLATION WITH ANTIBODIES

In 2013, the Garcia group reported 323 sites of Kme1, 127 sites of Kme2 and 102 sites of Kme3, and a total of 413 methylated proteins from HeLa-S3 cells [103]. In this work, Cao et al. labeled the methylome of HeLa-S3 cells with S-[methyl-$^{13}CD_3$]-methionine using the heavy methyl labeling technology (Figure 10.9). With Kme1/2/3-Ahx(aminocaproic acid)-Cys-conjugated KLH

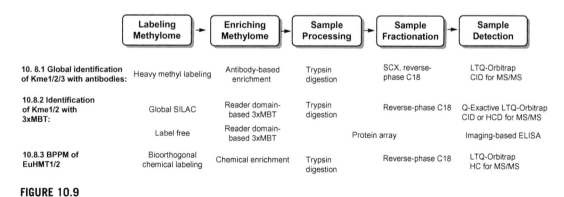

	Labeling Methylome	Enriching Methylome	Sample Processing	Sample Fractionation	Sample Detection
10. 8.1 Global identification of Kme1/2/3 with antibodies:	Heavy methyl labeling	Antibody-based enrichment	Trypsin digestion	SCX, reverse-phase C18	LTQ-Orbitrap CID for MS/MS
10.8.2 Identification of Kme1/2 with 3xMBT:	Global SILAC	Reader domain-based 3xMBT	Trypsin digestion	Reverse-phase C18	Q-Exactive LTQ-Orbitrap CID or HCD for MS/MS
	Label free	Reader domain-based 3xMBT		Protein array	Imaging-based ELISA
10.8.3 BPPM of EuHMT1/2	Bioorthogonal chemical labeling	Chemical enrichment	Trypsin digestion	Reverse-phase C18	LTQ-Orbitrap HC for MS/MS

FIGURE 10.9

Technological components of integrative application for methylome profiling. Three representative examples are chosen to highlight the current capacity for methylome profiling with respective technological modules dissected for direct comparison.

(keyhole limpet hemocyanin) as antigens, they developed three rabbit polyclonal antibodies against αKme1/2/3-containing peptides. After digesting the labeled proteome with trypsin, the resulting [methyl-$^{13}CD_3$]-labeled peptides were enriched with the polyclonal antibodies and subjected to SCX fractionation. After desalting with a homemade C_{18} tip, the resulting peptide sample was subjected to NanoLC followed by LTQ-Orbitrap MS equipped with CID mode for MS/MS. This single work presented the most extensive list of lysine methylation known at that time. Because many prior known Lys methylation sites are not included in this work, it remains to examine whether the additional methylation targets can be identified under the same context with other pan-αKme antibodies or by introducing more sophisticated fractionation steps such as WCX-HILIC developed by the same group.

10.8.2 IDENTIFICATION AND PROFILING OF PROTEIN LYSINE METHYLATION WITH 3×MBT READER DOMAIN

Given the low quality of many commercial pan-αKme antibodies, the Gozani group pursued a different path by developing pan-αKme enrichment reagents from naturally occurring Kme reader domains (Figure 10.9) [41]. They confirmed that the native 3 × MBT domain of L3MBTL1 but not its D355N mutant recognizes Kme1/2 via a single Kme binding pocket without significant dependence on Kme1/2's neighboring residues. To reveal human methylome with 3 × MBT and its D355N mutant as a pair of positive-control reporters, they carried out the two-way global SILAC experiment to label 293T cells with *L*-lysine/*L*-arginine or D_4-*L*-lysine/$^{13}C_6$-*L*-arginine ("light" and "heavy"), respectively. The resulting nucleoplasmic extracts were subject to the pulldown using GST-3 × MBT and GST-3 × MBTD355N, respectively. They also developed a three-way labeling experiment by culturing cells in light, medium, and heavy media [34]. This approach allows them to compare the 3 × MBT pull-down samples under two experimental settings (e.g. in the presence or absence of a PMT inhibitor) and one MBTD355N pull-down sample as a negative control. The bound proteins were then eluted out, resolved by SDS-PAGE and subject to trypsin digestion and HPLC packed with reverse-phase C_{18} resin. The identities of the methylome and methylation sites were characterized by MS. They have used Thermo Scientific Q-Exactive™, LTQ Orbitrap Elite™ and Velos instruments and obtained comparable results by performing MS/MS with CID or HCD fragmentation. A total of 544 proteins were identified in multiple repeats. Meanwhile, the Gozani group generated 3 × Flag-3 × MBT as a far Western blot reagent against Kme1/2-containing proteins (Figure 10.9). With 9,000 proteins on a protein chip (ProtoArrays) as arrayed candidates [41], the substrates of EuHMT1 were identified *in vitro* with the 3 × Flag-3 × MBT reporter, combined with mouse αFlag M2 antibody and an α-mouse fluorescent antibody. The robustness of using the 3 × MBT construct as a reporter of Kme1/2-containing proteins was also supported by subsequent success in validating multiple proteins as bona fide substrates of EuHMT1. The limitation of the current reader domain approach is that 3 × MBT only recognizes proteins containing Kme1/2 but not methylated peptides, making it challenging to map methylation sites. It will be interesting to explore whether other native or engineered reader domains can be used to overcome the limitation and to examine whether a similar approach can be applied to Rme-containing proteins, given the existence of Rme reader domains [67,68,104].

10.8.3 PROFILING SUBSTRATES OF EuHMT1 AND EuHMT2 WITH BPPM

With EuHMT1 and EuHMT2 as examples, our laboratory demonstrated the BPPM technology to dissect the methylome of specific PMTs from complicated cell lysates [45]. Here we engineered EuHMT1/2 and demonstrated that their Y1211A/Y1154A mutants are active toward Hey-SAM (Figure 10.9) [44,45,58]. The Hey-SAM cofactor was then added into the lysate of H293T cells that express the EuHMT1/2 mutants to label the methylome of EuHMT1/2. In parallel, we carried out a similar experiment with an empty vector as the negative control. A portion of the Hey-SAM-labeled methylome was subject to click chemistry with TAMRA-azide dye (Invitrogen), followed by in-gel fluorescence for visualization. This single experiment revealed hundreds of distinct fluorescent bands as potential substrates of EuHMT1/2. The methylome samples were then subject to click chemistry with an azide-azo-biotin probe. The resulting biotinylated methylome was enriched on streptavidin beads and cleaved off with sodium dithionite ($Na_2S_2O_4$), which targets the azo moiety of the click probe. The eluted protein samples were then subject to SDS-PAGE. After in-gel trypsin digestion, the resultant peptides were treated with respective TMT reagents. The products were further resolved by a nano-HPLC equipped with a homemade fused silica capillary column packed with reverse-phase C_{18} resin, and characterized by a Thermo LTQ-Orbitrap mass spectrometer equipped with HCD for MS/MS scanning. The BPPM approach allowed identification of >500 targets of EuHMT1/2 with representative targets validated with native EuHMT1/2 and SAM. This approach can be further improved upon by coupling the BPPM technology within living cells (see Subsection 10.2.3, "Chemical Labeling of Proteome-Wide PMT Substrates in Vitro and Within Living Cells") [36], the two/three-way global SILAC experiments (see Subsection 10.2.2, "Isotopic Labeling of Methylome Inside Living Cells") [34], and more sophisticated fractionation strategies as described under other settings (see Section 10.4, "Deconvolution of Methylome Sample") [33]. The limitation of the BPPM approach lies in its lack of generality. Here each PMT has to be engineered to utilize a bulky SAM analogue cofactor. Given the possibility that the bioorthogonal chemical labeling can be nonspecific for certain proteins or alter the substrate specificity of native enzyme-cofactor pairs to a small degree, BPPM-revealed targets need to be validated with native PMT-cofactor pairs [45,55].

10.9 PERSPECTIVE

Given the increased implications of PMTs under physiological and disease settings [2,8], more efforts have been made to elucidate PMT substrates at the proteomic level. The accompanying advancement of technology also allowed thousands of PMT substrates and methylation sites to be revealed at an unprecedented rate in the past several years. However, although individual approaches have respective merits for methylome profiling, their combination seems more powerful for addressing certain specific needs. Isotopic labeling of methylome inside living cells or during sample processing is always an ideal option for reliable sample quantification. It is also expected to be more efficient to enrich the methylome with a collection of antibodies, given the potential inability of a single reagent to recognize Kme/Rme epitopes independent of their neighboring sequences. More importantly, a significant margin is left to further leverage the overall quality of the antibodies for methylome enrichment. Proper sample processing with carefully selected proteolytic reagents and sophisticated sample fractionation with continuous chromatography will also improve the signal-to-noise ratios and thus uncover more targets during the subsequent MS analysis. Although only three PMTs were reported for BPPM application, the BPPM technology provides feasibility to assigning a single set of methylome to a single PMT. More efforts are thus expected to be made to engineer diverse PMTs for BPPM application. Additional efforts can be spent in developing bioinformatic algorithms to interrogate methylome of a specific PMT with more confidence. Given that methylome is context-dependent, it will be even more relevant to systematically annotate the targets and methylation sites of proteome-wide methylation events and elucidate the downstream roles in specific contexts. It will be interesting to dissect the methylation events of functional importance and examine whether the selective set of methylation events are sufficient to account for the observed phenotypes. In comparison with kinase-phosphatase biology, the complexity of the biology associated with PMTs is starting to be appreciated. Such recognition is expected to further inspire efforts to improve the current technology and develop new tools to address the needs of the emerging PMT field.

ACKNOWLEDGMENTS

The author is grateful for financial support from the National Institute of General Medical Sciences (1R01GM096056), the National Institute of Health (NIH) Director's New Innovator Award Program (1DP2-OD007335), the Starr Cancer Consortium, and Mr. William H. Goodwin and Mrs. Alice Goodwin, The Commonwealth Foundation for Cancer Research and The Experimental Therapeutics Center of Memorial Sloan Kettering Cancer Center.

REFERENCES

[1] Tuesta LM, Zhang Y. Mechanisms of epigenetic memory and addiction. EMBO J 2014;33(10):1091–103.
[2] Helin K, Dhanak D. Chromatin proteins and modifications as drug targets. Nature 2013;502(7472):480–8.
[3] Feng J, Nestler EJ. Epigenetic mechanisms of drug addiction. Curr Opin Neurobiol 2013;23(4):521–8.
[4] Jakovcevski M, Akbarian S. Epigenetic mechanisms in neurological disease. Nat Med 2012;18(8):1194–204.
[5] Afjehi-Sadat L, Garcia BA. Comprehending dynamic protein methylation with mass spectrometry. Curr Opin Chem Biol 2013;17(1):12–19.
[6] Kouzarides T. Chromatin modifications and their function. Cell 2007;128(4):693–705.
[7] Kouzarides T. SnapShot: histone-modifying enzymes. Cell 2007;128(4):802–3.
[8] Yang Y, Bedford MT. Protein arginine methyltransferases and cancer. Nat Rev Cancer 2013;13(1):37–50.
[9] Shimazu T, Barjau J, Sohtome Y, Sodeoka M, Shinkai Y. Selenium-based S-adenosylmethionine analog reveals the mammalian seven-beta-strand methyltransferase METTL10 to be an EF1A1 lysine methyltransferase. Plos One 2014;9(8):e105394.
[10] Kernstock S, Davydova E, Jakobsson M, Moen A, Pettersen S, Maelandsmo GM, et al. Lysine methylation of VCP by a member of a novel human protein methyltransferase family. Nat Commun 2012;3:1038.

[11] Jakobsson ME, Moen A, Bousset L, Egge-Jacobsen W, Kernstock S, Melki R, et al. Identification and characterization of a novel human methyltransferase modulating Hsp70 protein function through lysine methylation. J Biol Chem 2013;288(39):27752−63.

[12] Cloutier P, Lavallee-Adam M, Faubert D, Blanchette M, Coulombe B. A newly uncovered group of distantly related lysine methyltransferases preferentially interact with molecular chaperones to regulate their activity. PLoS Genet 2013;9(1):e1003210.

[13] Luo M. Current chemical biology approaches to interrogate protein methyltransferases. ACS Chem Biol 2012;7(3):443−63.

[14] Shinsky SA, Hu M, Vought VE, Ng SB, Bamshad MJ, Shendure J, et al. A non-active-site SET domain surface crucial for the interaction of MLL1 and the RbBP5/Ash2L heterodimer within MLL family core complexes. J Mol Biol 2014;426(12):2283−99.

[15] Patel A, Vought VE, Swatkoski S, Viggiano S, Howard B, Dharmarajan V, et al. Automethylation activities within the mixed lineage leukemia-1 (MLL1) core complex reveal evidence supporting a "two-active site" model for multiple histone H3 lysine 4 methylation. J Biol Chem 2014;289(2):868−84.

[16] Patel A, Vought VE, Dharmarajan V, Cosgrove MS. A novel non-SET domain multi-subunit methyl-transferase required for sequential nucleosomal histone H3 methylation by the mixed lineage leukemia protein-1 (MLL1) core complex. J Biol Chem 2011;286(5):3359−69.

[17] Wang R, Luo M. A journey toward Bioorthogonal Profiling of Protein Methylation inside living cells. Curr Opin Chem Biol 2013;17(5):729−37.

[18] Huang J, Dorsey J, Chuikov S, Zhang XY, Jenuwein T, Reinberg D, et al. G9a and Glp methylate lysine 373 in the tumor suppressor p53. J Biol Chem 2010;285(13):9636−41.

[19] Kurash JK, Lei H, Shen Q, Marston WL, Granda BW, Fan H, et al. Methylation of p53 by Set7/9 mediates p53 acetylation and activity in vivo. Mol Cell 2008;29(3):392−400.

[20] Saddic LA, West LE, Aslanian A, Yates JR, Rubin SM, Gozani O, et al. Methylation of the retinoblastoma tumor suppressor by SMYD2. J Biol Chem 2010;285(48):37733−40.

[21] Huang J, Perez-Burgos L, Placek BJ, Sengupta R, Richter M, Dorsey JA, et al. Repression of p53 activity by Smyd2-mediated methylation. Nature 2006;444(7119):629−32.

[22] Shi XB, Kachirskaia L, Yamaguchi H, West LE, Wen H, Wang EW, et al. Modulation of p53 function by SET8-mediated methylation at lysine 382. Mol Cell 2007;27(4):636−46.

[23] Mowen KA, Tang J, Zhu W, Schurter BT, Shuai K, Herschman HR, et al. Arginine methylation of STAT1 modulates IFNalpha/beta-induced transcription. Cell 2001;104(5):731−41.

[24] Vu LP, Perna F, Wang L, Voza F, Figueroa ME, Tempst P, et al. PRMT4 blocks myeloid differentiation by assembling a methyl-RUNX1-dependent repressor complex. Cell Rep 2013;5(6):1625−38.

[25] Yamagata K, Daitoku H, Takahashi Y, Namiki K, Hisatake K, Kako K, et al. Arginine methylation of FOXO transcription factors inhibits their phosphorylation by Akt. Mol Cell 2008;32(2):221−31.

[26] Lee JS, Kim Y, Kim IS, Kim B, Choi HJ, Lee JM, et al. Negative regulation of hypoxic responses via induced reptin methylation. Mol Cell 2010;39(1):71−85.

[27] Wu Q, Bruce AW, Jedrusik A, Ellis PD, Andrews RM, Langford CF, et al. CARM1 is required in embryonic stem cells to maintain pluripotency and resist differentiation. Stem Cells 2009;27(11):2637−45.

[28] Torres-Padilla ME, Parfitt DE, Kouzarides T, Zernicka-Goetz M. Histone arginine methylation regulates pluripotency in the early mouse embryo. Nature 2007;445(7124):214−18.

[29] Herz HM, Hu D, Shilatifard A. Enhancer malfunction in cancer. Mol Cell 2014;53(6):859−66.

[30] Liu Y, Liu K, Qin S, Xu C, Min J. Epigenetic targets and drug discovery: part 1: histone methylation. Pharmacol Ther 2014;143(3):275−94.

[31] Dhami GK, Liu H, Galka M, Voss C, Wei R, Muranko K, et al. Dynamic methylation of Numb by Set8 regulates its binding to p53 and apoptosis. Mol Cell 2013;50(4):565−76.

[32] Edwards A. Large-scale structural biology of the human proteome. Annu Rev Biochem 2009;78:541−68.

[33] Carlson SM, Gozani O. Emerging technologies to map the protein methylome. J Mol Biol 2014;426 (20):3350−62.

[34] Carlson SM, Moore KE, Green EM, Martin GM, Gozani O. Proteome-wide enrichment of proteins modified by lysine methylation. Nat Protoc 2014;9(1):37−50.

[35] Ong SE, Mittler G, Mann M. Identifying and quantifying in vivo methylation sites by heavy methyl SILAC. Nat Methods 2004;1(2):119−26.

[36] Wang R, Islam K, Liu Y, Zheng W, Tang H, Lailler N, et al. Profiling genome-wide chromatin methylation with engineered posttranslation apparatus within living cells. J Am Chem Soc 2013;135(3):1048−56.

[37] Cheng D, Vemulapalli V, Bedford MT. Methods applied to the study of protein arginine methylation. Methods Enzymol 2012;512:71−92.

[38] Levy D, Liu CL, Yang Z, Newman AM, Alizadeh AA, Utz PJ, et al. A proteomic approach for the identification of novel lysine methyltransferase substrates. Epigenetics Chromatin 2011;4(19):19.

[39] Lin Q, Jiang FY, Schultz PG, Gray NS. Design of allele-specific protein methyltransferase inhibitors. J Am Chem Soc 2001;123(47):11608−13.

[40] Chen DH, Wu KT, Hung CJ, Hsieh M, Li C. Effects of adenosine dialdehyde treatment on in vitro and in vivo stable protein methylation in HeLa cells. J Biochem 2004;136(3):371−6.

[41] Moore KE, Carlson SM, Camp ND, Cheung P, James RG, Chua KF, et al. A general molecular affinity strategy for global detection and proteomic analysis of lysine methylation. Mol Cell 2013;50(3):444−56.

[42] Wang R, Zheng W, Yu H, Deng H, Luo M. Labeling substrates of protein arginine methyltransferase with engineered enzymes and matched S-adenosyl-L-methionine analogues. J Am Chem Soc 2011;133 (20):7648−51.

[43] Wang R, Ibanez G, Islam K, Zheng W, Blum G, Sengelaub C, et al. Formulating a fluorogenic assay to evaluate S-adenosyl-L-methionine analogues as protein methyltransferase cofactors. Mol BioSyst 2011;7 (11):2970−81.

[44] Islam K, Zheng W, Yu H, Deng H, Luo M. Expanding cofactor repertoire of protein lysine methyltransferase for substrate labeling. ACS Chem Biol 2011;6:679−84.

[45] Islam K, Chen Y, Wu H, Bothwell IR, Blum GJ, Zeng H, et al. Defining efficient enzyme-cofactor pairs for bioorthogonal profiling of protein methylation. Proc Natl Acad Sci USA 2013;110(42):16778−83.

[46] Peters W, Willnow S, Duisken M, Kleine H, Macherey T, Duncan KE, et al. Enzymatic site-specific functionalization of protein methyltransferase substrates with alkynes for click labeling. Angew Chem Int Ed 2010;49(30):5170−3.

[47] Lukinavicius G, Lapiene V, Stasevskij Z, Dalhoff C, Weinhold E, Klimasauskas S. Targeted labeling of DNA by methyltransferase-directed transfer of activated groups (mTAG). J Am Chem Soc 2007;129 (10):2758−9.

[48] Klimasauskas S, Weinhold E. A new tool for biotechnology: AdoMet-dependent methyltransferases. Trends Biotechnol 2007;25(3):99−104.

[49] Dalhoff C, Lukinavicius G, Klimasauskas S, Weinhold E. Direct transfer of extended groups from synthetic cofactors by DNA methyltransferases. Nat Chem Biol 2006;2(1):31−2.

[50] Dalhoff C, Lukinavicius G, Klimasauakas S, Weinhold E. Synthesis of S-adenosyl-L-methionine analogs and their use for sequence-specific transalkylation of DNA by methyltransferases. Nat Protocols 2006;1:1879−86.

[51] Bothwell IR, Islam K, Chen Y, Zheng W, Blum G, Deng H, et al. Se-adenosyl-L-selenomethionine cofactor analogue as a reporter of protein methylation. J Am Chem Soc 2012;134(36):14905−12.

[52] Willnow S, Martin M, Luscher B, Weinhold E. A selenium-based click AdoMet analogue for versatile substrate labeling with wild-type protein methyltransferases. Chembiochem 2012;13(8):1167−73.

[53] Wu H, Min JR, Lunin VV, Antoshenko T, Dombrovski L, Zeng H, et al. Structural biology of human H3K9 methyltransferases. PloS One 2010;5(1):e8570.

[54] Qian C, Zhou MM. SET domain protein lysine methyltransferases: structure, specificity and catalysis. Cell Mol Life Sci 2006;63(23):2755−63.

[55] Guo H, Wang R, Zheng W, Chen Y, Blum G, Deng H, et al. Profiling substrates of protein arginine N-methyltransferase 3 with S-adenosyl-L-methionine analogues. ACS Chem Biol 2014;9(2):476−84.

[56] Blum G, Islam K, Luo M. Bioorthogonal profiling of protein methylation (BPPM) using an azido analog of S-adenosyl-L-methionine. Curr Protoc Chem Biol 2013;5(1):45−66.

[57] Blum G, Bothwell IR, Islam K, Luo M. Profiling protein methylation with cofactor analog containing terminal alkyne functionality. Curr Protoc Chem Biol 2013;5(1):67−88.

[58] Islam K, Bothwell I, Chen Y, Sengelaub C, Wang R, Deng H, et al. Bioorthogonal profiling of protein methylation using azido derivative of S-adenosyl-L-methionine. J Am Chem Soc 2012;134(13):5909−15.

[59] Wang L, Zhao Z, Meyer MB, Saha S, Yu M, Guo A, et al. CARM1 methylates chromatin remodeling factor BAF155 to enhance tumor progression and metastasis. Cancer Cell 2014;25(1):21−36.

[60] Liu H, Galka M, Mori E, Liu X, Lin YF, Wei R, et al. A method for systematic mapping of protein lysine methylation identifies functions for HP1beta in DNA damage response. Mol Cell 2013;50(5):723−35.

[61] Boisvert FM, Cote J, Boulanger MC, Richard S. A proteomic analysis of arginine-methylated protein complexes. Mol Cell Proteomics 2003;2(12):1319−30.

[62] Bremang M, Cuomo A, Agresta AM, Stugiewicz M, Spadotto V, Bonaldi T. Mass spectrometry-based identification and characterisation of lysine and arginine methylation in the human proteome. Mol Biosyst 2013;9(9):2231−47.

[63] Fisk JC, Li J, Wang H, Aletta JM, Qu J, Read LK. Proteomic analysis reveals diverse classes of arginine methylproteins in mitochondria of trypanosomes. Mol Cell Proteomics 2013;12(2):302−11.

[64] Lott K, Li J, Fisk JC, Wang H, Aletta JM, Qu J, et al. Global proteomic analysis in trypanosomes reveals unique proteins and conserved cellular processes impacted by arginine methylation. J Proteomics 2013;91:210−25.

[65] Guo A, Gu H, Zhou J, Mulhern D, Wang Y, Lee KA, et al. Immunoaffinity enrichment and mass spectrometry analysis of protein methylation. Mol Cell Proteomics 2014;13(1):372−87.

[66] Sylvestersen KB, Horn H, Jungmichel S, Jensen LJ, Nielsen ML. Proteomic analysis of arginine methylation sites in human cells reveals dynamic regulation during transcriptional arrest. Mol Cell Proteomics 2014;13(8):2072−88.

[67] Wagner T, Robaa D, Sippl W, Jung M. Mind the methyl: methyllysine binding proteins in epigenetic regulation. ChemMedChem 2014;9(3):466−83.

[68] Patel DJ, Wang Z. Readout of epigenetic modifications. Annu Rev Biochem 2013;82:81−118.

[69] Musselman CA, Lalonde ME, Cote J, Kutateladze TG. Perceiving the epigenetic landscape through histone readers. Nat Struct Mol Biol 2012;19(12):1218−27.

[70] Li H, Fischle W, Wang W, Duncan EM, Liang L, Murakami-Ishibe S, et al. Structural basis for lower lysine methylation state-specific readout by MBT repeats of L3MBTL1 and an engineered PHD finger. Mol Cell 2007;28(4):677−91.

[71] Min J, Allali-Hassani A, Nady N, Qi C, Ouyang H, Liu Y, et al. L3MBTL1 recognition of mono- and dimethylated histones. Nat Struct Mol Biol 2007;14(12):1229−30.

[72] West LE, Roy S, Lachmi-Weiner K, Hayashi R, Shi X, Appella E, et al. The MBT repeats of L3MBTL1 link SET8-mediated p53 methylation at lysine 382 to target gene repression. J Biol Chem 2010;285(48):37725−32.

[73] Garcia BA, Mollah S, Ueberheide BM, Busby SA, Muratore TL, Shabanowitz J, et al. Chemical derivatization of histones for facilitated analysis by mass spectrometry. Nat Protoc 2007;2(4):933−8.

[74] Plazas-Mayorca MD, Zee BM, Young NL, Fingerman IM, LeRoy G, Briggs SD, et al. One-pot shotgun quantitative mass spectrometry characterization of histones. J Proteome Res 2009;8(11):5367−74.

[75] Blair LP, Avaritt NL, Huang R, Cole PA, Taverna SD, Tackett AJ. MassSQUIRM: an assay for quantitative measurement of lysine demethylase activity. Epigenetics 2011;6(4):490−9.

[76] Dhayalan A, Kudithipudi S, Rathert P, Jeltsch A. Specificity analysis-based identification of new methylation targets of the SET7/9 protein lysine methyltransferase. Chem Biol 2011;18(1):111−20.

[77] Rathert P, Zhang X, Freund C, Cheng XD, Jeltsch A. Analysis of the substrate specificity of the dim-5 histone lysine methyltransferase using peptide arrays. Chem Biol 2008;15(1):5−11.

[78] Rathert P, Dhayalan A, Murakami M, Zhang X, Tamas R, Jurkowska R, et al. Protein lysine methyltransferase G9a acts on non-histone targets. Nat Chem Biol 2008;4(6):344−6.

[79] Rathert P, Dhayalan A, Ma HM, Jeltsch A. Specificity of protein lysine methyltransferases and methods for detection of lysine methylation of non-histone proteins. Mol BioSyst 2008;4(12):1186−90.

[80] Bicker KL, Obianyo O, Rust HL, Thompson PR. A combinatorial approach to characterize the substrate specificity of protein arginine methyltransferase 1. Mol Biosyst 2011;7(1):48−51.

[81] Lee J, Bedford MT. PABP1 identified as an arginine methyltransferase substrate using high-density protein arrays. EMBO Rep 2002;3(3):268−73.

[82] Chen CS, Korobkova E, Chen H, Zhu J, Jian X, Tao SC, et al. A proteome chip approach reveals new DNA damage recognition activities in *Escherichia coli*. Nat Methods 2008;5(1):69−74.

[83] Yu X, Woolery AR, Luong P, Hao YH, Grammel M, Westcott N, et al. Copper-catalyzed azide-alkyne cycloaddition (click chemistry)-based detection of global pathogen-host AMPylation on self-assembled human protein microarrays. Mol Cell Proteomics 2014;13(11):3164−76.

[84] Woolery AR, Yu X, LaBaer J, Orth K. AMPylation of Rho GTPases subverts multiple host signaling processes. J Biol Chem 2014;289(47):32977−88.

[85] Lu JY, Lin YY, Boeke JD, Zhu H. Using functional proteome microarrays to study protein lysine acetylation. Methods Mol Biol 2013;981:151−65.

[86] Lin YY, Lu JY, Zhang J, Walter W, Dang W, Wan J, et al. Protein acetylation microarray reveals that NuA4 controls key metabolic target regulating gluconeogenesis. Cell 2009;136(6):1073−784.

[87] Uhlmann T, Geoghegan VL, Thomas B, Ridlova G, Trudgian DC, Acuto O. A method for large-scale identification of protein arginine methylation. Mol Cell Proteomics 2012;11(11):1489−99.

[88] Young NL, DiMaggio PA, Plazas-Mayorca MD, Baliban RC, Floudas CA, Garcia BA. High throughput characterization of combinatorial histone codes. Mol Cell Proteomics 2009;8(10):2266−84.

[89] Bissinger E-M, Heinke R, Spannhoff A, Eberlin A, Metzger E, Cura V, et al. Acyl derivatives of p-aminosulfonamides and dapsone as new inhibitors of the arginine methyltransferase hPRMT1. Bioorg Med Chem Lett 2011;19(12):3717−31.

[90] Ragno R, Simeoni S, Castellano S, Vicidomini C, Mai A, Caroli A, et al. Small molecule inhibitors of histone arginine methyltransferases: homology modeling, molecular docking, binding mode analysis, and biological evaluations. J Med Chem 2007;50(6):1241−53.

[91] Erce MA, Pang CN, Hart-Smith G, Wilkins MR. The methylproteome and the intracellular methylation network. Proteomics 2012;12(4−5):564−86.

[92] Rappsilber J, Friesen WJ, Paushkin S, Dreyfuss G, Mann M. Detection of arginine dimethylated peptides by parallel precursor ion scanning mass spectrometry in positive ion mode. Anal Chem 2003;75(13):3107−14.

[93] Couttas TA, Raftery MJ, Bernardini G, Wilkins MR. Immonium ion scanning for the discovery of posttranslational modifications and its application to histones. J Proteome Res 2008;7(7):2632−41.

[94] Daily K., Radivojac P., Dunker A. Intrinsic disorder and protein modifications: building an SVM predictor for methylation. San Diego, California; 2005. p. 475−81.

[95] Chen H, Xue Y, Huang N, Yao X, Sun Z. MeMo: a web tool for prediction of protein methylation modifications. Nucleic Acids Res 2006;34(Web Server issue):249−53.

[96] Shao J, Xu D, Tsai SN, Wang Y, Ngai SM. Computational identification of protein methylation sites through bi-profile Bayes feature extraction. Plos One 2009;4(3):e4920.

[97] Shien DM, Lee TY, Chang WC, Hsu JB, Horng JT, Hsu PC, et al. Incorporating structural characteristics for identification of protein methylation sites. J Comput Chem 2009;30(9):1532−43.

[98] Hu LL, Li Z, Wang K, Niu S, Shi XH, Cai YD, et al. Prediction and analysis of protein methylarginine and methyllysine based on Multisequence features. Biopolymers 2011;95(11):763−71.

[99] Shi SP, Qiu JD, Sun XY, Suo SB, Huang SY, Liang RP. PLMLA: prediction of lysine methylation and lysine acetylation by combining multiple features. Mol Biosyst 2012;8(5):1520−7.

[100] Shi SP, Qiu JD, Sun XY, Suo SB, Huang SY, Liang RP. PMeS: prediction of methylation sites based on enhanced feature encoding scheme. Plos One 2012;7(6):e38772.

[101] Xu Y, Ding J, Huang Q, Deng NY. Prediction of protein methylation sites using conditional random field. Protein Pept Lett 2013;20(1):71−7.

[102] Qiu WR, Xiao X, Lin WZ, Chou KC. iMethyl-PseAAC: identification of protein methylation sites via a pseudo amino acid composition approach. Biomed Res Int 2014;2014:947416.

[103] Cao XJ, Arnaudo AM, Garcia BA. Large-scale global identification of protein lysine methylation in vivo. Epigenetics 2013;8(5):477−85.

[104] Taverna SD, Li H, Ruthenburg AJ, Allis CD, Patel DJ. How chromatin-binding modules interpret histone modifications: lessons from professional pocket pickers. Nat Struct Mol Biol 2007;14:1025−40.

BIOINFORMATICS AND BIOSTATISTICS IN MINING EPIGENETIC DISEASE MARKERS AND TARGETS

11

Junyan Lu, Hao Zhang, Liyi Zhang, and Cheng Luo

Drug Discovery and Design Center, State Key Laboratory of Drug Research, Shanghai Institute of Materia Medica, Chinese Academy of Sciences, Shanghai, China

CHAPTER OUTLINE

11.1 INTRODUCTION

Epigenetics refers to the study of meiotically and mitotically heritable changes in gene activity that do not involve the alteration of DNA sequence. Through covalent modifications on nucleotides and histones as well as other processes like RNA-associated silencing, epigenetic machinery orchestrates the distinct gene expression profiles in different cell lineages of an individual. The duty of epigenetic machinery is to activate or repress certain genes at the right time and the right place,

Y.G. Zheng (Ed): Epigenetic Technological Applications. DOI: http://dx.doi.org/10.1016/B978-0-12-801080-8.00011-9

which is crucial for the normal biological processes such as embryo development, stem cell self-renewal, and differentiation [1]. However, when this delicate balance fails, epigenetic machinery could also be the ringleader or accomplice in human diseases. A disrupted epigenetic regulatory system can lead to systematic genome instability as well as inappropriate expression or silencing of genes, resulting in "epigenetic diseases" such as cancers, autoimmune diseases, neural disorders, and other complex diseases [2]. Epigenetic malfunctions are now considered to be another driving force of Knudson's "two-hit" model in tumor development [3]. The strong link between aberrant epigenetic alterations and disease leads us to question whether epigenetic modifications on a single gene or the whole genome could be used as a criteria to distinguish healthy states and disease states, and whether epigenome-modifying enzymes could be regarded as therapeutic targets.

In contrast to genetic mutations, alteration of the epigenetic landscape is reversible. This motivates academic and industrial societies to put great effort into discovering and developing epigenetic drugs, which could reshape the disease epigenetic landscape toward a normal state through simply inhibiting enzymes or disrupting protein—protein interactions within the epigenetic machinery. To date, many epigenome-modifying agents have been discovered and some of them have shown promising therapeutic effect toward human diseases, especially cancers. The approval of histone deacetylase (HDAC) inhibitors — beginning with SAHA (suberanilohydroxamic acid) — as the first epigenetic drugs for the treatment of subtypes of leukemias and lymphomas, portends a bright future of targeting epigenetic enzymes as therapeutic strategies [4]. In addition, the tight association between epigenome states and cancer development or progression makes epigenetic modifications valuable biomarkers for cancer detection, tumor prognosis, and prediction of treatment responses (Figure 11.1). As a good example, CpG island hypermethylation has already been used as an epigenetic biomarker to detect cancer cells in several types of biological fluids and tissue biopsies [5,6]. Seligson et al. have shown that particular alterations of histone modifications were associated with recurrence of prostate cancer [7], indicating histone modification could also be used as disease markers. However, despite the promise of targeting epigenetic enzymes and using epigenetic modifications as disease markers, several concerns exist regarding their clinical applications. Although HDAC or DNMT (DNA methyltransferase) inhibitors exhibit promising antitumor effects, their therapeutic mechanism is still elusive. These agents mainly act through reducing the suppressive markers of gene expression in a genome-wide manner, which may lead to nonspecific activation of genes in normal cells, and therefore could be potentially mutagenic and carcinogenic. Intensive crosstalk has been found between epigenetic machinery and other cellular processes, such as signal transduction, metabolism, and gene regulatory networks. Perturbing epigenetic machinery without understanding its underlying mechanisms can be dangerous. Therefore, in order to mine more specific epigenetic targets and more indicative epigenetic biomarkers for diseases, the epigenetic machinery should be regarded as a part of the whole picture, and a more systematic approach should be adopted.

The rapid development of high-throughput techniques has led to an explosion of biological data and information, which gives us unprecedented opportunities to identify novel drug targets and biomarkers for complex disease. Computational tools, which are skilled at gathering, filtering, and analyzing huge amounts of hetero data and information, play indispensable roles in this target and biomarker mining process. In fact, bioinformatic and biostatistic methods have already greatly boosted the target identification and biomarker discovery for several complex diseases, such as Alzheimer's disease, autoimmune diseases, diabetes, and various cancers [8–10]. The identification of epigenetic disease markers and targets could also greatly benefit from these rapidly developing high-throughput technologies and computational tools. In this chapter, we review the application of state-of-the-art bioinformatic and biostatistic methods in mining epigenetic disease markers and targets. We will first introduce the

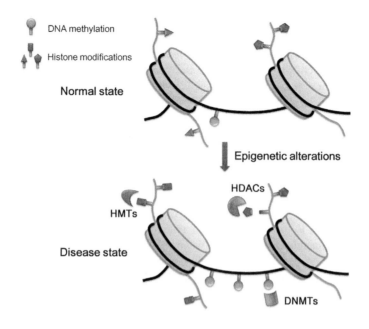

FIGURE 11.1

Epigenetic alterations and disease. Many human diseases correlate with aberrant DNA and histone modifications. On the one hand, the altered epigenomic profiles can be indicators of disease state. On the other hand, the epigenetic regulators that are responsible for these altered modifications can be therapeutically targeted. HMT, histone methyltransferase; HDAC, histone deacetylase; DNMT, DNA methyltransferase.

computational methods that process and interpret high-throughput epigenome data, whose reliability is of high importance for the subsequent data mining process. Then the recent development of bioinformatic and biostatistical methods, as well as their applications in discovering epigenetic disease markers and targets, will be described in detail. We will also briefly summarize several epigenome-related databases, software, and other packages that may be useful in conducting practical studies. Finally, the limits of current computational methods in epigenetic data mining and possible future directions will be discussed. This chapter is not a manual of the current computational tools and methods, but rather a guidebook for the researchers who intend to use bioinformatic and biostatistical methods in their epigenetic studies. We will mainly focus on epigenetic mechanisms involving DNA or histone modifications. Other epigenetic processes such as RNA processing are beyond the scope of this chapter.

11.2 ACQUISITION AND PROCESSING OF HIGH-THROUGHPUT EPIGENOMIC DATA

A wealth of homogeneous and reliable epigenomic data is a general prerequisite for mining drug targets or disease markers through bioinformatic or biostatistical approaches. Owing to the recent development of various high-throughput technologies, these data can be obtained quickly at a relatively low cost. Along with the power to generate massive amounts of data, these high-throughput technologies also bring greater challenges to data preprocess and quality-control techniques. Computational methods play irreplaceable roles at this stage to ensure the reliability of these high-throughput data, which greatly influences the accuracy and repeatability of the subsequent data mining results. In the following subsections, we will introduce some high-throughput technologies that are widely used in mapping epigenomic information and we will focus on the computational methods and tools for raw processing and quality control for the data generated by these high-throughput technologies.

11.2.1 BISULFITE SEQUENCING

Bisulfite sequencing is mainly used to detect DNA methylation patterns. As DNA methylation patterns are erased during PCR (polymerase chain reaction) amplification, current sequencing, and microarray technologies cannot distinguish between methylated and unmethylated cytosines. Thus, an extra step is needed to convert DNA methylation information into readily assayable DNA-sequencing information. Bisulfite-sequencing technology applies bisulfite to convert the methylation states of cytosines into methylation-dependent single-nucleotide polymorphisms (SNPs). Although this technology is restricted to DNA methylation, it is the gold standard due to its single-basepair resolution [11,12]. Analysis of sequences from bisulfite-treated DNA is based on single-nucleotide variant (SNV) detection methods, which identify variants from bisulfite conversion. Cytosines that have not been converted are assumed to be methylated. However, CpG dinucleotides are common sites of polymorphisms, which have to be distinguished from deaminations caused by bisulfite treatment. An error-prone approach is to exclude known polymorphisms, which might miss some potential methylations. Another approach utilizes the information from high-throughput sequencing. The SNV resulting from long-term deamination of cytosines to thymines will have been propagated on the opposing DNA strand as an adenine, while the deamination of unmethylated cytosines will leave guanines on the opposing strand unaffected. Thus sequencing both strands of bisulfite-treated DNA can discriminate between a CpG SNV and an unmethylated CpG [13].

So far, genome-wide high-throughput bisulfite-sequencing studies at single-basepair resolution have been performed mainly for small genomes. Major challenges for whole genome bisulfite sequencing are the limited sequencing capacities and high costs. Therefore, it is more practical to investigate parts of the genome to obtain the methylation patterns of mammals. The first approach to reduce the genomic complexity for bisulfite sequencing has been performed by PCR-amplification for targeted regions. One of the major innovations of this approach is the ESME (epigenetic sequencing methylation software) algorithm, which calculates methylation values directly from PCR-sequencing readouts and adjusts these values for incomplete conversion. This algorithm includes the following steps: entropy-based clipping, signal detection, alignment, trace correction, alignment-based clipping, signal normalization, compensation of incomplete conversion, and methylation estimation [14]. The second approach is the reduced representation bisulfite sequencing (RRBS), which involves restriction and window size selection, linker addition, bisulfite conversion, PCR amplification, and cloning. This approach is more efficient than the directed PCR approach, providing a comparative analysis of a defined proportion of a genome across multiple cell types without the need for targeted amplification. Nevertheless, the efficiency of this approach will decrease when applied to tissue samples that are inherently more heterogeneous than cell lines and thus higher sequencing coverage is required [13,15].

11.2.2 MICROARRAY AND CHIP-ON-CHIP

Although widely used in profiling transcriptomes, microarrays are seldom used alone to detect epigenetic modifications. The application of microarrays in profiling epigenomes usually requires a combination with other techniques such as bisulfite sequencing or chromatin immunoprecipitation (ChIP). A strategy to detect DNA methylation using microarrays involves the conventional bisulfite treatment of genomic DNA and PCR amplification for regions of interest. The products were then hybridized to microarrays that contain probes to discriminate converted from unconverted cytosines at the CpG site of interest, thus providing a readout for the methylation state at the CpG site. The major drawback of this approach is that most cytosines are converted to thymines after bisulfite treatment, which reduces the sequence complexity and makes it difficult to design unique probes to extend this approach for the whole genome [15,16]. Another strategy involves using a series of methylation sensitive restriction enzymes to discern methylated regions from unmethylated ones, which results in the differentiated enrichment of unmethylated and methylated DNA fragments. Subsequently, these fragments are interrogated by microarrays to decipher the methylation state on the original DNA sequence [17].

ChIP-on-chip combines chromatin immunoprecipitation (ChIP), which is used for specific DNA fragments enrichment, with microarray technology to detect differences between tested and control DNA samples. Initially, cells are treated with formaldehyde to cross-link any DNA-bound proteins to corresponding DNA. Subsequently, chromatin is extracted and sheared into small fragments of about 500 basepairs. Fragments are enriched using antibodies against histone modifications or chromatin proteins. Then DNA is released from these fragments and hybridized to a microarray. Regions that are significantly overrepresented in the immunoprecipitated DNA compared to control DNA are regarded as epigenetically modified or bound to proteins, according to the antibody used [11]. The key step of ChIP-on-chip data analysis is to derive a ranked list of overrepresented genomic regions from raw probe intensity data. Available algorithms are typically specifically targeted to peak finding for ChIP-on-chip data. The initial and still widely used strategy consists of three steps. First, microarrays are normalized to a common median intensity. Second, differential hybridization is tested locally on a sliding window and scores for each probe are derived. Third, significant probes are merged into regions of overrepresentation if sufficiently close to each other, and these regions are ranked by their combined scores [18]. Recent advances in this technology have made it possible to investigate chromatin structures on a genome-wide scale and have led to some interesting findings, such as the characteristic histone modification patterns marking gene promoters, broad domains of modifications at developmental loci, and modification patterns that are specific to embryonic stem cells [19–21].

11.2.3 NEXT-GENERATION SEQUENCING

The next-generation sequencing (NGS) platform performs massively parallel sequencing, which means that millions of small fragments of DNA are sequenced at the same time. NGS does not need prior information of DNA or RNA, compared with microarrays. Thus, NGS is able to detect additional SNPs, insertions or deletions, UTP regions, and splice junctions that are not detectable by common microarrays. The NGS platform was originally developed due to the demand for faster and cheaper sequencing of the entire genome. When combined with other experimental and computational techniques, NGS has been routinely used for high-throughput investigation of critical biological processes such as DNA methylation, histone modifications, the transcriptome, DNA—protein interactions, DNA accessibility, and long-range interactions on the genome-wide scale.

ChIP-sequencing, also known as ChIP-Seq, is one of the most widely used techniques to detect genome-wide epigenetic modifications. ChIP-Seq is a variant of ChIP-on-Chip that uses NGS technologies instead of microarrays to detect the differences between tested and control DNA samples. This method has two advantages over ChIP-on-Chip. First, data normalization is less of a concern in ChIP-Seq because sequencers produce absolute reading values rather than relative hybridization scores. Second, recent progress in sequencing technology makes ChIP-Seq highly cost-efficient. However, sharing the common dependence on high-quality antibodies with ChIP-on-Chip, ChIP-Seq has an additional disadvantage in that extra steps are required to restrict the data to specific regions on the chromatin. Nevertheless, ChIP-Seq has become the predominant means for investigating protein—DNA interactions in the human genome [11].

Similar to other high-throughput approaches that utilize NGS, ChIP-Seq relies heavily on bioinformatic techniques for data processing and analysis. The first step for ChIP-Seq data analysis is to match the sequenced short reads to a reference genome, which is called read mapping or alignment. Several flexible, memory efficient, and fast tools are available, including Bowtie, BWA, MAQ, Mosaik, Novoalign, SOAP, and ZOOM [22–24]. These programs achieve high speed by indexing the reference genome, which generates a look-up table that allows a fast search of subsequences contained in the reference genome. However, different algorithms have been adopted by these programs, which results in different performance. For example, MAQ uses a space-seeding approach that requires over 50 gigabytes while Bowtie and BWA use the Burrows-Wheeler transform method that requires only 2 gigabytes to store the human genome data. The Burrows-Wheeler transform algorithm is considerably more complex but much faster than a space-seeding algorithm [25,26].

After the readings have been mapped to the reference genome, the next step is peak calling to detect genomic loci that are highly enriched for sequenced reads. Bioinformatics analysis tools for detecting enriched regions usually take into account transcription factors when constructed. Transcription factors typically bind to short sequence elements of about ten basepairs, while histone marks are sometimes enriched over thousands of basepairs. To identify methylated histone regions, two different approaches have been applied in a pioneering study. One approach was based on fixed-size windows and an empirical background model, while the other approach was based on the hidden Markov model (HMM), which yields windows with variable length [27]. A major difficulty in detecting enriched regions is that there are three types of peaks: sharp, broad, and mixed. Sharp peaks are typically found for protein-DNA binding or histone modifications at regulatory elements, while broad peaks are usually associated with histone modifications that mark domains. Currently most algorithms have been developed for sharp peaks. A powerful method would be needed to detect peaks without previous knowledge of the type of enrichment, through combining approaches for both sharp peaks and broad ones [28,29].

The last step is to interpret the large amount of data obtained by ChIP-Seq. Many approaches can be applied to analyze the biological implications of ChIP-Seq data. For protein-DNA binding, the most common analysis is to discover binding motifs. Sequences of the top-ranking sites can be submitted to motif-finding programs such as MEME [30], MDScan [31], Weeder [32], and WebMOTIFS [33]. Afterward, potential binding motifs are reported along with corresponding statistical significance. The process of computing statistical significance is not straightforward, and available algorithms adopt different null models and multiple testing adjustments. Therefore, it is important to functionally validate any motifs that are found. Another common analysis with ChIP-Seq data is to annotate the peak locations on the genome in relation to known genomic features, such as the transcriptional site, exon-intron boundaries and 3′ ends of genes. More advanced analyses include the discovery of novel elements based on ChIP-Seq data and investigation of allele-specific binding and modification using ChIP-Seq data combined with SNP information [29].

A major application of ChIP-Seq in epigenetics is for mapping histone modifications. An approach called ChIPDiff was proposed to compare the genome-wide histone modification sites identified by ChIP-Seq, aiming at identifying different histone modification sites. This approach employs a HMM to infer the states of histone modification changes at each genomic site on the basis of ChIP fragment counts. ChIPDiff takes into account the correlation between consecutive bins, and automatically trains the transmission probabilities in an unsupervised manner, in order to avoid high variation caused by random sampling of ChIP fragments. Then the inference of the states of histone modification changes can be achieved using the trained HMM parameters [34]. An alternative way of deriving histone modifications from ChIP-Seq data is a spatial clustering approach for the identification of ChIP-enriched regions (SICER). This approach incorporates the tendency that histone modifications cluster to form domains. Through pooling together signals from all the nucleosomes located in the same modification state, this method improves the signal-to-noise ratio and deals with difficult cases where extended genomic regions are covered by diffuse enrichment. Moreover, this method identifies islands, rather than individual windows of fixed length, as clusters of enriched windows, which are the fundamental units of interest. Using both genome-scale analysis and datasets of genomic loci validated experimentally, SICER helps to identify ChIP-enriched domains in histone modification signals, especially those with diffuse profiles [35].

Application of NGS technology in DNA methylation profiling has also made mapping DNA methylation for the whole genome possible. Three approaches for genome-scale DNA methylation are commonly used. In the endonuclease restriction-based approach, fragmentation patterns generated through DNA digestion by methylation-sensitive enzymes are used to detect methylated regions. A version of this method in combination with NGS is called Methyl-seq. Another approach is affinity enrichment of methylated regions using immunoprecipitation with an antibody for either 5-methylcytosines or binding proteins. A version of this method in combination with NGS is MeDIP-seq. The third approach is bisulfite conversion coupled to sequencing (BS-seq), which exploits the fact that when DNA is treated with bisulfite, unmethylated cytosines (C) are converted into uracils (U). In the following replication, these uracils will become thymines (T). After the converted DNA is sequenced, methylated cytosines will appear in the obtained sequence reads as Cs, while unmethylated ones will appear as Ts. So far BS-seq is the only methylation mapping technique with single basepair resolution [28]. The biological complexity of DNA methylation is reflected in the multitude of available NGS protocols, which are classified into either the enrichment or the bisulfite conversion category. Enrichment methods yield a better cost to coverage ratio, while bisulfite conversion methods provide absolute readouts and single basepair resolution. However, currently no NGS protocol is available to specifically interrogate hmC at single basepair resolution.

11.2.4 **MASS SPECTROMETRY**

Due to the inherent difficulty of developing probe molecules to detect the quantitative levels of proteins, microarrays are used less in protein assays in comparison with nucleotide assays. Mass spectrometry (MS) is probably the most widely used technique for proteome assessment. It involves first separating a digested proteins mixture by liquid chromatography and then using online electrospray ionization to analyze the isolated peptides. MS is also a widely used high-throughput technique for characterizing histone modifications. Three main MS-based strategies exist for histone characterization. The most commonly used one is known as the *bottom-up approach*. In this workflow proteins are cleaved into small peptides, which are suitable for sequencing by MS/MS. However, the knowledge of global status of posttranslational modifications along the protein backbone is lost and mapping multivalent, coexisting posttranslational modification patterns is difficult when using this approach. Another strategy is known as the *top-down approach*, which characterizes the intact histone proteins by MS without cleaving the protein into peptides. This approach allows for mapping of coexisting posttranslational modifications, even those that are located far apart within the histone sequence. However, due to the higher charge states of intact protein ions, this approach often bring about complex MS and MS/MS spectra, as well as some difficulties in charge state determination, unless high-resolution tandem mass analyzers are used. The third strategy is known as the *middle-down approach*, which exploits the fact that the N-terminal tail of the histone can be cleaved off by specific proteases, producing a polypeptide of 40–50 residues that contain most posttranslational modifications of the histone. This approach reduces technical challenges of intact histone analysis and preserves a semiglobal overview of coexisting modifications [36]. MS is inherently not quantitative, but several developed methods add a quantitative abundance readout to the measurement. These methods include isotope-coded affinity tag (ICAT), isobaric tags for relative and absolute quantification (iTRAQ), stable isotope labelling by amino acids in cell culture (SILAC), and a variety of so-called label-free quantification methods. Enriched proteins can be easily distinguished from the background by their quantitative abundance ratio using a quantitative filter. In epigenetics this provides unique opportunities to investigate dynamic protein–protein interactions and their involvement in gene expression regulation [16,37].

11.3 **BIOINFORMATICS AND BIOSTATISTICS IN MINING EPIGENETIC BIOMARKERS**

A biomarker is an objectively measured characteristic that can be used as an indicator of a specific biological state. Broadly, biomarkers can be physiological indicators (such as body temperature, body weight, or blood pressure), medical images (such as B ultrasound, computed tomography [CT], or magnetic resonance imaging [MRI]), and biological molecules. Currently, the term "biomarker" usually refers to molecular biomarkers, which can be genes, peptides, proteins, and metabolites. Molecular biomarkers are usually measured in biological samples, including blood, urine, saliva, cerebrospinal fluid, or biopsy. According to their purposes, the clinical biomarkers can be categorized into five classes: (1) risk biomarkers, which are used to evaluate the risk of disease occurrence; (2) diagnostic biomarkers, which can help the physician to judge whether a patient has a certain disease; (3) staging biomarkers, which can be used to define the stage and severity of a disease; (4) prognostic biomarkers, which provide information about the patients' overall cancer outcome; (5) predictive biomarkers, which can be used to identify subpopulations of patients who are most likely to respond to a given therapy [38]. Molecular biomarkers, especially predictive biomarkers, can be extremely useful in helping physicians make individualized therapeutic decisions.

Genetic expression is regulated by epigenetic modifications, which should be stable enough to maintain patterns of genetic expression in different cell lineages during multiple rounds of division. However, these patterns are largely reshaped during the developmental processes and can be altered by environmental factors, such as lifestyle, stress or pharmacological intervention. Therefore, the epigenetic machinery should also be flexible enough to allow these changes. This dual nature of epigenetics between stability and plasticity renders them particularly useful for monitoring cell states, which leads to the concept of epigenetic biomarkers. "Epigenetic biomarkers" refers to the genomic regions that exhibit specific epigenetic alterations, which can lead to aberrant genetic expression profiles that correlate with the disease states. Epigenetic biomarkers can be based on either DNA methylation or histone modifications. Currently, most reported epigenetic biomarkers are DNA methylation biomarkers, because DNA methylation is stable and relatively easy to measure and its roles in disease development are more explicit. Although epigenetic biomarkers have not been used routinely in clinics, their effectiveness in diagnosis, and in predicting disease states and outcomes have been proven by a number of laboratory studies [39].

Molecular disease biomarkers can be discovered in two ways: a hypothesis-driven technique and a data-driven technique. As a more traditional way to identify biomarkers, the hypothesis-driven approach requires prior knowledge of a particular molecule in a disease model, and then a series of experimental studies are needed to validate its sensitivity, precision, and accuracy in indicating disease states. However, this approach can only find the biomarkers within the scope of current knowledge and cannot capture the whole complexity of human diseases. Alternatively, the ever-growing amount of experimental data and the development of high-throughput technologies have given rise to a more systematic and powerful way to discover novel disease biomarkers — the data-driven approach. Mining disease biomarkers through data-driven approaches involves three fundamental steps: the experimental design step, the data acquisition step, and the data mining step. Various experimental designs, such as simple case-control designs and more complex cohort study designs, can be used in biomarker studies. Among them, retrospective case-control design is the most frequently used, which compares the patients with a specific medical condition (case), for example a disease, with others who do not have this condition (control). In the data acquisition step, accurate measurements of the genomic, transcriptomic, proteomic, or epigenomic alterations in a given patient sample by experimental assays are required. In addition to advanced high-throughput technologies, these data can also be gathered from existing experimental results by using text mining or database. In the subsequent data mining step, various bioinformatic and biostatistical approaches provide an in-depth analysis of these data, in order to identify certain characteristics, such as the methylation state on one or several genes, that could separate the patient sample from normal samples (a diagnostic biomarker) or indicate the severity of a disease (a staging biomarker). Similar to the hypothesis-driven technique, biomarker candidates identified by the data-driven method also need follow-up experimental validation to judge whether they meet the stringent criteria of biomarkers for clinical application.

Except for techniques used in the data acquisition step (described in the previous section), the protocols for mining epigenetic biomarkers are not much different from those for genetic or proteomic biomarker development. Various computational methods can be used to mine disease biomarkers from a large dataset. The simplest and perhaps the most frequently used methods are based on classical statistical hypothesis tests, such as t-test, Chi-squared test, analysis of variance (ANOVA), Wilcoxon signed-rank test, and so on. In mining disease biomarkers, statistical hypothesis tests are usually used to select biological entities — usually genes, proteins, or metabolites — whose properties (genetic expression level, DNA methylation level, metabolite concentrations, etc.) are significantly different in two sample groups, such as a normal group and a disease group. Theoretically, the different traits of these biological entities in existing data sets can be used to classify new samples, and therefore they can be considered as candidates for disease biomarkers. Once the data have been well processed and normalized, the statistical hypothesis test can be performed straightforwardly by using statistical software such as R, SAS, and SPSS. Due to its simplicity, statistical hypothesis tests have been widely used in mining epigenetic or nonepigenetic biomarker candidates. Based on high-throughput methylation-specific oligonucleotide microarray data, Model et al. identified 20 most discriminating markers that were highly methylated in colorectal neoplasia through simple statistical hypothesis test methods. These markers showed high specificity for colorectal cancer and have potential as blood-based screening markers [40]. In another study, Lai et al. used a CpG island (CGI) microarray to uncover methylated genes in squamous cell carcinoma (SCC) of the uterine cervix. Through a subsequent hypothesis testing on these data, they discovered six genes that were more frequently methylated in SCC tissues than in the normal control, indicating these genes could be used for molecular screening of cervical cancer, although further validations are still needed [41]. However, statistical hypothesis test methods have many limitations. Aside from the large amount of criticism drawn toward the concept of the statistical hypothesis test itself [42], statistical hypothesis tests tend to oversimplify the disease model. The informative hidden structures within the sample data, such as gene—gene interactions, are lost when pair-wise hypothesis tests are performed between samples. In addition, hypothesis tests are weak at dealing with multiple groups, therefore unsuitable for discovering predictive and staging biomarkers.

The limit of statistical hypothesis test methods leads to the development of more sophisticated data mining and statistical bioinformatic methods, such as statistical learning methods. Statistical learning methods can be used to discover unknown properties in the data or to make novel predictions based on existing data (training data). There are two main categories of statistical learning methods: supervised learning and unsupervised learning. The major goal of supervised learning is to infer a function from labeled training data, and therefore supervised learning can be regarded as a classification process. The training data for supervised learning are composed by a set of training examples, and each training example is a pair consisting of an input object and a desired output value. In biomarker mining tasks, the input object is usually a vector that contains the organized sample data (or features), such as gene expression levels, DNA methylation levels, or enrichment of histone modifications, whereas the desired output value can be predefined sample categories, such as disease subtypes or disease stages. For biomarker discovering, most of the data mining tasks are "supervised" since the samples used for data mining are usually well phenotyped. Various supervised learning methods have been developed, such as linear regression, logistic regression, support vector machine (SVM), artificial neural networks (ANN), K-nearest neighbor (KNN), decision trees, and naïve Bayes. Detailed description of these methods can be found elsewhere [43]. Among them, SVM may be one of the most popular supervised learning methods and is frequently used in biomarker mining tasks. SVM tries to construct a hyperplane or a set of hyperplanes in a high-dimensional space, which can separate the samples from different categories with a clear gap that is as wide as possible. Wei et al. used a microarray-based approach to identify methylated loci prognostic for reduced progression-free survival (PFS) in advanced ovarian cancer patients. By applying SVM classifiers in combination with statistical hypothesis tests, they successfully identified 220 candidate PFS discriminatory methylated loci and demonstrated that 112 of them were capable of predicting PFS with 95% accuracy [44]. In an early study, Model et al. demonstrated that SVM in combination with microarray-based methylation analysis could reliably predict known tumor classes. In their study, they also emphasized the importance of feature selection in applying the supervised learning method to classify tumor samples, especially when the sample sizes were small [45]. Later, Zhang et al. developed an SVM based on the recursive feature elimination and cross validation (SVM-RFE-CV) algorithm for early detection of breast cancer in peripheral blood. Using the SVM-FRE-CV method, they successfully extracted 15 biomarkers with zero cross validation score and showed that this method outperformed other SVM-based approaches in both prediction performance and robustness to noisy data. In addition to SVM, other supervised learning methods have also been applied in biomarker mining tasks. In a study of characterizing the phenotypes of two main diffuse large B-cell lymphoma (DLBCL) subtypes − germinal center B-cell-like (GCB) and activated B-cell-like (ABC) − a Bayesian expression classifier was performed on DNA methylation data sets involving 50,000 CpG motifs among more than 14,000 promoter regions in these two DLBCL subtypes. This supervised analysis initially identified 311 differentially methylated probe sets (263 unique genes) between ABC and GCL. Further pathway enrichment and network analysis reduced this gene set to 16 genes, which could serve as an accurate predictor of ABC and GCB subtypes [46]. Rhee et al. conducted an integrated analysis of MethylCap-sequencing data and Affymetrix gene expression microarray data for 30 breast cancer cell lines. They used decision tree learning and Pearson's correlation analysis to investigate the effect of DNA methylation in various regions, including CGIs, promoter regions, first exons, etc., on the breast cancer subtype differential gene expression, and showed that the 30 breast cancer cell lines can be categorized into three molecular phenotypes − luminal, basal A and basal B − based on the differentially methylated regions [47]. Kisiel applied logistic regression in a study to identify noninvasive biomarkers for pancreatic cancer. They have demonstrated that the aberrant methylation state on several genes could be indicators of pancreatic cancer and BMP3 was the most discriminant methylation marker in stool [48].

Unsupervised learning tries to find hidden structure in unlabeled data, and therefore can be considered as a clustering problem. What differentiates supervised learning from unsupervised learning is that the data fed to the learner are unlabeled, and therefore there is no error or reward signal to evaluate a potential solution. Unsupervised learning methods include hierarchical clustering, K-means clustering, principal component analysis (PCA), self-organization maps, and HMM. Compared with supervised learning, unsupervised learning methods are less commonly used in biomarker mining tasks. However, unsupervised learning is better at discovering hidden correlations within sample data, such as nonlinear interactions between genes, and therefore the adoption of unsupervised methods can improve the success rate of biomarker development, when combined with other data mining methods. Pal et al. proposed a scheme that integrated neural network classifiers and fuzzy clustering for identifying a small set of biomarkers to distinguish four usually misdiagnosed childhood cancers. They have shown that their approach was capable of identifying a very small set of genes as biomarkers with high performance, by taking the interaction between genes into consideration [49]. In another study, Chambwe et al. performed unsupervised hierarchical clustering on genome-wide DNA methylation data of 140 DLBCL samples and ten normal germinal center B cells. Through this approach, they identified six novel epigenetic clusters that tightly correlated with survival outcomes. Further pathway enrichment analysis revealed these clusters are characterized by disruption of several biological pathways, such as cytokine-mediated signaling, ephrin signaling, apoptosis, and cell-cycle regulation. As a combination between supervised learning and unsupervised learning, semi-supervised learning can make use of unlabeled data for training and it has several applications in biomarker mining tasks [50−52].

During the process of biomarker discovery, the predictive value and generalizable power of selected candidates needs to be rigorously assessed by several objective statistical measures, such as sensitivity, specificity, the F-score, or the area under the receiver-operating characteristics (ROC) [53−55]. Most importantly, it is necessary to validate the identified biomarker candidates using larger sample sets, covering a broad cross-section of patients or populations. However, most laboratory identified epigenetic biomarkers fail to undergo this validation process, because the clinical epigenetic data samples are still costly. Alternatively, when the sample sizes are small and no independent cohort for validation is available, other statistical strategies such as cross-validation, bootstrapping, or permutation tests can be used for biomarker validation [56−58]. Overall, bioinformatics and biostatistics play key roles in almost every stage in the biomarker development pipeline, including data preparation, sample separation, data mining, and biomarker validation. Moreover, bioinformatic and biostatistical tools can help us make further biological and biochemical interpretations of identified epigenetic biomarker candidates in the context of a biological network, which will greatly enhance our current understanding of epigenetic machinery.

11.4 BIOINFORMATICS AND BIOSTATISTICS IN MINING EPIGENETIC DISEASE TARGETS

The success of drug discovery today is highly dependent on the selection of proper targets. Drug targets can be certain biological entities (proteins/genes/miRNA) that are relevant to a specific disease and whose activity can be modulated through interaction with small molecules, peptides, or proteins. Although disease biomarkers and drug targets share some similarities, they are intrinsically different in two ways. First, the activity of a drug target should be able to be modulated, which also is referred to as "druggable," while a biomarker does not need to be. For example, the concentration of certain metabolites or the methylation levels on certain genes can be biomarkers, but only the enzyme responsible for these changes can be drug targets. Second, the change of the activity of a drug target should have a clear impact on the biological state, while a biomarker is just an indicator of a biological state. Therefore, only the disease-driven genes or proteins can become drug targets, whereas many passengers can also be good biomarkers. These two differences lead to more strict criteria for target selection. The target discovery tasks require more knowledge about the disease mechanisms, the cellular interactions underlying disease phenotype and the complex biological networks.

Because of the stringent criteria to define a good drug target, current drug targets are distributed only among very limited families of biological components; most of them are G-protein-coupled receptors (GPCRs), nuclear receptors, ion channels, and protein kinases [59]. As previously described, epigenetic machinery plays a key role in human pathologies. Almost all epigenetic modifications are reversible, which renders them amenable to manipulation by extrinsic factors. Both the epigenome-modifying enzymes (writers) and the proteins that specifically recognize these modifications (readers) can be targeted to regulate epigenetic machinery. In addition, more and more evidence has supported a casual role of epigenetic alterations in the onset and progression of human diseases, which means the epigenetic states not only can be indicators of the disease states, but also can be therapeutically targeted. Despite the successes of several epigenetic drugs, such as DNMT inhibitors and HDAC inhibitors, many epigenetic modulators suffer from their promiscuous targets and elusive mechanism of action, which may bring potential safety issues into their clinical usages. Therefore, more specific epigenetic targets as well as a more comprehensive and systematic understanding of the epigenetic mechanisms are necessary.

Similar to biomarker discovery, drug target identification can also benefit from a wealth of literature-based and high-throughput technology-based data sources. Text mining of literate databases is one of the most popular approaches in identifying drug targets. Text mining is a computational way of uncovering novel information by automatically extracting information from literature sources, such as PubMed [60]. Text mining first retrieves literature or abstracts on a particular topic by using general search engines or specifically designed tools and then identifies or tabulates the relevant entities or facts from the retrieved information. The candidate genes or proteins for drug target can be inferred from the co-occurrence between the candidates and a certain disease or other known disease-related genes or proteins. Text mining can also be used to reconstruct a gene–gene or protein–protein interaction network, which can be further analyzed through network-based or statistical learning approaches to identify new drug target candidates. Pospisil and coworkers combined text mining and database searching to identify enzymes with hydrolytic activities that are expressed in the tumor microenvironment. Through this method, they discovered several cancer-related hydrolases, such as prostatic acid phosphatase (ACPP), prostate-specific antigen (PSA), and sulfatase 1 (SULF1), which could be suitable targets for cancer imaging and therapy [61]. Data generated by high-throughput technologies, such as microarrays, ChIP-Seq, bisulfite sequencing, MS, etc., are also frequently used for target mining. Owing to their similarities, drug targets and disease biomarkers can be discovered by using the same bioinformatic and biostatistical tools, such as statistical hypothesis testing, supervised learning, and unsupervised learning. In a study led by Perry et al., a series of statistical hypothesis tests, including t-test, ANOVA test, and Wilcoxon-matched pairs test, were used to globally identify potential targets of promoter hypermethylation in prostate cancer. In combination with experimental validations, they finally identified IGFBP3 as a novel methylation target in prostate cancer [62]. Ryu and colleagues compared gene expression profiles from melanoma cell lines in different stages of malignant progression to identify novel molecular signatures as well as therapeutic targets for aggressive melanoma. Using unsupervised hierarchical clustering, they initially categorized these cell lines into two distinctive molecular subclasses: aggressive metastatic tumor cell lines and less-aggressive primary tumor cell lines. Further supervised microarray data mining led them to identify a transcription factor – NF-κB – as being a potential "master regulator" of melanoma invasion, indicating it could be a novel drug target for metastatic melanoma [63].

In addition to these data mining techniques, network-based approaches are also widely used in the study of epigenetic regulatory mechanisms as well as target identification. Network biology is based on the principle that the behavior of most complex systems, such as cell and social networks, emerges from the orchestrated activity of pairwise interactions between many individuals. Therefore, the components in a complex system can be reduced to a series of nodes that are interconnected by links, with each link representing interaction between two components. The nodes and links together form a network. In a biological network, the nodes can represent biological entities such as genes, proteins, small molecules, etc., and the links can be the regulatory relationships, such as protein-protein interaction, transcription regulation, modification, etc. Useful information such as the hub nodes of the network, which is usually the most important component in certain biological processes, can be directly obtained by analyzing the network structure. In addition, by using various mathematical modeling techniques, the static network topology can be converted to dynamic and predictive models to gain deeper insight into the dynamic behavior of biological systems. Epigenetic machinery is tightly connected to other biological processes such as the gene regulatory network and the signaling transduction pathway, and therefore, network-based approaches are especially useful for studying epigenetic regulatory mechanisms and identifying key components that could be drug targets. Through a network-based approach, West and colleagues studied the mechanism of age-associated epigenetic drift. They first mapped genes with age-associated methylation changes onto a comprehensive and highly curated human protein-interaction network and then extensively studied the topological features of this integrated network. The results indicated that epigenetic drift preferentially occurs in parts of the cellular network that are not vital to the survival of a cell but that may lead to epigenetic deregulations of a few key transcription factors that correlated with longevity, stem cell function and disease predisposition [64]. In another study, Roznovat et al. constructed a dynamic network model for genetic and epigenetic changes observed at different stages of colon cancer, aiming to provide improved insight into the relationship between aberrant modifications and cancer initiation and progression. Through network dynamic simulation and analysis, they captured several key molecular events that served as potential "triggers" in cancer initiation and malignant transformation. These results could provide useful information on target identification of colon cancers [65].

To date, mining epigenetic drug target through systematic approaches is still in its infancy and very few successful examples have been reported. However, considering the diversity of epigenetic modifications and their close connections with various diseases, there are still many treasures to be mined in this field. The ever-increasing epigenomic data and quickly developing computational techniques will also greatly facilitate this process. In addition, other emerging new techniques and concepts such as chemogenomic [66] and network pharmacology [67] will push this field to a brighter future.

11.5 USEFUL BIOINFORMATIC RESOURCES

11.5.1 DATABASES

The rapid advances in epigenetics lead to a large body of experimental findings on this topic being published every year. Moreover, the application of high-throughput technology in epigenetic studies has resulted in an explosion of epigenomic data. While these biological "big data" can greatly facilitate the target or biomarker mining processes, they also bring more challenges in data collection, storage, curation, and presentation. Biological databases are aimed at solving this problem by integrating and presenting available data in a clear and easily accessible way. Bioinformaticians and experimentalists have already benefited greatly from several successful genomic databases, such as Gene Expression Omnibus (GEO) [68], the Ensembl genome database [69], Expression Atlas [70] and the Cancer Genome Atlas (TCGA) [71]. Compared to genomic data, epigenomic data are more scattered and inhomogeneous due to the technical limitations and the diversity of epigenetic modifications. This makes it more important to systemize epigenomic data through databases.

The Encyclopedia of DNA Elements (ENCODE) is a public research project launched by the US National Human Genome Research Institute (NHGRI) in September 2003 [72]. As a follow-up to the Human Genome Project, the ENCODE project aims at identifying all functional elements in the human genome sequence, including elements that act at the protein and RNA levels, and regulatory elements that control cells and circumstances in which a gene is active. A Data Coordination Center (genome.ucsc.edu) was built to track, store, integrate, and display ENCODE data. Different from many public genomic databases, the ENCODE Data Coordination Center performs a series of checks to ensure that the data meets specific quality standards when it is released to the public, making the ENCODE data highly reliable. The experimental data stored in ENCODE Data Coordination Center are mainly derived from ChIP-Seq, DNase I Hypersensitivity, RNA-Seq, and assays of DNA methylation. All ENCODE data is freely available for download and analysis. To this point, ENCODE has sampled 119 different DNA-binding proteins and a number of RNA polymerase components in 72 cell types as well as up to 12 histone modifications and DNA methylation in 46 cell types. The ENCODE project provides an unprecedented opportunity to study the epigenome in a systematic way and is an excellent source for data mining studies.

The target and biomarker mining approaches can also benefit from other more specific epigenetic databases. (These are summarized in Table 11.1, along with urls and reference info for each.) The majority of these databases are constructed to archive DNA methylation data and link this information with genotypic and phenotypic data. MethDB is a database that collects and attempts to unify the methylation data at different levels of resolution, ranging from ^{5m}C content of the genome to the methylation states of single nucleotides. It archives DNA methylation data

Table 11.1 Biological Databases for Epigenetic Studies

Database	Focus	URL	Reference
MethDB	Comprehensive methylation data	http://www.methdb.de	[73]
MethPrimerDB	PCR-based methylation data	http://medgen.ugent.be/methprimerdb	[74]
MethyCancer	DNA methylation in cancers	http://methycancer.psych.ac.cn	[76]
PubMeth	DNA methylation in cancers	http://www.pubmeth.org	[77]
MethyCancerDB	DNA methylation in cancers	http://www.methcancerdb.net	[78]
MENT	DNA methylation in cancers	http://mgrc.kribb.re.kr:8080/MENT/	[79]
DiseaseMeth	Human disease methylation data	http://bioinfo.hrbmu.edu.cn/diseasemeth	[80]
NGSmethDB	Methylation data from sequencing	http://bioinfo2.ugr.es/NGSmethDB	[75]
HHMD	Histone modifications in human	http://bioinfo.hrbmu.edu.cn/hhmd	[81]
HEMD	Epigenetic enzyme and modulator	http://mdl.shsmu.edu.cn/HEMD/	[82]

and information on the origins of investigated samples and experimental procedures. Queries are submitted based on ^{5m}C content, species, tissue, gene, sex, phenotype, sequence ID, and DNA type. The output lists all available information including relative gene expression levels. DNA methylation patterns and profiles can be visualized as a graph and as nucleotide sequences or tables containing the information of sequence positions and methylation levels [73]. MethPrimerDB is mainly focused on the storage and retrieval of validated PCR-based methylation assays. Records in the database can be searched by gene symbol, nucleotide sequence, analytical methods used, Entrez Gene or MethPrimerDB identifier, and name of the submitter. Each record contains a link to Entrez Gene and PubMed to retrieve additional information on the gene, where its genomic context and articles about methylation assays were described. Currently, MethPrimerDB contains primer records for the most popular PCR-based methylation analysis methods to investigate human, mouse, and rat epigenetic modifications [74]. NGSmethDB is a database archiving methylation data obtained from NGS. With a web interface linked to a MySQL database and several tools for data mining, users can retrieve methylation data for a set of tissues in a given chromosomal region, or display the methylation states of promoters among different tissues. NGSmethDB currently contains data mainly from human, mouse, and *Arabidopsis*, but methylomes from other species will be incorporated through an automatic pipeline as soon as new data is available [75].

Over the last 15 years, intensive studies have been conducted to reveal the epigenetic changes in cancer, which yield large amounts of data on cancer epigenetics. MethyCancer is a database that integrates data from public resources and data produced from the Cancer Epigenome Project in China. It archives highly integrated data of DNA methylation, cancer-related genes, mutation, and cancer information from public resources large-scale sequencing results. Moreover, along with MethyCancer comes a powerful graphical tool, MethView, which has been developed to provide user-friendly access to the data. MethView is able to display DNA methylation in the context of the genome, which facilitates the understanding of genetic and epigenetic mechanisms in cancer research [76]. PubMeth is a cancer methylation database, which is mainly based on literature information. A query can be either a gene name or a cancer type to investigate the correlation between hypermethylated genes and certain cancer types. PubMeth is based on automatic text mining of PubMed abstracts in combination with manual reading and annotation of preselected abstracts. The text-mining approach leads to higher speed and selectivity, while the manual reading raises the specificity and quality of the database [77]. MethCancerDB is a database that collects and annotates genes and sequences from published methylation studies and interlinks them to all relevant methylation resources. It provides a summary of preexisting information regarding DNA methylation in various cancers. A novelty of this database is that its entries are based on entire studies instead of single observations or patients. With experimental designs documented, users can quickly overview all studies and their results for a specific biological or clinical proposal [78]. MENT is an integrated database of DNA methylation and gene expression for diverse cancers, which contains data in paired samples and clinicopathological conditions gathered from the GEO and TCGA (The Cancer Genome Atlas) [79]. Databases for diseases other than cancer also exist. DiseaseMeth is a human disease methylation database, which focuses on the storage and statistical analysis of DNA methylation datasets from various diseases. The latest release covers the gene-centric methylation data from a variety of technologies and platforms. DiseaseMeth supports multiple search options such as gene ID and disease name to facilitate data extraction. It also provides integrated gene methylation data based on statistical analysis between disease and normal datasets. This information can be used to identify differentially methylated genes and to investigate the relationship between gene activity and diseases [80].

The diversity of histone modifications makes them superior targets or markers for complex diseases. However, compared to DNA methylation, the available experimental data and sources for histone modifications and related enzymes are much scarcer. The Human Histone Modifications Database (HHMD) is a comprehensive database for human histone modifications, aiming at integrating histone modification information from various experimental data [81]. So far, 43 location-specific histone modifications in human and a comprehensive resource of histone modification regulation in nine human cancer types have been incorporated into HHMD. This database also has a graphical interface, HisModView, to facilitate user browsing of histone modifications in the context of existing human genomic annotations [81]. As for the epigenome modifying enzymes, the Human Epigenetic Enzyme and Modulator Database (HEMD) provides a central resource for the display, search, and analysis of the structures, functions, and annotations for human epigenetic enzymes and their modulators. This database can be used to facilitate investigation of epigenetic targets for the query compound and to assist chemists in designing the structures of novel epigenetic drugs [82].

11.5.2 SOFTWARE AND PACKAGES

Gathering, analyzing, and interpreting high-throughput data requires both skill and patience. Luckily, various bioinformatic tools have been developed for researchers to assist their studies (summarized in Table 11.2). For bioinformaticians, several tools have been developed to help them process low-level data generated from high-throughput technologies. BRAT has been developed for mapping short reads generated from the Illumina Genome Analyzer following bisulfite treatment, supporting single and paired-end reads and handling input files containing reads and mates with different lengths [83]. BSMAP combines genome hashing and bitwise masking to achieve fast and accurate bisulfite mapping and runs faster and is more sensitive and more flexible than existing bisulfite mapping tools [84]. Bismark is a flexible tool for efficient analysis of bisulfite sequencing data, performing both read mapping and methylation calling in a single convenient step. Its output discriminates the cytosines in different contexts and enables visualization and interpretation of the methylation data soon after the sequencing run is completed [85]. MethBLAST is a sequence similarity search program derived from the original BLAST algorithm, but it queries *in silico* bisulfite-converted genome sequences to evaluate oligonucleotide sequence similarities. It also allows computational assessment of primer specificity in PCR-based methylation assays from the MethPrimerDB database, which provides a search portal for validated methylation assays [74].

For experimentalists without programming experience, there are also some handy tools to help them investigate the epigenome profiles or patterns from raw data. DBCAT (database of CpG islands and analytical tools) is a web-based database with several convenient tools to characterize DNA methylation profiles in human cancers. It consists of three parts: a CpG island finder, a genome query browser, and an analysis tool for microarray methylation data. These analytical tools facilitate identification of methylated genes from microarray data, comparing methylation status changes between different datasets, and analyzing related functions in addition to co-localizing the binding sites of transcription factors, which would be helpful in guiding further experimental designs [86]. MetMap is a statistical method that produces corrected site-specific methylation states

Table 11.2 Software and Packages for Epigenetic Studies

Tool	Functionality	URL	Reference
BRAT	Three-letter bisulfite alignment	http://compbio.cs.ucr.edu/brat/	[83]
BSMAP	Short reads mapping for bisulfite sequencing	http://code.google.com/p/bsmap/	[84]
Bismark	Three-letter bisulfite alignment	http://www.bioinformatics.babraham.ac.uk/projects/bismark/	[85]
MethBLAST	Sequence similarity search	http://medgen.ugent.be/methblast	[74]
DBCAT	CpG islands identification, Genome query	http://dbcat.cgm.ntu.edu.tw	[86]
	Microarray methylation data analysis		
MetMap	Site-specific methylation states inference	http://math.mcb.berkeley.edu/~meromit/MetMap	[87]
MethTools	DNA methylation data visualization and analysis	http://genome.imb-jena.de/methtools	[88]
ChIPseeqer	Comprehensive tool for ChIP-Seq data analysis	http://physiology.med.cornell.edu/faculty/elemento/lab/chipseq.shtml	[89]
EpiRegNet	Epigenetic regulatory networks construction	http://jjwanglab.org/EpiRegNet	[90]
Repitools	Epigenomic data interpretation and visualization	http://repitools.r-forge.r-project.org	[91]
EpiGRAPH	Web-based tool for enrichment analysis	http://epigraph.mpi-inf.mpg.de/WebGRAPH/	[92]

from experiments and annotated unmethylated islands across the genome. It integrates sequence information of the genome with experimental data, via a statistically sound and cohesive Bayesian network approach. It infers the extent of methylation at individual CGs and across regions, and serves as a framework for comparative methylation analysis within or among species [87]. MethTools is a collection of software that generates graphical outputs of methylation patterns and density, estimates the systematic error of the experiment and searches for conserved methylated nucleotide patterns. The programs are written in Perl with C, and the source code is freely available for download. The entire analysis procedure can be performed automatically by using some pre-written scripts with specific objects [88].

Besides the tools for processing and interpreting raw data from single experiments, advanced tools have been developed to integrate epigenomic data with other genomic or cellular pathway data, which is important to gain a holistic view of the epigenetic machinery. ChIPseeqer is a comprehensive computational framework to conduct broad and deep analysis of ChIP-Seq data. This framework includes various tools to perform gene-level peak annotation, nongenic peak annotation, pathways enrichment analysis, regulatory element analysis, conservation analysis, and clustering analysis. It can be used to integrate and compare data across different ChIP-Seq experiments and the results can be directly visualized through visualization tools incorporated in this framework [89]. EpiRegNet is a web tool to study the epigenetic factors that are responsible for gene expression changes. Provided with different categories of genes, the server will automatically find epigenetic factors responsible for the difference among the categories and construct an epigenetic regulatory network. To analyze the dynamic change of a histone modification mark under different cell conditions, the server will pinpoint the direct and indirect target genes of this mark by integrative analysis of experimental data and computational prediction, and then present a regulatory network regarding this mark. Currently, it covers genome-wide histone modification profiles for 12 cell lines in human and will continue to grow when new data is available. The server also provides network visualization via a user-friendly interface and data can be downloaded in batch mode. The web site is implemented in Perl, PHP, and MySQL and it is compatible with all major browsers [90]. Repitools is a toolbox for interrogating and visualizing epigenomic data. Developed for the R environment, this software package focuses on the analysis of enrichment-based epigenomic data. It includes useful functions for quality checking, analysis, and comparison of trends of both array data and sequencing data. The routines have been tested for Affymetrix and NimbleGen microarrays and Illumina Genome Analyzer sequencing data, while other platforms can be easily supported using generic data types [91]. EpiGRAPH was developed to assist researchers interpreting large-scale datasets, generated from ChIP-on-chip, tiling microarrays, and resequencing [92]. The original goal of EpiGRAPH was to predict DNA methylation from various genome properties, including DNA sequence and structure, gene and repeat distribution, SNPs and transcription factor binding sites [93]. In the follow-up project, it was improved for genome-wide predictive analysis across different datasets, cell lines, and tissues [94]. EpiGRAPH has already been successfully used in building the human genomic DNA melting map [95]. EpiGRAPH can also be used to identify promising cancer biomarkers from genome-wide DNA methylation profiles in a large patient sample.

11.6 CONCLUSIONS AND FUTURE PERSPECTIVES

In this chapter, we discussed the power of modern bioinformatic and biostatistical approaches in assisting epigenetic disease marker and target mining tasks. The diversity, complexity, and close connection with disease make epigenetic modifications and their modulators promising disease biomarkers as well as therapeutic targets. Many studies have proved the efficacy of epigenetic biomarkers in diagnosis, classification, and disease outcome prediction. More excitingly, a substantial number of drugs targeting epigenetic mechanisms are approved by FDA or under clinical trials. However, there is still urgent need to identify more indicative epigenetic disease markers and more specific epigenetic targets. Bioinformatic and biostatistics approaches can greatly facilitate this process by dealing with the massive amount of data and providing a more systematic perspective. Almost every stage in the biomarker or target mining process, including data acquisition and preprocessing, data integration, data mining and the final validation of candidates, can use the help of bioinformatic and biostatistical methods (Figure 11.2). Both the biomarker identification and drug target mining tasks can benefit greatly from the rich set of homogeneous data generated by high-throughput technologies, while these high-throughput technologies are highly dependent on bioinformatic methods to guarantee the reliability of the data they produce. In the actual data mining process, various computational tools, such as statistical hypothesis testing, statistical learning and network modeling, can be used to infer disease markers or drug targets from buried information in

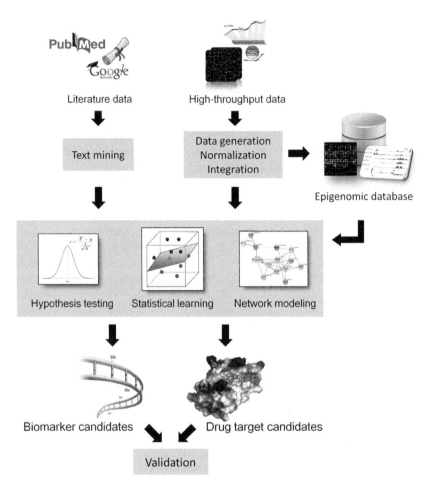

Literature data

High-throughput data

Text mining

Data generation
Normalization
Integration

Epigenomic database

Hypothesis testing   Statistical learning   Network modeling

Biomarker candidates   Drug target candidates

Validation

FIGURE 11.2

Epigenetic disease biomarker and target discovery through a systematic approach.

literature or high-throughput data. After the candidates for biomarkers or targets have been identified, biostatistical tools are also needed in their validation. In addition, bioinformatic resources such as biological databases can help the mining process by presenting the huge amount of epigenome data in a more accessible and organized way.

In spite of a number of successful applications of bioinformatic and biostatistical approaches in mining epigenetic disease markers and targets, there is still much uncharted territory in this field. Currently, most of the reported epigenetic biomarker candidates are based on DNA methylations. This mainly comes from two aspects: a practical consideration that DNA methylation is relatively stable and easy to detect and a more defined role of DNA methylations in human diseases. However, histone modifications are more diverse and therefore they may possess more discriminant power when used as biomarkers. In addition, histone modifications regulate gene transcriptional activity in a more sophisticated and subtle way and the number of histone-modifying enzymes far exceeds that of DNA-modifying enzymes. Therefore, targeting histone modifications may present an easier way to achieve specific and controlled downstream effects. The continuing development of high-throughput technologies will eventually lead to more convenient and cheaper ways to detect genome-wide histone modification profiles, therefore enabling the discovery and development of histone modification-based biomarkers or targets in a systematic way.

In addition to focusing on epigenetic modifications, future epigenome data analysis will take the proteins more into account, not only the epigenetic writers, but also the epigenetic readers as well as their interaction partners and regulatory networks. With the help of advanced bioinformatic and systems biology approaches, the whole epigenetic machinery can be seamlessly incorporated into protein–protein interaction networks, gene regulatory networks, drug-target networks, or other more complex networks. This integrated framework will improve our understanding of epigenetic mechanisms at multiple levels of complexity, from a single epigenetic marker to a regulatory pathway, a cell or even a whole living body. Further, it is interesting to explore the dynamic interplay between epigenetic machinery and other cellular processes through mathematical modeling techniques. The dynamic properties of this epigenetic network may be the key to some long-lasting mysteries, such as how the limited species of epigenetic modifying enzymes leads to such diverse epigenetic profiles in different cells. As reported by a recent study, the dynamic properties of a cellular network can also be targeted to achieve selected downstream effects and this finding will shed light on the identification of more specific epigenetic targets.

Another exciting possibility lies in the combinatory usage of both genetic and epigenetic alterations in monitoring disease states. As previously reported, epigenetic and genetic alterations may provide distinct routes for tumor cells to acquire the hallmarks of cancer [96]. Thus, the combined biomarkers may be substantially more accurate than exclusively epigenetic biomarkers. Similarly, simultaneously modulating epigenetic targets and classical antitumor targets, such as protein kinases, through drug combinations or multitarget drugs may have more potent antitumor efficacy while reducing side effects. As a matter of fact, several studies have already reported the promising results of combined epigenetic therapies [97–99]. However, a more rational way to design these combinations through bioinformatic methods awaits further characterization of the epigenetic mechanisms.

Although bioinformatic and biostatistical methods have already facilitated the biomarker and target identification pipeline, there are some remaining challenges. One major challenge in this field lies in the efficient processing of high-throughput data, especially the normalization and interpretation of data across various experiments from different laboratories. In addition, while the epigenomic data are increasing exponentially, a comprehensive and centralized repository, like GEO, for epigenetics-related data is still lacking. An integrated platform for effective data curation, annotation, and management will greatly facilitate the data mining process for epigenetic studies. In addition, more easy-to-use analysis software and more friendly front-ends of online resources will enable experimental biologists to analyze and visualize complex epigenome data without the requirement of strong statistical or programming skills. This will bring bioinformaticians and experimentalists much closer and make the biomarker and target mining process more efficient.

In conclusion, there is a bright future for developing epigenetic disease markers and targets, and bioinformatic and biostatistical techniques will greatly boost this process.

REFERENCES

[1] Li E. Chromatin modification and epigenetic reprogramming in mammalian development. Nat Rev Genet 2002;3(9):662−73.

[2] Egger G, Liang G, Aparicio A, Jones PA. Epigenetics in human disease and prospects for epigenetic therapy. Nature 2004;429(6990):457−63.

[3] Jones PA, Laird PW. Cancer epigenetics comes of age. Nat Genet 1999;21(2):163−7.

[4] Duvic M, Vu J. Vorinostat in cutaneous T-cell lymphoma. Drugs Today (Barc) 2007;43(9):585−99.

[5] Laird PW. The power and the promise of DNA methylation markers. Nat Rev Cancer 2003;3 (4):253−66.

[6] Shivapurkar N, Gazdar AF. DNA methylation based biomarkers in non-invasive cancer screening. Curr Mol Med 2010;10(2):123−32.

[7] Seligson DB, Horvath S, Shi T, Yu H, Tze S, Grunstein M, et al. Global histone modification patterns predict risk of prostate cancer recurrence. Nature 2005;435(7046):1262−6.

[8] Krauthammer M, Kaufmann CA, Gilliam TC, Rzhetsky A. Molecular triangulation: bridging linkage and molecular-network information for identifying candidate genes in Alzheimer's disease. Proc Natl Acad Sci USA 2004;101(42):15148−53.

[9] Gorr SU, Wennblom TJ, Horvath S, Wong DT, Michie SA. Text-mining applied to autoimmune disease research: the Sjogren's syndrome knowledge base. BMC Musculoskelet Disord 2012;13:119.

[10] Khan AR, Awan FR. Mining of protein based biomarkers for type 2 diabetes mellitus. Pak J Pharm Sci 2012;25(4):889−901.

[11] Bock C, Lengauer T. Computational epigenetics. Bioinformatics 2008;24(1):1−10.

[12] Bock C. Analysing and interpreting DNA methylation data. Nat Rev Genet 2012;13(10):705−19.

[13] Kerick M, Fischer A, Schweiger M-R. Generation and analysis of genome-wide DNA methylation maps. In: Rodríguez-Ezpeleta N, Hackenberg M, Aransay AM, editors. Bioinformatics for high throughput sequencing. New York: Springer; 2012. p. 151−67.

[14] Lewin J, Schmitt AO, Adorján P, Hildmann T, Piepenbrock C. Quantitative DNA methylation analysis based on four-dye trace data from direct sequencing of PCR amplificates. Bioinformatics 2004;20(17):3005−12.

[15] Beck S, Rakyan VK. The methylome: approaches for global DNA methylation profiling. Trends Genet 2008;24(5):231−7.

[16] Myers AJ. The age of the "ome": genome, transcriptome and proteome data set collection and analysis. Brain Res Bull 2012;88(4):294−301.

[17] Schumacher A, Kapranov P, Kaminsky Z, Flanagan J, Assadzadeh A, Yau P, et al. Microarray-based DNA methylation profiling: technology and applications. Nucleic Acids Res 2006;34(2):528−42.

[18] Cawley S, Bekiranov S, Ng HH, Kapranov P, Sekinger EA, Kampa D, et al. Unbiased mapping of transcription factor binding sites along human chromosomes 21 and 22 points to widespread regulation of noncoding RNAs. Cell 2004;116(4):499−509.

[19] Bernstein BE, Kamal M, Lindblad-Toh K, Bekiranov S, Bailey DK, Huebert DJ, et al. Genomic maps and comparative analysis of histone modifications in human and mouse. Cell 2005;120(2):169−81.

[20] Roh TY, Cuddapah S, Zhao K. Active chromatin domains are defined by acetylation islands revealed by genome-wide mapping. Genes Dev 2005;19(5):542−52.

[21] Huebert DJ, Kamal M, O'Donovan A, Bernstein BE. Genome-wide analysis of histone modifications by ChIP-on-chip. Methods 2006;40(4):365−9.

[22] Langmead B, Trapnell C, Pop M, Salzberg S. Ultrafast and memory-efficient alignment of short DNA sequences to the human genome. Genome Biol 2009;10(3):1−10.

[23] Li H, Ruan J, Durbin R. Mapping short DNA sequencing reads and calling variants using mapping quality scores. Genome Res 2008;18(11):1851−8.

[24] Li H, Durbin R. Fast and accurate long-read alignment with Burrows–Wheeler transform. Bioinformatics 2010;26(5):589–95.

[25] Flicek P, Birney E. Sense from sequence reads: methods for alignment and assembly. Nat Meth 2009;6 (11s):S6–12.

[26] Stolzenberg DS, Grant PA, Bekiranov S. Epigenetic methodologies for behavioral scientists. Horm Behav 2011;59(3):407–16.

[27] Mikkelsen TS, Ku M, Jaffe DB, Issac B, Lieberman E, Giannoukos G, et al. Genome-wide maps of chromatin state in pluripotent and lineage-committed cells. Nature 2007;448(7153):553–60.

[28] Huss M. Introduction into the analysis of high-throughput-sequencing based epigenome data. Brief Bioinform 2010;11(5):512–23.

[29] Park PJ. ChIP-Seq: advantages and challenges of a maturing technology. Nat Rev Genet 2009;10 (10):669–80.

[30] Bailey TL, Williams N, Misleh C, Li WW. MEME: discovering and analyzing DNA and protein sequence motifs. Nucleic Acids Res 2006;34(suppl. 2):W369–73.

[31] Liu XS, Brutlag DL, Liu JS. An algorithm for finding protein-DNA binding sites with applications to chromatin-immunoprecipitation microarray experiments. Nat Biotech 2002;20(8):835–9.

[32] Pavesi G, Mereghetti P, Mauri G, Pesole G. Weeder Web: discovery of transcription factor binding sites in a set of sequences from co-regulated genes. Nucleic Acids Res 2004;32(suppl. 2):W199–203.

[33] Romer KA, Kayombya G-R, Fraenkel E. WebMOTIFS: automated discovery, filtering and scoring of DNA sequence motifs using multiple programs and Bayesian approaches. Nucleic Acids Res 2007;35 (suppl. 2):W217–20.

[34] Xu H, Wei C-L, Lin F, Sung W-K. An HMM approach to genome-wide identification of differential histone modification sites from ChIP-seq data. Bioinformatics 2008;24(20):2344–9.

[35] Zang C, Schones DE, Zeng C, Cui K, Zhao K, Peng W. A clustering approach for identification of enriched domains from histone modification ChIP-Seq data. Bioinformatics 2009;25(15):1952–8.

[36] Sidoli S, Cheng L, Jensen ON. Proteomics in chromatin biology and epigenetics: elucidation of post-translational modifications of histone proteins by mass spectrometry. J Proteomics 2012;75 (12):3419–33.

[37] Stunnenberg HG, Vermeulen M. Towards cracking the epigenetic code using a combination of high-throughput epigenomics and quantitative mass spectrometry-based proteomics. BioEssays 2011;33 (7):547–51.

[38] Majkić-Singh N. What is a biomarker? From its discovery to clinical application. J Med Biochem 2011;30(3):186–92.

[39] Mulero-Navarro S, Esteller M. Epigenetic biomarkers for human cancer: the time is now. Crit Rev Oncol Hematol 2008;68(1):1–11.

[40] Model F, Osborn N, Ahlquist D, Gruetzmann R, Molnar B, Sipos F, et al. Identification and validation of colorectal neoplasia-specific methylation markers for accurate classification of disease. Mol Cancer Res 2007;5(2):153–63.

[41] Lai HC, Lin YW, Huang TH, Yan P, Huang RL, Wang HC, et al. Identification of novel DNA methylation markers in cervical cancer. Int J Cancer 2008;123(1):161–7.

[42] Harlow LL, Mulaik SA, Steiger JH. What if there were no significance tests? Psychology Press; 2013.

[43] Mohri M, Rostamizadeh A, Talwalkar A. Foundations of machine learning. MIT Press; 2012.

[44] Wei SH, Balch C, Paik HH, Kim YS, Baldwin RL, Liyanarachchi S, et al. Prognostic DNA methylation biomarkers in ovarian cancer. Clin Cancer Res 2006;12(9):2788–94.

[45] Model F, Adorjan P, Olek A, Piepenbrock C. Feature selection for DNA methylation based cancer classification. Bioinformatics 2001;17(Suppl. 1):S157–64.

[46] Shaknovich R, Geng H, Johnson NA, Tsikitas L, Cerchietti L, Greally JM, et al. DNA methylation signatures define molecular subtypes of diffuse large B-cell lymphoma. Blood 2010;116(20):e81−9.

[47] Rhee JK, Kim K, Chae H, Evans J, Yan P, Zhang BT, et al. Integrated analysis of genome-wide DNA methylation and gene expression profiles in molecular subtypes of breast cancer. Nucleic Acids Res 2013;41 (18):8464−74.

[48] Kisiel JB, Yab TC, Taylor WR, Chari ST, Petersen GM, Mahoney DW, et al. Stool DNA testing for the detection of pancreatic cancer: assessment of methylation marker candidates. Cancer 2012;118 (10):2623−31.

[49] Pal NR, Aguan K, Sharma A, Amari S. Discovering biomarkers from gene expression data for predicting cancer subgroups using neural networks and relational fuzzy clustering. BMC Bioinformatics 2007;8:5.

[50] Harris C, Ghaffari N. Biomarker discovery across annotated and unannotated microarray datasets using semi-supervised learning. BMC Genomics 2008;9(Suppl. 2):S7.

[51] Kamalakaran S, Varadan V, Giercksky Russnes HE, Levy D, Kendall J, et al. DNA methylation patterns in luminal breast cancers differ from non-luminal subtypes and can identify relapse risk independent of other clinical variables. Mol Oncol 2011;5(1):77−92.

[52] Ehrich M, Bullinger L, Dohner K, Schlenk RF, Nelson MR, Dohner H, et al. Survival prediction of Acute Myeloid Leukemia (AML) using a combination of DNA methylation analysis and gene expression profiling. Proc Am Assoc Cancer Res 2006;2006(1):561.

[53] Altman DG, Bland JM. Diagnostic tests. 1: sensitivity and specificity. BMJ 1994;308(6943):1552.

[54] Swets JA. Signal detection theory and ROC analysis in psychology and diagnostics: collected papers. Lawrence Erlbaum Associates, Inc; 1996.

[55] Powers DM. Evaluation: from precision, recall and F-measure to ROC, informedness, markedness & correlation. J Mach Learn Technol 2011;2(1):37−63.

[56. Kohavi R. A study of cross-validation and bootstrap for accuracy estimation and model selection. In: Proceedings of the fourteenth international joint conference on artificial intelligence (IJCAI). Morgan Kaufmann; 1995. p. 1137−45.

[57] Mooney CZ, Duval RD, Duval R. Bootstrapping: a nonparametric approach to statistical inference. Sage; 1993.

[58] Higgins JJ. An introduction to modern nonparametric statistics. Pacific Grove, CA: Brooks/Cole; 2004.

[59] Imming P, Sinning C, Meyer A. Drugs, their targets and the nature and number of drug targets. Nat Rev Drug Discov 2006;5(10):821−34.

[60] Krallinger M, Leitner F, Valencia A. Analysis of biological processes and diseases using text mining approaches. Methods Mol Biol 2010;593:341−82.

[61] Pospisil P, Iyer LK, Adelstein SJ, Kassis AI. A combined approach to data mining of textual and structured data to identify cancer-related targets. BMC Bioinformatics 2006;7:354.

[62] Perry AS, Loftus B, Moroose R, Lynch TH, Hollywood D, Watson RW, et al. In silico mining identifies IGFBP3 as a novel target of methylation in prostate cancer. Br J Cancer 2007;96(10):1587−94.

[63] Ryu B, Kim DS, Deluca AM, Alani RM. Comprehensive expression profiling of tumor cell lines identifies molecular signatures of melanoma progression. PloS One 2007;2(7):e594.

[64] West J, Widschwendter M, Teschendorff AE. Distinctive topology of age-associated epigenetic drift in the human interactome. Proc Natl Acad Sci USA 2013;110(35):14138−43.

[65] Roznovat IA, Ruskin HJ. A computational model for genetic and epigenetic signals in colon cancer. Interdiscip Sci 2013;5(3):175−86.

[66] Spring DR. Chemical genetics to chemical genomics: small molecules offer big insights. Chem Soc Rev 2005;34(6):472−82.

[67] Hopkins AL. Network pharmacology: the next paradigm in drug discovery. Nat Chem Biol 2008;4 (11):682−90.

[68] Edgar R, Domrachev M, Lash AE. Gene Expression Omnibus: NCBI gene expression and hybridization array data repository. Nucleic Acids Res 2002;30(1):207−10.

[69] Hubbard T, Barker D, Birney E, Cameron G, Chen Y, Clark L, et al. The Ensembl genome database project. Nucleic Acids Res 2002;30(1):38−41.

[70] Petryszak R, Burdett T, Fiorelli B, Fonseca NA, Gonzalez-Porta M, Hastings E, et al. Expression Atlas update − a database of gene and transcript expression from microarray- and sequencing-based functional genomics experiments. Nucleic Acids Res 2014;42(Database issue):D926−32.

[71] Weinstein JN, Collisson EA, Mills GB, Shaw KR, Ozenberger BA, Ellrott K, et al. The Cancer Genome Atlas Pan-Cancer analysis project. Nat Genet 2013;45(10):1113−20.

[72] The ENCODE (ENCyclopedia Of DNA Elements) Project. Science 2004;306(5696):636−40.

[73] Grunau C, Renault E, Rosenthal A, Roizes G. MethDB − a public database for DNA methylation data. Nucleic Acids Res 2001;29(1):270−4.

[74] Pattyn F, Hoebeeck J, Robbrecht P, Michels E, De Paepe A, Bottu G, et al. methBLAST and methPrimerDB: web-tools for PCR based methylation analysis. BMC Bioinformatics 2006;7(1):1−9.

[75] Hackenberg M, Barturen G, Oliver JL. NGSmethDB: a database for next-generation sequencing single-cytosine-resolution DNA methylation data. Nucleic Acids Res 2011;39(suppl. 1):D75−9.

[76] He X, Chang S, Zhang J, Zhao Q, Xiang H, Kusonmano K, et al. MethyCancer: the database of human DNA methylation and cancer. Nucleic Acids Res 2008;36(Database issue):D836−41.

[77] Ongenaert M, Van Neste L, De Meyer T, Menschaert G, Bekaert S, Van Criekinge W. PubMeth: a cancer methylation database combining text-mining and expert annotation. Nucleic Acids Res 2008;36 (suppl. 1):D842−6.

[78] Lauss M, Visne I, Weinhaeusel A, Vierlinger K, Noehammer C, Kriegner A. MethCancerDB − aberrant DNA methylation in human cancer. Br J Cancer 2008;98(4):816−17.

[79] Baek S−J, Yang S, Kang T−W, Park S-M, Kim YS, Kim S-Y. MENT: Methylation and expression database of normal and tumor tissues. Gene 2013;518(1):194−200.

[80] Lv J, Liu H, Su J, Wu X, Liu H, Li B, et al. DiseaseMeth: a human disease methylation database. Nucleic Acids Res 2012;40(Database issue):D1030−5.

[81] Li Y, Qiu C, Tu J, Geng B, Yang J, Jiang T, et al. HMDD v2.0: a database for experimentally supported human microRNA and disease associations. Nucleic Acids Res 2014;42(D1):D1070−4.

[82] Huang Z, Jiang H, Liu X, Chen Y, Wong J, Wang Q, et al. HEMD: an integrated tool of human epigenetic enzymes and chemical modulators for therapeutics. PloS One 2012;7(6):e39917.

[83] Harris EY, Ponts N, Levchuk A, Roch KL, Lonardi S. BRAT: bisulfite-treated reads analysis tool. Bioinformatics 2010;26(4):572−3.

[84] Xi Y, Li W. BSMAP: whole genome bisulfite sequence MAPping program. BMC Bioinformatics 2009;10(1):1−9.

[85] Krueger F, Andrews SR. Bismark: a flexible aligner and methylation caller for Bisulfite-Seq applications. Bioinformatics 2011;27(11):1571−2.

[86] Kuo HC, Lin PY, Chung TC, Chao CM, Lai LC, Tsai MH, et al. DBCAT: database of CpG islands and analytical tools for identifying comprehensive methylation profiles in cancer cells. J Comput Biol 2011;18(8):1013−17.

[87] Singer M, Boffelli D, Dhahbi J, Schönhuth A, Schroth GP, Martin DI, et al. MetMap enables genome-scale Methyltyping for determining methylation states in populations. PLoS Comput Biol 2010;6(8): e1000888.

[88] Grunau C, Schattevoy R, Mache N, Rosenthal A. MethTools − a toolbox to visualize and analyze DNA methylation data. Nucleic Acids Res 2000;28(5):1053−8.

[89] Giannopoulou E, Elemento O. An integrated ChIP-seq analysis platform with customizable workflows. BMC Bioinformatics 2011;12(1):1−17.

[90] Wang LY, Wang P, Li MJ, Qin J, Wang X, Zhang MQ, et al. EpiRegNet: constructing epigenetic regulatory network from high throughput gene expression data for humans. Epigenetics 2011;6(12):1505−12.

[91] Statham AL, Strbenac D, Coolen MW, Stirzaker C, Clark SJ, Robinson MD. Repitools: an R package for the analysis of enrichment-based epigenomic data. Bioinformatics 2010;26(13):1662−3.

[92] Bock C, Halachev K, Büch J, Lengauer T. EpiGRAPH: user-friendly software for statistical analysis and prediction of (epi)genomic data. Genome Biol 2009;10(2):1−14.

[93] Bock C, Paulsen M, Tierling S, Mikeska T, Lengauer T, Walter J. CpG island methylation in human lymphocytes is highly correlated with DNA sequence, repeats, and predicted DNA structure. PLoS Genet 2006;2(3):e26.

[94] Bock C, Walter J, Paulsen M, Lengauer T. CpG island mapping by epigenome prediction. PLoS Comput Biol 2007;3(6):e110.

[95] Liu F, Tostesen E, Sundet JK, Jenssen TK, Bock C, Jerstad GI, et al. The human genomic melting map. PLoS Comput Biol 2007;3(5):e93.

[96] Jones PA, Baylin SB. The fundamental role of epigenetic events in cancer. Nat Rev Genet 2002;3 (6):415−28.

[97] Matei DE, Nephew KP. Epigenetic therapies for chemoresensitization of epithelial ovarian cancer. Gynecol Oncol 2010;116(2):195−201.

[98] Kalac M, Scotto L, Marchi E, Amengual J, Seshan VE, Bhagat G, et al. HDAC inhibitors and decitabine are highly synergistic and associated with unique gene-expression and epigenetic profiles in models of DLBCL. Blood 2011;118(20):5506−16.

[99] Juergens RA, Wrangle J, Vendetti FP, Murphy SC, Zhao M, Coleman B, et al. Combination epigenetic therapy has efficacy in patients with refractory advanced non-small cell lung cancer. Cancer Discov 2011;1(7):598−607.

CHAPTER

COMPUTATIONAL MODELING TO ELUCIDATE MOLECULAR MECHANISMS OF EPIGENETIC MEMORY

12

Jianhua Xing[1,3], Jin Yu[1], Hang Zhang[1,2], and Xiao-Jun Tian[3]

[1]*Beijing Computational Science Research Center, Beijing, China* [2]*Department of Biological Sciences, Virginia Polytechnic Institute and State University, Blacksburg, VA, USA* [3]*Department of Computational and Systems Biology, School of Medicine, University of Pittsburgh, Pittsburgh, PA, USA*

CHAPTER OUTLINE

12.1 INTRODUCTION

There are more than 200 different cell types in a human body. These cells have drastically different shapes, physical properties, and physiological properties. Amazingly, all these cells (except reproductive cells) share the same set of genomes, and are developed from a single fertilized egg.

Y.G. Zheng (Ed): Epigenetic Technological Applications. DOI: http://dx.doi.org/10.1016/B978-0-12-801080-8.00012-0

Therefore, a fundamental and intriguing question in developmental biology is how a fertilized egg can develop into so many different types, in a controlled manner. Furthermore, a cell can preserve its identity after division. That is, a fibroblast cell divides into fibroblast cells. Recent studies show that it is possible, but difficult, to reprogram the identity of a terminally differentiated cell [1]. Then how can a cell remember its identity? Current accumulating evidence suggests that some heritable changes of gene activities are not caused by changes in the DNA sequence. Specifically in this chapter we will focus on heritable histone covalent modifications.

To form an organized and compact chromatin structure, a DNA molecule wraps around histone octamers to form nucleosomes. Covalent modifications such as methylation, acetylation, phosphorylation, and sumoylation can take place on a number of side residues of the histone proteins. Through changing the interactions between DNA and histone proteins, and between nucleosome and other regulatory elements such as histone modification enzymes, transcription factors, and regulatory noncoding RNAs, these covalent modifications affect higher-order packing of the nucleosomes and gene transcription efficiency [2]. Experimental studies reveal that at least some of the histone posttranslational modification patterns are inheritable, which is called histone epigenetic memory [3,4]. In the past few years, different groups have discovered multiple enzymes regulating the histone modification dynamics [3,5]. The so-called "histone code" proposal, although still under debate, has drawn extensive attention from the field [6]. Revealing the molecular mechanism of this histone modification memory has been a focused research subject for many years.

In recent years mathematical modeling has contributed significantly to our understanding of how histone epigenetic patterns are produced and maintained. In a seminal paper, Dodd et al. used a rule-based model to analyze the silenced mating-type locus of the fission yeast *Schizosaccharomyces pombe* (*S. Pombe*) [7]. *S. Pombe* has two mating-type cassettes that are normally in an epigenetically silent state. A mutant has been constructed with a portion of the silenced region removed and a ura4 + reporter gene inserted. Experimental studies on the mutant revealed that the DNA region (~ 60 nucleosomes) can exist in an inheritable epigenetic active or silent state, with a very low probability of stochastic transition between the two states of about 5×10^{-4} per cell division [8,9]. Furthermore, the two copies of the chromosomal region within one cell can exist in different epigenetic states. That is, cells can exist in a bistable region. The mathematical analysis of Dodd et al. showed that cooperativity among neighboring and beyond-neighboring nucleosomes are necessary and sufficient to generate robust bistable epigenetic states. Subsequently this pioneering study has been generalized to analyze systems such as vernalization in *Arabidopsis thaliana* [10], epigenetic switching at the genetic locus of Oct4 (also known as Pou5f1), a transcription factor essential for maintaining the embryonic stem cell state [11,12], and olfactory neuron differentiation [13]. Meanwhile studies using alternative approaches have also been developed to analyze various problems [14–24]. Especially, quantitative measurements on nucleosome covalent modification dynamics allow incorporation of molecular details in modeling studies. Steffen et al. [25] and Rohlf et al. [26] provided timely reviews on the experimental and mathematical modeling efforts to extract quantitative information on epigenetic regulation. In the remainder of the chapter, we discuss in detail the generic procedure of performing a mathematical modeling study. We will use a model of Zhang et al., for which all components are based on experimental information [27], as an example. The model's structure is similar to the well-studied Potts model in physics describing cooperative phenomenon. For simplicity we will call it the CoPE model, standing for coupled-Potts model of epigenetic dynamics of histone modifications.

12.2 IDENTIFY PUZZLE FROM EXPERIMENTAL STUDIES

The first step for a modeling study is to identify a problem that is both significant and suitable for theoretical studies. Modeling is not intended and is not capable of answering every question. For example, modeling studies can examine whether a proposed mechanism is consistent with available experimental observations, and the laws of physics and chemistry, but cannot decide whether the mechanism is actually what is assumed by the system. The confirmation must come from experimental studies. Similarly modeling studies may suggest whether a missing component is needed to reconcile existing data, but cannot determine the identity of the component.

For information inheritance from mother to daughter cells, the puzzle is how the information is transmitted and maintained. We can identify three types of heritable information: the DNA, whose double helix structure allows faithful reproduction; the abundance of proteins and other molecules (i.e. the transcriptome, proteome, etc.), which partition into two daughter cells either equally or asymmetrically; the covalent modification patterns on DNA molecules and on histones, whose inheritance mechanisms are less understood. For concreteness in this chapter we will focus on the problem of histone pattern inheritance, while the procedure can be easily generalized to DNA methylation as well.

A closer examination of the histone inheritance problem reveals that it is a highly nontrivial question. First, within a nucleus, there are constantly opposing histone modification enzymes attempting to add or remove each covalent mark and modify the histone pattern; thus instead of being static, the histone modification pattern is a consequence of a dynamic "tug-of-war." Second, although the interactions between a histone complex and DNA are not weak, the histone can stochastically detach from the DNA; then either the same histone or a new one, which likely bears no or covalent mark(s) different from the old one, quickly incorporates onto the DNA. This process is termed *histone turnover*. Furthermore, when cell division takes place, each histone is likely partitioned into one of the two daughter cells with equal probability; that is, each daughter cell needs to incorporate about half of the total DNA-binding histones with nascent unmarked histones. Amazingly, with all these large perturbations, cells can maintain at least some of the histone covalent patterns for generations.

Extensive biochemistry and biophysics studies reveal two prominent properties of the modification enzymes. First, the enzymes can recognize histone marks and thus have different free energy of binding. For example, Jacobs and Khorasanizadeh reported the structural basis for the chromodomain of *Drosophila* HP1 to recognize the trimethylated H3K9 residue [28]; Raphael Margueron et al. discovered that H3K27me3 propagation and maintenance require specific recognition of H3K27me3 by the polycomb protein complex [29]. Generally speaking, an enzyme has higher binding affinity to nucleosomes bearing the corresponding marks than those without mark or with different marks [30]. We want to point out that this property is typical for enzymes, i.e. an enzyme usually binds stronger to the substrate than to the product or to a nonsubstrate. Second, enzymes bound to neighboring nucleosomes can interact laterally. Canzio et al. showed that the HP1 proteins can form oligomers through chromodomain and chromoshadow domain lateral interactions, and enhanced lateral interactions lead to higher percentage of H3K9me3 [31,32]. Interestingly, mutations related to the histone modification enzyme lateral interactions have been reported in cancer cells [33].

Therefore the puzzle, or the question we want to address, is whether one can use the previously discussed molecular level information to explain the epigenetic histone memory. The process is complex, with many molecular species, and broad time scales involved. For the latter it ranges from subseconds for enzyme-binding/unbinding events, to months or longer for histone memory duration. For example, epigenetic state switches for the above-mentioned *S. Pombe* mutant take place about every 200 days on average [8,9]. Therefore mathematical modeling is necessary to fill in the huge gaps between the experimentally explored molecular events and collective epigenetic dynamics.

12.3 FORMULATE MATHEMATICAL MODEL

With the problem identified, one next needs to translate it into a mathematical model. Here we use the word "translate" literally. That is, each term in the mathematical model corresponds to a process identified as important for understanding histone memory. One does, however, need to consider carefully what level of detail to be included. In physics, a common criterion is based on the following famous quote from Einstein: a theory "should be made as simple as possible, but not simpler." That is, the model should contain just the amount of detail sufficient to explain the underlying phenomenon, but not more to distract one from the essential physics. For example, if one only wants to know the dimensions of a box, then information about the box color is irrelevant. To keep a model necessarily simple, abstraction is often needed.

Figure 12.1 summarizes the CoPE model, which includes a collection of N nucleosomes aligned as a one-dimensional array. A nucleosome i has three possible covalent states: bearing a repressive mark, unmarked, or bearing an active mark. For bookkeeping purposes, let us denote them as $s_i = -1, 0, 1$, respectively. In addition, four classes of covalent modification enzymes can bind to each nucleosome to catalyze adding or removing the marks. Thus, each nucleosome can have five possible enzyme-binding states: empty or one type of the enzymes bound. We denote these as σ_i $(= 1-5)$, indicating no enzyme bound ($\sigma_i = 1$), repressive modification addition enzyme bound ($\sigma_i = 2$), repressive modification

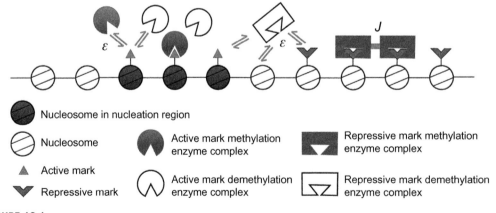

FIGURE 12.1

Schematic illustration of the CoPE model of Zhang et al. ε denotes enzyme-binding energy, J denotes enzyme lateral interaction energy.

Adapted from [27].

removal enzyme bound ($\sigma_i = 3$), active modification addition enzyme bound ($\sigma_i = 4$), and active modification removal enzyme bound ($\sigma_i = 5$). The σ state can change through enzyme binding and unbinding. The state of the system is thus denoted by the set of nucleosome indices $\{s_i\, \sigma_i, i = 1, \ldots, N\}$.

The overall s-σ state of the system evolves according to a Markovian dynamics. That is, the evolution depends only on the state in a previous time step. Enzyme binding/unbinding results in a σ-state change. The s state can change through histone turnover or enzyme catalyzed chemical reactions. For the latter, each of such reactions clearly requires that the corresponding enzyme binds to the nucleosome. Another relevant process is cell division. After each cell division, histones from a mother cell partition into two daughter cells. Current evidence suggests that this partition is random with equal probability to the daughter cells. Then nascent unmarked histones need to be incorporated to the DNA. In the language of modeling, for any given DNA-bound histone, during cell division its s state randomly either keeps its current value or resets to 0 with equal probability.

The covalent modification enzymes have no DNA sequence specificity. That is, they do not know which genome region to modify. Accumulating evidence suggests that some regulatory elements, such as transcription factors and noncoding RNAs, may recruit certain enzymes to specific DNA regions [34]. For example, the transcription factor SNAIL1 recruits to the E-cadherin promoter region histone demethylase LSD1 (lysine-specific demethylase 1) that removes H3K4me2 [35], histone deacetylase 1 (HDAC1) and HDAC2 [36], and PRC2, an H3K27me3 methyltransferase [37]. In addition, some enzymes, for example, MLL1 (Mixed Lineage Leukemia 1), KDM2A, and PRC2 (Polycomb Repressive Complex 2), have higher binding affinity at some DNA sequence elements, for example, CpG islands [38−41]. To reflect these observations, we follow the treatment of Angel et al. [10], and Hodge and Crabtree [11], to denote a "nucleation region" for a small number of nucleosomes, on which the enzymes have higher binding affinity compared to the nonspecific background-binding affinity on other nucleosomes. Existence of the nucleation regions can be inferred from the peaked distribution of histone modifications centered around many transcription-factor-binding sites [42].

Clearly the model is rather generic, and has neglected many details; a few are listed here:

1. Many residues can exist in multiple modification states. For example, a lysine can be mono-, di-, and tri-methylated, with different enzymes catalyzing each methylation and demethylation step. Also covalent modification enzymes may function redundantly and act on different substrates. For example, LSD1 (*lysine-specific demethylase 1*) can remove both mono- and di-methylation on H3K9 and H3K4. Both PRC1 (*Polycomb Repressive Complex 1*) and PRC2 (*Polycomb Repressive Complex 2*) can catalyze trimethylation on H3K27.

2. Each histone can have a large number of potential modification sites, leading to an even larger combinatory number of epigenetic states. According to the epigenetic code hypothesis, the covalent states of some sites may mutually affect each other and lead to different regulation on the gene activity.

3. A histone modification enzyme complex is usually bulky, and can interact with more than one nucleosome simultaneously.

4. The three-dimensional structure of chromatin affects the histone modification dynamics, for example, accessibility to the enzymes. In return, histone modifications may affect the three-dimensional packing of the chromatin.

These details likely have various biological implications. It is straightforward to expand the CoPE model to incorporate these details. However, the main purpose of that work of Zhang et al. [27] is to uncover the most essential molecular interactions and properties for histone memory. Therefore, these complexities are not explicitly considered. As we emphasized above, simplification is a key step for modeling.

12.4 CHOOSE APPROPRIATE MODELING TECHNIQUES

The described model is straightforward in terms of describing the relevant biological processes. However, technical difficulties exist upon studying it more closely. Even with this simplified model, each nucleosome has 3 of s states and 5 of σ states. With N nucleosomes, the total number of states is 15^N, which grows quickly with N. Furthermore, the possible dynamic processes, including enzyme binding and unbinding, chemical reactions, histone turnover, and cell cycles, span broad time scales, from subsecond-binding/unbinding events to the epigenetic state switching on the order of days to years. This large number of states and the broad time scale distribution make it computationally very expensive to simulate the system. Fortunately, the time scale of enzyme binding/unbinding is well separated from that of other processes, which suggests a quasi-equilibrium approximation.

One may remember the quasi-equilibrium approximation on deriving the Michaelis-Menten equation for enzymatic dynamics. One assumes that an enzymatic reaction follows this scheme:

$$E + S \underset{k_{-1}}{\overset{k_1}{\rightleftharpoons}} ES \overset{v}{\longrightarrow} E + P$$

That is, enzyme E and substrate S first form a complex ES, which then proceeds to form the product P and release the enzyme for the next enzymatic cycle. The quasi-equilibrium approximation assumes that the first step of forming ES is fast compared to the covalent bond breaking/forming step, so that E, S and ES concentrations reach an equilibrium distribution, $\frac{[ES]}{[E]} = \frac{K_1[S]}{k_{-1}} = e^{-\varepsilon/(k_B T)}$, where ε is the free energy of S binding to E at concentration $[S]$, (notice that $-\varepsilon$ is the binding affinity), k_B is Boltzmann's constant, T the temperature, and 1 $k_B T$ is ~ 0.6 kcal/mol at room temperature. If the total enzyme concentration is conserved, $[E] + [ES] = [E]_{tot}$, one has

$$[ES] = \frac{k_1[S]}{k_{-1} + k_1[S]}[E]_{tot} = \frac{e^{-\varepsilon/(k_B T)}}{1 + e^{-\varepsilon/(k_B T)}}[E]_{tot} \tag{12.1}$$

Here, for convenience of the following discussions, we have written the expression in the form of the Boltzmann distribution. Notice that an enzyme molecule can exist either in a free form E, or a bound form ES. If we set the E state, which we number as state 1, with a free energy level $\epsilon_1 = 0$, then the ES state, which we number as state 2, has a free energy level $\epsilon_2 = \varepsilon$. The Boltzmann distribution states that the probability of finding an enzyme in the ES state is given by

$$p_1 = \frac{1}{Z}, \ p_2 = \frac{e^{-\varepsilon/(k_B T)}}{Z} \tag{12.2}$$

where $Z = \sum_{i=1}^{2} e^{-\epsilon_i/(K_B T)} = 1 + e^{-\varepsilon/(K_B T)}$ is called the partition function in statistical physics, and is defined as summation over the Boltzmann factors of all states. Then, from $[ES] = p_2[E]_{tot}$, one recovers Equation 12.1. With Equation 12.1 one obtains the familiar Michaelis-Menten kinetics equation (under the quasi-equilibrium approximation),

$$\frac{d[P]}{dt} = k_2[ES] = \frac{k_1 v[S]}{k_{-1} + k_1[S]}[E]_{tot} = \frac{v e^{-\varepsilon/(k_B T)}}{1 + e^{-\varepsilon/(k_B T)}}[E]_{tot} = v p_2 [E]_{tot} \tag{12.3}$$

In the CoPE model, Zhang et al. adopts a similar approximation, although it is a little more complicated since enzymes can bind on any of the N nucleosomes and catalyze chemical reactions. For simplicity, let's consider a case with two nucleosomes. There are nine possible covalent states specified by $\{s_1, s_2\}$. For each of them, there are 25 possible enzyme-binding states specified by $\{\sigma_1, \sigma_2\}$. Again we can assign each enzyme-binding state a free energy level $\varepsilon_{s_1 s_2; \sigma_1 \sigma_2} = \varepsilon_{s_1 \sigma_1} + \varepsilon_{s_1 \sigma_1} - J_{\sigma_1 \sigma_2}$. Notice that the free energy of binding ε is s-dependent, and a term $-J_{\sigma_1 \sigma_2}$ represents the lateral interactions between two enzymes bound to the two neighboring nucleosomes. The Boltzmann distribution gives the probability of finding the system, that is, the two nucleosomes in state $\{s_1, s_2; \sigma_1, \sigma_2\}$ is $p_{s_1 s_2; \sigma_1 \sigma_2} = e^{-\varepsilon_{s_1 s_2; \sigma_1 \sigma_2}/(k_B T)}/Z_{s_1 s_2}$. The partition function $Z_{s_1 s_2}$ is obtained by summing over all the 25 enzyme-binding states with fixed covalent state $\{s_1, s_2\}$. The expression, $\bar{p}_{\sigma_1} = \sum_{\sigma_2 = 1}^{5} e^{-\frac{\varepsilon_{s_1 s_2; \sigma_1 \sigma_2}}{k_B T}}/Z_{s_1 s_2}$, gives the probability of finding one nucleosome, for example, nucleosome 1, with enzyme-binding state σ_1, irrespective of the enzyme-binding state of nucleosome 2. Then we can obtain the enzymatic reaction rate for a specific reaction using an expression similar to Equation 12.3. For example, suppose that the two nucleosomes are in state $\{s_1 = -1, s_2 = 0\}$. Nucleosome 1 can have its repressive mark removed by a bound corresponding removal enzyme E_{Rr}, and the rate is given by $k_1 = v_{-1 \to 0} \bar{p}_{\sigma_1 = 3}[E_{Rr}]_{\text{eff}} = v'_{-1 \to 0} \bar{p}_{\sigma_1 = 3}$, where $[E_{Rr}]_{\text{eff}}$ is the effective repressive mark removal enzyme concentration in the nucleus. With the quasi-equilibrium approximation, we separate the 3^N s states and 5^N σ states, and remove the necessity of treating the binding/unbinding processes explicitly, thus greatly reducing the computational cost.

With the enzyme-binding/unbinding processes treated by this quasi-equilibrium approximation, the following events can take place:

1. An enzymatic reaction or a process of histone turnover at site i with rate
$k_i = \delta_{s_i,0}(v'_{0 \to -1} \bar{p}_2 + v'_{0 \to 1} \bar{p}_4) + \delta_{s_i,-1} v'_{-1 \to 0}(\bar{p}_3 + d) + \delta_{s_i,1} v'_{1 \to 0}(\bar{p}_5 + d)$. Here δ_{ij} is the Kronecker delta function, which assumes a value 1 when $i = j$, and 0 when $i \neq j$. Notice here we take into account the fact that for an enzymatic reaction to take place, the corresponding enzyme has to bind to the nucleosome. The term d is the histone replacement rate due to stochastic turnover ($s_i \to 0$).
2. Every time when cell division takes place, each histone has 50% probability to be partitioned to one of the daughter cells.

Therefore the overall simulation procedure is as follows:
For each step with covalent state $\{s_i\}$,

1. Calculate $\bar{p}_{\sigma_i}$ and $\{k_i\}$.
2. Define the transition rate array $k = [k_1, \ldots, k_N]$. Then at a given simulation step, define elements of an accumulative reaction rate array as $\alpha_m = \sum_{i=1}^{m} k_i$.
3. Generate two random numbers r_1 and r_2 from a uniform distribution within $[0,1]$. The next time that an event will take place is given by $dt = \frac{1}{\alpha_N} \ln\left(\frac{1}{r_1}\right)$, so $t \to t + dt$, and the reaction channel

taking place is the smallest integer m satisfying $a_m \geq r_2 \alpha_N$. If $s_m \neq 0$, update s_m to 0. If $s_m = 0$, generate another random number r_3 from a uniform distribution within [0,1], update s_m to -1 if $r_3 \leq \left(v'_{0 \rightarrow -1}\bar{p}_2\right) / \left(v'_{0 \rightarrow -1}\bar{p}_2 + v'_{0 \rightarrow 1}\bar{p}_4\right)$, otherwise update s_m to 1.

4. Repeat Steps 1–3.

5. When it reaches the cell division time, for each nucleosome i generate a random number r_4 from a uniform distribution within [0,1]. If $r_4 \leq 0.5$ then $s_i = 0$, meaning that the histone is replaced by a nascent unmarked one; otherwise keep the original value of s_i, meaning the original histone is partitioned to this daughter cell being monitored. Here for simplicity we assume that the cell cycle time is fixed, which can be easily modified if variation of cell cycle time needs to be considered.

One can translate the above pseudo-code into any programming language, such as Matlab, Python or C.

12.5 DETERMINE MODEL PARAMETERS

To perform the numerical simulations discussed here, we need to determine the model parameters. A generally adopted strategy is to first determine or estimate the model parameters from experimental measurements.

Some parameters can be determined easily. If one assumes that some insulating elements constrain the histone modification patterns [43], one can estimate the number N from the DNA length within the constraints. For a gene length ~10 k bp including the promoter regions, the nucleosome length $N = 40$. Without insulating elements, the model studies of Hodges and Crabtree show that an inherently bound histone pattern domain can be formed when the mark addition and removal enzymes have comparable catalytic activities [11]. In that case the length of the domain is determined by the relative ratio between the addition and removal enzyme activities.

In the following discussions, we investigate how to determine other parameters.

12.5.1 NONSPECIFIC BACKGROUND FREE ENERGY OF BINDING OF ENZYMES

Several experimental techniques, such as fluorescence recovery after photobleaching (FRAP) and fluorescence correlation spectroscopy (FCS), can provide quantitative information about protein-chromatin binding [25]. In the literature what is usually reported is the fraction of enzymes bound to the histones. Below we discuss how to roughly estimate the free energy of binding from the data. Since these measurements are genome wide, therefore they reflect nonspecific protein–chromatin bindings instead of specific bindings facilitated by DNA-sequence specific elements.

Experimental data reveals that nonspecific protein-chromatin bindings are weak. Therefore we assume that the probability of having two neighboring nucleosomes occupied (from nonspecific background binding) at the same time is negligible. That is, for parameter estimation purposes we can neglect possible effects of the lateral interaction J, and treat each nucleosome as independent. Each histone can have two states: empty or occupied. Then with respect to an arbitrary reference state with binding energy ε_0 and free enzyme concentration c_0, the binding energy with free enzyme concentration c_{free} is $\varepsilon = \varepsilon_0 - \ln(c_{free}/c_0)$. From the Boltzmann distribution, the probability of observing a histone in the bound state is

$$p_H = \frac{exp\,(-\varepsilon/k_B T)}{1 + exp\,(-\varepsilon/k_B T)}, \tag{12.4}$$

then,

$$\varepsilon = -k_B T\ln\frac{p_H}{1 - p_H}, \; \varepsilon_0 = -k_B T\ln\frac{p_H c_0}{(1 - p_H)c_{free}} \tag{12.5}$$

From the cell volume and enzyme concentrations, we can estimate the total number of enzymes. Then from the measured fraction of bound enzymes, we obtain the total number of enzymes bound, noting that this number is also the total number of histones in the bound state. Next we can estimate the total number of nucleosomes from the genome size, assuming $\sim$200 base pairs per nucleosome. The total number of (nucleosome) H3 proteins is twice the number of nucleosomes (since each nucleosome contains two copies of H3 proteins). From all these numbers we can estimate p_H.

Table 12.1 summarizes our estimations based on available experimental data, using 1 μM as the reference free enzyme concentration c_0. Clearly our estimation is very rough. For example, we do not consider competition of binding from different types of enzymes. We also assume that every

Table 12.1 Estimation of Nonspecific Binding Energy from Experimental Data

	H3K9me3	H3K27me3	References
Enzyme	HP1α	Polycomb group (PcG) proteins	[25,44,49]
Cell source	Mouse L cells	*Drosophila* Neuroblasts/ Embryo	[18,46,51]
Nuclear volume (μM^3)	435	200	[25]
Estimated nucleosome number	21,120,000 (L cells)	960,000 (Embryo (cycle 14))	[25]
Nucleosome concentration	80.6 mM (L cells)	7.97 μM (Embryo (cycle 14))	[25]
Measured-enzyme-bound fraction	65% (Mouse NIH 3T3/iMEFs)	18.93% (*Drosophila* Neuroblasts cells)	[44,45]
Total enzyme concentration	1 μM	380 nM (*Drosophila* Neuroblasts cells)	[44,45]
Number of bound enzymes	149477	10350	Derived
P_H	0.004	0.0045	Derived based on Eqn. (12.4)
c_{free}	0.35 μM	0.308 μM	Derived
ε	4.5 $k_B T$	4.2 $k_B T$	Derived based on Eqn. (12.5)
Reproduced from [27].			

200 base pairs form a nucleosome. This is clearly an overestimation of the total number of nucleosomes since there are nucleosome-free regions. Including these corrections reduces the number of free nucleosomes, and leads to a lower binding energy.

Notice that the estimated values of free energy of binding are positive. That is, nonspecific binding of enzymes on DNA is very weak at physiological histone and enzyme concentrations. Mechanistically this weak binding is reasonable. From the above table, the total number of nucleosomes is far more than that of the enzymes. That is, the number of substrates is much larger than the number of enzymes. Strong nonspecific binding would not allow a binding enzyme to move and interact with other nucleosomes, and seriously deplete the pool of free enzymes.

12.5.2 FREE ENERGY OF BINDING OF ENZYMES WITHIN THE NUCLEATION REGION

There is no quantitative information on the enzyme free energy of binding at specific genome regions. The values are also affected by concentrations of the elements recruiting these enzymes. One piece of experimental information that can be used is the peaked distribution of the histone marks along the genome. The ratio between the peak value and that of the background value (for regions far away from the nucleation region) can be used to determine the specific binding affinities. That is, we require the ratio calculated from the model to match the experimental value (of Oct4 in the work of Zhang et al. [27]).

12.5.3 ENZYME LATERAL INTERACTIONS

The values of $J_{\alpha\alpha}$ are chosen to reproduce the bell-like shaped histone methylation pattern centered around the nucleation region with a half-height width of about 10 nucleosomes, to represent the histone modification distribution pattern of Oct4 gene [11]. In the work of Zhang et al. [27], for simplicity the same value of $J_{\alpha\alpha}$ is used for all enzymes. For interactions between different enzymes $J_{\alpha\beta}$ we simply assume that they may either be absent, or the enzyme interact unfavorably with several values examined to explore their effects on the epigenetic dynamics.

12.5.4 ENZYME RATE CONSTANTS

Without much direct experimental information, for simplicity we use the same rate constants for all four enzymes, and choose the value that reproduces the experimental observation that it takes about five cell cycles to switch Oct4 [11].

12.5.5 HISTONE EXCHANGE

The reported value of the histone exchange rate varies over a broad range and shows cell-type dependence. In reality one may also expect dependence of the histone exchange rate on the covalent marks. Active transcriptions can lead to higher histone exchange rate [38,48], and thus different histone exchange rates may exist for euchromatins and heterochromatins. For simplicity though, Angel et al. uses a single value estimated from measurements on *Drosophila* cells [10]. Zhang et al. adopt this value as well, and examine how changing the value affects the model behavior [27].

12.6 PERFORM COMPUTATIONAL STUDIES

Figure 12.2A shows a typical simulated trajectory using parameters roughly representing the gene Oct4. Clearly the s state of each nucleosome changes randomly and frequently. However, the system can exist in one collective epigenetic state, dominated by either repressive or active marks, for many cell cycles before stochastically switching to another state. A zoom-in of the trajectory (shown in Figure 12.2B) shows that a transition usually starts at one place, often within the nucleation region, then propagates outward. Statistically the system still spends most of the time around either the repressive or active mark-dominated states. That is, if one plots the fraction of time the system has n nucleosomes bearing repressive marks out of the N nucleosomes, one obtains a histogram with a bimodal distribution. In other words, the system exists as a bistable system.

Experimental studies reveal two essential molecular properties: enzymes can recognize the nucleosome marks and have mark-dependent free energy of binding, and enzymes bound to neighboring nucleosomes can interact laterally. Mathematically we use a quantity $\Delta\varepsilon$ to reflect the mark-dependent free energy of binding, assuming that the binding energies for the addition or removal enzymes to a nucleosome bearing the corresponding (or antagonizing) mark are $\Delta\varepsilon$ lower (or higher) than those binding to an unmodified nucleosome. That is, $\Delta\varepsilon$ is an energetic penalty for mismatched binding between an enzyme and a nucleosome. The parameter $J_{\alpha\alpha}$ specifies the strength of lateral interactions between two neighboring enzymes of the same type. Figure 12.2C shows the calculated bistable region in the $\Delta\varepsilon$ - $J_{\alpha\alpha}$ plane. Clearly a broad range of combinations of $\Delta\varepsilon$ and $J_{\alpha\alpha}$ lead to bimodal distributions. A finite value of $J_{\alpha\alpha}$, with a critical minimum value $\sim 2\ k_BT$, is necessary for generating bimodal distributions of the fraction of histones with repressive marks. Below this value of $J_{\alpha\alpha}$, increasing $\Delta\varepsilon$ values does not lead to a bimodal distribution. The required value of $J_{\alpha\alpha}$ also increases sharply upon decreasing $\Delta\varepsilon$. With $\Delta\varepsilon \to 0$, the value of $J_{\alpha\alpha}$ needed for generating a bimodal distribution increases sharply. Within intermediate values, a decrease of $\Delta\varepsilon$ can be compensated by an increase of $J_{\alpha\alpha}$. Therefore, these results demonstrate that $\Delta\varepsilon$ and $J_{\alpha\alpha}$, representing the two observed molecular properties, are both sufficient and necessary to generate the epigenetic histone memory. This is an essential result and the working mechanism obtained from analyzing the CoPE model.

As mentioned above, a major and typical concern for modeling complex biological systems is that many parameters cannot be well determined experimentally. Therefore a key concept arising in quantitative biology studies is that if it holds for a broad range of model parameters, a mechanism is robust, and one has higher confidence that it reflects the true biology of the system; on the other hand, one should be skeptical and cautious of a mechanism that requires fine tuning model parameters. To show that the above-discussed physical mechanism is not a result of fine-tuning the model parameters, Zhang et al. performed simulations using 4096 sets of parameters in a six-dimensional parameter space, with each dimension divided into four equally distributed grid points within a physically reasonable range. The six parameters are the free energy of binding and lateral interactions. They also used a *more* stringent criterion for the bistable region compared to what was used to generate Figure 12.2D: clear separation between the epigenetic states with high and low average number of nucleosomes with repressive marks (>4.5), significant epigenetic memory with the average dwelling time on each epigenetic state >2 cell cycles. It turns out that 1238 (30%) parameter sets satisfy the above requirement. Therefore, the mechanism is robust against parameter choices.

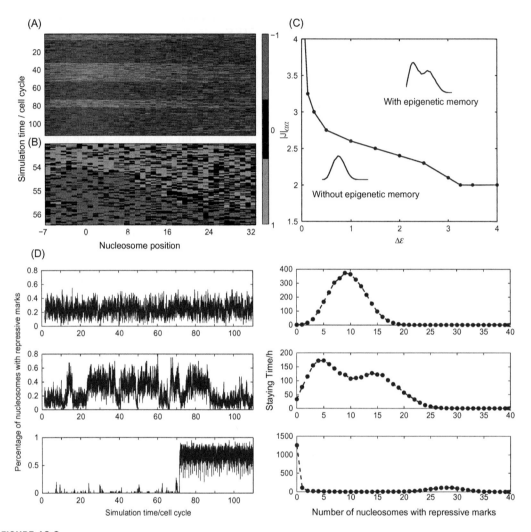

FIGURE 12.2

Simulation results using model parameters corresponding to Oct4. (A) Heat map representation of a typical simulation trajectory. (B) Zoom-in of the heat map in panel (A) showing epigenetic state transition. (C) Phase diagram on the $\Delta\varepsilon$ - J plane to illustrate bistability mechanism. (D) Typical trajectories of the fraction of nucleosomes with repressive marks (left) and the corresponding probability distribution of observing a given number of nucleosomes with repressive marks (right). All simulations are performed with $\Delta\varepsilon = 2$, but different $J_{\alpha\alpha}$ values, Upper panel, $J_{\alpha\alpha} = 0$; middle panel, $J_{\alpha\alpha} = 2{:}5$; lower panel, $J_{\alpha\alpha} = 3{:}5$. The dwelling time distribution is obtained by averaging over 100 trajectories, each started with a randomly selected initial histone modification configuration, simulated for 10^3 Gillespie steps, then followed by another 2×10^3 Gillespie steps for sampling. (*This figure is reproduced in color in the color plate section.*)

Adapted from [27].

12.7 IDENTIFY INSIGHTS FROM MODEL STUDIES AND MAKE TESTABLE PREDICTIONS

The previous model simulations reveal a simple molecular mechanism for generating epigenetic histone memory. Let's first consider an analogous situation. Suppose that there is a set of jigsaw puzzles (Figure 12.3A). A naughty kid randomly takes away pieces of the puzzle. You have two tasks:

1. Figure out what piece is missing. For more reliable reasoning of the original pattern it is better to examine not only the slots of missing pieces, but also a larger region.
2. Put back a piece of puzzle the same as the missing one from a reservoir of spare puzzle pieces. The process should be faster than the process of the puzzle pieces being taken away. Otherwise there would quickly be an accumulation of missing pieces, which makes it more and more difficult for the reasoning in step 1.

Cells essentially have the same tasks, and the molecular properties of the involved molecular species ensure robust completeness of the tasks. Let's consider a collection of nucleosomes dominated by repressive marks (Figure 12.3B). After cell division, some of the nucleosomes are replaced by unmarked ones. The remaining nucleosomes with repressive marks preferentially recruit

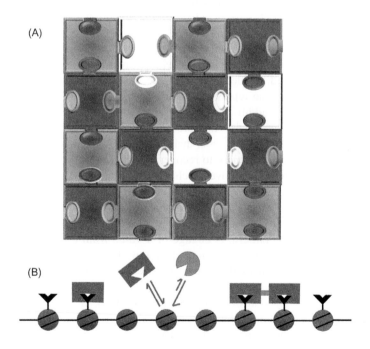

(A)

(B)

FIGURE 12.3

Schematic illustration of the reader/writer mechanism. (A) An analogous jigsaw puzzle reconstruction problem. (B) Reader-and-writer mechanism for epigenetic pattern reconstruction. (*This figure is reproduced in color in the color plate section.*)

Table 12.2 Calculated Enzyme-Binding Probabilities of a Two-Nucleosome System with $s_1 = 0$, and $s_2 = -1$

	$\sigma_1 = 1$	$\sigma_1 = 2$	$\sigma_1 = 3$	$\sigma_1 = 4$	$\sigma_1 = 5$
$\sigma_2 = 1$	0.285	0.0685	0.0685	0.010	0.010
$\sigma_2 = 2$	0.177	0.042	0.042	0.0062	0.0062
$\sigma_2 = 3$	0.177	0.042	0.042	0.0062	0.0062
$\sigma_2 = 4$	0.0032	0.0008	0.0008	0.0001	0.0001
$\sigma_2 = 5$	0.0032	0.0008	0.0008	0.0001	0.0001
$\bar{p}_{\sigma_1}$	0.645	0.155	0.155	0.0227	0.0227

All model parameters are taken from Table 12.1 of Zhang et al. [27]. Specifically, $J_{\alpha\alpha} = 3\,k_BT$, $\Delta\varepsilon = 2\,k_BT$.

repressive mark enzymes relative to active mark enzymes — a "reading" process. Because of enzyme lateral interactions, these bound enzymes help the unmarked nucleosomes preferentially also recruit repressive mark enzymes, and add the repressive marks — a "writing" process. Unlike genome inheritance, an epigenetic histone pattern, that is specific pattern of a given nucleosome, cannot be exactly inherited, but the overall pattern, repressive or active mark domination, can be rather faithfully maintained and inherited.

It may be easier to understand this molecular mechanism using a two-nucleosome system. Suppose that originally both of the two nucleosomes bear repressive marks. After cell division, nucleosome 1 becomes unmarked. Table 12.2 gives the enzyme-binding probabilities calculated from the Boltzmann distribution. The repressive-mark-bearing nucleosome 2 has higher probabilities of having the repressive mark addition or removal enzymes bound. Consequently, when it has these enzymes bound, nucleosome 1 also has higher probabilities of having the same enzyme bound. Overall nucleosome 1 has higher $\bar{p}_{\sigma_1} = 2$ than $\bar{p}_{\sigma_1} = 4$. That is, nucleosome 1 is more likely to add a repressive mark than an active mark to recover the original epigenetic pattern.

An immediate conjecture from the above mechanism is that the system needs to reconstruct the epigenetic pattern faster than the perturbations coming from histone turnover, enzymatic reactions, and cell division. Indeed Figure 12.4A shows that the model predicts sensitive dependence of the epigenetic state stability on the histone turnover rate d. Histone turnover is a major source of perturbations to the epigenetic pattern. A change of d value from 0.6 h^{-1} to 1.2 h^{-1} results in the average epigenetic state dwelling time changing from ~250 hours to 20 hours. Experimentally the value of d is difficult to measure accurately, and it varies over an order of magnitude [48–51]. The value also depends on the cell types. Embryonic stem cells have a histone turnover rate higher than that of differentiated cells [51]. It may be because embryonic stem cells only exist transiently during the developmental process, and thus there is no selection pressure to maintain the epigenetic memory long. On the other hand, for cells like neurons for which maintaining epigenetic information is crucial for physiological functions, we predict that the value of d should be kept small. The model results in Figure 12.4A also show that increasing the enzyme rate constant ν can compensate an increased value of d. Increasing ν allows faster

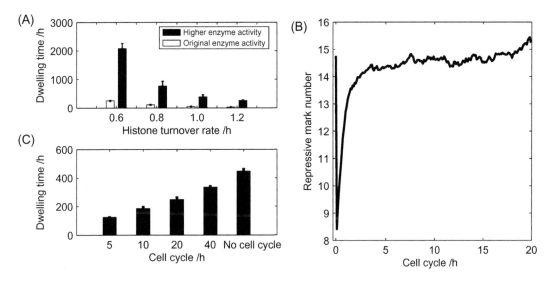

FIGURE 12.4

Different parameters affect histone epigenetic memory and dynamics. (A) Dependence of average dwelling time on histone exchange rate and different enzymatic activity. Higher enzymatic activity $v = 3.0$, original enzymatic activity $v = 1.5$. (B) After cell-cycle relaxation dynamics of the total number of nucleosomes with repressive marks. (C) Average state dwelling time as a function of the cell cycle.

Adapted from [27].

recovery of missed marks on the nucleosome due to histone turnover. The 60-kDa HIV-Tat inter-active protein (Tip60), a key member of the MYST family of histone acetyltransferases, can auto-acetylate its lysine residue K327. Yang et al. report that a K327 deacetylated Tip60 only loses its catalytic activity by less than one fold [52]. However, Yuan et al. introduced this mutant into yeast and found this lack of autoacetylation is fatal for the survival of the organisms [53]. Therefore, our model studies suggest that the enzyme activities (including concentrations) should be tightly regulated. Quantitative measurements are needed to test this prediction. If it is vali-dated, then how are the enzyme activities robustly regulated?

Cell division is another major source of perturbations. Figure 12.4B shows that a mammalian cell is capable of quickly recovering (within a few hours) the original epigenetic pattern after losing about half of the histones due to cell division. A direct conjecture is that if one reduces cell cycle time so the cell has less time to recover from this perturbation, there is higher probability that the perturbation may accumulate over cell cycles and lead to faster switching of the epigenetic state. This conjecture is numerically proven by the results in Figure 12.4C. This model result may help understand the experimental observation of Hanna et al. [54]. These authors show that decreasing cell cycle time can accelerate the process reprogramming somatic cells to induced pluripotent stem cells. The result in Figure 12.4C suggests that reduced cell cycle time may facilitate some genes to switch their epige-netic states and the cell could overcome the epigenetic barrier to achieve phenotypic transition.

12.8 CONCLUSION

Let's come back to the question we asked in the introduction. To understand how cells regulate and maintain phenotypes, a key step is to study how gene activities are regulated. Epigenetic histone modification is an essential part of the regulatory network. The mammalian cell reprogramming experiments reveal that epigenetic state switching is a rate-limiting step during the process [55−57]. Recent advances in techniques such as CRISPR open the possibility of easily editing the epigenome of a cell to artificially turn on or off selected genes. Therefore, understanding the molecular mechanism(s) of epigenetic regulation is of both theoretical and practical importance.

In this chapter we use the CoPE model of epigenetic memory from Zhang et al. as an example to illustrate how one constructs and analyzes a mathematical model. We argue that even for a system with many unknowns, one can still perform a certain level of mathematical modeling, and provide useful insights. One should be able to simplify and abstract the real system for modeling purposes, but in a well-controlled way so connections to the real physical quantities are transparent. Often there are a large number of model parameters that cannot be reliably constrained by available experimental data. One can still make qualitative and quantitative predictions through analyzing an ensemble of models with different parameter values. Last but not least, modeling is not the end but the starting point of another cycle of studies. While studying a complex biological system, modeling has its own strengths and limitations, and an effort cohesively integrating modeling and experiments is always desirable.

The CoPE model should be viewed as an initial step to model the complex process of epigenetic regulation. In the above we mentioned a number of limitations of the model. In their review Rohlf et al. have a detailed discussion on the additional features future modeling efforts should take into account [18]. For further development, more quantitative data and more molecular details would be needed. One may also adopt a multiscale modeling approach: using atomistic and coarse-grained modeling to explicitly include chromosome structure and provide inputs for the more coarse-grained modeling approach as described in this chapter.

Molecular dynamics (MD) simulation studies can provide down to atomistic level insight on some critical questions. For example, the histone turnover rate is an important parameter affecting the epigenetic dynamics. How is the rate affected by the histone covalent marks or DNA methylation? Does it show any sequence dependence? A single histone has many modifiable residues. How could different covalent marks crosstalk to each other through affecting binding affinities of various enzymes? In recent years, large-scale simulations based on structural details of biomolecules have advanced tremendously in terms of both simulation time scale [58] and simulation system size [59,60]. Since high-resolution crystal structures of nucleosomes have been made available [61], systematic computational studies on the nucleosome histone modifications, starting from the atomistic level, have become one of the important developments in the field of epigenetic research [62−64]. To support the atomistic scale simulation of the modified histone tails, commensurate efforts have been devoted to the force field developments to allow highly specific structural and energetic determination. For that purpose, *ab initio* quantum mechanics (QM) calculations, molecular mechanics (MM) or MD refinements, and experimental validation are all integrated [65]. For example, a user-friendly and freely available platform for automated introduction of posttranslocation modifications of choices to a protein 3D structure is presented by the Vienna-PTM web server [66]. Furthermore, the *ab initio* QM/MM techniques have also been implemented to study histone modifying enzymes on their reaction mechanisms [67].

Besides the atomistic level approach, coarse-grained modeling from nucleosome toward chromatin level [68], with a more or less structural basis and empirical interaction potentials, has also been developed accordingly. These types of models are quite adaptable to solve practical issues, without being restricted by time and spatial scales. For example, in a previous Monte Carlo simulation of a "mesoscale" chromatin model, histone tail flexibility, linker-histone electrostatic and orientation, magnesium ion induced electrostatic screening, and linker-DNA bending at physiological conditions, as well as thermal fluctuations and entropy effects, are all considered [46]. In a recent Brownian dynamics simulation study of DNA unrolling from the nucleosome, the mechanical forces from the histone core and effective electrostatic and site-specific binding of the DNA to the histone are considered, giving an estimation of the DNA-histone attraction at $\sim 2.7 \; k_B T$ per base pair [47]. In another bead-spring model of chromatin, the flexible histone tails are made available for temporary electrostatic interaction with nucleosomes; the inter-nucleosomal interactions are thus mediated by the histone tails to allow distant communication in chromatin [69]. A DNA lattice model in the framework of Ising-Markov approaches was developed as well, to describe transcription factor access to nucleosome DNA, taking into account the intermediate protein-binding state in which DNA is partially unwrapped from the histone octamer [70]. Although these models cannot deal with the chemical nature of histone modification, they can be combined with atomistic or *ab initio* types of simulation studies to reveal how local histone modifications impact on global properties of nucleosome-nucleosome interactions and chromatic structures.

In summary, structure-based modeling efforts, both at the atomistic and coarse-grained levels, will continue to help in analyzing existing experimental results and guiding new experimental studies toward elucidating the molecular mechanism of epigenetic regulation and how it is coupled to other regulatory schemes such as transcription and translation.

REFERENCES

[1] Takahashi K, Yamanaka S. Induction of pluripotent stem cells from mouse embryonic and adult fibroblast cultures by defined factors. Cell 2006;126(4):663−76.
[2] Bannister AJ, Kouzarides T. Regulation of chromatin by histone modifications. Cell Res 2011;21(3): 381−95.
[3] Greer EL, Shi Y. Histone methylation: a dynamic mark in health, disease and inheritance. Nat Rev Genet 2012;13(5):343−57.
[4] Beisel C, Paro R. Silencing chromatin: comparing modes and mechanisms. Nat Rev Genet 2011;12(2): 123−35.
[5] Black JC, Van Rechem C, Whetstine JR. Histone lysine methylation dynamics: establishment, regulation, and biological impact. Mol Cell 2012;48(4):491−507.
[6] Henikoff S, Shilatifard A. Histone modification: cause or cog? Trends Genet 2011;27(10):389−96.
[7] Dodd IB, Micheelsen MA, Sneppen K, Thon G. Theoretical analysis of epigenetic cell memory by nucleosome modification. Cell 2007;129(4):813−22.
[8] Grewal SI, Klar AJ. Chromosomal inheritance of epigenetic states in fission yeast during mitosis and meiosis. Cell 1996;86(1):95−101.
[9] Thon G, Friis T. Epigenetic inheritance of transcriptional silencing and switching competence in fission yeast. Genetics 1997;145(3):685−96.

[10] Angel A, Song J, Dean C, Howard M. A Polycomb-based switch underlying quantitative epigenetic memory. Nature 2011;476(7358):105−8.

[11] Hodges C, Crabtree GR. Dynamics of inherently bounded histone modification domains. Proc Natl Acad Sci 2012;109(33):13296−301.

[12] Hathaway NA, Bell O, Hodges C, Miller EL, Neel DS, Crabtree GR. Dynamics and memory of heterochromatin in living cells. Cell 2012;149(7):1447−60.

[13] Alsing AK, Sneppen K. Differentiation of developing olfactory neurons analysed in terms of coupled epigenetic landscapes. Nucleic Acids Res 2013;41(9):4755−64.

[14] Schwammle V, Jensen ON. A computational model for histone mark propagation reproduces the distribution of heterochromatin in different human cell types. PloS One 2013;8(9):e73818.

[15] Sedighi M, Sengupta AM. Epigenetic chromatin silencing: bistability and front propagation. Phys Biol 2007; 4(4):246−55.

[16] Dayarian A, Sengupta AM. Titration and hysteresis in epigenetic chromatin silencing. Phys Biol 2013; 10(3).

[17] Arnold C, Stadler PF, Prohaska SJ. Chromatin computation: epigenetic inheritance as a pattern reconstruction problem. J Theor Biol 2013;336:61−74.

[18] Binder H, Steiner L, Przybilla J, Rohlf T, Prohaska S, Galle J. Transcriptional regulation by histone modifications: towards a theory of chromatin re-organization during stem cell differentiation. Phys Biol 2013;10(2):026006.

[19] Schwab DJ, Bruinsma RF, Rudnick J, Widom J. Nucleosome switches. Phys Rev Lett 2008;100(22): 228105.

[20] Benecke A. Chromatin code, local non-equilibrium dynamics, and the emergence of transcription regulatory programs. Eur Phys J E 2006;19(3):353−66.

[21] Prohaska SJ, Stadler PF, Krakauer DC. Innovation in gene regulation: the case of chromatin computation. J Theor Biol 2010;265(1):27−44.

[22] David-Rus D, Mukhopadhyay S, Lebowitz JL, Sengupta AM. Inheritance of epigenetic chromatin silencing. J Theor Biol 2009;258(1):112−20.

[23] Raghavan K, Ruskin HJ, Perrin D, Goasmat F, Burns J. Computational micromodel for epigenetic mechanisms. PloS One 2010;5(11):e14031.

[24] Sontag LB, Lorincz MC, Georg LE. Dynamics, stability and inheritance of somatic DNA methylation imprints. J Theor Biol 2006;242(4):890−9.

[25] Steffen PA, Fonseca JP, Ringrose L. Epigenetics meets mathematics: towards a quantitative understanding of chromatin biology. Bioessays 2012;34(10):901−13.

[26] Rohlf T, Steiner L, Przybilla J, Prohaska S, Binder H, Galle J. Modeling the dynamic epigenome: from histone modifications towards self-organizing chromatin. Epigenomics 2012;4(2):205−19.

[27] Zhang H, Tian X-J, Mukhopadhyay A, Kim KS, Xing J. Statistical mechanics model for the dynamics of collective epigenetic histone modification. Phys Rev Lett 2014;112(6):068101.

[28] Jacobs SA, Khorasanizadeh S. Structure of HP1 chromodomain bound to a lysine 9-methylated histone H3 tail. Science 2002;295(5562):2080−3.

[29] Margueron R, Justin N, Ohno K, Sharpe ML, Son J, Drury III WJ, et al. Role of the polycomb protein EED in the propagation of repressive histone marks. Nature 2009;461(7265):762−7.

[30] Kouzarides T. Chromatin modifications and their function. Cell 2007;128(4):693−705.

[31] Cowieson NP, Partridge JF, Allshire RC, McLaughlin PJ. Dimerisation of a chromo shadow domain and distinctions from the chromodomain as revealed by structural analysis. Curr Biol 2000;10(9):517−25.

[32] Canzio D, Chang EY, Shankar S, Kuchenbecker KM, Simon MD, Madhani HD, et al. Chromodomain-mediated oligomerization of HP1 suggests a nucleosome-bridging mechanism for heterochromatin assembly. Mol Cell 2011;41(1):67−81.

[33] So CW, Lin M, Ayton PM, Chen EH, Cleary ML. Dimerization contributes to oncogenic activation of MLL chimeras in acute leukemias. Cancer Cell 2003;4(2):99−110.

[34] Buscaino A, Lejeune E, Audergon P, Hamilton G, Pidoux A, Allshire RC. Distinct roles for Sir2 and RNAi in centromeric heterochromatin nucleation, spreading and maintenance. Embo J 2013.

[35] Lin T, Ponn A, Hu X, Law BK, Lu J. Requirement of the histone demethylase LSD1 in Snail-mediated transcriptional repression during epithelial-mesenchymal transition. Oncogene 2010;29(35):4896−904.

[36] Peinado H, Ballestar E, Esteller M, Cano A. Snail Mediates E-Cadherin Repression by the Recruitment of the Sin3A/Histone Deacetylase 1 (HDAC1)/HDAC2 Complex. Mol Cell Biol 2004;24(1):306−19.

[37] Herranz N, Pasini D, Diaz VM, Franci C, Gutierrez A, Dave N, et al. Polycomb complex 2 is required for E-cadherin repression by the snail1 transcription factor. Mol Cell Biol 2008;28(15):4772−81.

[38] Gaffney DJ, McVicker G, Pai AA, Fondufe-Mittendorf YN, Lewellen N, Michelini K, et al. Controls of nucleosome positioning in the human genome. PLoS Genet 2012;8(11):e1003036.

[39] Mendenhall EM, Koche RP, Truong T, Zhou VW, Issac B, Chi AS, et al. GC-rich sequence elements recruit PRC2 in mammalian ES cells. PLoS Genet 2010;6(12):e1001244.

[40] Ku M, Koche RP, Rheinbay E, Mendenhall EM, Endoh M, Mikkelsen TS, et al. Genomewide analysis of PRC1 and PRC2 occupancy identifies two classes of bivalent domains. PLoS Genet 2008;4(10):e1000242.

[41] Xu C, Bian C, Lam R, Dong A, Min J. The structural basis for selective binding of non-methylated CpG islands by the CFP1 CXXC domain. Nat Commun 2011;2:227.

[42] Encode Project Consortium. An integrated encyclopedia of DNA elements in the human genome. Nature 2012; 489(7414):57−74.

[43] Bushey AM, Dorman ER, Corces VG. Chromatin insulators: regulatory mechanisms and epigenetic inheritance. Mol Cell 2008;32(1):1−9.

[44] Fonseca JP, Steffen PA, Muller S, Lu J, Sawicka A, Seiser C, et al. In vivo polycomb kinetics and mitotic chromatin binding distinguish stem cells from differentiated cells. Genes Dev 2012;26(8): 857−71.

[45] Muller KP, Erdel F, Caudron-Herger M, Marth C, Fodor BD, Richter M, et al. Multiscale analysis of dynamics and interactions of heterochromatin protein 1 by fluorescence fluctuation microscopy. Biophys J 2009;97(11):2876−85.

[46] Grigoryev SA, Arya G, Correll S, Woodcock CL, Schlick T. Evidence for heteromorphic chromatin fibers from analysis of nucleosome interactions. Proc Natl Acad Sci 2009.

[47] Wocjan T, Klenin K, Langowski J. Brownian dynamics simulation of DNA unrolling from the nucleosome. J Phys Chem B 2009;113(9):2639−46.

[48] Deal RB, Henikoff JG, Henikoff S. Genome-wide kinetics of nucleosome turnover determined by metabolic labeling of histones. Science 2010;328(5982):1161−4.

[49] Bhattacharya D, Talwar S, Mazumder A, Shivashankar GV. Spatio-temporal plasticity in chromatin organization in mouse cell differentiation and during *Drosophila* embryogenesis. Biophys J 2009;96 (9):3832−9.

[50] Zee BM, Levin RS, Dimaggio PA, Garcia BA. Global turnover of histone post-translational modifications and variants in human cells. Epigenetics Chromatin 2010;3(1):22.

[51] Barth TK, Imhof A. Fast signals and slow marks: the dynamics of histone modifications. Trends Biochem Sci 2010;35(11):618−26.

[52] Yang C, Wu J, Zheng YG. Function of the active site lysine autoacetylation in Tip60 catalysis. PLoS ONE 2012;7(3):e32886.

[53] Yuan H, Rossetto D, Mellert H, Dang W, Srinivasan M, Johnson J, et al. MYST protein acetyltransferase activity requires active site lysine autoacetylation. EMBO J 2012;31(1):58−70.

[54] Hanna J, Saha K, Pando B, van Zon J, Lengner CJ, Creyghton MP, et al. Direct cell reprogramming is a stochastic process amenable to acceleration. Nature 2009;462(7273):595−601.

[55] Pasque V, Jullien J, Miyamoto K, Halley-Stott RP, Gurdon JB. Epigenetic factors influencing resistance to nuclear reprogramming. Trends Genet 2011;27(12):516−25.

[56] Ang YS, Gaspar-Maia A, Lemischka IR, Bernstein E. Stem cells and reprogramming: breaking the epigenetic barrier? Trends Pharmacol Sci 2011;32(7):394−401.

[57] Papp B, Plath K. Epigenetics of reprogramming to induced pluripotency. Cell 2013;152(6):1324−43.

[58] Piana S, Klepeis JL, Shaw DE. Assessing the accuracy of physical models used in protein-folding simulations: quantitative evidence from long molecular dynamics simulations. Curr Opin Struct Biol 2014;24(0): 98−105.

[59] Freddolino PL, Arkhipov AS, Larson SB, McPherson A, Schulten K. Molecular dynamics simulations of the complete satellite tobacco mosaic virus. Structure 2006;14(3):437−49.

[60] Sanbonmatsu K, Blanchard S, Whitford P. Molecular dynamics simulations of the ribosome. In: Dinman JD, editor. Biophysical approaches to translational control of gene expression. Biophysics for the life sciences. 1. New York: Springer; 2013. p. 51−68.

[61] Andrews A, Luge K. Nucleosome structure(s) and stability: variations on a theme. Annu Rev Biophys 2011; 40:99−117.

[62] Potoyan DA, Papoian GA. Regulation of the H4 tail binding and folding landscapes via Lys-16 acetylation. Proc Natl Acad Sci 2012;109(44):17857−62.

[63] Sanli D, Keskin O, Gursoy A, Erman B. Structural cooperativity in histone H3 tail modifications. Protein Sci 2011;20(12):1982−90.

[64] Korolev N, Yu H, Lyubartsev AP, Nordenskiold L. Molecular dynamics simulations demonstrate the regulation of DNA-DNA attraction by H4 histone tail acetylations and mutations. Biopolymers 2014.

[65] Grauffel C, Stote RH, Dejaegere A. Force field parameters for the simulation of modified histone tails. J Comput Chem 2010;31(13):2434−51.

[66] Margreitter C, Petrov D, Zagrovic B. Vienna-PTM web server: a toolkit for MD simulations of protein post-translational modifications. Nucleic Acids Res 2013;41(W1):W422−6.

[67] Zhang Y. Ab initio quantum mechanical/molecular mechanical studies of histone modifying enzymes. In: York D, Lee T-S, editors. Multi-scale quantum models for biocatalysis. Challenges and advances in computational chemistry and physics. 7. Netherlands: Springer; 2009. p. 341−50.

[68] Korolev N, Fan Y, Lyubartsev AP, Nordenskiöld L. Modelling chromatin structure and dynamics: status and prospects. Curr Opin Struct Biol 2012;22(2):151−9.

[69] Kulaeva OI, Zheng G, Polikanov YS, Colasanti AV, Clauvelin N, Mukhopadhyay S, et al. Internucleosomal interactions mediated by histone tails allow distant communication in chromatin. J Biol Chem 2012; 287:20248−57.

[70] Teif VB, Ettig R, Rippe K. A lattice model for transcription factor access to nucleosomal DNA. Biophys J 2010;99(8):2597−607.

DNA METHYLTRANSFERASE INHIBITORS FOR CANCER THERAPY

José L. Medina-Franco[1], Jakyung Yoo[2], and Alfonso Dueñas-González[3]

[1]*Facultad de Química, Departamento de Farmacia, Universidad Nacional Autónoma de México, Mexico City, Mexico* [2]*Life Science Research Institute, Daewoong Pharmaceutical Co., Ltd, Gyeonggi-do, Republic of Korea* [3]*Instituto de Investigaciones Biomédicas, Universidad Nacional Autónoma de México and Unidad de Investigación Biomédica en Cáncer, Instituto Nacional de Cancerología, México*

CHAPTER OUTLINE

Y.G. Zheng (Ed): Epigenetic Technological Applications. DOI: http://dx.doi.org/10.1016/B978-0-12-801080-8.00013-2

13.1 INTRODUCTION

13.1.1 EPIGENETICS

The nucleosome can be considered to be the functional unit of DNA, as its status standing alone and within the nuclear context determines whether or not a gene is expressed [1]. Epigenetics, therefore, can be referred to as the study of all elements that participate in nucleosome-chromatin regulation as determinants of gene expression. In the 1970s it was discovered that the addition of a methyl group at the fifth position of the cytosine in a CpG dinucleotide could inactivate the expression of genes [2]. Thus, DNA methylation driven by DNA methyltransferase (DNMT) enzymes gained momentum as the most important epigenetic factor [3]. Afterward, the regulatory effect of histone acetylases and deacetylases added another level of complexity by regulating the acetylation status at the lysine tails at the histone core of the nucleosomes [4]. In addition, it was uncovered that these enzymes were also able to bind and regulate DNMTs and other proteins having a DNA methylation-binding domain [5]. The recognition that histone proteins can undergo methylation, phosphorylation and other modifications in addition to acetylation/deacetylation led to the discovery of histone methyltransferases, histone demethylases, and their numerous interactions with components of nucleosomes, histones, and DNA itself, as well as a number of protein complexes having nucleosome-remodeling activities [6]. The discovery of noncoding RNAs as another set of epigenetic players that are themselves regulated by epigenetic machinery elements, and the discovery of DNA demethylases [7] — the ten-eleven translocation (TET) enzymes, which convert 5-methylcytosine to 5-hydroxymethylcytosine — have resulted in the realization of an unanticipated higher complexity in how epigenetics drives the fine-tuning regulation of gene expression [8].

This increasing complexity of epigenetics, with new players and interactions participating in normal and disease processes such as cancer, has made it almost impossible to arrive at educated guesses on how to use this cascade of events: that is, how to best make use of pharmacological interventions to modify in a beneficial way the processes of cancer without at the same time producing epigenetic changes that could favor disease development. Years ago, DNA methylation alterations were thought to be the only or at least the most responsible for cancer development and progression; this allowed, in a simplistic way, the development of DNA methylation inhibitors (DNMTi) for cancer treatment.

13.1.2 DNA HYPOMETHYLATION IN CANCER

Paradoxically, and counterintuitively for development of DNMTi as cancer therapy, the first observation regarding the participation of DNA methylation in cancer was that tumor cells had global DNA hypomethylation that could be as much as 60% less than their normal counterparts [9]. This hypomethylation was observed to occur mainly in the body of genes (coding regions and introns), as well as in pericentromeric regions of chromosomes rich in repetitive DNA sequences [10]. Interestingly, hypomethylation is progressive from premalignant conditions to fully developed malignancies. The main mechanisms put forward in attempting to explain cancer causation by hypomethylation have included chromosome instability and reactivation of transposable elements and/or inappropriate gene activation [11]. There are two pieces of convincing evidence linking hypomethylation with chromosomal instability. The congenital disorder ICF syndrome (immunodeficiency, chromosomal instability, and facial anomalies), caused by mutations at *DNMT3b*, demonstrates loss of methylation in classical satellite DNA and mitogen-inducible formation of bizarre multiradial chromosomes that contain arms from chromosomes 1 and 16 [12]. This disorder, however, is not associated with cancer, but common somatic tumors such as breast, ovarian, and other epithelial tumors commonly have unbalanced chromosomal translocations with breakpoints in the pericentromeric DNA of chromosomes 1 and 16 [13]. In mouse models with an inactivated allele of *NF1* and *p53* genes, introduction of a hypomorphic *DNMT1* allele caused a 2.2-fold increase in LOH frequency [14]. DNA hypomethylation in cancer was further supported by the fact that many CpG islands are normally methylated in somatic tissues [15], and that demethylation could lead to activation of nearby genes such as *HRAS*. Indeed, experimental demonstration exists that hypomethylation leads to activation of genes important for cancer development, including promoter CpG demethylation and overexpression of *14-3-3sigma*, *maspin*, *heparanase*, and *S100A4* in several tumor types [16]. The question here is whether overexpression is indeed caused by hypomethylation or whether promoters are hypomethylated secondary to their high transcriptional activity. There are data showing that the sole hypomethylation as achieved by pharmacologic means is not sufficient to activate gene expression. In this context, some genes are not permissive for expression; this means that, despite the fact that methylation is relieved, the necessary ancillary factors to activate transcription are not present. Others are permissive and therefore reactivated by demethylation, whereas for others, hypomethylation does not affect their levels of expression but can be overexpressed due to activation of signalling pathways known to activate them [17].

13.1.3 DNA HYPERMETHYLATION AND GENE SILENCING IN CANCER

Despite the solid observations on DNA hypomethylation in cancer, the demonstration that tumor suppressor genes can be inactivated not only through structural changes (loss-of-function mutations, deletions) but also by lack of expression due to promoter hypermethylation positioned tumor suppressor gene epigenetic silencing as a well-established oncogenic process [18]. The first suppressor gene known to be hypermethylated and silenced was *RB* [19], which was followed by multiple publications describing similar findings for a variety of tumor suppressor genes, among them *p16*, *MLH1*, *VHL*, and *E-cadherin* [20]. Whether gene promoter hypermethylation is the cause or consequence for the tumor suppressor gene silencing is still a matter of controversy; nevertheless, these views are not mutually exclusive. That DNA methylation is causal has been shown by the ability of diverse pharmacologic compounds and molecular techniques to reactivate gene expression upon inhibition of DNA methylation in cancer cells [21]. On the other hand, other findings suggest that hypermethylation-induced gene silencing could be secondary to changes that determine gene expression, such as chromatin modification, so that methylation helps to maintain the silenced status of the gene. Strong support for the second view came from experiments showing that methylation of histone H3 lysine 9 — that is, chromatin modification — occurred, along with resilencing of *p16* in absence of DNA methylation in cells in which *p16* had previously been activated by being knocked out of DNA methyltransferase [22] and by data demonstrating *p16* silencing in mammary epithelial cells that had escaped senescence and had demethylated the promoter [23].

13.1.4 DNA METHYLTRANSFERASES

DNA methylation is a major epigenetic modification that regulates gene expression in the genome of higher eukaryotes by DNMTs. DNMTs promote the covalent addition of a methyl group from S-adenosyl-L-methionine (AdoMet or SAM) to the 5-position of cytosine, mostly within CpG dinucleotides [24]. In the mammals, four members of the DNMT family have been identified, including DNMT1, DNMT3A, DNMT3B, and DNMT3L. DNMT1 is the most abundant DNA methyltransferase involved in duplicating the pattern of DNA methylation during replication and is essential for mammalian development and cancer cell growth [25]. DNMT3A and DNMT3B which are highly expressed in early embryonic cells act as a *de novo* methyltransferase associated with DNMT3L [26]. DNMT3L lacks the residues required for DNMT activity in the C-terminal domain despite high sequence similarity with DNMT3A [27]. The protein DNMT2, now termed TRDMT1, is a methyltransferase homolog that can be found in mammalian cells. Recent experiments suggest that DNMT2 activity is not limited to tRNA^{Asp} [28].

DNMTs can be divided into a C-terminal catalytic domain and an N-terminal regulatory domain of variable size as shown in Table 13.1 [29]. The C-terminal catalytic domains of all DNMTs share a common core structure named the "AdoMet (SAM)-dependent MTase Fold," while N-terminal ones play a role in distinguishing hemi- and unmethylated DNA. The catalytic domain consists of substrate catalysis (motifs IV, VI, and VIII), cofactor binding (motif I and X), and a target recognition domain (TRD) that is involved in DNA recognition and specificity (motifs VIII and IX). The available three-dimensional (3D) structures of DNMTs with different domains are summarized in Table 13.2 [26,30−40].

Crystal structures of human DNMT1 containing methyltransferase domain bound to DNA-containing unmethylated CpG sites and different regions of the N-terminal domain (autoinhibitory linker, CXXC, BAH1/2) have been recently published (Protein Data Bank [PDB] IDs: 3SWR and 3PTA). The conformation of these crystal structures reflects the prevention of de novo methylation mechanisms by an autoinhibitory linker between the DNA and the active site of DNMT1. Also, crystal structures are available for the mouse DNMT bound to duplex DNA containing flipped out cytosine (PDB ID: 4DA4), DNMT3A-DNMT3L C-terminal domain complex (PDB ID: 2QRV) and bacterial DNMTs bound to duplex DNA and zebularine (PBD IDs: 6MHT and 1M0E, respectively).

DNMTs are key regulators of gene transcription, and their roles in carcinogenesis have been the subject of considerable interest over the last decade. It has been reported that the expression levels of DNMTs are elevated in cancers of the colon, prostate, breast, liver, and in leukemia. Subramaniam et al. recently reviewed and summarized the importance of each DNMT in specific cancers [41]. Since inhibition of DNMTs is correlated with reduction of tumorigenicity and increased expression of tumor suppressor genes, DNMTs are attractive targets for the development of anticancer agents [41]. Gros et al. has recently reviewed pathologies associated with DNA methylation [42]. In addition to cancer, a number of other diseases have been associated with epigenetic alterations, particularly those influenced by the environment. DNA methylation has shown importance in Alzheimer's disease and psychiatric diseases such as depression, bipolar disorder, and schizophrenia. DNA methylation is also involved in autoimmune diseases and genetic disorders [42].

Table 13.1 Graphical View of Human and Mouse DNMT1 (Equivalent residue numbers in parentheses correspond to the mouse DNMT1)

Domain	Position	Graphical View
DNMT1	1–1616 [1–1620]	
AdoMet-dependent MTase	1139–1599 (1142–1620)	
BAH1	755–880 [758–884]	
BAH2	972–1100 [976–1103]	
CXXC	646–692 [649–695]	
Autoinhibitory linker	693–754 [696–757]	
Required for activity	651–697	
Interaction with the PRC2/EED-EZH2 complex	308–606 [305–609]	
Homodimerization	310–502	
DNA replication foci-targeting sequence	331–550 [328–556]	
Interaction with the PRC2/EED-EZH2 complex	1–336 [1–343]	
Interaction with DNMT3A	1–148 [1–145]	
Interaction with DMAP1	1–120 [1–120]	
Interaction with DNMT3B	149–217 [147–217]	
Interaction with PCNA	163–174 [161–172]	

This chapter is organized in seven sections. After this introduction, Section 13.2 discusses the development of DNMTi for cancer therapy including DNMTi approved for clinical use. This section also introduces different classes of compounds that have been in clinical trials. The next section presents advances on the development of experimental assays to measure DNA methylation and to screen compound collections to identify DNMTi. Section 13.4 discusses the progress on the identification of distinct DNMTi following diverse approaches such as chemical synthesis, high-throughput screening (HTS), and drug repurposing. In Section 13.5 we present advances on computational studies to better understand the activity of known inhibitors and to uncover novel DNMTi in a systematic manner. Section 13.6 describes representative examples of data resources currently

Table 13.2 Selected Crystal Structures of DNMTs

PDB Code	Type	Sequence	Region	Resolution (Å)	Cofactor	Ligand	Polymer	Reference
3SWR	hDNMT1	601–1600	Autoinhibitory linker, CXXC, BAH1/2, methyltransferase domain	2.49	SFG (sinefungin)			(Hashimoto et al., to be published)
4DA4	mDNMT1	731–1602	BAH1/2, methyltransferase domain	2.60	SAH	5-Methyl-2′-deoxycytidine	DNA	[30]
3PTA	hDNMT1	646–1600	Autoinhibitory linker, CXXC, BAH1/2, methyltransferase domain	3.60	SAH		DNA	[30]
3PT6	mDNMT1	650–1602	Autoinhibitory linker, CXXC, BAH1/2, methyltransferase domain	3.00	SAH		DNA	
3PT9	mDNMT1	731–1602	BAH1/2, methyltransferase domain	2.50	SAH			
3AV6	mDNMT1	357–1608	RFD, Autoinhibitory linker, BAH1/2, methyltransferase domain	3.09	SAM			[31]
3EPZ	hDNMT1	351–600	RFTS	2.31				[32]
1MHT	M.HhaI	1–328	Methyltransferase domain	2.60	SAH	5-Methyl-5-fluoro-2′-deoxycytidine	DNA	[33]
6MHT	M.HhaI	1–327	Methyltransferase domain	2.05	SAM	5-Methyl-2′-deoxycytidine	DNA	[34]
10MH	M.HhaI	1–328	Methyltransferase domain	2.55	SAH	5,6-Dihydro-5-azacytosine	DNA	[35]
1M0E	M.HhaI	1–328	Methyltransferase domain	2.50	SAH	Zebularine	DNA	[36]
1DCT	M.HaeIII	1–324	Methyltransferase domain	2.80	SAH	5-Methyl-2′-deoxycytidine	DNA	[37]
1G55	hDNMT2	2–391	Methyltransferase domain	1.80	SAH			[38]
3A1A	DNMT3A	474–609	ATRX-DNMT3-DNMT3L (ADD) with histone H3	2.30				[39]
3A1B	DNMT3A	476–614	DNMT3L-Histone H3 tail	2.29				
2PV0	DNMT3L	34–380		3.30				[40]
2QRV	DNMT3A-DNMT3L	623–908 178–379	C-terminal domain complex	2.89	SAH			[26]

available for DNMTi which are intended to rapidly translate basic science into clinical practice. Section 13.7 presents summary conclusions and future directions.

13.2 DEVELOPMENT OF DNMTi AS CANCER THERAPY

Reactivation of tumor suppressor genes that have been silenced by an epigenetic mechanism such as gene promoter methylation continues to be an attractive molecular target for cancer therapy. In experimental systems, DNMTi have demonstrated their ability to inhibit hypermethylation, restore suppressor gene expression, and exert antitumor effects in *in vitro* and *in vivo* laboratory models. There are several demethylating agents currently being evaluated in preclinical and clinical studies.

13.2.1 DNMTi IN CLINICAL USE: NUCLEOSIDE ANALOGS

The classical demethylating agents comprise the analogs of deoxycytidine: 5-azacytidine, 5-aza-2-deoxycytidine, 1-β-D-arabinofuranosil-5-azacytosine, and dihydro-5-azacytidine. 5-Azacytidine (azacitidine, Vidaza®) and 5-aza-2′deoxycytidine (decitabine, Dacogen®) (Figure 13.1) were developed over 30 years ago as classical cytotoxic agents, but were subsequently discovered to be effective DNA methylation inhibitors [43]; these were tested as such in several phase II studies against solid tumors, demonstrating very modest activity [44]. Currently, both are Federal Drug Administration (FDA)-approved for use against myelodysplastic syndrome (MDS). Despite the poor activity against solid tumors of these nucleoside analogs, it is remarkable that the proof of the concept with regard to the ability of demethylating compounds to demethylate and reactivate tumor suppressor gene expression has been demonstrated in solid tumors [45]. Whether or not the reactivating effect can translate into a clinical response on its own or in combination with classical cytotoxic therapies remains to be demonstrated [46]. Azacitidine and decitabine are part of the four epigenome-targeted anticancer drugs approved by the FDA (the other two drugs are vorinostat and romidepsin which are inhibitors of histone deacetylases, HDACs) [47]. As reviewed below, other non-oncology approved drugs, including hydralazine, olsalazine, procaine, and procainamide, are also likely to interfere with the epigenome.

Azacitidine was the first hypomethylating agent, approved on May 19, 2004, for the treatment of patients with the following subtypes of MDS: refractory anemia or refractory anemia with ringed sideroblasts (if accompanied by neutropenia or thrombocytopenia and requiring transfusions), refractory anemia with excess blasts, refractory anemia with excess blasts in transformation, and chronic myelomonocytic leukemia. Decitabine was approved by the FDA on May 2, 2006 for the treatment of patients with MDS. Both agents which are pro-drugs and suicide inhibitors demonstrate significant, although usually transient, improvement in patient survival and are currently being tested in many solid cancers [48]. Since both agents are cytidine analogs they are used at low doses in order to deliver the demethylation effect with as low a cytotoxicity as possible [42].

Indeed, aza nucleosides have relatively low specificity and are characterized by substantial cellular and clinical toxicity [49]. As reviewed by Mummaneni and Shord [47], azacitidine enters the cell through a facilitative transport mechanism and is activated to a triphosphate form, which competes with cytidine triphosphate for incorporation into RNA. The drug may also be incorporated

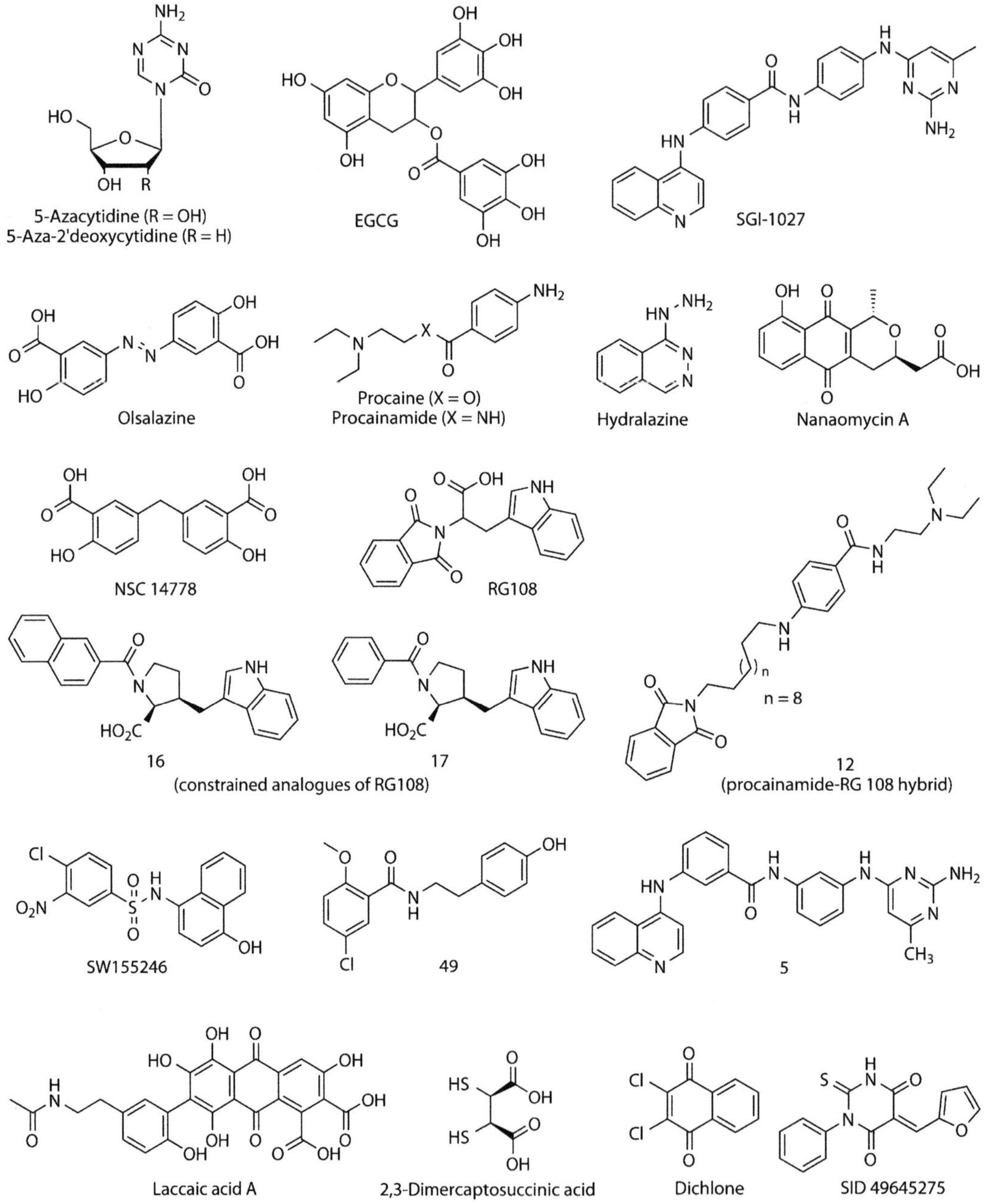

5-Azacytidine (R = OH)
5-Aza-2'deoxycytidine (R = H)

EGCG

SGI-1027

Olsalazine

Procaine (X = O)
Procainamide (X = NH)

Hydralazine

Nanaomycin A

NSC 14778

RG108

16
17
(constrained analogues of RG108)

12
(procainamide-RG 108 hybrid)

SW155246

49

5

Laccaic acid A

2,3-Dimercaptosuccinic acid

Dichlone

SID 49645275

FIGURE 13.1

Chemical structures of representative DNMTi and other compounds identified recently with demethylating properties.

into DNA. Similarly, decitabine undergoes a series of mono-, di-, and tri-phosphorylation steps. Then the drugs are incorporated into DNA via the formation of a covalent bond at position 6 of the 5-aza-cytosine ring. Under normal conditions, with a cytosine, DNMT would transfer the methyl group of SAM to the carbon 5 of the cytosine ring. This transfer enables the release of the enzyme from its covalent bond with the normal substrate cytosine. In contrast, in the case of aza nucleosides, when a 5′-aza-cytosine ring replaces cytosine into the DNA, the methyl transfer from SAM cannot occur, causing covalent trapping of the drugs and subsequent depletion of DNMTs [41].

Despite the high efficiency of azacitidine and decitabine, the relatively low specificity of these two drugs, which target all DNMT isoforms, along with their substantial cellular and clinical toxicity, their poor bioavailability, and their instability in physiological media, prompts a desire to identify novel and more specific DNMTi that do not function via incorporation into DNA [50].

In addition to azacitidine and decitabine there are other DNMTi that have been in clinical trials for various types of cancer [41].

13.2.2 NUCLEIC ACID-BASED

As a second category of demethylating agents, we note the antisense oligonucleotide MG98 against the 3′ untranslated region of *DNMT1* mRNA, which codes for the enzyme DNA methyltransferase 1 that is responsible for maintenance of DNA methylation [51]. This agent has shown an ability to inhibit DNMT1 expression without affecting other *DNMTs*, and to cause demethylation with re-expression of *p16* in bladder and colon cancer cell lines as well as to produce tumor growth inhibition in nude-mice-bearing human lung and colon xenografts. MG98 has been evaluated in a phase I trial in patients with advanced or refractory solid tumors and has demonstrated its tolerability despite the fact that dose-limiting toxicities of transaminitis, thrombocytopenia, and fatigue prohibited higher doses.

13.2.3 NON-NUCLEOSIDE ANALOGS

The fact that nucleoside analogs are not only carcinogenic but also exhibit neutropenia as their dose-limiting toxicity even when used at doses required for demethylation [52] has renewed interest in finding effective and less toxic demethylating agents. Zebularine is a new oral cytidine analog originally synthesized as a cytidine deaminase inhibitor that has been shown to cause demethylation and reactivation of a silenced and hypermethylated *p16* gene in human bladder tumor cells grown in nude mice. Zebularine was also demonstrated as minimally cytotoxic *in vitro* and *in vivo* and can be given continuously at a lower dose to maintain demethylation for a prolonged period, only possible because of its low-toxicity profile; to date, there are no results of clinical trials with this agent.

Within this class of so-called "nontoxic and orally administered agents" there is the green tea major polyphenol (-)-epigallocatechin-3-gallate (EGCG), which resulted as an effective inhibitor of DNMT activity at micromolar concentrations and that was able to demethylate and reactivate expression of several tumor suppressor genes such as *p16*, RAR-β2, and MGMT in cancer cell lines [53].

There is another class of so-called "old drugs" whose demethylating activity upon gene promoters of tumor suppressor genes was recently highlighted. Procainamide, a nonnucleoside inhibitor of DNA methyltransferases approved for treatment of cardiac arrhythmias, can demethylate the *GSTP1* promoter, a common somatic genome change in human prostate cancer and reactivates *in vitro* and in nude mice the expression of the gene [54]. A related drug, procaine, also has the ability of demethylating and reactivating tumor suppressor gene expression, such as the *RARβ2* gene in a breast cancer cell line effect that is accompanied by growth-inhibitory actions [55]. Among these non-nucleosides, hydralazine has been the most studied so far. Its DNA demethy-lating and its ability to restore expression of tumor suppressor genes silenced by promoter hypermethylation in cancer cell lines and primary tumors was uncovered in 2003 [56]. Singh et al. found in a validated model of human DNMT1 enzyme that hydralazine has inhibitory activity upon this enzyme [57]. A number of preclinical studies have further confirmed the DNA methylation inhibitory activity of hydralazine upon genes such as *APC, stanniocalcin-2, WWOX, SERCA2a,* and *Tead4* [58]. Hydralazine-induced DNA demethylation reverses doxorubicin resistance in a MCF-7 model that is associated with increased DNA methyltransferases, global DNA methylation and upregulation of the MDR gene. The results of this study in the MCF-7/Dox model demonstrate that global DNA hypermethylation participates in development of doxorubicin resistance and that the demethylating agent hydralazine can revert the resistant phenotype [59]. Indeed, hydralazine (in combination with the histone deacetylase inhibitor valproic acid) has undergone early clinical eval-uation in a number of solid tumors including cervical, ovarian, lung, breast, testicular, melanoma, colon, sarcomas, as well as in hematological neoplasias including myelodysplastic syndrome, cuta-neous T-cell lymphoma, and chronic myeloid leukemia. This DNMTi (in combination with valproic acid) either as a single agent or in combination with chemotherapy has been shown to be very well tolerated and has shown encouraging clinical activity. This combination was tested in a phase III placebo-controlled trial of advanced cervical cancer receiving standard cisplatin topotecan; how-ever, the study was closed early and no firm conclusions on its efficacy could be drawn despite showing a statistically significant difference on PFS [60].

13.3 ADVANCES IN EXPERIMENTAL METHODS

The promising activity of DNMTi for the therapeutic treatment of cancer warrants the development of assays for detecting this class of inhibitors.

13.3.1 ASSESSMENT OF DNA METHYLATION

The earliest approaches for DNA methylation analysis focused on the analysis of single or few CpG sequences within gene promoters under the candidate-gene approach, but more recently, approaches aim to analyze whole genome methylation. For either approach, methylation analysis has relied mainly on three experimental tools:

1. Restriction enzymes: Restriction endonucleases are used to enrich methylated from unmethylated DNA. Isoschizomers enzymes HpaII-MspI and SmaI-XmaI recognize CCGG and CCCGGG, respectively, but HpaII and SmaI lack activity when a methyl group is present in their recognition site [61]. McrBc restriction can also be used, as this enzyme only cuts methylated DNA including CpG islands.

2. Affinity enrichment (immunoprecipitation-based): These methods are based on enrichment of the methylated DNA using antibodies specific to ^{5m}C (MeDIP), or against the methyl-binding domain containing proteins such as MeCP2 and MBD2. Once methylated genomic fragments are captured these are analyzed following hybridization into microarrays [62].

3. Sodium bisulfite: Bisulfite treatment of DNA converts cytosine to uracil while ^{5m}C remains unchanged [63]. DNA methylation analysis methods that use bisulfite treatment include MSP (Methylation-Specific PCR), BSP (Bisulfite Sequencing PCR) and Combined Bisulfite Restriction Analysis (COBRA). These three methods are limited for the analysis of a single locus. However, more recently, bisulfite converted DNA can be submitted to high-throughput sequencing at the whole genome level (Bs-Seq) [64].

13.3.2 BENCHMARKING OF METHODS FOR METHYLOME ANALYSIS

The field of methylation analysis has shown considerable progress and development over the past decades; today, methylome analysis is a key in ongoing efforts to elucidate the epigenomes of healthy and diseased cell types and organisms. Currently, massively parallel sequencing and microarray platforms are the most utilized, but until recently it was not known how these methods perform when compared to each other in regard to resolution, coverage and accuracy — issues which are especially important for generating the so-called reference methylomes. A milestone in this important issue was the comparison on performance of six main technologies for methylome analysis, of which five are sequencing-based and one array-based [65]. Three of the methods — MethylC-seq, reduced representation bisulfite sequencing (RRBS) and the Infinium-27K bead-array — are sodium bisulfite-based, while the other three — methylated DNA immunoprecipitation sequencing (MeDIPseq), methylated DNA capture by affinity purification (MethylCap-seq), and methylated DNA-binding domain sequencing (MBDseq) — are affinity enrichment (immunoprecipitation-based). All of the evaluated methods are capable of producing accurate data for producing reference methylomes [66].

13.3.3 METHODS TO SCREEN COMPOUND DATA SETS

Gros et al. recently reviewed assays available for identifying DNMTi [42]. Both enzymatic and cell-based assays have been developed. Enzymatic assays can be classified into several types following different criteria; for example, based on the (A) phase test: into homogeneous and heterogeneous; (B) detection format: luminescence, restriction-based, or radiometric; (C) measurement of the product: if the assays are based on the quantification of either SAH or methylated DNA. Details of each method including advantages and disadvantages are reviewed elsewhere [42].

Cell-based assays are being developed and applied in HTS format. These are based on cell clones selected for their ability to express a reporter gene after treatment with compounds that target epigenetic drugs. Despite the fact that it is more difficult to deconvolute the precise mechanism of action of the compound in the cell, an advantage of cell-based assays over enzymatic assays is that the former provide readout more relevant to the development of compounds as therapeutic agents. This is because activity in a cell-based assay requires additional features of the compound, not only as an effective inhibitor of DNMTs, but also the compound will require appropriate solubility and stability [42]. Representative applications of screening assays are described in the next section.

13.4 PROGRESS ON THE IDENTIFICATION OF NOVEL DNMTi

Nowadays, there are an important number of known inhibitors but with still limited applicability due to high toxicity, weak activity, nonspecificity, or chemical instability, to name just a few limitations. Reviews on the progress and status of DNMTi have been published covering advances on the development of nucleoside, non-nucleoside inhibitors, natural products [67], current patents of the chemistry and biological activities of novel DNMTi and clinical studies [68], and advances on the development of inhibitors using computational approaches [29].

Initially, several DNMTi were identified at random with a few systematic studies. This is particularly true for the identification of new inhibitors with non-nucleoside scaffolds. However, the increasing evidence that DNMTi represent promising compounds for therapeutic treatment of cancer has boosted the number of rational approaches to identify or develop novel and improved DNMTi. Examples of major approaches utilized in the development of DNMTi include optimization of previously identified hits or lead compounds (guided by structure-based methods), HTS, and drug repurposing [69]. These approaches are commented on in the following sections. Figure 13.1 shows the chemical structures of representative DNMTi with emphasis on recently identified compounds.

13.4.1 CHEMICAL SYNTHESIS

N-phthaloyl-L-tryptophan 1, RG108 (Figure 13.1) was one of the first non-nucleoside inhibitors of DNMT identified using a systematic approach. Structure-based virtual screening of a subset of the National Cancer Institute (NCI) database led to the identification of RG108 that showed inhibition of DNMT in a noncovalent manner and caused demethylation of genomic DNA in cells without any detectable toxicity [70]. This well-validated DNMTi has been subject to several recent medicinal chemistry optimization programs through different strategies that include adding a maleimide substructure [71], and structure-activity relationship (SAR) studies through the systematic modification of the indole, carboxylate, and phthalimide moieties of RG108 [72]. In the latter study five compounds (including **16** and **17** in Figure 13.1) had at least ten fold increased potency over the reference showing also cytotoxicity on the prostate cancer DU145 cell line [72].

The quinoline derivative SGI-1027 (Figure 13.1) was originally synthesized for its antitumor activities and was found to be an efficient DNMTi both *in vitro* and in cells. SGI-1027 was shown to selectively decrease the levels of DNMT1 over DNMT3A or DNMT3B in cells by enhancing its rate of proteasomal degradation [73]. Recently, Gamage et al. reported the synthesis and SAR studies of 4-aminoquinolones to elucidate the structural features associated with the reduction of levels of DNMT1 in HCT116 human colon carcinoma cells [74]. Rilova et al. also conducted SAR studies of 25 SGI-1025 derivatives identifying four compounds with comparable activities to the parent molecule. The most active compounds were more potent against DNMT3A than DNMT1 and induced the re-expression of a reporter gene in leukemia KG-1 cells. In this study, the analogs were designed following a structure-based strategy starting from a docking model of SGI-1027 with *Haemophilus haemolyticus* cytosine-5 DNMT (MHhaI C5 DNMT) [75]. Also recently, Valente et al. followed a hit-to-lead strategy and optimized the structure of SGI-1027 to yield a potent DNMTi that was selective toward other AdoMet-dependent protein methyltransferases (compound no. **5** in original paper, Figure 13.1) [69].

Synthesis of RG108-procainamide conjugates led to the identification of six compounds with potent inhibition of the murine catalytic Dnmt3A/3L complex and of human DNMT1. It was found that the length of the linker between procainamide and RG108 had an important role in the activity. One of the best compounds (no. **12** in the original paper, Figure 13.1) showed good selectivity for C5 DNMTs over bacterial DNMTs. This and other four molecules had micromolar cytotoxicity toward the cancer cell lines DU145 and HCT116 [76]. Castellano et al. also reported the optimization of procaine/procainamide using a medicinal chemistry approach. Authors of that work developed a Δ^2-isoxazoline constrained analog of procaine/procainamide showing activity *in vitro* against HCT116 human colon carcinoma cells [77].

A further example is the optimization of compound NSC319745, initially identified by a docking-based virtual screening of a larger set of the NCI database using a homology model of the catalytic domain of human DNMT1 [78]. Kabro et al. recently reported the synthesis and DNMT inhibitory activity of 29 analogs of NSC319745. Seven compounds were identified as inhibitors DNMT3A. The most promising compound (no. **49** in the original paper, Figure 13.1) showed an apparent EC_{50} of 36 μM [79].

13.4.2 HIGH-THROUGHPUT SCREENING

Ceccaldi et al. developed a fluorescence-based HTS using a short DNA duplex immobilized on 96-well plates. Using this technology, the authors screened a database of 114 flavones and flavanones for the inhibition of the murine catalytic Dnmt3a/3L complex identifying compounds with IC_{50} in the nanomolar range [80].

Public chemical databases annotated with biological activity are frequently used in drug discovery. However, the biological data for DNMTi were limited in these public databases. About 2 years ago (March, 2012) the results of the screening data obtained using a novel nonradioactive HTS for human DNMT1 [81] was made publicly available in the ChEMBL and PubChem databases. The assay uses a fluorescent readout based on the ability of the DNMT1 methyltransferase to methylate hemi-methylated duplex DNA oligo substrate, thus protecting it from endonuclease cleavage [81]. The most active molecule in the confirmatory assay (PubChem Assay ID: 602382) was SID49645275 (Figure 13.1) with an $IC_{50} = 811$ nM.

Kilgore et al. developed a novel high-throughput scintillation proximity assay (SPA) to identify DNMTi. The SPA was used to evaluate ~180,000 molecules, including compounds from commercial sources and compounds synthesized at The University of Texas Southwestern [82]. In this assay, the hit confirmation rate was low (0.03%) and most of the hit compounds were found to be active due to the generation of reactive oxygen species. However, the sulfonamide inhibitor SW155246 (Figure 13.1) was selective toward human DNMT1 without affecting protein levels or generating reactive oxygen species [82]. This compound showed a 30-fold preference for human DNMT1 over murine DNMT3A. SAR studies indicated that the hydroxyl group of SW155246 was essential for its activity: loss of the hydroxyl group or addition of a methylated oxygen on the 1-position of the naphthyl ring completely abolished the ability of this compound to inhibit human DNMT1 activity *in vitro* and reduced the cell-based cytotoxicity [82].

More recently, a medium-throughput screening of a random database of 1,120 small molecules with low molecular weight (average molecular weight <200) identified dichlone as a nonselective inhibitor of DNA methyltransferases [83].

Fagan et al. developed a high throughput pipeline for identification of direct DNMTi [84]. The restriction enzyme-coupled fluorigenic assay uses an activated form of DNMT1 and it is performed in 384 well plates. The screening pipeline also includes a counter screen against the restriction enzyme, a screen to eliminate DNA intercalators, and a differential scanning fluorimetry assay to validate direct binders. Authors of that work employed their assay to screen a compound library with 2,320 compounds. Nine molecules were identified with dose responses ranging from 300 nM (2,3-dimer-captosuccinic acid, Figure 13.1) to 11 μM. Seven of nine inhibitors identified had two- to four fold selectivity for DNMT1 versus DNMT3A [84]. Most of the hit compounds were polycyclic aromatic molecules. Indeed, five molecules contain anthracene or anthraquinone-related structures such as the natural product laccaic acid A (*vide infra*). Despite the fact that polycyclic aromatic compounds are typical DNA intercalators, Fagan et al. showed experimentally that their hit molecules failed to compete with ethidium bromide in DNA intercalation assays. It is worth noting the similarity of the core scaffolds of laccaic acid A, dichlone, and nanaomycin A (Figure 13.1), all identified independently.

13.4.3 DRUG REPURPOSING

Drug repositioning or drug repurposing is an approach to accelerate the drug discovery process through the identification of a novel clinical use for an existing drug approved for a different indication. The increased success and applications of drug repurposing can be considered as one of the consequences of polypharmacology and it represents a manifestation of the shift from a single to multitarget paradigm in drug discovery [85]. Dueñas-González et al. have commented in detail on the important role of drug repurposing in cancer therapy [86]. Classical examples of drug repurposing in epigenetic compounds targeting DNMTs are the antihypertensive drug hydralazine or the local anesthetic procainamide (Figure 13.1) (*vide supra*). As reviewed above, procainamide has been a recent subject of optimization programs.

As part of an effort to identify approved drugs with potential for epigenetic therapeutic applications, Mendez-Lucio et al. implemented a computer-guided drug repurposing approach [87]. In that work, a public database of 1,582 approved drugs was subject to a systematic and fast computational search using as a reference molecule NSC14778 (Figure 13.1), a DNMTi previously identified from virtual screening [78]. The computational search was conducted using 3D similarity searching based on the principles of the molecular similarity that states that similar structures will have similar activity. Among the ten most similar compounds to the reference structure was olsalazine (Figure 13.1) which is an anti-inflammatory drug used in the treatment of inflammatory bowel disease and ulcerative colitis. It was approved by the FDA in 1990. In order to examine the ability of olsalazine to inhibit DNMT activity, a novel DNA methylation reprogramming system that operates in the context of living cells [88] was used. The cell-based screen used in that work is highly tractable, internally controlled and well-suited for a drug repurposing strategy in epigenetics. The authors concluded that olsalazine very closely mimics the action of decitabine with minimal cytotoxicity at the concentrations tested.

13.4.4 NATURAL PRODUCTS AND DIETARY SOURCES

Natural products have a long history in drug identification and compounds for health-related benefits. However, one needs to consider drawbacks of natural products for drug discovery applications such as the isolation and purification procedures, very small available amounts of lead compounds, the difficulty of synthesizing natural products with high structural complexity and the associated synthesis scale-up issues. Despite these drawbacks, the large success of natural products in producing bioactive compounds or bioactive mixtures has inspired the preparation of synthetic molecules that are drugs. Moreover, the unique structural features of natural products represent a promising opportunity to identify active compounds for emerging and "tough targets" difficult to address with classical organic small molecules. Because environmental exposures are commonly assumed to play a major role in the establishment of abnormal DNA methylation patterns, a constant uptake of DNA demethylating agents is believed to have a chemopreventive effect. This could be most conveniently achieved through the dietary uptake of natural product DNMT inhibitors and bioactive food components [89]. Indeed, "nutri-epigenomics" or "nutri-epigenetics," conceptualized as "the influence of dietary components on mechanisms influencing epigenomics," has emerged in the last few years as an exciting new field in modern epigenetic research [90].

A number of natural products are associated with direct or indirect inhibition of DNA methylation [91]. The main polyphenol compound from green tea, EGCG (Figure 13.1), has been proposed to inhibit DNMT1 by blocking the active site of the enzyme leading to reactivation of methylation-silenced genes in cancer cells. An alternative mode of action is the indirect inhibition of DNMTs through the interaction with cathechol-O-methyltransferase (COMT). Since COMT is a SAM-dependent enzyme, as the flavonoid becomes methylated, the SAH concentration increases. Since SAH is a potent DNMT inhibitor and the intracellular SAM/SAH equilibrium is displaced, there are no DNMTs available to methylate [42]. Other tea polyphenols such as catechin and epicatechin, and the bioflavonoids quercetin, fisetin, and myricetin have also been implicated in DNA methylation inhibition. Similarly, the catechol-containing dietary epicatechin has been suggested to indirectly inhibit DNA methylation as a consequence of COMT-mediated O-methylation of epicatechin. Other dietary catechols that may inhibit DNA methylation by a similar mechanism are the polyphenols from coffee: caffeic acid and chlorogenic acid. Apple polyphenols, the major isoflavone from soy bean genistein, and two other isoflavones, biochanin A and daidzein, have also been reported to possess demethylating activity. Mahanine, a plant-derived carbazole alkaloid, has been reported to induce the *RASSF1* gene in human prostate cancer cells, presumably by inhibiting DNMT activity. Psammaplin A and other disulfide bromotyrosine derivatives isolated from the marine sponge *Pseudoceratina purpurea* have been described as potent inhibitors of DNMT1. Other marine natural products with reported DNMT1 inhibitory activity are the peyssonenynes A and B isolated from the red alga *Peyssonnelia caulifera.* Curcumin, the major component of the Indian curry spice turmeric, and parthenolide, the principal sesquiterpene lactone of feverfew, have also been reported to inhibit DNMT1. Other natural products with reported inhibition of DNMTs are kazinol Q, withaferin A, resveratrol, and guggulsterone [41].

As discussed above, the natural product laccaic acid A (Figure 13.1) was recently identified from an HTS as a direct inhibitor of DNMT1. The highly substituted anthraquinone compound showed approximately four-fold selectivity for DNMT1 versus DNMT3A/DNMT3L. Additional characterization of laccaic acid A showed that it is a DNA-competitive inhibitor and reactivates the expression of a set of methylation-silenced genes in MCF-7 breast cancer cells [84]. Competitive biochemical assays showed that the natural product was not competitive with AdoMet but it was competitive with the oligonucleotide substrate. Authors of that work did not report a binding model of the natural product with human DNMT1. Figure 13.2 shows a binding model of laccaic acid A

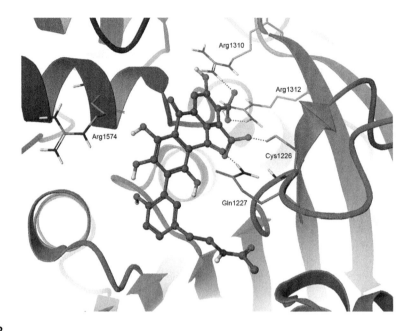

FIGURE 13.2

Induced-fit docking pose of laccaic acid A (Figure 13.1) within the substrate-binding site of human DNMT1. Selected amino acid residues of the binding pocket are shown. Hydrogen bonds between the ligand and side chains of DNMT1 are shown as dashed pink lines. (*This figure is reproduced in color in the color plate section.*)

with the substrate binding site of DNMT1 using the crystallographic structure 3SWR. In order to develop this model, we used an induced-fit docking (IFD) protocol very similar to the one we used before to generate the reported binding models of SW155246 and other inhibitors [92,93]. Similar to these works, IFD was used to consider, at least in part, the protein flexibility. According to the binding model, one of the carboxylate groups of the natural product forms hydrogen bonds with the side chains of Arg1310 and Arg1312. Of note, the binding modes obtained for a number of DNMTi including azacitydine and decitabine also suggest the formation of hydrogen bond interactions with Arg1310 and Arg1312 [50]. The second carboxylate group makes a hydrogen bond with the catalytic Cys1226. A third hydrogen bond is predicted between one of the carbonyl oxygen atoms with the side chain of Gln1227 (Figure 13.2).

13.5 ADVANCES IN COMPUTATIONAL STUDIES

An important number of computational approaches have been applied to help understand the activity of known compounds at the molecular level and to uncover or design novel DNMTi as potential leads. These methods include, but are not limited to, molecular docking, virtual screening, pharmacophore modeling, molecular dynamics, and similarity searching. Of note, before the crystallographic structure of the catalytic domain of human DNMT1 was available, virtual screening

Table 13.3 Summary of Representative Computational Studies Recently Conducted to Investigate DNMTi

Study	Outcome	Reference
Docking and IDF of the sulfonamide SW155246 with DNMT1 (PDB ID: 3SWR)	IFD provides structural basis to rationalize activity cliffs of a novel DNMTi of the sulfonamide class	[93]
Docking of **49** with DNMT3A/3L (PDB ID: 2QRV)	Docking helps to hypothesize that the inhibitor binds at the cofactor-binding site	[79]
IFD of constrained analogs of RG108 (**16** and **17**) with murine Dnmt1 (PDB ID: 4DA4)	IFD supports the hypothesis that the two compounds may bind in the catalytic-binding site	[72]
Docking of **5** with DNMT1 in the presence and absence of a 19 bp DNA duplex (PDB IDs: 3SWR and 3PTA)	Docking model with unbound DNA suggest that the DNMT1 occupies both the AdoMet-binding site and the CXXC region; the model suggests that **5** could be recognized by the DNA-unbound enzyme conformation	[69]
Docking of SID49645275 with DNMT1 (structure modeled into an active conformation from PDB ID: 3PTA)	Docking model proposed for a novel DNMTi obtained from HTS. The docked compound was used as query to conduct virtual-based similarity searching of different compound databases	[95]

conducted with homology models led to the identification of active compounds, including RG108 and NSC14778 (Figure 13.1, see above). Medina-Franco et al. have reviewed computational approaches for DNMTi performed in his and other groups [94]. The recent release of crystallo-graphic structures of the catalytic domain of humans has promoted the application of molecular docking in the research on DNMTi. Representative examples of recent computational studies are summarized in Table 13.3.

13.5.1 CHEMOINFORMATIC ANALYSIS

Chemoinformatic characterization of the chemical diversity and coverage of the chemical space of screening libraries is a useful practice that is being incorporated as part of the screening effort for DNMTi and other epidrugs. In order to increase the hit rates, the selection of screening libraries with drug-like properties and diversity is an important first step in library screening. Yoo et al. implemented a chemoinformatic analysis of four screening collections: a DNMT-focused library, an NCI diversity set, and two natural products collections using diverse criteria including physico-chemical properties, molecular fingerprints, and molecular scaffolds [96]. The DNMT focused library and the two natural products collections have molecules with properties similar to approved drugs. The natural products databases which have different chemical structures from approved drugs and synthetic databases have unique characteristics attractive for identifying DNMTi with novel molecular scaffolds. The scaffold analysis revealed that the DNMT focused library had the largest scaffold diversity with scaffolds not found in the other collections. Using these four screening collections, Yoo et al. reported the first chemoinformatic-based similarity approach applied to the systematic identification of DNMTi [97]. The similarity-based profile of four different screening libraries to the active reference compound SGI-1027 (Figure 13.1) was computed and classified using a total of 22 two-dimensional (2D) and three-dimensional (3D) representations and two similarity metrics including fingerprint- and shape-based methods. The DNMT focused library with the highest similarity profile represented is the most suitable candidate collection to identify additional DNMTi. The hit compounds with high similarity to SGI-1027 were identified as promising candidates for experimental validation.

13.5.2 MOLECULAR DOCKING OF SMALL MOLECULES

Most of the docking studies of DNMT inhibitors with similar size have been conducted at the substrate binding site of the MTase domain of DNMTs. Yoo et al. recently reported the first docking study of SGI-1027 and **12** (Figure 13.1) with "long" scaffolds in the MTase domain of human DNMT1 and DNMT3A in the presence and absence of other domains [92]. To explore the binding site and propose docking models for SGI-1027 and **12**, IFD considering receptor flexibility of DNMT1 and DNMT3A was carried out. IFD with DNMT1 in the MTase domain only revealed that SGI-1027 and **12** occupied the cofactor and substrate binding sites. In contrast, these compounds only occupied the cofactor site in the presence of other domains. The predicted binding models suggest that "long" scaffolds such as SGI-1027 and **12** mimic the function of the autoinhibitory linker or stabilize the position of the autoinhibitory linker. These hypotheses are in agreement with the reported autoinhibitory mechanism and provide insights for the design of novel DNMT inhibitors [30].

Medina-Franco et al. suggested the binding model of SID49645275 (Figure 13.1) which was the most potent DNMTi obtained by HTS and deposited in PubChem with Assay ID: 602382 (*vide supra*) [95]. The predicted binding models indicate that small molecule binds into the substrate-binding site. Based on the chemical structure of SID49645275, similarity searching followed by docking with human DNMT1 from five chemical databases provide 108 virtual screening hits including approved drugs for clinical use as potential candidates for drug repositioning.

Medina-Franco et al. reported a structure-based rationalization of the activity of SW155246 (Figure 13.1) using previously reported IFD with the human DNMT1 by Yoo et al. [92]. The predicted binding models of active compound SW155246 and inactive analogs with high structural similarity provided a structural explanation of 3D activity cliffs. SW155246 showed a significantly different binding mode as compared to the structural analogs in the cofactor binding site [93].

Kabro et al. reported the docking of **49** (Figure 13.1) with a crystal structure of the human catalytic DNMT3A/3L complex (PDB ID: 2QRV) concluding that this small molecule may bind in the cofactor binding site [79].

Asgatay et al. conducted IFD of two constrained analogs of RG108 **16** and **17** (Figure 13.1) using a crystallographic structure of murine catalytic Dnmt1 (residues 732−1600) (PDB ID: 4DA4) [72]. The predicted binding models indicate that these compounds may bind in the cofactor and catalytic pockets [72]. In order to explore the putative binding mode of compound **5** (Figure 13.1), Valente et al. conducted the docking of this compound with the crystal structure of DNMT1 in the presence and absence of the CXXC domain region [69]. A similar strategy (docking in the presence of binding domains other than the catalytic site) was previously published by Yoo et al. to elucidate the binding mode of compound **12** [92].

13.5.3 **VIRTUAL SCREENING**

Virtual screening, also called *in silico* screening, can be viewed as a funnel approach where one or more computational methods are used to select, from a pull of candidate molecules (usually in a molecular database), a subset of compounds for experimental validation. One of the major goals is to increase the probabilities of identifying active compounds. Depending on the information available for the system, the search can be performed using structure-based methods such as molecular docking, or ligand-based methods such as similarity searching.

Thus far, virtual screening has proven to be quite successful in identifying small molecule DNMTi. A primary example is the discovery of RG108 (Figure 13.1) by means of docking-based virtual screening of the NCI database (*vide supra*) [98]. A second example is the identification of 5,5′-methylenedisalicylic acid (NSC14778, Figure 13.1) also using docking-based screening of the NCI database [78]. Of note as discussed above, homology models of the catalytic domain of human DNMT1 were used in both cases because the crystallographic structures were not available at the time of those works. More recently, a ligand-based similarity searching of a database of approved drugs identified the anti-inflammatory drug olsalazine as a novel hypomethylating compound [87]. In that study, NSC14778 (5,5′-methylenedisalicylic acid) was used as the query. Authors of that work concluded that a similar computational approach coupled with experimental validation can be implemented to use other DNMTi as query molecules to search other molecular databases.

In addition to the hit compounds experimentally validated that have been obtained from virtual screening of the NCI database and approved drugs, there are reports of *in silico* screening of DNMT-focused and natural product databases. The chemical structures of the computational hits have been disclosed but it remains to conduct the experimental characterization [91,95].

13.6 **DATA RESOURCES FOR DNMTi**

Compound databases form part of an integrated effort of drug discovery projects. Depending on the goals of the project drug discovery endeavors frequently involve chemical libraries from different sources including commercial vendor compounds, natural products, combinatorial libraries, and focused libraries. These collections may contain hundreds, thousands, or millions of molecules and can be used for a wide variety of purposes, for example, SAR exploration and identification of polypharmacology.

One of the recent trends in drug discovery is to balance chemical novelty within a confined chemical space. One approach is developing and screening libraries focused on a specific target or target families [99]. In recent years, chemical companies have been developing epigenetic libraries that are available for purchase and screening. The chemical structures of the libraries are deposited on the company's web sites and can be requested directly from the chemical vendors. Table 13.4 summarizes examples of data sets with experimentally validated compounds hitting epigenetic targets (including DNMTs), and focused libraries designed using computational approaches.

Results of biological evaluations of compounds tested as DNMTi and/or hypomethylating agents are constantly collected in large public repositories such as PubChem and ChEMBL. For example, as discussed previously, PubChem contains the results of the HTS of a large collection of more than 359,000 compounds tested as inhibitors of DNMT1 (PubChem Assay ID 588458). At

Table 13.4 Examples of Epigenetic-Related Libraries Available for Screening

Library	Size	Description	URL
Epigenetics Compound Library	24	Small molecule modulators with biological activity used for epigenitc research.	www.selleckchem.com
Epigenetics Screening Library	>90	Molecules that are known to modulate the activity of methyltransferases, demethylases, histone acetyltransferases, histone deacetylases, and acetylated histone-binding proteins. The composition of this screening library will always vary somewhat depending upon our inventory.	www.caymanchem.com
Screen-Well® Epigenetics library	43	Compounds with defined activity against enzymes which carry out epigenetic modification of lysine. It also includes DNA methylation inhibitors.	www.enzolifesciences.com
DNMT1 and DNMT3b focused libraries	503 (DNMT1) and 1320 (DNMT3b)	Collections obtained from virtual screen of drug-like compounds against DNMT3b and DNMT1. Final selection of compounds was made with inspection of active site's critical structural determinants for ligand binding, docking score cut-off filtering, intermolecular hydrogen bonds with crucial active site's residues, and exclusion of predicted DNMT1 and DNMT3b cross-binders. A 3D pharmacophore model was applied during the selection.	www.otavachemicals.com

the time of writing (June 2014), ChEMBL collects the bioactivity information of 411, 64, and 12 compounds screened with human DNMT1, 3B, and 3A, respectively (IDs CHEMBL1993, 6095 and 1992, respectively). ChEMBL contains links to the literature reporting details of the methods of the biological evaluations.

In addition to public databases of screening data, another valuable repository of information in the public domain is the Human Epigenetic Enzyme and Modulator Database (HEMD) recently developed by Huang et al. This is an integrated repository curated from the literature that contains 269 epigenetic enzymes and 4,377 chemical modulators for therapeutics [100,101]. The modulators include regulators, activators and inhibitors. HEMD has been designed to conduct visual representations of the data, conduct online searches and analyze the structure and function. The enzymes include 19 proteins from DNA methylation, 219 from histone modification, 28 from chromatin remodeling, and 3 from noncoding RNA. It is expected that HEMD will be an interface between medicinal chemists and biologists to advance epigenetic therapeutics [100].

13.7 CONCLUSIONS

DNMTi are attractive candidates for development of single or combined therapies for the treatment of cancer and other diseases. Indeed, combination therapies with other epigenetic drugs such as HDAC inhibitors are attractive strategies to produce demethylation and reactivation of tumor suppressor genes [42]. The approval of two DNMTi for the treatment of MDS and the ongoing clinical development of investigational anticancer drugs as modulators of DNMTs highlight the potential translational applications of this family of epigenetic enzymes in hematologic malignancies and solid tumors. However, continued clinical testing of this agent class is clearly needed to determine their place in the therapeutical armamentarium against cancer. It is quite important to remark that despite the success of DNMTi in cancer therapy, no definitive evidence has shown that they indeed work by reactivating the expression of tumor suppressor genes as predicted by the rationale for its use. Thus, more effort is needed to find out how these agents lead to antitumor effects and also to develop biomarkers for response and toxicity in this era of personalized therapy. The availability of whole genome methylome analysis certainly would facilitate this endeavor.

A blend of well-established along with novel experimental and computational technologies continues to accelerate the discovery of DNMTi. In addition to therapeutic candidates, the development of chemical probes that show selectivity and specificity for each DNMT are required to investigate the specific roles of each isoform, the molecular basis of the enzymatic mechanism, and the DNA methylation cascade in cells [42].

Public repositories of information collecting the results of screening campaigns for DNMTi and online resources assembling the associations of epigenetic targets are clear examples of epigenetic technological applications that advance the field of DNMTi as promising epidrugs. Drug repurposing guided by experimental and/or computational approaches is an emerging approach to uncover DNMTi and hypomethylating agents. The first report of a computer-guided drug repurposing effort has been published and it is expected that an increased number of successful applications will follow this approach. Another future direction in this field is to augment the integration of *in silico* design and screening of chemical databases with experimental medium and HTS assays for DNMTi. In light of emerging and attractive areas such as *nutri-epigenomics*, it will be particularly relevant to screen databases of natural products, dietary supplements, and food components to identify candidate compounds for potential therapeutic treatment or chemoprevention of diseases associated with the irregular expression of DNMTi. One of the first examples in this direction is the computational-guided mining for bioactive compounds of the generally recognized as safe (GRAS) flavor chemicals [102]. Other examples of the increased application of computational approaches to identify novel modulators of DNMTi and other epigenetic targets are collected in the emerging area of *Epi-Informatics*.

ACKNOWLEDGMENTS

We thank M. Sc. Oscar Méndez-Lucio for insightful discussions.

REFERENCES

[1] You JS, Jones PA. Cancer genetics and epigenetics: two sides of the same coin? Cancer Cell 2012;22:9—20.
[2] Jurkowska RZ, Jurkowski TP, Jeltsch A. Structure and function of mammalian DNA methyltransferases. ChemBioChem 2011;12:206—22.

[3] Davie JR, Hendzel MJ. Multiple functions of dynamic histone acetylation. J Cell Biochem 1994;55:98–105.

[4] Heyn H, Esteller M. DNA methylation profiling in the clinic: applications and challenges. Nat Rev Genet 2012;13:679–92.

[5] Esteller M. Cancer epigenomics: DNA methylomes and histone-modification maps. Nat Rev Genet 2007; 8:286–98.

[6] Iorio MV, Piovan C, Croce CM. Interplay between microRNAs and the epigenetic machinery: an intricate network. Biochim Biophys Acta 2010;1799:694–701.

[7] Ito S, Shen L, Dai Q, Wu SC, Collins LB, Swenberg JA, et al. TET proteins can convert 5-methylcytosine to 5-formylcytosine and 5-carboxylcytosine. Science 2011;333:1300–3.

[8] Seisenberger S, Peat JR, Hore TA, Santos F, Dean W, Reik W. Reprogramming DNA methylation in the mammalian life cycle: building and breaking epigenetic barriers. Philos Trans R Soc Lond B Biol Sci 2013;368:20110330.

[9] Goelz SE, Vogelstein B, Hamilton SR, Feinberg AP. Hypomethylation of DNA from benign and malignant human-colon neoplasms. Science 1985;228:187–90.

[10] Ehrlich M. DNA hypomethylation, cancer, the immunodeficiency, centromeric region instability, facial anomalies syndrome and chromosomal rearrangements. J Nutr 2002;132:2424S–2429SS.

[11] Gamasosa MA, Slagel VA, Trewyn RW, Oxenhandler R, Kuo KC, Gehrke CW, et al. The 5-methylcytosine content of DNA from human-tumors. Nucl Acids Res 1983;11:6883–94.

[12] Eden A, Gaudet F, Waghmare A, Jaenisch R. Chromosomal instability and tumors promoted by DNA hypomethylation. Science 2003;300:455.

[13] Hansen RS, Wijmenga C, Luo P, Stanek AM, Canfield TK, Weemaes CMR, et al. The DNMT3b DNA methyltransferase gene is mutated in the ICF immunodeficiency syndrome. Proc Natl Acad Sci USA 1999;96:14412–17.

[14] Narayan A, Ji WZ, Zhang XY, Marrogi A, Graff JR, Baylin SB, et al. Hypomethylation of pericentromeric DNA in breast adenocarcinomas. Int J Cancer 1998;77:833–8.

[15] Strichman-Almashanu LZ, Lee RS, Onyango PO, Perlman E, Flam F, Frieman MB, et al. A genome-wide screen for normally methylated human CpG islands that can identify novel imprinted genes. Genome Res 2002;12:543–54.

[16] Ogishima T, Shiina H, Breault JE, Tabatabai L, Bassett WW, Enokida H, et al. Increased heparanase expression is caused by promoter hypomethylation and up-regulation of transcriptional factor early growth response-1 in human prostate cancer. Clin Cancer Res 2005;11:1028–36.

[17] Karpf AR, Lasek AW, Ririe TO, Hanks AN, Grossman D, Jones DA. Limited gene activation in tumor and normal epithelial cells treated with the DNA methyltransferase inhibitor 5-aza-2′-deoxycytidine. Mol Pharmacol 2004;65:18–27.

[18] Laird PW, Jaenisch R. DNA methylation and cancer. Hum Mol Genet 1994;3:1487–95.

[19] Greger V, Passarge E, Hopping W, Messmer E, Horsthemke B. Epigenetic changes may contribute to the formation and spontaneous regression of retinoblastoma. Hum Genet 1989;83:155–8.

[20] Santini V, Kantarjian HM, Issa JP. Changes in DNA methylation in neoplasia: pathophysiology and therapeutic implications. Ann Intern Med 2001;134:573–86.

[21] Szyf M. Targeting DNA methylation in cancer. Ageing Res Rev 2003;2:299–328.

[22] Bachman KE, Park BH, Rhee I, Rajagopalan H, Herman JG, Baylin SB, et al. Histone modifications and silencing prior to DNA methylation of a tumor suppressor gene. Cancer Cell 2003;3:89–95.

[23] Clark SJ, Melki J. DNA methylation and gene silencing in cancer: which is the guilty party? Oncogene 2002;21:5380–7.

[24] Rius M, Lyko F. Epigenetic cancer therapy: rationales, targets and drugs. Oncogene 2012;31:4257–65.

[25] Chen T, Hevi S, Gay F, Tsujimoto N, He T, Zhang B, et al. Complete inactivation of dnmt1 leads to mitotic catastrophe in human cancer cells. Nat Genet 2007;39:391–6.

[26] Jia D, Jurkowska RZ, Zhang X, Jeltsch A, Cheng X. Structure of Dnmt3a bound to Dnmt3L suggests a model for de novo DNA methylation. Nature 2007;449:248−51.

[27] Deplus R, Brenner C, Burgers WA, Putmans P, Kouzarides T, Launoit Yd, et al. Dnmt3L is a transcriptional repressor that recruits histone deacetylase. Nucl Acids Res 2002;30:3831−8.

[28] Schaefer M, Lyko F. Solving the Dnmt2 enigma. Chromosoma 2010;119:35−40.

[29] Yoo J, Medina-Franco JL. Inhibitors of DNA methyltransferases: insights from computational studies. Curr Med Chem 2012;19:3475−87.

[30] Song J, Rechkoblit O, Bestor TH, Patel DJ. Structure of dnmt1-DNA complex reveals a role for autoinhibition in maintenance DNA methylation. Science 2011;331:1036−40.

[31] Takeshita K, Suetake I, Yamashita E, Suga M, Narita H, Nakagawa A, et al. Structural insight into maintenance methylation by mouse DNA methyltransferase 1 (Dnmt1). Proc Natl Acad Sci USA 2011; 108:9055−9.

[32] Syeda F, Fagan RL, Wean M, Avvakumov GV, Walker JR, Xue S, et al. The replication focus targeting sequence (rfts) domain is a DNA-competitive inhibitor of Dnmt1. J Biol Chem 2011;286:15344−51.

[33] Klimasauskas S, Kumar S, Roberts RJ, Cheng XD. HhaI methyltransferase flips its target base out of the DNA helix. Cell 1994;76:357−69.

[34] Kumar S, Horton JR, Jones GD, Walker RT, Roberts RJ, Cheng X. DNA containing 4′-thio-2′-deoxycytidine inhibits methylation by HhaI methyltransferase. Nucl Acids Res 1997;25:2773−83.

[35] Sheikhnejad G, Brank A, Christman JK, Goddard A, Alvarez E, Ford H, et al. Mechanism of inhibition of DNA (cytosine C5)-methyltransferases by oligodeoxyribonucleotides containing 5,6-dihydro-5-azacytosine. J Mol Biol 1999;285:2021−34.

[36] Zhou L, Cheng X, Connolly BA, Dickman MJ, Hurd PJ, Hornby DP. Zebularine: a novel DNA methylation inhibitor that forms a covalent complex with DNA methyltransferases. J Mol Biol 2002;321:591−9.

[37] Reinisch KM, Chen L, Verdine GL, Lipscomb WN. The crystal-structure of HaeIII methyltransferase covalently complexed to DNA − an extrahelical cytosine and rearranged base-pairing. Cell 1995;82:143−53.

[38] Dong AP, Yoder JA, Zhang X, Zhou L, Bestor TH, Cheng XD. Structure of human dnmt2, an enigmatic DNA methyltransferase homolog that displays denaturant-resistant binding to DNA. Nucl Acids Res 2001;29:439−48.

[39] Otani J, Nankumo T, Arita K, Inamoto S, Ariyoshi M, Shirakawa M. Structural basis for recognition of H3K4 methylation status by the DNA methyltransferase 3A atrx-dnmt3-dnmt3l domain. EMBO Rep 2009;10:1235−41.

[40] Ooi SKT, Qiu C, Bernstein E, Li KQ, Jia D, Yang Z, et al. Dnmt3l connects unmethylated lysine 4 of histone H3 to de novo methylation of DNA. Nature 2007;448:714−17.

[41] Subramaniam D, Thombre R, Dhar A, Anant S. DNA methyl transferases: a novel target for prevention and therapy. Front Oncol 2014;4:80.

[42] Gros C, Fahy J, Halby L, Dufau I, Erdmann A, Gregoire J-M, et al. DNA methylation inhibitors in cancer: recent and future approaches. Biochimie 2012;94:2280−96.

[43] Jones PA, Taylor SM. Cellular-differentiation, cytidine analogs and DNA methylation. Cell 1980;20: 85−93.

[44] Abele R, Clavel M, Dodion P, Bruntsch U, Gundersen S, Smyth J, et al. The eortc early clinical-trials cooperative group experience with 5-aza-2′-deoxycytidine (NSC-127716) in patients with colo-rectal, head and neck, renal carcinomas and malignant melanomas. Eur J Cancer Clin Oncol 1987;23:1921−4.

[45] Zambrano P, Segura-Pachero B, Perez-Cardenas E, Cetina L, Revilla-Vazquez AR, Taja-Chayeb LAT, et al. A phase I study of hydralazine to demethylate and reactivate the expression of tumor suppressor genes. BMC Cancer 2005;5:44.

[46] Fandy TE, Herman JG, Kerns P, Jiemjit A, Sugar EA, Choi SH, et al. Early epigenetic changes and DNA damage do not predict clinical response in an overlapping schedule of 5-azacytidine and entinostat in patients with myeloid malignancies. Blood 2009;114:2764−73.

[47] Mummaneni P, Shord SS. Epigenetics and oncology. Pharmacotherapy 2014;34:495−505.

[48] Schrump DS, Fischette MR, Nguyen DM, Zhao M, Li XM, Kunst TF, et al. Phase I study of decitabine-mediated gene expression in patients with cancers involving the lungs, esophagus, or pleura. Clin Cancer Res 2006;12:5777−85.

[49] Stresemann C, Lyko F. Modes of action of the DNA methyltransferase inhibitors azacytidine and decitabine. Int J Cancer 2008;123:8−13.

[50] Yoo J, Kim JH, Robertson KD, Medina-Franco JL. Molecular modeling of inhibitors of human DNA methyltransferase with a crystal structure: discovery of a novel DNMT1 inhibitor. Adv Protein Chem Struct Biol 2012;87:219−47.

[51] Goffin J, Eisenhauer E. DNA methyltransferase inhibitors − state of the art. Ann Oncol 2002;13:1699−716.

[52] Stewart DJ, Issa JP, Kurzrock R, Nunez MI, Jelinek J, Hong D, et al. Decitabine effect on tumor global DNA methylation and other parameters in a phase I trial in refractory solid tumors and lymphomas. Clin Cancer Res 2009;15:3881−8.

[53] Fang MZ, Wang YM, Ai N, Hou Z, Sun Y, Lu H, et al. Tea polyphenol (−) − epigallocatechin-3-gallate inhibits DNA methyltransferase and reactivates methylation-silenced genes in cancer cell lines. Cancer Res 2003;63:7563−70.

[54] Lin XH, Asgari K, Putzi MJ, Gage WR, Yu X, Cornblatt BS, et al. Reversal of GSTP1 CpG island hypermethylation and reactivation of pi-class glutathione S-transferase (GSTP1) expression in human prostate cancer cells by treatment with procainamide. Cancer Res 2001;61:8611−16.

[55] Villar-Garea A, Fraga MF, Espada J, Esteller M. Procaine is a DNA-demethylating agent with growth-inhibitory effects in human cancer cells. Cancer Res 2003;63:4984−9.

[56] Segura-Pacheco B, Trejo-Becerril C, Perez-Cardenas E, Taja-Chayeb L, Mariscal I, Chavez A, et al. Reactivation of tumor suppressor genes by the cardiovascular drugs hydralazine and procainamide and their potential use in cancer therapy. Clin Cancer Res 2003;9:1596−603.

[57] Singh N, Dueñas-González A, Lyko F, Medina-Franco JL. Molecular modeling and dynamics studies of hydralazine with human DNA methyltransferase 1. ChemMedChem 2009;4:792−9.

[58] O'Driscoll CM, Coulter JB, Bressler JP. Induction of a trophoblast-like phenotype by hydralazine in the p19 embryonic carcinoma cell line. Biochim Biophys Acta 2013;1833:460−7.

[59] Segura-Pacheco B, Perez-Cardenas E, Taja-Chayeb L, Chavez-Blanco A, Revilla-Vazquez A, Benitez-Bribiesca L, et al. Global DNA hypermethylation-associated cancer chemotherapy resistance and its reversion with the demethylating agent hydralazine. J Transl Med 2006;4:32.

[60] Dueñas-Gonzalez A, Coronel J, Cetina L, González-Fierro A, Chavez-Blanco A, Taja-Chayeb L. Hydralazine-valproate, a repositioned drug combination for the epigenetic therapy of cancer. Exp Opin Drug Metab Toxicol 2014;10(10):1433−44.

[61] Costello JF, Fruhwald MC, Smiraglia DJ, Rush LJ, Robertson GP, Gao X, et al. Aberrant CpG-island methylation has non-random and tumour-type-specific patterns. Nat Genet 2000;24:132−8.

[62] Seifert M, Cortijo S, Colome-Tatche M, Johannes F, Roudier F, Colot V. MeDIP-HMM: genome-wide identification of distinct DNA methylation states from high-density tiling arrays. Bioinformatics 2012;28:2930−9.

[63] Frommer M, McDonald LE, Millar DS, Collis CM, Watt F, Grigg GW, et al. A genomic sequencing protocol that yields a positive display of 5-methylcytosine residues in individual DNA strands. Proc Natl Acad Sci USA 1992;89:1827−31.

[64] Krueger F, Kreck B, Franke A, Andrews SR. DNA methylome analysis using short bisulfite sequencing data. Nat Methods 2012;9:145−51.

[65] Beck S. Taking the measure of the methylome. Nat Biotech 2010;28:1026−8.

[66] Bock C, Tomazou EM, Brinkman AB, Muller F, Simmer F, Gu H, et al. Quantitative comparison of genome-wide DNA methylation mapping technologies. Nat Biotech 2010;28:1106−14.

[67] Singh V, Sharma P, Capalash N. DNA methyltransferase-1 inhibitors as epigenetic therapy for cancer. Curr Cancer Drug Targets 2013;13:379−99.

[68] Fahy J, Jeltsch A, Arimondo PB. DNA methyltransferase inhibitors in cancer: a chemical and therapeutic patent overview and selected clinical studies. Expert Opin Ther Pat 2012;22:1427−42.

[69] Valente S, Liu YW, Schnekenburger M, Zwergel C, Cosconati S, Gros C, et al. Selective non-nucleoside inhibitors of human DNA methyltransferases active in cancer including in cancer stem cells. J Med Chem 2014;57:701−13.

[70] Brueckner B, Boy RG, Siedlecki P, Musch T, Kliem HC, Zielenkiewicz P, et al. Epigenetic reactivation of tumor suppressor genes by a novel small-molecule inhibitor of human DNA methyltransferases. Cancer Res 2005;65:6305−11.

[71] Suzuki T, Tanaka R, Hamada S, Nakagawa H, Miyata N. Design, synthesis, inhibitory activity, and binding mode study of novel DNA methyltransferase 1 inhibitors. Bioorg Med Chem Lett 2010;20:1124−7.

[72] Asgatay S, Champion C, Marloie G, Drujon T, Senamaud-Beaufort C, Ceccaldi A, et al. Synthesis and evaluation of analogues of N-phthaloyl-L-tryptophan (RG108) as inhibitors of DNA methyltransferase 1. J Med Chem 2014;57:421−34.

[73] Datta J, Ghoshal K, Denny WA, Gamage SA, Brooke DG, Phiasivongsa P, et al. A new class of quinoline-based DNA hypomethylating agents reactivates tumor suppressor genes by blocking DNA methyltransferase 1 activity and inducing its degradation. Cancer Res 2009;69:4277−85.

[74] Gamage SA, Brooke DG, Redkar S, Datta J, Jacob ST, Denny WA. Structure−activity relationships for 4-anilinoquinoline derivatives as inhibitors of the DNA methyltransferase enzyme DNMT1. Bioorg Med Chem 2013;21:3147−53.

[75] Rilova E, Erdmann A, Gros C, Masson V, Aussagues Y, Poughon-Cassabois V, et al. Design, synthesis and biological evaluation of 4-amino-n-(4-aminophenyl)benzamide analogues of quinoline-based SGI-1027 as inhibitors of DNA methylation. ChemMedChem 2014;9:590−601.

[76] Halby L, Champion C, Sénamaud-Beaufort C, Ajjan S, Drujon T, Rajavelu A, et al. Rapid synthesis of new DNMT inhibitors derivatives of procainamide. ChemBioChem 2012;13:157−65.

[77] Castellano S, Kuck D, Viviano M, Yoo J, López-Vallejo F, Conti P, et al. Synthesis and biochemical evaluation of δ2-isoxazoline derivatives as DNA methyltransferase 1 inhibitors. J Med Chem 2011; 54:7663−77.

[78] Kuck D, Singh N, Lyko F, Medina-Franco JL. Novel and selective DNA methyltransferase inhibitors: docking-based virtual screening and experimental evaluation. Bioorg Med Chem 2010;18:822−9.

[79] Kabro A, Lachance H, Marcoux-Archambault I, Perrier V, Dore V, Gros C, et al. Preparation of phenylethylbenzamide derivatives as modulators of DNMT3 activity. MedChemComm 2013;4:1562−70.

[80] Ceccaldi A, Rajavelu A, Champion C, Rampon C, Jurkowska R, Jankevicius G, et al. C5-DNA methyltransferase inhibitors: from screening to effects on zebrafish embryo development. ChemBioChem 2011;12:1337−45.

[81] Ye Y, Stivers JT. Fluorescence-based high-throughput assay for human DNA (cytosine-5)-methyltransferase 1. Anal Biochem 2010;401:168−72.

[82] Kilgore JA, Du XL, Melito L, Wei SG, Wang CG, Chin HG, et al. Identification of DNMT1 selective antagonists using a novel scintillation proximity assay. J Biol Chem 2013;288:19673−84.

[83] Ceccaldi A, Rajavelu A, Ragozin S, Sénamaud-Beaufort C, Bashtrykov P, Testa N, et al. Identification of novel inhibitors of DNA methylation by screening of a chemical library. ACS Chem Biol 2013; 8:543−8.

[84] Fagan RL, Wu M, Chedin F, Brenner C. An ultrasensitive high throughput screen for DNA methyltransferase 1-targeted molecular probes. PLoS One 2013;8:e78752.

[85] Medina-Franco JL, Giulianotti MA, Welmaker GS, Houghten RA. Shifting from the single to the multi-target paradigm in drug discovery. Drug Discov Today 2013;18:495−501.

[86] Dueñas-González A, Garcia-Lopez P, Herrera LA, Medina-Franco JL, Gonzalez-Fierro A, Candelaria M. The prince and the pauper. A tale of anticancer targeted agents. Mol Cancer 2008;7:82.

[87] Méndez-Lucio O, Tran J, Medina-Franco JL, Meurice N, Muller M. Towards drug repurposing in epigenetics: olsalazine as a novel hypomethylating compound active in a cellular context. ChemMedChem 2014;9:560–5.

[88] Morano A, Angrisano T, Russo G, Landi R, Pezone A, Bartollino S, et al. Targeted DNA methylation by homology-directed repair in mammalian cells. Transcription reshapes methylation on the repaired gene. Nucl Acids Res 2014;42:804–21.

[89] Li YY, Saldanha SN, Tollefsbol TO. Impact of epigenetic dietary compounds on transgenerational prevention of human diseases. AAPS J 2014;16:27–36.

[90] Gerhauser C. Cancer chemoprevention and nutri-epigenetics: state of the art and future challenges. In: Pezzuto J, Suh N, editors. Natural products in cancer prevention and therapy. Berlin, Germany: Springer-Verlag; 2013. p. 73–132.

[91] Medina-Franco JL, López-Vallejo F, Kuck D, Lyko F. Natural products as DNA methyltransferase inhibitors: a computer-aided discovery approach. Mol Divers 2011;15:293–304.

[92] Yoo J, Choi S, Medina-Franco JL. Molecular modeling studies of the novel inhibitors of DNA methyltransferases SGI-1027 and CBC12: implications for the mechanism of inhibition of DNMTs. PLoS One 2013;8:e62152.

[93] Medina-Franco J, Méndez-Lucio O, Yoo J. Rationalization of activity cliffs of a sulfonamide inhibitor of DNA methyltransferases with induced-fit docking. Int J Mol Sci 2014;15:3253–61.

[94] Medina-Franco JL, Yoo J. Molecular modeling and virtual screening of DNA methyltransferase inhibitors. Curr Pharm Des 2013;19:2138–47.

[95] Medina-Franco JL, Yoo J. Docking of a novel DNA methyltransferase inhibitor identified from high-throughput screening: insights to unveil inhibitors in chemical databases. Mol Divers 2013;17:337–44.

[96] Yoo J, Medina-Franco JL. Chemoinformatic approaches for inhibitors of DNA methyltransferases: comprehensive characterization of screening libraries. Comp Mol Biosci 2011;1:7–16.

[97] Yoo J, Medina-Franco JL. Towards the chemoinformatic-based identification of DNA methyltransferase inhibitors: 2D- and 3D-similarity profile of screening libraries. Curr Comput-Aided Drug Des 2012;8:317–29.

[98] Siedlecki P, Boy RG, Musch T, Brueckner B, Suhai S, Lyko F, et al. Discovery of two novel, small-molecule inhibitors of DNA methylation. J Med Chem 2006;49:678–83.

[99] Medina-Franco JL, Martinez-Mayorga K, Meurice N. Balancing novelty with confined chemical space in modern drug discovery. Expert Opin Drug Discov 2014;9:151–65.

[100] Huang Z, Jiang H, Liu X, Chen Y, Wong J, Wang Q, et al. HEMD: an integrated tool of human epigenetic enzymes and chemical modulators for therapeutics. PLoS One 2012;7:e39917.

[101] Shanghai Jiaotong University. Human epigenetic enzyme and modulator database (HEMD). 2015. Available at: <http://mdl.shsmu.edu.cn/HEMD/>.

[102] Martinez-Mayorga K, Peppard TL, López-Vallejo F, Yongye AB, Medina-Franco JL. Systematic mining of generally recognized as safe (GRAS) flavor chemicals for bioactive compounds. J Agric Food Chem 2013;61:7507–14.

HISTONE ACETYLTRANSFERASES: ENZYMES, ASSAYS, AND INHIBITORS

14

Yepeng Luan[1], Liza Ngo[1], Zhen Han[1], Xuejian Wang[2], Meihua Qu[2], and Y. George Zheng[1]

[1]*Department of Pharmaceutical and Biochemical Sciences, The University of Georgia, Athens, Georgia, USA*
[2]*College of Pharmaceutical and Biological Sciences, Weifang Medical University, Weifang, Shandong, P.R. China*

CHAPTER OUTLINE

Y.G. Zheng (Ed): Epigenetic Technological Applications. DOI: http://dx.doi.org/10.1016/B978-0-12-801080-8.00014-4

291

14.1 **HISTONE ACETYLTRANSFERASES**

Lysine acetylation of nucleosomal histones is one of the most studied epigenetic modifications in the epigenetics field [1]. Although it was reported as early as the 1960s that histone acetylation impacts gene transcription [2], the field of histone acetylation research was not launched until the discovery of the underlying enzymes in the mid-1990s. It is now clear that reversible lysine acetylation is dynamically mediated by two classes of enzymes with opposing activities, histone acetyltransferase (HAT) and histone deacetylase (HDAC). While HAT enzymes add the acetyl group to the ε-amino group of lysine residues in a protein substrate, HDAC enzymes conversely catalyze the removal of this acetyl group from the lysine. The majority of important acetylated lysine residues on core histones are located in the amino-terminal region [1,3]. In general, histone acetylation loosens the nucleosome structure and promotes the accessibility of transcription factors to local and global genetic loci [1,4]. Thus, HATs and histone acetylation are functionally linked with transcription activation, whereas HDACs and histone deacetylation are associated with gene repression (Figure 14.1). The effect of histone acetylation in chromatin-templated processes is generally thought to be induced via two major mechanisms [5,6]. First, histone acetylation directly modulates the packaging of chromatin by removing the positive charge of histone lysine residues, which weakens the nucleosomal packing and alters internucleosomal interactions. This leads to the facile access of DNA-binding proteins such as transcription factors to the targeted genes, signaling for transcription activation. Second, lysine acetylation regulates chromatin structure and function by recruiting modification-specific binding proteins, which recognize acetylated histones via specialized structural folds such as the bromodomain [7,8]. Further, histone acetylation may compete with

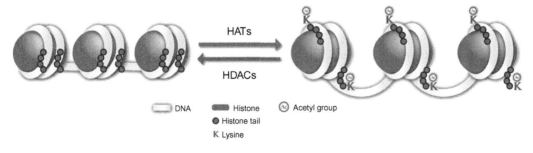

FIGURE 14.1

Acetylation and deacetylation catalyzed by HATs and HDACs, respectively.

modifications on the same residue or crosstalk with modifications on neighboring residues [9]. With all possible mechanisms acting either alone or in concert, acetylation-induced changes in histone−DNA and histone−protein interactions are in turn translated into biological readouts [10].

In recent years, protein lysine acetylation has been recognized as more of a general posttranslational modification, far beyond the chromatin realm, regulating diverse biological cascades such as transcriptional regulation and signal transduction. Thus, in a number of recent publications, HAT and HDAC are also referred to as KAT (lysine acetyltransferase) and KDAC (lysine deacetylase), respectively. These two types of enzymes cooperate to control dynamic homeostasis of lysine acetylation. Mass-spectrometry-based proteomic screening has identified thousands of nonhistone substrates, suggesting acetylation is a posttranslational modification of general significance in biology [11−17]. A notable observation is that the metabolic pathways could be regulated by the acetylation of metabolism-mediating enzymes [18−21]. Zhao et al. [19] uncovered that virtually every enzyme involved in glycolysis, gluconeogenesis, the tricarboxylic acid (TCA) cycle, the urea cycle, fatty acid metabolism, and glycogen metabolism were acetylated in human liver tissue.

A number of proteins possessing intrinsic HAT activity were identified. The first discovered HATs were *Tetrahymena* Gcn5 (general control nonderepressible 5) and yeast HAT1. So far, about 20 proteins that encode HAT activities have been characterized. These HAT proteins show remarkable diversity in primary amino acid sequence. Table 14.1 summarizes the list of HAT enzymes that are studied, which include HAT1, GCN5/PCAF, yeast Hpa2/3, Elp3, p300/CBP, the MYST family (composed of MOF, MORF, MOZ, Ybf2/Sas3, Sas2, and Tip60), and yeast

Table 14.1 Major HAT Enzymes		
Family	**Representative Members**	**Histone Substrates**
GNAT superfamily	HAT1/KAT1	H4(K5,12), H2A(K5)
	Gcn5/KAT2A	H2B(K16), H3(K14,18,23)
	PCAF/KAT2B	H3(K14)
	yHpa2/KAT10	H3(K4, 14); H4(K5, 12 > K8)
	yHpa3	H4(K8 > K5, 12)
	Elp3/KAT9	H3(K14), H4(8)
p300/CBP	CBP/KAT3A	H2A(K5), H2B (K12, 15), H3(K14, 18, 23), H4(K5, 8, 12, 16)
	p300/KAT3B	H2A(K5), H2B(K12, 15), H3(K14, 18, 23), H4(K5, 8, 12, 16)
MYST	Tip60/KAT5	H2A(K5), H3(K14), H4(5, 8,12, 16)
	MOZ/MYST3/KAT6A	H3, H4
	MORF/MYST4/KAT6B	H2A, H3, H4
	HBO1/MYST2/KAT7	H4
	MOF/MYST1/KAT8	H4(K16)
	yEsa1 (KAT5)	H2AZ, H4
	ySas2 (KAT8)	H4(K16)
	ySas3 (KAT6)	H3(K14)
Rtt109	yRtt109 (KAT11)	H3(K56)

RTT109 (Table 14.1). HAT1, GCN5/PCAF, Elp3, and MYST have homologs from yeast to human. p300/CBP is metazoan specific. Rtt109 and Hpa2/3 are fungal specific. Human GCN5 and PCAF (p300/CBP-associated factor) have 89% sequence identity in their HAT domains. Yeast Hpa2 and Hpa3 have nearly 50% identity and 68% similarity over 156 amino acids. The p300 and CBP HAT domains are 88% identical. The MYST members have a highly conserved catalytic domain, containing a characteristic C2HC zinc finger motif and three key residues in the active site (i.e. K274, C316, and E350 in human MOF).

Crystal structures for several HAT proteins have been determined. The structural elucidation reveals that, despite the diversity in the primary amino acid sequence, the majority of HAT structures contain a remarkable conserved structural fold that binds acetyl-CoA (acetyl-coenzyme A) cofactor, which is characteristic to the superfamily of GCN5-related *N*-acetyltransferase (GNAT) enzymes [22]. This characteristic GNAT structural fold is apparent in HAT1, PCAF/GCN5, MYST enzymes, and yeast Hpa2. Unlike the GNAT structural fold, the HAT domains of p300 and Rtt109 contain a unique, approximately 25 residue substrate-binding loop, which appears to participate in the binding of both Ac-CoA and histone substrates. Most HATs contain varied domains that flank their catalytic core region, and are likely responsible for protein—protein interaction and substrate targeting.

The search for HAT enzymes is still ongoing, and it is likely that many other proteins possessing HAT activity have not yet been discovered. For instance, *Salmonella enterica* aminoglycoside 6′-*N*-acetyltransferases have high structural homology with the yeast Hpa2 [23]. This raises the question of whether the bacterial aminoglycoside *N*-acetyltransferase could catalyze protein acetylation [22]. Recently, Karmodiya et al. [24] found that a Camello family protein, CMLO3, exhibited specific activity toward histone H4. Knockdown of this perinuclear protein in zebrafish embryos exhibited defects in zebrafish development; however, further work is needed to prove whether Camello proteins truly represent an independent HAT family.

14.2 THE PHARMACOLOGIC SIGNIFICANCE OF HATs

The acetylation level of proteins is orchestrated by HATs and HDACs. Disturbing the coordination between these two enzymes could give rise to diseased phenotypes [25,26]. Indeed, many studies show that anomalous expression of HATs associates with the occurrence of such diseases as neurodegenerative amyotrophic lateral sclerosis, inflammation, diabetes, asthma, chronic obstructive pulmonary disease, cystic fibrosis, and cancer. p300 and CBP are the best-known HATs involved in multiple cellular processes, including proliferation, differentiation, growth arrest, apoptosis, angiogenesis, and cellular response to stress environments [27]. Expression of p300/CBP decreases during chemical hepatocarcinogenesis and mutations in *p300/CBP* are associated with different cancers [28]. In acute myeloid leukemia, the *CBP* gene is translocated and fused to either *MOZ* and *MORF* or to the *MLL* gene [29]. The p300 and CBP genes are mutated in >85% of microsatellite instable (MSI)$^{+}$ colon cancer cell lines and primary tumors. p300 null MSI + cancer cell lines that reexpress exogenous p300 showed slower growth and a more flattened morphology. p300/CBP HAT activity is important for the G1/S transition and downregulation of p300 activity resulting in growth inhibition and activation of a senescence checkpoint in human melanocytes. It was also reported that CBP could interact with catenin to regulate the activity of the WNT/β-catenin pathway, which is at an increased level in colorectal cancers. In prostate cancer, p300 is involved in the transactivation with DHT (dihydrotestosterone) in an androgen independent mode. It was also reported that the cysteine/histidine-rich domain 1 of p300/CBP has a tight interaction with the C-terminal activation domain of HIF-1α whose high expression leads to tumor angiogenesis. The cut-off of this interaction could be an effective way to inhibit the angiogenesis of tumor [30,31]. All these data raise the idea that chromosomal trans-locations to HAT genes could lead to diseases like cancer and that small-molecule HAT inhibitors could be exploited therapeutically [32].

PCAF and GCN5 perform global histone acetylation, locus-specific histone acetylation, as well as acetylation of nonhistone proteins. These enzymes are often part of large multiprotein complexes, which regulate their activity and specificity. GCN5 plays a key role in EGF-mediated gene transcription and provides a target for the downregulation of oncogenic EGF signaling [33]. Furthermore, it has been shown that GCN5 is crucial for cell cycle progression [34,35]. PCAF can directly acetylate one of the specific CDK inhibitors, p27, whose downregulation has been proven to be involved in tumorigenesis. The acetylation on Lys100 mediated by PCAF could degrade p27 [36]. In breast cancer, resistance to chemotherapy is usually caused by the ATP-binding cassette (ABC) transporter proteins, such as PgP and MDR, which can efflux anticancer drugs out of the tumor cell. It has been found that PCAF and GCN5-mediated H3K9 acetylation is elevated in the major promoter and first exon of the MDR1 gene in a drug-resistant cell line compared to a drug-sensitive cell line [37].

Tip60, a member of the MYST HAT family, was first identified as a HIV tat interactive protein and later was found to be a HAT [38]. In the tumor suppressor p53, Tip60 can acetylate Lys120 [39]. In prostate cancer, the AR has been shown to be regulated by posttranscriptional modification such as acetylation by Tip60 [40] in an androgen independent way. The proliferation and progression of prostate cancer has been correlated to the unregulated activity of Tip60 [41]. Besides p53 and AR, Tip60 can also acetylate other nonhistone proteins like NF-κB and ATM (ataxia telangiectasia mutated) kinase, many of which are important proteins involved in tumorigenesis [42]. Inhibition of Tip60 activity could suppress the tumorigenesis directly, or make tumor cells susceptible to anticancer agents. Another important role of Tip60 is its ability to repair DNA double-strand breaks (DSBs) [43] by acetylating and activating ATM, which is involved in DNA repair. Therefore, the inhibition of Tip60 activity using inhibitors could be a novel therapeutic and radio-sensitizing option for cancer therapy.

HBO1 (histone acetyltransferase binding to ORC1), another MYST member and a key regulator of DNA replication and cell proliferation [44], was identified as a subunit of the origin recognition complex (ORC) with HAT activity [45]. This HAT enzyme has interactions with ORC1 [45], MCM2 [46], AR [47], progesterone receptor [48], Jad-1 [49], ING4 and ING5 [50], and is involved in a wide range of regulatory functions [51,52]. HBO1 also interacts with CDK11[p58], which is closely related to cell cycle, apoptotic signaling, and tumorigenesis [53]. An overexpression of HBO1 causes an increase in colony formation on soft agar in breast cancer cell lines [54] and in a specific subset of human primary cancers [44]; however, the tumor suppressor p53 can inhibit the HAT activity of HBO1 [51]. HBO1 might coordinately alter the expression profile through changes in histone acetylation and promotion of estrogen receptor α (ERα) degradation through ubiquitination [55]. HBO1 also suppressed NF-κB-induced transcription in an AR-independent manner [56].

14.3 HAT BIOCHEMICAL ASSAYS

Assays have direct and central roles in HAT activity validation and drug discovery. Given the significance of protein lysine acetylation in normal physiology and pathology, effective biochemical assays are required to detect and quantitate HAT activities both *in vitro* and *in vivo* [57,58]. For basic molecular and cellular biology research that typically does not require a very large scale screening, the requirement of a HAT assay is not so strict as long as the assay allows for faithful and quantitative demonstration of HAT enzymatic activities, either by directly measuring the acetylated protein amount or by quantifying the side product CoA. For medicinal chemistry discovery, nevertheless, there is a particular need for assay speed and throughput. Screening HAT inhibitors, especially in an HTS fashion, would require appropriate assay systems that ideally combine robustness, sensitivity, speed, simplicity, cost-effectiveness, and readiness for automation [59]. Major HAT assays are reviewed in the following sections.

14.3.1 RADIOMETRIC ASSAYS

The radiometric assays take the leading role and represent the gold standard in studying HAT activities [57,60]. These assays rely on [3H]- or [14C]-labeled Ac-CoA as the acetyl donor (Figure 14.2A). Following a HAT-catalyzed acetyl transfer reaction, the acetylated peptide or protein product, which becomes radioactive owing to the added isotopic acetyl group, is then detected and quantitated with radiometry. The radiometric HAT assay can be implemented in two different experimental protocols, that is, radiographic gel imaging and filter binding [57,61]. In the first method, the HAT reaction mixture is resolved by sodium dodecyl sulfate-polyacrylamide gel electrophoresis (SDS-PAGE). Next, the proteins on the polyacrylamide gel can be either transferred to a nitrocellulose membrane using a standard Western blot transfer device or the electrophoretic gel is directly dried using a Bio-Rad vacuum drier. Then the nitrocellulose membrane or the dried gel is exposed to a fluorographic film or a storage phosphor screen for signal readout. During the exposure process, ionizing radiations emanating from the 14C or 3H expose the photographic emulsion, causing reduction of silver ions in the emulsion formulation to become reduced to metallic silver grains intimately above the position of the radioactive source (acetylated proteins). Because 3H radioactivity is much weaker than 14C, fluorographic films are particularly needed for 3H exposure. Following radiation exposure, the radiographic film is processed in a darkroom to develop the radiogram. The storage phosphor screen, following exposure, does not need further chemical development in the darkroom and is directly scanned on a storage phosphor imaging system (e.g. GE Typhoon Scanner and Molecular Dynamics Phosphorimager).

The second radiometric method quantifies the amount of radioactivity incorporated into the peptide or protein substrates with filter binding followed by liquid scintillation counting [57,60,62]. The acetylation reaction mixture is typically loaded onto a Waterman P81 filter paper disc, which is composed of an anionic phosphocellulose material. Charged HAT substrates, such as the highly positively charged histones, allow the substrate to bind to the paper disc by electrostatic interaction. Washing with 50 mM sodium bicarbonate solution (pH 9) removes any unreacted radioisotopic acetyl-CoA. After air drying, the radioactive products on the paper disc can be quantified by liquid scintillation on a liquid scintillation counter.

The filter-binding assay is convenient to implement and is less time-consuming than the fluorography method because it does not need a SDS PAGE gel and the development of a radiographic image. This assay simply relies on the ability of the acetylated peptide/protein product to electrostatically bind to the phosphocellulose disc and also depends upon accurate determination of the [^{3}H or ^{14}C]-acetyl-CoA-specific activity. However, substrates in this assay are limited to basic peptides and proteins. It can be erroneous to assume that all of the radiolabeled products produced in the assay are bound to the phosphocellulose disk. If binding is less than complete, this can lead to an underestimation of the calculated rate constants. In particular, certain buffer conditions, for example, high concentrations of phosphate, may cause failure of peptides to bind to the P81 filters efficiently [63]. Also, if the specific radioactivity of [^{3}H or ^{14}C]-acetyl-CoA is overestimated (due to partial hydrolysis of [^{3}H or ^{14}C]-acetyl-CoA), the amount of product formed will be underestimated in the final calculations [64]. The radiometric assays have been widely used to measure HAT activities from various sources, including kinetic characterization of HAT enzymology [57,65], and study of HAT inhibitors [66].

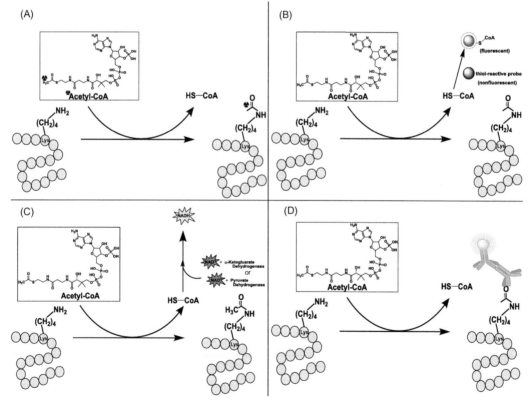

FIGURE 14.2

Some common formats of HAT assays. (A) Radioactive method to detect HAT activity. (B) Fluorogenic method to detect HAT activity. (C) An enzyme-coupled NADH assay. (D) Antibody-based method to detect HAT activity.

14.3.2 COUPLED SPECTROPHOTOMETRIC AND SPECTROFLUORIMETRIC ASSAYS

Given the limitation of the radioactivity-based assays, a number of efforts were investigated in the development of nonradioactive HAT assays. Nonradioactive strategies are particularly desired for HAT inhibitor screening. The design of nonradioactive HAT assays, however, is technically challenging because the substrates and products of HATs exhibit little spectroscopic difference. Several nonradioactive assays are developed based on detection and quantification of the side product CoA using either chemical- or enzyme-coupled spectrophotometric methods [64,67,68]. For example, the HAT reaction has been directly monitored with thiol reactive chemical DTNB (Ellman's reagent, 5,5'-dithiobis-(2-nitrobenzoic acid)) [68]. CoA reacts with this compound, cleaving the disulfide bond to give 3-thio-6-nitrobenzoate (TNB^-), which ionizes to the TNB^{2-} dianion at neutral and alkaline pH. This TNB^{2-} ion has a yellow color, with absorbance at 412 nm. This reaction is rapid and stoichiometric, offering a good linear correlation of the amount of TNB^{2-} with the production of CoA.

Another spectroscopic method for HAT assay relies on the chemical reaction of CoA with fluorogenic compounds (Figure 14.2B). Upon conjugation to the thiol group of CoA, these molecules gain enhanced quantum yields and strong fluorescence, permitting facile quantitation of HAT activities. We evaluated and compared the assay performances of a series of thiol-sensitive fluorogenic compounds for the detection of the enzymatic activities of different HAT enzymes, including kinetics of reaction with CoA, fluorescence amplification factors, and influence on HAT activity. Our data showed that 7-diethylamino-3-(4'-maleimidylphenyl)-4-methylcoumarin (CPM) and 3-(7-hydroxy-2-oxo-2H-chromen-3-ylcarbamoyl)-acrylic acid methyl ester (CME) are particularly excellent HAT probes owing to their fast kinetics of reacting with CoA and strong fluorogenicity. Both compounds have very weak fluorescence, but give strong fluorescence upon reacting with CoA, with maximum emission at 482 nm and 465 nm, respectively. CPM is commercially available, and has been used in several cases to study enzyme-catalyzed CoA formation [67]. This fluorescent method is robust and amenable to minimization and automation in multiwalled plates, thereby suited for adaption to screen small molecule inhibitors of HATs in the HTS format, highlighting the value of this assay strategy in new drug discovery. Because free sulfhydryls on the enzyme or substrate or in the assay buffers can react with the fluorogenic probe to produce fluorescent signals, appropriate controls should be used to exclude the background fluorescence. Further, some fluorogenic compounds, such as CPM, are detrimental to the intrinsic activity of certain HATs [69]. In such cases, it is necessary to add the fluorogenic probe after the HAT reaction is over, which means that carrying out the assay in the continuous mode is prohibited.

Denu et al. [64] developed an enzyme-coupled strategy to measure HAT activity (Figure 14.2C). In that approach, the CoA generated in the HAT reaction was continuously measured by using a coupled enzyme system with either α-ketoglutarate dehydrogenase or pyruvate dehydrogenase. The CoA-dependent oxidation of α-ketoglutarate or pyruvate is accompanied by the reduction of NAD to NADH, which was monitored spectrophotometrically at 340 nm. The continuous, nonradioactive assay is less labor intensive, less expensive, and is a more accurate way of measuring kinetic parameters for HATs than the commonly used radioactive filter-binding assay. The coupled assay is not limited by the kinds of basic peptide/protein substrates that are required for efficient binding to the negatively charged phosphocellulose disks. These assays can be used in microplate format to increase the capacity of kinetic analyses. Due to the limited molecular absorptivity of NADH ($\varepsilon_{340} = 6230$ $M^{-1}cm^{-1}$), the enzyme-coupled assay requires more product formation (low μM) to produce significant changes in absorbance at 340 nm. It is worth noting that the coupled spectrophotometric methods can be used to measure activities of other enzyme systems that produce CoA as a product, such as protein *N*-myristoyltransferase.

14.3.3 IMMUNOSORBENT ASSAYS

From the beginning of acetylation biology, antibody-based immunosorbent assays are a basic molecular biology tool used in studying protein acetylation and HAT function (Figure 14.2D). Many antiacetyl antibodies are commercially available for either sequence-independent pan lysine acetylation recognition or protein sequence-specific acetylation site. The antibody-based assays can be implemented in different formats depending on experimental needs. For immunoblotting applications, HAT reaction with a protein substrate is quenched and then is resolved by polyacrylamide gel electrophoresis and transferred to a nitrocellulose or polyvinylidene fluoride (PVDF) membrane for Western blotting using the appropriate antiacetyl antibody. Interestingly, Kuninger et al. [70] expressed histone tail domain in fusion with maltose-binding protein in soluble form (note: bacteria-expressed recombinant histone proteins are in inclusion body) and used it as a substrate to establish robust immunoblotting assays for HATs.

Immunosorbent acetyltransferase assays were used by Jung and coworkers to study HAT inhibitors [71,72]. In a typical platform, this heterogeneous assay is performed in streptavidin-coated 96-well plates. First, biotinylated histone peptide is bound to the wells. Then, the bound substrate is turned over in an enzymatic reaction catalyzed by HATs in the presence of different concentrations of inhibitor; the enzymatic reaction is initiated by adding acetyl-CoA solution to each well. The amount of the turnover is detected by a primary rabbit IgG antibody against acetyl-lysine. Next is an incubation step with a horseradish-peroxidase-conjugated secondary antibodies for enzymatic bioluminescence readout (i.e. ELISA, enzyme-linked immunosorbent assay), or europium (N1-Eu)-labeled secondary antibody. The europium label produces time-resolved fluorescence: an enhancement solution is added to cleave off the europium label, followed by time-resolved fluorescence measurement at λex/λem: 340/615 nm. Data from the readout were plotted against the logarithm of compound concentration in order to obtain IC_{50} values. Similar immuno assays can be set up to evaluate the acetylation status of histones and nonhistone proteins in cells [58,73]. Key experimental procedures include cell culturing, fixation, membrane permeabilization, probing with primary anti-acetyl-lysine antibody, and then with secondary antibody. Cell immunosorbent assays of this type can be carried out on microtiter plates, thus possessing a higher throughput capacity than Western blotting, which is particularly useful for screening and validating cell-permeable acetylation regulators.

14.4 HAT INHIBITORS

The multifold evidence for disease relevance of lysine acetylation highlights the importance of HATs as potential drug targets. Relative to HDAC inhibitors, however, development of HAT inhibitors in both academics and pharmaceutics lags greatly behind. Numerous HDAC inhibitors with diverse pharmacophores have been developed and entered clinics such as Romidepsin (FK228) and

Vorinostat (SAHA, suberoylanilide hydroxamic acid) which has been approved by the Food and Drug Administration (FDA) to be clinically used for the therapy of cutaneous T-cell lymphoma, and many others are under different phases of clinical trials. In contrast with HDAC, many fewer HAT inhibitors were obtained, and no inhibitor is in preclinical phase [74]. The reasons underlying the greater emphasis on HDAC inhibitors may reflect our incomplete understanding of acetylation biology, limitations of assay technologies, or purely biased conceptions. At any rate, the predominance of HDAC inhibitors greatly overshadows the significance of HAT inhibitors. As the biology of HATs in cancer and other pathological pathways becomes better elucidated, there is no doubt that HAT inhibitors as drug molecules will have great pharmaceutical potential. As a matter of fact, HATs are as important as HDACs in maintaining proliferation and metastasis of cancer cells. In a number of studies, HAT inhibitors have been shown to act as equally potent blockers of tumorigenesis as the HDAC inhibitors. One technical difficulty is that the assays for screening HAT inhibitors are not as easy as for HDAC inhibitors. HDAC reactions can be measured with a single-step fluorogenic assay, which is less costly and can be implemented in standard biochemistry labs. On the other hand, HAT reactions most commonly rely on radiometric methods in which safety regulations often restrict access for many labs.

Nevertheless, with more and more HAT biology and pathology being elucidated, HAT inhibitor development has attracted increasing attention in recent years [75−77]. Just as in the application of other types of enzyme inhibitors, selective HAT inhibitors have great utility in proving the HAT function in particular biological pathways. The developed HAT inhibitors can be classified into two groups: substrate-based bivalent inhibitors and small molecule inhibitors [78,79]. In the following, we briefly summarize the HAT inhibitors discovered or synthesized to date.

14.4.1 BISUBSTRATE HAT INHIBITORS

HATs transfer the acetyl group from acetyl-CoA to the ε-amino group of lysine within protein substrates. During this catalytic process, a ternary complex is formed that involves spatial proximity of acetyl-CoA and lysine residue in the enzyme active, which is required for almost all HAT enzymes with the exception of Esa1 [80,81]. Thus, covalently linking acetyl-CoA with lysine-residue of a substrate peptide will be an effective design strategy for creating potent HAT inhibitors [82]. This feature is particularly true for the GNAT superfamily of acetyltransferases, for which many members have been reported to be effectively inhibited by covalent bisubstrate analogs. The substrate-specific sequence provides a great advantage for achieving selectivity for a particular HAT. Design of the CoA-substrate type of bisubstrate inhibitors evolves over time. In the 1960s Chase and Tubbs proposed one of the earliest bisubstrate analogs, a carnitine-CoA adduct that was involved in the self-catalyzed inactivation of carnitine O-acetyltransferase [83]. In the early 1980s, several polyamine-CoA conjugates were found to block histone/polyamine acetyltransferase activities in cellular and nuclear extracts [84,85]. The bisubstrate strategy was employed for inhibition of several GNAT superfamily members, such as serotonin *N*-acetyltransferase [86,87], gentamicin *N*-acetyltransferase [88], and *Enterococcus faecium* aminoglycoside 6′-*N*-acetyltransferase type Ii [89,90].

Cole and coworkers are among the first to utilize the bisubstrate strategy in their design of specific HAT inhibitors [91]. In their original work, two substrate-based analogs, Lys-CoA (entry **1**) and H3K14CoA (entry **2**) were found to be potent inhibitors for p300 ($IC_{50} = 0.5\ \mu M$) and PCAF ($IC_{50} = 0.3\ \mu M$) [91], respectively. The compound H3K14CoA is very potent for PCAF, due to the fact that H3K14 is a favorable acetylation site for PCAF [92]. Of unique interest is that Lys-CoA, in which CoA is linked to a single lysine amino acid, shows particular potency and selectivity for p300, as compared to PCAF and MSYT HATs. Such a selection likely is associated with the atypical Theorell-Chance kinetic mechanism of p300/CBP catalysis, which involves a fleeting interaction of the histone peptide substrate. Inspection of the X-ray structure of p300 HAT complexed to Lys-CoA suggested it could be possible to find a bisubstrate analog with a longer linker that stretches to cover the presumed peptide-binding pocket. Thus, Cole et al. [93] synthesized and evaluated several histone H4 peptide-CoA conjugates with different length linkers as inhibitors of p300 HAT. The results showed that longer linker length tends to present higher potency for p300 inhibition. The authors deduced that this linker allowed the inhibitor to interact with both the Lys-CoA tunnel as well as the peptide-binding pocket. Kwie et al. [94] recently designed a set of bisubstrate-like inhibitors for p300 in which the lysine part is replaced with various organic groups (e.g. entry **3**). Impressively, some of the compounds showed similar or enhanced potency in comparison to Lys-CoA. Similar effect on potency was also observed when the lysine region was modified in Cole's work [95]. These results highlight the great potential of optimizing the substrate portion in order to obtain more potent Lys-CoA type of bisubstrate p300 inhibitors.

Although shown to be efficacious for inhibiting HATs, CoA-containing bisubstrate HAT inhibitors generally suffer poor pharmacokinetic performance *in vivo*, due to low cell permeability attributed to the existence of highly polar phosphate groups in the CoA moiety. For example, Lys-CoA alone could not be used in cell tests, unless there was aid from microinjection or cell permeability agents; however, Lys-CoA is not metabolically stable [96]. Although CoA is a negatively charged motif of high polarity, incorporation of cell-permeable motif in the substrate portion could make it cell permeable. Therefore, efforts have been investigated to improve the chemical structures of bisubstrate HAT inhibitors for enhanced cell permeability. For example, the HIV-1 TAT transduction domain Tat sequence [97,98] was conjugated to Lys-CoA and H3K14CoA, thus generating LysCoA-TAT and H3K14CoA-TAT. Indeed, both Tat-tagged inhibitors showed cellular activity. Zheng et al. designed a prodrug strategy for cell delivery of Lys-CoA inhibitor [99]. In another example, Bandyopadhyay et al. [100] showed that spermidine(N^1)−CoA (entry **4**) was capable of being internalized into the cell, and inhibit histone acetylation, DNA synthesis and DNA repair. Spermidine(N^1)−CoA also makes tumor cells susceptible to DNA destroying anticancer drugs such as camptothecin. Cole's group also tried to truncate the CoA structures to reduce the polarity [101]. Unfortunately, in the assay of anti-p300 activity, these compounds did not show desirable activities; the IC_{50} values of Lys-phosphopantetheine and Lys-pantetheine were greater than 6.4 μM, while the IC_{50} value of positive control Lys-CoA was 50 nM. Compound 3′-dephospho-Lys-CoA, showed a mild activity with an IC_{50} value of 1.6 μM. The result demonstrated the intact Lys-CoA should be indispensable for p300 inhibition, and the phosphate acid group in the 3′ position had important interaction with the p300.

We designed a series of bisubstrate analogs with H4 peptide linked to CoA at K5, K8, K12 or K16 sites, and tested their inhibition potency for the MYST family HAT members Esa1 and Tip60 [66]. Enzymatic assays showed that these bisubstrate analogs are much more potent than small molecules, curcumin and anacardic acid. In particular, H4K16CoA was tested as one of the most potent inhibitors for both Esa1 and Tip60, and the inhibition is competitive versus acetyl-CoA and noncompetitive versus the histone H4 substrate, which supports an ordered substrate-binding mechanism in Esa1. Since H3 Lys-14 is also a substrate of Esa1 and Tip60, not surprisingly we tested that K3K14CoA is indeed a low micromolar inhibitor of both Esa1 and Tip60. In our recent work, we incorporated mono- and trimethylated lysine residues at Lys-4 and/or Lys-9 sites of H3 in the H3K14CoA inhibitor. The enzymatic assay results showed that the presence of methyl group(s) on the substrate resulted in more potent inhibitors for Tip60, relative to the parent H3K14CoA ligand. Importantly, by comparing the inhibitory properties of the ligands against full-length Tip60 versus the Tip60 catalytic domain, we determined that the K4me1 and K9me3 marks contributed to the potency augmentation by interacting with the catalytic region of the enzyme (entry **5**), instead of the chromodomain [102]. These results highlight that distal amino acid in the context of the acetylation site can influence the affinity of the substrate to enzyme and optimizing peptide sequence can significantly increase the potency of bisubstrate HAT inhibitors.

Bisubstrate enzyme inhibitors can be applied as powerful mechanistic tools for defining kinetic mechanisms of HATs, dissecting the relative roles of HATs in protein acetylation and transcription, and serving as structural biology tools [82,103]. In particular, bisubstrate analogs have demonstrated to be binding ligands for the structural characterization of target enzymes, which allow for the examination of key amino acid residues involved in substrate recognition and catalysis [104]. For instance, the first crystal structure of p300 HAT was determined in complex with Lys-CoA [105]. This structure reveals that, in addition to the pocket binding with the lysine, there is an extra pocket with highly negative electricity also playing an important role in the substrate binding, and a narrow shallow electronegative groove that connects the two pockets together. A high-ordered ternary complex could not be detected during the progress of transferring the acetyl group from acetyl-CoA to protein substrates [106]. The structural information was utilized later as a critical insight for the design of p300 HAT inhibitors with enhanced properties [93,94].

14.4.2 SMALL MOLECULE HAT MODULATORS

Low molecular-weight compounds are of great significance for pharmacological modulation of HATs. Compared to peptide-CoA conjugates, small molecules are smaller in size, more biostable and biocompatible, easily able to penetrate cellular membrane, and have a better chance of developing into medicinal drugs, due to their desired pharmacokinetic properties. In recent years, a number of efforts have investigated the development of small molecule HAT modulators, including both naturally derived compounds and synthetic organic molecules (Figure 14.3).

14.4.2.1 Anacardic acid and its analogs

The naturally occurring compound, 6-nonadecyl salicylic acid (anacardic acid, AA, entry 6), which is rich in the liquid of cashew nut shells, was reported in 2003 to inhibit HAT activity [107]. In the original report, AA worked as a nonspecific inhibitor of p300/CBP and PCAF ($IC_{50} = 8.5\ \mu M$ and $5\ \mu M$, respectively) and was capable of permeating the cells in culture. Interestingly, a recent study showed that AA inhibited the MYST member Tip60 with better specificity than PCAF and p300 under similar experimental conditions [108]. CTPB (entry 7), the amide derivative of AA, acts as an activator of p300 but not of PCAF [107]. This is consistent with Ghizzoni's report that some AA analogs acted as HAT activators [108]. The cause for the agonist and antagonist actions of AA compounds is worth further investigation. A co-crystal structure of AA or AA analog in complex with HATs will be valuable for understanding the mechanism of AA action and for further structural optimization. It should be noted that a number of studies have been performed using AA for

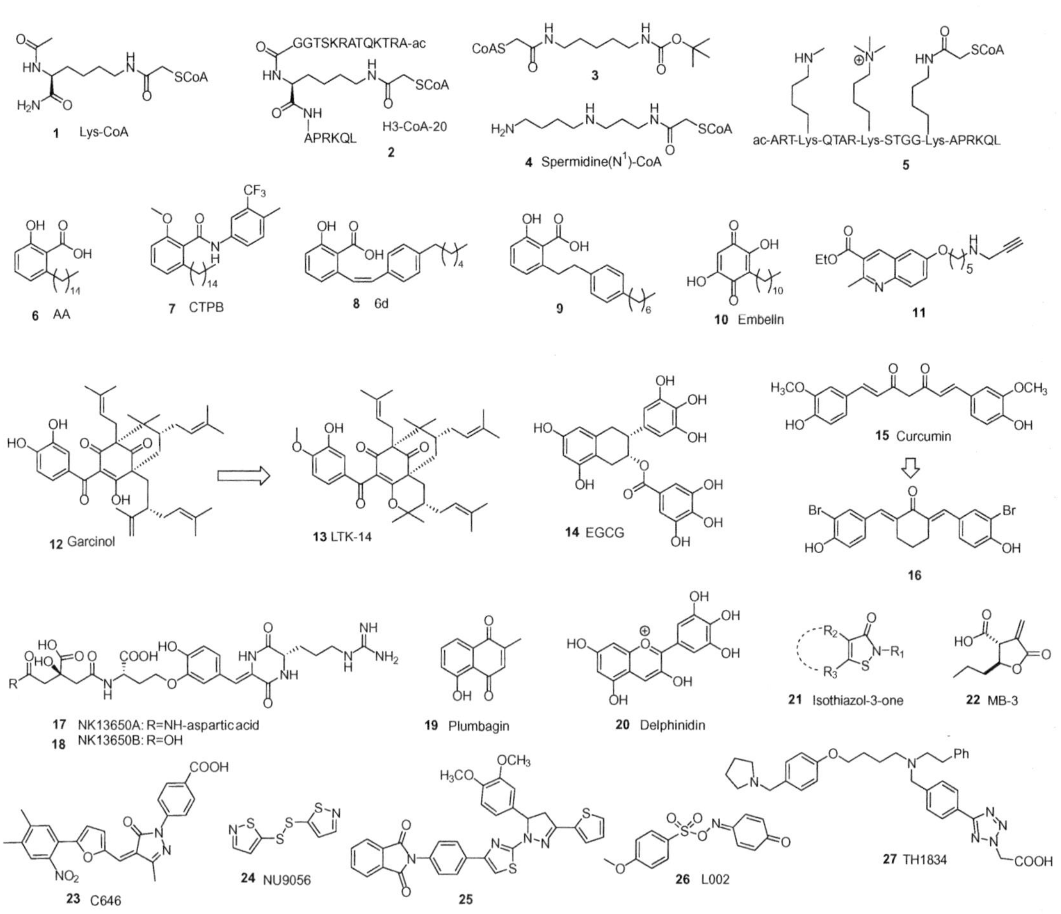

FIGURE 14.3

Structures of chemical modulators for HATs.

treatment of diseases such as cancer, inflammation, obesity, and oxidative damage, likely because of the pleiotropic capability of AA to affect multiple enzyme targets such as NF-κB kinase, lipoxygenase (LOX-1), xanthine oxidase, tyrosinase, ureases, matrix metalloproteinase-2 (MMP-2), and matrix metalloproteinase-9 (MMP-9) [109].

Efforts have been made by several groups to design and synthesize AA analogs or derivatives to improve cell permeability, physiochemical properties, and inhibitory efficacy of HATs [72,110–114]. For example, Ghizzoni et al. [108] synthesized several AA analogs by substituting the long alkyl chain of AA to different hydrophobic groups with a key Sonogashira coupling reaction. Compound 6d (entry **8**) showed slightly better activity than AA for PCAF inhibition (IC$_{50}$ 662 μM) [115]. Subsequently, these and other AA analogs were comparatively studied for inhibiting Tip60, p300, and PCAF. Interestingly, several compounds, including AA and compound **9**, showed better activity for the MYST HATs than p300 and PACF. According to the docking result, compound **9** binds to the same binding site as Ac-CoA, and the long chain on position 6 could mimic that of Ac-CoA. The esterification of the carboxyl group of salicylic acid was detrimental to the activity.

The key structural components of AA and AA analog compounds responsible for HAT inhibition are the aromatic head group and the alkyl chain tail. The hydrophobicity and length of the chain seem to be substantial for the HAT inhibition [108]. Analogous to the structural framework of AA compounds, several other HAT inhibitors also possess the aromatic head and an alkyl side chain, for example embelin (entry **10**) [116]. This small molecule has a hydroxybenzoquinone moiety that played an important role in selectively inhibiting PCAF *in vitro* and *in vivo*. Quinoline derivatives were another set of compounds containing head and tail structures; however, in this case, the benzene ring is substituted with quinoline [114,117]. Lenoci et al. recently reported a series of quinoline HAT inhibitors for p300 inhibition [118] which incorporated various large alkyl substitutions on the C2 position, and alkoxyl groups on the C6 position. Biochemical assays showed that anti-HAT activity is fairly tolerant to C2 substitution but is affected drastically by variation at C6 substitution. Of these inhibitors, compound **11** was found to show activities with IC$_{50}$ 57.5 μM for p300 and induced G2/M phase arrest and apoptosis in U937 cells.

14.4.2.2 Garcinol and its analogs

This tri-isoprenylated chalcone, garcinol (entry **12**), is harvested from *Garcinia indica*, which has been traditionally used and appreciated in tropical regions for centuries. The core structure of this polyphenol compound consists of a polyisoprenylated benzophenone, and has been found to possess many biological properties such as antioxidant, anti-inflammatory, anti-HIV, and anticancer activity. Garcinol is another natural product that displays inhibition activity toward p300 with IC$_{50}$ value of 7 μM and PCAF with IC$_{50}$ value of 5 μM [119]. Treatment of garcinol to human cancer cells induces apoptosis [120] while garcinol treatment to HeLa cells inhibits histone acetylation and HAT dependent chromatin transcription [121].

A structure and activity relationship study showed that the catechol hydroxyl groups were proposed to interact with the acetyl-CoA-binding domain, while the isoprenyl moieties interacted with the substrate-binding site of enzyme. Mantelingu and colleagues developed an isogarcinol derivative, LTK-14 (entry **13**), which is a p300-specific inhibitor. Even at a concentration of 89 μM, LTK-14 did not show conspicuous inhibition activity to PCAF. Due to the methoxy group on the aromaticring, LTK-14 has lower toxicity than garcinol to T-cells. The SAR of garcinol, isogarcinol, and LTK-14 was studied by Arif et al. [122] Garcinol and its derivatives supply a promising scaffold for seeking more effective and specific small molecular inhibitors of HATs.

14.4.2.3 Epigallocatechin-3-gallate

Rich in green tea, epigallocatechin-3-gallate (EGCG) (entry **14**) is another polyphenol natural product that shows inhibitory activity for HAT. At a dose of 100 μM, EGCG abrogated p300-induced p65 acetylation *in vitro* and *in vivo*, and suppressed tumor necrosis factor α (TNFα) induced NF-κB activation [123]. EGCG was also found to inhibit the TNFα-induced p65 translocation to the nucleus. The pharmacologic merit of EGCG is clear because acetylation of p65 can activate the NF-κB pathway, which is an appealing target for anti-inflammation and anticancer drug discovery. It is noteworthy to mention that EGCG has a wide spectrum of biological affects; for instance, it is also an inhibitor of DNMT1 [124] and topoisomerases [125].

14.4.2.4 Curcumin and cinnamoyl analogs

As a major component of *Curcuma longa rhizome* and commonly used as a yellow coloring and flavoring agent in food, curcumin (entry **15**) is another natural product found to show inhibitory activity to HAT, particularly p300/CPB [126]. Curcumin shows specific inhibition activity toward p300/CBP (IC_{50} 25 μM) but not to PCAF and other enzymes involved in chromatic epigenetic modulation such as HDACs and DNA methyltransferases. It can also promote the degradation of p300 through the proteasome-dependent way. The result from the enzyme kinetics study suggests that curcumin does not bind to the catalytic domain of p300, but to a special site exclusively for curcumin. The anti-p300 activity of curcumin is consistent with its anti-acetylation of the histone proteins and p53, a nonhistone protein substrate of p300. Another feature of curcumin is its smooth cell entry, which is attributed to the anti-acetylation activity *in vivo*. The SAR study showed that the two double bonds in the curcumin structure, which form two Michael reaction acceptors, are pivotal for binding with p300. The reduction of the two double bonds that destroys the Michael reaction system leads to the total disappearance of the anti-p300 activity [127]. The interaction mode of curcumin with p300 was illustrated recently [128]. Two cinnamoyl curcumin analogs, 1,7-Bis-(3-bromo-4-hydroxy-phenyl)-5-hydroxy-hepta-1,4,6-trien-3-one and 2,6-Bis-(3-bromo-4-hydroxy-benzylidene)-cyclohexanone (entry **16**) display inhibitory activities against p300. Docking results showed that these two cinnamoyl analogs have lower binding energies to p300 which means that they should have better activity than the parent curcumin [129]. It is worth mentioning that a large number of studies have proven curcumin's wide range of pharmacological properties, such as antibacterial, antifungal, antiviral, antioxidative, anti-inflammatory, antitumorigenesis, anti-angiogenesis, and antimetastasis [130]. Further structural improvement of curcumin would be an effective way to enhance p300 inhibitor discovery.

14.4.2.5 Peptide mimics

Tohyama et al. [131] identified two natural compounds (NK13650A − entry **17** and NK13650B − entry **18**) during the screening of 19,320 samples for HAT activity inhibition. The structures were identified with the aid of HRMS, UV, IR, and two-dimensional NMR. The two compounds have a peptide skeleton containing aspartic acid, glycine, arginine, and uncommon amino acid homoserine. Citric acid acted as the bridge to link the amino acids together, which was a very novel and interesting structure. In the biological activity evaluation, these two compounds showed low IC_{50} values to p300 (11 nM and 22 nM, respectively) and high selectivity in contrast with Tip60, another HAT playing key roles in AR activation and in androgen-independent prostate cancer cell growth. Based on these two natural products, p300 selective inhibitors with better activity and physiochemical characters could be designed according to the structure and activity relationship study.

14.4.2.6 Plumbagin

Plumbagin (entry **19**), isolated from *Plumbago rosea* root extract, possesses many biological activities and is another specific natural inhibitor of p300 in the p300-mediated p53 acetylation pathway [132,133]. As a noncompetitive inhibitor of p300, the free 5-hydroxyl proved to be indispensable for p300 inhibitory activity. Methylation of the hydroxyl group or derivatization conjugated with chemical groups damped the activity. The docking study with plumbagin in the p300 HAT domain

(PDB 3BIY) showed that the single hydroxyl group had an interaction with Lys1358, a key residue in p300 acetyltransferase activity. The discovery of plumbagin supplies a new source for designing novel and active p300 selective inhibitors.

Besides the natural HAT inhibitors mentioned above, there exist other natural products also possessing inhibitory activity to HATs, such as triptolide [134] and delphinidin [135] (entry **20**); however, an in-depth investigation of these compounds for HAT inhibition is still needed.

14.4.2.7 Isothiazol-3-ones

Isothiazol-3-one is another class of cell permeable HAT inhibitors (scaffold shown in entry **21**). Aherne and coworkers performed a high-throughput screening using the FlashPlate assay and identified a set of isothiazol-3-one compounds for PCAF inhibition [136]. Some of the hits are capable of blocking histone acetylation and colon tumor cell proliferation. N-arylated compounds showed stronger potency than N-alkylated ones. The inhibition mechanism appears to be suicide inhibition because of the covalent disulfide bond that is formed between the inhibitor and a cysteine residue in PCAF. Testing has also shown that DDT can interrupt this interaction and diminish inhibitory activity.

Since the first discovery of isothiazol-3-one for HAT inhibition, it was conjectured that the substitution group on the parent skeleton has deep influence on the inhibitor activity. Indeed several efforts have been devoted to exploring structural optimization of isothiazol-3-one inhibitors. Haisma et al. made a set of 5-chloro-isothiazolone with IC_{50} ranging from $2-3\,\mu M$ for PCAF [137]. The presence of the 5-chloro appears to be enhancing the potency. Gorsuch et al. studied isothiazolones derivatives with substitution at N2, 4, and 5 positions [138]. N-pyridyl, -4-morpholino-phenyl, and -4-dimethylaminophenyl substituted compounds all showed potent activity to PCAF with the IC_{50} values from 1.5 to 19 μM. The substitution on the isothiazolone ring affects activity as well. By introducing large groups there is steric hindrance, which affects the nucleophilic attack on the isothiazol ring and decreases the activity to PCAF HAT. Based on the parent structure of isothiazolone and isothiazolidinone rings, Furdas [71] designed a scaffold of novel pyridoisothiazolones that showed low micromolar potency to PCAF. Oxidation of the sulfur atom to sulfoxide or sulfone significantly reduced the inhibitory activity, which emphasizes the importance of the free active S-N bond. Overall, substituted isothiazolone molecules for HAT are attracting more and more attention. Given their irreversible covalent nature of inhibition, attention to nonselectivity and off-target effects should be carefully given.

14.4.2.8 γ-butyrolactone

Biel et al. [139] reported small molecule compounds containing a scaffold of α-methylene-γ-butyrolactone for GCN5 inhibition. The inspiration for designing γ-butyrolactone is based on the theory that the desired structure should present a possible hydrogen-bond acceptor for the backbone amide of Cys177 and a polar group for interaction with Glu173 in GCN5. Moreover, the molecule should possess an aliphatic side chain analogous to that of lysine. The best hit, MB-3 (entry **22**), inhibited the GCN5 activity with IC_{50} value of 100 μM, which is comparable to that of the natural substrate H3 and showed selectivity to GCN5 compared with CBP ($IC_{50} = 500\,\mu M$). This compound provides a promising starting point for further study of structure−activity relationships and exploring new inhibitors to GCN5.

14.4.2.9 C646

Cole and coworkers [140] conducted a virtual ligand screening based on the Lys-CoA/p300-binding structure. After several steps of screening and validation, three distinct hits were finally found. The most potent hit was C646 (entry **23**), which contains a pyrazolone conjugated to a terminal benzoic acid. The compound showed high potency for p300 HAT (K_i of 400 nM) and more than eight-fold selectivity compared with other HATs including PCAF, GCN5, and MYST families. C646 was proved to be a reversible competitive inhibitor of p300 HAT. SAR studies showed that the free acid group ($-COOH$ or $-SO_3H$) plays an important role in the interaction between C646 and p300 for the carboxyl acid methylated derivative largely loses activity. Another derivative in which the double bond was reduced did not show detectable inhibition activity, which means that keeping the planarity and maintaining the conjugation system is critical. Finally substituting the $-NO_2$ to $-CH_2OH$ group also reduced the activity heavily, which showed the importance of the nitro group that could form hydrogen bonds with p300. Studies using C646 as a molecular tool for p300 inhibition show that the compound has antitumor efficacy in several cancer types, including leukemia and prostate cancer [141,142].

14.4.2.10 (Iso)thiazoles

Aiming to search for a Tip60 inhibitor for prostate cancer therapy, Coffey et al. [143] screened a series of small molecule compounds containing a disulfide bond. Out of the compounds screened, compound NU9056 (entry **24**) with a skeleton of isothiazole was identified to have IC_{50} of 2 μM. Research showed that both the disulfides bond and the structure of side rings proved to be crucial to the inhibition of Tip60. This information could guide further inhibitor design. In biological studies, NU9056 inhibited proliferation of prostate cancer cell lines and induced apoptosis via activation of caspase-3 and -9 in a concentration and time-dependent manner. Also, decreased protein levels of AR, prostate-specific antigen, p53 and p21, were observed in response to the treatment with NU9056. Furthermore, pretreatment with NU9056 could inhibit both ATM kinase (a substrate of Tip60) phosphorylation and Tip60 stabilization in response to IR (ionizing radiation). Based on the activity of NU9056 and its specificity toward Tip60, it could be a potential auxiliary therapeutic agent for prostate cancer.

We performed a virtual screening by using the crystal structure of Esa1 (the yeast homolog of Tip60) on a small molecule library database with the aim to discover Tip60 inhibitors with novel scaffolds [144]. Three thiazole-containing compounds (one structure is shown in entry **25**) were identified with micromolar inhibition potency for HATs from the biochemical studies. Computer modeling and kinetic assays suggest that these molecules target the acetyl-CoA binding site in Tip60. Further structural modification may improve the potency.

14.4.2.11 L002

Liao et al. [145] performed a high-throughput screening of a library of compounds targeting p300 HAT, in search for a therapeutic compound against the triple-negative breast cancer. A small molecule inhibitor (compound L002, entry **26**) for p300 was discovered with a scaffold of iminoquinone and methyoxyphenyl rings connected by a sulfonyl linker. The IC_{50} value for p300 was measured to be 1.98 μM using the fluorescent-coupled assay, whereas the IC_{50} obtained from the radioactive filter binding was 128 μM. The compound was smoothly docked into the

AcCoA-binding pocket of p300/CBP. The authors showed that L002-inhibited acetylation of histones and p53, and suppressed STAT3 activation. This compound suppressed tumor growth and histone acetylation in MDA-MB-468 xenograft animal model and decreased growth in leukemia, lymphoma, and breast cancer cell lines.

14.4.2.12 TH1834

Recently, a Tip60 inhibitor TH1834 (entry **27**) was reported [146]. The author developed a Tip60 model portraying the active-binding pocket as a positive electrostatic surface potential on one end and negative on the other end. Structure-based drug design was then used to develop inhibitors to fit this specific pocket. Based on this model, they synthesized TH1834 possessing two oppositely charged ends to match the electrostatic surface potential of the Tip60 pocket. The *in vitro* activity of TH1834 seemed to be weak, with an IC_{50} around 500 μM. The compound did not seem to affect the activity of MOF, a related MYST member. In cellular assays, TH1834 caused increased DNA double-strand break following ionizing radiation treatment in several cancer cell lines (MCF7, PC-3, and DU-145), resulting in apoptosis.

14.5 CONCLUSION AND PERSPECTIVE

Since the identification of the first set of HAT and HDAC enzymes in the mid-1990s, there has been significant development in our understanding of the mechanism, structure, and function of lysine acetylation. Discovery of HAT modulators as tool probes is a continuing voyage. Nevertheless, our current knowledge of HAT chemistry and biology is still quite incomplete and there is much that we do not understand which will require vigorous future studies.

14.5.1 HAT ENZYME DISCOVERY

Bioinformatic analysis suggests that there are likely other lysine acetyltransferases existing in cells, with sequence and structure different from the currently known HATs [24,147]. Extensive biochemical studies of these proteins and genetic analysis of the corresponding genes are needed to approve or disapprove whether they are bona fide HAT enzymes *in vitro* and *in vivo*. Also, some known HATs may possess novel acyltransferase activities. For instance, p300 was discovered to catalyze not only lysine acetylation, but also lysine propionylation [148]. These novel activities are largely unexplored. On the other hand, there are many novel acylations, for example, lysine succinylation, which cannot be catalyzed by known HATs, so it is highly possible that new acyltransferases remain to be discovered. The protein members of the GNAT superfamily could be promising candidates toward this investigation. Beyond the histone realm, hundreds and thousands of acetylated proteins have been identified by mass-spectrometry-based proteomics screening in both eukaryotes and prokaryotes [17]. These acetylation reactions occur in multiple subcellular compartments such as nucleus, ribosome, and mitochondria, and are associated with diverse cellular processes from chromatin remodeling, protein translation, to mitochondrial metabolism [11,149].

Efforts will be needed to investigate the enzymes responsible for establishing particular sub-acetylomes and biological consequences of the individual acetylation pathways [150]. Elucidation of the matching relationship of individual acetyltransferases and their corresponding acetylomes is also closely aligned with the key question of how the substrate specificity of different HAT enzymes is regulated in the cell.

14.5.2 CHALLENGES IN HAT ASSAYS

Several assay formats are available for characterizing HAT activities. However, there seems to be no ideal assay method suited for high-throughput screening of HAT inhibitors. The CPM-coupled fluorogenic assay, although it is straightforward and does not require any washing-steps, generates a fluorescent signal in the blue light range, which is readily interfered with by autofluorescent compounds producing false negatives. The criteria to consider in developing assays for HAT inhibitor screening and characterization are multi-fold. A facile assay should be robust, reproducible, and implementable in 96-well plate or higher density plate with superior performance. High signal stability and high signal-to-background ratios are desired, which is typically manifested from the assay parameters Z and Z′. The assay protocol should be simple enough for automated handling. Ideally, one wishes to conduct an assay which is a homogeneous, continuous, single-step measurement of HAT activity. "Mix-and-read" approaches without need of time-consuming separation and wash-steps are highly desired. Therefore, current assays need to be improved and new assay formats need to be invented, in order to meet the demand of HAT inhibitor discovery on large scales.

14.5.3 DEVELOPMENT OF HAT MODULATORS

Although HAT and HDAC proteins were discovered at around the same time in the 1990s, development of HAT inhibitors has progressed much slower than that of HDAC inhibitors. Among the HAT inhibitors reported so far, natural products and bisubstrate inhibitors dominate as the most potent compounds. Drawbacks exist in terms of low potency, low selectivity, low solubility, poor cell permeability, metabolic instability, and multitarget actions, etc. Improvement in potency and selectivity are particularly required. Because different assay systems were adopted, often there is inconsistency about IC_{50} data for reported HAT inhibitors. Most HAT inhibitors were discovered and characterized by *in vitro* assays, but *in vivo* target validation is of great necessity. It is critical to determine whether a cellular phenotype response is caused by HAT inhibition or by off-target effects. The mechanism of HAT activation by certain small molecules is very interesting and worth further investigation. The majority of currently reported small molecule HAT inhibitors targets p300/CBP rather than PCAF/GCN5 and MYST HATs. The reason is unclear. Perhaps, p300/CBP has pockets on its surface that are more amenable to target by small molecules. Because HAT inhibitors reported to date have relatively modest potency and selectivity profiles. It is also important to obtain structural information of HAT/inhibitor complexes, which will facilitate the development of more potent and selective inhibitors with possible therapeutic applications.

ACKNOWLEDGMENTS

We acknowledge the National Institutes of Health and the American Heart Association for funding our work. We apologize to those colleagues whose work cannot be cited due to the limited space.

REFERENCES

[1] Sterner DE, Berger SL. Acetylation of histones and transcription-related factors. Microbiol Mol Biol Rev 2000;64(2):435−59.

[2] Allfrey VG, Faulkner R, Mirsky AE. Acetylation and methylation of histones and their possible role in the regulation of RNA synthesis. Proc Natl Acad Sci USA 1964;51:786−94.

[3] Bannister AJM, Miska EA. Regulation of gene expression by transcription factor acetylation. Cell Mol Life Sci 2000;57:1184−92.

[4] Roth SY, Denu JM, Allis CD. Histone acetyltransferases. Annu Rev Biochem 2001;70:81−120.

[5] Kouzarides T. Chromatin modifications and their function. Cell 2007;128:693−705.

[6] Ruthenburg AJ, Li H, Patel DJ, Allis CD. Multivalent engagement of chromatin modifications by linked binding modules. Nat Rev Mol Cell Biol 2007;8(12):983−94.

[7] Zeng L, Zhou MM. Bromodomain: an acetyl-lysine binding domain. FEBS Lett 2002;513(1):124−8.

[8] Marmorstein R, Zhou MM. Writers and readers of histone acetylation: structure, mechanism, and inhibition. Cold Spring Harb Perspect Biol 2014;6(7):a018762.

[9] Feng Y, Wang J, Asher S, Hoang L, Guardiani C, Ivanov I, et al. Histone H4 acetylation differentially modulates arginine methylation by an in Cis mechanism. J Biol Chem 2011;286(23):20323−34.

[10] Margueron R, Trojer P, Reinberg D. The key to development: interpreting the histone code? Curr Opin Genet Dev 2005;15(2):163−76.

[11] Spange S, Wagner T, Heinzel T, Kramer OH. Acetylation of non-histone proteins modulates cellular signalling at multiple levels. Int J Biochem Cell Biol 2009;41(1):185−98.

[12] Polevoda B, Sherman F. The diversity of acetylated proteins. Genome Biol 2002;3(5): reviews0006.

[13] Nallamilli BR, Edelmann MJ, Zhong X, Tan F, Mujahid H, Zhang J, et al. Global analysis of lysine acetylation suggests the involvement of protein acetylation in diverse biological processes in rice (*Oryza sativa*). PLoS One 2014;9(2):e89283.

[14] Li T, Du Y, Wang L, Huang L, Li W, Lu M, et al. Characterization and prediction of lysine (k)-acetyltransferase specific acetylation sites. Mol Cell Proteomics 2012;11(1): M111.011080.

[15] Zhang J, Sprung R, Pei J, Tan X, Kim S, Zhu H, et al. Lysine acetylation is a highly abundant and evolutionarily conserved modification in *Escherichia coli*. Mol Cell Proteomics 2009;8(2):215−25.

[16] Basu A, Rose KL, Zhang J, Beavis RC, Ueberheide B, Garcia BA, et al. Proteome-wide prediction of acetylation substrates. Proc Natl Acad Sci USA 2009;106(33):13785−90.

[17] Choudhary C, Kumar C, Gnad F, Nielsen ML, Rehman M, Walther TC, et al. Lysine acetylation targets protein complexes and co-regulates major cellular functions. Science 2009;325(5942):834−40.

[18] Starai VJ, Celic I, Cole RN, Boeke JD, Escalante-Semerena JC. Sir2-dependent activation of acetyl-CoA synthetase by deacetylation of active lysine. Science 2002;298(5602):2390−2.

[19] Zhao S, Xu W, Jiang W, Yu W, Lin Y, Zhang T, et al. Regulation of cellular metabolism by protein lysine acetylation. Science 2010;327(5968):1000−4.

[20] Patel J, Pathak RR, Mujtaba S. The biology of lysine acetylation integrates transcriptional programming and metabolism. Nutr Metab 2011;8:12.

[21] Wellen KE, Hatzivassiliou G, Sachdeva UM, Bui TV, Cross JR, Thompson CB. ATP-citrate lyase links cellular metabolism to histone acetylation. Science 2009;324(5930):1076–80.

[22] Vetting MW, S de Carvalho LP, Yu M, Hegde SS, Magnet S, Roderick SL, et al. Structure and functions of the GNAT superfamily of acetyltransferases. Arch Biochem Biophys 2005;433(1):212–26.

[23] Angus-Hill ML, Dutnall RN, Tafrov ST, Sternglanz R, Ramakrishnan V. Crystal structure of the histone acetyltransferase Hpa2: a tetrameric member of the Gcn5-related N-acetyltransferase superfamily. J Mol Biol 1999;294(5):1311–25.

[24] Karmodiya K, Anamika K, Muley V, Pradhan SJ, Bhide Y, Galande S. Camello, a novel family of histone acetyltransferases that acetylate histone H4 and is essential for zebrafish development. Sci Rep 2014;4:6076.

[25] Yang XJ. The diverse superfamily of lysine acetyltransferases and their roles in leukemia and other diseases. Nucleic Acids Res 2004;32(3):959–76.

[26] Heery DM, Fischer PM. Pharmacological targeting of lysine acetyltransferases in human disease: a progress report. Drug Discov Today 2007;12(1–2):88–99.

[27] Stiehl DP, Fath DM, Liang D, Jiang Y, Sang N. Histone deacetylase inhibitors synergize p300 autoacetylation that regulates its transactivation activity and complex formation. Cancer Res 2007;67(5):2256–64.

[28] Giles RH, Peters DJ, Breuning MH. Conjunction dysfunction: CBP/p300 in human disease. Trends Genet 1998;14(5):178–83.

[29] Rowley JD, Reshmi S, Sobulo O, Musvee T, Anastasi J, Raimondi S, et al. All patients with the T(11;16)(q23;p13.3) that involves MLL and CBP have treatment-related hematologic disorders. Blood 1997;90(2):535–41.

[30] Kwon HS, Kim DR, Yang EG, Park YK, Ahn HC, Min SJ, et al. Inhibition of VEGF transcription through blockade of the hypoxia inducible factor-1alpha-p300 interaction by a small molecule. Bioorg Med Chem Lett 2012;22(16):5249–52.

[31] Yin S, Kaluz S, Devi NS, Jabbar AA, de Noronha RG, Mun J, et al. Arylsulfonamide KCN1 inhibits *in vivo* glioma growth and interferes with HIF signaling by disrupting HIF-1alpha interaction with cofactors p300/CBP. Clin Cancer Res 2012;18(24):6623–33.

[32] Dekker FJ, Haisma HJ. Histone acetyl transferases as emerging drug targets. Drug Discov Today 2009; 14(19–20):942–8.

[33] Cheung P, Tanner KG, Cheung WL, Sassone-Corsi P, Denu JM, Allis CD. Synergistic coupling of histone H3 phosphorylation and acetylation in response to epidermal growth factor stimulation. Mol Cell 2000;5(6):905–15.

[34] Howe L, Auston D, Grant P, John S, Cook RG, Workman JL, et al. Histone H3 specific acetyltransferases are essential for cell cycle progression. Genes Dev 2001;15(23):3144–54.

[35] Kikuchi H, Takami Y, Nakayama T. GCN5: a supervisor in all-inclusive control of vertebrate cell cycle progression through transcription regulation of various cell cycle-related genes. Gene 2005;347(1):83–97.

[36] Pérez-Luna M, Aguasca M, Perearnau A, Serratosa J, Martínez-Balbas M, Jesús Pujol M, et al. PCAF regulates the stability of the transcriptional regulator and cyclin-dependent kinase inhibitor p27 Kip1. Nucleic Acids Res 2012;40(14):6520–33.

[37] Toth M, Boros IM, Balint E. Elevated level of lysine 9-acetylated histone H3 at the MDR1 promoter in multidrug-resistant cells. Cancer Sci 2012;103(4):659–69.

[38] Yamamoto T, Horikoshi M. Novel substrate specificity of the histone acetyltransferase activity of HIV-1-Tat interactive protein Tip60. J Biol Chem 1997;272(49):30595–8.

[39] Tang Y, Luo J, Zhang W, Gu W. Tip60-dependent acetylation of p53 modulates the decision between cell-cycle arrest and apoptosis. Mol Cell 2006;24(6):827–39.

[40] Shiota M, Yokomizo A, Masubuchi D, Tada Y, Inokuchi J, Eto M, et al. Tip60 promotes prostate cancer cell proliferation by translocation of androgen receptor into the nucleus. Prostate 2010;70(5):540–54.

[41] Halkidou K, Gnanapragasam VJ, Mehta PB, Logan IR, Brady ME, Cook S, et al. Expression of Tip60, an androgen receptor coactivator, and its role in prostate cancer development. Oncogene 2003;22(16): 2466−77.

[42] Avvakumov N, Cote J. The MYST family of histone acetyltransferases and their intimate links to cancer. Oncogene 2007;26(37):5395−407.

[43] Squatrito M, Gorrini C, Amati B. Tip60 in DNA damage response and growth control: many tricks in one HAT. Trends Cell Biol 2006;16(9):433−42.

[44] Iizuka M, Takahashi Y, Mizzen CA, Cook RG, Fujita M, Allis CD, et al. Histone acetyltransferase Hbo1: catalytic activity, cellular abundance, and links to primary cancers. Gene 2009;436(1−2):108−14.

[45] Iizuka M, Stillman B. Histone acetyltransferase HBO1 interacts with the ORC1 subunit of the human initiator protein. J Biol Chem 1999;274(33):23027−34.

[46] Burke TW, Cook JG, Asano M, Nevins JR. Replication factors MCM2 and ORC1 interact with the histone acetyltransferase HBO1. J Biol Chem 2001;276(18):15397−408.

[47] Sharma M, Zarnegar M, Li X, Lim B, Sun Z. Androgen receptor interacts with a novel MYST protein, HBO1. J Biol Chem 2000;275(45):35200−8.

[48] Georgiakaki M, Chabbert-Buffet N, Dasen B, Meduri G, Wenk S, Rajhi L, et al. Ligand-controlled interaction of histone acetyltransferase binding to ORC-1 (HBO1) with the N-terminal transactivating domain of progesterone receptor induces steroid receptor coactivator 1-dependent coactivation of transcription. Mol Endocrinol 2006;20(9):2122−40.

[49] Foy RL, Song IY, Chitalia VC, Cohen HT, Saksouk N, Cayrou C, et al. Role of Jade-1 in the histone acetyltransferase (HAT) HBO1 complex. J Biol Chem 2008;283(43):28817−26.

[50] Doyon Y, Cayrou C, Ullah M, Landry AJ, Côté V, Selleck W, et al. ING tumor suppressor proteins are critical regulators of chromatin acetylation required for genome expression and perpetuation. Mol Cell 2006;21(1):51−64.

[51] Iizuka M, Sarmento OF, Sekiya T, Scrable H, Allis CD, Smith MM. Hbo1 Links p53-dependent stress signaling to DNA replication licensing. Mol Cell Biol 2008;28(1):140−53.

[52] Miotto B, Struhl K. HBO1 histone acetylase activity is essential for DNA replication licensing and inhibited by Geminin. Mol Cell 2010;37(1):57−66.

[53] Zong H, Li Z, Liu L, Hong Y, Yun X, Jiang J, et al. Cyclin-dependent kinase 11(p58) interacts with HBO1 and enhances its histone acetyltransferase activity. FEBS Lett 2005;579(17):3579−88.

[54] Hu X, Stern HM, Ge L, O'Brien C, Haydu L, Honchell CD, et al. Genetic alterations and oncogenic pathways associated with breast cancer subtypes. Mol Cancer Res 2009;7(4):511−22.

[55] Iizuka M, Susa T, Takahashi Y, Tamamori-Adachi M, Kajitani T, Okinaga H, et al. Histone acetyltransferase Hbo1 destabilizes estrogen receptor alpha by ubiquitination and modulates proliferation of breast cancers. Cancer Sci 2013;104(12):1647−55.

[56] Contzler R, Regamey A, Favre B, Roger T, Hohl D, Huber M. Histone acetyltransferase HBO1 inhibits NF-kappaB activity by coactivator sequestration. Biochem Biophys Res Commun 2006;350(1):208−13.

[57] Berndsen CE, Denu JM. Assays for mechanistic investigations of protein/histone acetyltransferases. Methods 2005;36(4):321−31.

[58] Wynne Aherne G, Rowlands MG, Stimson L, Workman P. Assays for the identification and evaluation of histone acetyltransferase inhibitors. Methods 2002;26(3):245−53.

[59] Wegener D, Hildmann C, Schwienhorst A. Recent progress in the development of assays suited for histone deacetylase inhibitor screening. Mol Genet Metab 2003;80(1−2):138−47.

[60] Aherne GW, Rowlands MG, Stimson L, Workman P. Assays for the identification and evaluation of histone acetyltransferase inhibitors. Methods 2002;26(3):245−53.

[61] Wiley EA, Mizzen CA, Allis CD. Isolation and characterization of in vivo modified histones and an activity gel assay for identification of histone acetyltransferases. Methods Cell Biol 2000;62:379−94.

[62] Mizzen CA, Brownell JE, Cook RG, Allis CD. Histone acetyltransferases: preparation of substrates and assay procedures. Methods Enzymol 1999;304:675–96.

[63] Benson LJ, Annunziato AT. In vitro analysis of histone acetyltransferase activity. Methods 2004;33(1): 45–52.

[64] Kim Y, Tanner KG, Denu JM. A continuous, nonradioactive assay for histone acetyltransferases. Anal Biochem 2000;280(2):308–14.

[65] Yang C, Wu J, Sinha SH, Neveu JM, Zheng YG. Autoacetylation of the MYST Lysine Acetyltransferase MOF. J Biol Chem 2012;287:34917–26.

[66] Wu J, Xie N, Wu Z, Zhang Y, Zheng YG. Bisubstrate Inhibitors of the MYST HATs Esa1 and Tip60. Bioorg Med Chem 2009;17(3):1381–6.

[67] Trievel RC, Li FY, Marmorstein R. Application of a fluorescent histone acetyltransferase assay to probe the substrate specificity of the human p300/CBP-associated factor. Anal Biochem 2000;287(2): 319–28.

[68] Thompson PR, Wang D, Wang L, Fulco M, Pediconi N, Zhang D, et al. Regulation of the p300 HAT domain via a novel activation loop. Nat Struct Mol Biol 2004;11(4):308–15.

[69] Gao T, Yang C, Zheng YG. Comparative studies of thiol-sensitive fluorogenic probes for HAT assays. Anal Bioanal Chem 2013;405(4):1361–71.

[70] Kuninger D, Lundblad J, Semirale A, Rotwein P. A non-isotopic in vitro assay for histone acetylation. J Biotechnol 2007;131(3):253–60.

[71] Furdas SD, Shekfeh S, Bissinger EM, Wagner JM, Schlimme S, Valkov V, et al. Synthesis and biological testing of novel pyridoisothiazolones as histone acetyltransferase inhibitors. Biorg Med Chem 2011; 19(12):3678–89.

[72] Eliseeva ED, Valkov V, Jung M, Jung MO. Characterization of novel inhibitors of histone acetyltransferases. Mol Cancer Ther 2007;6(9):2391–8.

[73] Koeller KM, Haggarty SJ, Perkins BD, Leykin I, Wong JC, Kao MC, et al. Chemical genetic modifier screens: small molecule trichostatin suppressors as probes of intracellular histone and tubulin acetylation. Chem Biol 2003;10(5):397–410.

[74] Chen CC, Tyler J. Selective sensitization of cancer cells to DNA damage by a HAT inhibitor. Cell Cycle 2009;8(18):2867.

[75] Suzuki T, Miyata N. Epigenetic control using natural products and synthetic molecules. Curr Med Chem 2006;13(8):935–58.

[76] Biel M, Wascholowski V, Giannis A. Epigenetics – an epicenter of gene regulation: histones and histone-modifying enzymes. Angew Chem Int Ed Engl 2005;44(21):3186–216.

[77] Zheng YG, Wu J, Chen Z, Goodman M. Chemical regulation of epigenetic modifications: opportunities for new cancer therapy. Med Res Rev 2008;28(5):645–87.

[78] Furdas SD, Kannan S, Sippl W, Jung M. Small molecule inhibitors of histone acetyltransferases as epigenetic tools and drug candidates. Arch Pharm (Weinheim) 2012;345(1):7–21.

[79] Selvi BR, Mohankrishna DV, Ostwal YB, Kundu TK. Small molecule modulators of histone acetylation and methylation: a disease perspective. Biochim Biophys Acta 2010;1799(10–12):810–28.

[80] Berndsen CE, Albaugh BN, Tan S, Denu JM. Catalytic mechanism of a MYST family histone acetyltransferase. Biochemistry 2007;46(3):623–9.

[81] Yan Y, Harper S, Speicher DW, Marmorstein R. The catalytic mechanism of the ESA1 histone acetyltransferase involves a self-acetylated intermediate. Nat Struct Biol 2002;9(11):862–9.

[82] Yu M, Magalhaes ML, Cook PF, Blanchard JS. Bisubstrate inhibition: theory and application to N-acetyltransferases. Biochemistry 2006;45(49):14788–94.

[83] Chase JF, Tubbs PK. Conditions for the self-catalysed inactivation of carnitine acetyltransferase. A novel form of enzyme inhibition. Biochem J 1969;111(2):225–35.

[84] Cullis PM, Wolfenden R, Cousens LS, Alberts BM. Inhibition of histone acetylation by N-[2-(S-coenzyme A) acetyl] spermidine amide, a multisubstrate analog. J Biol Chem 1982;257(20):12165−9.

[85] Erwin BG, Persson L, Pegg AE. Differential inhibition of histone and polyamine acetylases by multi-substrate analogues. Biochemistry 1984;23(18):4250−5.

[86] Khalil EM, Cole PA. A potent inhibitor of the melatonin rhythm enzyme. J Am Chem Soc 1998;120(24):6195−6.

[87] Kim CM, Cole PA. Bisubstrate ketone analogues as serotonin N-acetyltransferase inhibitors. J Med Chem 2001;44(15):2479−85.

[88] Williams JW, Northrop DB. Synthesis of a tight-binding, multisubstrate analog inhibitor of gentamicin acetyltransferase I. J Antibiot (Tokyo) 1979;32(11):1147−54.

[89] Magalhaes ML, Vetting MW, Gao F, Freiburger L, Auclair K, Blanchard JS. Kinetic and structural analysis of bisubstrate inhibition of the *Salmonella enterica* aminoglycoside 6′-N-acetyltransferase. Biochemistry 2008;47(2):579−84.

[90] Gao F, Yan XX, Baettig OM, Berghuis AM, Auclair K. Regio- and chemoselective 6′-N-derivatization of aminoglycosides: bisubstrate inhibitors as probes to study aminoglycoside 6′-N-acetyltransferases. Angew Chem Int Ed Engl 2005;44(42):6859−62.

[91] Lau OD, Kundu TK, Soccio RE, Ait-Si-Ali S, Khalil EM, Vassilev A, et al. HATs off: selective synthetic inhibitors of the histone acetyltransferases p300 and PCAF. Mol Cell 2000;5(3):589−95.

[92] Schiltz RL, Mizzen CA, Vassilev A, Cook RG, Allis CD, Nakatani Y. Overlapping but distinct patterns of histone acetylation by the human coactivators p300 and PCAF within nucleosomal substrates. J Biol Chem 1999;274(3):1189−92.

[93] Karukurichi KR, Cole PA. Probing the reaction coordinate of the p300/CBP histone acetyltransferase with bisubstrate analogs. Bioorg Chem 2011;39(1):42−7.

[94] Kwie FH, Briet M, Soupaya D, Hoffmann P, Maturano M, Rodriguez F, et al. New potent bisubstrate inhibitors of histone acetyltransferase p300: design, synthesis and biological evaluation. Chem Biol Drug Des 2011;77(1):86−92.

[95] Sagar V, Zheng W, Thompson PR, Cole PA. Bisubstrate analogue structure-activity relationships for p300 histone acetyltransferase inhibitors. Biorg Med Chem 2004;12(12):3383−90.

[96] Castellano S, Milite C, Feoli A, Viviano M, Mai A, Novellino E, et al. Identification of structural features of 2-alkylidene-1,3-dicarbonyl derivatives that induce inhibition and/or activation of histone acetyltransferases KAT3B/p300 and KAT2B/PCAF. ChemMedChem 2015;10(1):144−57.

[97] Wadia JS, Dowdy SF. Transmembrane delivery of protein and peptide drugs by TAT-mediated trans-duction in the treatment of cancer. Adv Drug Deliv Rev 2005;57(4):579−96.

[98] Fuchs SM, Raines RT. Polyarginine as a multifunctional fusion tag. Protein Sci 2005;14(6):1538−44.

[99] Zheng Y, Balasubramanyam K, Cebrat M, Buck D, Guidez F, Zelent A, et al. Synthesis and evaluation of a potent and selective cell-permeable p300 histone acetyltransferase inhibitor. J Am Chem Soc 2005;127(49):17182−3.

[100] Bandyopadhyay K, Baneres JL, Martin A, Blonski C, Parello J, Gjerset RA. Spermidinyl-CoA-based HAT inhibitors block DNA repair and provide cancer-specific chemo- and radiosensitization. Cell Cycle 2009;8(17):2779−88.

[101] Cebrat M, Kim CM, Thompson PR, Daugherty M, Cole PA. Synthesis and analysis of potential prodrugs of coenzyme A analogues for the inhibition of the histone acetyltransferase p300. Biorg Med Chem 2003;11(15):3307−13.

[102] Yang C, Ngo L, Zheng YG. Rational design of substrate-based multivalent inhibitors of the histone acetyltransferase Tip60. ChemMedChem 2014;9(3):537−41.

[103] Zheng Y, Thompson PR, Cebrat M, Wang L, Devlin MK, Alani RM, et al. Selective HAT inhibitors as mechanistic tools for protein acetylation. Methods Enzymol 2004;376:188−99.

[104] Poux AN, Cebrat M, Kim CM, Cole PA, Marmorstein R. Structure of the GCN5 histone acetyltransferase bound to a bisubstrate inhibitor. Proc Natl Acad Sci USA 2002;99(22):14065—70.

[105] Liu X, Wang L, Zhao K, Thompson PR, Hwang Y, Marmorstein R, et al. The structural basis of protein acetylation by the p300/CBP transcriptional coactivator. Nature 2008;451(7180):846—50.

[106] Wang L, Tang Y, Cole PA, Marmorstein R. Structure and chemistry of the p300/CBP and Rtt109 histone acetyltransferases: implications for histone acetyltransferase evolution and function. Curr Opin Struct Biol 2008;18(6):741—7.

[107] Balasubramanyam K, Swaminathan V, Ranganathan A, Kundu TK. Small molecule modulators of histone acetyltransferase p300. J Biol Chem 2003;278(21):19134—40.

[108] Ghizzoni M, Wu J, Gao T, Haisma HJ, Dekker FJ, George Zheng Y. 6-alkylsalicylates are selective Tip60 inhibitors and target the acetyl-CoA binding site. Eur J Med Chem 2012;47(1):337—44.

[109] Hemshekhar MS, Kemparaju K, Girish KS. Emerging roles of anacardic acid and its derivatives: a pharmacological overview. Basic Clin Pharmacol Toxicol 2011;110:122—32.

[110] Souto JA, Conte M, Alvarez R, Nebbioso A, Carafa V, Altucci L, et al. Synthesis of benzamides related to anacardic acid and their histone acetyltransferase (HAT) inhibitory activities. ChemMedChem 2008;3(9): 1435—42.

[111] Sbardella G, Castellano S, Vicidomini C, Rotili D, Nebbioso A, Miceli M, et al. Identification of long chain alkylidenemalonates as novel small molecule modulators of histone acetyltransferases. Bioorg Med Chem Lett 2008;18(9):2788—92.

[112] Milite C, Castellano S, Benedetti R, Tosco A, Ciliberti C, Vicidomini C, et al. Modulation of the activity of histone acetyltransferases by long chain alkylidenemalonates (LoCAMs). Bioorg Med Chem 2011;19(12):3690—701.

[113] Park WJ, Ma E. Inhibition of PCAF histone acetyltransferase and cytotoxic effect of N-acylanthranilic acids. Arch Pharm Res 2012;35(8):1379—86.

[114] Mai A, Rotili D, Tarantino D, Ornaghi P, Tosi F, Vicidomini C, et al. Small-molecule inhibitors of histone acetyltransferase activity: identification and biological properties. J Med Chem 2006;49(23): 6897—907.

[115] Ghizzoni M, Boltjes A, Graaf C, Haisma HJ, Dekker FJ. Improved inhibition of the histone acetyltransferase PCAF by an anacardic acid derivative. Bioorg Med Chem 2010;18(16):5826—34.

[116] Modak R, Basha J, Bharathy N, Maity K, Mizar P, Bhat AV, et al. Probing p300/CBP associated factor (PCAF)-dependent pathways with a small molecule inhibitor. ACS Chem Biol 2013;8(6): 1311—23.

[117] Mai A, Rotili D, Tarantino D, Nebbioso A, Castellano S, Sbardella G, et al. Identification of 4-hydroxyquinolines inhibitors of p300/CBP histone acetyltransferases. Bioorg Med Chem Lett 2009; 19(4):1132—5.

[118] Lenoci A, Tomassi S, Conte M, Benedetti R, Rodriguez V, Carradori S, et al. Quinoline-based p300 histone acetyltransferase inhibitors with pro-apoptotic activity in human leukemia U937 cells. ChemMedChem 2014;9(3):542—8.

[119] Sarli V, Giannis A. Selective inhibition of CBP/p300 HAT. Chem Biol 2007;14(6):605—6.

[120] Balasubramanyam K, Altaf M, Varier RA, Swaminathan V, Ravindran A, Sadhale PP, et al. Polyisoprenylated benzophenone, garcinol, a natural histone acetyltransferase inhibitor, represses chromatin transcription and alters global gene expression. J Biol Chem 2004;279(32):33716—26.

[121] Varier RA, Swaminathan V, Balasubramanyam K, Kundu TK. Implications of small molecule activators and inhibitors of histone acetyltransferases in chromatin therapy. Biochem Pharmacol 2004;68(6): 1215—20.

[122] Arif M, Pradhan SK, Thanuja GR, Vedamurthy BM, Agrawal S, Dasgupta D, et al. Mechanism of p300 specific histone acetyltransferase inhibition by small molecules. J Med Chem 2009;52:267—77.

[123] Choi KC, Jung MG, Lee YH, Yoon JC, Kwon SH, Kang HB, et al. Epigallocatechin-3-gallate, a histone acetyltransferase inhibitor, inhibits EBV-induced B lymphocyte transformation via suppression of RelA acetylation. Cancer Res 2009;69(2):583−92.

[124] Fang MZ, Wang Y, Ai N, Hou Z, Sun Y, Lu H, et al. Tea polyphenol (−) − epigallocatechin-3-gallate inhibits DNA methyltransferase and reactivates methylation-silenced genes in cancer cell lines. Cancer Res 2003;63(22):7563−70.

[125] Suzuki K, Yahara S, Hashimoto F, Uyeda M. Inhibitory activities of (−) − epigallocatechin-3-O-gallate against topoisomerases I and II. Biol Pharm Bull 2001;24(9):1088−90.

[126] Balasubramanyam K, Varier RA, Altaf M, Swaminathan V, Siddappa NB, Ranga U, et al. Curcumin, a novel p300/CREB-binding protein-specific inhibitor of acetyltransferase, represses the acetylation of histone/nonhistone proteins and histone acetyltransferase-dependent chromatin transcription. J Biol Chem 2004;279(49):51163−71.

[127] Marcu MG, Jung YJ, Lee S, Chung EJ, Lee MJ, Trepel J, et al. Curcumin is an inhibitor of p300 histone acetyltransferase. Med Chem 2006;2:169−74.

[128] Devipriya B, Kumaradhas P. Molecular flexibility and the electrostatic moments of curcumin and its derivatives in the active site of p300: a theoretical charge density study. Chem Biol Interact 2013;204(3):153−65.

[129] Costi R, Di Santo R, Artico M, Miele G, Valentini P, Novellino E, et al. Cinnamoyl compounds as simple molecules that inhibit p300 histone acetyltransferase. J Med Chem 2007;50(8):1973−7.

[130] Sunagawa Y, Morimoto T, Wada H, Takaya T, Katanasaka Y, Kawamura T, et al. A natural p300-specific histone acetyltransferase inhibitor, curcumin, in addition to angiotensin-converting enzyme inhibitor, exerts beneficial effects on left ventricular systolic function after myocardial infarction in rats. Circ J 2011;75(9):2151−9.

[131] Tohyama S, Tomura A, Ikeda N, Hatano M, Odanaka J, Kubota Y, et al. Discovery and characterization of NK13650s, naturally occurring p300-selective histone acetyltransferase inhibitors. J Org Chem 2012;77(20):9044−52.

[132] Dalvoy Vasudevarao M, Dhanasekaran K, Selvi RB, Kundu TK. Inhibition of acetyltransferase alters different histone modifications: probed by small molecule inhibitor plumbagin. J Biochem 2012;152(5):453−62.

[133] Ravindra KC, Selvi BR, Arif M, Reddy BA, Thanuja GR, Agrawal S, et al. Inhibition of lysine acetyltransferase KAT3B/p300 activity by a naturally occurring hydroxynaphthoquinone, plumbagin. J Biol Chem 2009;284(36):24453−64.

[134] Park B, Sung B, Yadav VR, Chaturvedi MM, Aggarwal BB. Triptolide, histone acetyltransferase inhibitor, suppresses growth and chemosensitizes leukemic cells through inhibition of gene expression regulated by TNF-TNFR1-TRADD-TRAF2-NIK-TAK1-IKK pathway. Biochem Pharmacol 2011;82(9):1134−44.

[135] Seong AR, Yoo JY, Choi K, Lee MH, Lee YH, Lee J, et al. Delphinidin, a specific inhibitor of histone acetyltransferase, suppresses inflammatory signaling via prevention of NF-kappaB acetylation in fibroblast-like synoviocyte MH7A cells. Biochem Biophys Res Commun 2011;410(3):581−6.

[136] Stimson L, Rowlands MG, Newbatt YM, Smith NF, Raynaud FI, Rogers P, et al. Isothiazolones as inhibitors of PCAF and p300 histone acetyltransferase activity. Mol Cancer Ther 2005;4(10):1521−32.

[137] Dekker FJ, Ghizzoni M, van der Meer N, Wisastra R, Haisma HJ. Inhibition of the PCAF histone acetyl transferase and cell proliferation by isothiazolones. Bioorg Med Chem 2009;17(2):460−6.

[138] Gorsuch S, Bavetsias V, Rowlands MG, Aherne GW, Workman P, Jarman M, et al. Synthesis of isothiazol-3-one derivatives as inhibitors of histone acetyltransferases (HATs). Bioorg Med Chem 2009;17(2):467−74.

[139] Biel M, Kretsovali A, Karatzali E, Papamatheakis J, Giannis A. Design, synthesis, and biological evaluation of a small-molecule inhibitor of the histone acetyltransferase Gcn5. Angew Chem Int Ed Engl 2004;43(30):3974−6.

[140] Bowers EM, Yan G, Mukherjee C, Orry A, Wang L, Holbert MA, et al. Virtual ligand screening of the p300/CBP histone acetyltransferase: identification of a selective small molecule inhibitor. Chem Biol 2010;17(5):471−82.

[141] Santer FR, Höschele PP, Oh SJ, Erb HH, Bouchal J, Cavarretta IT, et al. Inhibition of the acetyltransferases p300 and CBP reveals a targetable function for p300 in the survival and invasion pathways of prostate cancer cell lines. Mol Cancer Ther 2011;10(9):1644−55.

[142] Gao XN, Lin J, Ning QY, Gao L, Yao YS, Zhou JH, et al. A histone acetyltransferase p300 inhibitor C646 induces cell cycle arrest and apoptosis selectively in AML1-ETO-positive AML cells. PLoS One 2013;8(2):e55481.

[143] Coffey K, Blackburn TJ, Cook S, Golding BT, Griffin RJ, Hardcastle IR, et al. Characterisation of a Tip60 specific inhibitor, NU9056, in prostate cancer. PLoS One 2012;7(10):e45539.

[144] Wu J, Wang J, Li M, Yang Y, Wang B, Zheng YG. Small molecule inhibitors of histone acetyltransferase Tip60. Bioorg Chem 2011;39(1):53−8.

[145] Yang H, Pinello CE, Luo J, Li D, Wang Y, Zhao LY, et al. Small-molecule inhibitors of acetyltransferase p300 identified by high-throughput screening are potent anticancer agents. Mol Cancer Ther 2013;12(5):610−20.

[146] Gao C, Bourke E, Scobie M, Famme MA, Koolmeister T, Helleday T, et al. Rational design and validation of a Tip60 histone acetyltransferase inhibitor. Sci Rep 2014;4:5372.

[147] Arrowsmith CH, Bountra C, Fish PV, Lee K, Schapira M. Epigenetic protein families: a new frontier for drug discovery. Nat Rev Drug Discov 2012;11(5):384−400.

[148] Zhang K, Chen Y, Zhang Z, Zhao Y. Identification and verification of lysine propionylation and butyrylation in yeast core histones using PTMap software. J Proteome Res 2009;8(2):900−6.

[149] Smith KT, Workman JL. Introducing the acetylome. Nat Biotechnol 2009;27(10):917−19.

[150] Yang C, Mi J, Feng Y, Ngo L, Gao T, Yan L, et al. Labeling lysine acetyltransferase substrates with engineered enzymes and functionalized cofactor surrogates. J Am Chem Soc 2013;135(21):7791−4.

IN VITRO HISTONE DEACETYLASE ACTIVITY SCREENING: MAKING A CASE FOR BETTER ASSAYS

15

Quaovi H. Sodji[1], James R. Kornacki[3], Milan Mrksich[3], and Adegboyega K. Oyelere[1,2]

[1]*School of Chemistry and Biochemistry, Georgia Institute of Technology, Atlanta, GA, USA* [2]*Parker H. Petit Institute for Bioengineering and Bioscience, Georgia Institute of Technology, Atlanta, GA, USA* [3]*Departments of Chemistry and Biomedical Engineering, Northwestern University, Evanston, IL, USA*

CHAPTER OUTLINE

15.1 INTRODUCTION

Protein acetylation and deacetylation are key regulatory mechanisms involved in the modulation of nearly 2,000 targets [1]. Such pervasive activity bolsters the role of histone acetyl transferase (HAT) and histone deacetylase (HDAC) enzymes, which mediate protein acetylation and deacetylation, respectively. There are 11 zinc-dependent HDAC isoforms subdivided into three classes (I, II, and IV), and 7 nicotinamide adenine dinucleotide (NAD +)-dependent isoforms (class III) called Sirtuins [2,3]. Aberrant HDAC activity has been linked to the pathogenesis of various illnesses including inflammatory, parasitic, neurodegenerative diseases, and cancer [2–6], prompting the development of various HDAC inhibitors (HDACi) as potential therapeutics (Figure 15.1).

Y.G. Zheng (Ed): Epigenetic Technological Applications. DOI: http://dx.doi.org/10.1016/B978-0-12-801080-8.00015-6

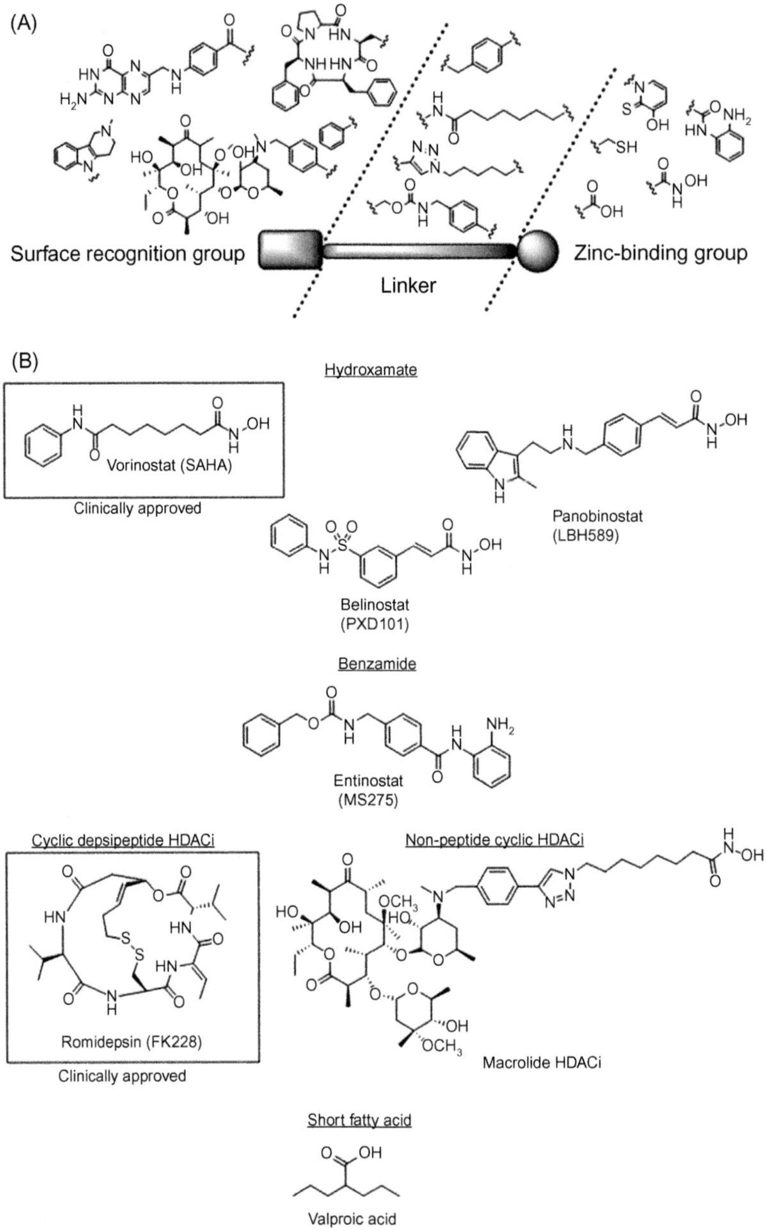

(A)

Surface recognition group   Zinc-binding group

Linker

(B)

Hydroxamate

Vorinostat (SAHA)

Clinically approved

Panobinostat
(LBH589)

Belinostat
(PXD101)

Benzamide

Entinostat
(MS275)

Cyclic depsipeptide HDACi   Non-peptide cyclic HDACi

Romidepsin (FK228)

Clinically approved

Macrolide HDACi

Short fatty acid

Valproic acid

FIGURE 15.1

HDAC inhibitors. (A) HDACi pharmacophoric model with various chemical moieties used to represent each segment. (B) Structures of selected HDACi including the FDA-approved Vorinostat (SAHA) and Romidepsin (FK228).

HDACi share structural features that interact with enzyme surface motifs and regions of catalytic importance. The pharmacophoric model for active site inhibition includes molecular structures bearing a surface recognition group that interacts with amino acids surrounding the vicinity of the HDAC channel entrance, a linker that spans the hydrophobic channel leading to the active site, and a zinc-binding group (ZBG) necessary for chelation of an active site zinc ion (Figure 15.1A) [2]. Among the multitude of HDACi potential indications, anticancer activity has been the most explored, and various small molecules have been investigated [7−9]. This endeavor led to many clinical trials culminating in FDA approvals of Vorinostat (SAHA) and Romidepsin (FK228) for the treatment of cutaneous T-cell lymphoma (CTCL) in 2006 and 2009, respectively and Panobinostat (Farydak) in 2015 for patients with multiple myeloma (Figure 15.1B) [10−15].

Although significant progress has been made in understanding the roles of the 18 human HDAC isoforms in cellular function and disease involvement, obstacles are still to be overcome. One pressing challenge is the development of an HDAC activity assay that can selectively report the activity of a specific isoform in a cellular extract or intracellular milieu [16,17].

Successfully pinpointing individual isoform activities will more clearly distinguish the mechanisms of action governing individual functional roles. Furthermore, an assay that can be automated, and thereby compatible with high-throughput screening (HTS), would be an invaluable tool in the drug discovery field as HDACi have already proved useful as drugs. Here, we discuss various assays for zinc-dependent HDACs and touch briefly on sirtuin ($NAD+$-dependent HDAC) assays. We promote substrate-matched, label-free *in vitro* assays as desirable alternatives to the commonly used fluorescence-based HDAC activity assays.

15.2 OVERVIEW OF HDAC ASSAYS

15.2.1 SIRTUIN ACTIVITY ASSAYS

Class III histone deacetylases, or sirtuins, catalyze the deacetylation of histone and nonhistone proteins using nicotinamide adenine dinucleotide (NAD^+) as a cofactor (Figure 15.2) [18,19].

Assays to measure the activity of these enzymes are based on direct or indirect detection of the products generated from the reaction. In one version of the assay, HPLC is used to separate the deacetylated product from the acetylated precursor in order to quantify their relative intensities. As an inherently label-free technique, the method does not require radio-labeling or potentially disruptive tags and is therefore safer and less prone to artifactual results. However, limitations of the HPLC-based assay include low throughput, arising from the long duration of the assay and the difficulty of automation to handle large numbers of samples, and the requirement for larger volume injection for samples with low peptide concentration [20].

Sirtuin deacetylase activities are also measured using radiolabeled substrates. During the deacetylase reaction, the acetyl group is transferred from the substrate's lysine residue onto NAD+, resulting in O-acetyl-ADP-ribose (Figure 15.2). Incubation of sirtuins with [3H]acetylated substrate results in O-[3H]acetyl-ADP-ribose whose acetyl moiety is easily hydrolyzed by treatment with NaOH and heat. Unlike the substrates and other metabolites which are sequestered by the activated charcoal, the released [3H]acetate is not, enabling its extraction from the aqueous reaction mixture with ethyl acetate followed by quantification using scintillation counting (Figure 15.3A). The quantity of the

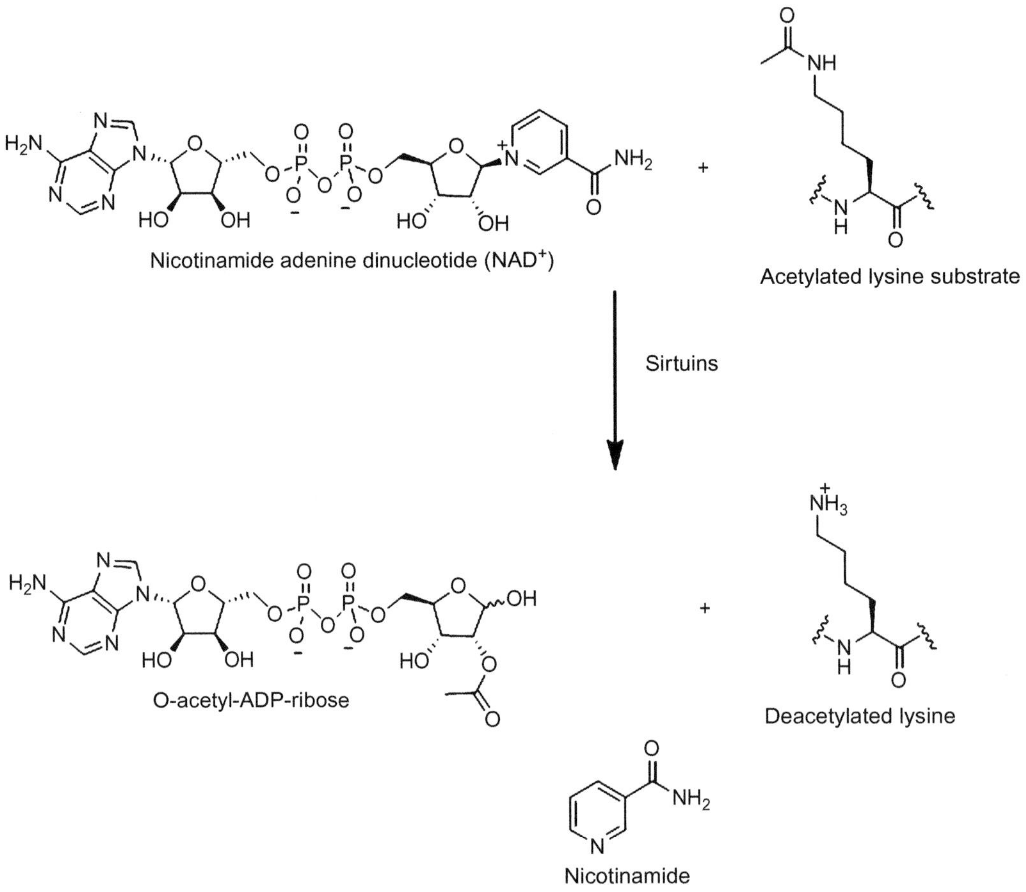

FIGURE 15.2

Schematic of sirtuins-mediated deacetylation [24].

released [3H]acetate is a direct measure of sirtuin activity [20]. Alternatively, starting with [14C] NAD+ or [32P]NAD+ as cofactor for the sirtuin reaction, the radio-labeled analogs of the other byproducts of the deacetylation — nicotinamide and O-acetyl-ADP-ribose — can also be monitored and quantified using a densitometer (Figure 15.3B) [20,21]. The use of radioisotopes is less convenient due to lab personnel safety concerns and the generation of highly toxic wastes, not to mention their high cost. A less direct assay for sirtuin activity relies on an enzyme-coupled system in which the nicotinamide generated from the deacetylation reaction is metabolized by nicotinamidase into ammonia and nicotinic acid. Glutamate dehydrogenase transfers the product ammonia to α-ketoglutarate in the presence of NAD(P)H yielding glutamate and NAD(P)+ which is monitored by absorbance at 340 nm (Figure 15.3C) [22]. We direct the readers to reviews by Landry and Sternglanz, and Schutkowski et al. for more information on sirtuin activity assays [23,24].

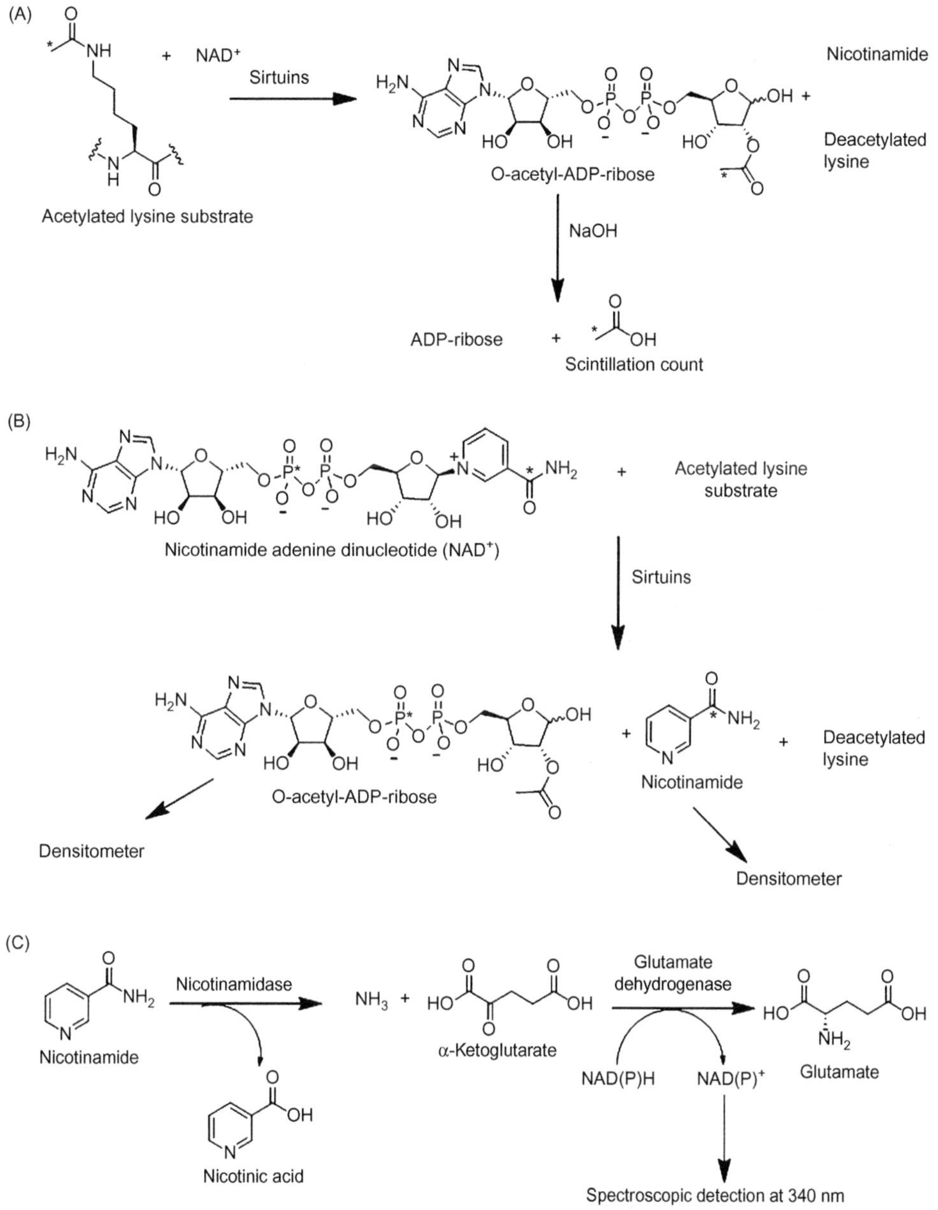

FIGURE 15.3

Schematic of selected sirtuins assays. *symbolizes radioactive labeling. (A) Charcoal-binding sirtuins assay: radioactively labeled [3H]-acetate-lysine is used and [3H]-acetate quantified via scintillation count. (B) TLC-based assay: For O-acetyl-ADP-ribose monitoring, [32P]NAD+ is used, whereas for nicotinamide [14C]NAD+ is used. (C) Enzyme-coupled detection of nicotinamide based on NH3 release by nicotinamidase [20,22].

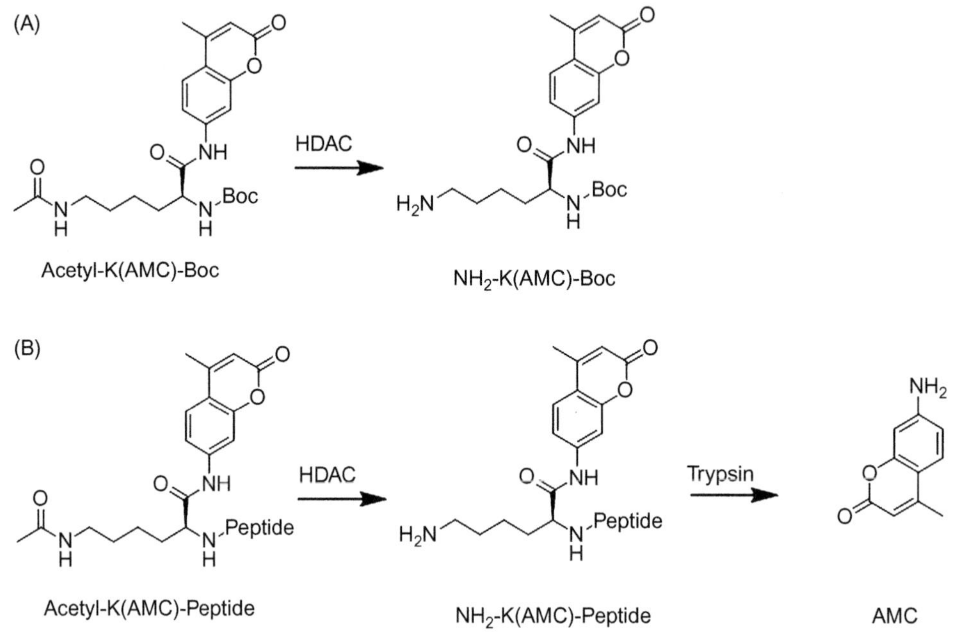

(A)

Acetyl-K(AMC)-Boc   NH$_2$-K(AMC)-Boc

(B)

Acetyl-K(AMC)-Peptide   NH$_2$-K(AMC)-Peptide   AMC

FIGURE 15.4

Schematic of the nonradioactive HDAC assay. (A) HPLC monitoring of AMC as reporter of HDAC activity: Acetyl-K(AMC)-Boc (t_{ret} = 6.99 min), NH$_2$-K(AMC)-Boc (t_{ret} = 5.54 min), flow rate = 1.2 mL/min, Multospher 100 RP-18, 250 × 4 mm, Phenomenex, acetonitrile/water (40:60), 0.01% trifluoroacetica acid [29]. (B) HDAC fluorogenic assay: fluorescence of AMC (fluorophore) as indirect reporter of HDAC activity. Step 1: incubation of the substrate (ε-acetyl-Lys(AMC)-peptide) containing a quenched fluorophore (AMC) with HDAC; Step 2: Incubation with trypsin releases the fluorophore from the peptide sequence. Free AMC has an excitation wavelength (λ_{ex}) of 390 nm and an emission (λ_{em}) of 460 nm [30].

15.2.2 ZINC-DEPENDENT HDAC ACTIVITY ASSAYS

Analogous to the Sirtuin activity assay, the original assay widely used to measure HDAC activity relied on radio-labeled acetylated histones as substrate. Following incubation of the [^{3}H]-acetyl-histone with HDAC, quantification of the released [^{3}H]-acetyl by scintillation counting correlates the HDAC activity [25]. In addition to the inconvenience of radioisotope-based assays, obtaining the radioactively labeled substrates can be labor intensive [26], and separating the released [^{3}H]-acetyl from the substrate involves an organic phase extraction, which precludes an easy transition to high-throughput capability. The development of the scintillation proximity assay has obviated the need for organic solvent extraction; however this improved protocol still relies on histone labeling and isolation of radioactive materials [27].

Immunoblotting of acetylated histone served as one of the early nonradioactive alternatives used to determine HDAC activity. Immunoblotting directly measures HDAC activities with an antibody that recognizes a particular deactetylated lysine target of a protein substrate. Deacetylation activity is then quantified by the signal generated from a secondary antibody application. While inherently sensitive and specific, immunoblotting requires antibodies raised against particular lysine motifs. Further attention must then ensure that the antibodies are not cross-reactive with other lysine residues, or the acetylated target residue. Given the difficulty in raising robust antibodies, immunoblotting is ill-suited for drug discovery, or other applications that required high-throughput capability [28].

The first nonradioactive HDAC assay was developed by Hoffmann et al. using an HPLC with a fluorescence detector to indirectly report HDAC activity (Figure 15.4). A 7-amino-4-methyl-coumarin (AMC)-tagged ε-acetylated lysine (acetyl-K(AMC)-Boc) was used as a substrate, and the deacetylated product is easily separated (Figure 15.4A) [29]. However, the acetyl-K(AMC)-Boc substrate does not resemble the natural histone and possesses a higher affinity for HDAC (K_M value of 0.68 μM for rat liver HDACs) relative to the natural substrate histone (K_M value of 20 μM for rat liver HDACs) [29]. While the HPLC requirement precludes HTS capability, the assay is greatly advantaged by fluorescence detection, a feature that dominates subsequent assay development.

15.3 FLUOROGENIC HDAC ACTIVITY ASSAY

The fluorogenic HDAC activity assay revolutionized the study of HDAC and HDACi discovery by allowing automation and adaptability with HTS. It built on advances made by Hoffmann and cow-orkers by retaining the use of a fluorophore (AMC), but established the monitoring of the fluorophore as an indirect measurement of HDAC activity. A short peptide sequence, containing an ε-acetylated lysine residue and a C-terminus conjugated to a fluorophore such as AMC, is used as substrate. Upon incubation of this substrate (ε-acetyl- Lys(AMC)-peptide) with the HDAC enzyme, the lysine residue is deacetylated and subsequent addition of trypsin to the deacetylated substrate (NH$_2$-Lys(AMC)-peptide) releases AMC, whose fluorescence can be measured distinctly from substrate-linked AMC (Figure 15.4B) [30].

15.4 DISADVANTAGES OF HDAC FLUOROGENIC ASSAYS

The strategy of using a fluorescent reporter for various assays has been popular, as it can be adapted to various systems, detected easily, and used for HTS [31]. However, such labeling methods can perturb enzymatic activity, yielding false positive or false negative results [17]. The alleged activation of sirtuins (SIRT) by resveratrol epitomized such effect. Resveratrol was found to activate SIRT1 activity on the standard p53 fluorogenic substrate, and this activation was credited to explaining resveratrol's role in the extension of yeast lifespan [32]. However, subsequent studies proved that the observed activation was merely an artifact stemming from use of a tagged substrate [33]. Native substrates, lacking the fluorophore, did not indicate a direct activation of SIRT1 by resveratrol [33]. Furthermore, incorporation of the fluorophore increased the affinity of the substrate for SIRT1 and resveratrol caused a tighter binding between the fluorophore-containing substrate and the enzyme [34]. The use of label-free substrates would surely have prevented such confusion [31].

HDAC fluorogenic assays also require the addition of trypsin to release the fluorophore. This added step increases the likelihood of false positives as inhibitors of trypsin would show up as hits, just as inhibitors of the target HDAC would. A mandatory trypsin inhibition assay on potential HDACi has been suggested prior to running the HDAC fluorogenic assay to discriminate true HDAC inhibition from false positives that resulted from trypsin inhibition [30]. An idealized format would eliminate the necessity of a secondary enzyme, like trypsin, altogether. Indeed, this second step can be eliminated by instead relying on electrophoretic separation. Automated systems can directly detect the shift in mobility of the deacetylated product by monitoring fluorescence in microfluidic arrays [35]. Such technology makes HTS possible, but it cannot avoid the uncertainty imbued by the required fluorophore.

15.5 HDAC ASSAY USING THE SELF-ASSEMBLED MONO-LAYERS FOR MATRIX-ASSISTED LASER

15.5.1 DESORPTION IONIZATION TIME-OF-FLIGHT MASS SPECTROMETRY

Based on the preceding analysis, the fluorogenic HDAC assay, which is to date the most commonly used assay, has two main challenges: identification of oligopeptide sequences resembling the natural substrate of each HDAC isoform and direct measurement of the enzymatic activity. The self-assembled mono-layers for matrix assisted laser desorption ionization (SAMDI-MS) appears to have solved the latter challenge. As a label-free method, SAMDI-MS can be used to measure the activity of various enzymes, including HDAC, caspases, and other enzymes [36,37]. In the context of HDAC activity assay, a label-free peptide substrate with an acetylated lysine residue is immobilized onto a plate. Upon exposure to the HDAC, the loss of the acetyl group is detected by mass spectrometry as a change in the mass to charge ratio (m/z) of 42 (Figure 15.5), thus providing a direct measure of HDAC activity [31]. An alternative protocol involves conducting a solution-phase deacetylation reaction and immobilizing the product onto the plate followed by mass spectrometry [38].

The fluorogenic assay, due to its use of a relative few substrates not optimized for all HDAC isoforms, suffers from a lack of selectivity. In a milieu such as a cell lysate which contains multiple isozymes, one may not objectively measure the contribution of each isoform to the overall HDAC activity. The SAMDI-MS assay comes with specific substrates (amino acid sequences) for selected HDACs, thus potentially addressing the problem of lack of specificity observed with the fluorogenic assay substrates (Table 15.1) [39].

15.6 COMPARISON OF THE FLUOROGENIC AND SAMDI-MS HDAC ASSAYS — OUR EXPERIENCE

We first observed the dissimilarity between the fluorogenic and the SAMDI-MS assays by measuring the HDAC8 inhibition profile of selected ketolide-derived macrocyclic HDACi **1a−e** (Figure 15.6). Although we observed a similar inhibition pattern between the two assays for compounds **1a−e**,

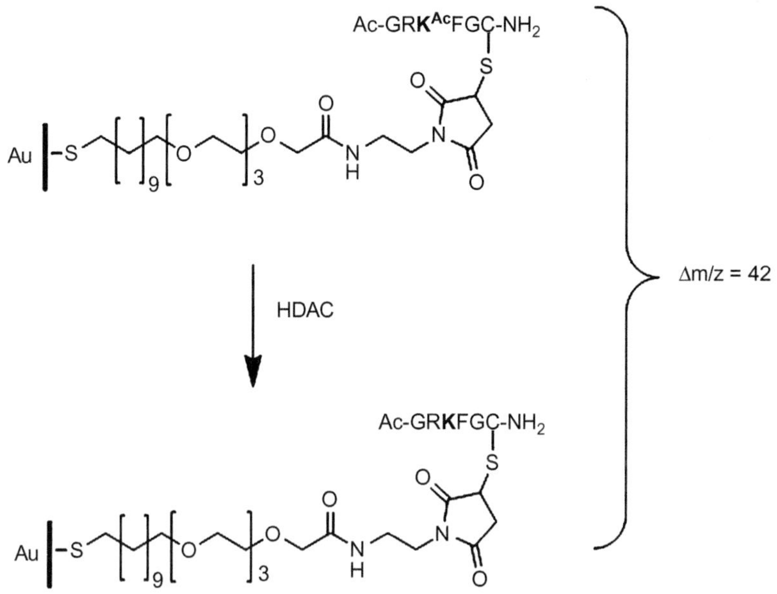

FIGURE 15.5

HDAC activity assay using SAMDI-MS. The acetylated substrate (Ac-GRK^{Ac}FGC-NH$_2$) is immobilized on a gold plate. Exposure of this substrate to HDAC results in the loss of the acetyl group indicated by a $\Delta m/z = 42$ (deacetylated substrate Ac-GRKFGC-NH$_2$) [31].

Table 15.1 Specific Substrate of Selected Deacetylase Enzymes [39]	
Enzyme	**Substrate Sequence**
HDAC2	Ac-GRK^{Ac}YWC-NH$_2$
HDAC8	Ac-GRK^{Ac}FPC-NH$_2$
SIRT1	Ac-GRK^{Ac}RVC-NH$_2$

there is a disparity in the absolute magnitude of the inhibition constants with the fluorogenic assay overestimating HDAC inhibition activity [8]. In another published study, we found that the fluorogenic assay was inconsistent in evaluating the HDAC activity of 3-hydroxypyridin-2-thione-based HDACi, whereas no such inconsistency was seen using the SAMDI-MS assay [38,40].

Informed by these previous observations, we have conducted a detailed analysis of the anti-HDAC activities of a different set of HDACi, against HDAC1, 6 and 8, using the SAMDI-MS and commercial fluorogenic assays. Specifically, we investigated pteroate-derived HDACi and measured their HDAC inhibition activity in both Fluor-de-lys® and the SAMDI-MS assays. In concordance with our previous observation, the HDAC inhibition IC$_{50}$ values obtained from both

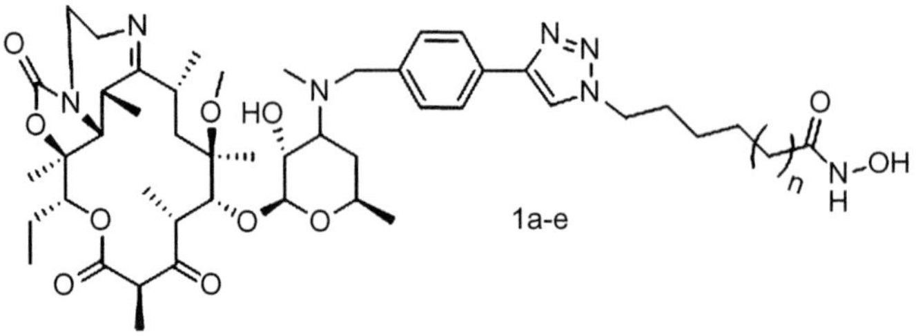

FIGURE 15.6

Structure of Ketolide-derived macrocyclic HDACi **1a−e** used for comparative HDAC8 inhibition study in fluorogenic and SAMDI-MS assays [8].

Table 15.2 HDAC Inhibition Profile of Pteroate-Derived HDACi Fluorogenic Assay vs SAMDI-MS

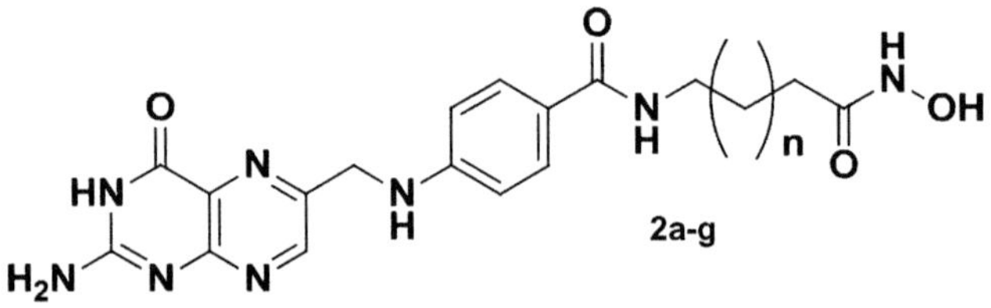

Compound	n	HDAC 1IC_{50}(nM) Fluor-de-lys[†]	HDAC 1IC_{50}(nM) SAMDI	HDAC 6IC_{50}(nM) Fluor-de-lys	HDAC 6IC_{50}(nM) SAMDI	HDAC 8IC_{50}(nM) Fluor-de-lys	HDAC 8IC_{50}(nM) SAMDI
2a	1	8000 ± 2000	$22.1 \pm 1.7\%*$	4000	1460 ± 310	NI	$25 \pm 18\%$
2b	2	190 ± 30	6150 ± 1320	52 ± 6	212 ± 15	7000 ± 2000	$39 \pm 6\%$
2c	3	70 ± 8	79.6 ± 12.6	50 ± 5	43.6 ± 8.4	5400 ± 800	$41 \pm 17\%$
2d	4	53 ± 7	23.4 ± 4.2	22 ± 3	12.2 ± 0.2	NI	1060 ± 320
2e	5	11 ± 1	16.1 ± 2.8	76 ± 9	55.4 ± 2.2	NI	9380 ± 1720
2f	6	104 ± 9	524 ± 36	4.9 ± 0.4	10.2 ± 1.0	NI	7590 ± 2120
2g	7	270 ± 20	$53 \pm 7\%$	70 ± 10	476 ± 48	NI	$40 \pm 10\%$
SAHA	–	42 ± 3	38 ± 2	34 ± 2	144 ± 23	2800 ± 200	232 ± 19

*For IC_{50} greater than 10 μM, % inhibition at 10 μM is indicated; NI, No inhibition at 10 μM;
[†]HDAC assay performed through a contractual agreement with BPS Biosciences.

assays are not identical. In this case, however, similar HDAC inhibition trends are observable only for HDAC1 and 6 (Table 15.2).

All the compounds seem to be more active when their HDAC1 inhibition inhibitory profile is measured using the fluorogenic assay except for the 6-methylene compound **2d** (n = 4). Of particular concern is the apparent gross overestimation of the activities of compounds **2b** and **2g** by the fluorogenic assay, showing almost a 33-fold increase in the potency of these compounds against HDAC1 relative to the SAMDI-MS. Despite the discrepancy between the absolute IC_{50} values, the observed trend for the HDAC1 inhibition between both assays is similar. An increase in the linker length improves the anti-HDAC1 activity starting from compound **2a** (3-methylene linkers between the amide and the hydroxamate moieties) with the optimum methylene linker length being the 7 as in compound **2e**.

Against HDAC6, both assays revealed that compound **2a** is the least potent in the series; however, the fluorogenic assay gave an IC_{50} that is four fold lower than that obtained from the SAMDI-MS. Subsequent increase in methylene linker length resulted in enhanced HDAC6 inhibition activity with no discernible trend in the $IC_{50}s$ values obtained from both assays. The only silver lining is that both assays designated compound **2f** as the most potent HDAC6 inhibitor followed by compounds **2d** and **2c**, respectively. Furthermore, the $IC_{50}s$ of the three most active HDAC6 inhibitors obtained from both assays are very similar.

The dissimilarity between the two assays is made more glaring by the measurement of the HDAC8 inhibition activities. The fluorogenic assay revealed that only two compounds have measurable, though weak, $IC_{50}s$. The inability to estimate the HDAC8 inhibitory activity of most of these compounds may be due to fluorophore label-induced change in the affinity of the substrate for the HDAC8 enzyme, analogous to the previous observation on HDAC8 and SIRT1 [36,41]. If the fluorophore label increases the affinity of substrate for HDAC8, the pteroate-derived HDACi (weak HDAC8 inhibitors) may not be able to effectively compete against the said substrate [34]. Since the substrates for SAMDI-MS assay are not labeled, such artificial alteration in the substrates' HDAC-binding affinities is quite unlikely. This may partly explain the determination of the HDAC8 inhibitory activities of these pteroate-derived HDACi in the SAMDI-MS assay (Table 15.2). Based on the foregoing, fluorogenic assays, such as the one used here, may not be effective in gauging the HDAC8 inhibitory profile of these and possibly many HDACi. Moreover, Riester and coworkers demonstrated that HDAC8 is the most selective of all HDACs for its substrate [42], Hence, the substrate optimization performed prior to the SAMDI-MS may have been beneficial relative to the fluorogenic assay where such optimization was not done.

The importance of substrate optimization was further emphasized with studies evaluating HDAC isoform inhibition of Tubastatin A, an HDAC6 selective HDACi. Using different substrates, Tubastatin A was found by Butler et al. to have an $IC_{50} > 30\,\mu M$ against HDAC 2, whereas Wagner and coworkers reported an IC_{50} of $0.360\,\mu M$ against the same isoform [43,44].

Such discrepancies can also be addressed by the standardization of isoform optimized substrates for all HDAC assays described in the literature. This will not only result in an increased reproducibility and accuracy of the assays, but also will allow an objective comparison between HDACi developed by different laboratories.

The SAMDI-MS assay has addressed some of the problems seen with commonly used HDAC assays. These include fluorophore tag-induced perturbation of substrate-binding affinities, false positive as seen with resveratrol, and the indirect measurement of HDAC activity which can result in data inconsistency. However, a concern that it has not completely addressed is the development of isoform matched or specific substrates, though significant efforts have been made in that direction. Isoform specific substrates may not be needed for closely related HDAC isoforms such as HDACs 1, 2, and 3. In fact attainment of substrate specificity among such closely related isoforms may not be feasible as they all have the same natural substrates, histones. On the other hand, substrate specificity may be achievable between the class I isoform (HDAC1, 2, 3) and the class IIb HDAC6, as these two classes have very distinct substrates: histones and cytoplasmic proteins, respectively [45]. Furthermore, the highly conserved regions within classes I and II seem to be solely at the active site of these enzymes. The area surrounding the active site channel entrance, which may mediate binding interactions between the HDAC and the substrates, are highly different, suggesting interactions with different amino acid sequences by class I and II HDACs [46]. Obtaining a subclass selective substrate can constitute an important step toward the discovery of a new generation of class I selective HDACi devoid of any activity against class IIa HDAC isoforms. Such small molecules may prove to be advantageous for HDACi therapy as class I isoforms are overexpressed in many malignancies as opposed to class IIa, which are cardioprotective [47,48].

15.7 CONCLUSION

The fluorogenic HDAC assays have been one of the most commonly used to evaluate HDAC activity due to their ease and compatibility with automation screening platforms such as HTS. However, fluorophore labeled substrates have been shown to interfere with the HDAC enzymatic activities. Also, the limited substrate peptide sequences may not allow for distinction between the activities of various HDAC isozymes in a cellular lysate. Furthermore, the indirect measurement of HDAC activity heightens concern about data accuracy and the likelihood of false positive. On the other hand, the SAMDI-MS, a label-free technique, does not require any modifications to the substrate and enables a direct monitoring of HDAC activity. For selected isoforms, the substrate sequences have been optimized relative to the fluorogenic assay. These attributes confer a higher level of specificity and consistency to the SAMDI-MS data. In the examples cited, the major discrepancy between SAMDI-MS and fluorogenic assays was seen with HDAC8, possibly due to the influence of the fluorophore label on substrate affinity. It is therefore possible that, for compounds with relatively lower anti-HDAC activity, the increased affinity of the substrate ensuing from the presence of the fluorophore can hinder the measurement of the HDAC activity using the fluorogenic assay.

REFERENCES

[1] Choudhary C, Kumar C, Gnad F, Nielsen ML, Rehman M, Walther TC, et al. Lysine acetylation targets protein complexes and co-regulates major cellular functions. Science 2009;325(5942):834−40.
[2] Gryder BE, Sodji QH, Oyelere AK. Targeted cancer therapy: giving histone deacetylase inhibitors all they need to succeed. Future Med Chem 2012;4(4):505−24.
[3] Haigis MC, Sinclair DA. Mammalian sirtuins: biological insights and disease relevance. Annu Rev Pathol 2010;5(1):253−95.
[4] Yang XJ, Seto E. HATs and HDACs: from structure, function and regulation to novel strategies for therapy and prevention. Oncogene 2007;26(37):5310−18.
[5] Kazantsev AG, Thompson LM. Therapeutic application of histone deacetylase inhibitors for central nervous system disorders. Nat Rev Drug Discov 2008;7(10):854−68.

[6] Rotili D, Simonetti G, Savarino A, Palamara AT, Migliaccio AR, Mai A. Non-cancer uses of histone deacetylase inhibitors: effects on infectious diseases and beta-hemoglobinopathies. Curr Top Med Chem 2009;9(3):272−91.

[7] Beckers T, Burkhardt C, Wieland H, Gimmnich P, Ciossek T, Maier T, et al. Distinct pharmacological properties of second generation HDAC inhibitors with the benzamide or hydroxamate head group. Int J Cancer 2007;121(5):1138−48.

[8] Mwakwari SC, Guerrant W, Patil V, Khan SI, Tekwani BL, Gurard-Levin ZA, et al. Non-peptide macrocyclic histone deacetylase inhibitors derived from tricyclic ketolide skeleton. J Med Chem 2010;53 (16):6100−11.

[9] Stimson L, Wood V, Khan O, Fotheringham S, La Thangue NB. HDAC inhibitor-based therapies and haematological malignancy. Ann Oncol 2009;20(8):1293−302.

[10] Tan JH, Cang SD, Ma YH, Petrillo RL, Liu DL. Novel histone deacetylase inhibitors in clinical trials as anti-cancer agents. J Hematol Oncol 2010;3.

[11] Qiu TZ, Zhou L, Zhu W, Wang TS, Wang J, Shu YQ, et al. Effects of treatment with histone deacetylase inhibitors in solid tumors: a review based on 30 clinical trials. Future Oncol 2013;9(2):255−69.

[12] Duvic M, Talpur R, Ni X, Zhang C, Hazarika P, Kelly C, et al. Phase 2 trial of oral vorinostat (suberoylanilide hydroxamic acid, SAHA) for refractory cutaneous T-cell lymphoma (CTCL). Blood 2007;109(1):31−9.

[13] Piekarz RL, Frye R, Turner M, Wright JJ, Allen SL, Kirschbaum MH, et al. Phase II multi-institutional trial of the histone deacetylase inhibitor romidepsin as monotherapy for patients with cutaneous T-cell lymphoma. J Clin Oncol 2009;27(32):5410−17.

[14] Mann BS, Johnson JR, Cohen MH, Justice R, Pazdur R. FDA approval summary: vorinostat for treatment of advanced primary cutaneous T-cell lymphoma. Oncologist 2007;12(10):1247−52.

[15] VanderMolen KM, McCulloch W, Pearce CJ, Oberlies NH. Romidepsin (Istodax, NSC630176, FR901228, FK228, depsipeptide): a natural product recently approved for cutaneous T-cell lymphoma. J Antibiot 2011;64(8):525−31.

[16] Bradner JE, West N, Grachan ML, Greenberg EF, Haggarty SJ, Warnow T, et al. Chemical phylogenetics of histone deacetylases. Nat Chem Biol 2010;6(3):238−43.

[17] Kaeberlein M, McDonagh T, Heltweg B, Hixon J, Westman EA, Caldwell SD, et al. Substrate-specific activation of sirtuins by resveratrol. J Biol Chem 2005;280(17):17038−45.

[18] Sanders BD, Jackson B, Marmorstein R. Structural basis for sirtuin function: what we know and what we don't. Biochim Biophys Acta 2010;1804(8):1604−16.

[19] Sauve AA. Sirtuin chemical mechanisms. Biochim Biophys Acta Waltham, Massachusetts 2010;1804(8): 1591−603.

[20] Borra MT, Denu JM. Quantitative assays for characterization of the sir2 family of NAD+ - dependent deacetylases. In: Allis CD, Carl W, editors. Methods in enzymology, Vol. 376. Academic Press; 2003. p. 171−87.

[21] Landry J, Sutton A, Tafrov ST, Heller RC, Stebbins J, Pillus L, et al. The silencing protein SIR2 and its homologs are NAD-dependent protein deacetylases. Proc Natl Acad Sci 2000;97(11):5807−11.

[22] Smith BC, Hallows WC, Denu JM. A continuous microplate assay for sirtuins and nicotinamide-producing enzymes. Anal Biochem 2009;394(1):101−9.

[23] Landry J, Sternglanz R. Enzymatic assays for NAD-dependent deacetylase activities. Methods 2003;31 (1):33−9.

[24] Schutkowski M, Fischer F, Roessler C, Steegborn C. New assays and approaches for discovery and design of Sirtuin modulators. Expert Opinion on Drug Discov 2014;9(2):183−99.

[25] Hay CW, Candido EP. Histone deacetylase. Association with a nuclease resistant, high molecular weight fraction of HeLa cell chromatin. J Biol Chem 1983;258(6):3726−34.

[26] Kölle D, Brosch G, Lechner T, Lusser A, Loidl P. Biochemical methods for analysis of histone deacetylases. Methods 1998;15(4):323−31.

[27] Nare B, Allocco JJ, Kuningas R, Galuska S, Myers RW, Bednarek MA, et al. Development of a scintillation proximity assay for histone deacetylase using a biotinylated peptide derived from histone-H4. Anal Biochem 1999;267(2):390−6.

[28] Zhang Y, LeRoy G, Seelig H-P, Lane WS, Reinberg D. The dermatomyositis-specific autoantigen Mi2 is a component of a complex containing histone deacetylase and nucleosome remodeling activities. Cell 1998;95(2):279−89.

[29] Hoffmann K, Jung M, Brosch G, Loidl P. A non-isotopic assay for histone deacetylase activity. Nucleic Acids Res 1999;27(9):2057−8.

[30] Wegener D, Wirsching F, Riester D, Schwienhorst A. A fluorogenic histone deacetylase assay well suited for high-throughput activity screening. Chem Biol 2003;10(1):61−8.

[31] Gurard-Levin ZA, Scholle MD, Eisenberg AH, Mrksich M. High-throughput screening of small molecule libraries using SAMDI mass spectrometry. ACS Comb Sci 2011;13(4):347−50.

[32] Howitz KT, Bitterman KJ, Cohen HY, Lamming DW, Lavu S, Wood JG, et al. Small molecule activators of sirtuins extend Saccharomyces cerevisiae lifespan. Nature 2003;425(6954):191−6.

[33] Pacholec M, Bleasdale JE, Chrunyk B, Cunningham D, Flynn D, Garofalo RS, et al. SRT1720, SRT2183, SRT1460, and resveratrol are not Direct activators of SIRT1. J Biol Chem 2010;285(11):8340−51.

[34] Borra MT, Smith BC, Denu JM. Mechanism of human SIRT1 activation by resveratrol. J Biol Chem 2005;280(17):17187−95.

[35] Schroeder FA, Lewis MC, Fass DM, Wagner FF, Zhang Y-L, Hennig KM, et al. A selective HDAC 1/2 inhibitor modulates chromatin and gene expression in brain and alters mouse behavior in two mood-related tests. Plos One 2013;8(8).

[36] Gurard-Levin ZA, Kim J, Mrksich M. Combining mass spectrometry and peptide arrays to profile the specificities of histone deacetylases. ChemBioChem 2009;10(13):2159−61.

[37] Su J, Rajapaksha TW, Peter ME, Mrksich M. Assays of endogenous caspase activities: a comparison of mass spectrometry and fluorescence formats. Anal Chem 2006;78(14):4945−51.

[38] Sodji QH, Patil V, Kornacki JR, Mrksich M, Oyelere AK. Synthesis and structure−activity relationship of 3-hydroxypyridine-2-thione-based histone deacetylase inhibitors. J Med Chem 2013;56(24):9969−81.

[39] Gurard-Levin A, Kilian KA, Kim J, Bähr K, Mrksich M. Peptide arrays identify isoform-selective substrates for profiling endogenous lysine deacetylase activity. ACS Chem Biol 2010;5(9):863−73.

[40] Patil V, Sodji QH, Kornacki JR, Mrksich M, Oyelere AK. 3-Hydroxypyridin-2-thione as novel zinc binding group for selective histone deacetylase inhibition. J Med Chem 2013;56(9):3492−506.

[41] Beher D, Wu J, Cumine S, Kim KW, Lu S-C, Atangan L, et al. Resveratrol is not a direct activator of SIRT1 enzyme activity. Chem Biol Drug Des 2009;74(6):619−24.

[42] Riester D, Hildmann C, Grünewald S, Beckers T, Schwienhorst A. Factors affecting the substrate specificity of histone deacetylases. Biochem Biophys Res Commun 2007;357(2):439−45.

[43] Butler KV, Kalin J, Brochier C, Vistoli G, Langley B, Kozikowski AP. Rational design and simple chemistry yield a superior, neuroprotective HDAC6 inhibitor, tubastatin A. J Am Chem Soc 2010;132(31):10842−6.

[44] Wagner FF, Olson DE, Gale JP, Kaya T, Weïwer M, Aidoud N, et al. Potent and selective inhibition of histone deacetylase 6 (HDAC6) does not require a surface-binding motif. J Med Chem 2013;56(4):1772−6.

[45] Dokmanovic M, Clarke C, Marks PA. Histone deacetylase inhibitors: overview and perspectives. Mol Cancer Res 2007;5(10):981−9.

[46] Marmorstein R. Structure of histone deacetylases: insights into substrate recognition and catalysis. Structure 2001;9(12):1127−33.

[47] Weichert W. HDAC expression and clinical prognosis in human malignancies. Cancer Lett 2009;280 (2):168−76.

[48] Backs J, Olson EN. Control of cardiac growth by histone acetylation/deacetylation. Circ Res 2006;98 (1):15−24.

ENZYMATIC ASSAYS OF HISTONE METHYLTRANSFERASE ENZYMES 16

Hao Zeng and Wei Xu

McArdle Laboratory for Cancer Research, University of Wisconsin-Madison, Madison, WI, USA

CHAPTER OUTLINE

Y.G. Zheng (Ed): Epigenetic Technological Applications. DOI: http://dx.doi.org/10.1016/B978-0-12-801080-8.00016-8

16.1 INTRODUCTION

Histone methyltransferases can be classified into two families, histone lysine methyltransferases (HKMTs) and protein arginine methyltransferases (PRMTs). HKMTs and PRMTs share the common mechanism of methylation that the methyl moiety of the universal methyl donor *S*-adenosyl-L-methionine (SAM; AdoMet) is enzymatically transferred to a nitrogen atom of lysine or arginine side chains of protein substrate, leading to the formation of methylated substrate and a by-product *S*-adenosyl-homocysteine (SAH; AdoHcy) [1]. To date, more than 50 HKMTs and 10 PRMTs have been identified and many of them have been implicated in a variety of human diseases including cancers [2]. Epizyme Inc. published the phylogenetic trees of HKMTs and PRMTs based on their amino acid similarity in *Nature Chemical Biology* in 2011. Some cancer-relevant enzymes are discussed in the following sections.

16.1.1 HISTONE LYSINE METHYLTRANSFERASES

The majority of histone lysine methylation occurs on histone H3 (K4, K9, K27, K36, and K79), with additional methylation events on histone H1K26, H4K20, and H2BK5 [3]. The extent of methylation on a single lysine residue can be mono-, di-, or tri-methylated, which are linked to distinct biological consequences. For example, mono-methylation of H3K27 has been linked to gene activation, while tri-methylation of H3K27 is linked to transcriptional repression [4]. All but one of the HKMTs contain a common SET (Su(var), Enhancer of zeste, Trithorax) domain which is essential for AdoMet binding and catalysis [5]. Based on the sequence homology in the SET domain and the adjacent domain features, HKMTs can be classified into four main families: the SUV39 family, the SET1 family, the SET2 family, and the RIZ family [5].

The SUV39 family features a unique pre-SET domain that is necessary for the SET domain to exert its catalytic function toward histone H3K9 [5]. The family members include SUV39H1, SUV39H2, G9a, GLP, ESET, and CLLL8, among which SUV39H1 was the first lysine methyltransferase identified in human [5]. SUV39H1 has been shown to be overexpressed in colorectal tumors and recently has been shown to interact with retinoblastoma protein, Rb, for the transcriptional regulation of the cyclin E gene [6]. Another frequently studied member of the SUV39 family is G9a methyltransferase, which predominantly methylates H3K9 as well as H3K27. G9a is elevated in hepatocellular carcinoma and also is required for cell growth of PC3 prostate cancer cells [7].

The SET1 family is structurally defined by the localization of the SET domain at the very carboxyl terminus with a following post-SET domain [5]. However, the most often studied SET1 family members, EZH1 and EZH2, do not possess the post-SET domain and may therefore be classified as an individual family [5,8]. EZH2 (Enhancer of Zeste Homologue 2) is a crucial component of the polycomb repressive complex 2 (PRC2), which is responsible for the tri-methylation of histone H3K27, leading to the repression of target genes involved in a variety of cellular processes [9]. EZH2 is amplified or overexpressed in a variety of cancer types, including prostate, breast, bladder, colon, lung, and gastric cancers [9]. A recent study further demonstrated that EZH2 acts as a driving factor in prostate cancer metastasis [10]. As for breast cancer, higher EZH2 expression is correlated with aggressive basal-like subtype and poorer metastasis-free survival [11].

Moreover, EZH2 has been shown to support the expansion of breast-tumor-initiating cells [12]. In addition to EZH proteins, the other members of the SET1 family include the human homologues of yeast SET1 (hSET1A and hSET1B) and human trithorax proteins (MLL1-5), with the capacity to methylate histone H3K4 [8]. The MLL methyltransferases have been extensively studied in leukemia due to the frequent translocations and fusion with other proteins [6,7,13].

The SET2 family is characterized by the presence of SET domain between a cysteine-rich AWS domain and a post-SET domain [5,8]. This family includes the mammalian nuclear receptor-binding SET-domain-containing proteins (NSD1-3), the huntingtin-interacting protein SETD2, and the SMYD family proteins [5,8]. NSD1, the enzyme responsible for methylation of histone H3K36 and H4K20, is a coregulator of the androgen receptor, which is able to enhance transcriptional activation in prostate cancer cells [14]. NSD2 is well known as the Multiple Myeloma SET domain (MMSET) protein or Wolf-Hirschhorn Syndrome Candidate 1 (WHSC1), which is responsible for di-methylation of histone H3K36 and is involved in the t(4;14) translocation present in about 15% of multiple myeloma patients [15]. The NSD3 gene has been shown to be amplified in human breast cancer cell lines [16]. SETD2 is a H3K36-specific methyltransferase and its methyltransferase activity has been shown required for embryonic vascular remodeling [17]. Within the SMYD subfamily, *SMYD3* gene was found overexpressed in the majority of colorectal and hepatocellular carcinomas. Moreover, the H3K4-specific methyltransferase SMYD3 is required for proliferation of cancer cells by forming a complex with RNA polymerase II [18].

The RIZ family is characterized by the localization of the SET domain at the amino terminus, without pre- or post-SET domain. This family is composed of three members, RIZ1, BLIMP1, and PFM1, among which RIZ1 is the founding member [5,8]. RIZ1 is a tumor suppressor protein with methyltransferase activity toward histone H3K9. It was initially identified as a retinoblastoma-interacting partner [19] and further was shown to be a coactivator of estrogen receptor (ER) in breast cancer cells [20].

In addition to these four main families characterized by different domain features, SET-domain-containing HKMTs also contain a number of unclassified members, such as SET7/9, SET8, SUV4-20H1, and SUV4-20H2 [8]. SET7/9 is responsible for methylation of histone H3K4, which is associated with transcriptional activation. Besides, SET7/9 can also methylate a number of nonhistone proteins. For example, SET7/9 can specifically methylate ER at lysine 302 and therefore stabilizing ER protein and activating the transcription of ER-target genes in breast cancer cells [21]. In addition, the well-known tumor suppressor p53 can be methylated by SET7/9 at lysine 369, which is required for p53 activation *in vivo* [22].

Distinct from the SET-domain-containing HKMTs, DOT1L is unique in that it does not contain the SET domain and it is structurally similar to PRMTs [23]. DOT1L methyltransferase is specific for histone H3K79 methylation which is associated with active gene transcription [24]. DOT1L as well as its methyltransferase activity have been demonstrated to be required for proliferation of *MLL*-rearranged leukemia cells by activating a number of leukemogenic genes through H3K79 methylation [25,26].

Taken together, the HMKTs can specifically methylate histone substrates and play important roles in regulation of gene transcription as well as disease development and progression [7,13]. Additionally, more and more nonhistone proteins, including nuclear receptors, tumor suppressor proteins, and transcription factors, have been identified as substrates of some HMKTs, indicating that HMKTs also function through methylation of nonhistone proteins [21,22,27].

16.1.2 PROTEIN ARGININE METHYLTRANSFERASES

Histone arginine methylation is catalyzed by protein arginine methyltransferases (PRMTs), and methylated arginine species can exist in three different types in mammalian cells: ω-N^G-monomethylarginine (MMA), ω-N^G,N^G-asymmetric dimethylarginine (ADMA), and ω-N^G,N'^G-symmetric dimethylarginine (SDMA) [28]. Ten mammalian PRMTs (PRMT1-10) have been identified to date, which can be classified into three distinct types according to their corresponding methylarginine products. Type I PRMTs (PRMT1, 2, 3, 4, 6, and 8) catalyze the formation of MMA and ADMA, whereas type II PRMTs (PRMT5, 7, and 9) catalyze the production of MMA and SDMA [29]. In addition to type II activity, PRMT7 also possesses type III activity, which only generates MMA products [29]. No methyltransferase activity has been characterized for PRMT10. The PRMTs harbor four conserved motifs (I, post-I, II, and III) and a THW loop, among which motifs I, post-I, and the THW loop form part of the AdoMet-binding pocket [30]. Like HMKTs, PRMTs methylate histones as well as nonhistone proteins, making them multifunctional proteins engaged in diverse cellular processes, such as mRNA splicing, signal transduction, transcriptional regulation, and DNA repair, among others [28–33].

The dysregulation of PRMTs has been linked to various human diseases including cancers [28,29]. PRMT1 is the predominant PRMT in mammalian cells, and the dimethylation of histone H4R3 catalyzed by PRMT1 has been associated with transcriptional activation and positively correlates with prostate tumor grade [28]. PRMT1 has also been shown aberrantly expressed in breast, prostate, lung, colon, and bladder cancer as well as leukemia [28]. In human breast cancer cells, PRMT1 specifically methylates ER at R260 which localizes in the DNA-binding domain of ER, activating the SRC-PI3K-FAK cascade and AKT and thus accelerating cell proliferation [34]. PRMT2 has weak methyltransferase activity, yet it is still found to be a coactivator of ER and overexpressed in breast cancer [35]. PRMT3 has been shown to interact with DAL-1/4.1B tumor suppressor protein, and this interaction significantly inhibits the methyltransferase activity of PRMT3 [36]. PRMT4 or coactivator associated arginine methyltransferase 1 (CARM1) was the first PRMT characterized as a transcriptional activator by methylating histone H3 [37]. CARM1 functions as a transcriptional coactivator for a series of cancer-relevant transcription factors, including NF-kB, p53, E2F1, β-catenin, and steroid receptors, among which coactivation of ER has been best characterized [28,30–33]. Our lab demonstrated that CARM1 activates ER-target genes via H3R17 dimethylation and subsequent recruitment of human RNA polymerase-II-associated factor complex (hPAFc) [38]. More recently, we identified BAF155 as a novel substrate of CARM1 and validated that methylation of BAF155 by CARM1 contributes to breast cancer metastasis [39]. PRMT5 is the major type II PRMT, which catalyzes the methylation of histone H3R8 and H4R3, associating with transcriptional repression [28]. PRMT5 has been shown overexpressed in lung cancer and glioblastoma, and also required for human embryonic stem cell proliferation [28,29]. PRMT6 has been reported to be elevated in lung and bladder cancer cells [28]. It also regulates DNA base excision repair by methylating DNA polymerase β [40]. PRMT7 catalyzes the methylation of histone H4R3 and plays a role in male germline imprinted gene methylation through interaction with testis-specific factor CTCFL [41]. Moreover, decreasing levels of PRMT7 could sensitize tumor cells to topoisomerase I inhibitor camptothecin [42]. PRMT8 is a plasma membrane bound and neuron-specific PRMT, which has a high degree of sequence identity with PRMT1 [43]. Collectively, PRMTs work in conjunction with their various substrates to control a variety of cellular processes in normal and disease conditions, rendering them as potential therapeutic targets for cancer treatment.

16.2 ENZYMATIC ASSAYS FOR HISTONE METHYLTRANSFERASES

16.2.1 RADIOACTIVE ASSAYS

16.2.1.1 *Standard* in vitro *methylation assays*

Standard *in vitro* radioactive assay or radioisotope-labeled assay has been the gold standard approach for characterizing *in vitro* methyltransferase activity of HKMTs and PRMTs due to its sensitivity and accuracy [27]. In general, the tritiated methyl groups from [^{3}H]-SAM are first enzymatically transferred into the peptide or protein substrates of methyltransferases. Subsequently, the methylated substrates can be precipitated with trichloroacetic acid, leaving unreacted [^{3}H]-SAM in supernatant, followed by scintillation counting or SDS-PAGE to separate proteins on the gel for autoradiography or gel extraction and subsequent scintillation counting [44]. The standard *in vitro* methylation assays have been well applied to determine the specificity of HKMTs and PRMTs, to identify novel substrates, and to study the potency and selectivity of small molecule inhibitors [45]. For example, our laboratory utilized the standard *in vitro* methylation assays to validate that BAF155 is a novel CARM1 substrate, which can be specifically methylated at R1064 [39]. The Bedford laboratory utilized the standard *in vitro* methylation assays to demonstrate that some xenoestrogens are able to modulate the enzymatic activity of arginine methyltransferases [46]. Although the standard approaches to measuring histone methyltransferase activity are sensitive and applicable to most histone methyltransferases, the process is time-consuming, which significantly impedes the rapid analysis of the methylation reaction. Moreover, the separation steps significantly limit the number of assays that could be analyzed simultaneously.

16.2.1.2 *Filter-binding assays*

To decrease the analysis time of the standard approach, radioactive filter-binding assay was developed to monitor methyltransferase activity (Figure 16.1A). The filter-binding assay takes advantage of the property of phosphocellulose filter paper to bind peptide or protein products with higher affinity than free SAM. Following the standard *in vitro* methylation reaction, the unreacted [^{3}H]-SAM can be filtered off and the methylated products will be retained on the phosphocellulose filters assembled in 96- or 384-well vacuum plates, which can be easily analyzed by subsequent scintillation counting (Figure 16.1A). Using this technique, the Imhof group identified the fungal mycotoxin chaetocin as a specific inhibitor of histone lysine methyltransferase SU(VAR)3-9 after screening 2,976 compounds [47]. This success demonstrates the application of radioactive filter-binding assay in the characterization of methyltransferase activity and in small molecule inhibitor screening. The Jayaraman group also utilized the filter-binding assay to monitor CARM1 methyltransferase activity and identified pyrazole amide 7b as a potent and selective inhibitor of CARM1 [48]. The Reaction Biology Corporation commercialized a HotSpot^{SM} platform initially to characterize protein kinase catalytic activity and profile kinase inhibitors based on the radioactive filter-binding assay technology [49]. Recently, they further adapted the kinase assay platform to monitor histone methyltransferase activity and proved that it can be utilized effectively for determination of substrate specificity, compound profiling, enzyme kinetic analysis, and high-throughput screening (HTS) [50]. After the identification of GSK126 as a potential EZH2 inhibitor from their initial biochemical screen, GlaxoSmithKline further assessed and confirmed the selectivity of GSK126 in a

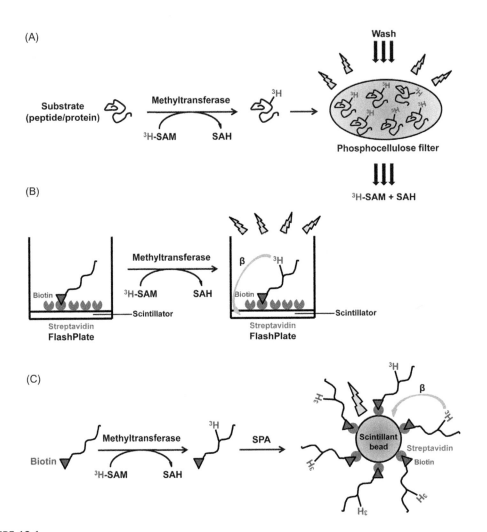

FIGURE 16.1

Schematic representation of radioactive assays. (A) Filter-binding assay: Standard *in vitro* methylation assay is performed with substrate, methyltransferase, and ^{3}H-SAM, leading to the incorporation of tritiated methyl group into the lysine/arginine residue of substrate. Unreacted ^{3}H-SAM is filtered off and the methylated products are retained on the phosphocellulose filters assembled in a vacuum plate. Scintillation fluid is then added and radiolabeled products are quantified by scintillation counting. (B) FlashPlate-based scintillation proximity assay: Biotinylated histone peptide or protein substrates are coated onto the wells of a streptavidin FlashPlate with scintillator embedded into the walls. Following the methylation reaction, β-particles generated by the radioactive decay of plate-bound substrates excite the scintillator to emit light that can be quantified. (C) Bead-based scintillation proximity assay: Streptavidin-coated microscopic beads containing scintillator are employed instead of the FlashPlate to capture the methylated biotinylated substrates followed by the methylation reaction. β-particles generated by the radioactive decay of bead-bound substrates excite the scintillator to emit light that can be quantified.

panel of histone methyltransferases using the HotSpot technology, highlighting the application of HotSpot technology in compound profiling [51].

To further reduce the process time of radioactive assay, the Hevel group developed an assay with the utilization of commercially available ZipTip$_{c4}$ pipette tips (Millipore) with 0.6 μL of reverse-phase chromatography resins (C4, silica, 15 μm, 300 Å pore size) filling the end of the tips [52]. Reverse-phase resins have been extensively utilized for the separation of peptides and proteins [52]. Herein the ZipTip$_{c4}$ tips were used to rapidly separate unreacted [^{3}H]-SAM from radiolabeled protein products, and the bound radiolabeled protein products were eluted directly into scintillation solution and counted. The separation process is extremely fast with the ZipTip protocol, which can be completed within 2−14 min [52]. Using the ZipTip$_{c4}$ tips, they generated a Michaelis-Menten curve for the methylation of heterogeneous nuclear ribonucleoprotein K (hnRNP K) catalyzed by PRMT1, variant 1, and also demonstrated that glutathione S-transferase (GST) tag fused to the N terminus of PRMT6 largely reduced its methyltransferase activity [52]. The use of ZipTip$_{c4}$ tips decreased the volume of radiolabeled waste by more than 3000-fold over the traditional SDS-PAGE separation. However, the ZipTip$_{c4}$ tip has a binding capacity of approximately 3.3 μg proteins, which should be taken into consideration when designing the experiment [52]. Although multichannel pipettes can be used to handle more samples at the same time, the pipetting handling and the requirement of washing steps are not suitable for automation and thus restrict its application in HTS.

16.2.1.3 Scintillation proximity assays

To eliminate the washing and pipetting steps required for the previously described assays, the Jeltsch group established a continuous scintillation proximity assay (SPA) for histone lysine methyltransferase activity quantification and inhibitor screening [53,54]. Their assay uses streptavidin FlashPlates with polystyrene-based scintillator embedded into the walls of the microplate (PerkinElmer, Waltham, MA, USA) (Figure 16.1B). During the assay performance, the wells of a FlashPlate are first coated with biotinylated histone peptide or protein substrates followed by washing off the unbound excess. After the addition of methyltransferase/[^{3}H]-SAM mixture, the transfer of the radiolabeled methyl groups to the target substrates leads to a close approximation of β-particles generated by the radioactive decay and the scintillation fluid, which results in a scintillation signal (Figure 16.1B) [53,54]. One major advantage of this assay is that there is no need to remove the unreacted [^{3}H]-SAM, mainly because the β-particles emitted from the free [^{3}H]-SAM in solution will not produce a strong scintillation signal. The solvent quenches the majority of β-particles from free [^{3}H]-SAM before they reach the bottom of the wells, where the scintillant is embedded. Therefore the SPA platform allows continuous monitoring of the methylation reaction. The FlashPlates-based SPA has been applied to the enzymatic activity measurement of histone lysine methyltransferases Dim-5 and G9a, using histone H3 peptide and WIZ (widely interspaced zinc finger motifs) protein as substrate, respectively [53,54]. The initial rate of each methylation reaction was measured and the Michaelis constant K_m of Dim-5 for AdoMet was determined to be 0.68 (-0.16, $+0.25$) μM [53]. In addition, the authors also determined the IC$_{50}$ values of AdoHcy and BIX-01294, a G9a selective inhibitor, for G9a-catalyzed WIZ protein methylation to be 25.7 ± 1.5 μM and 20 ± 1 μM, respectively, indicating the capability of SPA for small molecule inhibitor profiling [54]. One major limitation of the FlashPlates-based SPA platform is that the effective substrate concentration in the methylation reaction is unknown due to the substrate binding and subsequent washing steps. Thus the K_m for substrate cannot be determined using this assay.

To resolve this issue, the Zheng group modified the SPA platform, in which streptavidin-coated microscopic beads containing scintillant were employed instead of the FlashPlates to capture the methylated biotinylated histone peptides after the methylation reaction (Figure 16.1C) [55]. Using this bead-based SPA, the authors successfully measured the enzymatic activities of type I arginine methyltranferase PRMT1 and type II arginine methyltransferase PRMT5. The merit of this assay was further shown in a dose-dependent inhibition experiment for PRMT1 by AdoHcy and NS-1, two previously reported arginine methyltransferase inhibitors, with IC_{50} values 0.3 μM and 14.7 μM, respectively. Moreover, this assay was demonstrated suitable for HTS of PRMT inhibitors, with a Z' value of 0.80 [55]. The bead-based SPA assay was also utilized by Luo and colleagues to monitor methyltransferase activities of SET7/9, SET8, and SETD2 and identified NF279, a suramin analog, as an inhibitor of SET7/9 and SETD2 with IC_{50} values of 1.9 μM and 1.1 μM, respectively [56].

Generally speaking, the radioactive assays provide sensitive and accurate means to measure the enzymatic activities of histone methyltransferases from various sources. However, dependence on the radioisotope-labeled materials, the cost of $[^3H]$-SAM, and radiolabeled waste disposal in the radioactive assays limit the broad application of *in vitro* methylation assays. Moreover, the radioactive assays cannot distinguish the degree of methylation such as mono-/di-/tri-methylation for HKMTs and MMA/ADMA/SDMA for PRMTs. This problem can be partially resolved in antibody-based assays (to be discussed later) with the availability of highly specific antibodies [1]; and the methylation extent can be accurately determined by mass spectrometry based methods. Consequently, the design and utilization of nonradioactive assays are highly desired to further evaluate protein methyltransferase activities.

16.2.2 FLUORESCENCE POLARIZATION ASSAYS

16.2.2.1 Direct fluorescence polarization assay for methyltransferases

Fluorescence polarization (FP) assays are based on the principle that the degree of polarization of a fluorophore is inversely related to its molecular rotation, where the fluorophore rotates more quickly when unbound by a small molecule in solution (Figure 16.2) [57]. Quantitatively, FP is defined as the difference of the emission light intensity parallel ($I||$) and perpendicular ($I\perp$) to the excitation light plane normalized by the total fluorescence emission intensity [58]. $FP = (I|| - I\perp)/(I|| + I\perp)$. Given the enzyme/substrate interaction in the methylation reaction, the Zheng group synthesized fluorescein-labeled peptides based on two different substrates of PRMT1, glycine and arginine-rich (GAR) motifs and the N-terminal 20-aa sequence of histone H4, as fluorescent reporters for monitoring PRMT1 methyltransferase activity in the FP assays [59]. Using this platform, they unraveled the inhibitory mechanism of AIM-1, the first reported PRMT inhibitor [60], as it competes with the peptide substrate for the binding pocket in PRMT1 [59]. They also utilized the same FP assay to study the transient kinetics of PRMT1 catalysis, which cannot be easily achieved by the classic radioactive assays [61]. Similarly, the Jin group utilized the direct FP assay to monitor the binding of a fluorescein-labeled histone H3 peptide to methyltransferase G9a in the presence of the selective G9a inhibitor UNC0224. The FP signal intensity reflects the inhibitory potency of this compound [62].

On the other hand, given the enzyme/co-factor (SAM) interaction in the methylation reaction, the Cayman Chemical Company commercialized a SAM-Screener™ Assay Kit for screening SAM-binding

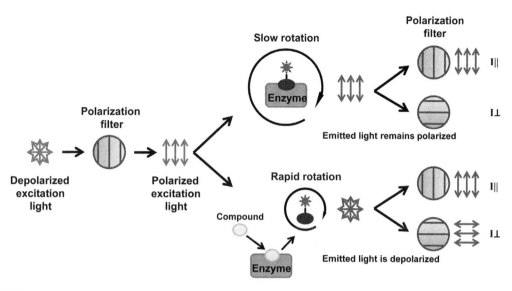

FIGURE 16.2

Principle of fluorescence polarization assay. The fluorophore labeled on the substrate is excited by light that is linearly polarized through a polarization filter. When the substrate is enzyme-bound, the rotation of fluorophore is slow, leading to the polarized light emission. When a compound competes with the fluorophore-labeled substrate for the binding site, the fluorophore is released, leading to the depolarized light emission. Fluorescence polarization is then measured as the difference of the emission light intensity parallel $(I_{\parallel})$ and perpendicular $(I_{\perp})$ to the excitation light plane normalized by the total fluorescence emission intensity.

site inhibitors of HKMTs (SET7/9, MLL1, and G9a). In the SAM-Screener assay, a fluorescent small molecule probe can bind with nanomolar affinity to the SAM-binding site on methyltransferase and generate a strong FP signal. Compounds able to compete with SAM for the SAM-binding site on methyltransferase will release the fluorescent probe from methyltransferase, leading to a decrease of the FP signal. With regard to arginine methylation, the Mowen laboratory established a similar FP assay based on the SAM-PRMT1 interaction for screening PRMT1-specific inhibitors [63]. Interestingly, PRMT1 has been validated to contain a hyperreactive cysteine 101 (Cys101) which directly contacts SAM during PRMT1-catalyzed methylation. However, all other PRMTs except PRMT8 are free of the homologous cysteine [64]. Based on this property, the Mowen laboratory synthesized a thiol-reactive maleimide probe conjugated to Alexa Fluor488 which can covalently bind to Cys101 of PRMT1 as the indicator for FP [63]. Using this FP assay, the authors screened 16,000 compounds in the Maybridge Hitfinder Collection and identified compounds 5380390 and 2818500 as specific inhibitors for the SAM-binding cysteine containing PRMTs (PRMT1 and PRMT8) but not the other PRMTs or HKMTs [63].

To this end, these direct FP strategies are homogeneous, sensitive, and particularly useful for monitoring interactions between methyltransferases and their respective substrates or the cofactor SAM. Therefore they are applicable for the identification and characterization of chemical inhibitors that target the substrate-binding pocket or SAM-binding pocket of methyltransferases [27].

However, the large amounts of enzyme required for the direct FP assay restricts its application in the large-scale discovery of methyltransferase inhibitors [57,58].

16.2.2.2 Competitive fluorescence polarization immunoassay

To reduce the amount of enzyme used in the direct FP assay, the Scott group developed a high-throughput, competitive fluorescence polarization immunoassay (FPIA) to monitor the AdoMet-utilizing methyltransferase reactions [65]. Although they used catechol-O-methyltransferase (COMT) as a model, this assay is applicable to HKMTs and PRMTs, since AdoHcy is measured in this assay. Technically, the Scott group employed commercially available anti-AdoHcy antibody and fluorescein-AdoHcy conjugate tracer to measure the level of AdoHcy generated as a by-product of the methylation reaction [65]. Before the methylation reaction occurs, the fluorescein-AdoHcy conjugate tracer is antibody bound, resulting in the deceleration of its rotation. When the methylation reaction happens, the produced AdoHcy will compete with the tracer to bind anti-AdoHcy antibody, leading to the release of tracer and consequent decrease in the fluorescence polarization signal. The decrease of the FP signal is proportional to the level of AdoHcy generated by the methylation reaction and thus to the methyltransferase activity. With this technology, the Scott group characterized the methyltransferase activity of COMT in the presence of Ro 41-0960, a well-known selective inhibitor of COMT activity [65]. The IC_{50} value obtained with competitive FPIA was consistent with the previously reported data. In addition, the authors demonstrated that the competitive FPIA is highly sensitive to initial reaction product based on their result that it can detect as low as 0.15 pmol of AdoHcy [65]. Although the authors stated that the anti-AdoHcy antibody had more than a 150-fold preference for binding AdoHcy relative to AdoMet, the antibody cross-reactivity still limits the AdoMet concentration to 3 μM, which is not well suited for some methyltransferases possessing K_m values of $1-25\,\mu$M for AdoMet [65]. The same competitive FPIA has also been applied to monitor the methyltransferase activity of EZH2 and determine the K_m for histone H3 peptide substrate [66].

16.2.3 MICROFLUIDIC CAPILLARY ELECTROPHORESIS

Microfluidic capillary electrophoresis (MCE) separates nanoliter-sized small biomolecules in a capillary quartz channel based on their charge-to-mass ratio. However, lysine methylation, unlike acetylation, does not alter the electrophoretic properties of peptide or protein substrates, prohibiting direct electrophoretic separation. To overcome this hurdle, the Janzen laboratory took advantage of the methylation-sensitive property of endoproteinase-LysC (Endo-LysC), which selectively cleaves peptide bonds on the carboxyl side of the unmethylated lysine residues, but not methylated ones [67]. Consequently, the cleaved versus noncleaved peptides can be separated on a Caliper Life Sciences EZ Reader II based on their different charge-to-mass ratios. 5-carboxylfluorescein tracer (5-FAM) was incorporated at the C-terminus of synthetic peptide substrate for detection at the end of the capillary channel (Figure 16.3). With G9a as a model methyltransferase, the authors monitored the initial velocities of G9a-catalyzed methylation reaction and measured the kinetic parameters K_m (31.7 μM for AdoMet and 36.2 μM for substrate) and k_{cat} (13 $\pm$ 3 min^{-1}), indicating that the MCE assay is highly quantitative and suitable for the kinetic analysis [67]. Moreover, the assay was configured for small molecule inhibitor screening, with a Z$'$ factor 0.92 [67]. As a proof-of-principle experiment, the authors confirmed the inhibitory mechanism of UNC0224, a potent G9a

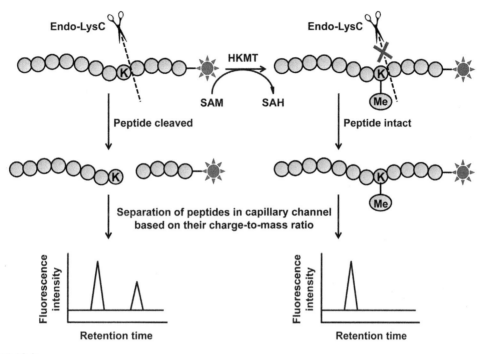

FIGURE 16.3

Microfluidic capillary electrophoresis assay concept. 5-carboxylfluorescein tracer (5-FAM) is incorporated at the C-terminus of synthetic substrate peptide for fluorescent detection. Endoproteinase-LysC (Endo-LysC) cleaves peptide bonds on the carboxyl side of unmethylated lysine residues, but not methylated ones. Then the cleaved versus intact peptides are separated in a capillary channel based on their charge-to-mass ratio and the fluorescent intensity is measured at the end of the capillary channel.

inhibitor, to be competitive with peptide substrate and determined the K_i of UNC0224 to be 1.6 nM [67]. Similarly, the Jin group utilized the MCE assay to determine the Morrison K_i of G9a inhibitor UNC0321 to be 63 pM [68]. In addition, endoproteinase-ArgC (Endo-ArgC) was shown able to distinguish unmethylated and methylated arginine catalyzed by PRMT3, indicating the possible application of MCE assay in the characterization of PRMT methyltransferase activity [67].

16.2.4 FLUORESCENCE LIFETIME ASSAY

Similar to the MCE assay, the properties of Endo-LysC and Endo-ArgC were employed by the Almac Group (UK) to develop and commercialize the *FLEXYTE*™ Fluorescence Lifetime (FLT) assays for characterization of enzymatic activities of G9a, SET7/9, and PRMT5 [69]. In principle, FLT is defined as the average time taken for the fluorescence intensity of a fluorophore to decay from the excited state to the ground state [70]. In their *FLEXYTE* FLT assay, the corresponding histone peptide substrate is labeled at amino terminus with a proprietary long lifetime fluorophore, 9-aminoacridine (9AA), and an aromatic moiety is incorporated several amino acids apart from the methylated residue [71]. The proximity between 9AA and the aromatic moiety significantly restricts the FLT of 9AA. Endo-LysC and Endo-ArgC can cleave the unmethylated peptide, releasing 9AA from quenching of the aromatic moiety and thus increasing the FLT signal. Peptide is protected from cleavage when specific lysine or arginine residue is methylated. Hence, action of histone methyltransferase can be monitored as a change in the FLT signal. The Almac Group has characterized the homogeneous *FLEXYTE* FLT assay to be compatible with a wide range of AdoMet concentrations and suitable for compound profiling as well as HTS applications with Z′ factor >0.7. Moreover, FLT is not affected by autofluorescence from compound or light scattering, therefore minimizing the background interference when compound libraries are screened [70]. One drawback for the assay is that the modified peptide substrate should be designed specifically and individually for different methyltransferases, due to the requirement of localization of the methylation site between 9AA and the aromatic moiety.

16.2.5 COUPLED ENZYME ASSAYS

Since all PMT-catalyzed methylation reactions produce a common by-product, *S*-adenosyl-L-homocysteine (AdoHcy; SAH), monitoring AdoHcy production stands out as a promising approach for establishing general enzymatic assays of all AdoMet-utilizing methyltransferases, including HKMTs and PRMTs. However, the little spectral difference between AdoHcy and AdoMet excludes the possibility of directly monitoring AdoHcy production [72]. In order to overcome this hurdle, several laboratories have carried out enzymatic assays with additional enzymes to generate AdoHcy derivatives that can be easily detected via spectrophotometric, fluorescent, or luminescent methods (Figure 16.4).

16.2.5.1 Enzyme-coupled colorimetric/spectrophotometric assays

The Zhou group reported an enzyme-coupled colorimetric assay for monitoring salicylic acid carboxyl methyltransferase (SAMT) activity [73]. In their assay, AdoHcy is first hydrolyzed by recombinant AdoHcy nucleosidase (SAHN) into adenine and *S*-ribosyl-homocysteine, the latter of which is further cleaved by recombinant *S*-ribosyl-homocysteinase (LuxS) to form homocysteine (Hcy). Subsequently, Hcy is reacted with Ellman's reagent (5,5′-dithiobis-2-nitrobenzoic acid; DTNB) to form the yellow-colored TNB (5-thio-2-nitrobenzoic acid), which can be measured by the absorption change at 412 nm [73]. With the aid of this three-step enzyme-coupled colorimetric assay, the authors determined the precise kinetic parameters of SAMT activity and demonstrated that the conversion from AdoHcy to TNB is linear from 5 μM to at least 200 μM product formed [73]. This procedure is amenable to batch assay in a larger scale because the recombinant SAHN and LuxS can be easily expressed and purified in large amounts [73]. In addition, AdoHcy has been validated as a potent feedback inhibitor for nearly all methyltransferases [74]; therefore the utilization of SAHN greatly reduces the accumulation of AdoHcy in the reaction mixture and consequently alleviates the feedback inhibitory effect. However, several limitations are noted in the DTNB-based assay. First, the conversion from DTNB to TNB is pH and temperature sensitive because the conversion only happens optimally in neutral and alkali pHs and around 27 degrees [75]. Second, DTNB is light sensitive; hence the assay should be performed in a dark environment to protect from light interference [73,75].

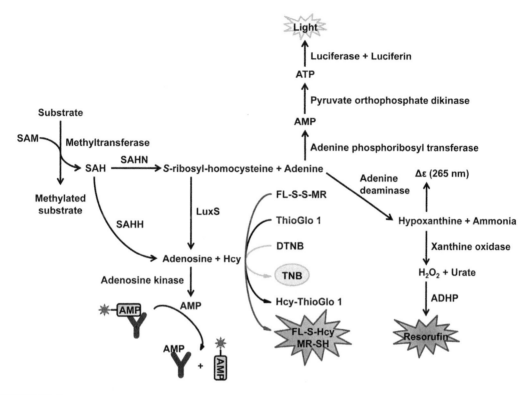

FIGURE 16.4

Coupled enzyme assays overview. All protein methyltransferase-catalyzed methylation reactions produce a common byproduct, S-adenosyl-L-homocysteine (SAH). Coupled enzyme assays utilize various coupling enzymes to generate SAH derivatives that can be easily detected via spectrophotometric, fluorescent, or luminescent methods.

Similarly, the Hevel group adapted the enzyme-coupled colorimetric assay in a continuous format, in which the adenine hydrolyzed from AdoHcy by SAHN is converted to hypoxanthine and ammonia by recombinant adenine deaminase, leading to an absorbance decrease at 265 nm that can be easily and continuously detected by UV spectrometry [72]. Using this continuous assay, the authors successfully determined the kinetic parameters of rat PRMT1 and confirmed the previously reported finding that RGG-repeats-containing substrates can be more efficiently catalyzed by PRMT1 than histone H4 [72,76]. Akin to SAHN and LuxS, recombinant adenine deaminase can be easily purified in large amounts, providing an adequate resource for the scale-up of this assay [72]. Although the continuous assay format allows more accurate kinetic analysis, some potential limitations of this assay still exist, including the interference from some protein substrates with a strong UV absorbance and the background absorbance from AdoMet [72].

16.2.5.2 Enzyme-coupled fluorescent assays

G-Biosciences further extended Hevel's enzyme-coupled continuous assay to commercialize the SAMfluoro™: SAM Methyltransferase Assay Kit [77], in which an additional coupling enzyme xanthine oxidase is employed to rapidly convert hypoxanthine to urate and hydrogen peroxide (H_2O_2). The production of H_2O_2 is subsequently measured by the reaction with 10-acetyl-3, 7-dihydroxyphenoxazine (ADHP), leading to the formation of a highly fluorescent compound resorufin ($\lambda_{ex} = 530-540$ nm and $\lambda_{em} = 585-595$ nm). Similarly, the Cayman Chemical Company also provides this fluorometric assay platform for monitoring methyltransferase activity.

The Hrycyna laboratory developed a simple and sensitive fluorescence-based two-step enzyme-coupled assay to monitor porcine COMT activity [78]. In their assay, the methylation reaction is carried out with the addition of S-adenosyl-homocysteine hydrolase (SAHH) to convert AdoHcy to adenosine and thiol homocysteine (Hcy) [78]. They also synthesized a fluorescence resonance energy transfer (FRET)-based fluorescein-cystamine-methyl red (FL-S-S-MR) reporter molecule, in which the disulfide bond within the cystamine tether can be cleaved by the thiol Hcy, leading to the activation of fluorescent signal. Meanwhile, the fluorescence intensity can be easily measured at $\lambda_{ex} = 493$ nm and $\lambda_{em} = 516$ nm [78]. The authors demonstrated that this fluorescence-based enzyme-coupled assay can detect $\leq 1 \mu M$ concentration of Hcy, which is more sensitive than the previously discussed enzyme-coupled colorimetric assay [73,78].

In a similar manner, the Trievel laboratory developed a fluorescence-based SAHH-coupled assay for monitoring enzymatic activity of *Schizosaccharomyces pombe* CLR4, a histone H3 Lys-9-specific methyltransferase, toward the synthetic histone H3 peptide substrate [79]. In general, the enzyme-coupled steps are the same as Hrycyna's FL-S-S-MR reporter assay [78,79]. The free sulfhydryl group of Hcy is then reacted with the maleimide moiety of a thiol-sensitive fluorophore, ThioGlo 1 [10-(2,5-dihydro-2,5-dioxo-1H-pyrrol-1-yl)-9-methoxy-3-oxo-, methyl ester 3H-naphthol (2,1-β)pyran-S-carboxylic acid], leading to a strong fluorescent emission. The fluorescence intensity can be measured at $\lambda_{ex} = 400$ nm and $\lambda_{em} = 515$ nm [79]. Using this fluorescent assay, the authors measured the initial velocities and determined the kinetic parameters K_{cat} (0.0099 ± 0.0003 s^{-1}) and K_m ($101 \pm 9 \mu M$) values for CLR4-catalyzed methylation of histone H3 peptide [79]. Moreover, this fluorescence-based SAHH-coupled assay can detect as low as approximately 50 pmol of AdoHcy, indicating that it is a sensitive approach to monitor methyltransferase activity [79]. However, these two enzymatic assays as well as the previously discussed DTNB-based assay are based on the reactive thiol, and thus can be interfered with by the thiol-containing reducing reagents such as dithiothreitol, β-mercaptoethanol, and TCEP that may be used to preserve optimal methyltransferase activities. In addition, the free cysteines of methyltransferases may interfere because of the reactive thiol of cysteine.

The BellBrook Labs developed another generic enzymatic assay using fluorescent immunodetection of adenosine monophosphate (AMP) [80]. In their assay, AdoHcy generated in a methylation reaction is converted to AMP by two consecutive enzymatic reactions catalyzed by SAHH and then adenosine kinase [80]. The resultant AMP is detected using the commercialized Transcreener® AMP/GMP assay kit, in which the produced AMP disrupts the interaction between AMP-Alexa® Fluor633 tracer and anti-AMP monoclonal antibody with extremely high selectivity, leading to the decrease of FP signal [81]. The authors demonstrated the merit of this assay by measuring the initial velocity of methylation reaction catalyzed by G9a and performing a pilot screen of 8800 small molecules for G9a inhibitors in the miniaturized 1536-well plate format. The Z′ was determined to be 0.59, indicating its adaptability for high throughput screening [80]. Furthermore, the authors validated that this assay can measure the initial velocity within the physiological range of AdoMet concentrations ($0.20-50 \mu M$) [80]. Another advantage of this assay is that the tracer used in the Transcreener AMP/GMP assay kit emits in the far red, which minimizes possible interference from fluorescent compounds in the HTS libraries and light scattering [81].

16.2.5.3 Enzyme-coupled luminescent assay

To increase the assay sensitivity, the Luo laboratory developed an enzyme-coupled ultrasensitive luminescence assay for monitoring protein methyltransferase activity [82]. In this assay, AdoHcy is sequentially converted to adenine, AMP, and then adenosine 5′-triphosphate (ATP) by SAHN, adenine phosphoribosyl transferase, and pyruvate orthophosphatedikinase, respectively. Then the final product ATP is quantified by a commercially available luciferase kit (ATPlite, Perkin Elmer) through the detection of luminescence signals [82]. The assay is particularly featured by its linear response to AdoHcy concentration and ultrasensitivity with the capability to trace 0.3 pmol of AdoHcy in the methylation reaction mixture [82]. The authors demonstrated the sensitivity of this assay by characterizing the steady state kinetics of G9a and SET7/9 methyltransferases. The kinetic parameters of SET7/9 were determined (K_{cat} of 0.31 min^{-1}, K_m for histone H3 peptide substrate of 29 μM, and the K_m for AdoMet of 17 μM), which were consistent with the previously reported data [82]. With the aid of this assay, the authors confirmed the ordered mechanism for SET7/9-catalyzed methylation in which AdoMet binding takes place before substrate binding [82]. Moreover, this assay has been miniaturized to 384-well plate format, which is amenable to automation and HTS. Unlike the previously discussed thiol Hcy-based fluorescent assays, this luciferase-based assay well tolerates the reducing reagents [82].

In summary, the coupled enzyme assays provide valuable tools to characterize methyltransferase activities and screen small-molecule modulators of methyltransferases in a homogenous format [1]. The capability to detect a broad range of methylated substrates through AdoHcy derivatives makes the coupled enzyme assays well suited for the kinetic analysis of methyltransferases. However, the coupled enzyme assays involve multiple coupling enzymes, so meticulous optimization is required for the assay development and validation. More importantly, when applied to HTS of methyltransferase inhibitors, the coupled enzyme assays are expected to yield high false-positive rates because the coupling enzymes may be inhibited by some compounds.

16.2.6 ANTIBODY-BASED ASSAYS

16.2.6.1 Enzyme-linked immunosorbent assay

To measure methyltransferase activity, the enzyme-linked immunosorbent assay (ELISA) employs the immobilization of peptide or protein substrate on the microplate, followed by the methylation reaction to generate methyl-lysine or methyl-arginine mark. Methyl-specific primary antibody and horseradish peroxidase (HRP)-conjugated secondary antibody are sequentially captured on the plate and then the chemiluminescent signal is detected by autoradiography (Figure 16.5A) [60]. With the support of this principle, the Bedford laboratory immobilized GST-Npl3p (*Saccharomyces cerevisiae* RNA-binding protein; substrate of yeast arginine methyltransferase Hmt1p) on the 384-well plates and successfully monitored Hmt1p enzymatic activity [60]. The merit of ELISA was highlighted by the identification of the first protein arginine methyltransferase inhibitor AMI-1, a symmetrical sulfonated urea, by screening 9,000 individual synthetic compounds from the DIVETSet library of Chembridge (San Diego, CA) [60]. The authors further validated AMI-1 to be

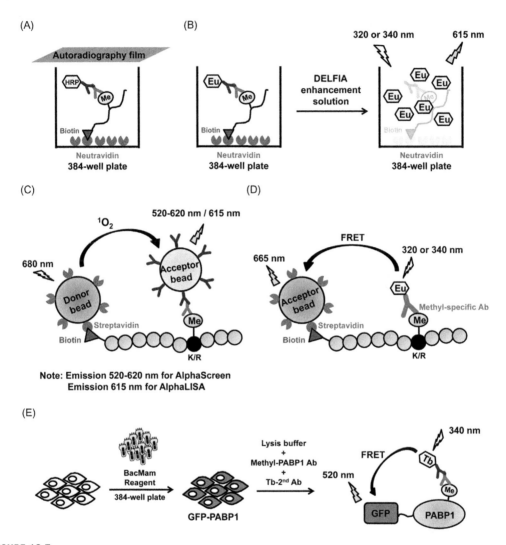

FIGURE 16.5

Schematic representation of antibody-based assays. (A) Enzyme-linked immunosorbent assay (ELISA): ELISA employs the immobilization of peptide or protein substrate on the microplate, followed by the methylation reaction to generate methyl-lysine or methyl-arginine mark. Methyl-specific primary antibody and horseradish peroxidase (HRP)-conjugated secondary antibody are sequentially captured on the plate and then the chemiluminescent signal is detected by autoradiography. (B) Dissociation-enhanced lanthanide fluoroimmunoassay (DELFIA): DELFIA is an evolved version of ELISA, in which biotinylated peptide is bound to neutravidin-coated 384-well plate and the methylation reaction is then carried out. The resultant methylated products are detected with the methyl-specific primary antibody and europium (Eu)-conjugated secondary antibody. The europium is then dissociated from the plate by addition of DELFIA enhancement solution (Perkin cell-permeable and demonstrated that AMI-1 specifically inhibits arginine, but not lysine, methyl-transferase activity [60]. Despite the identification of AMI-1, the ELISA process requires multiple washing steps that hamper automation and the chemiluminescence detection method could not generate precise quantitative data, thus limiting its broad application for characterizing methyltransferase activity.

16.2.6.2 Dissociation-enhanced lanthanide fluoroimmunoassay

The Jenuwein group adapted a dissociation-enhanced lanthanide fluoroimmunoassay (DELFIA) [83], which is an evolved version of ELISA, to monitor G9a methyltransferase activity and discover G9a specific inhibitors (Figure 16.5B) [84]. In this assay, the biotinylated histone H3 peptides were bound to neutravidin-coated 384-well microplates and dimethylated at lysine-9 by recombinant G9a. The resultant H3K9me2 products were detected with an H3K9me2-specific primary antibody along with a europium-conjugated secondary antibody [84]. The europium was then dissociated from the plate by the addition of an enhancement solution (Perkin Elmer) to form a homogeneous micellar chelate solution which has fluorescence emission at 615 nm upon excitation at 340 nm (Figure 16.5B) [83]. The signal was measured in the time-resolved fluorescence (TRF) manner, which could greatly increase signal-to-noise ratios [85]. With the aid of DELFIA, the authors identified BIX-01294 (diazepin-quinazolin-amine derivative) as a potent and selective G9a methyltransferase inhibitor from a screen of 125,000 preselected compounds [84]. The IC_{50} of BIX-01294 for G9a was also determined to be 2.7 μM using DELFIA [84]. In a similar manner, the Jung laboratory employed the DELFIA approach to validate the initial hits from the virtual screening using the

---

◄ Elmer) to form a homogeneous micellar chelate solution which has fluorescence emission at 615 nm upon excitation at 320 or 340 nm. (C) Amplified luminescence proximity homogeneous assays (AlphaScreen and AlphaLISA): Methylation of a biotinylated peptide substrate is recognized by the methyl-specific primary antibody and secondary antibody conjugated with acceptor beads (incorporating thioxene, anthracene, and rubrene for AlphaScreen or europium for AlphaLISA). Streptavidin-coated phthalocyanine photosensitizer-containing donor beads are used to capture the biotinylated peptide/antibody complex, leading to the close proximity between donor and acceptor beads. The donor beads are then excited by a laser at 680 nm to convert ambient oxygen to singlet oxygen, which is able to reach the acceptor beads within a distance of about 200 nm in solution, leading to the excitation of acceptor beads to emit light at 520−620 nm for AlphaScreen or 615 nm for AlphaLISA. (D) The LANCE *Ultra* assay is based on the principle of time-resolved fluorescence resonance energy transfer (TR-FRET), in which the methylated biotinylated peptide is captured by europium-labeled methyl-specific antibody and U*Light*™-conjugated streptavidin beads. Excitation of europium at 320 or 340 nm allows TR-FRET to occur between europium and U*Light* dye when they are in close proximity, giving rise to light emission at 665 nm. (E) Methyl-GFP-PABP1 TR-FRET assay for detecting CARM1 cellular activity. GFP-PABP1 substrate protein is transiently expressed in MCF7 human breast cancer cells plated in 384-well plate via the BacMam gene delivery system. The methylation event is captured by methyl-PABP1-specific primary antibody and terbium (Tb)-labeled secondary antibody in the cell lysate, leading to the formation of the antibody/substrate complex. Upon excitation of terbium by light at 340 nm, the proximity between terbium and GFP enables energy transfer to occur. The TR-FRET ratio of the GFP-specific signal (measured at 520 nm) to the terbium-specific signal (measured at 495 nm) reflects the level of GFP-PABP1 methylation, and thus the CARM1 methyltransferase activity.

National Cancer Institute (NCI) diversity set for inhibitors of PRMT1, with histone H4 peptide as substrate and H4R3me2-specific antibody as a detection tool [86]. Two compounds, allantodapsone and stilbamidine, were identified and validated as inhibitors of PRMT1-mediated H4R3 di-methylation [86].

The GlaxoSmithKline Corporation further reported a modified cell-based DELFIA assay to measure cellular histone H3K27 tri-methylation [87]. Briefly, PC3 prostate cancer cells are cultured in 384-well plate and treated with screening compounds. Afterward, the 384-well Nunc high-binding MaxiSorp plate (eBioscience, San Diego, CA) is used to capture endogenous histone proteins from PC3 cell lysate. After multiple washing steps, the traditional antibody-binding and signal detection procedures are performed to obtain a DELFIA signal, representing the level of cellular H3K27 tri-methylation [87].

Despite several successful cases, the ELISA and DELFIA assays are nonhomogeneous assays and require multiple washing steps, so are not ideal for rapid HTS [88]. In addition, the peptide or protein binding prior to the methylation reaction can generate large variations for quantitative measurement due to the well-to-well variation of binding efficiency.

16.2.6.3 Amplified luminescence proximity homogeneous assays

To transform the antibody-based assays into a homogeneous format, the amplified luminescence proximity homogeneous assay (AlphaScreen) was developed (Figure 16.5C). AlphaScreen is a bead-based proximity assay adapted from a diagnostic luminescent oxygen channeling assay (LOCI) [88,89]. In terms of histone methyltransferases, the Simeonov group utilized AlphaScreen to characterize G9a methyltransferase activity [90]. Technically, methylation of a biotinylated histone H3 peptide substrate is recognized by the polyclonal anti-monomethyl histone H3K9 antibody and secondary antibody conjugated with acceptor beads (incorporating thioxene, anthracene, and rubrene). Streptavidin-coated phthalocyanine photosensitizer-containing donor beads are used to capture the biotinylated peptide/antibody complex, leading to the close proximity between donor and acceptor beads. The donor beads are then excited by a laser at 680 nm to convert ambient oxygen to singlet oxygen, which is able to reach the acceptor beads within a distance of about 200 nm in solution. The singlet oxygen is then reacted with thioxene incorporated in the acceptor beads to produce light energy, which initiates the energy transfer cascade involving anthracene and rubrene, thus leading to the excitation of acceptor beads to emit light at 520−620 nm (Figure 16.5C) [88,90]. The AlphaScreen has been demonstrated to respond appropriately to BIX-01294, a methyltransferase inhibitor of G9a [90]. The same AlphaScreen approach was also employed by the Jin group to determine the selectivity of UNC0321, which is the most potent G9a inhibitor to date [68]. Using AlphaScreen, the IC_{50} of UNC0321 for G9a was determined to be 6.0 nM [68].

AlphaLISA is an evolution of the AlphaScreen technology in that the acceptor beads are chemically modified to contain europium instead of thioxene, anthracene, and rubrene (Figure 16.5C). The excitation of europium provides an emission that is more intense and spectrally defined (615 nm), thus markedly reducing the background interference [88]. The Rodriguez-Suarez group at PerkinElmer utilized the AlphaLISA technology to monitor the enzymatic activity of SET7/9 methyltransferase, using both biotinylated histone H3 peptide and full length histone H3 protein as substrates [91]. They optimized the assay in 384-well plate format and demonstrated that only nanomolar concentrations of enzyme and substrate are required to generate a robust assay to profile small-molecule inhibitors, with $Z' \geq 0.7$ [91]. The Amgen Corporation also employs AlphaLISA technology to monitor EZH2 methyltransferase activity and identify EZH2 inhibitors [66]. In a similar manner, the Jung laboratory further adapted the AlphaLISA platform for the characterization of PRMT1 methyltransferase activity, using native core histones as a substrate which better resembled the native conditions as compared to using histone peptide as the substrate [92]. In their modified AlphaLISA assay, biotinylated secondary antibody and anti-histone H3 (C-terminus) primary antibody were used to capture core histones, in place of directly using biotinylated histone H3 peptides [92]. With the aid of AlphaLISA, the authors measured the dose-dependent inhibition of PRMT1 methyltransferase activity by the PRMT1 inhibitor allantodapsone [92].

Collectively, AlphaScreen and AlphaLISA technologies provide a rapid and homogeneous assay platform as compared to the classical ELISA and DELFIA assays, in the sense that no washing steps are required [88]. They can also be miniaturized to even a 1536-well plate format, which well enables the application of HTS in large scale. In addition, the AlphaScreen and AlphaLISA assays require smaller sample volumes than the ELISA and DELFIA assays [88]. Moreover, the Perkin Elmer Incorporation provides a series of commercially available methyl mark specific antibodies directly conjugated to acceptor beads, which can be used in place of the combination of primary antibody and secondary antibody conjugated acceptor beads. Nonetheless, the AlphaScreen/AlphaLISA signal is subjected to interference from intense light and compounds that are able to quench singlet oxygen, which should be taken into consideration when designing assays for monitoring methyltransferase activity and identifying small-molecule inhibitors [88].

16.2.6.4 *Time-resolved fluorescence resonance energy transfer assays*

The Perkin Elmer Corporation developed the LANCE *Ultra* assay to measure the enzymatic activity of SET7/9 methyltransferase [91]. The LANCE *Ultra* assay is based on the principle of time-resolved fluorescence resonance energy transfer (TR-FRET), in which the methylated biotinylated histone peptide was captured by europium-labeled anti-H3K4me1-2 antibody and ULight™-conjugated streptavidin beads [91,93]. Excitation of europium at 320 or 340 nm allows TR-FRET to occur between europium and ULight dye when they are in close proximity, giving rise to light emission at 665 nm (Figure 16.5D) [91]. Taking advantage of the LANCE *Ultra* technology, the authors successfully measured the equivalent apparent K_m for AdoMet of 75 nM and profiled the inhibitory effects by AdoHcy and sinefungin, a nonselective AdoMet analog inhibitor for methyltransferases, with IC_{50} values of 93 μM and 769 nM, respectively [91]. This assay has also been miniaturized to 384-well plate format and optimized for HTS, with $Z' \geq 0.7$ [91].

Analogous to the LANCE *Ultra* assay, the Life Technologies Corporation developed the first cellular TR-FRET-based LanthaScreen assay, in which a terbium (the long-lifetime donor fluorophore)-labeled antibody is employed to detect specific modification on a GFP (the short-lifetime acceptor fluorophore)-substrate fusion protein expressed in cells [94]. Using this cellular assay, the authors interrogated a series of lysine methylations on histone H3, catalyzed by different HKMTs [94]. Our laboratory further adapted this technology to characterize the methyltransferase activity of arginine methyltransferase CARM1 in human breast cancer cells [95]. It is the first time to monitor lysine or arginine methyltransferase activity respectively in the cellular context, preserving the intracellular complexity and pathway specificity [94,95]. Taking our TR-FRET assay (Figure 16.5E) as an example, GFP-PABP1 substrate protein is transiently expressed in MCF7 human breast cancer cells plated in 384-well plate via the BacMam gene delivery system, which has been applied for drug discovery in the pharmaceutical industry [96]. The methylation event is captured by the methyl-PABP1-specific primary antibody and terbium-labeled secondary antibody in the cell lysate, leading to the formation of the antibody/substrate complex. Upon excitation of terbium by light at 340 nm, the proximity between terbium and GFP enables energy transfer to occur. The TR-FRET ratio of the GFP-specific signal (measured at 520 nm) to the terbium-specific signal (measured at 495 nm) reflects the level of GFP-PABP1 methylation, and thus the CARM1 methyltransferase activity [95]. The merit of our TR-FRET assay was demonstrated by showing the dose-dependent response to the inhibition by adenosine dialdehyde (Adox), an indirect methyltransferase inhibitor, or activation by synthetic CARM1 activators 1k, 3c, or 3g (aryl ureido acetamido indole carboxylates) [95]. Moreover, our assay system proved suitable and robust for screening allosteric activators of CARM1, as shown in a pilot screen of the NIH Clinical Collection Library, with the Z' factor determined to be 0.76 [95]. Similarly, Cisbio utilized the TR-FRET technology to commercialize the EPIgeneous™ HTRF assay products for reliable, high-throughput screening and investigation of epigenetic targets such as H3K4me2, H3K27me3, and H3K36me2 [97].

Collectively, the TR-FRET technology provides a fully homogeneous and rapid platform, without any washing, lysate transfer, or separation procedures. Importantly, the time-resolved detection method remarkably circumvents the issues of background fluorescence resulting from absorption of the excitation light by the acceptor fluorophore and auto-fluorescence emitted by some compounds [85,93]. Technically, the long-lifetime donor fluorophore and the short-lifetime fluorophore were utilized together with the pulse excitation and electronic gating detection to distinguish the temporal difference between FRET signal and the background fluorescence signal [85,93]. Particularly, the LanthaScreen cellular assay has the advantage of using the biologically relevant cellular milieu and focusing on cell-permeable hits when applied to HTS. The utilization of the BacMam gene delivery system allows for efficient and rapid substrate expression in a variety of cell lines [94,95]. Moreover, the TR-FRET ratio calculation circumvents the compound quenching problem in HTS and produces more accurate data [93].

In general, the antibody-based enzymatic assays have been broadly used for quantifying the methyltransferase activities of a variety of HKMTs and PRMTs, as well as screening small-molecule modulators. However, the antibody-based assays have some shortcomings. The major limitation is that the antibody-based assays greatly depend on the specificity and performance of the primary antibodies directed toward specific methylation marks [98]. Although many primary antibodies have been raised against different posttranslational modifications (including lysine and arginine methylation) on histone and nonhistone protein substrates, at least 25% have substantial problems of cross-reactivity or nonspecificity when more than 200 antibodies are examined [99]. Hence the specificity and performance of the individual antibody must be well characterized before applying to the antibody-based assays for monitoring methyltransferase activity. In addition, the cost of the antibodies further limits the scale-up of the antibody-based assays for HTS. Moreover, the antibody-based assays are not suitable for kinetic analysis of histone methyltransferases, as compared to the coupled enzyme assays discussed previously [1].

16.2.7 OTHER REPORTER ASSAYS FOR HISTONE METHYLTRANSFERASES

16.2.7.1 Chemical labeling of methyltransferase substrates

Although AdoMet/SAM is the universal cofactor for histone methyltransferases, emerging evidence implies the capacity of a number of SAM analogs as alternative cofactors, for example several alkyne/azido-containing SAM analogs [100,101]. However, these SAM surrogates are too bulky for the native methyltransferases to exert fast and efficient methyltransferase activity [101]. To resolve this issue, the Luo laboratory recently synthesized a novel SAM analog, propargylic *Se*-adenosyl-L-selenomethionine (ProSeAM) [102]. ProSeAM has advantages over previously reported SAM analogs due to its small and stable propargylic moiety that does not interfere with adaption by multiple native histone methyltransferases [102]. During the methyltransferase reaction, an alkyne moiety in ProSeAM is transferred onto the lysine or arginine methylation site specifically, which can be subsequently labeled and visualized by using an azido fluorescent dye (Az-Rho) or isolated by an azido-biotin tag, via copper-catalyzed azide-alkyne cycloaddition (CuAAC) chemistry [100,102]. With the aid of ProSeAM, Luo and colleagues characterized the methyltransferase activities of G9a, GLP1, and SUV39H2 in diverse cellular contexts [102]. In summary, ProSeAM serves as a powerful reporter to interrogate enzymatic activity of various methyltransferases by detecting labeled substrate.

16.2.7.2 Cell-based reporter assays

As discussed in Section 16.2.6, "Antibody-Based Assays," a number of methyl-specific antibodies have been raised against methylated lysine or arginine at specific sites. In addition to utilization in these antibody-based HTS assays, the site- and methyl-specific antibodies can also be used to detect endogenous methylated substrates by Western blot assay, in which the methylation level specifically reflects the corresponding methyltransferase activity [103]. Chromatin immunoprecipitation (ChIP) assay can also be performed to measure the change of specific histone methylation at defined genomic regions upon alteration of the corresponding histone methyltransferase activity [27]. When examining methylation by Western blot assay, one should keep in mind that, unlike lysine methylation, arginine methylation is extremely stable since no arginine demethylase has been hitherto identified. For example, our lab had shown that knocking down 90% of endogenous CARM1 in MCF7 human breast cancer cells only slightly decreases the methylation of PABP1 substrate [95]. In order to gauge CARM1 methyltransferase activity, the Bedford laboratory developed a PABP1 inducible cell line, in which the expression of Flag-tagged PABP1 can be induced by treatment of tetracycline. Therefore cells can be treated with tetracycline and small molecule modulators of CARM1 at the same time and then CARM1 methyltransferase activity can be reflected by Western blot analysis of Flag-PABP1 methylation [104]. In addition to the site- and methyl-specific antibodies, pan methyl-specific antibodies, such as ab7315 and ab23367 from Abcam for pan lysine methylation, ASYM24 and SYM11 from Millipore and D5A12 and CST#13522 from Cell Signaling Technology for pan arginine methylation, have been developed in an effort to identify methylated cellular proteins on a large scale [45]. In order to identify cancer-relevant substrates and probe the oncogenic functions of CARM1 in breast cancer, our lab took advantage of the Zinc-Finger Nuclease technology to knock out *CARM1* from several breast cancer cell lines [39]. Anti-asymmetrically dimethylated arginine (α-ADMA) antibody was utilized to pull down endogenous proteins from either parental MCF7 cells or *CARM1* knockout cells. Using these methods, we successfully identified several novel CARM1 substrates in breast cancer cells [39].

Collectively, those cell-based reporter assays in combination with methyl-specific antibodies provide invaluable tools to characterize enzymatic activity of specific histone methyltransferase. These assays, however, are semiquantitative, and are less automatic and high throughput as compared with the previously discussed assays. Hence, these cell-based reporter assays would best be used as a validation tool following the previously discussed assays.

16.3 CONCLUSIONS

Emerging evidence implies the causal effects of histone methyltransferases in human diseases and cancer. Targeting those enzymes with specific small molecule modulators has become appealing. For example, Epizyme is developing EPZ-6438, a small molecule inhibitor of EZH2, for the treatment of patients with non-Hodgkin lymphomas (PMBCL) and follicular lymphoma. In June 2013, Epizyme initiated a Phase 1/2 clinical trial of EPZ-6438 in patients with advanced solid tumors or with relapsed or refractory B-cell lymphoma [105]. Being able to study the regulation of histone methyltransferase activity comprehensively and to identify small molecule modulators has necessitated the development of various enzymatic assays. In this chapter, we have focused on the background of histone methyltransferases as well as the achievements and progress in development of enzymatic assays to measure histone methyltransferase activity. The overall strategy and "pro et contra" and application of each assay have been discussed in detail in this chapter (Table 16.1). Radioactive assays are the gold standard for characterization of methyltransferase activity, especially for validation of new substrates or new enzymes. Fluorescence polarization assays are useful

Table 16.1 Comparison of Assays

Enzymatic Assays	Applications	Advantages	Disadvantages
Radioactive Assays			
In vitro Methylation	Novel substrate identification, determination of small molecule selectivity	Sensitive, applicable to most HMTs	Time-consuming, high-cost, radiolabeled waste disposal
Filter Binding	Determination of substrate specificity, compound profiling, HTS		
SPA	Determination of substrate and compound specificity, continuous monitor of reaction, HTS		
Fluorescence Polarization (FP) Assays			
Direct FP	Monitoring interactions between methyltransferases and their respective substrates or SAM, compound profiling	Homogeneous, applicable for identification of inhibitors that target the substrate-binding pocket or SAM-binding pocket of methyltransferases	Low sensitivity, interfereable with fluorescent compounds, requirement of large amount of purified protein
FPIA	Determination of compound potency and selectivity	High sensitivity	Antibody cross-reactivity
Microfluidic Capillary Electrophoresis (MCE)			
	Kinetic analysis, small molecule screening	Highly quantitative	Good for few peptide substrates

Table 16.1 Comparison of Assays *Continued*

Enzymatic Assays	Applications	Advantages	Disadvantages
Fluorescence Lifetime (FLT) Assay			
	Compound profiling, HTS	Minimized background interference from compound auto-fluorescence or light scattering	Good for few peptide substrates
Coupled Enzyme Assays			
Colorimetric Assay	Monitoring AdoHcy production, kinetic analysis, compound profiling, HTS	Homogeneous, amendable to batch assay, feedback inhibition alleviation	Background interference, sensitive to pH, temperature and light, high false-positive rate
Fluorescent Assay		High sensitivity, feedback inhibition alleviation, homogeneous	Interference from thiol-containing reagents, high false-positive rate
Luminescent Assay		High sensitivity, feedback inhibition alleviation, reducing reagent tolerance, homogeneous	High false-positive rate
Antibody-Based Assays			
ELISA	Determination of compound potency and selectivity	Directly visible result	Semi-quantitative, high variation, non-homogeneous
DELFIA	HTS, compound profiling	Time-resolved measurement	
AlphaScreen/ AlphaLISA	HTS, compound profiling, determination of compound potency and selectivity	Rapid, homogeneous, miniaturizable, less sample volume	Subject to interference from intense light and compounds that are able to quench singlet oxygen
TR-FRET		Rapid, homogeneous, miniaturizable, less sample volume, time-resolved measurement, low background interference	Not suitable for kinetic analysis
Reporter Assays			
Chemical Labeling	Compound profiling	Sensitive	Not applicable for HTS, low throughput, semi-quantitative, lack of automation
Cell-Based Reporter Assay	Determination of compound potency and selectivity	Specific for individual methyltransferase	

for studying enzyme/substrate/cofactor interaction and the competitive inhibitors. Coupled enzyme assays are universal for all AdoMet-utilizing methyltransferases via the detection of AdoHcy derivatives and suitable for enzyme kinetic measurement as well as HTS for small molecule modulators. Antibody-based assays are sensitive and specific for individual methylation events. In general, these assays can largely benefit both the academic and pharmaceutical companies in pursuit of the epi-drugs that target specific histone methyltransferases. Herein we do not discuss the mass spectrometry technologies that are widely used to identify novel substrates and map methylation sites. Although various assay platforms are accessible to investigators, continuous development, and optimization in this area are still necessary to identify selective methylation inhibitors to be used in translational epigenetics research.

REFERENCES

[1] Luo M. Current chemical biology approaches to interrogate protein methyltransferases. ACS Chem Biol 2012;7(3):443—63.

[2] Copeland RA, Solomon ME, Richon VM. Protein methyltransferases as a target class for drug discovery. Nat Rev Drug Discov 2009;8(9):724—32.

[3] Martin C, Zhang Y. The diverse functions of histone lysine methylation. Nat Rev Mol Cell Biol 2005;6 (11):838—49.

[4] Barski A, Cuddapah S, Cui K, Roh TY, Schones DE, Wang Z, et al. High-resolution profiling of histone methylations in the human genome. Cell 2007;129(4):823—37.

[5] Kouzarides T. Histone methylation in transcriptional control. Curr Opin Genet Dev 2002;12 (2):198—209.

[6] Schneider R, Bannister AJ, Kouzarides T. Unsafe SETs: histone lysine methyltransferases and cancer. Trends Biochem Sci 2002;27(8):396—402.

[7] Varier RA, Timmers HT. Histone lysine methylation and demethylation pathways in cancer. Biochim Biophys Acta 2011;1815(1):75—89.

[8] Dillon SC, Zhang X, Trievel RC, Cheng X. The SET-domain protein superfamily: protein lysine methyltransferases. Genome Biol 2005;6(8):227.

[9] Chase A, Cross NC. Aberrations of EZH2 in cancer. Clin Cancer Res 2011;17(9):2613—18.

[10] Ren G, Baritaki S, Marathe H, Feng J, Park S, Beach S, et al. Polycomb protein EZH2 regulates tumor invasion via the transcriptional repression of the metastasis suppressor RKIP in breast and prostate cancer. Cancer Res 2012;72(12):3091—104.

[11] Kleer CG, Cao Q, Varambally S, Shen R, Ota I, Tomlins SA, et al. EZH2 is a marker of aggressive breast cancer and promotes neoplastic transformation of breast epithelial cells. Proc Natl Acad Sci USA 2003;100(20):11606—11.

[12] Chang CJ, Yang JY, Xia W, Chen CT, Xie X, Chao CH, et al. EZH2 promotes expansion of breast tumor initiating cells through activation of RAF1-beta-catenin signaling. Cancer Cell 2011;19 (1):86—100.

[13] Albert M, Helin K. Histone methyltransferases in cancer. Semin Cell Dev Biol 2010;21(2):209—20.

[14] Wang X, Yeh S, Wu G, Hsu CL, Wang L, Chiang T, et al. Identification and characterization of a novel androgen receptor coregulator ARA267-alpha in prostate cancer cells. J Biol Chem 2001;276 (44):40417—23.

[15] Chesi M, Nardini E, Lim RS, Smith KD, Kuehl WM, Bergsagel PL. The t(4;14) translocation in myeloma dysregulates both FGFR3 and a novel gene, MMSET, resulting in IgH/MMSET hybrid transcripts. Blood 1998;92(9):3025—34.

[16] Angrand PO, Apiou F, Stewart AF, Dutrillaux B, Losson R, Chambon P. NSD3, a new SET domain-containing gene, maps to 8p12 and is amplified in human breast cancer cell lines. Genomics 2001;74 (1):79—88.

[17] Hu M, Sun XJ, Zhang YL, Kuang Y, Hu CQ, Wu WL, et al. Histone H3 lysine 36 methyltransferase Hypb/Setd2 is required for embryonic vascular remodeling. Proc Natl Acad Sci USA 2010;107 (7):2956−61.

[18] Hamamoto R, Furukawa Y, Morita M, Iimura Y, Silva FP, Li M, et al. SMYD3 encodes a histone methyltransferase involved in the proliferation of cancer cells. Nat Cell Biol 2004;6(8):731−40.

[19] Buyse IM, Shao G, Huang S. The retinoblastoma protein binds to RIZ, a zinc-finger protein that shares an epitope with the adenovirus E1A protein. Proc Natl Acad Sci USA 1995;92(10):4467−71.

[20] Abbondanza C, Medici N, Nigro V, Rossi V, Gallo L, Piluso G, et al. The retinoblastoma-interacting zinc-finger protein RIZ is a downstream effector of estrogen action. Proc Natl Acad Sci USA 2000;97 (7):3130−5.

[21] Subramanian K, Jia D, Kapoor-Vazirani P, Powell DR, Collins RE, Sharma D, et al. Regulation of estrogen receptor alpha by the SET7 lysine methyltransferase. Mol Cell 2008;30(3):336−47.

[22] Kurash JK, Lei H, Shen Q, Marston WL, Granda BW, Fan H, et al. Methylation of p53 by Set7/9 mediates p53 acetylation and activity in vivo. Mol Cell 2008;29(3):392−400.

[23] Min J, Feng Q, Li Z, Zhang Y, Xu RM. Structure of the catalytic domain of human DOT1L, a non-SET domain nucleosomal histone methyltransferase. Cell 2003;112(5):711−23.

[24] Kim SK, Jung I, Lee H, Kang K, Kim M, Jeong K, et al. Human histone H3K79 methyltransferase DOT1L protein [corrected] binds actively transcribing RNA polymerase II to regulate gene expression. J Biol Chem 2012;287(47):39698−709.

[25] Bernt KM, Zhu N, Sinha AU, Vempati S, Faber J, Krivtsov AV, et al. MLL-rearranged leukemia is dependent on aberrant H3K79 methylation by DOT1L. Cancer Cell 2011;20(1):66−78.

[26] Bernt KM, Armstrong SA. A role for DOT1L in MLL-rearranged leukemias. Epigenomics 2011;3 (6):667−70.

[27] Li KK, Luo C, Wang D, Jiang H, Zheng YG. Chemical and biochemical approaches in the study of histone methylation and demethylation. Med Res Rev 2012;32(4):815−67.

[28] Yang Y, Bedford MT. Protein arginine methyltransferases and cancer. Nat Rev Cancer 2013;13 (1):37−50.

[29] Wei H, Mundade R, Lange KC, Lu T. Protein arginine methylation of non-histone proteins and its role in diseases. Cell Cycle 2014;13(1):32−41.

[30] Bedford MT. Arginine methylation at a glance. J Cell Sci 2007;120(Pt 24):4243−6.

[31] Di Lorenzo A, Bedford MT. Histone arginine methylation. FEBS Lett 2011;585(13):2024−31.

[32] Bedford MT, Clarke SG. Protein arginine methylation in mammals: who, what, and why. Mol Cell 2009;33(1):1−13.

[33] Bedford MT, Richard S. Arginine methylation an emerging regulator of protein function. Mol Cell 2005;18(3):263−72.

[34] Le Romancer M, Treilleux I, Leconte N, Robin-Lespinasse Y, Sentis S, Bouchekioua-Bouzaghou K, et al. Regulation of estrogen rapid signaling through arginine methylation by PRMT1. Mol Cell 2008;31(2):212−21.

[35] Qi C, Chang J, Zhu Y, Yeldandi AV, Rao SM, Zhu YJ. Identification of protein arginine methyltransferase 2 as a coactivator for estrogen receptor alpha. J Biol Chem 2002;277(32):28624−30.

[36] Singh V, Miranda TB, Jiang W, Frankel A, Roemer ME, Robb VA, et al. DAL-1/4.1B tumor suppressor interacts with protein arginine N-methyltransferase 3 (PRMT3) and inhibits its ability to methylate substrates in vitro and in vivo. Oncogene 2004;23(47):7761−71.

[37] Chen D, Ma H, Hong H, Koh SS, Huang SM, Schurter BT, et al. Regulation of transcription by a protein methyltransferase. Science 1999;284(5423):2174−7.

[38] Wu J, Xu W. Histone H3R17me2a mark recruits human RNA polymerase-associated factor 1 complex to activate transcription. Proc Natl Acad Sci USA 2012;109(15):5675−80.

[39] Wang L, Zhao Z, Meyer MB, Saha S, Yu M, Guo A, et al. CARM1 methylates chromatin remodeling factor BAF155 to enhance tumor progression and metastasis. Cancer Cell 2014;25(1):21−36.

[40] El-Andaloussi N, Valovka T, Toueille M, Steinacher R, Focke F, Gehrig P, et al. Arginine methylation regulates DNA polymerase beta. Mol Cell 2006;22(1):51−62.

[41] Jelinic P, Stehle JC, Shaw P. The testis-specific factor CTCFL cooperates with the protein methyltransferase PRMT7 in H19 imprinting control region methylation. PLoS Biol 2006;4(11):e355.

[42] Verbiest V, Montaudon D, Tautu MT, Moukarzel J, Portail JP, Markovits J, et al. Protein arginine (N)-methyl transferase 7 (PRMT7) as a potential target for the sensitization of tumor cells to camptothecins. FEBS Lett 2008;582(10):1483−9.

[43] Lee J, Sayegh J, Daniel J, Clarke S, Bedford MT. PRMT8, a new membrane-bound tissue-specific member of the protein arginine methyltransferase family. J Biol Chem 2005;280(38):32890−6.

[44] Sayegh J, Webb K, Cheng D, Bedford MT, Clarke SG. Regulation of protein arginine methyltransferase 8 (PRMT8) activity by its N-terminal domain. J Biol Chem 2007;282(50):36444−53.

[45] Cheng D, Vemulapalli V, Bedford MT. Methods applied to the study of protein arginine methylation. Methods Enzymol 2012;512:71−92.

[46] Cheng D, Bedford MT. Xenoestrogens regulate the activity of arginine methyltransferases. Chembiochem 2011;12(2):323−9.

[47] Greiner D, Bonaldi T, Eskeland R, Roemer E, Imhof A. Identification of a specific inhibitor of the histone methyltransferase SU(VAR)3-9. Nat Chem Biol 2005;1(3):143−5.

[48] Purandare AV, Chen Z, Huynh T, Pang S, Geng J, Vaccaro W, et al. Pyrazole inhibitors of coactivator associated arginine methyltransferase 1 (CARM1). Bioorg Med Chem Lett 2008;18(15):4438−41.

[49] Anastassiadis T, Deacon SW, Devarajan K, Ma H, Peterson JR. Comprehensive assay of kinase catalytic activity reveals features of kinase inhibitor selectivity. Nat Biotechnol 2011;29(11):1039−45.

[50] Horiuchi KY, Eason MM, Ferry JJ, Planck JL, Walsh CP, Smith RF, et al. Assay development for histone methyltransferases. Assay Drug Dev Technol 2013;11(4):227−36.

[51] McCabe MT, Ott HM, Ganji G, Korenchuk S, Thompson C, Van Aller GS, et al. EZH2 inhibition as a therapeutic strategy for lymphoma with EZH2-activating mutations. Nature 2012;492(7427):108−12.

[52] Suh-Lailam BB, Hevel JM. A fast and efficient method for quantitative measurement of S-adenosyl-L-methionine-dependent methyltransferase activity with protein substrates. Anal Biochem 2010;398(2):218−24.

[53] Rathert P, Cheng X, Jeltsch A. Continuous enzymatic assay for histone lysine methyltransferases. BioTechniques 2007;43(5):602−6.

[54] Dhayalan A, Dimitrova E, Rathert P, Jeltsch A. A continuous protein methyltransferase (G9a) assay for enzyme activity measurement and inhibitor screening. J Biomol Screen 2009;14(9):1129−33.

[55] Wu J, Xie N, Feng Y, Zheng YG. Scintillation proximity assay of arginine methylation. J Biomol Screen 2012;17(2):237−44.

[56] Ibanez G, Shum D, Blum G, Bhinder B, Radu C, Antczak C, et al. A high throughput scintillation proximity imaging assay for protein methyltransferases. Comb Chem High Throughput Screen 2012;15(5):359−71.

[57] Owicki JC. Fluorescence polarization and anisotropy in high throughput screening: perspectives and primer. J Biomol Screen 2000;5(5):297−306.

[58] Lea WA, Simeonov A. Fluorescence polarization assays in small molecule screening. Expert Opin Drug Discov 2011;6(1):17−32.

[59] Feng Y, Xie N, Wu J, Yang C, Zheng YG. Inhibitory study of protein arginine methyltransferase 1 using a fluorescent approach. Biochem Biophys Res Commun 2009;379(2):567−72.

[60] Cheng D, Yadav N, King RW, Swanson MS, Weinstein EJ, Bedford MT. Small molecule regulators of protein arginine methyltransferases. J Biol Chem 2004;279(23):23892−9.

[61] Feng Y, Xie N, Jin M, Stahley MR, Stivers JT, Zheng YG. A transient kinetic analysis of PRMT1 catalysis. Biochemistry 2011;50(32):7033−44.

[62] Liu F, Chen X, Allali-Hassani A, Quinn AM, Wasney GA, Dong A, et al. Discovery of a 2,4-diamino-7-aminoalkoxyquinazoline as a potent and selective inhibitor of histone lysine methyltransferase G9a. J Med Chem 2009;52(24):7950−3.

[63] Dillon MB, Bachovchin DA, Brown SJ, Finn MG, Rosen H, Cravatt BF, et al. Novel inhibitors for PRMT1 discovered by high-throughput screening using activity-based fluorescence polarization. ACS Chem Biol 2012;7(7):1198−204.

[64] Weerapana E, Wang C, Simon GM, Richter F, Khare S, Dillon MB, et al. Quantitative reactivity profiling predicts functional cysteines in proteomes. Nature 2010;468(7325):790−5.

[65] Graves TL, Zhang Y, Scott JE. A universal competitive fluorescence polarization activity assay for S-adenosylmethionine utilizing methyltransferases. Anal Biochem 2008;373(2):296−306.

[66] Simard JR, Plant M, Emkey R, Yu V. Development and implementation of a high-throughput AlphaLISA assay for identifying inhibitors of EZH2 methyltransferase. Assay Drug Dev Technol 2013;11(3):152−62.

[67] Wigle TJ, Provencher LM, Norris JL, Jin J, Brown PJ, Frye SV, et al. Accessing protein methyltransferase and demethylase enzymology using microfluidic capillary electrophoresis. Chem Biol 2010;17 (7):695−704.

[68] Liu F, Chen X, Allali-Hassani A, Quinn AM, Wigle TJ, Wasney GA, et al. Protein lysine methyltransferase G9a inhibitors: design, synthesis, and structure activity relationships of 2,4-diamino-7-aminoalkoxy-quinazolines. J Med Chem 2010;53(15):5844−57.

[69] Almac Group Ltd. 2015. Flexyte™ fluorescence lifetime assays. Website available at: <http://www.almacgroup.com/api-services-chemical-development/flexyte-assays/epigenetics/hmts/>.

[70] Paterson MJ, Dunsmore CJ, Hurteaux R, Maltman BA, Cotton GJ, Gray A. A fluorescence lifetime-based assay for serine and threonine kinases that is suitable for high-throughput screening. Anal Biochem 2010;402(1):54−64.

[71] Maltman BA, Dunsmore CJ, Couturier SC, Tirnaveanu AE, Delbederi Z, McMordie RA, et al. 9-Aminoacridine peptide derivatives as versatile reporter systems for use in fluorescence lifetime assays. Chem Commun 2010;46(37):6929−31.

[72] Dorgan KM, Wooderchak WL, Wynn DP, Karschner EL, Alfaro JF, Cui Y, et al. An enzyme-coupled continuous spectrophotometric assay for S-adenosylmethionine-dependent methyltransferases. Anal Biochem 2006;350(2):249−55.

[73] Hendricks CL, Ross JR, Pichersky E, Noel JP, Zhou ZS. An enzyme-coupled colorimetric assay for S-adenosylmethionine-dependent methyltransferases. Anal Biochem 2004;326(1):100−5.

[74] Graham I. Homocysteine in health and disease. Ann Intern Med 1999;131(5):387−8.

[75] Schaberle TF, Siba C, Hover T, Konig GM. An easy-to-perform photometric assay for methyltransferase activity measurements. Anal Biochem 2013;432(1):38−40.

[76] Rajpurohit R, Lee SO, Park JO, Paik WK, Kim S. Enzymatic methylation of recombinant heterogeneous nuclear RNP protein A1. Dual substrate specificity for S-adenosylmethionine: histone-arginine N-methyltransferase. J Biol Chem 1994;269(2):1075−82.

[77] Geno Technology Inc. 2014. SAM-fluoro: SAM Methyltransferase Assay. G-Biosciences website. Available at: <http://www.gbiosciences.com/ResearchProducts/samfluoro.aspx>.

[78] Wang C, Leffler S, Thompson DH, Hrycyna CA. A general fluorescence-based coupled assay for S-adenosylmethionine-dependent methyltransferases. Biochem Biophys Res Commun 2005;331 (1):351−6.

[79] Collazo E, Couture JF, Bulfer S, Trievel RC. A coupled fluorescent assay for histone methyltransferases. Anal Biochem 2005;342(1):86−92.

[80] Klink TA, Staeben M, Twesten K, Kopp AL, Kumar M, Dunn RS, et al. Development and validation of a generic fluorescent methyltransferase activity assay based on the transcreener AMP/GMP assay. J Biomol Screen 2012;17(1):59−70.

[81] Staeben M, Kleman-Leyer KM, Kopp AL, Westermeyer TA, Lowery RG. Development and validation of a transcreener assay for detection of AMP- and GMP-producing enzymes. Assay Drug Dev Technol 2010;8(3):344−55.

[82] Ibanez G, McBean JL, Astudillo YM, Luo M. An enzyme-coupled ultrasensitive luminescence assay for protein methyltransferases. Anal Biochem 2010;401(2):203−10.

[83] Garcia-Martinez LF, Bilter GK, Wu J, O'Neill J, Barbosa MS, Kovelman R. In vitro high-throughput screening assay for modulators of transcription. Anal Biochem 2002;301(1):103−10.

[84] Kubicek S, O'Sullivan RJ, August EM, Hickey ER, Zhang Q, Teodoro ML, et al. Reversal of H3K9me2 by a small-molecule inhibitor for the G9a histone methyltransferase. Mol Cell 2007;25 (3):473−81.

[85] Morrison LE. Time-resolved detection of energy transfer: theory and application to immunoassays. Anal Biochem 1988;174(1):101−20.

[86] Spannhoff A, Heinke R, Bauer I, Trojer P, Metzger E, Gust R, et al. Target-based approach to inhibitors of histone arginine methyltransferases. J Med Chem 2007;50(10):2319−25.

[87] Xie W, Ames RS, Li H. A cell-based high-throughput screening assay to measure cellular histone h3 lys27 trimethylation with a modified dissociation-enhanced lanthanide fluorescent immunoassay. J Biomol Screen 2012;17(1):99−107.

[88] Eglen RM, Reisine T, Roby P, Rouleau N, Illy C, Bosse R, et al. The use of AlphaScreen technology in HTS: current status. Curr Chem Genomics 2008;1:2−10.

[89] Ullman EF, Kirakossian H, Switchenko AC, Ishkanian J, Ericson M, Wartchow CA, et al. Luminescent oxygen channeling assay (LOCI): sensitive, broadly applicable homogeneous immunoassay method. Clin Chem 1996;42(9):1518−26.

[90] Quinn AM, Allali-Hassani A, Vedadi M, Simeonov A. A chemiluminescence-based method for identification of histone lysine methyltransferase inhibitors. Mol Biosyst 2010;6(5):782−8.

[91] Gauthier N, Caron M, Pedro L, Arcand M, Blouin J, Labonte A, et al. Development of homogeneous nonradioactive methyltransferase and demethylase assays targeting histone H3 lysine 4. J Biomol Screen 2012;17(1):49−58.

[92] Hauser AT, Bissinger EM, Metzger E, Repenning A, Bauer UM, Mai A, et al. Screening assays for epigenetic targets using native histones as substrates. J Biomol Screen 2012;17(1):18−26.

[93] Hemmila I. LANCETM: homogeneous assay platform for HTS. J Biomol Screen 1999;4(6):303−7.

[94] Machleidt T, Robers MB, Hermanson SB, Dudek JM, Bi K. TR-FRET cellular assays for interrogating posttranslational modifications of histone H3. J Biomol Screen 2011;16(10):1236−46.

[95] Zeng H, Wu J, Bedford MT, Sbardella G, Hoffmann FM, Bi K, et al. A TR-FRET-based functional assay for screening activators of CARM1. Chembiochem 2013;14(7):827−35.

[96] Kost TA, Condreay JP, Ames RS, Rees S, Romanos MA. Implementation of BacMam virus gene delivery technology in a drug discovery setting. Drug Discov Today 2007;12(9−10):396−403.

[97] Cisbio.com. 2015. Epigenetic screening. Cisbio Assays website. Available at: <http://www.htrf.com/usa/epigenetic-screening>.

[98] Quinn AM, Simeonov A. Methods for activity analysis of the proteins that regulate histone methylation. Curr Chem Genomics 2011;5(Suppl. 1):95−105.

[99] Egelhofer TA, Minoda A, Klugman S, Lee K, Kolasinska-Zwierz P, Alekseyenko AA, et al. An assessment of histone-modification antibody quality. Nat Struct Mol Biol 2011;18(1):91−3.

[100] Slade DJ, Subramanian V, Fuhrmann J, Thompson PR. Chemical and biological methods to detect post-translational modifications of arginine. Biopolymers 2013.

[101] Wang R, Zheng W, Yu H, Deng H, Luo M. Labeling substrates of protein arginine methyltransferase with engineered enzymes and matched S-adenosyl-L-methionine analogues. J Am Chem Soc 2011;133 (20):7648−51.

[102] Bothwell IR, Islam K, Chen Y, Zheng W, Blum G, Deng H, et al. Se-adenosyl-L-selenomethionine cofactor analogue as a reporter of protein methylation. J Am Chem Soc 2012;134(36):14905−12.

[103] Qian J, Lu L, Wu J, Ma H. Development of multiple cell-based assays for the detection of histone H3 Lys27 trimethylation (H3K27me3). Assay Drug Dev Technol 2013;11(7):449−56.

[104] Castellano S, Spannhoff A, Milite C, Dal Piaz F, Cheng D, Tosco A, et al. Identification of small-molecule enhancers of arginine methylation catalyzed by coactivator-associated arginine methyltransferase 1. J Med Chem 2012;55(22):9875−90.

[105] Epizyme Inc. 2013. EZH2 inhibitor − EPZ-6438 for genetically defined non-Hodgkin lymphoma and INI1-deficient tumors. Epizyme website available at: <http://www.epizyme.com/programs/ezh2-inhibitor/>.

HISTONE METHYLTRANSFERASE INHIBITORS FOR CANCER THERAPY

Keqin Kathy Li[1,2], Kenneth Huang[1], Shukkoor Kondengaden[1], Jonathan Wooten[1], Hamed Reyhanfard[1], Zhang Qing[1], Bingxue Chris Zhai[3], and Peng George Wang[1]

[1]Department of Chemistry, Georgia State University, Atlanta, GA, USA [2]State Key Lab of Medical Genomic, Ruijin Hospital Affiliated to Medical School of Shanghai Jiaotong University, Shanghai, P.R. China [3]Department of Biology, Georgia State University, Atlanta, GA, USA

CHAPTER OUTLINE

Y.G. Zheng (Ed): Epigenetic Technological Applications. DOI: http://dx.doi.org/10.1016/B978-0-12-801080-8.00017-X

17.1 **INTRODUCTION**

Epigenetics is the study of alterations to gene expression that occur without entailing any modifications to the DNA sequence itself [1−3]. Efforts over the last 15 years have provided insight into the various roles that epigenetic regulation plays in cell and organism models, ranging from genome imprinting and establishment, gene transcription and silencing, and cell differentiation [4]. The primary modifications for epigenetic regulation are DNA methylation, histone modifications, and expression of noncoding RNAs.

Eukaryotic organisms tightly pack genomes into chromatin; the basic chromatin units are nucleosomes, comprised of 146 DNA base pairs wrapped around an octamer core of histones. This structure provides the basis on which all nuclear processes linked to genetic activity is based. Within each nucleosome, there are two copies of histones H2A, H2B, H3, and H4. The core histones can be characterized by a distinguishing globular fold domain that is responsible for engaging in histone−histone and histone−DNA interactions. Interactions by this fold domain make up the principles of how the nucleosome disc forms. From this disc, the N-terminals of the histones protrude out in an unstructured manner [5]. Chromatin modifications manifest through two means: methylation of a cytosine nucleotide on the DNA, or posttranslational modifications (PTMs) on the histones. It should be noted that, despite the positioning of the N-terminal, modifications to histones are not restricted and may occur elsewhere along the histone itself. Likewise, different histone PTMs are able to interact with each other; interplays resulting from PTM patterns have been suggested to serve as regulatory motifs [6]. Known as histone codes, these motifs are written by chromatin-modifying enzymes and then read downstream by effectors, which in turn signal the transcription status of target genes [7]. The exact impact of histone marks is dependent upon the context of where methylation occurs, making deciphering the epigenetic network a daunting task.

Within the number of genes regulated by DNA methylation and histone modifications, many of the target genes are tumor-specific [8,9]. As with genetic alterations, mutations, and translocations, aberrant alterations of the epigenetic landscape are fundamental processes in cancer development, and correlate with all stages of cancer development: initiation, progress, invasion, and metastasis [10]. Two methods by which cancer cells escape the cell checkpoint machinery are the anomalous silencing of tumor suppressor genes and magnification of oncogenes [11,12]. Notably, these changes can occur in the initial stages of tumorigenesis and are required to maintain genomic methylation.

Epigenetic alterations occur gradually, providing an opportunity to develop strategies aimed toward the preliminary stages of diseases. Since epigenetic alterations are distinct from genetic mutations by virtue of being reversible, the enzymes that are responsible for establishing and maintaining epigenetic marks are now considered a new class of targets for treatments. The development of epigenetic therapeutics has been heralded by small molecule inhibitors targeting enzymes that have been shown to contribute to tumorigenesis and cancer cell renewal [13]. The study of histone methylation has been revitalized in recent years with the advancement and application of new analytical methods and technologies. Crystal structures of HMTs provide insight into the biological roles that epigenetic marks have as structural modifiers and how methylation alters biological function (Figure 17.1). Here, we discuss recent developments in histone methyltransferases (HMTs) related to cancer and the inhibitors targeting them.

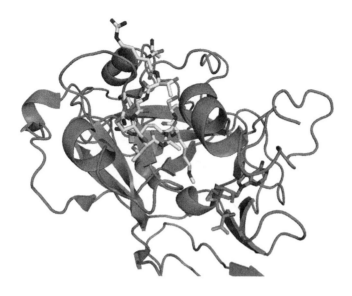

FIGURE 17.1

Crystal structure of GLP catalytic domain in complex with SAH and H3K9me1 (PDB 3HNA). The whole structure of GLP is colored in cyan. The H3K9me1 peptide is colored yellow and SAH is colored in cyan. (*This figure is reproduced in color in the color plate section.*)

17.2 HISTONE LYSINE METHYLTRANSFERASES

Presently, more than 50 lysine methyltransferases have been identified in humans, many of which display substrate specificity. All but one histone lysine methyltransferase (HKMT) possess a characteristic SET (Su(var)3−9, enhancer-of-Zeste, trihorax) domain that is evolutionarily conserved and the source of the methylation activity. The ε-amino group of lysine residues can be mono-, di-, and trimethylated. The functional diversity of lysine methylation is due to the differential methylation patterns, and this range of genetic alterations through HKMT activity has been tied to human cancers.

17.2.1 DOT1L

DOT1L methylates histone 3 lysine 79 (H3K79) and is principally responsible for gene regulation [14]. DOT1L catalyzes the mono-, di-, and trimethylation of the ε-amino group on H3K79 by using S-adenosyl-L-methionine (SAM) as the source of the methyl groups (Figure 17.2).

Composed of 1,537 highly conserved amino acids, the C-terminal of mammalian DOT1L interacts with transcriptionally relevant proteins [14]. DOT1L is a class I methyltransferase; other known HKMTs are class V methyltransferases and possess a conserved SET domain [14]. The differences are further highlighted based on target location − H3K79 lies within histone H3's ordered core, as DOT1L can only methylate nucleosomes, while other HMKT substrates are situated in unordered histone tails.

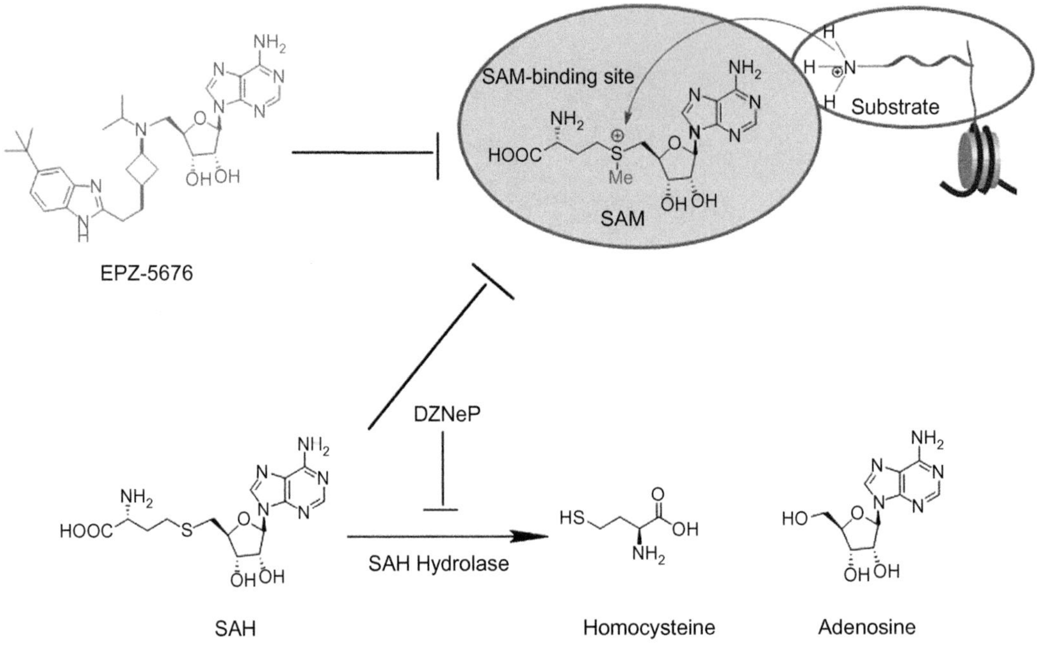

FIGURE 17.2

Schematic representation of the DOT1L activity; DOT1L brings the substrate (H3 peptide) and the cofactor (SAM) together and transfers a methyl group from SAM to H3 lysine.

In humans, part of the DOT1L amino acid sequence (1−351) is conserved in H3K79 methyl-transferases. However, the recombinant proteins are inactive enzymatically − 1−416 is the minimal length for HKMT activity. This is because the C-terminal residues 391−416 are critical to the electrostatic interactions in substrate binding and recognition; negatively charged DNA in nucleosomes binds to the positively charged C-terminal [14]. The regions responsible for cofactor binding are residues 161−169, 186−191, 239−245, 133−139, and 221−224. Regions 161−169, 186−191, and 239−245 are motifs found in methyltransferases that utilize SAM as a cofactor, while residues 133−139 and 221−224 are unique to DOT1L. Residues 133−139 are noteworthy for being part of a flexible loop that regulates SAM/SAH entry and exit, respectively.

When SAM is bound to DOT1L, the majority of the cofactor is buried inside the protein, correlating to its strong binding affinity. The adenosine ring of SAM constitutes five hydrogen bonds with DOT1L with Glu 186, Lys 187, Asp 222, and Phe 223, as well as contributing π−π stacking and hydrophobic interactions [14]. Additional hydrogen bonds can be formed between SAM and Thr 139, Asp 161, Gly 163, and Gln 168 [14]. The contributed methyl group of SAM is allocated to the lysine channel, which is large enough for a trimethylated lysine.

17.2.1.1 DOT1L and diseases

In translocated acute leukemia, DOT1L is required to initiate tumorigenesis and maintain the malignant phenotype of mixed-lineage leukemia (MLL) with gene translocations [15]. MLL-rearranged leukemia of either the acute lymphocytic or acute myeloid phenotype is reported in about 10% of child/adult and 80% of infant acute leukemia; this form of leukemia has remarkably poor prognosis [14,16]. The enzymatic activity of DOT1L is known to drive cellular proliferation in MLL-rearranged leukemia. Four common fusion partners of MLL (AF4, AF9, AF10, and ENL) have displayed the capacity to bind to DOT1L to form a transcription complex [14,17].

H3K79 hypermethylation has also been observed in lung cancer, and DOT1L deficiency has been tied to the inhibition of proliferation in lung cancer cells. Induced cell senescence and growth inhibition of lung cancer cells can be achieved through siRNA knockdown; overexpression of DOT1L can similarly reverse changes induced by siRNA [14]. Inhibition of DOT1L has been shown to induce pluripotency in somatic cells, suggesting a link to cellular reprogramming. This is further evidenced by disruption of genes along DOT1L methylation pathways reducing mRNA and promoting reprogramming [14].

17.2.1.2 Inhibitors of DOT1L

Due to the critical role DOT1L has in MLL-rearranged leukemias, there have been numerous attempts to synthesize selective inhibitors. A number of potent and selective DOT1L inhibitors have been discovered, which can be divided into four categories based on their mode of action: SAH mimetic compounds (Table 17.1, entries 1−3), mechanism-based compounds (Table 17.1, entries 4,5), compounds with either benzimidazole or urea (Table 17.1, entries 6−13), and carbamate-containing compounds (Table 17.1, entries 14,15).

S-adenosyl homocysteine (SAH), the resulting product of the methyl transfer, can also act as a potent inhibitor of DOT1L. However, its amino acid composition severely limits its therapeutic uses, as it is unable to enter cells. Carbamate containing compounds were found to be inhibitors of DOT1L (Ki = 20 μM), but lacked specificity. However, this class of compounds resulted in the discovery of potent, urea-containing DOT1L inhibitors (Ki = 13 nM) and EPZ04777 (Ki = 0.3 nM) [14,18]. EPZ004777 was the first selective DOT1L inhibitor discovered, and was reported as having selective activity in treating MLL-rearranged leukemias (Table 17.1, entry 11) [18]. However, EPZ004777 has a poor half-life in the body; the poor pharmacokinetics are due to the inclusion of ribose. Since multiple enzymes can recognize the adenosine moiety in ribose, adenine tends to be rapidly cleaved, causing metabolic instability [14].

Structure-guided modifications led to the SAM competitive derivative EPZ-5676 (Table 17.1, entry 12). The provision of more hydrogen bond donors and a reduction in conformational flexibility reduces clearance from plasma to increase bioavailability to improve enzyme (Ki = 0.08 nM) and biological activity [18]. EPZ-5676 is superior to previously described inhibitors of DOT1L, including EPZ004777. This is attributed to how it occupies the SAM pocket and induces conformational changes in DOT1L; a hydrophobic pocket beyond the amino acid portion of SAM opens to contribute to binding (Figure 17.3A). Crystallographic studies have confirmed the high affinity and potent inhibition of DOT1L by EPZ-5676 due to this conformational change.

EPZ5676 exhibits high selectivity for DOT1L over other HMTs, and only weakly associates with PRMT5; it is currently in phase I clinical trials to treat patients with MLL-rearranged

Table 17.1 Molecular Structures of Known HMT Inhibitors

G9a/GLP Inhibitors

SAH-like

SAH
Ki=160 nM
less selective

1

Ki= 290 nM
Selective

2

Ki= 12 mM

3

Mechanism Based

IC$_{50}$= 16 mM

4

IC$_{50}$= 120 nM
Selective

5

Urea- and Benzimidazole containing

SYC-522 (R=H), Ki= 0.5 nM
SYC-534 (R=Me), Ki = 0.8 nM
Selective

6

R=Cl, Ki= 82 nM
R=t-Bu, Ki = 70 nm
R= CF$_3$, Ki 58 nm

7

SGC 0946
IC$_{50}$= 0.3 nM

8

SYC 687, Ki= 1.1 nM
Selective

9

SYC 687, Ki= 1.3 nM
Selective

10

EPZ 004777
Ki = 0.3 nm

11

EPZ-5676
Ki= 80 pM
37000 fold selective over other PMTs
Plasma half life 0.25 to 1.5 hr in rats
Phase I Clinical trial

12

Ki= 13 nM
Selective

13

Carbamate containing

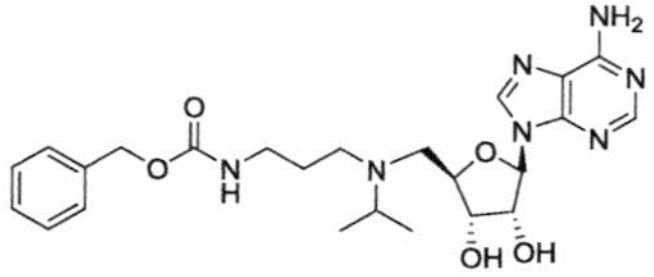

Ki= 22 mM
14

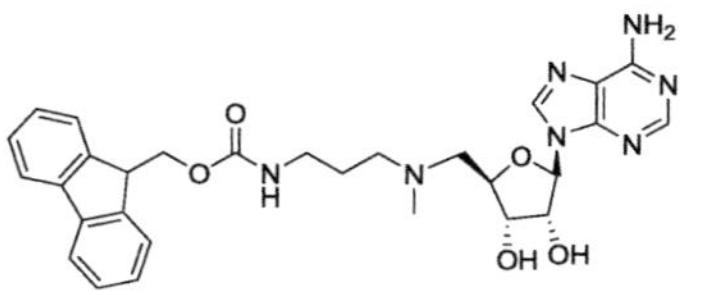

Ki= 20 mM
15

EZH2 Inhibitors

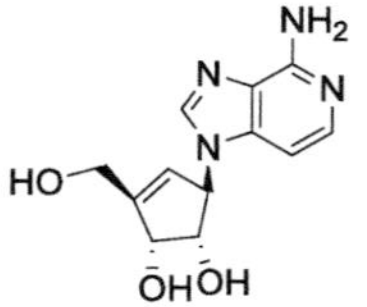

DZNeP
16

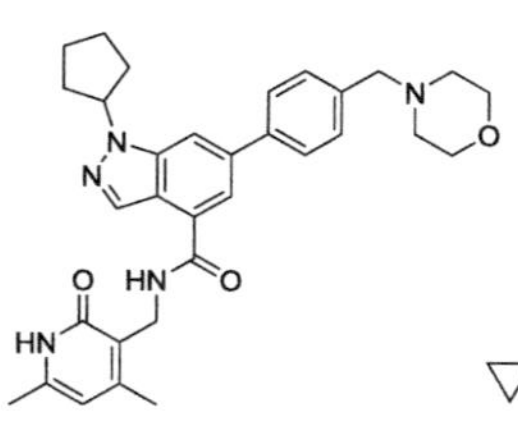

EPZ005687
17

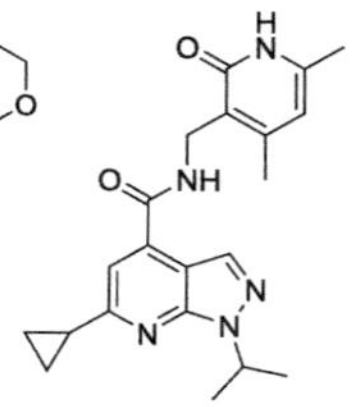

GSK-A
18

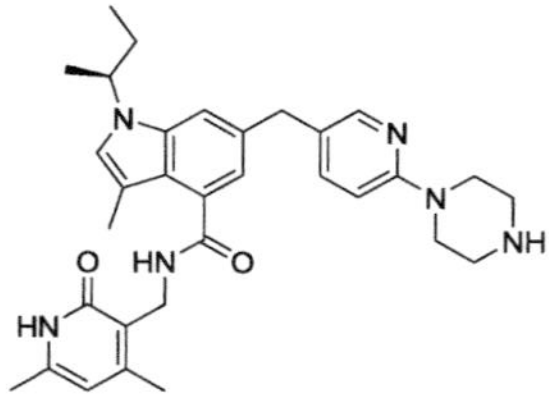

GSK 126
19

G9a/GLP Inhibitors

Type I

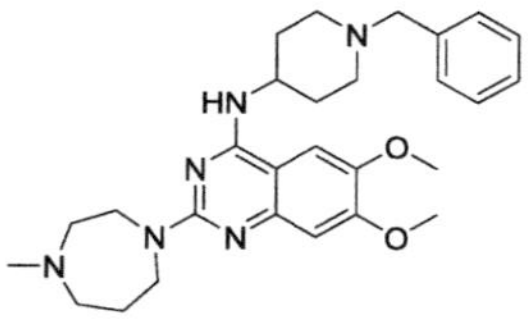

BIX01294
20

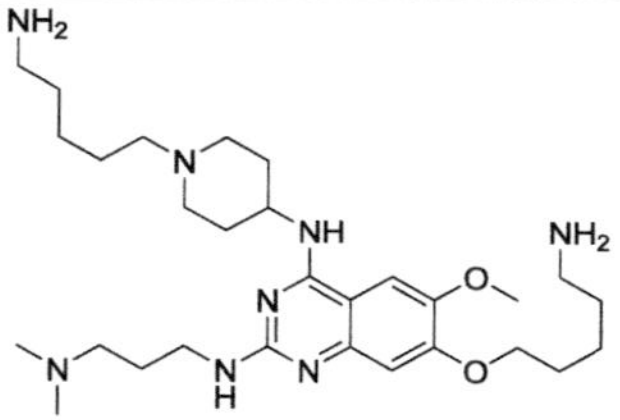

E72
21

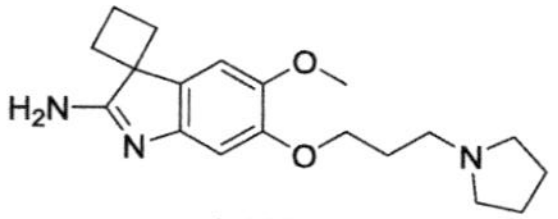

A-366
22

(Continued)

Table 17.1 Molecular Structures of Known HMT Inhibitors *Continued*

UNC0224
23

UNC0321
24

UNC0638
25

UNC0642
26

BIX01338
27

BRD9539
28

Chaetocin
29

Type II

Type III

MML Inhibitors

WDR5-0101
30

WDR5-0102
31

PRMT Inhibitors

Methylgene compound
CARM1 inhibitor IC50 = 60 nM
32

CARM1 inhibitor IC50 = 40 nM (BMS)
33

AMI I
34

AMI-6
35

AMI-9
36

stilbamidine
37

Table 17.1 Molecular Structures of Known HMT Inhibitors *Continued*

DB75

38

allantodapsone

39

RM-65

40

trimethine cyanine dye

41

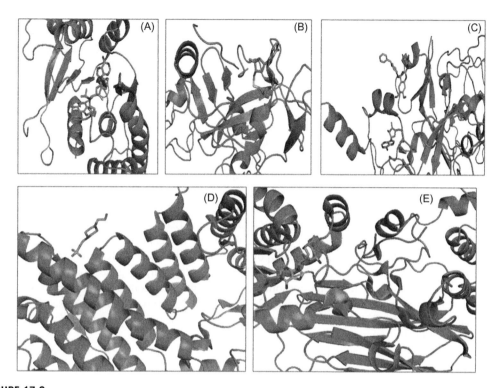

FIGURE 17.3

Co-crystal structures of inhibitors and the important cancer-related histone methyltransferases. (A) Crystal structure of DOT1L with EPZ-5676 [PDB 4HRA]. (B) Crystal structure of EZH2 SET domain [PDB 4MI5] aligned to domain of crystal structure of G9a with SAM [PDB 3HNA]; peptide residue from EED complex, molecule present is cofactor SAM. (C) Crystal structure of GLP [PDB 3FPD] in complex with SAM (green) and BIX-01294 (orange). (D) Crystal structure of menin-MLL complex [PDB 4GPQ] dissociated by small molecule MI-2. (E) Crystal structure of PRMT5:MEP50P complex [PDB 4GQB]. (*This figure is reproduced in color in the color plate section.*)

leukemia [14]. Treatment of leukemia cells with EPZ-5676 results in concentration and time-dependent reduction of H3K79 methylation without effects on the methylation status of other histone sites. The reduction of H3K79 methylation leads to inhibition of MLL target genes and selective apoptosis in MLL-rearranged leukemia cells, but minimal impact on nonrearranged cells. No obvious toxicity was observed in rats treated with EPZ-5676 in a rat xenograft model.

The University of Toronto, Canada, has discovered that 7-substituted adenosine derivatives may also act as potent inhibitors of DOT1L [14,19]. The SGC series of compounds, analogs of SAH with a 7-bromo group (Figure 17.3), appear to be potent inhibitors of DOT1L ($IC_{50} = 77$ nM) with high selectivity with a much higher activity than SAH. This suggests that 7-ring substituents are favored in DOT1L selectivity and inhibition (Table 17.1, entry 8). Moreover, the SGC series are inactive against SET-domain HMTs.

17.2.2 ENHANCER OF ZESTE (EZH2)

Enhancer of Zeste (EZH2) is a subunit in the Polycomb repressor complex 2 (PRC2), a protein complex belonging to the family of Polycomb group (PcG) proteins. EZH2 is the site of catalysis in the complex. Proteins in the PcG family are transcriptional repressors with two functionally distinct complexes — Polycomb repressive complex 1 (PRC1) and Polycomb repressive complex 2 (PRC2). PcG proteins act as transcriptional repressors and are involved in cellular memory via genomic imprinting, X chromosome inactivation, cell proliferation, cell differentiation, and identity via epigenetic modifications of histones [20]. PcG proteins have been shown to display activity as a HMT in methylating core histones. PRC1 and PRC2 achieve gene silencing through distinct methods: PRC2 adds up to three methyl groups to lysine 27 on histone 3(H3K27) and PRC1 mono-ubiquitylates histone H2A lysine 119 (H2AK119). Though PRC2 can also mono- and dimethylate H3K27, only the trimethylated form conveys biological function, while the monoubiquitylation of H2AK119 is a collaborative effort between PRC1 and PRC2 [21]. H2AK119 monoubiquitylation occurs after PRC2 is recruited to the target gene; the exact mechanism for PRC2 recruitment is unknown but likely varies across cell types. Upon PRC2 trimethylating H3K27, PCR1 is recruited by a network of regulatory micro-RNAs (miRNAs) to the target gene for ubiquitylation. Subsequent ubiquitylation of the gene consolidates transcriptional repression by preventing transcriptional elongation by RNA polymerase II [21]. As miRNAs involved are repressed by EZH2 and regulate PRC1 expression, PRC1 is upregulated to create a positive feedback loop between PRC1 and PRC2 [22].

PRC2 consists of several essential subunits: enhancer of Zeste 2 (EZH2), embryonic ectoderm development isoform 1 (EED1), retinoblastoma-binding protein 7/4 (RBBP7/RBBP4), and the suppressor of Zeste 12 (SUZ12), with EZH2 and EED forming the EED−EZH2 complex responsible for methylation [23]. In addition to the four subunits, PRC2 may contain other proteins, like AEBP2, PHF1 (PCL1), MTF2 (PCL2), PHF19 (PCL3), and JARID2, which help regulate the activity of PRC2 and aid in localizing the complex to target genes. EZH2 is comprised of a conserved SET domain that provides the active site for the covalent methylation reaction, acting as a multimeric effector to catalyze the trimethylation (me3) of histone 3 lysine 27 (H3K27).

Despite the fact that EZH2 is the catalytic subunit of PRC2, EZH2 by itself is unable to display enzymatic activity as the lysine and methyl donor cofactor-binding pockets are on opposite sides (Figure 17.3B). In order for trimethylation of H3K27 to take place, EED1, RBBP7/RBBP4 and SUZ12 are required to be present in addition to cofactor SAH; both EED1 and SUZ12 are essential to complex formation [22]. Even within the EED−EZH2 complex, EED1 is the subunit that contributes affinity for nucleosomes by binding to the lysine [22].

PRC2 is able to interact with other histone modifiers and histones. Minor methylation activity has been detected *in vitro* for H3K9; however, whether this activity is intrinsic to EZH2 is debated. Physical and functional links between EZH2 and histone deacetylase (HDAC) are well established, as PRC2-mediated repression of gene activity involves histone deacetylation. HDAC activity is required prior to EZH2-mediated H3K27 methylation. PRC2 can physically associate with HDAC1 and HDAC2, which deacetylase H3K27, H3K9, H3K14, or H4K811 [21]. However, HDACs should not be considered subunits of EZH2.

There is evidence that suggests the stable repression of specific genes, including many tumor suppressors, requires EZH2 to be present. Furthermore, EZH2 is capable of interacting with multiple DNA methyltransferases, such as DNMT1, DNMT3A, and DNMT3B which requires EZH. Trimethylation of H3K27 and DNA hypermethylation have been considered independent epigenetic systems as both H3K27 and DNA hypermethylation can be linked for transcriptional repression; not all genes suppressed by PcG methylation are maintained by DNA methylation. The discovery that EZH2 is capable of recruiting DNMTs to genes that promote gene silencing through DNA methylation has changed this view and it is now accepted that, prior to DNA methylation, histone methylation occurs; once PCR2 hypermethylates a gene, DNA methylation maintains it [24]. Genes with H3K27 trimethylated by PRC2 containing EZH2 in normal development stem cells are predisposed to *de novo* DNA hypermethylation, leading to long-term silencing in the presence of oncogenic cues [25]. EZH2 also preordains genes toward hypermethylation during cellular transformation [25].

17.2.2.1 *EZH2 in disease*

Histone modifications have been implicated in oncogenesis with the methylation/acetylation of histones H3 and H4 being associated with multiple cancer types. PRC1 and PRC2 tag histones with specific marks that allow the silencing of tumor suppressor genes and have been tied to carcinogenesis [26]. Initially tied to prostate cancer, EZH2 is known to be involved in cellular proliferation and is linked to the metastasis of various other cancers. Forced expression of EZH2 results in increased cell proliferation and cells with overexpressed EZH2 are tumorigenic [27]. Breast and colon cancers with EZH2-positive tumors have a much higher proliferative index than those with EZH2-negative tumors [28]. EZH2 expression increases in order from benign, organ confined, and metastatic tumors. Increased levels of EZH2 may precede metastasis, which could be useful in predicting prognosis and determining potential changes. High EZH2 levels correlate to poor prognosis, tumor progression and risk of recurrence, particularly in prostate and breast cancer.

The current view of EZH2 is that it acts as a transcriptional repressor that silences multiple discrete genes, particularly tumor suppression and cell proliferation genes, to aid the development of the metastatic phenotype and can be considered as an oncogene due to its protumorigenic properties. EZH2 is crucial in tumor development, growth, differentiation, and spread. Inhibition of EZH2 decreases DNA synthesis and knockdown of EZH2 leads to inhibition of tumor growth and cell migration, suggesting a link to continued proliferation [29]. The fate of the accumulated EZH2 in cancer cells is still debated. The dominant views are that the EZH2 assembles to form high levels of PCR2 or that the imbalance of PCR2 subunits leads to unassembled EZH2, which then triggers aberrant PCR2 subcomplexes.

However, the mechanisms of EZH2 in metastasis are well documented. EZH2 is able to silence the Ras GAP gene and Disabled Homolog2-Interacting Protein (DAB2IP) to trigger metastasis [30]. Increased EZH2 drives the metastatic phenotype by promoting anchorage-independent growth, cell invasion and silencing genes responsible for inhibiting tumor angiogenesis [31]. Vascular endothelial growth factor (VEGF), which stimulates angiogenesis, can transactivate EZH2 expression to silence expression of negative regulator of angiogenesis Vasohibin 1 (VASH1) to enhance angiogenesis [31]. Knockout/silencing of EZH2 in cancer-derived endothelial cells leads to inhibition of cell migration and tube formation, suggesting EZH2 is fundamental in angiogenesis [31]. EZH2 also regulates apoptosis in cancer cells by antagonizing the activity of pro-apoptotic factors [32].

17.2.2.2 Inhibitors of EZH2

As one of the first PKMTs to be implicated in human cancers, EZH2 presents a favorable target for cancer treatments. EZH2 inhibitors differ based on specificity, modes of action and toxicity profiles against normal human cells. Blockage of PRC2 function might inhibit multiple signaling pathways to yield dramatic antitumorigenic effects, as shown in animal studies. Since EZH2 is highly expressed in advanced stages of cancer and is causal in driving metastasis, EZH2 inhibitors may have significant therapeutic effect on metastatic cancers for which no effective curative treatment is available.

The most studied inhibitor of EZH2 is a carbocyclic adenosine analog 3-deazaneplanocin (DZNep), a derivative of the naturally occurring antibiotic neplanocin-A (Table 17.1, entry 16). Found through library-based screening, DZNep inhibits SAH hydrolase, a component of the methionine cycle, resulting in accumulation of SAH while disrupting methylation of substrates by EZH2 [33]. DZNep has been shown to deplete PCR2 subunits, reactivate PCR2 silenced genes, and reduce proliferation/invasiveness in breast, lung and prostate cancer. DZNep also reduced tumor angiogenesis in a glioblastoma xenograft model [34]. *In vivo* studies have shown that DZNep is able to induce apoptosis specifically in cancerous cells with low cytotoxicity in normal cells. Due to its hydrophilic nature, DZNep has a very short half-life *in vivo* which is overcome by the formulation of DNZep encapsulated inside pegylated liposomes. The combined treatment by DZNep and HDAC inhibitor panobinostat was able to induce an apoptotic effect upon primary leukemia cells compared with normal cells [35]. However, the effect of DZNep upon histone methylation is as a blanket histone methyltransferase inhibitor rather than EZH2 specific.

A number of other molecules have been found with varying degrees of success. EPZ005687, a SAM-competitive, has shown potent inhibition in lymphoma cell lines with mutant EZH2 (Table 17.1, entry 17). Another promising compound is GSK-A from the GlaxoSmithKline (GSK) collection. GSK-A exhibits inhibitory activity toward EZH2 by competing with the endogenous substrate, reducing H3K27 trimethylation [36]. *In vivo* studies with the cousin compound GSK126 found GSK126 to selectively inhibit EZH2 in lymphoma cells, supporting the evidence that EZH2 could be inhibited in a selective manner (Table 17.1, entries 18,19). A combinatorial approach with DZNep-like inhibitors, DNA demethylating agents, and HDAC inhibitors should also be considered [37].

Downstream or upstream disruption targets are also viable, as EZH2 expression is regulated at transcriptional, posttranscriptional, and posttranslational levels. Several examples are AZD6244, 3,3′-diindolylmethane (BR-DIM), and curcumin; all three compounds use distinct pathways to affect EZH2 expression: BR-DIM upregulates regulatory miRNAs to enhance the downregulation of EZH2 downstream and AZD6244 inhibits cancer progression through downstream targeting RAF1-ERK signaling and eliminating BTICs [36]. Curcumin, isolated from turmeric spice and a well-known anti-carcinogenic agent, downregulates EZH2 by affecting the mitogen-activated protein kinase (MAPK) pathway [38]. Disruption of HDAC and DNMTs can also inhibit EZH2 activity through downstream effects [39]. HDAC-targeting drugs and green tea polyphenols have both been found to be able to inhibit EZH2-activity at micromolar concentrations [39].

While PcG silencing is an attractive candidate for cancer therapy, care must be applied in developing a treatment due to the complexity of the interactions of EZH2 to avoid adverse side effects. Conditional inactivation of EZH2 in adult mouse hematopoietic/pancreatic stem cells yields minor defects in normal organ development, suggesting that a balance of H3K27me3 and H3K27 methylation could be critical in both maintaining pluripotency in developmental cells and normal cell growth. EZH2 inhibitors may require tumor-specific delivery systems to avoid broad cytotoxicity. Further characterization of oncogenic targets and pathways along with EZH2 specificity can improve treatments.

17.2.3 **G9A**

Excepting DOT1L, all currently known HKMTs can be characterized by a conserved catalytic SET domain from/to the Su(var)3−9 family of proteins. Two members, G9a (EuHMT2 or KMT1C) and G9a-like protein GLP (EuHMT1 or KMT1D), are two HKMTs responsible for catalyzing the transfer of one to three methyl groups from the cofactor SAM to the ε-amino group of a target lysine to achieve mono-, di-, and trimethylation; G9a and GLP display a similar substrate affinity on histones. The second HKMT discovered, G9a, primarily methylates histone 3 lysine 9 (H3K9), in addition to histone H1 and histone 3 lysine 27 (H3K27) *in vivo* [40].

Highly expressed in multiple tissue types, including skeletal muscles, peripheral blood leukocytes and bone marrow, G9a and GLP are corepressors that silence genes involved in genomic imprinting, pluripotency, and other important roles in cell development. Knockout of G9a and GLP leads to embryonic death through a loss of silencing and growth retardation through apoptosis instead of growth arrest [41]. G9a-deficient embryonic stem cells will exhibit defects upon differentiating, but lack any defects present during growth. G9a/GLP is a controller for differentiation; overexpression of G9a will negatively regulate the process. This is done through H3K9 dimethylation (H3K9me2), a conserved marker found along transcriptionally silenced regions and integral for recruiting and condensing chromatin domains [42]. Distributed on long stretches of DNA, H3K9me2 is spatially longer compared to other histone methyl modifications [43]. When H3K9me2 levels decline during differentiation, the absence of repressive H3K9me2 marks allows genes expressed only in stem cells to be induced, reverting cells to pluripotent states.

In order to function properly, G9a forms heteromeric and homomeric complexes with GLP through their SET domains; GLP has an 80% sequence identity to G9a in their respective SET domains. G9a cannot compensate for the loss of GLP, and vice versa. The G9a-GLP complex is primarily heteromeric and acts as a functional H3K9 methyltransferase to mono- (H3K9me1) and dimethylate H3K9 (H3K9me2). H3K9 trimethylation (H3K9me3) occurs mainly from HMTs SUV39H1 and SUV39H2 [41,43]. This is due to a domain containing ankyrin repeating at the non-catalytic N-terminal region in G9a, responsible for G9a's nuclear import, and protein−protein interactions. The hydrophobic cages of the ankyrin repeat domain bind H3K9me1/2 and are not required for histone methyltransferase activity, but are required for DNA methylation. This ankyrin repeat domain is too narrow for H3K9me3, explaining why H3K9me3 occurs through other HMTs. Likewise, the catalytic SET domain is required for proper G9a activity − catalytically inactive G9a also generates a phenotype similar to that of G9a knockout [41]. However, the isolated catalytic domain targeted to DNA is sufficient for transcriptional silencing.

G9a is notable due to its ability to methylate other nonhistone proteins, such as p53 (K372), WIZ (K305), CDYL1 (K135), ACINUS (K654), and Reptin (K67) [44]. Many of the nonhistone proteins are repressive chromatin proteins and multizinc finger molecules, but the biological roles of these nonhistone methylations are not well understood [42]. G9a is capable of self-methylation; self-methylated G9a resembles trimethylated H3K9. G9a may also act as an internal regulator of other methyltransferases, as it is also capable of methylating other methyltransferases, such as mAM, GLP, and nonhistone substrates like HDAC1 and DNMT1.

As demonstrated in its interactions with histone 3, a partiality toward hydrophobic and hydrophilic amino acids gives rise to G9a's ability to methylate nonhistone proteins. Specifically targeting residues 6−11, arginine 8 adjacent to lysine 9 is required for activity; changes at these positions will halt methylation. This is due to a G9a recognition motif made up of alanine-arginine-lysine-threonine (ARKT). Nonhistone proteins methylated by G9a have this motif present; this motif is present at the N-terminus of G9a, allowing it to self-methylate. However, G9a methylated nonhistone protein functions have not been well established, though multiple transcription factors appear to be negatively affected.

There are also functional links between H3K9 and DNA methylation in mammals. Knocking out GLP or G9a reduces DNA methylation activity, and knockdown of DNA methyltransferase 1 (DNMT1) induces a reduction in H3K9me2 levels and G9a/GLP expression [45]. It is known that during replication, G9a in complex with DNMT1 at specific loci can coordinate DNA and H3K9 methylation during lineage commitment and differentiation to methylate DNA and silence genes. DNMT1 can also regulate H3K9 methylation by directing the chromatin loading of G9a, although the degree of interaction is still debated [45].

There has been evidence suggesting that H3K9 and H3K27 methylations cooperate to ensure epigenetic gene silencing at certain chromatin regions, but evidence of a functional interaction between PRC2 and H3K9 KMTs is lacking.

17.2.3.1 G9a in disease

It should not be surprising that G9a expression is found to be upregulated in multiple forms of cancer, such as colon, prostate, lung and lymphocytic leukemia, due to the role G9a plays in cellular replication and proliferation. G9a knockdown studies in lung, leukemia and prostate cancer cell lines caused both growth suppression and apoptosis [46]. Inhibition of G9a by pharmacological agents causes a reduction in proliferation that is followed by G1 cell cycle arrest, mitigation in cellular migration, lowering global methylation levels and a reprogramming of cells into an induced pluripotent state [47].

H3K9 methylation, which is tied to transcriptional silencing, is located at promoter regions of silenced tumor suppressor genes in cancer cells. An example is tumor suppressor p53 − G9a dimethylates lysine 373 of p53 to inactivate the transcriptional activity [48]. The exact pathology of how G9a triggers tumorigenesis is not well understood or characterized. Access to regulatory factors and complexes that modulate biological processes seem to be determined through modifications to the terminal tails of histones. Knockdown of G9a decreases H3K9 methylation and generates morphological changes with shortened telomeres and a loss of telomerase activity [49]. This suggests that perpetuation and continued proliferation of malignant phenotypes is governed by G9a, possibly through transcriptional silencing and centrosome duplication via changes to the chromatin structure.

G9a is also involved in silencing viral genomes. Viral genomes are introduced into the host genome by a provirus, which then proceeds to disrupt normal cellular activity. Silencing proviruses is in the interests of the cell, as in the case of HIV. G9a has displayed the ability to regulate HIV1 gene expression through transcriptional silencing by H3K9me2 on the HIV1 promoter, suggesting that G9a is critical for viral genome and provirus silencing [50]. Furthermore, G9a has been tied to cocaine addiction, mental retardation, and DNA methylation, presenting a viable therapeutic target.

17.2.3.2 Inhibitors of G9a

Current G9a inhibitors may be categorized by their mode of action or their parent compound. Current G9a inhibitors act either as SAM competitive inhibitors or peptide competitive inhibitors; BIX-01294 (Table 17.1, entry 20) and its derivative family of compounds represent G9a peptide substrate mimetic, while BIX-0138 (Table 17.1, entry 27) and its derivative family along with chaetocin (Table 17.1, entry 29) represent SAM competitive inhibitors.

Fungal metabolite chaetocin is a competitive inhibitor of cofactor SAM, isolated from fermentation of *Chaetomium minutum*, and is the first HKMT inhibitor discovered to target the enzymatic activities of HKMTs belonging to members of the Suv39 family [51,52]. A histone lysine methyltransferase Su(var)3−9 selective inhibitor in *Drosophila melanogaster*, the compound displayed weaker inhibition for G9a at low concentrations without affecting other HKMTs [53]. However, chaetocin has a broad range of off-target proteins that are functionally and structurally unrelated, contributing to its broad cytotoxicity [51,54]. Chaetocin belongs to a family of fungal toxins, which are a subset of the epipolythiodioxopiperazine (ETP) class of compounds. These compounds have a distinct disulfide bridge across a diketopiperazine scaffold [55]. The disulfide bridge is responsible for the inhibitory activity and contributes to its nonspecific mechanism through rendering adduct formation reversible. Analogs of chaetocin without the disulfide bridge present are inactive; conversely, this disulfide functionality also confers the broad nonselective cytotoxicity [56,57].

The BIX-0138 family is also comprised of cofactor-competitive inhibitors that are based on parent BIX-0138, including all its derivatives, such as BRD9539 (Table 17.1, entry 28). Discovered in the same screening as BIX-01294 and chaetocin, the BIX-0138 family is a SAM mimetic that inhibits HKMTs such as G9a, and is able to induce cell-cycle arrest and inhibit both anchorage-dependent and independent cell proliferation. However, BIX-0138 is unable to modulate cellular H3K9 methylation and its derivatives are similarly nonselective [52]. SAM mimetic compounds are nonspecific and result not only in the inhibition of HMTs, but also DNMTs and other enzymes that use SAM as a cofactor, preventing them from being of much therapeutic use.

A more specific H3K9 HKMT inhibitor, BIX-01294, was found in 2007 as the first potent and selective small-molecule inhibitor of G9a and GLP [52]. BIX-01294 and its derivative family of small molecules act as competitive inhibitors, occupying the histone peptide-binding site, binding to the acidic surface of the peptide-binding groove of the N-terminal from H3K4 to H3R8 or H3K9 [58]. Thus the peptide-binding pocket is inaccessible to other substrates, while the lysine-binding channel remains open, with minimal contact with BIX-01294 (Figure 17.3C). BIX-01294 has since been used successfully as a probe of G9a in cellular reprogramming, gene reactivation, and reduced H3K9me2 levels at G9a target genes. However, BIX-01294 has poor separation between concentrations necessary for functional activity and concentrations that are toxic to cells, limiting its usefulness [52]. This off-target toxicity is not linked to HMT inhibition.

A wide variety of other selective and potent substrate competitive G9a inhibitors have also emerged, such as E72, A-366, and the UNC series (UNC0224, UNC0638, UNC0642, UNC0321, UNC0638) (Table 17.1, entries 21−26). These compounds, such as UNC0638, have a much better toxicity to function ratio compared to the parent BIX-01294, performing well in modulating G9a regulated genes and reducing H3K9me2 levels [55]. However, some of these, like UNC0321 and

E72, are less potent than BIX-01294 in cellular assays [58]. As these compounds are generated through structure-activity relationship (SAR) transformations to BIX-01294, the quinazoline core is a dominant feature and the biological activity tends to be similar. This is caused by BIX-01294 owing most of the activity and interactions it has with G9a to the quinazoline core, while the branched benzene moiety has little to no direct contact with the enzyme. As such, derivative compounds in the BIX-01294 family retain the 2,4-diamino-6,7-dimethoxyquinazoline template.

17.2.4 MIXED LINEAGE LEUKEMIA

In mammals, there are at least ten histone H3 lysine 4 (H3K4) methyltransferases (HMTs). In humans, there are at least eight HMTs specific toward H3K4, with six (MLL1−MML5, SET1B) from the MLL family. The MLL HMTs are a family of large multidomain proteins that catalyze the methyl group transfer from SAM to the ε-amino acid of H3K4 via a conserved SET (Suppressor of variegation, Enhancer of Zeste, Trithorax) domain to mono-, di-, or trimethylate H3K4. Also called human trithorax (HRX) or acute lymphocytic leukemia-1 (ALL-1), the MLL gene is located on human chromosome 11q23. Alterations on 11q23 are associated with converting MLL to an active oncogene associated with acute leukemias and the source of the gene's name [59].

The MLL gene encodes a protein with multiple domains, including a cysteine rich CxxC domain that acts as a homolog of DNMTs, DNA-binding AT hooks, plant homeodomain finger (PHD) and Win motifs, a bromodomain (BD), a transactivation domain (TAD), a nuclear receptor interaction motif (NR box), and SET domain at the C-terminal that confers MLL its histone methyltransferase (HMT) activity [60]. To become active, the MLL protein must first be cleaved proteolytically by taspase1 into a N-terminal fragment containing the DNA-targeting sequence involved in gene repression and the C-terminal fragment with transcriptional activation activity [61]. Following association between the N-terminal and the C-terminal fragments, they recruit and form a multicomponent super-complex for transcriptional activation of target genes by mediating chromatin remodeling and modifications (methylation and acetylation) [61,62]. PTMs can then adjust the enzymatic activity of MLL.

The MLL HMTs belong to a larger family of proteins responsible for regulating homeobox (HOX) genes and transcription through H3K4 methylation. H3K4 methylation is a conserved mark that predisposes gene transcriptional activation found at the start site of all transcribed genes and only occurs after acetylation of K9 and K14 on H3 peptides. MLL is tied to embryonic skeletal formation and required for hematopoietic development and maintenance of adult hematopoietic stem cells (HSCs) through the regulation of HOX genes [63]. MLL proteins can also target non-HOX genes and are involved in cellular memory and heritable changes through chromatin modification.

MLL proteins are incorporated into larger complexes with various cell cycle regulators, including Polycomb group (PcG) proteins, histone acetylases (HACs) and deacetylases (HDACs), and other HMTs [64]. The MLL complexes all have a core required for enzymatic activity comprised of three common subunits − WDR5 (WD repeat domain 5), RBPB5 (retinoblastoma-binding protein 5) and ASH2L (*Drosophila* ASH2-like) [65]. MLL is only active as an HMT when it associates with the core complex; however, the mechanism of how the target gene recruits MLL or the multiunit complex is not well understood.

MLL HMTs are unusual due to their ability to generate MLL-fusion proteins through chromosomal translocations. MLL fusions function as general transcription factors able to indiscriminately activate many different genes, determined by their fusion partner. Most MLL-fusion genes preserve their ability to regulate HOX genes, while the SET domain responsible for MLL HMT activity is lost in MLL fusions. In particular, MLL proteins seem to serve as an integrative platform for histone acetylation, methylation, and remodeling to achieve heritable activation of a target gene. This is seen in MLL HMT activity displaying a preference for pre-acetylated H3 peptides, and its ability to recruit other HMTs like DOT1L. However, the nature of how MLL-introduced histone modifications are translated or affect transcriptional activation is not well characterized [66].

17.2.4.1 Disease and MLL

Myeloid-lymphoid leukemia, also known as MLL, is an aggressive blood cancer that occurs in pediatric patients characterized by translocations and rearrangements at chromosome 11q23 of the MLL gene and the presence of MLL-fusion proteins. MLL-rearranged leukemia (MLLr) can be acute myeloid leukemia (AML), MLL, or acute lymphoid leukemia (ALL) that shares phenotypic characteristics of both ALL and AML. MLL rearrangements occur in about 70% of infant ALL and 8−10% for ALL; about 10% of human acute leukemias harbor MLL translocations [67]. Leukemias with MLL rearrangements display an increased resistance to standard and targeted therapies, with MLL having poor prognosis in contrast to other pediatric acute leukemias [61]. Pediatric ALL without an MLL translocation has an overall survival above 80% [68].

Four distinct types of deregulated MLL have been thus far: MLL amplification, MLL-partial tandem duplication (PTD), MLL fusion, and internal deletion of MLL exons coding the first PHD finger; all four cause structural alternations of the MLL sequence [61]. Due to the structural and functional disparities of the resultant MLL proteins, it is probable that each form emerges from a distinct transformation mechanism. The poor prognosis of MLL, MLLr AML, and MLLr ALL is largely due to the presence of MLL-fusion proteins, which can induce leukemia via unregulated transcriptional regulation at MLL-fusion target genes. Leukemogenesis results through fusion products activating aberrant self-renewal and sustaining the expression programs that confer aberrant self-renewal in hematopoietic stem cells.

As a fusion partner, MLL-fusion proteins are involved in translocations along the chromosome in acute *de novo* and therapy-related leukemias, primarily seen in hematopoietic myeloid and lymphoid cells [67]. Chromosomal translocations of 11q23 or other abnormalities in turn alter the MLL gene structure, generating a fusion between MLL and a translocated gene from another chromosomal region. When the resulting fusion gene is read for transcription and translation, the fusion protein is produced; this occurs when the oncogenic MLL protein loses its C-terminal sequence in order to fuse to the N-terminus of a fusion partner protein [66]. In return, the fusion product loses transactivation, SET domains and normal HMT function, gaining heterologous functions to act as a transcriptional activator via transcriptional effectors or homodimerization domains from its fusion partner [61,66].

While upward of 60 distinct MLL-fusion proteins are currently known, the most frequent partner proteins for MLL are AF4, AF9, ENL, AF10, AF6, ELL, AF1P, AF17, and SEPT6, constituting nearly 90% of MLL rearrangements [61]. AF4 is the most common MLL partner, followed closely by AF9 and ENL [67]. Resulting fusions can then act as transcriptional regulators targeting genes normally controlled by MLL; transcriptional complexes are critical for oncogenic functions of MLL fusions. Despite the diversity of fusion partners, all fusion proteins follow a similar structural scheme with clear separation into two groups. All of the most common partner proteins, with the exception of AF6, are nuclear proteins. Therefore, MLL-fusion proteins can be separated by their functional properties and localizations – nuclear proteins with transcription effector domains and cytoplasmic proteins with dimerization domains [61,69].

Likewise, there appear to be two distinct mechanisms involved in MLL fusions activating transcription. Most of the fusion partners are either chromatin modifiers that acetylate histones or recruit HMTs. The link between MLL and histone acetylation is not well known. HDACs have been seen to bind to MLL proteins, so it is possible the loss of HDAC silencing of genes targeted by MLL contribute to MLL-fusion leukemogenesis. Histone methylation, particularly methylation of H3K79 by DOT1L, has been linked to positive transcriptional regulation [70]. H3K79 methylation is a histone mark associated with actively transcribed genes, linked to chromatin activated by MLL-fusion proteins and associated with MLL-fusion target genes. As several common MLL fusion partners can coordinate DOT1L activity to stimulate the elongation phase of transcription, it is possible DOT1L has a role in maintaining MLL-fusion gene expression. This aberrant recruitment of H3K79 HMTs or arginine HMTs likely influences gene expression, suggesting that MLL-fusion proteins rely on association with specific HMTs to achieve leukemogenic transformation.

17.2.4.2 Inhibitors targeted to MML

MLL treatments have stalled, with only about 40% of infants surviving 5 years after diagnosis. Most treatments for MLL or MLLr leukemias are with hypomethylating agents that reactivate tumor suppressor gene expression and the inhibition of cell growth and proliferation. MLLr leukemias have been found to be more sensitive to hypomethylating agents *in vitro* than some other lines of leukemia; profiling of MLLr ALL has found an association between DNA methylation and therapeutic outcome [71]. Likewise, inactivation of DNA methyltransferase 1 (DNMT1) was able to delay the onset of MLLr AML and suppressed development of induced leukemia. These studies suggest DNA methylation is critical for hematopoiesis and leukemia development, indicating that approaches targeting DNA methylation in leukemia may be successful.

Inhibition of MLL-fusion proteins and recruitment of co-activators to target genes have also seen some success; small molecule inhibitors have seen some success by blocking MLL and menin, for a critical co-activator in MLL-fusion protein transformation (Figure 17.3D) [67]. Other domains in MLL have also been studied as potential targets – the CxxC domain of MLL, a vital component in MLL-DNA interactions, is being studied as a potential target to block DNA binding and transcription [61,72]. Presently, two classes of compounds are in clinical trials; the GSK3β inhibitors and flavonoids such as flavopiridol. Flavopiridol and other flavonoids have potent and specific inhibition of P-TEFb, a transcriptional elongation factor involved in several common fusion partners (AF4, AF9, and ENL) and required for the onset of leukemia [73]. Currently in phase II trials, these compounds suffer from limited efficacy and high toxicity [73]. The GSK3β inhibitors work through blocking association of transcription complexes with MLL with co-activator CREB1 [74]. Disruption of MLL partner proteins like WRD5 is also possible by small molecule inhibitors, such as WRD5-0101 and WRD5-0102 (Table 17.1, entries 30,31) [75].

Due to the involvement of DOT1L and HDACs in MLL transcription, both DOT1L and HDAC inhibitors have been researched as potential therapeutic approaches for MLLr leukemias [76].

Several HDAC inhibitors have entered clinical trials for hematologic malignancies and displayed the capacity to halt proliferation and induce apoptosis in MLLr ALL [77]. Suppression or inactivation of DOT1L expression has been shown to suppress MLL-fusion target gene expression, leading to growth arrest and apoptosis in human MLLr ALL and MLLr AML. Inactivation of DOT1L in mouse MLLr AML inhibited leukemia development [78]. DOT1L inhibitors as targeted therapeutics for MLLr leukemias is promising. EPZ004777, a small molecule inhibitor specific for DOT1L, has shown anti-proliferative activity and was selective for cell lines with MLL rearrangements [79].

However, the largest concern with these strategies is the vital function of MLL will also be affected, leading to cytotoxicity. It is also possible that modifying the DNA methylation state in MLLr leukemias could lead to unwanted alterations in gene expression.

17.3 PROTEIN ARGININE METHYLTRANSFERASES

Protein arginine methyltransferases (PRMTs) are enzymes that mediate posttranslational methylation of arginine on histones and other proteins. It is estimated that about 0.5% arginine moieties are susceptible to these modifications, and primarily occur at glycine, alanine and arginine rich portions (GAR motifs) [80]. Acting in conjunction with other PMTs, PRMTs regulate activation and suppression of gene expression in accordance with the arginine residue involved and the methylation mark. Across the core histones, over 30 arginine methylation sites have been found wherein arginine methylation occurs and this methylation is facilitated by the several potential hydrogen bond donors in the guanidine group [81]. Consequently, any modifications on these guanidine groups will affect the interaction of the arginine containing protein with the corresponding hydrogen bond acceptors like DNA, RNA, and proteins; which eventually leads to altered physiological functions [81].

More than 60 PMTs — including 11 known PRMTs and more than 50 protein lysine methyltransferases (PKMTs) — are encoded by the human genome. PRMTs methylate the guanidine nitrogen of arginine using the universal methyl donor SAM as a cofactor yielding SAH and a methylated arginine (Figure 17.4). The methylated arginine exists in three forms in mammals: monomethylarginine (MMA), asymmetric dimethylarginine (aDMA), and symmetric dimethylarginine (sDMA). Apart from this a monomethylation of the internal δ-guanidino nitrogen atom (δMMA) is also seen, but it has been reported only in yeast proteins [80,81].

17.3.1 PRMT STRUCTURE AND ACTIVE DOMAIN

The number of amino acid residues in PRMTs varies from 316 to 956 amino acids, but all of them possess a catalytic core region consisting of around 300 amino acid residues which is highly conserved among mammals (Figure 17.5) [81]. Each PRMT differs in their function and selectivity because of the difference in the unique N-terminal region. Like PRMT3 with a zinc finger (ZnF) and PRMT8 with a myristoylation domain, the N-terminals of PRMTS are of variable length and distinct domain motifs [81]. A recent X-ray crystallography study of PRMT1, PRMT3, and PRMT5 gave much needed insight to the highly conserved catalytic core which consists of three structural regions, namely a methyltransferase (MTase) domain, a β-barrel domain unique to the PRMT family and a dimerization arm. The N-terminal region of the catalytic core consists of an α/β Rossmann fold (containing the conserved motifs I, post-I, II, and III) and act as the MTase domain [81].

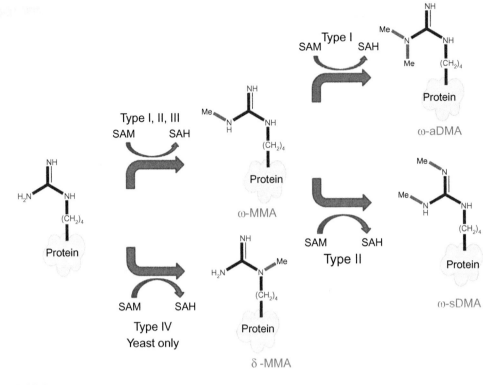

FIGURE 17.4

Methylation of the arginine side chain in proteins by protein arginine methyltransferases (PRMTs). The arginine residue holds five potential hydrogen bond donors. Mammalian PRMTs use the methyl group from a molecule of SAM to form ω-NG-monomethylarginine (MMA). Subsequently, Type I PRMTs add a methyl group to the same nitrogen atom forming ω-NG,NG-dimethylarginine (aDMA), whereas Type II PRMTs generate ω-NG,N′G-dimethylarginine (sDMA).

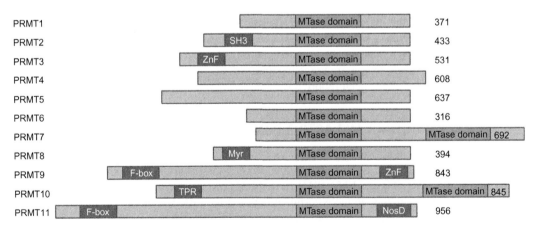

FIGURE 17.5

Schematic representation of the human protein arginine methyltransferase (PRMT) family. For each member, the length of the longest isoform is indicated on the right. All members of the family possess at least one conserved MTase domain with signature motifs I, post-I, II, and III and a THW loop. Additional domains are marked in maroon boxes: SH3, ZnF (zinc finger), Myr (myristoylation), F-box, TPR (tetratricopeptide), and NosD (nitrous oxidase accessory protein). (*This figure is reproduced in color in the color plate section.*)

Most of the PRMTs have a MTases domain with specific sequences (GxGxG) in its C-terminal part, which is dedicated to SAM binding. MTase domains have a localized structural element containing two invariant glutamate residues (double-E loop). Adjacent to this is a highly conserved THW (threonine-histidine-tryptophan) loop and altogether they form the active site. Along with this, strands 6 and 7 of the β-barrel domain have a three-helix segment inserted in between which is responsible for dimerization, essential for enzyme activity [81,82].

17.3.2 PRMT CLASSIFICATIONS

As mentioned earlier, PRMTs are able to generate MMA, aDMA, and sDMA, and are classified according to the methylation they catalyze. The first three Types (I, II, and III) generate MMA initially, while Type I PRMTs (1−4, 6, 8, and CARM1) − perform a second methylation. This second step generates the aDMA mark. In a similar fashion, Type II PRMT5 generates the sDMA mark. Type III PRMT7 can generate only the MMA mark. The MMA mark acts as a marker for the subsequent methylation by Type I and II PRMTs. However, certain proteins can still exist in a heavily monomethylated state. Type IV is seen only in yeasts and catalyzes the monomethylation of the internal guanidine nitrogen atom [81].

17.3.2.1 Mammalian arginine methyltransferases
17.3.2.1.1 PRMT1

Protein arginine methyltransferase 1 (PRMT1) is the major arginine methylation enzyme, strictly generating MMA and aDMA and responsible for 85% of arginine methylation activity. PRMT1 displays wide substrate specificity, as it preferentially methylates arginine residues with the GAR motif present. PRMT1 mediated methylation of histone H4 arginine 3 (H4R3me2a mark) functions as a transcriptional activation mark, which can be read by other methyl-binding proteins. As a transcriptional co-activator, PRMT1 is recruited to promoters by a number of different transcription factors. Information on how PRMT1 is regulated is limited, but we can assume that because of these factors, PRMT1 is likely to be regulated by a complex sequence. Recent studies correlate the role of PRMT1 methylation in many diseases; PRMT1 is significantly overexpressed in various types of cancer, notably bladder and prostate cancer, and it was found to be required for cell proliferation [83].

17.3.2.1.2 PRMT2

PRMT2 methylates histone H4, although the modification sites have not yet been well mapped [84]. Tissue distribution of PRMT2 is highest in the heart, prostate, ovary and neuronal system and cell distribution was found predominantly in the nucleus and partly to the cytoplasm [81], Though it was believed that PRMT2 does not possess methyltransferase activity, recent studies report very weak Type I activity on histone H4, the retinoblastoma gene product, STAT3 (signal transducer and activator of transcription 3) protein, the androgen receptor, the estrogen receptor alpha, and the heterogeneous nuclear ribonucleoprotein (hnRNP) E1B-AP5 [81,84−86]. PRMT2 has an SH3 domain that can bind the N-terminal domain of the PRMT8 and act together [87].

17.3.2.1.3 PRMT3

PRMT3, widely expressed in human tissues, was the first identified PRMT1-binding partner and has a predominantly cytosolic subcellular localization [81]. Asymmetric dimethylation of the 40S ribosomal protein S2 by PRMT3 plays a critical role in ribosomal biosynthesis; PRMT3's actions are also implicated in cancer through interaction with the DAL-1 tumor suppressor protein [88]. The N-terminus of PRMT3 harbors a zinc-finger domain responsible for the substrate recognition and helping to bind with the 40S ribosomal protein S2 (rpS2) [84]. Several studies suggest that PRMT3s may not directly impact epigenetic pathways since its cellular distribution is almost exclusively limited in the cytosol [81,84].

17.3.2.1.4 PRMT4

PRMT4 also function as a steroid receptor co-activator (also referred to as CARM1) and was the first PRMT showing that arginine methylation regulates transcription [84]. Apart from a steroid receptor co-activator, it also reinforces other transcription factor pathways; PRMT4 methylates histone H3 (H3R17me2a and H3R26me2a) and regulates the transcriptional activity [81,84,89]. CARM1 mediated methylation of splicing factors regulates the coupling of transcription and splicing and plays an important role in a number of biological processes including muscle cell differentiation, T cell development, and adipocyte differentiation. PRMT4 maintains embryonic stem cell (ESC) pluripotency and inhibits ESC differentiation [89].

17.3.2.1.5 PRMT5

PRMT5 accounts for the largest share of Type II arginine methylation in mammals and generates ω-NG-monomethyl and ω-NG,N'G-symmetric dimethylarginine residues [90]. PRMT5 is widely expressed in mammals and is located both in the nucleus and in the cytoplasm of tissues like the heart, muscle and testis. It was initially identified as a Jak2-binding protein and shown to methylate histones H2A, H3, and H4. PRMT5 plays a role in cell signaling, chromatin remodeling, and control of gene expression through its association with MEP50 and the subsequent methylation of many cytoplasmic and nuclear substrates, including histone, nucleolin, p53, eIF4E, EGFR, and C-RAF (Figure 17.3E) [81,90,91]. PRMT5 is emerging as an important enzyme involved in tumorigenesis and stem cell maintenance. In the nucleus, PRMT5 binds to its cooperator COPR5 and exert its transcriptional corepressor activities [91]. Major cytoplasmic activity of PRMT5 is in snRNP biogenesis, PRMT5 methylate a number of Sm proteins and Piwi proteins which regulate a class of small noncoding RNAs [87].

17.3.2.1.6 PRMT6

PRMT6 is the smallest protein in its class with 316 amino acids and its cellular distribution is mostly limited to the nucleus. It catalyzes the formation of asymmetric ω-NG, NG-dimethyl arginine and ω-NG-monomethyl arginine on both histone and nonhistone targets. PRMT6 is the major enzyme responsible for H3R2 methylation in mammalian cells, counteracting the H3K4me3 activation mark, making it a transcriptional repressor [84]. H3R2 dimethylation antagonizes the binding of effector proteins sensitive to H3K4 methylation, such as the MLL complex methyltransferase. Even though PRMT6s transcriptional repressor activities are not clear cut yet, its functions as a co-activator with nuclear receptors have been demonstrated recently. The co-activator ability of

PRMT6 may be attributed to its capacity to methylate the R3 position of the N-terminal tails of histones H4 and H2A [84].

17.3.2.1.7 PRMT7

PRMT7 is a Type III PRMT that is 692-amino acids long and produces only MMA from its substrates and is one of the two PRMTs to harbor two SAM-binding motifs, both of which are required for activity [92,93]. PRMT7 is predominantly distributed in thymus, dendritic cells and testis with nuclear and cytosolic localizations [81]. Whether PRMT7 was capable of forming sDMA as a Type II enzyme was debated until recently. Studies have confirmed its Type III activity of monomethylation of H2B (Arg-29, Arg-31, Arg-33 and H4 Arg-17, Arg-19) as the best substrate among recombinant human core histones [93]. PRMT7 activity has been correlated with either resistance or sensitivity to DNA damaging agents [87].

17.3.2.1.8 PRMT8

PRMT8 is more than 84% identical to PRMT1 and is unique among the PRMTs due to its restricted expression, which is found only in the central nervous system. The N-terminal end of PRMT8 features a unique functional myristoylation motif which forms many favorable electrostatic interactions with membrane lipids and further localization of PRMT8 in plasma membrane [87]. Two proline-rich sequences, contained in the N-terminal region, bind a number of SH3 domains, including PRMT2 [87]. PRMT8 targets the pro-oncoprotein encoded by the Ewing sarcoma (EWS) gene, which contains a GAR motif in its C-terminal terminus [81].

17.3.2.1.9 PRMT9 (4q31)

PRMT9 is a Type II PRMT capable of catalyzing the formation of MMA and aDMA [81]. However, the PRMT9 sequence varies considerably from other human PRMTs and has both nuclear and cytoplasmic localization [81]. PRMT9, like PRMT7, has two putative SAM-binding motifs and two tetratricopeptide repeats (TPR) at its N-terminal, which mediate protein–protein interactions [87]. PRMT9 targets histones, the maltose-binding protein and several peptides [81].

17.3.2.1.10 PRMT10 and PRMT11

PRMT10 and PRMT11 were identified by their homology with PRMT7 and PRMT9, respectively, but their PRMT activity or biological roles are yet to be identified [81].

17.3.3 PRMTS AS THERAPEUTIC TARGETS FOR DISEASES

Apart from playing diverse roles in normal physiological processes including signal transduction, transcription, DNA repair, and RNA processing, PRMT dysregulation is a main cause of concern in many diseases. Through the aberrant upregulation of oncogenes, PRMTs are found to have links to cancer [94]. For downstream signals of the MLL transcriptional complex, the enzymatic activities of DOT1L and PRMT1 were shown to be essential [94]. Beyond this, PRMT1 and PRMT6 activity were associated with bladder and lung cancer cells while PRMT4 has shown to be overexpressed in prostrate adenocarcinomas, human grade-III breast tumors and colorectal cancer.

PRMT5, a type-II PRMT, also plays a major role in promoting cell growth and transformation. PRMT5 is overexpressed in many cancers including colon, liver, kidney, lung, pancreas, bladder, ovary, breast, prostate, cervix, and skin; the overexpression in colon cancer is particularly noticeable. This may be attributed to the activation of PRMT5 through the nuclear cyclin D1/CDK4 kinase-dependent cullin4 (CUL4) expression.

PRMTs have a role in cardiovascular diseases as well. It has been reported that plasma levels of ω-aDMA are elevated with endothelial dysfunction-related cardiovascular and pulmonary diseases, including coronary heart disease (CHD) and pulmonary hypertension, ω-aDMA interacts with endothelial nitric oxide (NO)-producing enzyme dimethylarginine dimethylaminohydrolase-1 and -2 (DDAH), which may increase susceptibility to cardiovascular disease [81].

PRMTs are also found to be involved in virus-related diseases as well, such as endothelial cell (EC) inflammatory response through PRMT5-induced methylation of HOXA9 (Homeobox protein Hox-A9) on R140 [95]. PRMT1-mediated methylation of FOXO transcription factors have a critical role in the homoeostasis of cell apoptosis. A methylated FOXO protein cannot be further phosphorylated by Akt, thus increasing oxidative-stress-induced apoptosis from the PI3K Akt signaling pathway [81]. These findings may help in developing inhibitors of PRMT1 for the treatment of neurodegenerative diseases and cell damage due to oxidative stress.

Due to their characteristic co-activation of nuclear receptors, arginine methyltransferases have become an attractive therapeutic target for hormone-dependent cancers. PRMT1, 2, 4, and 5 have been identified as co-activators of either androgen or estrogen receptors [81]. In particular, PRMT5, along with being a regulator of p53 transcription, can lead to apoptosis in the context of DNA damage control. Alternatively, co-repressor activity of PRMT5 on tumor suppressor genes ST7 and NM23 is linked to proliferation and the regulation of cell growth [91]. Though these findings suggest PRMT inhibitors as potential drugs, the biological impact of inhibiting arginine methylation is a major concern.

Another aspect to be addressed is the crosstalk between various PTMs. Recent studies suggest the existence of multiple instances of crosstalk between arginine methylation and other PTMs. For example serine phosphorylation and lysine methylation on histone tails are corelated [84]. In a similar fashion, there is crosstalk between arginine and lysine methylation which has been termed an arginine/lysine-methyl/methyl switch [84]. Examples of such pairs in histone tails are H3R2/H3K4; H3R8/H3K9;H3R26/H3K27, where both residues are methylated [84].

17.3.4 PRMT INHIBITORS

In 2004, a screening of 9,000 compounds by Bedford et al. led to the discovery of the first non-nucleoside specific inhibitors of PRMTs − nine compounds (Table 17.1, entries 34−36) with a low micromolar inhibition of PRMT1 (0.19−16.3 μM) named arginine methyltransferase inhibitors (AMIs) [96]. These compounds were used as lead molecules in the design and further optimization of AMIs, generating many potent and selective analogs. Later, Purandare et al. identified a series of pyrazole amide containing compounds as potential inhibitors through high-throughput screening; preliminary optimization of these compounds led to Methylgene compound 7a (Table 17.1, entry 32, IC_{50} hPRMT4 = 60 nM) and Bristol Myers Squibb compound 7f (Table 17.1, entry 33, IC_{50} hPRMT4 = 40 nM), each leading to a variety of active analogs [94,96]. Target-based virtual screening for PRMTs by Spannhoff et al. in 2007 produced two compounds, allantodapsone (Table 17.1, entry 39, IC_{50} hPRMT1 = 1.7 μM) and stilbamidine (Table 17.1, entry 37, IC_{50} hPRMT1 = 57 μM). A later fragment-based virtual screening by Spannhoff et al. led to the discovery of RM65

(Table 17.1, entry 40, IC$_{50}$ hPRMT155 μM) [95,96]. A recent study by Leilei Yan et al. shows that 2,5-bis(4-amidinophenyl)furan (DB75, Table 17.1, entry 38) is also a selective PRMT1 inhibitor (IC$_{50}$ = 9.4 ± 1.1) [97]. It is also worth mentioning the carbocyanine inhibitors developed by Sinha et al., which are not only selective PRMT1 inhibitors (Table 17.1, entry 41, IC50 = 4.1 ± 0.1) but also a photoactive chemical probe for the further study and characterization of PRMT inhibitors [98]. Developing potent and selective inhibitors of PRMTs is a large field in medicinal chemistry currently, with multiple groups working with reasonable success in finding potent and selective inhibitors.

Protein arginine methylation has received widespread interest in the scientific community. It is now clear that PTM impacts crucial biological functions such as transcriptional regulation, RNA processing, and DNA repair and signal transduction. It is also important to note that the metabolism of arginine and protein arginine methylation are biochemically linked since the methyl group marking of arginine residues originates from SAM. Unfortunately, information regarding important protein arginine methylation is scarce. There is still much to discover and learn about its regulation by other PTMs, how it affects protein-ligand interactions, what factors govern the specificity of the different PRMTs toward the target protein, as well as the pathological implications of arginine methylation disruption [81].

17.4 CONCLUSION

Epigenetics is a science with far-reaching potential. As the development and maintenance of any organism is orchestrated by a set of chemical reactions that regulate parts of the genome at specific intervals, understanding this process is critical. The remodeling of chromatin can selectively activate or inactivate genes. This process determines their expression; which in turn may subsequently influence pathogenesis and the outcome of many diseases. Studies show that these chromatin modifications can be tied to the onset of inflammatory diseases and the development of aberrant proteins found frequently in many cancers. Epigenetics offers insight into the genetic process, the biological role of chromatin-associated proteins, and the roles these proteins have on development.

As the initiation and progression of cancer is controlled by both genetic and epigenetic events and epigenetic modifications are potentially reversible, the development of specific and potent small molecule inhibitors is a promising therapeutic option for cancers. Targeting enzymes tied to histone methylation involved in cancer has shown potential as a treatment. PKMTs and PRMTs in particular constitute an attractive and diverse selection of targets, as many genetic modifications can be traced back to these writers, erasers, and readers. Small molecules can exploit this relationship by blocking enzymatic activity or chromatin recognition.

This approach is well validated, stemming from the success of inhibiting DOT1L and EZH2 in cancerous cell lines and xenograft models. Companies such as Epizyme Inc. and Glaxo Smith Kline have independently reported small molecule inhibitors and chemical probes for PRC2 and EZH2, DOT1L, PRC2, and G9a/GLP demonstrating, in a concentration-dependent fashion, the ability to reduce intracellular levels of the relevant histone methyl marks. Some of these inhibitors, such as DOT1L inhibitor EPZ-5676, have already begun clinical trials. These discoveries have built momentum, leading to the discovery of several more potent, selective inhibitors of methyltransferases this year. Likewise, future efforts should be focused on the continued discovery of potent inhibitors; however, the impact and consequences of inhibiting histone methylation is not well understood and requires more research.

Methylation of histones has long been associated with chromatin, and in turn tied solely to transcriptional activation and repression. Now, with the discovery that methylation can occur across a variety of nonchromatin histone proteins, histone methylation cannot be considered the sole domain of chromatin biology. However, the exact function of nonchromatin methylation in proteins is not well understood. As methylation has no impact on residue charge, addition of methyl groups is unlikely to create significant conformational shifts within the substrate protein. Alternatively, bimolecular interactions may be impeded by the steric hindrance of the methyl groups, or provide an epitopic motif to introduce interactions with downstream effector proteins. Several protein domains have shown recognition toward certain methylation marks, suggesting that methylation of proteins may mediate protein—nucleic acid or protein—protein interactions [99,100]. A systematic review of the specific interactions and roles of these new domains, using novel inhibitors and chemical probes, should be undertaken to understand the relationship of aberrant methylation in cancer.

REFERENCES

[1] Altucci L, Stunnenberg HG. Time for epigenetics. Int J Biochem Cell Biol 2009;41:2−3.
[2] Baylin SB, Schuebel KE. Genomic biology: the epigenomic era opens. Nature 2007;448:548−9.
[3] Goldberg AD, Allis CD, Bernstein E. Epigenetics: a landscape takes shape. Cell 2007;128:635−8.
[4] Kim JK, Samaranayake M, Pradhan S. Epigenetic mechanisms in mammals. Cell Mol Life Sci 2009;66:596−612.
[5] Luger K, Mader AW, Richmond RK, Sargent DF, Richmond TJ. Crystal structure of the nucleosome core particle at 2.8 Å resolution. Nature 1997;389:251−60.
[6] Strahl BD, Allis CD. The language of covalent histone modifications. Nature 2000;403:41−5.
[7] Taverna SD, Li H, Ruthenburg AJ, Allis CD, Patel DJ. How chromatin-binding modules interpret histone modifications: lessons from professional pocket pickers. Nat Struct Mol Biol 2007;14 (11):1025−40.
[8] Iacobuzio-Donahue CA. Epigenetic changes in cancer. Annu Rev Pathol 2009;4:229−49.
[9] Wang GG, Allis CD, Chi P. Chromatin remodeling and cancer, Part I: covalent histone modifications. Trends Mol Med 2007;13:363−72.
[10] Marmorstein R, Trievel RC. Histone modifying enzymes: structures, mechanisms, and specificities. Biochim Biophys Acta 2009;1789:58−68.
[11] Altucci L, Minucci S. Epigenetic therapies in haematological malignancies: searching for true targets. Eur J Cancer 2009;45(7):1137−45.
[12] Zhang K, Dent SY. Histone modifying enzymes and cancer: going beyond histones. J Cell Biochem 2005;96:1137−48.
[13] Cole PA. Chemical probes for histone-modifying enzymes. Nat Chem Biol 2008;4:590−7.
[14] Anglin JL, Song Y. A medicinal chemistry perspective for targeting histone H3 Lysine-79 methyltransferase DOT1L: miniperspective. J Med Chem 2013;56(22):8972−83.
[15] Okada Y, Feng Q, Lin Y, Jiang Q, Li Y, Coffield VM, et al. Cell 2005;121:167.
[16] Krivtsov AV, Armstrong SA. MLL translocations, histone modifications and leukaemia stem-cell development. Nat Rev Cancer 2007;7:823−33.
[17] Krivtsov AV, Feng Z, Lemieux ME, Faber J, Vempati S, Sinha AU, et al. H3K79 methylation profiles define murine and human MLL-AF4 leukemias. Cancer Cell 2008;14:355−68.

[18] Daigle SR, Olhava EJ, Therkelsen CA, Majer CR, Sneeringer CJ, Song J, et al. Selective killing of mixed lineage leukemia cells by a potent small-molecule DOT1L inhibitor. Cancer Cell 2011;20:53–65.

[19] Yu W, Smil D, Li F, Tempel W, Fedorov O, Nguyen KT, et al. Bromo-deaza-SAH: a potent and selective DOT1L inhibitor. Bioorg Med Chem 2013;21:1787–94.

[20] Lund AH, van Lohuizen M. Polycomb complexes and silencing mechanisms. Curr Opin Cell Biol 2004;16:239–46.

[21] Moss TJ, Wallrath LL. Connections between epigenetic gene silencing and human disease. Mutat Res 2007;618:163–74.

[22] Cao R, Zhang Y. SUZ12 is required for both the histone methyltransferase activity and the silencing function of the EED-EZH2 complex. Mol Cell 2004;15:57–67.

[23] Kuzmichev A, Jenuwein T, Tempst P, Reinberg D. Different EZH2-containing complexes target methylation of histone H1 or nucleosomal histone H3. Mol Cell 2004;14:183–93.

[24] Tamaru H, Selker EU. A histone H3 methyltransferase controls DNA methylation in *Neurosporacrassa*. Nature 2001;414:277–83.

[25] Schlesinger Y, Straussman R, Keshet I, Farkash S, Hecht M, Zimmerman J, et al. Polycomb-mediated methylation on Lys27 of histone H3 pre-marks genes for de novo methylation in cancer. Nat Genet 2007;39:232–6.

[26] Yu J, Mani RS, Cao Q, Brenner CJ, Cao X, Wang X, et al. An integrated network of androgen receptor, polycomb, and TMPRSS2-ERG gene fusions in prostate cancer progression. Cancer Cell 2010;17:443–54.

[27] Cha TL, Zhou BP, Xia W, Wu Y, Yang CC, Chen CT, et al. Akt-mediated phosphorylation of EZH2 suppresses methylation of lysine 27 in histone H3. Science 2005;310:306–10.

[28] Fluge Ø, Gravdal K, Carlsen E, Vonen B, Kjellevold K, Refsum S, et al. Expression of EZH2 and Ki-67 in colorectal cancer and associations with treatment response and prognosis. Br J Cancer 2009;101:128–9.

[29] Zheng F, Liao YJ, Cai MY, Liu YH, Liu TH, Chen SP, et al. The putative tumor suppressor microRNA-124 modulates hepatocellular carcinoma cell aggressiveness by repressing ROCK2 and EZH2. Gut 2011;61:278–89.

[30] Min J, Zaslavsky A, Fedele G, McLaughlin SK, Reczek EE, De Raedt T, et al. An oncogene-tumor suppressor cascade drives metastatic prostate cancer by coordinately activating Ras and nuclear factor-kappaB. Nat Med 2010;16:286–94.

[31] Lu C, Han HD, Mangala LS, Ali-Fehmi R, Newton CS, Ozbun L, et al. Regulation of tumor angiogenesis by EZH2. Cancer Cell 2010;18:185–97.

[32] Wu ZL, Zheng SS, Li ZM, Qiao YY, Aau MY, Yu Q. Polycomb protein EZH2 regulates E2F1-dependent apoptosis through epigenetically modulating Bim expression. Cell Death Diff 2010;17:801–10.

[33] Glazer RI, Hartman KD, Knode MC, Richard MM, Chiang PK, Tseng CK, et al. 3-Deazaneplanocin: a new and potent inhibitor of S-adenosylhomocysteine hydrolase and its effects on human promyelocytic leukemia cell line HL-60. Biochem Biophys Res Commun 1986;135:688–94.

[34] Suva ML, Riggi N, Janiszewska M, Radovanovic I, Provero P, Stehle JC, et al. EZH2 is essential for glioblastoma cancer stem cell maintenance. Cancer Res 2009;69:9211–18.

[35] Fiskus W, Wang Y, Sreekumar A, Buckley KM, Shi H, Jillella A, et al. Combined epigenetic therapy with the histone methyltransferase EZH2 inhibitor 3-deazaneplanocin A and the histone deacetylase inhibitor panobinostat against human AML cells. Blood 2009;114:273–343.

[36] Kong D, Heath E, Chen W, Cher ML, Powell I, Heilbrun L, et al. Loss of let-7 upregulates EZH2 in prostate cancer consistent with the acquisition of cancer stem cell signatures that are attenuated by BR-DIM. PloS One 2012;7(3):e33729.

[37] Diaz E, Machutta CA, Chen S, Jiang Y, Nixon C, Hofmann G, et al. Development and validation of reagents and assays for EZH2 peptide and nucleosome high-throughput screens. J Biomol Screen 2012;17:1279–92.

[38] Hua WF, Fu YS, Liao YJ, Xia WJ, Chen YC, Zeng YX, et al. Curcumin induces down-regulation of EZH2 expression through the MAPK pathway in MDA-MB-435 human breast cancer cells. Eur J Pharmacol 2010;637:16–21.

[39] Fiskus W, Buckley K, Rao R, Mandawat A, Yang Y, Joshi R, et al. Panobinostat treatment depletes EZH2 and DNMT1 levels and enhances decitabine mediated de-repression of JunB and loss of survival of human acute leukemia cells. Cancer Biol Ther 2009;8:939–50.

[40] Tachibana M, Sugimoto K, Fukushima T, Shinkai Y. Set domain-containing protein, G9a, is a novel lysine-preferring mammalian histone methyltransferase with hyperactivity and specific selectivity to lysines 9 and 27 of histone H3. J Biol Chem 2001;276:25309–17.

[41] Tachibana M, Ueda J, Fukuda M, Takeda N, Ohta T, Iwanari H, et al. Histone methyltransferases G9a and GLP form heteromeric complexes and are both crucial for methylation of euchromatin at H3-K9. Genes Dev 2005;19:815–26.

[42] Bannister AJ, Zegerman P, Partridge JF, Miska EA, Thomas JO, Allshire RC, et al. Selective recognition of methylated lysine 9 on histone H3 by the HP1 chromo domain. Nature 2001;410:120–4.

[43] Rice JC, Briggs SD, Ueberheide B, Barber CM, Shabanowitz J, Hunt DF, et al. Mol Cell 2003;12:1591–8.

[44] Allali-Hassani A, Wasney GA, Siarheyeva A, Hajian T, Arrowsmith CH, Vedadi M. Fluorescence-based methods for screening writers and readers of histone methyl marks. J Biomol Screen 2012;17(1):71–84.

[45] Estève PO, Chin HG, Smallwood A, Feehery GR, Gangisetty O, Karpf AR, et al. Direct interaction between DNMT1 and G9a coordinates DNA and histone methylation during replication. Genes Dev 2006;20:3089–103.

[46] Goyama S, Nitta E, Yoshino T, Kako S, Watanabe-Okochi N, Shimabe M, et al. EVI-1 interacts with histone methyltransferases SUV39H1 and G9a for transcriptional repression and bone marrow immortalization. Leukemia 2010;24:81–8.

[47] Shi Y, Desponts C, Do JT, Hahm HS, Schöler HR, Ding S, et al. Induction of pluripotent stem cells from mouse embryonic fibroblasts by Oct4 and Klf4 with small-molecule compounds. Cell Stem Cell 2008;3:568–74.

[48] Huang J, Desponts C, Do JT, Hahm HS, Schöler HR, Ding S, et al. G9A and GLP methylate lysine 373 in the tumor suppressor p53. J Biol Chem 2010;285:9636–41.

[49] Kondo Y, Shen L, Ahmed S, Boumber Y, Sekido Y, Haddad BR, et al. Downregulation of histone H3 lysine 9 methyltransferase G9a induces centrosome disruption and chromosome instability in cancer cells. PLoS One 2008;3:e2037.

[50] Leung DC, Dong KB, Maksakova IA, Goyal P, Appanah R, Lee S, et al. Lysine methyltransferase G9a is required for de novo DNA methylation and the establishment, but not the maintenance, of proviral silencing. Proc Natl Acad Sci 2011;108(14):5718–23.

[51] Tibodeau JD, Benson LM, Isham CR, Owen WG, Bible KC. Antioxid Redox Signal 2009;11:1097–106.

[52] Kubicek S, O'Sullivan RJ, August EM, Hickey ER, Zhang Q, Teodoro ML, et al. Reversal of H3K9me2 by a small-molecule inhibitor for the G9a histone methyltransferase. Mol Cell 2007;25:473–81.

[53] Greiner D, Bonaldi T, Eskeland R, Roemer E, Imhof A. Identification of a specific inhibitor of the histone methyltransferase SU(VAR)3–9. Nat Chem Biol 2005;1:143–5.

[54] Cook KM, Hilton ST, Mecinović J, Motherwell WB, Figg WD, Schofield CJ. Epidithiodiketopiperazines block the interaction between hypoxia-inducible factor-1α (HIF-1α) and p300 by a zinc ejection mechanism. J Biol Chem 2009;284(39):26831–8.

[55] Jenuwein T, Allis CD. Translating the histone code. Science 2001;293:1074–80.

[56] Gardiner DM, Waring P, Howlett BJ. The epipolythiodioxopiperazine (ETP) class of fungal toxins: distribution, mode of action, functions and biosynthesis. Microbiology 2005;151:1021–32.

[57] Saito T, Suzuki Y, Koyama K, Natori S, Iitaka Y, Kinoshita T. Chetracin A and Chaetocins B and C: three new epipolythiodioxopiperazines from *Chaetomium* spp. Chem Pharm Bull 1988;36:1942–56.

[58] Chang Y, Ganesh T, Horton JR, Spannhoff A, Liu J, Sun A, et al. Adding a lysine mimic in the design of potent inhibitors of histone lysine methyltransferases. J Mol Biol 2010;400:1–7.

[59] Domer PH, Fakharzadeh SS, Chen CS, Jockel J, Johansen L, Silverman GA, et al. Acute mixed lineage leukemia t(4;11)(q21;q23) generates an MLL-AF4 fusion product. Proc Natl Acad Sci U S A 1993;90 (16):7884–8.

[60] Cosgrove MS, Patel A. Mixed lineage leukemia: a structure function perspective of the MLL1 protein. FEBS J 2010;277:1832–42.

[61] Yip BH, So CW. Mixed lineage leukemia protein in normal and leukemic stem cells. Exp Biol Med (Maywood) 2013;238(3):315–23.

[62] Takahashi YH, Westfield GH, Oleskie AN, Trievel RC, Shilatifard A, Skiniotis G. Structural analysis of the core COMPASS family of histone H3K4 methylases from yeast to human. Proc Natl Acad Sci U S A 2011;108:20526–31.

[63] Ansari KI, Mandal SS. Mixed lineage leukemia: roles in gene expression, hormone signaling and mRNA processing. FEBS J 2010;277:1790–804.

[64] Liu H, Cheng EH, Hsieh JJ. MLL fusions: pathways to leukemia. Cancer Biol Ther 2009;8:1204–11.

[65] Malik S, Bhaumik SR. Mixed lineage leukemia: histone H3 lysine 4 methyltransferases from yeast to human. FEBS J 2010;277(8):1805–21.

[66] Ayton PM, Cleary ML. Molecular mechanisms of leukemogenesis mediated by MLL fusion proteins. Oncogene 2001;20:5695–707.

[67] Muntean AG, Hess JL. The pathogenesis of mixed lineage leukemia. Annu Rev Pathol 2012;7:283–301.

[68] Yeung J, Esposito MT, Gandillet A, Zeisig BB, Griessinger E, Bonnet D, et al. B-catenin mediates the establishment and drug resistance of MLL leukemic stem cells. Cancer Cell 2010;18:606–18.

[69] Meyer C, Kowarz E, Hofmann J, Renneville A, Zuna J, Trka J, et al. New insights to the MLL recombinome of acute leukemias. Leukemia 2009;23:1490–9. [PubMed: 19262598].

[70] Krivtsov AV, Feng Z, Lemieux ME, Faber J, Vempati S, Sinha AU, et al. H3K79 methylation profiles define murine and human MLL-AF4 leukemias. Cancer Cell 2008;14(5):355–68.

[71] Stumpel DJ, Schneider P, van Roon EH, Boer JM, de Lorenzo P, Valsecchi MG, et al. Specific promoter methylation identifies different subgroups of MLL rearranged infant acute lymphoblastic leukemia, influences clinical outcome, and provides therapeutic options. Blood 2009;114(27):5490–8.

[72] Muntean AG, Tan J, Sitwala K, Huang Y, Bronstein J, Connelly JA, et al. The PAF complex synergizes with MLL fusion proteins at HOX loci to promote leukemogenesis. Cancer Cell 2010;17:609–21.

[73] Rizzolio F, Tuccinardi T, Caligiuri I, Lucchetti C, Giordano A. CDK inhibitors: from the bench to clinical trials. Curr Drug Targets 2010;11:279–90.

[74] Wang Z, Iwasaki M, Ficara F, Lin C, Matheny C, Wong SH, et al. GSK-3 promotes conditional association of CREB and its coactivators with MEIS1 to facilitate HOX mediated transcription and oncogenesis. Cancer Cell 2010;17:597–608.

[75] Senisterra G, Wi H, Allali-Hassani A, Wasney GA, Barstye-Lovejoy D, Dombrovski L, et al. Small-molecule inhibition of MLL activity by disruption of its interaction with WDR5. Biochem J 2013;449:151–9.

[76] Chang MJ, Wu H, Achille NJ, Reisenauer MR, Chou CW, Zeleznik-Le NJ, et al. Histone H3 lysine 79 methyltransferase Dot1 is required for immortalization by MLL oncogenes. Cancer Res 2010;70:10234–42.

[77] Lane AA, Chabner BA. Histone deacetylase inhibitors in cancer therapy. J Clin Oncol 2009;27 (32):5459—68.

[78] Nguyen AT, Taranova O, He J, Zhang Y. DOT1L, the H3K79 methyltransferase, is required for MLL-AF9-mediated leukemogenesis. Blood 2011;117(25):6912—22.

[79] Daigle SR, Olhava EJ, Therkelsen CA, Majer CR, Sneeringer CJ, Song J, et al. Selective killing of mixed lineage leukemia cells by a potent small molecule DOT1L inhibitor. Cancer Cell 2011;20 (1):53—65.

[80] Gayatri S, Bedford M. Readers of histone methylarginine marks. Biochim Biophys Acta 2014;1839 (8):702—10.

[81] Ruben E, Leandro P, Rivera I, Tavares de Almeida I, Blom HJ, Castro R, et al. Deciphering protein arginine methylation in mammals. Molecular themes in DNA replication. In: Dricu A, editor. Methylation: from DNA, RNA and histones to diseases and treatment. Cambridge: Royal Society of Chemistry; 2012. p. 91—116.

[82] Rust HL, Subramanian V, West GM, Young DD, Schultz PG, Thompson PR. Using unnatural amino acid mutagenesis to probe the regulation of PRMT1. ACS Chem Biol 2014;9(3):649—55.

[83] Elakoum R, Gauchotte G, Oussalah A, Wissler MP, Clément-Duchêne C, Vignaud JM, et al. CARM1 and PRMT1 are dysregulated in lung cancer without hierarchical features. Biochimie 2014;97:210—18.

[84] Bedford MT. Histone arginine methylation. FEBS Lett 2011;585(13):2024—31.

[85] Meyer R, Wolf SS, Obendorf M. PRMT2, a member of the protein arginine methyl-transferase family, is a coactivator of the androgen receptor. J Steroid Biochem Mol Biol 2007;107:1—14.

[86] Lakowski TM, Frankel A. Kinetic analysis of human protein arginine N-methyltrans-ferase 2: formation of monomethyl- and asymmetric dimethyl-arginine residues on histone H4. Biochem J 2009;421:253—61.

[87] Bedford M, Clarke SG. Protein arginine methylation in mammals: who, what, and why. Mol Cell 2009;33(1):1—13.

[88] Liu F, Li F, Ma A, Dobrovetsky E, Dong A, Gao C, et al. Exploiting an allosteric binding site of PRMT3 yields potent and selective inhibitors. J Med Chem 2013;56(5):2110—24.

[89] Vu LP, Perna F, Wang L, Voza F, Figueroa ME, Tempst P, et al. PRMT4 blocks myeloid differentiation by assembling a methyl-RUNX1-dependent repressor complex. Cell Rep 2013;5(6):1625—38.

[90] Han X, Li R, Zhang W, Yang X, Wheeler CG, Friedman GK, et al. Expression of PRMT5 correlates with malignant grade in gliomas and plays a pivotal role in tumor growth in vitro. J Neurooncol 2014;118(1):61—72.

[91] Karkhanis V, Hu U-J, Baiocchi RA, Imbalzano AN, Sif S. Versatility of PRMT5-induced methylation in growth control and development. Trends Biochem Sci 2011;36(12):633—41.

[92] Hu Y, Sif S, Imbalzano A. Prmt7 is dispensable in tissue culture models for adipogenic differentiation. F1000Res 2013;2:279.

[93] Feng Y, Maity R, Whitelegge JP, Hadjikyriacou A, Li Z, Zurita-Lopez C, et al. Mammalian protein arginine methyltransferase 7 (PRMT7) specifically targets RXR sites in lysine- and arginine-rich regions. J Biol Chem 2013;288(52):37010—25.

[94] Wigle TJ, Copeland RA. Drugging the human methylome: an emerging modality for reversible control of aberrant gene transcription. Curr Opin Chem Biol 2013;17(3):369—78.

[95] Heinke R, Spannhoff A, Meier R, Trojer P, Bauer I, Jung M, et al. Virtual screening and biological characterization of novel histone arginine methyltransferase PRMT1 inhibitors. Chem Med Chem 2009;4 (1):69—77.

[96] Yost JM, Korboukh I, Liu F, Gao C, Jin J. Targets in epigenetics: inhibiting the methyl writers of the histone code. Curr Chem Genomics 2011;5(Suppl. 1):72.

[97] Yan L, Yan C, Qian K, Su H, Kofsky-Wofford SA, Lee WC, et al. Diamidine compounds for selective inhibition of protein arginine methyltransferase 1. J Med Chem 2014;57(6):2611−22.

[98] Sinha SH, Owens EA, Feng Y, Yang Y, Xie Y, Tu Y, et al. Synthesis and evaluation of carbocyanine dyes as PRMT inhibitors and imaging agents. Eur J Med Chem 2012;54:647−59.

[99] Cosgrove MS, Boeke JD, Wolberger C. Regulated nucleosome mobility and the histone code. Nat Struct Mol Biol 2004;11:1037−43.

[100] Min J, Zhang Y, Xu RM. Structural basis for specific binding of Polycomb chromodomain to histone H3 methylated at Lys 27. Genes Dev 2003;17:1823−8.

DISCOVERY OF HISTONE DEMETHYLASE INHIBITORS

Alexander-Thomas Hauser[1], Martin Roatsch[1], Johannes Schulz-Fincke[1,2,3], Dina Robaa[4], Wolfgang Sippl[4], and Manfred Jung[1]

[1]Institute of Pharmaceutical Sciences, Albert-Ludwigs-University of Freiburg, Germany [2]German Cancer Consortium, Heidelberg, Germany [3]German Cancer Research Center, Heidelberg, Germany [4]Department of Pharmaceutical Chemistry, Martin Luther University of Halle-Wittenberg, Germany

CHAPTER OUTLINE

Y.G. Zheng (Ed): Epigenetic Technological Applications. DOI: http://dx.doi.org/10.1016/B978-0-12-801080-8.00018-1

18.1 INTRODUCTION

It is well accepted that the methylation of histones is involved in major regulatory functions such as the formation and maintenance of heterochromatin, X-chromosome inactivation, transcriptional regulation, and genomic imprinting [1]. Histone proteins can be methylated either on arginine or on lysine residues by the action of histone methyltransferases (HMTs). While arginines can be mono- or dimethylated, lysine residues can be mono-, di-, or trimethylated [2]. Furthermore, depending on the site and even on the degree of methylation, histone methylation can either be a repressive or an activating mark [2].

For a long period of time, histone methylation has been considered to be an irreversible biochemical modification [3], but starting from the first publication on a histone lysine demethylating enzyme in 2004, it became clear that enzymes exist that are able to actively remove methyl marks from histone lysine residues [4−7]. Subsequently, the demethylation of other proteins has also been described [8−10].

In the following, we will focus on histone lysine demethylases, the structural aspects of inhibitor binding and mainly on the screening technologies that are used to identify and characterize new lysine demethylase inhibitors.

18.2 HISTONE DEMETHYLASES AND THEIR INVOLVEMENT IN DISEASE

18.2.1 JUMONJIC DOMAIN-CONTAINING HISTONE DEMETHYLASES

So far, about 30 members of JumonjiC domain-containing proteins have been identified, 18 of which have shown to possess histone demethylase activity [11]. Based on phylogenetic analysis as well as on domain architecture, these proteins have been divided into seven subfamilies that consist of several enzyme subtypes [12]. Table 18.1 gives a brief overview over the different subtypes and their respective target sites. For a detailed description of the phylogenetic tree and the differences in domain architecture, see reference [13].

The JumonjiC domain-containing demethylases belong to the cupin superfamily of oxygenases [14] and remove the methyl marks from histone lysines in a totally different manner than the flavin-dependent demethylase LSD1 (see below). The Jumonji demethylases contain a Fe^{2+}-ion in their catalytic pocket and use α-ketoglutarate as a co-substrate. During the demethylation reaction, the methyl group is hydroxylated, resulting in an unstable heminaminal, which spontaneously decomposes to result in the formation of the demethylated lysine residue under the release of formaldehyde [7,11,15,16]. In contrast to LSD1, this mechanism also enables the removal of a trimethyl mark, as no free electron pair on the nitrogen atom is needed (see below). All JumonjiC domain-containing demethylases catalyze hydroxylation reactions, but they significantly differ regarding their substrate specificity and their ability to recognize the respective methylation status

Table 18.1 Classification and Substrate Specificity of JumonjiC Domain-Containing Demethylases

Subfamily	Subtypes	Histone Substrate
KDM2	FBXL10	H3K36me1/me2, H3K4me3
	FBXL11	H3K36me1/me2
KDM3	JMJD1A	H3K9me1/me2
	JMJD1B	H3K9me1/me2
	JMJD1C	Unknown
	HR	Unknown
KDM4	JMJD2A	H3K9me2/me3, H3K36me2/me3, H1.4K26me2/me3
	JMJD2B	H3K9me2/me3, H3K36me2/me3, H1.4K26me2/me3
	JMJD2C	H3K9me2/me3, H3K36me2/me3, H1.4K26me2/me3
	JMJD2D	H3K9me2/me3, H1.4K26me2/me3
	JMJD2E	H3K9me2/me3, H1.4K26me2/me3
	JMJD2F	Unknown
KDM5	JARID1A	H3K4me1/me2/me3
	JARID1B	H3K4me1/me2/me3
	JARID1C	H3K4me1/me2/me3
	JARID1D	H3K4me1/me2/me3
	JARID2	No demethylase activity
KDM6	UTX	H3K27me2/me3
	UTY	Unknown
	JMJD3	H3K27me2/me3
KDM7	PHF2	H3K9me2
	PHF8	H3K9me2/me1, H4K20me1
	KIAA1718	H3K9me2/me1, H3K27me1/me2
KDM8	JMJD5	H3K36me2

(me1, me2, or me3). This is determined by a methylammonium-binding pocket that enables the direct hydroxylation of the methyl group by bringing it in close proximity to the Fe^{2+}-binding site [17−19].

18.2.2 LYSINE-SPECIFIC DEMETHYLASE 1

Since the existence of dynamic regulation of the methylation state at histone tails has been proven by Shi et al., [4] there is much interest in the biochemical function of the lysine-specific demethylase (LSD1), also referred to as lysine demethylase 1 (KDM1). The demethylating mechanism of LSD1 depends on the cofactor flavin adenine dinucleotide (FAD) [20] and it shares a high sequence homology with other FAD-dependent amine oxidases [21]. It is known to be part of several corepressor complexes, especially Co-REST [22], and shows substrate specificity to H3K4me1/me2, which can change to H3K9me1/me2 upon association with the androgen receptor [5]. However, there is no strong preference observed for either H3K4me1 or H3K4me2 [23]. In contrast to JumonjiC domain-containing histone demethylases, LSD1 is not able to demethylate trimethyl-lysine residues (Kme$_3$) based on its mechanism [20]. In an oxidative process, LSD1 forms an imine intermediate which subsequently will be hydrolyzed to an aldehyde and its corresponding amine. During the last step, the reduced cofactor FAD is regenerated by oxygen concomitant with generation of hydrogen peroxide [4].

The recently discovered demethylase LSD2 (KDM1B) with activity toward H3K4me1/me2 [24,25] has not been validated as a target for drug discovery at this point.

Although the physiological role of the demethylation of histone lysines is poorly understood, it is known that lysine demethylases play an important role in embryonic development. So it has been shown that the knockdown of LSD1 and the subsequent increase of H3K4 methylation of target genes lead to differentiation of embryonic stem cells [26]. The knockout of the JumonjiC domain-containing histone demethylase JMJD5 upregulates the transcription of the tumor suppressor p53 in mice embryos [27] and the knockdown of JMJD2A leads to a significant decrease in the expression of neural crest specifier genes in embryonic chicken [28].

It is also well accepted that, in general, an aberrant histone methylation state seems to be correlated with tumorigenesis. Numerous examples underline the link between the overexpression of a histone demethylase and the progression of malignant tumors. The activity of LSD1 for example was demonstrated to be required for cell proliferation by promoting the G2/M transition [29] and the overexpression of LSD1 also is correlated with prostate cancer and its recurrence during therapy [30]. Other examples are the overexpression of the JumonjiC domain-containing histone demethylases JARID1B and JMJD2A in breast and testis cancer as well as in esophageal squamous carcinoma [31−33]. Lysine demethylases also could be linked to drug resistance, as a knockdown of JARID1A in the non−small cell lung carcinoma cell line PC9 reduces the levels of drug tolerant cells [34].

For JMJD2A, an involvement in the regulation of the replication of Kaposi's sarcoma-associated herpesvirus (KSHV) has also been shown [35]. Studies with mice lacking or overexpressing JMJD2A in the heart also give evidence that this demethylase promotes cardiac hypertrophy in response to cardiac stress [36]. Demethylases also were shown to play a role in metabolic diseases as JMJD1A-knockout mice display an adult obesity phenotype [37]. Several studies underline the involvement of lysine demethylases in inflammation. It could be shown that JMJD2D controls H3K9me3 levels of cell-type-specific enhancer regions in immune cells, thus promoting transcription of functional enhancers [38]. A role in inflammation has also been shown for JMJD3 as its expression in macrophages is regulated through the NF-κB pathway in LPS-stimulated macrophages [39]. A few other reports show the involvement of histone demethylases in other diseases like hyperglycemia [40] or X-linked mental retardation [41]. The emerging role of histone demethylases and their inhibition in the therapy of cancer and other diseases has been discussed in more detail in recent review articles [13,42−44].

18.3 ASSAYS FOR HISTONE DEMETHYLASES AND THEIR ROLE IN THE DISCOVERY OF INHIBITORS

18.3.1 JUMONJIC DOMAIN-CONTAINING HISTONE DEMETHYLASES

Structural information and insight into the catalytic mechanism and substrate specificity of JumonjiC domain-containing histone demethylases is increasingly becoming available and has spurred an interest in the development of novel inhibitors for this enzyme class. Quite a number of inhibitors have been described as recently reviewed [42,45,46].

Based on the mechanism of action of these enzymes [6,17], three main strategies can, in principle, be envisaged to measure their *in vitro* catalytic activity and the potency of potential inhibitory compounds:

- quantification of the consumption or formation of one cofactor or coupled reaction product;
- antibody-based measurements of the degree of methylation of a substrate, requiring highly specific antibodies that are able to distinguish; for example, dimethyl-lysine (Kme2) from trimethyl-lysine (Kme3) in peptide or histone substrates;
- mass-spectrometry-based techniques that measure and quantify the mass change of a peptidic substrate upon demethylation.

18.3.1.1 Quantification of the reaction product formaldehyde

As for the first option, the quantification of formaldehyde as one of the coupled products of the demethylation reaction constitutes one of the most commonly used assay systems for inhibitor screening and characterization for JumonjiC demethylases. Originally introduced by Couture et al. in 2007 [17], it uses a secondary enzyme, formaldehyde dehydrogenase (FDH), as the detection system. The basic principle is outlined in Figure 18.1A.

The JumonjiC demethylase in question is incubated with the substrate, usually a shortened synthetic peptide of the original histone sequence bearing the desired modification, along with the cofactors NAD^+, Fe^{2+}, and α-ketoglutarate as well as ascorbic acid and formaldehyde dehydrogenase, continuously releasing stoichiometric amounts of formaldehyde. This, in turn, is further oxidized by FDH under consumption of its cofactor NAD^+ and formation of NADH. The fluorescence intensity of NADH can be measured at $\lambda ex = 330\,nm$ and $\lambda em = 460\,nm$ and correlated to the amount of formaldehyde formed, allowing for the direct quantification of demethylase activity. One advantage is that this assay allows for a continuous monitoring of reaction progress, facilitating kinetic studies.

This assay system has been instrumental in the discovery and development of some of the first JumonjiC demethylase inhibitors [47]. In their seminal paper, Sakurai et al. redesigned, simplified, miniaturized, and optimized the FDH-coupled assay, making it amenable to high-throughput screening for inhibitors even in 384- and 1536-well format [48], which has allowed for many more lead discovery studies.

Using the FDH-coupled assay, the Schofield and Oppermann groups were able to present the first collection of JumonjiC demethylase inhibitors in 2008 based on similar compounds that were

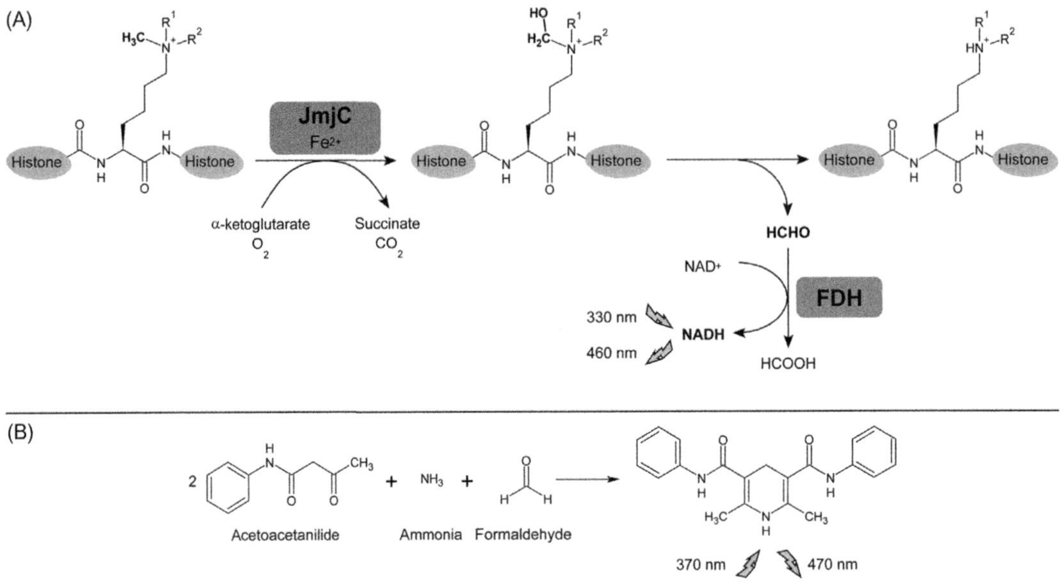

(A)

(B)

Acetoacetanilide   Ammonia Formaldehyde

FIGURE 18.1

(A) Basic principle of the formaldehyde dehydrogenase (FDH)-coupled enzymatic assay for Jumonji-type histone demethylases. See text for details. (B) Reaction for the *in situ* formation of a fluorescent dihydropyridine dye from formaldehyde and acetoacetanilide as in the commercial demethylase activity kit.

known to inhibit other iron(II) and α-ketoglutarate (α-KG)-dependent enzymes [47]. This includes *N*-oxalyl glycine (NOG, **1**, see Figure 18.2), an amide analog of the enzyme co-substrate α-ketoglutarate, which is a competitive inhibitor of JMJD2E with an IC_{50} value of 78 μM.

The original series of inhibitors also included pyridine 2,4-dicarboxylic acid (2,4-PDCA, **2**), which is now the most commonly used *in vitro* reference inhibitor. It exhibited an IC_{50} value of only 1.4 μM on JMJD2E, making it also one of the most potent inhibitors known to date. A co-crystal structure of JMJD2A in complex with **2** revealed insights into its mode of inhibition as a bidentate chelator of the central metal ion, while the other carboxylate is involved in an ionic interaction with the highly conserved lysine residue Lys206, which otherwise stabilizes the co-substrate α-ketoglutarate [47].

Furthermore, compounds bearing a hydroxamic acid moiety such as trichostatin A (TSA) and a simpler version suberoylanilide hydroxamic acid (SAHA, **3**), which are actually known to be quite potent inhibitors of zinc-dependent histone deacetylases, were also reported to inhibit JMJD2E in the FDH-coupled assay with an IC_{50} value of 14 μM for **3** [47]. For TSA, it was found that it also inhibits the coupled enzyme FDH, revealing the limitations of this assay system as it can be prone to false positives. However, its inhibitory effect could be verified in a MALDI-TOF (matrix assisted laser desorption ionization − time of flight) mass-spectrometry (MS) assay, as well as for other compounds in that study [47].

Nonetheless, the potential of active site iron (II) chelation by hydroxamic acids initiated a study by Hamada et al., who developed inhibitors with internal hydroxamic acids and longer aminoalkyl

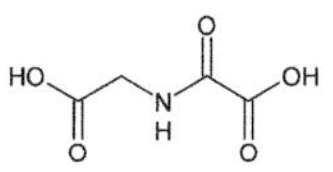

N-oxalyl glycine (NOG)
1

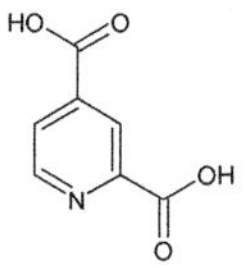

2,4-PDCA
2

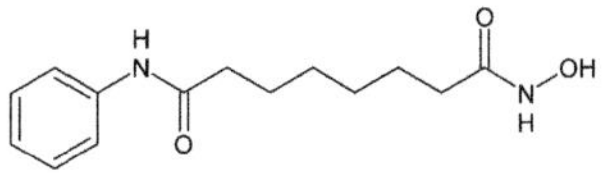

Suberoylanilide hydroxamic acid (SAHA)
3

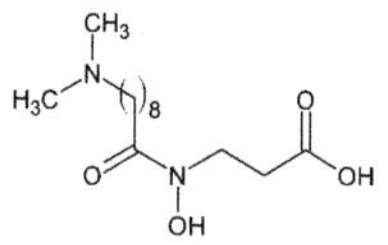

4

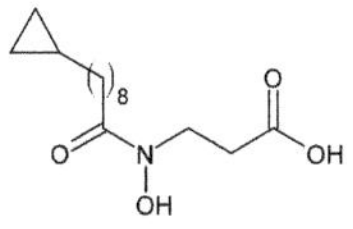

5

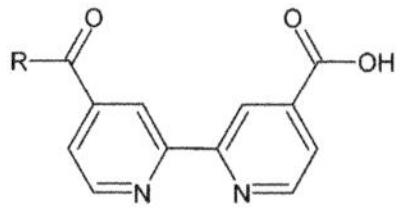

2,2′-bipyridine derivatives
6a R = OCH_3
6b R = $NHCH_2CH_2NH_2$

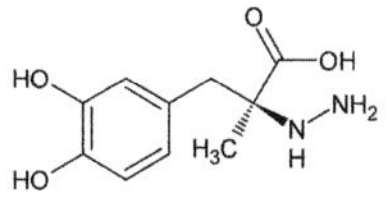

(S)-(-)-carbidopa
7

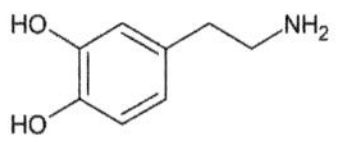

Dopamine
8

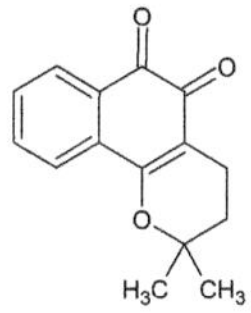

β-lapachone
9

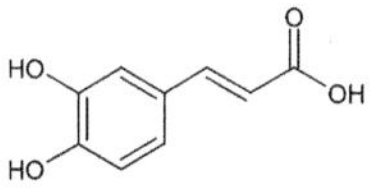

Caffeic acid
10

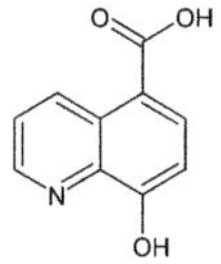

8-HQ-5-COOH
11

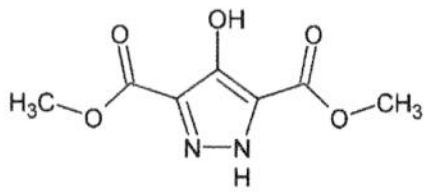

4-hydroxypyrazole
12

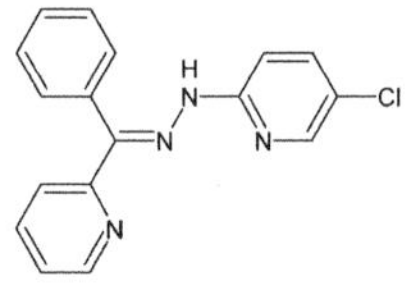

JIB-04
13

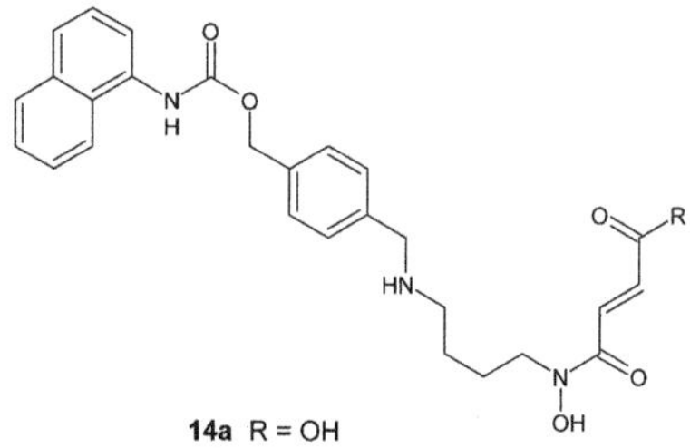

14a R = OH
14b R = OCH_3 methylstat

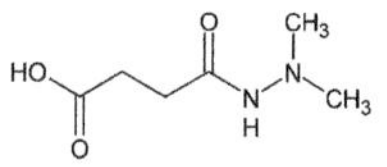

Daminozide
15

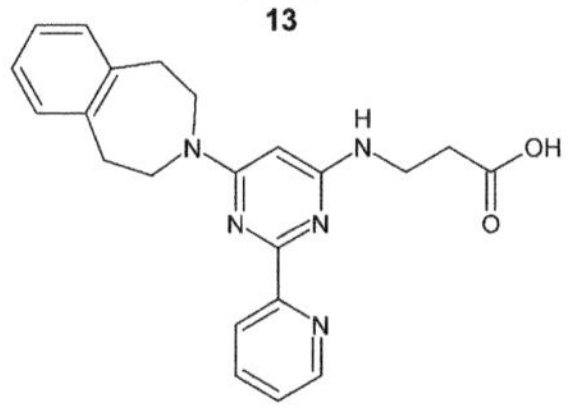

GSK-J1
16

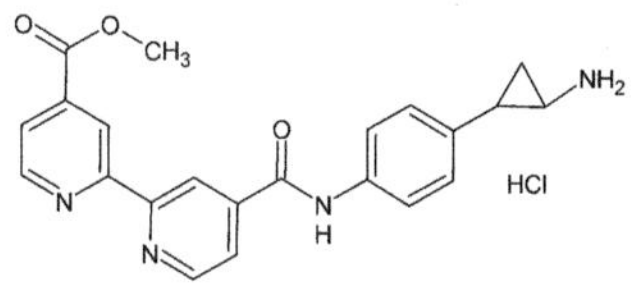

Hybrid pan-demethylase inhibitor **17**

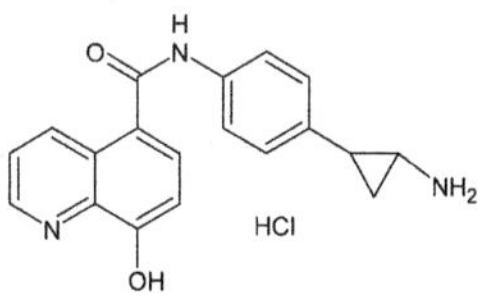

Hybrid pan-demethylase inhibitor **18**

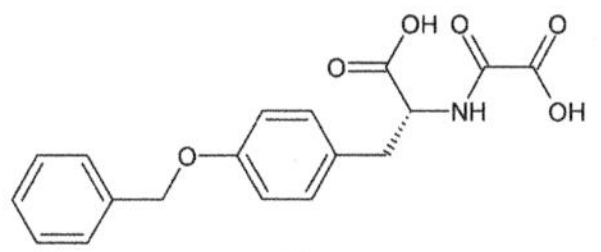

26

FIGURE 18.2

Chemical structures of reported *in vitro* JumonjiC histone demethylase inhibitors that were discovered using the assay techniques as outlined in the text.

chains. Their structure-guided design approach yielded a series of novel compounds that were tested in the FDH-coupled assay and the optimized compound **4** bearing a dimethylamino group connected to the hydroxamic acid via an octyl chain linker. **4** showed an IC_{50} value of $1.0\,\mu M$ on JMJD2C and $3.0\,\mu M$ on JMJD2A, but a greater than 100-fold selectivity on other iron-binding enzymes [49]. These compounds also bear a propanoate side chain on the hydroxamic acid, which can be thought of as a group mimicking the α-ketoglutarate co-substrate and interacting with the same conserved lysine residue as discussed previously.

This concept was further elaborated in a recent study [50], leading to inhibitors that are selective for the KDM2 and KDM7 demethylase subfamilies. The optimized cyclopropyl compound **5** exhibited IC_{50} values of $6.8\,\mu M$ and $0.20\,\mu M$ on KDM2A and KDM7A, respectively. While these values were measured in MS assays, counterscreens against other demethylase subtypes were performed in the FDH-coupled assay, revealing much higher IC_{50} values for those enzymes [50].

Another inhibitor published in the original series using the FDH-coupled assay [47] is the 2,2'-bipyridine derivative **6a** that exhibited an IC_{50} value of $6.6\,\mu M$ on JMJD2E. This structural template was investigated further with more analogs, albeit using different assay systems.

Upon optimizing the FDH assay and making it accessible to high-throughput screening, Sakurai et al. also discovered a number of catechols and other natural products to quite potently inhibit demethylation, including (S)-$(-)$-carbidopa **7**, dopamine **8**, and β-lapachone **9**, with IC_{50} values on JMJD2E of $8.0\,\mu M$, $7.1\,\mu M$, and $3.6\,\mu M$, respectively [48].

A larger collection of catechol-based *in vitro* inhibitors was recently discovered and characterized, including, among others, caffeic acid **10** with an IC_{50} value of $13.7\,\mu M$ on JMJD2C. This study used an enzyme-linked based immunoassay (ELISA) for screening of a natural compound library and verified inhibition for their hit compounds in the FDH-coupled assay [51].

In another high-throughput screening approach using the optimized FDH-coupled assay, the Simeonov group identified 8-hydroxyquinolines (8HQs) as a new compound class with inhibitory action against JumonjiC demethylases. Hits were validated in a secondary MALDI-TOF assay. Structure-activity relationships were studied and led to compound **11** bearing an additional carboxylic acid as the most potent inhibitor with an IC_{50} value of $0.2\,\mu M$ on JMJD2E, which even showed *in vivo* inhibitory potential without the need for prior derivatization or synthesis of a prodrug [52].

Again using the FDH-coupled assay, 4-hydroxypyrazoles such as compound **12** were identified as JMJD2C inhibitors, albeit with a relatively high IC_{50} value of $147\,\mu M$. In their study, Leurs et al. also investigated the structure-activity relationship, underlining the importance of the 4-hydroxy and 3-carbonyl groups, presumably because of their iron(II) chelating function [53].

Highlighting the broad usability of the FDH-coupled assay, substrate-based peptidic inhibitors of JumonjiC demethylases could also be developed. Lohse et al. investigated kinetics and binding affinities of different peptides and coupled them to a small molecule inhibitor such as bromouracil at the H3K9-lysine ε-amino group yielding the 5-mer peptide ARK(BrU)ST, an inhibitor of JMJD2C ($K_i = 27\,\mu M$) with a four fold selectivity over the related demethylase JMJD2A [54].

A different method of quantification of formaldehyde release by JumonjiC demethylases consists of the *in situ* formation of a fluorescent dihydropyridine dye from formaldehyde and acetoacetanilide in the presence of ammonia in a Hantzsch reaction, as illustrated in Figure 18.1B. It is based on the development of novel reagents for Hantzsch reactions to fluorimetrically detect formaldehyde [55] and marketed in an assay kit, for example by Cayman Chemical. However, it has not been used in the discovery of any published inhibitors to date.

18.3.1.2 Antibody-based assay systems

A conceptually completely different approach to assaying the demethylase activity of JumonjiC enzymes is the quantification of the degree of methylation on lysines in peptides by using highly specific antibodies directed against these modifications. The challenge here, thus, lies in the production of antibodies that are able to recognize the minute changes in the chemical structure of the peptides that they are binding. While with acetylation the lysine side chain alters its charge, methylation only leads to a gradual increase in size and lipophilicity. The assays used for inhibitor screens generally differ only in the way that antibody binding is quantified and eventually measured.

Simple approaches are basic Western blot or dot blot techniques, where binding of the antibody is usually detected by secondary antibodies that are equipped with radioactive or luminescent markers. In a similar way, standard ELISA techniques can be used for *in vitro* screening of demethylation activity and inhibitory potency. This was used in the screening of a natural compound library that led to the discovery of catechols such as caffeic acid **10** (see above), employing a horseradish peroxidase-coupled secondary antibody [51].

In the same way, Wang et al. used the ELISA technology as well as the FDH-coupled assay to characterize a novel lysine demethylase inhibitor, termed JIB-04 **13**, which was discovered from a phenotypic cell-based screen, with the subtype JARID1A being the most sensitive enzyme ($IC_{50} = 230$ nM). This compound was described as noncompetitive with the enzyme co-substrate α-ketoglutarate and also potently showed antiproliferative effects in human cells and a mouse tumor model. Effects on cellular methylation were not shown [56].

A number of other antibody-based assays have been developed that differ in their detection system. However, they usually contain proprietary components that are only marketed by the original developing company. One such system is the LANCE*Ultra* technology (lanthanide chelate excite) marketed by PerkinElmer. They have published its use for LSD1 in a recent paper [57] and applications of the technology for Jumonji-type demethylases are in a number of technical and application notes, available from the manufacturer's web site. In this system, the antibody that specifically recognizes the demethylated peptide after the reaction by JumonjiC enzymes is covalently modified with a europium-chelate complex. The detection system requires the peptide substrate to be biotinylated, as a second proprietary fluorescent dye called U*Light* is coupled to streptavidin and, thus, binds to the peptide. Only if both components, the antibody and streptavidin-bound U*light*, bind to the peptide and are thus in proximity, a fluorescence resonance energy transfer (FRET effect) will occur. Advantages of using lanthanide fluorescence are its long emission duration, which can be used to limit interference as organic molecules such as the test compounds usually have a much shorter fluorescence lifetime. Additionally, they possess a large Stokes shift, which also limits interference by autofluorescence. Due to the high sensitivity of the system, these assays typically require drastically less enzyme than does, for example, the FDH-coupled assay.

Similarly, time-resolved fluorescence (TRF) by europium ions can be employed in a technology termed DELFIA (dissociation-enhanced lanthanide fluorescence immunoassay), that is also marketed by PerkinElmer. In this heterogeneous assay, streptavidin-coated plates are used and a biotinylated peptide is immobilized on a microtiter plate. After incubation with the enzyme and the resulting demethylation, a primary unlabeled antibody specific for the demethylated peptide is added. A secondary antibody (modified covalently with a europium-chelate complex) against the constant fragment of the primary antibody is added. After washing off the unbound antibodies, a detection solution is added that releases europium from the complex only in wells, where the antibody has recognized the demethylation. A high fluorescence intensity caused by Eu^{3+} in the detection solution can, thus, be measured only if the demethylation has occurred — that is, the enzyme was not inhibited.

The DELFIA technology was used in the characterization of a novel rationally designed JumonjiC demethylase inhibitor **14a**, which inhibits JMJD2C with an *in vitro* IC_{50} value of 4.3 μM. However, this compound is relatively unselective with regard to other subtypes and unrelated iron(II)-dependent hydroxylases. Its methyl ester prodrug **14b** has become known as methylstat and inhibits demethylase activity *in vivo* as well as proliferation of an esophageal carcinoma cell line [58].

Third, specific antibody binding after histone peptide turnover can also be measured using the AlphaScreen technology (amplified luminescent proximity homogeneous assay) as recently published [59]. Here, biotin-labeled peptide substrates are recognized by streptavidin-coated beads ("donor beads"), carrying a photosensitizer, which, upon laser irradiation, emits singlet oxygen $^1O_2$. A methylation-state specific antibody recognizes and binds the demethylated peptide and is, in turn, bound by "acceptor beads" coated with protein A as well as a thioxene reagent. Only if the two beads are in proximity, can the $^1O_2$ molecule reach the acceptor bead and after reaction with the thioxene a luminescence signal is released. The assay was originally developed for JMJD2E, but can certainly also be applied to other subtypes given the right peptide substrates and specific antibodies. It was validated using reference inhibitors such as NOG **1**, 2,4-PDCA **2**, and the bipyridine derivative **6a** [59].

The AlphaScreen technology was also used in the recent discovery that the plant growth regulator daminozide **15** is a selective inhibitor of the KDM2 and KDM7 subfamilies of histone demethylases. It inhibited KDM2A with an IC_{50} value of 1.5 μM in an α-ketoglutarate-competitive manner, as demonstrated by kinetic analyses and a co-crystal structure [60].

Using a combination of AlphaScreen assays, protein crystallography, and MS techniques, GlaxoSmithKline recently reported the first small molecule inhibitor selective for the KDM6 subfamily of H3K27 demethylases. In their AlphaScreen assay, the lead compound GSK-J1 **16** exhibited an IC_{50} value of 60 nM against JMJD3. Selectivity against a large panel of other Jumonji-type demethylases and unrelated enzymes was demonstrated. A cell-permeable ethyl ester prodrug reduced pro-inflammatory cytokine production in human macrophages, a process dependent on both JMJD3 and UTX [61].

In a very recent paper, Rotili et al. reasoned that since there are a number of different mechanisms at play in cancer cells and since both LSD1 and Jumonji-type histone demethylases are involved in disease-initiating androgen receptor-mediated mechanisms, it might be advantageous to create hybrid inhibitors for both LSD1 as well as JmjC demethylases. The known LSD1 inhibitor tranylcypromine was, thus, coupled to JmjC inhibitors such as bipyridines of the type of **6a** and 8-hydroxyquinoline **11**, yielding hybrid inhibitors **17** and **18**, respectively, which showed pan-demethylase inhibitory potential as determined in AlphaScreen assays. The strongest inhibition was observed for JMJD2C with IC_{50} values of 2.7 μM for **17** and 1.2 μM for **18**. Notably, the inhibitory potential was not much changed as compared to the parent compounds. **17** and **18** also showed dose-dependent *in vivo* effects on histone methylation in LNCaP prostate cancer cells [62].

18.3.1.3 Mass-spectrometry-based assay systems

Yet another technologically different approach to assaying JumonjiC demethylase activity is the use of MS techniques. Here, synthetic peptides carrying the lysine modification of interest are incubated with the JmjC demethylase and potential inhibitors and then subjected to MS analysis. The loss of one or more methyl groups can easily be observed by a mass shift of multiples of 14 Da. When quantified, these mass spectra can be used to calculate enzyme activity or inhibitory potency of test compounds.

Matrix-assisted laser desorption ionization (MALDI) MS has been used in combination with other techniques, for example in the discovery of the very first inhibitors [47], the 8-hydroxyquinoline series [52], and of the substrate-based peptidic inhibitors as discussed above [54].

A slightly different approach lies in the use of nondenaturing MS techniques such as electrospray ionization (ESI-MS). Some types of noncovalent interactions between the enzyme and a substrate or inhibitors can survive the transition from solution to gas phase and, thus, novel peaks appear that correspond to the mass of enzyme plus bound small molecules. Notably, these assays reveal binding of test compounds to the enzyme rather than effects on catalytic activity and can also be misleading in case unspecific binding occurs elsewhere than in the catalytic site. Use of this method was first demonstrated in 2010 [63] and led to the discovery of some N-oxalyl D-tyrosine derivatives as inhibitors of JMJD2E, an elaboration of the original lead structure NOG **1**. A series of N-oxalyl amino acids was investigated using this binding assay and results correlated with enzymatic inhibition studies using the FDH-coupled assay. Notably, only compounds deriving from D-amino acids showed any inhibitory potential [63].

Similarly, thiol-containing test compounds could be studied with this technique, when they were allowed to react with thiols in the enzyme active site (from cysteine residues) to form disulfide bonds. The mass change due to the covalently bound inhibitor could be revealed via nondenaturing MS, even in a dynamic combinatorial library where thiol compounds of different mass were incubated simultaneously with JMJD2E [63].

This nondenaturing ESI-MS technology was also used in the advancement of the 2,2'-bipyridyl compound **6a** that was published in the original inhibitor series [47]. Upon inspection of co-crystal structures and guided by structure-activity relationships, the optimized compound **6b** was developed with an ethylenediamine residue. The positively charged amine residue was reasoned to be involved in an ionic interaction with an active site aspartate. Compounds were checked for enzyme inhibition in the FDH-coupled assay revealing an IC_{50} value of 180 nM for **6b** on JMJD2E, while binding was verified in the ESI-MS assay and from an X-ray structure of the enzyme-inhibitor co-crystal [64].

In a similar approach, a novel peptidic inhibitor was designed by investigating available crystal structures, which led to chemically connecting a mimic of the peptide substrate to an analog of the co-substrate α-ketoglutarate via a disulfide bridge. This enabled the compound to simultaneously bind to both binding sites in the enzyme with high affinity. The binding mode was confirmed through crystal structure analyses of the enzyme in complex with the two-component inhibitor and using the nondenaturing ESI-MS technology. A chemically stabilized analog potently inhibited JMJD2E ($IC_{50} = 90$ nM) and JMJD2A ($IC_{50} = 270$ nM) [65].

In an effort to overcome the inherently low throughput and long analysis times typically associated with MS techniques, a novel label-free RapidFire assay has been developed, using a 15 amino acid peptide as substrate for JMJD2 demethylases. Formation of the dimethylated H3K9me2 peptide from H3K9me3 can directly be monitored and quantified. With a sampling time of only 7 s per well and high assay robustness, analysis times were drastically reduced and MS assays became amenable to high-throughput screens. The assay was correlated with results from the FDH-coupled assay and validated with NOG **1** and 2,4-PDCA **2**, while a screen of a >100,000 compound library revealed several hits of the 8-hydroxyquinoline **11** structure class [66].

Very recently, throughput of the RapidFire-MS assay has been increased even further by using a multiplexing strategy. This relies on pooling several samples in one analysis and being able to later deconvolute their origin based on the fact that they stem from wells using peptide substrates of different sequence and mass yet equal binding affinity. This mass-tagging approach has made the assay suitable for screens of large compound libraries on the order of millions of compounds [67].

18.3.1.4 Other assay technologies

Besides these well-established assay technologies, other methods that are based on different physical principles were also described.

Using the reported *in vitro* inhibitor methylstat acid **14a** [58] as a scaffold, a fluorescently labeled version of this compound was synthesized and used as a tracer in a novel fluorescence-polarization assay. Its use for competition studies for some JumonjiC demethylase subtypes such as JHDM1A was demonstrated and it could be validated using reference inhibitors such as NOG **1** and 2,4-PDCA **2**. This assay, like nondenaturing MS techniques discussed previously, investigates binding of test compounds to the enzyme and does not rely on inhibition of catalytic activity [68].

Moreover, Leung et al. recently reported on the use of nuclear magnetic resonance (NMR) techniques for a binding assay of α-ketoglutarate-dependent oxygenases. Using $[1,2,3,4-^{13}C_4]$-labeled α-ketoglutarate as a reporter ligand, a competition binding assay was established. The NMR signals of ^{13}C-labeled α-ketoglutarate can only be measured for the free molecule in solution, while its resonances are broadened and nearly attenuated when bound to the protein. When an organic small molecule competitively binds to the active site and displaces α-ketoglutarate, the signal intensities increase again, as was demonstrated using standard reference inhibitors such as **1**, **2**, and **11** and the JumonjiC subtype JHDM1A (FBXL11). A clear advantage of this method is that only specific active site binders give a signal in this assay and fewer false positives are to be expected. However, the cost of the ^{13}C-labeled probe and measurement times likely prohibit its use in larger screening campaigns [69].

18.3.2 LYSINE-SPECIFIC DEMETHYLASE 1

Due to the fact that demethylation can actually mediate transcriptional repression and activation, it is complicated to distinguish the pathological dimension of inhibiting LSD1. However, LSD1 has proven to be an interesting target for several cancers, for example, prostate cancer, breast cancer, and acute myeloid leukaemia [30,70,71] and selective inhibitors for LSD1 of great potency in a lower nM range already exist and some are very promising in preclinical investigations [45,72].

The close homology of LSD1 to monoamine oxidases (MAO A, B) explains the inhibiting potency of several already approved drugs for the treatment of depression; for example, Tranylcypromine (PCPA, **19**, see Figure 18.3), Pargyline (**20**), and Clorgyline (**21**). PCPA, a nonselective irreversible inhibitor, is commonly used as reference inhibitor in *in vitro* assay systems. The difference of the published IC_{50} values (2−32 μM) is related to the individual assay configurations [73,74]. Since this discovery has taken place in 2006, the scaffold of Tranylcypromine acts still as a starting point toward new, more potent derivatives (**22,23**) [75,76].

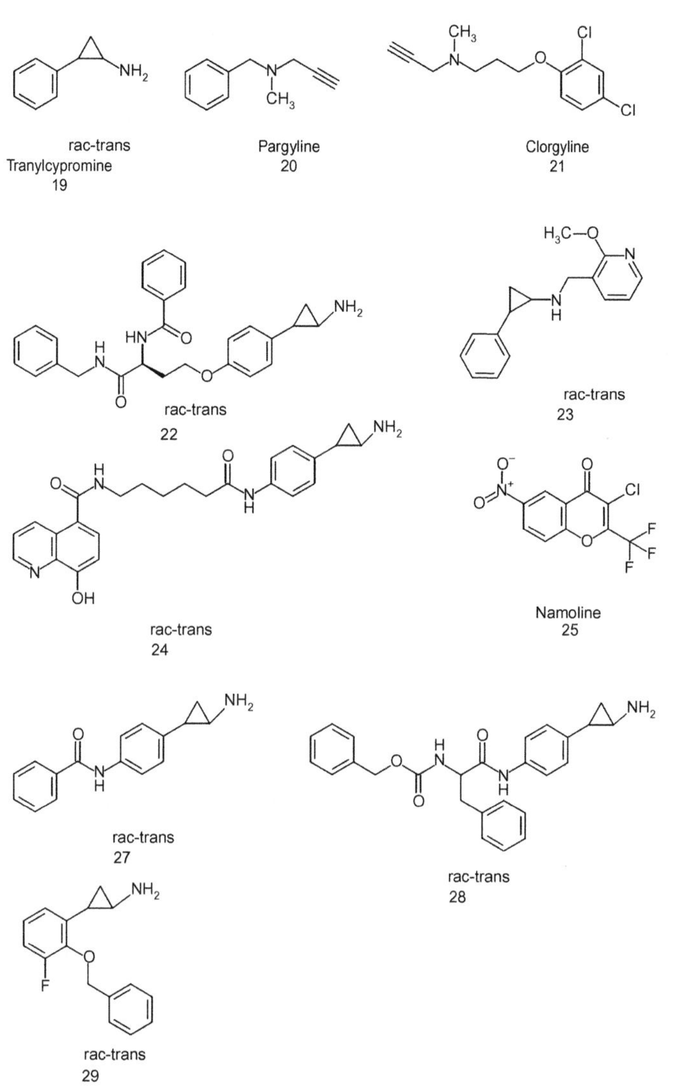

FIGURE 18.3

Chemical structures of reported inhibitors for the histone demethylase LSD1. Rac trans refers to the configuration at the cyclopropyl ring.

For the description of assay methodology and structural aspects of inhibition, we mainly want to focus on PCPA and its analogs. However, a variety of other scaffolds have already been presented as inhibitors of LSD1. Besides peptides [77–79], cyclic peptidomimetics [80] and polyamine analogs [81], also a number of small-molecule LSD1 inhibitors have been discovered. Among these are (bis)ureas and (bis)biguanides [82], the first small molecule lysine demethylase inhibitors that were described, and their isosteres [83–85] as well as amidoxines [86], phenyloxazoles [87] and dithiocarbamates [88,89]. The gamma-pyrone namoline (**25**) has been identified as a reversible inhibitor of LSD1 with a high selectivity over monoamine oxidases [90] and also propargylamines with a lysine or an aryloxypropanol scaffold have been found to inhibit LSD1 in the micromolar range [91].

Importantly, the general strategies to determine the *in vitro* potency of potential inhibitors for JumonjiC domain-containing demethylases can be applied to LSD1 as well:

- quantification of the consumption or formation of one cofactor or coupled reaction product;
- antibody-based measurements of the degree of methylation of a peptide substrate;
- MS-based techniques that measure and quantify the mass change of a peptidic substrate upon demethylation.

18.3.2.1 Quantification of the reaction product formaldehyde

For LSD1 (as wells as for JumonjiC domain-containing demethylases), the FDH-coupled assay can be used for activity testing, as both types of enzymes lead to a release of formaldehyde throughout the demethylation reaction. As described in more detail for the JumonjiC domain-containing demethylases (see above), finally the reduction of NAD^+ to NADH can be measured quantitatively by its absorbance at a wavelength of 330 nm or more sensitively by its fluorescence intensity at $\lambda ex = 330$ nm and $\lambda em = 460$ nm. In this range, the feasibility is often prevented by compound interferences, for example, caused by autofluorescence [92].

18.3.2.2 Quantification of the reaction coproduct hydrogen peroxide

To circumvent the absorption/fluorescence interferences one encounters using the FDH-coupled assay, commonly a horseradish peroxidase (HRP)-coupled assay is used [93].

The reaction of the enzyme and a dimethylated substrate, usually a commercially available histone derived synthetic peptide (aa 1-21), takes place at room temperature. Although the substrate depletion by LSD1 is very fast, kinetic measurements are still possible.

The detecting reagent contains horseradish peroxidase Type II and 10-acetyl-3,7-dihydroxyphenoxazine (e.g. Amplex Red®) [94]. By the addition of Amplex Red detecting reagent to the reaction mixture, the conversion to the fluorescence product Resorufin takes place [93]. Further mechanism details are lined out in Figure 18.4A. This assay principle is also convenient for high-throughput screening (HTS) [95]. However, depending on purification conditions, the enzyme solution can contain ingredients, which are converted by HRP [96]. This leads to a high background signal which correlates with the enzyme concentration. High concentrations of Resorufin can act as a substrate for HRP, as well, which can lead to a continuous increase of the fluorescence signal. Also the presence of NADH and/or GSH (glutathione) can generate noticeable amounts of hydrogen peroxide, thus complicating the clear determination of the demethylation reaction.

Commercial assay kits advise to use wavelengths of $\lambda ex = 530$ nm and $\lambda em = 590$ nm, but it is possible to enlarge the Stokes shift up to $\lambda em = 615$ nm when using the corresponding filter system [92].

IC_{50} values of the already mentioned "pan"-demethylase inhibitors (**17,18,24**) have been determined using this assay system. In cases of interferences with the detection reagent (e.g. **11**), it is necessary to confirm the results with an alternative assay system [62].

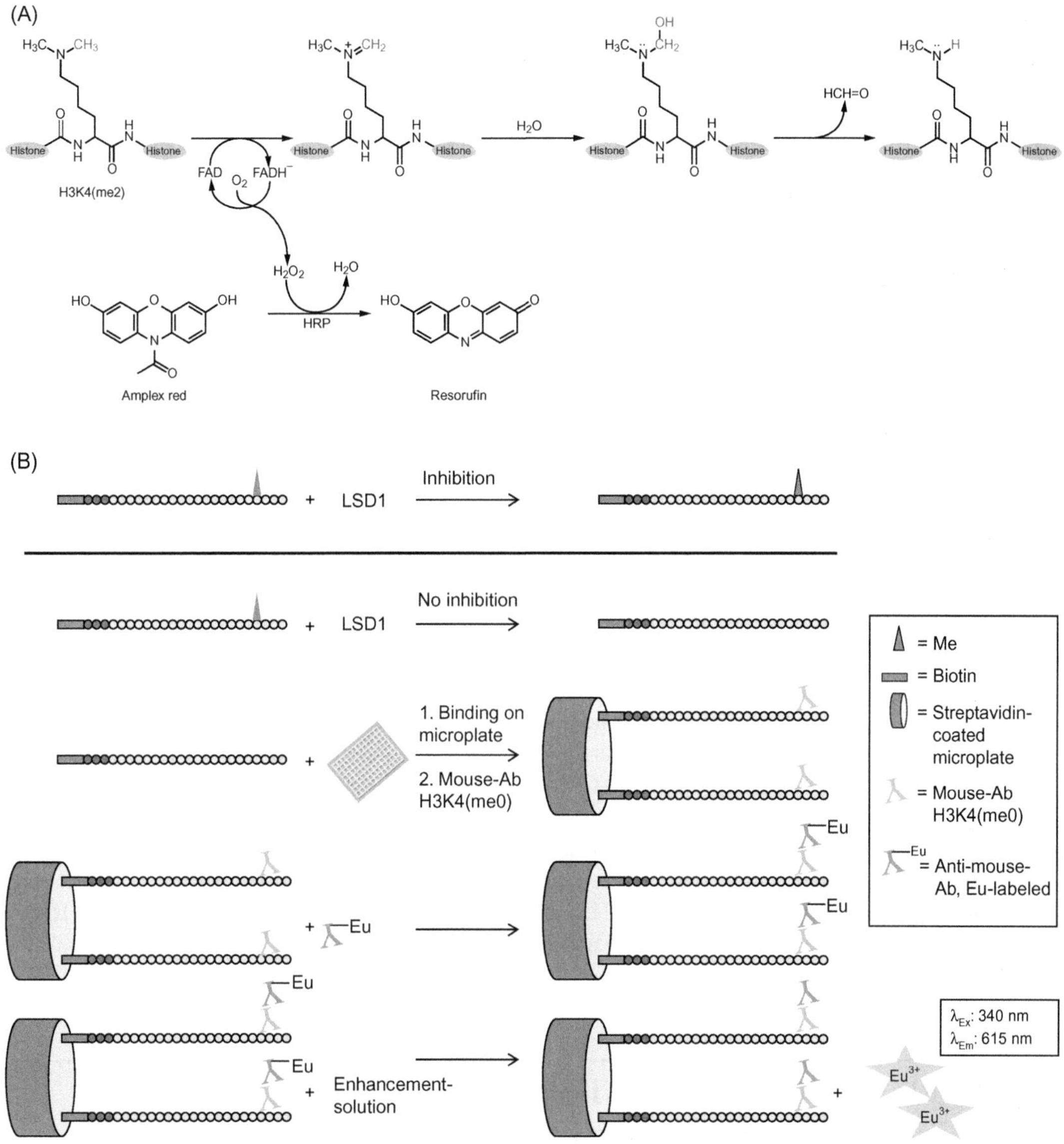

FIGURE 18.4

(A) Mechanism of FAD-dependent oxidative demethylation. The necessity of an imine intermediate prohibits Kme3 as a substrate of LSD1. (B) Principle of the heterogeneous DELFIA assay.

18.3.2.3 Quantification of the demethylated reaction product

If the determination of IC_{50} values is difficult in the peroxidase assay, for example, by quenching effects, it is possible to quantify substrate conversion. There are commercial assay systems available from PerkinElmer, such as AlphaLISA and DELFIA, that can be used for the readout. The well-established homogeneous AlphaLISA (amplified luminescent proximity homogeneous assay) is easy to use and already validated. The highly sensitive luminescence proximity technique needs less volume, is automation-ready and therefore is suitable for HTS. However, inhibitors may also lead to quenching in this proximity-based system.

The heterogeneous DELFIA (dissociation-enhanced lanthanide fluorescent immunoassay) is less susceptible to intrinsic fluorescence and quenching nor to interactions between compounds and co-products like H_2O_2 and formaldehyde. Its washing-steps and time-resolved fluorescence (TRF) read-out avoid interferences of the assay reagents but hinder the employment of HTS. All reagents are commercially available for both systems.

Usually for the DELFIA assay, a biotinylated peptide is used as LSD1 substrate. However, we could show that core histones that more resemble the native substrate can be used as well [97]. Probably due to steric hindrances inside the wells of the coated plates, the enzyme reaction with biotinylated peptides has to take place separately before coating the wells for quantitation. The reaction can be stopped by addition of a washing buffer (pH >9.5), so LSD1 becomes inactive. After that, the biotinylated substrate will be coated onto the surface of streptavidin-coated plates. The critical point for these LSD1 assays is the availability of sensitive and specific primary antibodies for selective recgonition of defined levels of methylation. Then the second europium-labeled antibody can detect the first antibody easily and quantitatively. The acidic enhancement solution releases Eu^{3+} from the chelate bound to the secondary antibody. Further details are outlined in Figure 18.4B. The large Stokes shift of nearly 300 nm ($\lambda_{ex} = 340$ nm and $\lambda_{em} = 615$ nm) and the high sensitivity result in a very good signal-to-noise ratio. A published assay using H3K4me1 as a substrate has a Z-factor of 0.7 and is thus considered to be excellent. IC_{50} values for the known reference inhibitor **25** have been determined using the HRP-coupled assay as well as the DELFIA assay. The differing values of $56.03\,\mu M \pm 1.62\,\mu M$ (HRP) and $3.08\,\mu M \pm 0.35\,\mu M$ (DELFIA) depend on the differences in assay design. The inhibitor **19** was also measured with differing IC_{50}-values of $16.2\,\mu M \pm 2.30\,\mu M$ (HRP) or $2.15\,\mu M \pm 0.70\,\mu M$ (DELFIA), respectively.

18.4 STRUCTURAL ASPECTS OF INHIBITOR BINDING

18.4.1 JUMONJIC DOMAIN-CONTAINING DEMETHYLASES

As already mentioned, the Jumonji-type demethylases belong to the cupin superfamily of enzymes [14]. They show a conserved β-barrel fold which harbors the binding site for Fe^{2+} and the cofactor α-ketoglutarate (α-KG) [18,45]. Binding of Fe^{2+} is achieved through the motif HXXD/E······H; where the presence of the three conserved residues (two histidines and an aspartate or glutamate) seems to be mandatory for catalytic activity [44]. Besides the coordination with the two histidine residues and the acidic amino acid (Glu or Asp) from the conserved motif, the octahedral coordination of Fe^{2+} is completed by the cofactor α-KG and a water molecule. The binding pocket of α-KG itself shows less conservation among the subfamilies and can hence be exploited for the development of selective inhibitors [42].

Numerous crystal structures of JmjC domain-containing proteins in complex with inhibitors have appeared in the last few years, which gives an important insight into the inhibitory mechanism of these compounds (for review see [42]). All inhibitors reported so far are metal chelators which bind competitively in the α-KG binding pocket. This is the main reason why many of these inhibitors show an inhibitory activity at different JmjC-protein subtypes as well as at other Fe(II)/α-KG-dependent enzymes, such as prolyl hydroxylase domain-containing proteins (PHDs) and factor-inhibiting hypoxia-inducible factor (FIH).

Crystal structure analysis showed that JMJDs possess a subpocket, adjacent to the α-KG-binding pocket, which is larger and more open than in the related enzymes FIH and PHD, in order to accommodate the peptide substrate. N-oxalyl-D-tyrosine derivatives were hence designed to address this subpocket, in an attempt to improve the selectivity toward the JMJD2 subfamily [63]. The most active compounds showed an inhibitory activity higher than NOG and selectivity for JMJD2A and JMJD2E over PHD2, but not over FIH. The crystal structure of derivative **26** in complex with JMJD2A (Figure 18.5A) showed that the compound is able to occupy the α-KG binding pocket as well as part of the H3K9me3 pocket. The oxalyl-carbonyl oxygen and the adjacent carboxylate group act as metal chelators; the other carboxylate group interacts with the side chain N-atom of Lys206 and the OH-group of Tyr132, while the benzyl group projects into the substrate binding pocket, specifically where Thr11 and Gly12 of H3K9me3 bind [63].

The classical inhibitor 2,4-pyridine dicarboxylic acid also binds to the α-KG binding pocket in a manner similar to α-KG and NOG (**1**) [47]. The pyridyl nitrogen and the 2-carboxylate oxygen are involved in the chelation of the metal ion, while the other carboxylate group undergoes H-bond interactions with Lys206 and Tyr132 in JMJD2a. The pyridine ring is additionally involved in π-π stacking with Phe185 (see Figure 18.5B).

A very similar binding mode could be observed for the analogous 8-hydroxy-quinoline-5-carboxylate (8HQ, **11**). The crystal structure of JMJD2A in complex with 8HQ (see Figure 18.5D) shows that the quinoline nitrogen and the phenolic OH coordinate the metal ion whereas the 5-carboxylate group forms interactions with Lys206 and Tyr132. π-π stacking is also observed with Phe185 [52]. In JMJD3 8HQ adopts a very similar binding mode (Figure 18.5C). However, due to the differences in the α-KG binding site, other residues are involved in the interaction with the 5-carboxylate group (paper not published). Lys206 and Tyr132 in JMJD2A are replaced by Cys1405 and Gly179 in JMJD3. The 5-carboxylate group interacts instead with Lys1378, Asn1397, Thr1384 and a water molecule. π-π stacking with the aromatic ring is missing, since Phe185 of JMJD2A is replaced by Thr1384. Both crystal structures reveal another interesting finding, namely that the binding of the inhibitor induces the translocation of the metal ion, thereby disrupting the interaction between the metal ion and one of the conserved histidine residues [52]. This is another factor that should be considered when designing new inhibitors. 8-Hydroxy-quinoline-5-carboxylate shows an improved selectivity profile with a preference for the JMJD2 family, JMJD3 and JMJD1A, albeit with still a considerable inhibitory activity against other 2-oxoglutarate oxygenases like PHD2 and FIH [98]. However, it can be foreseen that rational optimization of 8HQ could lead to more potent and/or selective drugs [52,98].

Bipyridyl derivatives are another class of pyridine-based analogs, which showed promising results as KDM inhibitors. The crystal structure of **6b**, the highly active bipyridyl derivative bearing

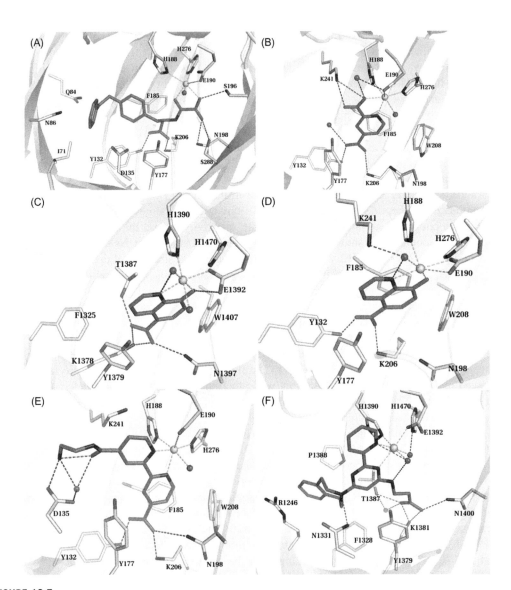

FIGURE 18.5

Crystal structures of JMJD2/3 inhibitor-complexes. (A) Compound **26** (cyan) in complex with JMJD2a (white) (PDB ID 2WWJ). (B) 2,4-Pyridine dicarboxylic acid (**2**, cyan) in complex with JMJD2a (white) (PDB ID 2VD7). (c) 8-Hydroxyquinoline-5-carboxylate (**11**, cyan) in complex with JMJD3 (C) and JMJD2A (D) (JMJD2: PDB ID 3NJY, JMJD3: PDB ID 2XXZ). (E) Compound **6b** (cyan) in complex with JMJD2a (white) (PDB ID 3PDQ). (F) GSK-J1 (**16**, cyan) in complex with JMJD3 (white) (PDB ID 4ASK). In all complexes the metal ion is shown in yellow, and the coordination interactions as yellow dashed lines. Only the side chains of the residues involved in the interaction with the ligand are shown as white sticks. H-bond interactions are shown as black dashed lines. (*This figure is reproduced in color in the color plate section.*)

an ethylenediamine moiety with JMJD2A (see Figure 18.5E) shows that both pyridyl nitrogens are involved in metal chelation, while the 3-carboxylate group undergoes the classical H-bond interactions with Lys206 and Tyr132. The additional pyridine ring with the ethylenediamine moiety projects into the peptide binding pocket, where K9me3 and Ser10 of H3K9me3 bind, and thereby disrupts the peptide binding to the enzyme. The second pyridine ring is involved in cation-π interaction with Lys241, whereas the terminal protonated nitrogen of the ethylenediamine moiety interacts with Asp135, which is otherwise involved in H-bond interaction with the backbone nitrogen of Thr11 of H3K9me3 [64].

Through analyzing the crystal structure of JMJD3 in complex with H3K27me3 and identifying the key interactions responsible for the substrate selectivity of the enzyme, the first selective inhibitor for the KDM6 family (JMJD3, UTX and UTY) was identified [61]. GSK−J1 (**16**) binds to both the α-KG and peptide binding sites as revealed by its crystal structures in complex with JMJD3 and UTY. The nitrogen atoms of the pyridine and pyrimidine rings chelate the metal ion, whereas the propanoate group mimics the second carboxylate function of α-KG and undergoes interactions with Lys1381, Thr1387 and Asn1480 of JMJD3 (Figure 18.5F). Again, here a ligand-induced translocation of the metal ion is observed. The key contributor to the selectivity is the tetrahydrobenzazepine moiety which is located in the same cleft as Pro30 of H3K27me3 and interacts with Pro1388 in JMJD3. The hydrophobic interaction between Pro30 of H3K27me3 and Pro1388 of JMJD3 was found to be crucial namely for the peptide substrate recognition by JMJD3 [61].

In recent years, several lead compounds have been disclosed as inhibitors of KDMs. Although most of these compounds lack selectivity and are expected to act at related and undesired targets, some breakthroughs have been made toward finding selective inhibitors. While targeting the metal ion is obviously necessary for the activity of the inhibitors, the simultaneous targeting of the substrate binding pocket, which shows less conservation among α-KG dependent oxygenases, seems to be profitable in achieving selectivity. It can hence be expected that further structure-based studies, based on the available crystal structures of JmjC-domains, would yield more potent and selective KDM inhibitors.

18.4.2 LYSINE-SPECIFIC DEMETHYLASE 1

As revealed by its crystal structure, LSD1 is composed of three domains: the tower domain (aa 419−520), the SWIRM domain (aa 166−260) and the amine oxidase domain (AOD, aa 520−852) [99,100]. The catalytic site is located in the AOD, where both the cofactor FAD and the peptide substrate bind [99−101].

The AOD of LSD1 shares a significant sequence similarity with other FAD-dependant amine oxidases. It shows a 26% homology with PAO (plant amine oxidase) and 20% homology with MAO A and MAO B enzymes [102]. While the FAD-binding pocket shares high similarity with the other amine oxidases, substantial differences are found in the subdomain which accommodates the natural substrate [44,101,102]. This subdomain is much more spacious and open in LSD1 and it harbors a deep negatively charged pocket in order to accommodate the peptide substrate [101−103]. Moreover, the classical aromatic cage found in other amine oxidases, which is responsible for the recognition of the positively charged methylated amine through cation-π interactions, is missing in LSD1. Only one aromatic amino acid (Tyr761) is preserved whereas the second one is replaced by Thr810 [101,102],

So far, only tranylcypromine (PCPA) derivatives have been cocrystallized with LSD1. PCPA acts as an irreversible inhibitor by a covalent addition to the cofactor FAD. Figure 18.6 shows its mechanism of inhibition as was suggested from the crystal structure and MS analysis [104].

Further crystal structures analysis [104–106] revealed that PCPA also seems to form an N(5)-adduct [106]. More recent studies [105] showed that the two enantiomers of PCPA form different adducts with FAD and adopt different positions in LSD1 (see Figure 18.7A–C).

PCPA derivatives with bulkier substituents were designed in order to increase the selectivity for LSD1 [105,107]. Crystal structures of LSD1 in complex with p-substituted PCPA derivatives

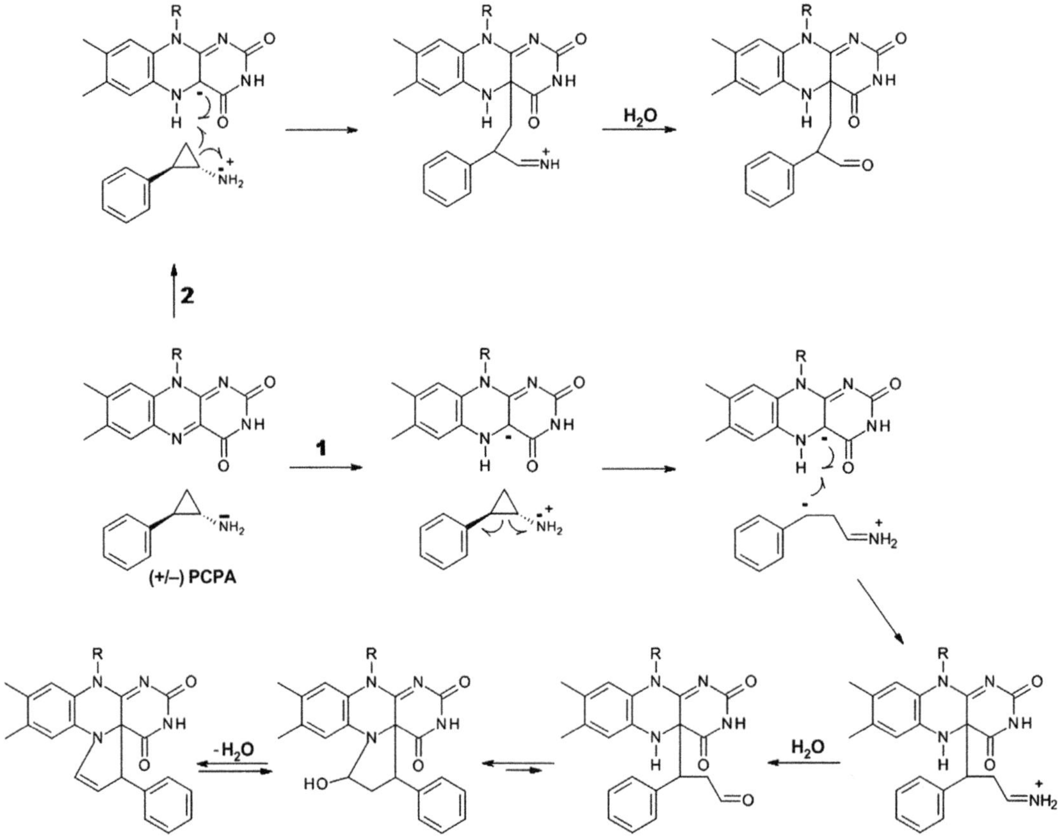

FIGURE 18.6

Mechanism of inhibition of PCPA. FAD extracts an electron from the nitrogen of PCPA. In pathway 1, the opening of the cyclopropyl ring leads to a benzylic carbon radical, which can react with C(4) of FAD to form a new bond. Subsequent hydrolysis of the imine intermediate and cyclization leads to the formation of a five-membered ring. In the second pathway, opening of the cyclopropyl ring and the formation of a new bond between FAD and PCPA occurs in a concerted action.

(Figure 18.7D), showed the warhead of the inhibitors forming an N(5) adduct with FAD, while the bulky substituent extends orthogonally to the flavin ring. In both cocrystallized derivatives (**27** and **28**) an aromatic group is embedded in a highly negatively charged region and is surrounded by three acidic amino acids. The second phenyl group of the branched derivative (**28**) projects outside

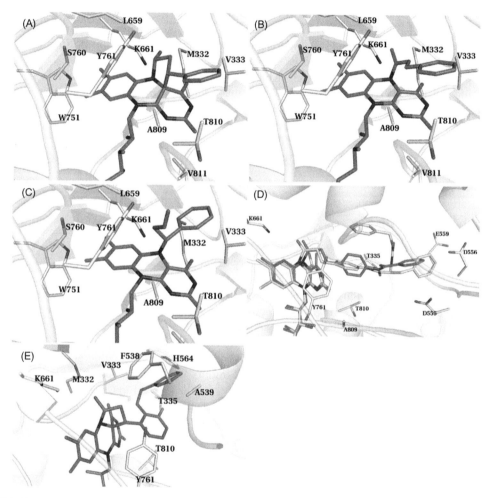

FIGURE 18.7

Crystal structures of LSD1-inhibitor complexes. (A) Five-membered ring adduct (cyan) of PCPA with FAD in LSD1 (white) (PDB ID 2Z5U (depicted) and 2UXX). (B) N(5)-adduct (cyan) of PCPA (PDB ID 2EJR (depicted), 2Z3Y, and 2XAJ). (C) The (+)-PCPA enantiomer forms a different adduct (cyan) with FAD and the phenyl group adopts a different orientation (PDB ID 2XAH). (D) LSD1 in complex with p-substituted PCPA derivatives. Compound **27**-FAD adduct is shown in cyan and the adduct of the branched derivative **28** in yellow (PDB ID 2XAS and 2XAQ). (E) LSD1 (white) in complex with ortho-benzyloxy-fluoro-PCPA − FAD adduct (cyan). (PDB ID 3ABU). In all complexes only side chains of LSD1 residues involved in the interaction are shown as white sticks for clarity. (*This figure is reproduced in color in the color plate section.*)

the pocket. This shows that the compounds are far from perfectly fitting to the pocket. As suggested by the authors, these inhibitors rather act as a plug that blocks the substrate binding pocket [105]; they are attached to the cofactor FAD, whereas few interactions are formed with the substrate-binding site.

Ortho-substituted PCPAs were designed based on the comparison of the crystal structures of LSD1 and MAO B with PCPA. A bulky ortho-substituent was expected to encounter steric hindrance in MAO B but not in LSD1 [107]. In the resolved crystal structure, compound **29** forms a five-membered ring adduct with FAD, whereas the phenyl group at the ortho position undergoes hydrophobic interactions with V333 and F538 of LSD1 (see Figure 18.7E).

Crystal structures of LSD1 with covalent inhibitors have given valuable insights into the mode of inhibition of these derivatives and have shown that addressing the peptide subpocket is a successful approach to obtaining more active and selective LSD1 inhibitors. However, the lack of crystal structures of LSD1 in complex with noncovalent small-molecule inhibitors limits our full understanding of the structural basis for LSD1 inhibition. New crystal structures with noncovalent inhibitors should significantly facilitate the structure-based design of active and selective LSD1 inhibitors.

18.5 CONCLUSION

In conclusion, one can see that a wide range of different assay formats for histone lysine demethylases have been established thus far, based on very different physical read-out principles. All of them have their own advantages and disadvantages and can suffer from system-inherent drawbacks such as high enzyme consumption, long measurement times and limited throughput, incompatibility of assay components with test compounds, reactivity or instability of compounds under assay conditions, interferences by autofluorescence and so forth, which makes them prone to false positive or negative results. It is, thus, highly advisable to always use combinations of physically and conceptually different orthogonal assay technologies as demonstrated in the screening and drug development studies presented here in order to validate the results for hit compounds in complementary systems. Together with computational methods and cocrystallization efforts, this can aid in the paramount objective of discovering and developing more potent and, particularly important, selective inhibitors targeting specific subtypes and isoforms in order to develop them into useful pharmacological tools.

ACKNOWLEDGMENTS

We thank the Deutsche Forschungsgemeinschaft (DFG) for funding within CRC992 (Medical Epigenetics) and the Studienstiftung des Deutschen Volkes for a doctoral scholarship to M. R.

REFERENCES

[1] Zhang Y, Reinberg D. Transcription regulation by histone methylation: interplay between different covalent modifications of the core histone tails. Genes Dev 2001;15(18):2343−60.

[2] Martin C, Zhang Y. The diverse functions of histone lysine methylation. Nat Rev Mol Cell Biol 2005;6 (11):838−49.

[3] Rice JC, Allis CD. Histone methylation versus histone acetylation: new insights into epigenetic regulation. Curr Opin Cell Biol 2001;13(3):263−73.

[4] Shi Y, Lan F, Matson C, Mulligan P, Whetstine JR, Cole PA, et al. Histone demethylation mediated by the nuclear amine oxidase homolog LSD1. Cell 2004;119(7):941−53.

[5] Metzger E, Wissmann M, Yin N, Muller JM, Schneider R, Peters AH, et al. LSD1 demethylates repressive histone marks to promote androgen-receptor-dependent transcription. Nature 2005;437 (7057):436−9.

[6] Tsukada Y, Fang J, Erdjument-Bromage H, Warren ME, Borchers CH, Tempst P, et al. Histone demethylation by a family of JmjC domain-containing proteins. Nature 2006;439(7078):811−16.

[7] Klose RJ, Zhang Y. Regulation of histone methylation by demethylimination and demethylation. Nat Rev Mol Cell Biol 2007;8(4):307−18.

[8] Huang J, Sengupta R, Espejo AB, Lee MG, Dorsey JA, Richter M, et al. p53 is regulated by the lysine demethylase LSD1. Nature 2007;449(7158):105−8.

[9] Nicholson TB, Chen T. LSD1 demethylates histone and non-histone proteins. Epigenetics 2009;4 (3):129−32.

[10] Agger K, Christensen J, Cloos PA, Helin K. The emerging functions of histone demethylases. Curr Opin Genet Dev 2008;18(2):159−68.

[11] Kooistra SM, Helin K. Molecular mechanisms and potential functions of histone demethylases. Nat Rev Mol Cell Biol 2012;13(5):297−311.

[12] Klose RJ, Kallin EM, Zhang Y. JmjC-domain-containing proteins and histone demethylation. Nat Rev Genet 2006;7(9):715−27.

[13] Johansson C, Tumber A, Che K, Cain P, Nowak R, Gileadi C, et al. The roles of Jumonji-type oxygenases in human disease. Epigenomics 2014;6(1):89−120.

[14] Clissold PM, Ponting CP. JmjC: cupin metalloenzyme-like domains in Jumonji, hairless and phospholipase A2beta. Trends Biochem Sci 2001;26(1):7−9.

[15] Whetstine JR, Nottke A, Lan F, Huarte M, Smolikov S, Chen Z, et al. Reversal of histone lysine trimethylation by the JMJD2 family of histone demethylases. Cell 2006;125(3):467−81.

[16] Arrowsmith CH, Bountra C, Fish PV, Lee K, Schapira M. Epigenetic protein families: a new frontier for drug discovery. Nat Rev Drug Discov 2012;11(5):384−400.

[17] Couture JF, Collazo E, Ortiz-Tello PA, Brunzelle JS, Trievel RC. Specificity and mechanism of JMJD2A, a trimethyllysine-specific histone demethylase. Nat Struct Mol Biol 2007;14(8):689−95.

[18] Hou H, Yu H. Structural insights into histone lysine demethylation. Curr Opin Struct Biol 2010;20 (6):739−48.

[19] Lee J, Thompson JR, Botuyan MV, Mer G. Distinct binding modes specify the recognition of methylated histones H3K4 and H4K20 by JMJD2A-tudor. Nat Struct Mol Biol 2008;15(1):109−11.

[20] Forneris F, Binda C, Vanoni MA, Mattevi A, Battaglioli E. Histone demethylation catalysed by LSD1 is a flavin-dependent oxidative process. FEBS Lett 2005;579(10):2203−7.

[21] Forneris F, Orru R, Bonivento D, Chiarelli LR, Mattevi A. ThermoFAD, a Thermofluor-adapted flavin ad hoc detection system for protein folding and ligand binding. FEBS J 2009;276(10):2833−40.

[22] You A, Tong JK, Grozinger CM, Schreiber SL. CoREST is an integral component of the CoREST-human histone deacetylase complex. Proc Natl Acad Sci USA 2001;98(4):1454−8.

[23] Forneris F, Binda C, Vanoni MA, Battaglioli E, Mattevi A. Human histone demethylase LSD1 reads the histone code. J Biol Chem 2005;280(50):41360−5.

[24] Ciccone DN, Su H, Hevi S, Gay F, Lei H, Bajko J, et al. KDM1B is a histone H3K4 demethylase required to establish maternal genomic imprints. Nature 2009;461(7262):415−18.

[25] Karytinos A, Forneris F, Profumo A, Ciossani G, Battaglioli E, Binda C, et al. A novel mammalian flavin-dependent histone demethylase. J Biol Chem 2009;284(26):17775−82.

[26] Adamo A, Sese B, Boue S, Castano J, Paramonov I, Barrero MJ, et al. LSD1 regulates the balance between self-renewal and differentiation in human embryonic stem cells. Nat Cell Biol 2011;13 (6):652−9.

[27] Oh S, Janknecht R. Histone demethylase JMJD5 is essential for embryonic development. Biochem Biophys Res Commun 2012;420(1):61−5.

[28] Strobl-Mazzulla PH, Sauka-Spengler T, Bronner-Fraser M. Histone demethylase JmjD2A regulates neural crest specification. Dev Cell 2012;19(3):460−8.

[29] Scoumanne A, Chen X. The lysine-specific demethylase 1 is required for cell proliferation in both p53-dependent and -independent manners. J Biol Chem 2007;282(21):15471−5.

[30] Kahl P, Gullotti L, Heukamp LC, Wolf S, Friedrichs N, Vorreuther R, et al. Androgen receptor coactivators lysine-specific histone demethylase 1 and four and a half LIM domain protein 2 predict risk of prostate cancer recurrence. Cancer Res 2006;66(23):11341−7.

[31] Barrett A, Madsen B, Copier J, Lu PJ, Cooper L, Scibetta AG, et al. PLU-1 nuclear protein, which is upregulated in breast cancer, shows restricted expression in normal human adult tissues: a new cancer/testis antigen? Int J Cancer 2002;101(6):581−8.

[32] Yang ZQ, Imoto I, Pimkhaokham A, Shimada Y, Sasaki K, Oka M, et al. A novel amplicon at 9p23−24 in squamous cell carcinoma of the esophagus that lies proximal to GASC1 and harbors NFIB. Jpn J Cancer Res 2001;92(4):423−8.

[33] Yamane K, Tateishi K, Klose RJ, Fang J, Fabrizio LA, Erdjument-Bromage H, et al. PLU-1 is an H3K4 demethylase involved in transcriptional repression and breast cancer cell proliferation. Mol Cell 2007;25 (6):801−12.

[34] Sharma SV, Lee DY, Li B, Quinlan MP, Takahashi F, Maheswaran S, et al. A chromatin-mediated reversible drug-tolerant state in cancer cell subpopulations. Cell 2010;141(1):69−80.

[35] Chang PC, Fitzgerald LD, Hsia DA, Izumiya Y, Wu CY, Hsieh WP, et al. Histone demethylase JMJD2A regulates Kaposi's sarcoma-associated herpesvirus replication and is targeted by a viral transcriptional factor. J Virol 2011;85(7):3283−93.

[36] Zhang QJ, Chen HZ, Wang L, Liu DP, Hill JA, Liu ZP. The histone trimethyllysine demethylase JMJD2A promotes cardiac hypertrophy in response to hypertrophic stimuli in mice. J Clin Invest 2011;121(6):2447−56.

[37] Inagaki T, Tachibana M, Magoori K, Kudo H, Tanaka T, Okamura M, et al. Obesity and metabolic syndrome in histone demethylase JHDM2a-deficient mice. Genes Cells 2009;14(8):991−1001.

[38] Zhu Y, van Essen D, Saccani S. Cell-type-specific control of enhancer activity by H3K9 trimethylation. Mol Cell 2012;46(4):408−23.

[39] De Santa F, Totaro MG, Prosperini E, Notarbartolo S, Testa G, Natoli G. The histone H3 lysine-27 demethylase Jmjd3 links inflammation to inhibition of polycomb-mediated gene silencing. Cell 2007;130(6):1083−94.

[40] Brasacchio D, Okabe J, Tikellis C, Balcerczyk A, George P, Baker EK, et al. Hyperglycemia induces a dynamic cooperativity of histone methylase and demethylase enzymes associated with gene-activating epigenetic marks that coexist on the lysine tail. Diabetes 2009;58(5):1229−36.

[41] Iwase S, Lan F, Bayliss P, de la Torre-Ubieta L, Huarte M, Qi HH, et al. The X-linked mental retardation gene SMCX/JARID1C defines a family of histone H3 lysine 4 demethylases. Cell 2007;128 (6):1077−88.

[42] Hoffmann I, Roatsch M, Schmitt ML, Carlino L, Pippel M, Sippl W, et al. The role of histone demethylases in cancer therapy. Mol Oncol 2012;6(6):683−703.

[43] Spannhoff A, Hauser AT, Heinke R, Sippl W, Jung M. The emerging therapeutic potential of histone methyltransferase and demethylase inhibitors. ChemMedChem 2009;4(10):1568–82.

[44] Hojfeldt JW, Agger K, Helin K. Histone lysine demethylases as targets for anticancer therapy. Nat Rev Drug Discov 2013;12(12):917–30.

[45] Lohse B, Kristensen JL, Kristensen LH, Agger K, Helin K, Gajhede M, et al. Inhibitors of histone demethylases. Bioorg Med Chem 2011;19(12):3625–36.

[46] Suzuki T, Miyata N. Lysine demethylases inhibitors. J Med Chem 2011;54(24):8236–50.

[47] Rose NR, Ng SS, Mecinovic J, Lienard BM, Bello SH, Sun Z, et al. Inhibitor scaffolds for 2-oxoglutarate-dependent histone lysine demethylases. J Med Chem 2008;51(22):7053–6.

[48] Sakurai M, Rose NR, Schultz L, Quinn AM, Jadhav A, Ng SS, et al. A miniaturized screen for inhibitors of Jumonji histone demethylases. Mol Biosyst 2010;6(2):357–64.

[49] Hamada S, Suzuki T, Mino K, Koseki K, Oehme F, Flamme I, et al. Design, synthesis, enzyme-inhibitory activity, and effect on human cancer cells of a novel series of Jumonji domain-containing protein 2 histone demethylase inhibitors. J Med Chem 2010;53(15):5629–38.

[50] Suzuki T, Ozasa H, Itoh Y, Zhan P, Sawada H, Mino K, et al. Identification of the KDM2/7 histone lysine demethylase subfamily inhibitor and its antiproliferative activity. J Med Chem 2013;56 (18):7222–31.

[51] Nielsen AL, Kristensen LH, Stephansen KB, Kristensen JB, Helgstrand C, Lees M, et al. Identification of catechols as histone-lysine demethylase inhibitors. FEBS Lett 2012;586(8):1190–4.

[52] King ON, Li XS, Sakurai M, Kawamura A, Rose NR, Ng SS, et al. Quantitative high-throughput screening identifies 8-hydroxyquinolines as cell-active histone demethylase inhibitors. PLoS One 2010;5(11): e15535.

[53] Leurs U, Clausen RP, Kristensen JL, Lohse B. Inhibitor scaffold for the histone lysine demethylase KDM4C (JMJD2C). Bioorg Med Chem Lett 2012;22(18):5811–13.

[54] Lohse B, Nielsen AL, Kristensen JB, Helgstrand C, Cloos PA, Olsen L, et al. Targeting histone lysine demethylases by truncating the histone 3 tail to obtain selective substrate-based inhibitors. Angew Chem Int Ed Engl 2011;50(39):9100–3.

[55] Li Q, Sritharathikhun P, Motomizu S. Development of novel reagent for Hantzsch reaction for the determination of formaldehyde by spectrophotometry and fluorometry. Anal Sci 2007;23(4):413–17.

[56] Wang L, Chang J, Varghese D, Dellinger M, Kumar S, Best AM, et al. A small molecule modulates Jumonji histone demethylase activity and selectively inhibits cancer growth. Nat Commun 2013;4:2035.

[57] Gauthier N, Caron M, Pedro L, Arcand M, Blouin J, Labonte A, et al. Development of homogeneous nonradioactive methyltransferase and demethylase assays targeting histone H3 lysine 4. J Biomol Screen 2012;17(1):49–58.

[58] Luo X, Liu Y, Kubicek S, Myllyharju J, Tumber A, Ng S, et al. A selective inhibitor and probe of the cellular functions of Jumonji C domain-containing histone demethylases. J Am Chem Soc 2011;133 (24):9451–6.

[59] Kawamura A, Tumber A, Rose NR, King ON, Daniel M, Oppermann U, et al. Development of homogeneous luminescence assays for histone demethylase catalysis and binding. Anal Biochem 2010;404 (1):86–93.

[60] Rose NR, Woon EC, Tumber A, Walport LJ, Chowdhury R, Li XS, et al. Plant growth regulator daminozide is a selective inhibitor of human KDM2/7 histone demethylases. J Med Chem 2012;55 (14):6639–43.

[61] Kruidenier L, Chung CW, Cheng Z, Liddle J, Che K, Joberty G, et al. A selective jumonji H3K27 demethylase inhibitor modulates the proinflammatory macrophage response. Nature 2012;488 (7411):404–8.

[62] Rotili D, Tomassi S, Conte M, Benedetti R, Tortorici M, Ciossani G, et al. Pan-histone demethylase inhibitors simultaneously targeting Jumonji C and lysine-specific demethylases display high anticancer activities. J Med Chem 2014;57(1):42—55.

[63] Rose NR, Woon EC, Kingham GL, King ON, Mecinovic J, Clifton IJ, et al. Selective inhibitors of the JMJD2 histone demethylases: combined nondenaturing mass spectrometric screening and crystallographic approaches. J Med Chem 2010;53(4):1810—18.

[64] Chang KH, King ON, Tumber A, Woon EC, Heightman TD, McDonough MA, et al. Inhibition of histone demethylases by 4-carboxy-2,2′-bipyridyl compounds. ChemMedChem 2011;6(5):759—64.

[65] Woon EC, Tumber A, Kawamura A, Hillringhaus L, Ge W, Rose NR, et al. Linking of 2-oxoglutarate and substrate binding sites enables potent and highly selective inhibition of JmjC histone demethylases. Angew Chem Int Ed Engl 2012;51(7):1631—4.

[66] Hutchinson SE, Leveridge MV, Heathcote ML, Francis P, Williams L, Gee M, et al. Enabling lead discovery for histone lysine demethylases by high-throughput RapidFire mass spectrometry. J Biomol Screen 2012;17(1):39—48.

[67] Leveridge M, Buxton R, Argyrou A, Francis P, Leavens B, West A, et al. Demonstrating enhanced throughput of RapidFire mass spectrometry through multiplexing using the JmjD2d demethylase as a model system. J Biomol Screen 2014;19(2):278—86.

[68] Xu W, Podoll JD, Dong X, Tumber A, Oppermann U, Wang X. Quantitative analysis of histone demethylase probes using fluorescence polarization. J Med Chem 2013;56(12):5198—202.

[69] Leung IK, Demetriades M, Hardy AP, Lejeune C, Smart TJ, Szollossi A, et al. Reporter ligand NMR screening method for 2-oxoglutarate oxygenase inhibitors. J Med Chem 2013;56(2):547—55.

[70] Kauffman EC, Robinson BD, Downes MJ, Powell LG, Lee MM, Scherr DS, et al. Role of androgen receptor and associated lysine-demethylase coregulators, LSD1 and JMJD2A, in localized and advanced human bladder cancer. Mol Carcinog 2011;50(12):931—44.

[71] Harris WJ, Huang X, Lynch JT, Spencer GJ, Hitchin JR, Li Y, et al. The histone demethylase KDM1A sustains the oncogenic potential of MLL-AF9 leukemia stem cells. Cancer Cell 2012;21(4):473—87.

[72] Crea F, Sun L, Mai A, Chiang YT, Farrar WL, Danesi R, et al. The emerging role of histone lysine demethylases in prostate cancer. Mol Cancer 2012;11:52.

[73] Lee MG, Wynder C, Schmidt DM, McCafferty DG, Shiekhattar R. Histone H3 lysine 4 demethylation is a target of nonselective antidepressive medications. Chem Biol 2006;13(6):563—7.

[74] Schmidt DM, McCafferty DG. trans-2-Phenylcyclopropylamine is a mechanism-based inactivator of the histone demethylase LSD1. Biochemistry 2007;46(14):4408—16.

[75] Ueda R, Suzuki T, Mino K, Tsumoto H, Nakagawa H, Hasegawa M, et al. Identification of cell-active lysine specific demethylase 1-selective inhibitors. J Am Chem Soc 2009;131(48):17536—7.

[76] Guibourt N., Ortega M.A., Castro-Palomino L.J. Phenylcyclopropylamine derivatives and their medical use. PCT/EP2010/050697. 2010;WO2010084160 A1.

[77] Culhane JC, Szewczuk LM, Liu X, Da G, Marmorstein R, Cole PA. A mechanism-based inactivator for histone demethylase LSD1. J Am Chem Soc 2006;128(14):4536—7.

[78] Culhane JC, Wang D, Yen PM, Cole PA. Comparative analysis of small molecules and histone substrate analogues as LSD1 lysine demethylase inhibitors. J Am Chem Soc 2010;132(9):3164—76.

[79] Forneris F, Binda C, Adamo A, Battaglioli E, Mattevi A. Structural basis of LSD1-CoREST selectivity in histone H3 recognition. J Biol Chem 2007;282(28):20070—4.

[80] Kumarasinghe IR, Woster PM. Synthesis and evaluation of novel cyclic peptide inhibitors of lysine-specific demethylase 1. ACS Med Chem Lett 2014;5(1):29—33.

[81] Huang Y, Hager ER, Phillips DL, Dunn VR, Hacker A, Frydman B, et al. A novel polyamine analog inhibits growth and induces apoptosis in human breast cancer cells. Clin Cancer Res 2003;9(7):2769—77.

[82] Huang Y, Greene E, Murray Stewart T, Goodwin AC, Baylin SB, Woster PM, et al. Inhibition of lysine-specific demethylase 1 by polyamine analogues results in reexpression of aberrantly silenced genes. Proc Natl Acad Sci USA 2007;104(19):8023−8.

[83] Pachaiyappan B, Woster PM. Design of small molecule epigenetic modulators. Bioorg Med Chem Lett 2014;24(1):21−32.

[84] Sharma SK, Wu Y, Steinbergs N, Crowley ML, Hanson AS, Casero RA, et al. (Bis)urea and (bis)thiourea inhibitors of lysine-specific demethylase 1 as epigenetic modulators. J Med Chem 2010;53(14):5197−212.

[85] Sharma SK, Hazeldine S, Crowley ML, Hanson A, Beattie R, Varghese S, et al. Polyamine-based small molecule epigenetic modulators. Medchemcomm 2012;3(1):14−21.

[86] Hazeldine S, Pachaiyappan B, Steinbergs N, Nowotarski S, Hanson AS, Casero Jr. RA, et al. Low molecular weight amidoximes that act as potent inhibitors of lysine-specific demethylase 1. J Med Chem 2012;55(17):7378−91.

[87] Dulla B, Kirla KT, Rathore V, Deora GS, Kavela S, Maddika S, et al. Synthesis and evaluation of 3-amino/guanidine substituted phenyl oxazoles as a novel class of LSD1 inhibitors with anti-proliferative properties. Org Biomol Chem 2013;11(19):3103−7.

[88] Duan YC, Ma YC, Zhang E, Shi XJ, Wang MM, Ye XW, et al. Design and synthesis of novel 1,2,3-triazole-dithiocarbamate hybrids as potential anticancer agents. Eur J Med Chem 2013;62:11−19.

[89] Duan YC, Zheng YC, Li XC, Wang MM, Ye XW, Guan YY, et al. Design, synthesis and antiprolifera-tive activity studies of novel 1,2,3-triazole-dithiocarbamate-urea hybrids. Eur J Med Chem 2013;64:99−110.

[90] Willmann D, Lim S, Wetzel S, Metzger E, Jandausch A, Wilk W, et al. Impairment of prostate cancer cell growth by a selective and reversible lysine-specific demethylase 1 inhibitor. Int J Cancer 2012;131 (11):2704−9.

[91] Schmitt ML, Hauser AT, Carlino L, Pippel M, Schulz-Fincke J, Metzger E, et al. Nonpeptidic propar-gylamines as inhibitors of lysine specific demethylase 1 (LSD1) with cellular activity. J Med Chem 2013;56(18):7334−42.

[92] Schmitt ML, Ladwein KI, Carlino L, Schulz-Fincke J, Willmann D, Metzger E, et al. Heterogeneous antibody-based activity assay for lysine specific demethylase 1 (LSD1) on a histone peptide substrate. J Biomol Screen 2014.

[93] Mohanty JG, Jaffe JS, Schulman ES, Raible DG. A highly sensitive fluorescent micro-assay of H_2O_2 release from activated human leukocytes using a dihydroxyphenoxazine derivative. J Immunol Methods 1997;202(2):133−41.

[94] Forneris F, Binda C, Dall'Aglio A, Fraaije MW, Battaglioli E, Mattevi A. A highly specific mechanism of histone H3-K4 recognition by histone demethylase LSD1. J Biol Chem 2006;281(46):35289−95.

[95] Guang HM, Du GH. High-throughput screening for monoamine oxidase-A and monoamine oxidase-B inhibitors using one-step fluorescence assay. Acta Pharmacol Sin 2006;27(6):760−6.

[96] Votyakova TV, Reynolds IJ. Detection of hydrogen peroxide with Amplex Red: interference by NADH and reduced glutathione auto-oxidation. Arch Biochem Biophys 2004;431(1):138−44.

[97] Hauser AT, Bissinger EM, Metzger E, Repenning A, Bauer UM, Mai A, et al. Screening assays for epigenetic targets using native histones as substrates. J Biomol Screen 2012;17(1):18−26.

[98] Hopkinson RJ, Tumber A, Yapp C, Chowdhury R, Aik W, Che KH, et al. 5-Carboxy-8-hydroxyquinoline is a broad spectrum 2-oxoglutarate oxygenase inhibitor which causes iron transloca-tion. Chem Sci 2013;4(8):3110−17.

[99] Chen Y, Yang Y, Wang F, Wan K, Yamane K, Zhang Y, et al. Crystal structure of human histone lysine-specific demethylase 1 (LSD1). Proc Natl Acad Sci USA 2006;103(38):13956−61.

[100] Stavropoulos P, Blobel G, Hoelz A. Crystal structure and mechanism of human lysine-specific demethylase-1. Nat Struct Mol Biol 2006;13(7):626−32.

[101] Yang M, Culhane JC, Szewczuk LM, Gocke CB, Brautigam CA, Tomchick DR, et al. Structural basis of histone demethylation by LSD1 revealed by suicide inactivation. Nat Struct Mol Biol 2007;14 (6):535−9.

[102] Forneris F, Battaglioli E, Mattevi A, Binda C. New roles of flavoproteins in molecular cell biology: histone demethylase LSD1 and chromatin. FEBS J 2009;276(16):4304−12.

[103] Yang M, Gocke CB, Luo X, Borek D, Tomchick DR, Machius M, et al. Structural basis for CoREST-dependent demethylation of nucleosomes by the human LSD1 histone demethylase. Mol Cell 2006;23 (3):377−87.

[104] Yang MJ, Culhane JC, Szewczuk LM, Jalili P, Ball HL, Machius M, et al. Structural basis for the inhibition of the LSD1 histone demethylase by the antidepressant trans-2-phenylcyclopropylamine. Biochemistry 2007;46(27):8058−65.

[105] Binda C, Valente S, Romanenghi M, Pilotto S, Cirilli R, Karytinos A, et al. Biochemical, structural, and biological evaluation of tranylcypromine derivatives as inhibitors of histone demethylases LSD1 and LSD2. J Am Chem Soc 2010;132(19):6827−33.

[106] Mimasu S, Sengoku T, Fukuzawa S, Umehara T, Yokoyama S. Crystal structure of histone demethylase LSD1 and tranylcypromine at 2.25 Å. Biochem Biophys Res Commun 2008;366(1):15−22.

[107] Mimasu S, Umezawa N, Sato S, Higuchi T, Umehara T, Yokoyama S. Structurally designed trans-2-phenylcyclopropylamine derivatives potently inhibit histone demethylase LSD1/KDM1. Biochemistry 2010;49(30):6494−503.

HISTONE DEMETHYLASES: BACKGROUND, PURIFICATION, AND DETECTION

19

Kelly M. Biette, Joshua C. Black, and Johnathan R. Whetstine

Massachusetts General Hospital Cancer Center and Department of Medicine, Harvard Medical School,
Charlestown, MA, USA

CHAPTER OUTLINE

Y.G. Zheng (Ed): Epigenetic Technological Applications. DOI: http://dx.doi.org/10.1016/B978-0-12-801080-8.00019-3

19.1 INTRODUCTION

Histone lysine methylation is a highly regulated event in the eukaryotic genome. The addition of this modification is catalyzed by lysine methyltransferases (KMTs) and methyl groups are removed by lysine demethylases (KDMs). The degree of lysine methylation (mono-, di-, or tri-methylated) and the specific lysine modified within a histone tail are associated with various gene expression states, enhancer elements, structural features (telomeres and centromeres), and other chromatin states [1−3]. Generally, methylation of histone H3 lysine 4 (H3K4), H3K36, or H3K79 is associated with activation of gene expression, whereas methylation of H3K9, H3K27, and H4K20 is typically linked to gene silencing [4,5].

The impact of histone methylation extends beyond gene expression and genome structure. Recent studies clearly demonstrate important roles for lysine methylation dynamics in cell cycle, replication, and DNA damage response [4,6−11]. Recent studies have demonstrated the importance of KMTs and KDMs in determining cell fate and maintaining genomic stability [12−15]. Additionally, lysine methylation dynamics are implicated in diverse development and pathological states and this complex relationship is an exciting area of future research [16−22]. Understanding the enzymology and *in vivo* function of KMTs and KDMs will be critical for future biomarker studies as well as chemical targeting of disease states.

This book chapter focuses on methods to purify and assay the enzymatic activity of histone demethylases. We provide a brief history of the discovery of this class of enzymes, discuss original and recently developed enzymatic assays, and note important considerations that impact enzymatic activity both *in vitro* and *in vivo*.

19.1.1 HISTORICAL PERSPECTIVE

The first histone KDM, LSD1/KDM1A was discovered by Yang Shi and colleagues in late 2004 as part of the C-terminal binding protein-1 complex (CtBP1) co-repressor complex [23,24]. KDM1A removes methyl marks from H3K4 to regulate gene expression and is unable to oxidize polyamines despite containing a polyamine oxidase domain. KDM1A requires flavin adenine dinucleotide (FAD) as a cofactor and acts on protonated nitrogens, which restricts its activity to di-methylated (me2) and mono-methylated (me1) lysines. This reaction produces the by-products $FADH_2$ and formaldehyde (Figure 19.1).

Following the identification of KDM1A and its substrate, additional KDMs with specific activity toward other lysine modifications were speculated to exist [25,26]. A new class of histone KDM enzymes containing a JmjC domain was discovered in 2006 with the observation that KDM2 and KDM3 family members could demethylate mono- and di-methylated lysines [27,28]. Shortly after this discovery, the first tri-demethylase JmjC family, KDM4A-KDM4D, was identified by a number of groups [29−32]. The JmjC enzymes catalyze demethylation by methyl group oxidation and require α-ketoglutarate, molecular oxygen (O_2), and Fe(II) as reaction cofactors [33]. During the reaction, α-ketoglutarate is converted to succinate. The Fe(II) promotes hydroxylation and subsequent demethylation of the methylated lysine, producing formaldehyde. This enzymatic reaction is compatible with lysine mono- (me1), di- (me2) and tri- (me3), demethylation (Figure 19.1).

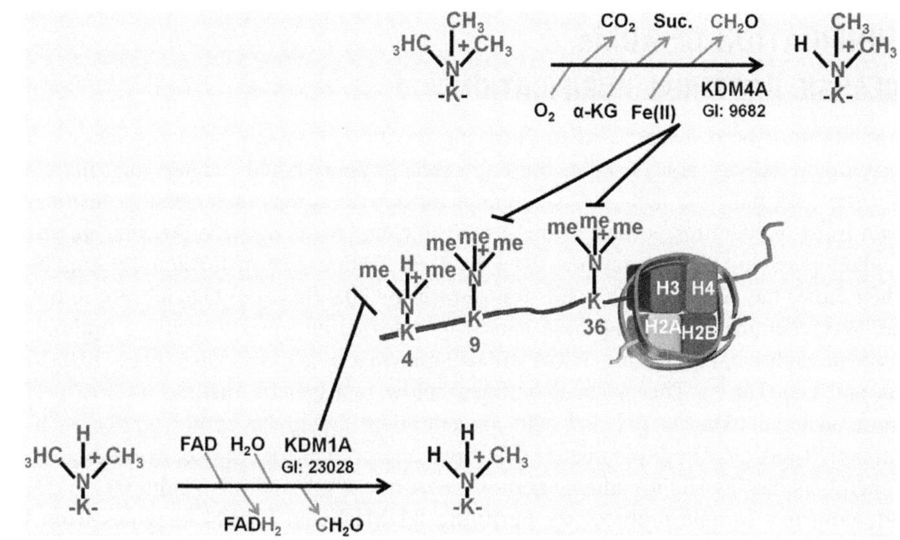

FIGURE 19.1

Schematic representation of the demethylase reaction. KDM4A demethylates di- and tri-methylated H3K9 and H3K36 by methyl group oxidation in the presence of cofactors α-ketoglutarate, molecular oxygen (O_2), and Fe (II). This reaction produces succinate, CO_2, and formaldehyde. KDM1A, with cofactors FAD and H_2O, demethylates mono- and di-methylated H3K4, producing $FADH_2$ and formaldehyde. (*This figure is reproduced in color in the color plate section.*)

Data generated from numerous groups suggests that KDMs act with high substrate specificity for both the degree of methylation and location of the lysine within the histone tail (reviewed in [4]). For example, KDM4A primarily catalyzes the tri- to di-demethylation of H3K9, H3K36, and H1.4K26, while KDM4D removes tri- and di-methylation on H3K9 and H1.4K26 (example shown in Figure 19.3) [32,34]. These findings are consistent with the fact that histone methyltransferases exhibit similar site and degree of methylation specificity during methyl mark deposition [4]. Through purification of both full-length and catalytic domains, important biochemical properties of KDMs have been elucidated. Crystal and co-crystal structures have been resolved and the impact of KDM inhibitors on the catalytic domains can be directly established or modeled based on previous structure and function studies [35–39].

The current state of the field emphasizes the importance of fine-tuning lysine methylation states in the genome, but many major questions remain for both methyltransferase and demethylase biology. Due to the specific nature of the methylation site, both for particular lysines in the histone tail and for larger genomic regions, there is a need for biochemical analysis to be coupled with *in vivo* studies of genome-wide methylation changes, gene expression changes, impact on cell cycle, and genome stability analysis. Future research will continue to require optimal purification, enzymology and cellular activity studies, many conditions for which will be discussed below.

19.2 PURIFICATION OF KDMs
19.2.1 CLASSIC BACTERIAL PURIFICATION

Some key points must be considered when purifying the JmjC domain-containing histone demethylases for enzymatic activity assays. First, the expressed protein should include the zinc finger that accompanies the enzymatic domain, and is essential for activity. JmjC enzymes that also contain the JmjN domain must include this portion of the protein for enzymatic activity. Another important consideration when purifying these enzymes from bacteria is that enzymatic activity can be greatly reduced when either the GST (glutathione S-transferase) or HIS (histone) tags are not removed from the enzyme. We recommend cleaving these tags from purified enzymes to achieve optimal activity. In recent years, additional tags that allow robust enzymatic activity have been identified (e.g. Streptavidin and HaloTag®). The protocol outlined below is a generic method used to purify these proteins from bacterial extracts. pH and salt concentration should be optimized for each purified enzyme and activity assay. These parameters are especially critical for JmjC enzymes because different family members display slightly altered activity in different salt and pH conditions.

The following is a sample protocol for purifying KDM1A and KDM4 enzymes from bacteria, which can be applied to other JmjC-containing KDMs:

1. In the case of KDM1A, the entire protein can be cloned into an expression vector used to express soluble enzyme for purification. For KDM4 JmjC-containing enzymes, the first 350 amino acids (or another region containing the enzymatic domain and zinc finger) are expressed from plasmids transformed into Rosetta cells. Rosetta cells are used for constructs that express GST-tagged (e.g. pGEX-4T1) or HIS-tagged (e.g. pET28a) enzymatic domains.

2. Inoculate a single colony in a 50 mL culture overnight with the appropriate antibiotic selection. Use a small fraction of this culture to inoculate a larger culture and grow to an optical density (O.D.) of 0.3 at 600 nm. Induce protein expression with 0.2 mM IPTG (isopropyl-beta-D-thiogalactopyranoside) for 3−6 hours at 37°C or overnight at 20°C. The induction time and temperature should be optimized to achieve high protein quality and solubility.

3. When the culture reaches an optimal O.D. (typically <1.0), the bacteria are pelleted by centrifugation at 4°C. Wash the pellet twice with 1X PBS (phosphate-buffered saline), pH 7.4. After the last wash, resuspend the pellet in lysis buffer (50 mM Tris-HCl, pH 8.0, 500 mM NaCl, 0.2% Triton-X 100, 3 mM β-mercaptoethanol, protease inhibitors). Add the protease inhibitors and β-mercaptoethanol fresh immediately before use. Lyse the bacteria by sonication on a dry ice bath to avoid overheating. Following sonication, spin the sample at 18,000 g for 20 minutes. Remove the supernatant and transfer it to a fresh tube. It is important to keep nonsonicated, sonicated, and centrifuged lysate to assess the quality and solubility of the protein isolation.

4. Glutathione beads are used to purify GST-tagged protein and nickel agarose beads are used for HIS-tagged protein. Wash the beads twice in lysis buffer before adding them to the lysate. Incubate HIS-tagged protein for 2−4 hours or GST-tagged protein for 30−45 minutes with the appropriate beads.

5. Following incubation, spin down lightly and save the supernatant to assess depletion and the amount of protein recovered. Wash the beads twice in 10 mL lysis buffer in batch and two

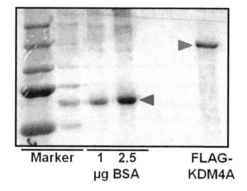

FIGURE 19.2

Quantification of KDM4A purified from Sf9 cells. FLAG-tagged KDM4A was expressed in Sf9 cells, purified as described, run on an SDS-PAGE gel and Coomassie stained. The resulting enzyme prep (green arrow) is of high purity and can be quantified by comparison to a known concentration of BSA protein standard (red arrow). (*This figure is reproduced in color in the color plate section.*)

additional times in a disposable chromatography column. For HIS-tagged protein and nickel beads, elute with 200−250 mM imidazole. For GST-tagged proteins, elute twice for 30 minutes in 20 mM reduced glutathione and 50 mM Tris-HCl, pH 7.4. Dialyze eluted samples three times into either KDM1A buffer (1 × PBS pH 7.4, 10% glycerol and 1 mM PMSF) or JmjC buffer (20 mM Tris-HCl pH 7.3, 150 mM NaCl, 8.0% glycerol, 1 mM DTT, and 1 mM PMSF) at 4°C.

6. Following purification and dialysis, estimate protein concentration by running on an SDS gel loaded with a titration of a protein standard, such as BSA (bovine serum albumin) (see Figure 19.2). After determining the approximate concentration of the purified protein, remove the tag by incubating with thrombin at 4°C for several hours.

19.2.2 ALTERNATIVE TAGS AND PURIFICATION APPROACHES FROM BACTERIA

Two additional tags have been developed to purify more robust and consistent JmjC-containing enzymes for demethylation assays. The use of these two alternative tags provides purification strategies that avoid oxidants, reductants, and metals, which may impede demethylase function. The HaloTag® from Promega Corporation is an engineered protein from a bacterial dehalogenase (MW 34 kDa) that covalently attaches to a set of chloroalkane ligands. These ligands can then be attached to solid surfaces through functional group interactions, which allow protein immobilization on a resin, similar to the streptavidin and biotin interaction. The high affinity association can be competed with the ligand to produce highly pure enzyme.

Another purification method that has been very important in evaluating enzyme kinetics for the KDM4 family of enzymes is streptavidin purification, which reduces protein and metal contaminants in the enzyme preparation. A recent study using this method demonstrated that metal contaminates are a major reason for variability and reduced demethylase activity *in vitro* of bacterial purified enzymes [40]. A complete description for these approaches and tags can be found in Krishnan et al. [40] and on the Promega Corporation website.

19.2.3 BACULOVIRAL PURIFICATION FROM SF9 CELLS

In order to study the impact of specific domains on demethylase activity, it is necessary to express pure, full-length protein with consistent enzyme kinetics. Full-length KDMs can be easily purified from bacteria (KDM1A family members) or *Spodoptera frugiperda* (Sf9) cells (primarily JmjC-containing enzymes) following transduction with baculovirus. Transduction of Sf9 cells allows for full-length protein purification and activity assays for a number of demethylases, and has facilitated the study of individual domains within the enzymes. For example, it has been observed that individuals with cognitive disorders contain multiple mutations in the JmjC-containing KDM5C/SmcX enzyme, which were shown to abrogate enzymatic activity [41]. We primarily use FLAG-tagged demethylases in this purification; however, other expression tags such as HaloTag or HA may also be used for purification and enzymatic analysis.

There are several different methods for the preparation of baculovirus containing a gene of interest. We have predominantly used the commercially available Bac-N-Blue™ kit from Life Technologies. A complete manual can be found at the url given in [42]. The modified method below was used to express and purify full-length proteins from the KDM4 enzyme family. For this reason, these proteins will be used as an example, but the procedure can be extended to other KDMs.

1. Seed Sf9 cells in Grace's Insect media supplemented with 10% FBS (fetal bovine serum) and 1% penicillin/streptomycin at 1×10^6 cells per well on a 6-well dish at least 30 minutes prior to transfection.
2. Add 2.5 μg of expression vector, 0.25 μg linear Bac-N-Blue™ viral DNA, and 12 μL of Cellfectin® reagent to 500 μL unsupplemented Grace's Insect medium (no antibiotic or FBS). Incubate for 15 minutes at room temperature.
3. While the DNA–lipid complex is forming, wash cells once with unsupplemented Grace's Insect medium.
4. After 15 min, aspirate the wash and add the 500 μL DNA–lipid complex dropwise to the cells. Place the dishes into a sealed container and leave the cells in the transfection media for 4 hours at 27°C.
5. After the incubation time, add 1.5 mL of complete insect medium to the cells. Place the dishes into a sealed container for at least 72 h and monitor the cells to determine optimal collection time. Typically, the virus is collected at Day 4 or 5. The cells should be larger, no longer dividing, and obviously vacuolated. Remove the viral supernatant (referred to as P_0), wrap in foil, and store at 4°C. This can be used to infect additional plates and scale up protein production. The virus can also be plaque purified (see below), which increases the number of viral particles containing the protein of interest.
6. Optional plaque purification: Seed 1.7×10^6 Sf9 cells in 3 mL complete Express medium in a 60 mm^2 dish. After they have attached, remove 1.8 mL of medium and add 0.3 mL of the P_0 (or P_1, second infection) to the cells for 1 hour at room temperature. Mix Express medium with 4% agarose and add this to the cells after the media is removed. Allow the gel to solidify at room temperature (about 30 minutes) and then move the plates to the incubator. Positive plaques will be milky white and negative plaques will be clear. The positive plaques can now be isolated and used for expansion and subsequent infections.

19.2.3.1 *Protein purification from transduced Sf9 cells*

1. After transducing the cells, keep the media for additional infections (store in foil at 4°C). Wash the cells twice with 1X PBS, pH 7.4. Resuspend the pellet in 1 volume the size of the cell pellet of lysis buffer (100 mM Tris-HCl, pH 7.3, 3 mM $MgCl_2$, 300 mM NaCl, 0.1% NP-40 (or substitute), 1 mM DTT and 20% glycerol). Dounce homogenize the cells 30 times and sonicate for 10 minutes before snap freezing in liquid nitrogen. Slowly thaw the samples and centrifuge at maximum speed at 4°C to clear the lysate.

2. Dilute the supernatant with equal volume of Buffer Zero (100 mM Tris-HCl, pH 7.3, 3 mM $MgCl_2$, 0.1% NP-40 (or substitute), 1 mM DTT and 20% glycerol). The lysate is now ready for an IgG preclear. Wash the IgG beads (either magnetic or sepharose) once with lysis buffer before adding to the lysate. Incubate for 1 hour at 4°C.

3. Discard the supernatant and wash FLAG-conjugated beads with lysis buffer. Incubate the precleared lysate with the FLAG-tagged beads for at least 4 hours, but preferably overnight at 4°C.

4. Save the supernatant to ensure immune-depletion of tagged protein and wash the beads three times with Wash Buffer (20 mM Tris-HCl pH 7.3, 150 mM NaCl, 8% glycerol, 1 mM DTT (added fresh) and 0.1% NP40 or substitute). For the fourth wash, use Wash Buffer lacking NP-40 (referred to as Buffer B), as NP-40 can interfere with MALDI-TOF.

5. To elute, incubate the beads with equal volume Buffer B and a 1:20 dilution of FLAG peptide (stock 5 μg/μL) twice for 45−60 minutes. Conduct this step with constant agitation at room temperature.

6. Quantify the protein by running it against a protein standard, such as BSA, on a polyacrylamide gel (Figure 19.2). Aliquot the protein and store at −80°C.

19.3 *IN VITRO* ASSAYS TO EVALUATE DEMETHYLASE ACTIVITY

19.3.1 ENZYMATIC REACTIONS

19.3.1.1 *KDM1A*

The following reaction is used to evaluate KDM1A activity *in vitro*. A few components (BSA, NAD^+ (nicotinamide adenine dinucleotide), and FAD) are not essential but are added to ensure robust demethylase activity. Similar to its addition in a restriction endonuclease digestion, BSA will improve the activity of some proteins but not be detrimental to others.

Purified histones are used at a range of 4−10 μg/reaction. In the case of histone peptides, 1−2 μg are included in the reactions. The buffer for the KDM1A demethylation assay is 1X PBS, pH 7.4. The following reaction is enough for five Western blots of histones and at least 50 MALDI-TOF spectrometry assays. Combine the reaction components in a final volume of 50 μL and incubate at 37°C for four to five hours.

> $1 \times$ PBS, pH 7.4
> 5 mM $MgCl_2$
> 0.2 mM FAD
> 0.2 mM NAD^+
> <u>1−2 μg KDM1A</u>
> 4−10 ug histones or 1−2 ug peptide
> Final volume 50 μL

19.3.1.1.1 JmjC-containing enzymes

The following *in vitro* reaction is used to evaluate JmjC-containing enzymes. Similar to the KDM1A assay, the nonessential component ascorbic acid (2 mM final concentration) is added to ensure robust demethylation activity. The reaction can be conducted in a Tris-HCl or HEPES (2-[4-(2-hydro-xyethyl)piperazin-1-yl]ethanesulfonic acid) buffer, but higher quality (less noise) MALDI-TOF spectrometry results from use of the Tris-HCl buffer. For *in vitro* formaldehyde assays, we recommend the HEPES buffer to avoid background formaldehyde production when Tris-HCl is left in plastic tubes at or above 37°C [43]. It is important to consider pH and salt concentration when conducting demethylase assays with JmjC-containing enzymes, as not all members of this enzyme class are equally soluble or active at a fixed salt concentration or pH. We recommend optimizing these conditions for each enzyme assayed, which will improve the consistency and robustness of the demethylation reaction. The sample protocol below uses conditions for the KDM4 family of enzymes.

These enzymatic assays require 5 μg of purified histones for analysis by Western blot (typically 1 μg per lane is sufficient). Histone peptides may also be used at $1-2$ μg or 10 μM and should be approximately 21 amino acids in length. This type of peptide reaction is adequate for at least 100 MALDI-TOF spectrometry analyses.

> 20 mM Tris-HCl, pH 7.3
> 150 mM NaCl
> 50 μM $(NH4)_2Fe(SO4)_2 + 6(H_2o)$
> 1 mM α-ketoglutarate
> 2 mM ascorbic acid
> Final volume 100 μL

This reaction is conducted at 37°C for $2-5$ hours. In this assay, MALDI-TOF can detect products within minutes (Figure 19.3). Competitive amounts of metals such as Zn^{2+}, Ni^{2+}, or divalent cation chelators (ethylenediaminetetraacetic acid − EDTA) can be used to quench the demethylase activity. Interestingly, Ni^{2+} was used to help crystallize the KDM4A enzyme with the H3 peptide substrates [38].

19.3.2 **MALDI-TOF**

Matrix-associated laser desorption/ionization coupled to time-of-flight mass spectrometry (MALDI-TOF) is highly sensitive and allows for the observation of subtle enzymatic activity. Since these enzymes demethylate by oxidation, a percentage of the enzyme could be inactivated and the increased sensitivity of this screening assay was important in identifying different demethylase classes (shown in Figure 19.3 and Figure 19.4 and in [32]). This method allowed us to observe activity for the KDM4B enzyme, which did not exhibit enzymatic activity in other *in vitro* assays.

The Western blot of the H3K36me3 peptide shows a modest reduction at 60 minutes in the presence of full-length FLAG-tagged KDM4A; whereas, the mass spectrometric analysis shows demethylation of H3K36me3 peptide by KDM4A within 1 minute after the enzyme was added to the reaction (Figure 19.4). MALDI-TOF against peptides is not a quantitative approach, but does give clear insight into the speed of the reaction and relative species abundance within samples (see Figures 19.3 and 19.4). A good example of this is seen when comparing KDM4A and KDM4D

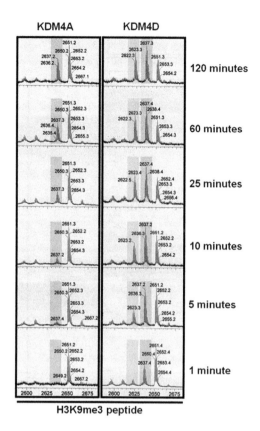

H3K9me3 peptide

FIGURE 19.3

Comparison of KDM4A and KDM4D demethylation activity. Purified enzymes were incubated with H3K9me3 peptide for 120 minutes and analyzed by MALDI-TOF at six time points. KDM4D generates di-methylated H3K9 (red) within 1 minute, but the same amount is not observed with KDM4A until 25 minutes. KDM4D also produces H3K9me1 (green) at 1 minute, but this activity is undetectable for KDM4A. (*This figure is reproduced in color in the color plate section.*)

from 1 minute to 120 minutes. KDM4D is able to generate dimethylated H3K9me2 (histone 3 lysine 9 dimethylation; red) in 1 minute; whereas, KDM4A does not generate a similar amount of dimethylated residue until 25 minutes of reaction time (Figure 19.3). These data also highlight the ability to visualize stepwise demethylation. KDM4D generates H3K9me1 within 1 minute, while activity is not observed with KDM4A under the same conditions (Figure 19.3; green). Because of this sensitivity, MALDI-TOF has considerable advantages for analyzing demethylase activity. However, this approach is less amenable to large-scale drug screens for KDM inhibitors and several higher throughput analyses are discussed below.

To conduct the assay, activate and equilibrate a C18 ZipTip (Millipore) by pipetting and ejecting 10 μL of 50% acetonitrile/0.1% trifluoroacetic acid (TFA) three times prior to sample loading. After the demethylase reactions are complete, desalt 1 μL of reaction by pipetting onto an activated

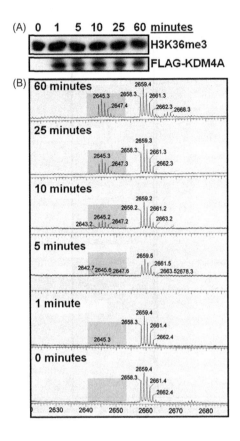

FIGURE 19.4

H3K36me3 demethylation by KDM4A. A. Western blot for H3K36me3 and the KDM4A peptide levels at each time point corresponding to the MALDI TOF spectra in panel B. B. Quantification of H3K36me3 demethylation reaction by MALDI-TOF spectrometry. Demethylation of the residue is observed beginning 1 minute after the addition of KDM4A to the reaction (red). (*This figure is reproduced in color in the color plate section.*)

and equilibrated ZipTip. After loading the sample onto the ZipTip, the tip is washed with 10 μL 0.1% TFA five independent times. Elute the bound material with 2 μL of 70% acetonitrile containing 1 mg/mL α-cyano-4-hydroxycinnamic acid MALDI matrix and 0.1% TFA. Spot the eluate onto the MALDI plate, allow it to evaporate and crystalize, then load the samples into the MALDI-TOF.

When using MALDI-TOF for studying demethylases or in efforts to isolate new enzymes, one should consider the length of the peptides being used. Approximately 21−23 amino acid peptides with a linker and conjugated biotin has resulted in identification of all known histone demethylases thus far. However, activity can be missed when screening is restricted to this method, as in the discovery of the H3K9 and H4K20 demethylase KDM7B/PHF8. This enzyme exhibits altered substrate affinity when incubated with peptides compared to more complex substrates, such as nucleosomes. For this reason, MALDI-TOF should be coupled to another approach (e.g. Western blotting and immunostaining, which are outlined below).

19.3.3 WESTERN BLOT ANALYSIS

In order to conduct reliable Western blots on histones or nucleosomes, we recommend loading at least 1 μg per lane. In order to detect histone modifications with less background, membranes should be blocked in 5% BSA in 1X PBS with 0.1% Tween-20 (PBST − 1X PBS with 0.1% Tween-20). Appropriate antibodies should be selected and optimized based on the residue, modification state, and species being tested. The unmodified version of the histone should be used as a loading control in immunoblots. The histones most commonly used in these assays are calf thymus type II-A histones (Sigma). For more complex substrates, one can purify mononucleosomes or oligonucleosomes or purchase them commercially. Lusser and Kadonaga have also published a method for nucleosome purifications [44].

Western blots have a more limited dynamic range than what is observed using other approaches (Figure 19.4). This assay can be variable at times and is greatly affected by the activity of the enzyme. For this reason, this method is not the best approach for discovery or high throughput analysis of demethylase activity, but serves as a reliable readout for well-optimized, low-throughput activity assays.

19.3.4 RADIOLABELED FORMALDEHYDE RELEASE ASSAY

The first KDM assay with moderately increased throughput was the radiolabeled formaldehyde release assay (Figure 19.5A). This assay measures the production of formaldehyde during demethylation and is used for both the amine oxidase and JmjC-dependent demethylase enzymes. Formaldehyde is converted to 3,5-diacethyl-1, 4-dihydrolutidine (DDL) through a NASH reaction and the DDL is recovered by organic phase extraction with 1-pentanol [27]. The amount of radiolabeled formaldehyde released is quantified by liquid scintillation. Because the radiolabel on the formaldehyde is derived from the original methyl group, the exact demethylated residue can be determined.

Histone methyltransferase reactions are performed with specific KMTs in the presence of S-adenosyl-L-methyl-[^{3}H]-methionine to generate specifically radiolabeled methyl groups (Figure 19.5A). This reaction can be used to generate specifically methylated peptides, histones, nucleosomes, or oligonucleosomes, which is advantageous for a number of reasons. First, it is possible to screen for demethylation of many different lysine residues relatively quickly compared to mass spectrometry or Western analysis. Second, this method is easily adaptable to different histone substrates. Finally, this method is also viable for assaying nonhistone substrates if the methyl groups in question can be radiolabeled.

The major disadvantage of this approach is the reliance on KMTs to supply the methyl group, which limits this technique to residues and methylation states for which a KMT has been identified and purified. It is also challenging to produce tri-methylation states *in vitro*. While it is possible to generate H3K9me3 using Suv39h1, consistent trimethylation of other residues has not been achieved. It is also difficult to achieve quantitative methylation *in vitro*, which renders this method unsuitable for kinetic determination of demethylase activity. Finally, because the assay measures formaldehyde production, it can only identify the demethylated residue, but cannot characterize the state of demethylation. Simply, it is difficult to determine if the radiolabeled formaldehyde was produced from a labeled mono-, di-, or tri-methylated state. For these reasons, it is most practical to use this method to identify candidate lysine residues and couple the approach to Western blotting or mass spectrometry for precise identification of regulated methylation states (as seen in [27]).

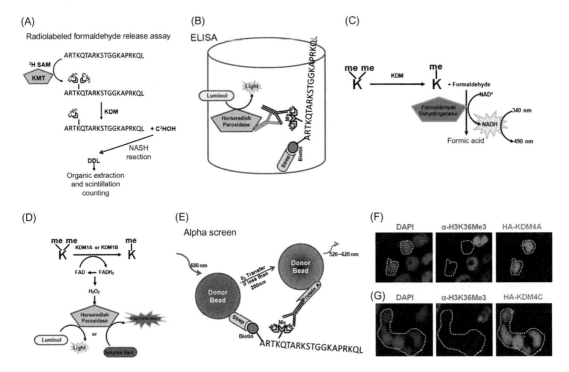

FIGURE 19.5

Methods to detect KDM activity. (A) Radiolabeled formaldehyde release assay. In the radiolabeled formaldehyde release assay a specific methyltransferase is used in conjunction with tritiated S-adenosyl-methionine to add tritiated methyl groups to a specific lysine residue on a histone tail peptide. Following demethylation the radioactivity is released as part of the formaldehyde by-product. The formaldehyde is converted to 3,5-diacethyl-1, 4-dihydrolutidine (DDL) through a NASH reaction. The DDL is organically extracted and the radioactivity is counted by a filter-binding assay followed by liquid scintillation. (B) ELISA. In the ELISA assay, biotinylated histone peptides with specific modifications (H3K4me3 is depicted) are affixed to streptavidin-coated microplate wells. Antibodies to specific histone modifications interact with the methylation and are subsequently bound by a secondary antibody coupled to horseradish peroxidase. The amount of methylated peptide is measured by the light produced from HRP conversion of luminol, with demethylation resulting in reduced signal. (C) Enzyme-coupled fluorescence assay. The JmjC-containing demethylases and FAD-dependent amine oxidases generate formaldehyde as a by-product of the demethylation reaction. Formaldehyde dehydrogenase uses NAD^+ to convert the formaldehyde to formic acid, which also reduces the NAD^+ to NADH. The amount of NADH can be assayed by absorbance at 490 nm. (D) Enzyme-coupled fluorescent and luminescent assays. The FAD-dependent amine oxidases generate hydrogen peroxide as a by-product of the demethylation reaction. The H_2O_2 allows horseradish peroxidase to convert luminescent substrates (Luminol) or fluorescent substrates (Amplex Red) into their light emitting compounds, which directly correlate with the amount of enzymatic activity. (E) AlphaScreen. In an AlphaScreen, donor beads coated with streptavidin are coupled to biotinylated histone peptides. The acceptor bead is coupled to antibodies directed against a specific histone modification, in this case H3K4me3. The binding of the antibody to the modification brings the acceptor bead within 200 nm of the donor bead, allowing an oxygen singlet to transfer to the acceptor bead causing emission of fluorescence. (F) Immunofluorescence. Cells are co-stained with an antibody specific to H3K36me3 and an antibody to HA for detection of tagged KDM4A (A) or KDM4C (B). Nuclei are visualized with DAPI. Cells expressing a high level of KDM (dashed line) have lower levels of H3K36me3 compared to cells not overexpressing the KDM. (*This figure is reproduced in color in the color plate section.*)

19.3.5 ELISA-BASED ASSAYS

Enzyme-linked immunosorbent assays (ELISAs) have been extensively utilized to characterize biological molecules including antibodies, chemokines, infectious agents, and allergens. This method has been used with histone peptides to identify an inhibitor of PRMT1 and has the potential for use in demethylase assays [44]. In an ELISA, a biotinylated histone peptide methylated at a specific residue is affixed to the surface of a streptavidin-coated multi-well plate (Figure 19.5B). The purified KDM of interest and essential cofactors are added to the well and peptide demethylation occurs. Antibodies to the specific methylation state are then added and detected with horseradish peroxidase (HRP) conjugated secondary antibodies. HRP substrate is added to generate a colorimetric or luminescent readout, which is quantified in a plate reader. While ELISAs are potentially better suited to KMT assays because of the gain in signal, loss of signal is also quantifiable with control wells lacking the KDM or containing catalytically inactive KDMs.

One potential drawback to the ELISA is the challenge of using nonpeptide substrates. It is possible to generate biotinylated histones or nucleosomes or to immobilize substrates using antibodies to the core domains of the histones, but screening peptides is much simpler [45,46]. Furthermore, the ELISA is limited by the availability of high-quality antibodies specific to the modifications of interest. However, ELISA assays allow for the screening of any methylated site or state for which antibodies are available and is semi-quantitative when run with appropriate controls. For this reason, ELISA analysis can be useful in high-throughput screens for modulators of demethylase activity.

19.3.6 FLUORESCENT AND LUMINESCENT ASSAYS

Fluorescent and luminescent assays for measuring KDM activity are gaining popularity due to their high-throughput nature. The most common of these assays measure the production of reaction by-products (formaldehyde and H_2O_2) or use an antibody-based approach similar to an ELISA to measure demethylation of specific residues (Figure 19.5C). This section will briefly describe both assay types.

Trievel and Schofield have developed a fluorescence-based system for measuring formaldehyde production during demethylation by KDM4 family enzymes and this methodology is applicable to all KDMs. In this assay, the formaldehyde produced by demethylation is oxidized by formaldehyde dehydrogenase (FDH) into formic acid, which requires the reduction of NAD^+ to NADH. Excitation of NADH at 340 nm causes fluorescence at 490 nm (Figure 19.5C). Production of NADH is measured, which correlates with formaldehyde levels and demethylation. NADH production is quantitated by comparison to a standard curve of NADH in reaction buffer, allowing precise determination of rates of formaldehyde production.

This assay is amenable to any KDM substrate (peptides, histones, nucleosomes, or oligonucleosomes) and is useful for high-throughput screening as all reaction components are added simultaneously and followed by single measurements. A major drawback to this method is its inability to determine which methylation state the enzyme is acting upon. However, kinetic activity against mono-, di- and tri-methylation can be measured using peptide substrates with specific methylation states. This method has been particularly useful in designing JmjC protein inhibitors and in determining the kinetic properties of the KDM4 family [40,47].

KDM1A and KDM1B activity can also be measured using fluorescent and luminescent assays for detection of the production of H_2O_2 (Figure 19.5D). The KDM1A demethylation reaction reduces FAD to $FADH_2$, which reacts with oxygen to produce H_2O_2. The addition of HRP to the reaction uses H_2O_2 by-product to oxidize AmplexRed (10-acetyl-3,7-dihydroxyphenoxazine) or luminol into fluorescent or luminescent signal [48]. Similar to the FDH conversion of formaldehyde, this assay is very amenable to high throughput screening and is most informative when the demethylated substrate is already known to aid in the identification of small molecule modulators of enzymatic activity.

Luminescent technology has also been adapted into an assay similar to the ELISA, called the amplified luminescent proximity homogenous assay (ALPHA) screen (Figure 19.5E) [49]. In an ALPHA screen, a biotinylated histone peptide is attached to a "donor bead" coated with a photosensitizer. When a laser excites the donor bead at 680 nm, it converts oxygen from the demethylase reaction into an excited singlet oxygen molecule. This singlet oxygen then transfers the energy to an "acceptor bead" in close proximity (approximately 200 nm), resulting in emission of a luminescent signal between 520 and 620 nm. The acceptor bead is coupled to an antibody for a specific histone methylation state. When using a methylated peptide, the acceptor will be close enough to generate luminescent signal. When demethylation of the peptide occurs, the antibody on the acceptor bead has nothing to recognize, the oxygen returns to its ground state, and no energy transfer occurs, resulting in a signal decreases (Figure 19.5E). Schofield and colleagues have used the ALPHA screen to identify small molecule inhibitors of KDM4E [49].

This methodology can be used to analyze nucleosomes by coating the "donor bead" with antibodies specific to the core histone domains or other histones not containing the modification of interest in order to immobilize the nucleosome on the donor bead. The ALPHA screen can also be modified to detect binding events by coupling the KDM or protein of interest to the acceptor bead. In this scenario, if the KDM binds a specific histone modification, it will bring the acceptor bead into proximity for energy transfer and luminescence will be observed. This technology has great potential for screening for inhibitors that interfere with both KDM activity and targeting.

The major drawback of the ALPHA screen technology is the inability to determine kinetic parameters, as the binding capacity and bead proximity are limited. However, this assay still facilitates the accurate determination of IC_{50} values for small molecule inhibitors. The major advantage of an ALPHA screen is the homogeneous nature of the assay. All of the reaction components are added in a single step and no washing steps are required prior to detection. Furthermore, this assay involves direct detection of the methylated substrates, which eliminates false positives and negatives associated with assaying reaction by-products. These strengths make ALPHA technology an excellent platform for high-throughput screening for small molecule modulators of KDM activity.

19.4 *IN VIVO* DETECTION OF KDM ACTIVITY IN CELL CULTURE MODELS

Determination of KDM activity and the validation of KDM targets in cells is essential following characterization *in vitro*. This is particularly important for KDMs, which may exhibit broader substrate activity *in vivo*, as exemplified by PHF8/KDM7 [50,51]. This altered *in vivo* demethylation may result from different associations on chromatin or interaction with additional proteins that can modulate activity to specific methylation states. In the case of KDM1A alone, no activity is observed toward nucleosomes, but when KDM1A overexpressed in cells or is purified with its associated proteins, H3K4me2 demethylation occurs [51]. These observations reiterate the importance of combining *in vitro* and *in vivo* analyses.

Characterization of *in vivo* demethylase activity has primarily involved three assays: immunofluorescence (IF), isolation of histones from cell lysates followed by immunoblotting, and chromatin immunoprecipitation (ChIP) with antibodies to specific methylation states. These techniques are usually coupled with cellular overexpression or depletion of the KDM of interest. Interpretation of these experimental results can be compounded by enzyme redundancy as well as by the effect of other KDMs or KMTs on the substrate being assayed. Thus, it is important to be careful when interpreting the cellular assays presented below and to combine them with the *in vitro* analyses described above.

19.4.1 ANALYSIS OF KDM ACTIVITY BY IMMUNOFLUORESCENCE

Immunofluorescent analysis (IF) of KDM activity typically involves modulating the expression levels of the KDM of interest and detecting the effect on cellular methylation states with specific antibodies (Figure 19.5F,G). Secondary antibodies are conjugated to a fluorescent label and visualized by microscopy. Immunofluorescence for methylation marks is typically conducted on fixed cells that have been grown on a coverslip or in a chambered microscope slide.

There are several important considerations in the design of IF experiments. First, the choice of cell line is very important. For example, the chromatin architecture in mouse cells is well defined and distinct patterns are often visible (e.g. heterochromatin foci), which is not necessarily the case for other organisms. The cell cycle phase is also an important variable, as several histone modifications are regulated across the cell cycle, including methylation of H4K20, H3K27, and H3K4 [52−54]. If this is the case, not all cells will demonstrate the same pattern of expression, which can make interpretation difficult. The choice of fixation conditions is also critical, as different fixation techniques will work for some antibodies but not others. This parameter needs to be determined empirically and optimized for each cell line and antibody selected. Similarly, permeabilization before, during, or after fixation may impact the activity observed or provide more information about a particular enzyme. Finally, the concentration of the methylation-state specific antibody is critical to ensure low background and high specificity. Two common approaches used in our laboratory and others are described below.

In addition to the fixation, permeabilization, and antibody conditions, the method of KDM overexpression or depletion is also important. For example, successful detection of globally decreased histone methylation following KDM expression requires high expression of the KDM for an optimized amount of time. High-level expression of certain KDMs can be cytotoxic, and thus requires completion of the assay before widespread cell death is observed. Historically, cell fixation and IF analysis has been conducted 24 hours following observation of high-level KDM expression. Because expression levels will vary between cells, different degrees of demethylation will be detected; but co-staining for an expression marker allows for the scoring of only high-expressing cells to better determine specificity.

In order to be confident that the effect being observed is specific for the enzyme being tested, catalytically inactive enzymes should also be overexpressed and scored. We recommend doing these experiments in a blinded fashion to improve the accuracy and decrease the subjectivity of the data. Finally, it is often easier to detect the loss of signal generated by KDM overexpression than to detect and measure gain of signal expected by KDM depletion.

19.4.1.1 Immunostaining protocol options
19.4.1.1.1 Method 1: Simultaneous permeabilization and fixation
This permeabilization method reduces cytoplasmic signal.

1. Cells are growing on coverslips in a six-well plate. Wash cells two times with $1 \times$ PBS, pH 7.4.
2. Add 2 mL Perm/Fix Buffer (0.5% Triton-X, 20 mM Hepes, pH 7.9, 50 mM NaCl, 3 mM KCl, 300 mM sucrose, 3% paraformaldehyde) per well. Leave dishes on the bench at room temperature.
3. Remove Perm/Fix Buffer after 30 minutes at room temperature and wash with Buffer A (0.5% NP40, 0.3% sodium azide in $1 \times$ PBS) three times for 10 minutes each.
4. After the washes, block with Buffer A + 10% FBS for at least 1 hour. Fixed cells can be left in block indefinitely as long as sodium azide was added to the block solution. Store blocked slides at 4°C.
5. Remove the block solution and add 35 µL primary antibody diluted in block solution for 3−4 hours at room temperature. Place a piece of parafilm cut to the size of the coverslips on top of the antibody to prevent evaporation. Because of the low volume of antibody, ensure that the slides are kept in a humid chamber (damp paper towels will suffice).
6. After incubation with the primary antibody, wash the coverslips three times with Buffer A for 10 minutes each. Increasing the length or number of washes can reduce the background fluorescence significantly.
7. After the washes, add the appropriate secondary antibody and co-stain nuclei with DAPI for 45−60 minutes. Follow this incubation with three 10-minute washes in Buffer A.
8. Rinse the coverslips once in 1X PBS, pH 7.4 before adding mounting media and affixing the coverslips to microscope slides. Coverslip edges can be sealed with nail polish.

19.4.1.1.2 Method 2: Standard fixation before permeabilization
This method allows observation of both cytoplasmic and nuclear signal.

1. Cells are growing on coverslips in a six-well dish. Wash cells two times $1 \times$ PBS, pH 7.4.
2. Add 2 mL PFA (paraformaldehyde) Fix Buffer (3.7% paraformaldehyde in $1 \times$ PBS, pH 7.4). Incubate on the bench for 30 minutes at room temperature.
3. Follow Method 1 from step 3 onward.

In order to visualize demethylation, the antibody recognizing the substrate of the enzyme needs to be titrated down so that the signal is within a range where decreased signal can be observed (Figure 19.5F,G). It is also important to note that for many of the demethylases, only cells expressing higher levels of the KDM actually have reduced signal for the modification being assayed.

19.4.2 ANALYSIS OF KDM ACTIVITY BY HISTONE PURIFICATION AND WESTERN BLOT

Analysis of changes in histone modifications following overexpression or depletion of KDMs should also be accompanied by Western blot analysis of acid extracted histones using methylation specific antibodies. This is important, as subtle differences could be missed in IF by cross reactivity or proximity of antibody epitopes.

For analyzing cellular histones by Western blot, it is crucial that all histones are equally extracted. This is not always achieved using standard, detergent-based lysis buffers, which often poorly extract histones enriched in heterochromatic regions. For this reason, an acid-extraction protocol should be used. These protocols typically isolate cell nuclei and precipitate acid-insoluble material using a low concentration of HCl ($0.2-0.4$ M) or H_2SO_4. The amino acid composition of histones allows them to remain soluble in the low acid concentration, whereas the majority of cellular proteins precipitate and are removed by centrifugation. Histones are then precipitated by the addition of trichloroacetic acid to at least 12.5% by volume. Wash the histones several times in ice-cold acetone and resuspend them in distilled water. Histones can be accurately quantified against a protein standard and loaded onto SDS-PAGE gels for Western blot analysis.

19.4.3 ANALYSIS OF KDM ACTIVITY BY CHROMATIN IMMUNOPRECIPITATION

Immunofluroescence and Western blotting techniques allow for analysis of global methylation levels. However, it is common for KDMs to function only in specific genomic contexts or at specific genes that may not be observed using more global methods. In these cases, ChIP is particularly useful for analyzing changes at a specific genomic locus.

Cells are cross-linked using formaldehyde to preserve histone−DNA interactions. ChIP immunoprecipitates modified histones and associated DNA using methylation-state specific antibodies. Following a number of washes to remove nonspecific interactions, the associated DNA is purified and assayed by quantitative polymerase chain reaction (qPCR) using primers specific for the regions of interest.

One major drawback to ChIP is that it requires prior knowledge of target genomic loci. It can also be difficult to prove causation between the enzyme of interest and the observed change in methylation. Additionally, different antibodies have different efficiencies, so the amount of immunoprecipitated DNA cannot be quantitatively compared between different methylation marks. It is also essential to assay the levels of the histone itself to ensure that changes in methylation do not reflect changes in the amount of histones present at the target region. ChIP does allow for temporal comparisons, resulting in establishment of correlations between KDM recruitment and the disappearance of particular methyl modifications *in vivo*. Finally, by coupling ChIP to next-generation sequencing technology, researchers are able to determine the majority of regions bound by KDMs, as well as regions where methyl marks change.

19.4.3.1 ChIP method

A. Preparation of chromatin:

1. Crosslink cells by adding 1% formaldehyde to the media for $10-15$ minutes at 37°C. Stop the fixation by adding Glycine to 0.125 M.
2. Wash the cells with cold 1 × PBS, pH 7.4. The cells are scraped and collected in a 15 mL tube and centrifuged for 2 minutes at 800 rpm, 4°C. The cellular pellet is resuspended in cellular

lysis buffer (5 mM PIPES, pH 8.0, 85 mM KCl, 0.5% NP40) with fresh protease inhibitors (1 mL for one 15 cm^2 dish). The cells are incubated 5 min on ice and then spun at 800 RPM for 2 minutes at 4°C. The supernatant is discarded and the pellet is resuspended into nuclear lysis buffer (50 mM Tris, pH 8.0, 10 mM EDTA, pH 8.0, 0.2% SDS) with fresh protease inhibitors (300 μL for one 15 cm^2 dishes). Samples are stored at −80°C prior to sonication.

B. Sonication of chromatin: Sonication can be performed with a probe, bath sonicator, such as Diagenode's Biorupter, or the cup horn sonicator system Q800R from QSonica.

1. Q800R cup horn sonicator: one 15 cm^2 dish equivalent is sonicated for 30 min at 70% in thin wall 0.5 mL PCR tubes. The sonication time can vary depending on the cell line used. This example is most compatible with RPE cells. The chromatin is cleared by centrifuging for 10 minutes at 14,000 rpm and 4°C.

2. Check the sonication at this point by reversing the crosslink, digesting associated protein and RNA, and running a 1% agarose gel of the chromatin. For most applications, the goal is to prepare a sample that is primarily the size of mononucleosomes (∼150 base pairs). If the majority of the sample does not run at this size, additional sonication is required.

C. Immunoprecipitation of chromatin:

1. The beads are prebound with the appropriate primary antibody. For example, incubate 900 μL dilution IP buffer (16.7 mM Tris, pH 8.0, 1.2 mM EDTA, pH 8.0, 167 mM NaCl, 0.01% SDS, 1.1% Triton-X100) with fresh protease inhibitors and 25 μL protein A or G magnetic beads (A for rabbit polyclonal antibodies, G for mouse monoclonal antibodies) and 2 μg of primary antibody in a 1.5 mL tube. The amount of antibody used can vary and needs to be determined empirically. The mix is incubated on a rotator for at least 6 hours at 4°C.

2. Add 100 μL of the sonicated chromatin (in nuclear lysis buffer) to the antibody/bead mix (usually 20−100 μg chromatin) and incubate on the rotator overnight at 4°C.

3. The next day, conduct the following washes: two washes with 1 mL dilution IP buffer, one wash with TSE buffer (20 mM Tris, pH 8.0, 2 mM EDTA, pH 8.0, 500 mM NaCl, 1% Triton-X100, 0.1% SDS), one wash with the LiCl buffer (100 mM Tris pH 8.0, 500 mM LiCl, 1% deoxycholic acid, 1% NP40), and two TE (10 mM Tris, pH 8.0, 1 mM EDTA, pH 8.0) washes. For each wash, vortex and incubate for a couple minutes at room temperature.

4. After the washes are complete, elute the DNA by adding 150 μL elution buffer (50 mM NaHCO$_3$, 140 mM NaCl, 1% SDS) and 2 μL RNase A (100 mg/mL). Incubate the elution for 30 minutes at 37°C with constant agitation. After 30 minutes, add 1 μL proteinase K (10 mg/mL) incubate at 55°C for 1 hour with constant agitation. The inputs need to be treated exactly the same from this step forward.

5. To reverse the crosslink, incubate for at least 4 hours at 65°C.

6. Purify the eluted DNA with a PCR clean up kit (such as the Promega Wizard system) and perform qPCR using primers specific to the regions of interest.

19.5 CONCLUSION

The focus of this chapter is to provide a historical context for the study of histone demethylases and to illustrate several methods for the purification and analysis of their enzymatic activity. We also highlighted alternative purification strategies and both traditional and more recent advances in assaying the enzymatic activity of histone demethylases. Finally, we have shared protocols that allow for the *in vivo* characterization of these enzymes. While this chapter documents current approaches in demethylase biology, the field is in need of fine-tuned, *in vivo* readouts specific to each family of enzymes. Improving assay technology will allow for increased understanding of the role of histone demethylases in specific biological contexts, such as chromosome and cell cycle changes, gene expression, and organism development [13,55,56]. Ultimately, the combined use of *in vitro* and *in vivo* approaches will provide the most complete picture of the function of histone demethylase enzymes.

ACKNOWLEDGMENTS

We would like to thank members of the Whetstine laboratory for critical reading and comments. We especially would like to thank Dr. Capucine Van Rechem for providing a detailed ChIP protocol. Dr. Johnathan R. Whetstine is supported by grants GM097360, CA059267 and the American Cancer Society Basic Research Scholar grant. Dr. Whetstine is also a Leukemia and Lymphoma Society Scholar and the Tepper Family MGH Scholar. Dr. Joshua C. Black is a fellow of the Jane Coffin Childs Memorial Fund.

REFERENCES

[1] Hnilicova J, Stanek D. Where splicing joins chromatin. Nucleus 2011;2(3):182−8.

[2] Kouzarides T. Chromatin modifications and their function. Cell 2007;128(4):693−705.

[3] Smith E, Shilatifard A. The chromatin signaling pathway: diverse mechanisms of recruitment of histone-modifying enzymes and varied biological outcomes. Mol Cell 2010;40(5):689−701.

[4] Black JC, Van Rechem C, Whetstine JR. Histone lysine methylation dynamics: establishment, regulation, and biological impact. Mol Cell 2012;48(4):491−507.

[5] Zhang X, Tamaru H, Khan SI, Horton JR, Keefe LJ, Selker EU, et al. Structure of the Neurospora SET domain protein DIM-5, a histone H3 lysine methyltransferase. Cell 2002;111(1):117−27.

[6] Jha DK, Strahl BD. An RNA polymerase II-coupled function for histone H3K36 methylation in checkpoint activation and DSB repair. Nat Commun 2014;5:3965.

[7] Li F, Mao G, Tong D, Huang J, Gu L, Yang W, et al. The histone mark H3K36me3 regulates human DNA mismatch repair through its interaction with MutSalpha. Cell 2013;153(3):590−600.

[8] MacAlpine DM, Almouzni G. Chromatin and DNA replication. Cold Spring Harb Perspect Biol 2013;5 (8):a010207.

[9] Mallette FA, Mattiroli F, Cui G, Young LC, Hendzel MJ, Mer G, et al. RNF8- and RNF168-dependent degradation of KDM4A/JMJD2A triggers 53BP1 recruitment to DNA damage sites. Embo J 2012;31(8):1865−78.

[10] Rivera C, Gurard-Levin ZA, Almouzni G, Loyola A. Histone lysine methylation and chromatin replication. Biochim Biophys Acta 2014;1839(12):1433−9.

[11] Young LC, McDonald DW, Hendzel MJ. Kdm4b histone demethylase is a DNA damage response protein and confers a survival advantage following gamma-irradiation. J Biol Chem 2013;288(29):21376−88.

[12] Black JC, Allen A, Van Rechem C, Forbes E, Longworth M, Tschop K, et al. Conserved antagonism between JMJD2A/KDM4A and HP1gamma during cell cycle progression. Mol Cell 2010;40(5):736−48.

[13] Black JC, Manning AL, Van Rechem C, Kim J, Ladd B, Cho J, et al. KDM4A lysine demethylase induces site-specific copy gain and rereplication of regions amplified in tumors. Cell 2013;154(3): 541−55.

[14] Cheedipudi S, Genolet O, Dobreva G. Epigenetic inheritance of cell fates during embryonic development. Front Genet 2014;5:19.

[15] Li X, Liu L, Yang S, Song N, Zhou X, Gao J, et al. Histone demethylase KDM5B is a key regulator of genome stability. Proc Natl Acad Sci USA 2014;111(19):7096−101.

[16] Berry WL, Janknecht R. KDM4/JMJD2 histone demethylases: epigenetic regulators in cancer cells. Cancer Res 2013;73(10):2936−42.

[17] Black JC, Whetstine JR. Tipping the lysine methylation balance in disease. Biopolymers 2013;99(2): 127−35.

[18] Cascante A, Klum S, Biswas M, Antolin-Fontes B, Barnabe-Heider F, Hermanson O. Gene-specific methylation control of H3K9 and H3K36 on neurotrophic BDNF versus astroglial GFAP genes by KDM4A/C regulates neural stem cell differentiation. J Mol Biol 2014;426(20):3467−77.

[19] Das PP, Shao Z, Beyaz S, Apostolou E, Pinello L, De Los Angeles A, et al. Distinct and combinatorial functions of Jmjd2b/Kdm4b and Jmjd2c/Kdm4c in mouse embryonic stem cell identity. Mol Cell 2014; 53(1):32−48.

[20] Greer EL, Shi Y. Histone methylation: a dynamic mark in health, disease and inheritance. Nat Rev Genet 2012;13(5):343−57.

[21] Roidl D, Hacker C. Histone methylation during neural development. Cell Tissue Res 2014;356(3): 539−52.

[22] Van Rechem C, Whetstine JR. Examining the impact of gene variants on histone lysine methylation. Biochim Biophys Acta 2014;1839(12):1463−76.

[23] Shi Y, Lan F, Matson C, Mulligan P, Whetstine JR, Cole PA, et al. Histone demethylation mediated by the nuclear amine oxidase homolog LSD1. Cell 2004;119(7):941−53.

[24] Shi Y, Sawada J, Sui G, Affar el B, Whetstine JR, Lan F, et al. Coordinated histone modifications mediated by a CtBP co-repressor complex. Nature 2003;422(6933):735−8.

[25] Bannister AJ, Schneider R, Kouzarides T. Histone methylation: dynamic or static? Cell 2002;109(7): 801−6.

[26] Trewick SC, McLaughlin PJ, Allshire RC. Methylation: lost in hydroxylation? EMBO Rep 2005;6(4): 315−20.

[27] Tsukada Y, Fang J, Erdjument-Bromage H, Warren ME, Borchers CH, Tempst P, et al. Histone demethylation by a family of JmjC domain-containing proteins. Nature 2006;439(7078):811−16.

[28] Yamane K, Toumazou C, Tsukada Y, Erdjument-Bromage H, Tempst P, Wong J, et al. JHDM2A, a JmjC-containing H3K9 demethylase, facilitates transcription activation by androgen receptor. Cell 2006;125 (3):483−95.

[29] Cloos PA, Christensen J, Agger K, Maiolica A, Rappsilber J, Antal T, et al. The putative oncogene GASC1 demethylates tri- and dimethylated lysine 9 on histone H3. Nature 2006;442(7100):307−11.

[30] Fodor BD, Kubicek S, Yonezawa M, O'Sullivan RJ, Sengupta R, Perez-Burgos L, et al. Jmjd2b antagonizes H3K9 trimethylation at pericentric heterochromatin in mammalian cells. Genes Dev 2006;20(12): 1557−62.

[31] Klose RJ, Yamane K, Bae Y, Zhang D, Erdjument-Bromage H, Tempst P, et al. The transcriptional repressor JHDM3A demethylates trimethyl histone H3 lysine 9 and lysine 36. Nature 2006;442(7100): 312−16.

[32] Whetstine JR, Nottke A, Lan F, Huarte M, Smolikov S, Chen Z, et al. Reversal of histone lysine trimethylation by the JMJD2 family of histone demethylases. Cell 2006;125(3):467−81.

[33] Shi Y, Whetstine JR. Dynamic regulation of histone lysine methylation by demethylases. Mol Cell 2007; 25(1):1−14.

[34] Trojer P, Zhang J, Yonezawa M, Schmidt A, Zheng H, Jenuwein T, et al. Dynamic histone H1 isotype 4 methylation and demethylation by histone lysine methyltransferase G9a/KMT1C and the Jumonji domain-containing JMJD2/KDM4 proteins. J Biol Chem 2009;284(13):8395−405.

[35] Chen Z, Zang J, Whetstine J, Hong X, Davrazou F, Kutateladze TG, et al. Structural insights into histone demethylation by JMJD2 family members. Cell 2006;125(4):691−702.

[36] Del Rizzo PA, Trievel RC. Molecular basis for substrate recognition by lysine methyltransferases and demethylases. Biochim Biophys Acta 2014.

[37] Lohse B, Kristensen JL, Kristensen LH, Agger K, Helin K, Gajhede M, et al. Inhibitors of histone demethylases. Bioorg Med Chem 2011;19(12):3625−36.

[38] Ng SS, Kavanagh KL, McDonough MA, Butler D, Pilka ES, Lienard BM, et al. Crystal structures of histone demethylase JMJD2A reveal basis for substrate specificity. Nature 2007;448(7149):87−91.

[39] Thinnes CC, England KS, Kawamura A, Chowdhury R, Schofield CJ, Hopkinson RJ. Targeting histone lysine demethylases − Progress, challenges, and the future. Biochim Biophys Acta 2014;1839(12): 1416−32.

[40] Krishnan S, Collazo E, Ortiz-Tello PA, Trievel RC. Purification and assay protocols for obtaining highly active Jumonji C demethylases. Anal Biochem 2012;420(1):48−53.

[41] Iwase S, Lan F, Bayliss P, de la Torre-Ubieta L, Huarte M, Qi HH, et al. The X-linked mental retardation gene SMCX/JARID1C defines a family of histone H3 lysine 4 demethylases. Cell 2007;128(6):1077−88.

[42] Life Technologies. 2012. Bac-N-Blue™ transfection and expression guide. Available at: <http://tools. lifetechnologies.com/content/sfs/manuals/bacnblue_man.pdf>.

[43] Shiraishi H, Kataoka M, Morita Y, Umemoto J. Interactions of hydroxyl radicals with tris (hydroxy-methyl) aminomethane and Good's buffers containing hydroxymethyl or hydroxyethyl residues produce formaldehyde. Free Radic Res Commun 1993;19(5):315−21.

[44] Lusser A, Kadonaga JT. Strategies for the reconstitution of chromatin. Nat Methods 2004;1(1):19−26.

[45] Arita A, Niu J, Qu Q, Zhao N, Ruan Y, Nadas A, et al. Global levels of histone modifications in peripheral blood mononuclear cells of subjects with exposure to nickel. Environ Health Perspect 2012;120(2): 198−203.

[46] Deal RB, Henikoff JG, Henikoff S. Genome-wide kinetics of nucleosome turnover determined by metabolic labeling of histones. Science 2010;328(5982):1161−4.

[47] Sakurai M, Rose NR, Schultz L, Quinn AM, Jadhav A, Ng SS, et al. A miniaturized screen for inhibitors of Jumonji histone demethylases. Mol Biosyst 2010;6(2):357−64.

[48] Quinn AM, Simeonov A. Methods for activity analysis of the proteins that regulate histone methylation. Curr Chem Genomics 2011;5(Suppl. 1):95−105.

[49] Kawamura A, Tumber A, Rose NR, King ON, Daniel M, Oppermann U, et al. Development of homogeneous luminescence assays for histone demethylase catalysis and binding. Anal Biochem 2010;404 (1):86−93.

[50] Kleine-Kohlbrecher D, Christensen J, Vandamme J, Abarrategui I, Bak M, Tommerup N, et al. A functional link between the histone demethylase PHF8 and the transcription factor ZNF711 in X-linked mental retardation. Mol Cell 2010;38(2):165−78.

[51] Liu W, Tanasa B, Tyurina OV, Zhou TY, Gassmann R, Liu WT, et al. PHF8 mediates histone H4 lysine 20 demethylation events involved in cell cycle progression. Nature 2010;466(7305):508−12.

[52] Aoto T, Saitoh N, Sakamoto Y, Watanabe S, Nakao M. Polycomb group protein-associated chromatin is reproduced in post-mitotic G1 phase and is required for S phase progression. J Biol Chem 2008;283(27): 18905−15.

[53] Beck DB, Oda H, Shen SS, Reinberg D. PR-Set7 and H4K20me1: at the crossroads of genome integrity, cell cycle, chromosome condensation, and transcription. Genes Dev 2012;26(4):325−37.

[54] Blobel GA, Kadauke S, Wang E, Lau AW, Zuber J, Chou MM, et al. A reconfigured pattern of MLL occupancy within mitotic chromatin promotes rapid transcriptional reactivation following mitotic exit. Mol Cell 2009;36(6):970−83.

[55] Eissenberg JC, Shilatifard A. Histone H3 lysine 4 (H3K4) methylation in development and differentiation. Dev Biol 2010;339(2):240−9.

[56] Wagner EJ, Carpenter PB. Understanding the language of Lys36 methylation at histone H3. Nat Rev Mol Cell Biol 2012;13(2):115−26.

ANIMAL MODEL STUDY OF EPIGENETIC INHIBITORS

20

Aili Chen[1,2,3] **and Gang Huang**[2,3]

[1]*Key Laboratory of Genomic and Precision Medicine, Beijing Institute of Genomics, Chinese Academy of Sciences, Beijing, China* [2]*Division of Pathology, Cincinnati Children's Hospital Medical Center, Cincinnati, Ohio, USA* [3]*Divisions of Experimental Hematology and Cancer Biology, Cincinnati Children's Hospital Medical Center, Cincinnati, Ohio, USA*

CHAPTER OUTLINE

Y.G. Zheng (Ed): *Epigenetic Technological Applications.* DOI: http://dx.doi.org/10.1016/B978-0-12-801080-8.00020-X

20.1 INTRODUCTION

Higher eukaryotic transcription is regulated by genetic/epigenetic regulators which mediate the fine tuning of genetic processes in a given context. Epigenetics is defined as heritable traits that are not linked to changes in the DNA sequence. To be more specific, epigenetics uses chromatin modification to regulate chromatin structure dynamically. Chromatin can be regulated by epigenetic factors that modify both DNA and histone which include DNA methylation, histone modifications, chromatin remodeling, and regulation of noncoding RNAs.

DNA methylation is executed by DNA methyltransferases (DNMTs), which transfer a methyl group from the universal methyl donor, S-adenosyl-L-methionine (SAM), to the 5-position of cytosine residues in DNA sequences, mainly in a CpG dinucleotide context [1]. Hypermethylation of DNA sequences often coincides with gene silencing. Four DNMT members are found in mammals: they are DNMT1, DNMT3A, DNMT3B, and DNMT3L. DNMT1 plays a key role in maintenance methylation while DNMT3A and DNMT3B are responsible for *de novo* methylation [2]. Unlike other DNMTs, DNMT3L has no inherent enzymatic activity [3]. DNA methylation has been shown to play an instrumental role in regulating embryonic development with regard to imprinting, X inactivation, and silencing pluripotent or tissue-specific gene expression [4]. DNA methylation is also crucial to the stability of chromatin in differentiated cells and helps to protect against mutations by insertion through the repression of transposons and repeated DNA elements [5].

Similar to DNA, histones can undergo different types of modifications which include methylation, acetylation, phosphorylation, ubiquitination, and SUMOylation [6]. Among them, methylation and acetylation have been well studied, and are both reversible dynamic processes [4]. The "histone code" was proposed by Jenuwein and Allis in 2001 and presents information that extends the limits of the genetic code. Their study suggested that the presence of histone modifications provides effector protein binding sites that change how the gene is translated [7]. These histone modifications are mediated through the processes of different types of enzymes. Histone methyltransferases (HMTs) and acetyltransferases are introduced by "writers." These can be removed by histone demethylases and deacetyelases, which are enzymes referred to as "erasers." "Readers" are enzymes or enzyme complexes that bind to modified proteins that are able to recognize histone modifications [8]. These enzymes collaborate to provide a transcriptional product, and mistakes in their activity have been related to cellular transformation and malignancies. The study of these chromatin regulators has become an important component of cancer pathology research [9].

In 1983, epigenetic changes were first detected in human disease by Feinberg and Vogelstein [10]. Currently, research has shown that epigenetic alterations share equal responsibility as genetic mutations for carcinogenesis [11]. Thus far, epigenetic aberrations have been well established in a diverse assortment of diseases such as cancer [12], diabetes [13], lupus [14], asthma [15] and a variety of neurological disorders [12,16,17]. With few exceptions, cancers possess hypomethylation in the global genome and hypermethylation in tumor suppressor genes or DNA repair genes [18,19].

Hypomethylation particularly occurs in gene bodies and intergenic regions (including repetitive elements), which leads to genomic instability. This results in stable gene silencing. In addition to changes in DNA methylation, histone modification changes are also common features in different diseases. The increasing use of a next-generation sequencing approach reveals a number of recurrent genetic mutations in the epigenetic "writers." Using myeloid malignancy as an example, mutations were found in the methylcytosine hydroxylase TET2, DNA methyltransferase DNMT3A, histone H3K27 methyltransferase EZH2, and Polycomb-related protein ASXL1, in patients with myeloproliferative neoplasms (MPNs), myelodysplastic syndromes (MDSs), and/or acute myeloid leukemia (AML) [20]. Mutations in IDH1 and IDH2 likely also affect the epigenome through neomorphic generation of 2-hydroxyglutarate, which inhibits TET2 and Jumonji-domain histone demethylases in blood diseases and brain tumors [21]. Moreover, epigenetic modification provides insight to clinicians, allowing them to provide individualized treatment due to the ability to discriminate between disease subtypes and varying severities that correlate to specified epigenetic changes [21,22]. For instance, malignant prostate tissue differs from benign tissue by the presence of dimethylated H3K9 [23].

The reversible property of these epigenetic modifiers makes them perfect candidates for identifying molecular targets to be used for drug production (Figure 20.1) [24]. Epigenetic inhibitors from preclinical study regularly enter phase I and II clinical trials. These preclinical trials generate data to indicate how effective the drug will be in phase I and II trials. Often *in vitro*, promising drugs proceed to be tested in animal models to provide data on dosing, toxicity, and other important aspects that allow for safe and efficient clinical trials (Figure 20.2). Because epigenetic inhibitors regularly find themselves surmounting obstacles that block many potential drugs from clinical trials, the drugs derived from epigenetic modifiers hold promising potential in the future treatment of disease [25].

In this chapter, we first discuss the latest epigenetic inhibitors according to their biological roles as well as their functions and importance in diseases. This is followed by a discussion of the importance of using animal models in preclinical studies and an examination of studies of some particular epigenetic inhibitors in mouse models.

20.2 EPIGENETIC INHIBITORS

Epigenetic therapy refers to altering epigenetic irregularities, usually through the use of epigenetic inhibitors. These therapies provide useful targets for treatment and prevention of disease because of

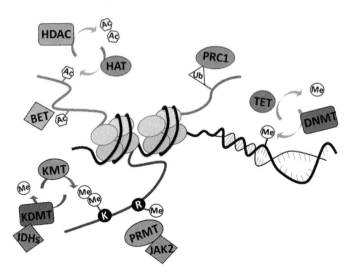

FIGURE 20.1 Schematic representation of the reversible property of epigenetic modifiers and their target sites.

The addition of a methyl group to convert the DNA base cytosine to 5-methylcytosine is catalyzed by DNA methyltransferase (DNMT). Hydroxymethylation of the DNA is catalyzed by TET family gene products, which are known to modulate hydroxymethylation by catalyzing the conversion of 5-methylcytosine to 5-hydroxymethylcytosine. Histone tail modifications include methylation, acetylation, phosphorylation, ADP-ribosylation, and ubiquitination. Of these modifications, methylation and acetylation have the most influence on chromatin structure. Histone acetyltransferases (HATs) catalyze acetylation of the histone tails, and histone deacetylases (HDACs) reverse acetylation. Histone methylation can involve mono-, di-, or trimethylation of arginine and lysine residues of one of the highly conserved histone units. Histone demethylation can remove mono-, di-, or trimethylation of arginine and lysine residues of one of the highly conserved histone units.

the reversibility of epigenetic modification [26–28]. There are limitations to epigenetic therapy, and thus corresponding specificity and off-target activation/inactivation must be well understood for a given disease [29]. The reversible potential for epigenetic inhibitors means the possibility of reverting to previous methylation patterns; therefore this must be addressed in preclinical trials [30]. Examples of drugs that have progressed to clinical trials are often associated with solid tumors and hematological malignancies and involve enzymes that are HATs, HDACs, DNMTs, and HMTs [27].

20.2.1 INHIBITORS OF DNMTS

DNMT inhibitors can be useful in cancer treatment in which hypermethylation of promoter regions occurs [31]. Generally, promoter regions are unmethylated in noncancerous cells. Virtually all human cancers contain hypermethylation of promoter regions, which represses transcription of the gene. These changes in tumor suppressor genes have been shown to be as common as genetic mutations in human cancer [31]. DNMT inhibitors provide more potential than cancer therapy alone; they are also being investigated as a means to reactivate the fetal hemoglobin gene in patients with

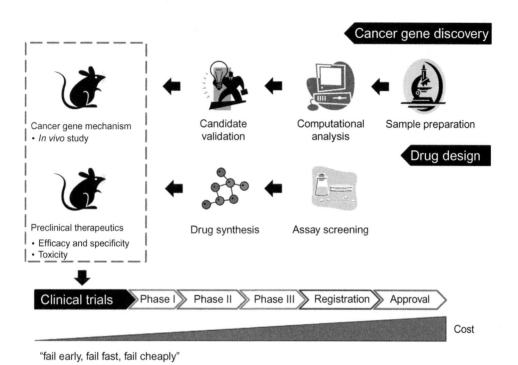

FIGURE 20.2 Preclinical uses of murine models for epigenetic inhibitors discovery.

Essential preclinical steps of targeted inhibitor design are shown; those clinical trials largely depend on studies in mouse models in the modern era. As lack of efficacy is a major reason for inhibitor attrition during compound development, better target validation is crucial to the process. Most importantly, target validation relies on the conditional ability to activate and inactivate the target of interest in a mouse model. This can be achieved through pharmacological inactivation of a target, which mimics genetic inactivation of that target. Numerous studies have indicated that poor pharmacokinetics and toxicity were causes of costly late-stage failure in drug development. The ability to detect issues with pharmacokinetics before the drug moves into clinical testing will ultimately save considerable resources in time and money for both academic researches and pharmaceutical/biotechnology companies, which is the "fail early, fail fast, fail cheaply" strategy.

thalassemia and sickle cell anemia [16]. There are many uses for DNMT inhibitors and they can be divided into three groups based on their structures, as discussed in the following subsections.

20.2.1.1 Nucleoside analog inhibitors

Nucleoside analog inhibitors are analogs of cytosine, the nucleotide base that is methylated by DNMTs. 5-azacytidine (5-aza-CR) [32] and decitabine (5-aza-2′-deoxycytidine or 5-aza-CdR) [33] are prototypes of these drugs, which were initially developed as cytotoxic drugs. Both of these drugs are S-phase specific drugs; they are phosphorylated to the deoxynucleotide triphosphate and then incorporated instead of cytosine into replicating DNA. However, both drugs are myelotoxic, which is thought to be due to their incorporation into DNA and not related to their DNA hypomethylation effects [34]. Zebularine is a more recent cytosine analog which was serendipitously discovered and is less toxic, but it is considerably less potent than 5-azacytidine and decitabine and needs to be administered at a higher dose [35,36]. Another inhibitor, 5-F-CdR, is already under clinical trials [37].

20.2.1.2 Non-nucleoside analog inhibitors

The search of non-nucleoside analogs is encouraged because of the myelotoxic effects of the nucleoside analogs (molecules that act like nucleosides in DNA synthesis) [18]. The widely used vasodilator and antiarrhythmic agents, hydralazine [38], and procainamide [39], have been demonstrated to be non-nucleoside analog DNMT inhibitors. These drugs exhibit less cytotoxicity because they do not require incorporation into DNA [18]. Other such inhibitors including RG108 [40], SGI-1027 [41], and EGCG [42] have been found and identified. Some of these types of drugs are currently undergoing preclinical trial [43]. Moreover, hydralazine is currently being tested in phase III of brain or ovary tumors [44].

20.2.1.3 Antisense oligonucleotides

Another way to inhibit DNMTs is to use antisense oligonucleotides and micro RNA. Their sequences are complementary to mRNAs for human DNMT1 and hybridize with them to make them inactive, thereby repressing translation. MG98 [45] and miR29b [46] are examples of this type of inhibitor, currently undergoing preclinical as well as clinical trials.

Table 20.1 lists examples of DNMT inhibitors in clinical development.

Table 20.1 DNMT Inhibitors in Clinical Development				
Drug Classification	**Compound Name**	**Indications**	**Developmental Phase**	**Reference**
Nucleoside analog	5-Azacytidine	MDS, solid tumors, leukemia	FDA approved for MDS (2005)	[32]
	Decitabine	MDS, CML	FDA approved (2006)	[33]
	Zebularine	Urinary bladder cancer	Preclinical	[36]
	5-F-CdR	Oncology	Phase I	[37]
Non-nucleoside analog	Procaine	Breast cancer, oncology	Preclinical	[18]
	Procainamide	Prostate cancer, oncology	Preclinical	[39]
	Hydralazine	Oncology	Phase III	[38,44]
	RG108	Oncology	Preclinical	[40]
	SGI-1027	Oncology	Preclinical	[41]
	EGGG	Oncology	Phase II	[42]
Antisense oligonucleotides	MG98	Oncology	Phase II	[45]
	miR29b	Oncology	Preclinical	[46]

20.2.2 INHIBITORS OF HDACs

Histone deacetylase (HDAC) inhibitors have great potential against human diseases; therefore many studies are currently working on identifying or synthesizing HDAC inhibitors. These inhibitors have different biochemical and biologic properties, from simple compounds such as valproate, to more elaborate designs, such as MS-275 [47]. Overall, HDAC inhibitors manifest a wide range of activity against all HDACs, with only a few exceptions [48]. These compounds can be classified according to their chemical nature and mechanism of inhibition, as follows.

20.2.2.1 Hydroxamates

Most of the available HDAC inhibitors belong to this group. Most of the chemicals in this group are very potent, but also reversible. The general structure of these substances consists of a hydrophobic linker that allows the hydroxamic acid moiety to chelate the cation at the bottom of the HDAC catalytic pocket, while the heavier part of the molecule acts as a cap for the tube [47]. (hydroxamic acid is a class of organic compounds bearing the functional group $RC(O)N(OH)R'$, with R and R' as organic residues and CO as a carbonyl group. They are amides ($RC(O)NHR'$) in which the NH center has been replaced by an OH.) TSA was one of the first HDAC inhibitors to be depicted [49] and is widely used as a reference in HDAC hydroximic acid research. Inspiring by TSA structure, SAHA (suberoylanilide hydroxamic acid) [50], Scriptaid [51], and the latest drugs NVP-LAQ-824 [52,53] and PXD-101 [54], are designed as HDAC inhibitors and have been tested *in vitro*, *in vivo*, and subsequently in clinical trials with acceptable degrees of success [55].

20.2.2.2 Carboxylic acid/short chain fatty acid inhibitors

This class of inhibitors is less potent but has pleiotropic effects. Valproic acid [56,57] and 4-phenylbutanoic [58] have already been approved for use in treating epilepsy and some cancers. Butanoic acid [59] is currently undergoing clinical trials [60].

20.2.2.3 Benzamides

MS-275 along with mocetinostat (MGCD0103) belongs to this group; they can both inhibit HDAC with an unknown mechanism. Preclinical studies showed that mocetinostat has a potent antiproliferative activity against a wide range of cancers [61]. MS-275 and mocetinostat are currently both undergoing clinical trials [62,63].

20.2.2.4 Epoxide moiety HDAC inhibitors

The HDAC inhibitors in this class of compounds are a number of natural products such as depeudecin, trapoxin A, etc. These chemicals are supposed to trap HDACs through the reaction of the epoxide moiety with the zinc cation or an amino acid (forming a covalent attachment) in the binding pocket. However, the reaction of the epoxide functionality prevents significant *in vivo* activity, which makes them of little pharmacologic interest [47].

20.2.2.5 Other HDAC inhibitors

There are also some natural HDAC inhibitors which do not belong to any of the above groups: for example, the fungal metabolite, FK228 and Apicidin A/B/C. Apicidin A is a cyclic tetrapeptide bearing an alkylketone residue. Apicidins B and C differ from Apicidin A by a single residue.

The epoxide moiety trapoxin has a similar structure to apicidins, the main difference between the two groups being that the former bear an epoxyketone functionality rather than an alkylketone functionality, which makes the compound much less stable under physiologic conditions [47,64,65].

20.2.3 INHIBITOR OF HATs

The identification of histone acetyltransferase (HAT) inhibitors has proved to be more challenging than others [66]. Researchers have identified a series of HAT inhibitors based on bisubstrate analogs or through screens [67−76]. One of the most potent and selective compounds, Lys-CoA, has been converted to a cell-permeable form with Tat peptide attachment and has been used in a variety of studies, but its complexity is somewhat limiting for pharmacologic applications [68−70]. High-throughput screening experiments have led to several small molecule synthetic agents and natural product derivatives of moderate potency as p300 HAT inhibitors micromolar Ki values, but their selectivity and mechanism of inhibition remain to be fully understood [71−76]. In 2010, Bowers et al. used a structure-based, *in silico* screening approach to identify a commercially available pyrazolone-containing small molecule p300 HAT inhibitor, C646 [77]. C646 is a competitive p300 inhibitor with a Ki of 400 nM and is very selective versus other acetyltransferases [77].

Table 20.2 lists HDAC and HAT inhibitors that are currently in clinical development.

Table 20.2 HDAC/HAT Inhibitors in Clinical Development

Drug Classification	Compound Name	Indications	Developmental Phase	Reference
HDAC inhibitors				
Hydroxamates	TSA	Oncology, inflammation	Preclinical	[49]
	SAHA	CTCL	FAD approved (2006)	[50]
	Scriptaid	Oncology	Preclinical	[51]
	LAQ824	Oncology	Phase II	[52,53]
	PDX-101	Oncology	Phase II	[54]
Carboxylic acid/short chain fatty acid	Valproate	Oncology	Phase III	[56,57]
	4-Phenylbutanoic	Oncology	Preclinical	[58]
	Butyrate	MDS, leukemia	Phase I	[59,60]
Benzamides	MS275	Oncology	Phase II	[62]
	MGCD0103	Oncology	Phase II	[63]
Epoxide moiety	Trapoxin A	Oncology	Preclinical	[47]
others	FK228	Oncology	Preclinical	[47]
	Apicidin A	Oncology	Preclinical	[64,65]
HAT inhibitors				
	Lys-CoA	Oncology	Preclinical	[68−70]
	C646	Oncology	Preclinical	[77]

20.2.4 INHIBITOR OF HISTONE LYSINE METHYLTRANSFERASES

Histone lysine methylation plays an important role in the chromatin organization and the regulation of gene expression [78]. Histone methyltransferases (HMTs) have more than 50 different SET-domains and differ in their specificity for a target lysine residue [79]. DOT1L is the only non SET-domain HMT currently known. Many HMTs such as SUV391H, EZH2, MLL, NSD1, RIZ, and others are implicated in tumor development. Therefore, in the last 10 years many epigenetics research has focused on histone lysine methyltransferase (HKMT) inhibitors (HMTi). As there are many different methyltransferases which respond for different histone modifications, we will introduce them based on their targets as opposed to their chemical structures.

20.2.4.1 G9a/GLP: Histone H3K9me1/2 methyltransferases

The G9a/GLP complex is a HMT that catalyzes mono- and di- methylation of H3K9, resulting in transcriptional repression [80]. This complex plays multiple biological roles in cells and its overexpression is observed in many different cancers. Therefore, it is highly desirable to develop G9a/GLP-specific inhibitors.

BIX-01294 is the first G9a/GLP-specific inhibitor, which was identified by screening a library of 125,000 preselected compounds [81]. On the basis of the GLP and BIX-01294 complex structure, highly potent compounds UNC0224 [82] and UNC0321 [83] have been designed and synthesized. A potent and less toxic compound E72 was also reported by another group almost at the same time [84]. Nevertheless, both UNC0321 and E72 are less potent than BIX-01294 in cellular activity [85], which was toxic at high concentrations. Therefore, UNC0638 [86] and its biotinylated version UNC0965 [87] was identified to improve the situation. More G9a/GLP inhibitors with high potency and less toxicity have been designed or discovered by screening [88,89].

20.2.4.2 Suv39H1/2: Histone H3K9me3 methyltransferases

The function of Su(var)3-9 is unclear until human SUV39H1 and murine Suv39h1 are found as histone H3K9 methyltransferases with SET domain (named after Suv39, E(z) and Trithorax) [90]. In 2005, fungal metabolite Chaetocin was identified as the first Su(var)3-9 histone H3K9 methyltransferase inhibitor by screening natural products [91]. The same screen also revealed a pair of synthetic compounds BIX-01338 and BIX-01337, these two compounds acted broadly on all enzymes tested, including SUV39H1, G9a, and PRMT1 [81].

20.2.4.3 DOT1L: A histone H3K79 methyltransferase

DOT1L is the only known enzyme responsible for H3K79 methylation, and carries out mono-, di-, and trimethylation in a nonconsecutive manner [92]. Human DOT1L contains a non-SET domain catalytic domain, which is the only non-SET HMT to date. Histone H3K79 methylation is generally associated with transcriptional activation. Most importantly, DOT1L was shown to interact with a subset of MLL translocation fusion proteins. AML includes translocations involving mixed lineage leukemia 1 (MLL1) occurring in at least 10% of adult AML patients and >70% of infant acute leukemias. So far, 104 different MLL rearrangements were found in adult and pediatric acute leukemia patients [93]. The four most frequent MLL translocations (MLL-AF4, MLL-AF9, MLL-AF10, and MLL-ENL) result in recruitment of DOT1L to the fusion protein and acquisition of histone 3 lysine

79 (H3K79) methyltransferase activity. This has led to the concept of developing targeted therapy for DOT1L inhibition in the therapy of MLL-translocated leukemias [20].

In 2011, Epizyme Inc. reported a compound, EPZ004777, that could specifically inhibit H3K79 methylation, and selectively killed leukemic cells bearing the MLL translocation genes while non-MLL-translocated cells survived [94]. Since then, over 20 DOT1L inhibitors have been reported by different groups [95−100]. The optimized version of EPZ004777, EPZ-5676, is now in phase I clinical trials for advanced hematologic malignancies, including acute leukemia with rearrangement of the MLL gene, which is a first-in-human study [88].

20.2.4.4 EZH2 and the PRC2 complex: A histone H3K27 methyltransferase

Polycomb group genes (PcGs) are epigenetic effectors, essential for stem cell self-renewal and pluripotency. Polycomb proteins are organized in two main polycomb repressive complexes (PRCs), PRC1 and PRC2 [101]. PRCs are involved in gene silencing through histone posttranslational modifications. PcGs have also been the focus of investigation in cancer research. Many cancer types show an overexpression of PcGs, predicting poor prognosis, metastasis, and chemoresistance.

PRC2 exhibits HMT activity on H3K27 via its catalytic subunit E(z) or mammalian homologs EZH1/EZH2 (Enhancer of zeste homolog 1 or 2) [102]. EZH2 is linked to various human cancers through distinct mechanisms, including overexpression, and both activating and inactivating mutations. In 2012, EPZ005687, GSK126, and El1were reported by three groups, respectively; they are all small-molecule inhibitors for EZH2, which could diminish histone H3K37 methylation and inhibit diffuse large B-cell lymphoma with EZH2-activating mutations [103−105]. An improved version of EZH2 inhibitors UNC1999 was then developed by Dr. Jin's group [106].

20.2.5 INHIBITOR OF KDMTs

Previously, methylation was considered to constitute a permanent and irreversible histone modification that defined epigenetic programs in concert with DNA methylation. However, the discovery of lysine-specific demethylase 1 (LSD1, also known as KDM1A, AOF2, BHC110, and KIAA0601), LSD2 and later the JmjC-domain-containing lysine demethylase family has completely changed this view [107,108].

20.2.5.1 LSD1: Mainly H3K4 demethylase

First-generation mechanism-based inhibitors of the flavin-dependent lysine-specific demethylases LSD1 and LSD2, such as tranylcypromine (TCP), lacked potency, and selectivity over their historical targets, the monoamine oxidases [109]. Now, the general monooxidase inhibitor TCP is more commonly used as an LSD1 inhibitor [110]. Recently, Oryzon Genomics has reported on the further development of a clinical compound, ORY-1001, which is more than 1,000 times more potent than TCP and highly selective compared to related enzymes, including LSD2 [111].

20.2.5.2 The JmjC-domain family

All current inhibitors of Jumonji domain-containing lysine demethylases compete with the cofactor 2-oxoglutarate and bind to the catalytic iron in the active site. The highly polarized compound 2,4-pyridine-dicarboxylate inhibits the Jumonji domain-containing demethylases as well as other 2-oxoglutarate-dependent oxygenases, so the inhibitors are not very specific. Two more

Table 20.3 HMT/KDMT Inhibitors
(All Are Under Preclinical Phase)

Drug Classification	Compound Name	IC_{50}(uM)/Targets	Reference
HMT inhibitors			
G9a/GLP inhibitors	BIX-01294	1.7/38	[81]
	UNC0224	0.015/0.05	[82]
	UNC0321	0.009/0.015	[83]
	E72	−/0.1	[84]
	UNC0638	0.015/0.019	[86]
	UNC0965	<0.0025/−	[87]
Suv39H1/2 inhibitors	chaetocin		[91]
	BIX-01338	4.7/−	[81]
DOT1L inhibitors	EPZ004777	0.0004	[94]
	EPZ-5676	8.00E − 05	[88]
EZH2 inhibitors	EPZ005687	0.054/1	[103]
	GSK126	0.0005/0.089	[105]
	Ell	0.0094/1.34	[104]
	UNC1999	0.0046/0.045	[106]
KDMT inhibitors			
LSD1 inhibitors	TCP	<0.002	[110]
	ORY-1001	<0.02/0.1	[111]
JmjC inhibitors	SID 85736331	0.2	[112]
	compound 15c	0.11	[113]
	GSK-J1	0.06/−	[114]

recent series of Jumonji domain-containing demethylase inhibitors include 8-hydroxyquinolines (for example, SID 85736331) and 2,2′-bipyridines (e.g. compound 15c), which are potent inhibitors with subtype selectivity and more drug-like properties [112,113]. Inhibitors that are competitive with 2-oxoglutarate and noncompetitive with a peptide substrate, shown to be specific for the JMJD3 (also known as KDM6B) and UTX H3K27 demethylases, have also been designed recently [114].

Table 20.3 contains a list of HMT/KDMT inhibitors.

20.2.6 INHIBITOR OF PRMTs

Arginine methylation is a very abundant covalent posttranslational modification which regulates diverse cellular processes, including transcriptional regulation, RNA processing, signal transduction, and DNA repair [115]. At least nine protein arginine methyltransferases (PRMTs) have been identified in the human genome, which can be grouped into three classes. Protein arginine methyltransferases can methylate a variety of targets, including histones, SM proteins, and transcription factors [115]. Overexpression, aberrant splicing, or mutations of these different PRMTs have been implicated in various types of cancer and have been recently reviewed [116]. Here we will focus on advances in developing PRMT inhibitors as chemical probes and therapeutic reagents.

The first series of arginine methyltransferase inhibitors were reported by the group of Dr. Bedford from the M.D. Anderson Cancer Center in 2004. One of these inhibitors, AMI-1 (arginine methyltransferase inhibitor-1), was shown to be non-SAM competitive and specifically inhibit arginine methylation, not lysine methylation. Since then, many groups have devoted a great deal of effort to developing arginine methyltransferase inhibitors, which can be classified as AM1-1-like [117,118], substrate-competitive inhibitors [119], bisubstrate inhibitors [120], substrate-targeting inhibitors (inhibitors targeting substrates directly, such as histone H3 or H4 other than PRMTs) [121] and allosteric inhibitors of PRMT3 [122,123]. Most of these inhibitors have modest inhibition activity and poor cell permeability with no selectivity among the PRMT members. It is worth noting that the indole and pyrazole inhibitors of CARM1, which occupy the substrate-binding groove, have relatively decent selectivity over PRMT1/3 (over 300 folds) [119]. Allosteric inhibitors are another set of inhibitors for PRMTs [122,123].

20.2.7 INHIBITOR OF PRC1

We mentioned PRCs and its ability to catalyze histone H3 Lys 27 trimethylation with EZH2 of PRC2. However, the other polycomb repressive complex, PRC1, is a histone H2A/H2AX E3 ubiquitin ligase which completes gene silencing through the histone H2A ubiquitylation. BMI1/RING1 are catalytic subunits of PRC1 [124]. PRC1 has a well-established role in the regulation of differentiation and the maintenance of stem cell populations [125].

Recently, 2-pyridine-3-yl-methylene-indan-1, 3-dione (PRT4165), an inhibitor of BMI1/RING1-mediated polyubiquitylation of topoisomerase II was identified [126]. PRT4165 is a potent inhibitor of PRC1-mediated H2A ubiquitylation *in vivo* and *in vitro*. The drug also inhibits the accumulation of all detectable ubiquitin at sites of DNA double-strand breaks (DSBs), the retention of several DNA damage response proteins in foci that form around DSBs, and the repair of the DSBs [127].

20.2.8 INHIBITOR OF JAK2, IDH1/2

JAK2 is a nonreceptor tyrosine kinase that regulates several cellular processes by inducing cytoplasmic signaling cascades. JAK2 signaling is implicated in various biological processes, including cell cycle progression, apoptosis, mitotic recombination, genetic instability, and alteration of heterochromatin [128]. The diverse roles of JAK2 in normal and leukemic hematopoiesis are believed to be mediated by cytoplasmic signaling pathways [129]. Nevertheless, JAK2 is present in the nucleus of hematopoietic cells and directly phosphorylates Tyr 41 (Y41) on histone H3, stimulating growth. Both JAK2 inhibitors, TG101209 and AT9283, can inhibit the phosphorylation of H3Y41 *in vitro* and *in vivo* [130].

The type II arginine methyltransferase PRMT5 was first identified as JAK-binding protein 1 (JBP1) in a yeast two-hybrid assay [131]. It mediates the symmetrical dimethylation of arginine residues within histones H2A, H3, and H4, and methylates other cellular proteins as well. The most common somatic alteration of JAK2 is a gain-of-function mutation (JAK2 V617F) associated with human myelo-proliferative diseases [129]. A recent study has determined that oncogenic mutations within the JAK2 tyrosine kinase (V617F and K539L) enhance its interaction with PRMT5, leading to PRMT5 phosphorylation *in vivo* [132]. Although both the wild-type and mutant forms of JAK2 proteins interact with PRMT5, phosphorylation of PRMT5 is a "gain-of-function" of the mutant JAK2 kinases, which reduces PRMT5 methyltransferase activity and decreases global histone H2A/H4 R3 methylation [132]. Thus, JAK2 inhibitor may have broad epigenetic activity.

Recurrent mutations in isocitrate dehydrogenase 1 (IDH1) and IDH2 have been identified in gliomas, acute myeloid leukaemias (AML) and chondrosarcomas, and share a novel enzymatic property of producing 2-hydroxyglutarate (2HG) from a-ketoglutarate [133]. 2HG-producing IDH mutants can prevent the histone demethylation that is required for lineage-specific progenitor cells to differentiate into terminally differentiated cells [134]. IDH1 mutation could increase H3K9me3 and H3K27me3 in tumors. This suggests IDH mutations might preferentially affect the regulation of repressive histone methylation marks *in vivo* [134]. Biochemical studies propose 2HG is a universal inhibitor of JHDM family members [135]. Future investigation of the sensitivity to 2HG inhibition among JHDM family members and/or cellular feedback mechanisms activated after defective histone demethylation will be needed.

20.2.9 INHIBITOR OF THE BROMODOMAIN-CONTAINING FAMILY

The bromodomain-containing family of proteins represents an important class of histone modification reader proteins that recognize acetylated lysine residues [136]. Bromodomains comprise a small family of proteins that recognize and bind to acetylated lysine residues on histone tails. Bromodomain-containing proteins are considered "readers" of the histone code. The current interest in targeting various bromodomains originated in the demonstration by GlaxoSmithKline (GSK), the SGC and the Dr. Bradner lab that the bromodomain and extra-terminal (BET) subfamily (Brd2, Brd3, Brd4, and BrdT) could be targeted by small molecule antagonists [137]. By directly binding to the BET proteins, such compounds prevent the interaction of the reader module to the acetylated histone thereby preventing assembly of an active gene transcriptional complex. The compounds JQ1 and IBET represent novel chemical templates that are distinct from the previously reported simple acetyl-containing templates, and they have a clear mode of action.

20.2.10 INHIBITOR OF SPLICOSOME

The spliceosome is a ribonucleoprotein complex involved in RNA splicing — that is, the removal of noncoding introns from precursor messenger RNA [138]. In general, most genes give rise to multiple spliced transcripts by alternative splicing. Recent research has revealed that alternatively spliced products can be linked to various (genetic) diseases and also may play a role in cancer development [139]. As alternative splicing affects most genes, it is likely that cell cycle control, signal transduction, angiogenesis, motility, and metastasis, apoptosis, and other processes that are often impaired in cancer will be affected [140].

Recently, two chemically different microbial products with profound cancer cell inhibiting potential were found to target the spliceosome. Both compounds, the pladienolide derivatives [141] and spliceostatin A [142], appear to bind to SF3b, a subcomplex of U2 snRNP, which is an essential component of the spliceosome. Clearly, binding of pladienolide and spliceostatin A to SF3b components interferes with the splicing process as well as surveillance mechanisms that are operational in the cell to prevent the synthesis of truncated proteins [138].

20.3 **USE OF ANIMAL MODEL TO STUDY THE INHIBITORS**

We briefly reviewed the most current epigenetic inhibitors according to their biological roles as well as their functions. As mentioned, not all substances found *in vitro* become suitable candidates for further *in vivo* study or clinical trials. Dating back to the 1990s, studies have indicated that poor pharmacokinetics and toxicity caused expensive failures in the late stages of drug development, accounting for 20—40% of all drug development. Being able to detect these pharmacokinetics and toxicity early on saves both time and money. This is the "fail early, fail cheaply strategy" (Figure 20.2). Using animal models has proved to be an instrumental practice in efficiently determining which drugs hold promise in clinical study.

There are a number of model organisms which have been used as animal models for *in vivo* drug tests, such as *Drosophila*, zebra fish, mouse, rat, hamster, and monkey. Each animal has its own advantages in different research disciplines. For example, *Drosophila*, the common laboratory fruit fly, has well-differentiated organ systems [143]. They are divided as males and females with functional gonads and complex reproductive behavior and physiology; their hearts and blood circulation typically weaken with age, as is the case with humans. They have distinguishable and integrated nervous systems; their lifespans are short enough to allow complete lifespan assays for side effects on both mortality and fertility; and their complex behavioral patterns are able to allow us to readily detect neurological impairment. There is still debate about the importance of murine models, which are the most experimentally tractable mammalian systems in advancing our basic understanding of cancer biology. The use of transgenic and knockout technology has enabled several of the most important basic scientific discoveries in cancer research during the past two decades [144].

Most researchers use mouse models to test drugs *in vivo*, which includes xenograft and genetically engineered mouse models (GEMMs) (Figure 20.3). Xenograft mouse models are created by injecting homogeneous human immortalized or tumor cell lines into immunodeficient (e.g. severe combined immunodeficiency) mice [144]. This technology became available in the 1980s as a means to inactivate genes in the germline through homologous recombination of embryonic stem cells (ESCs). Since advent of this technology, well-defined murine models have been produced with increasing sophistication, allowing researchers to augment their understanding of most human cancers.

In xenograft analysis, limited collections of human tumor cell lines are injected into immunodeficient (severe combined immunodeficient (SCID) or nude) mice before being treated with the compound of interest. This approach has extensively been used in academia and pharmaceutical companies because of its ease and cost-effectiveness, to the point that nearly every modern cancer therapy has undergone xenograft testing. Xenograft models are not without their imperfections. First, host mice have profound defects in their immune response which precludes the testing of immunomodulatory agents in the system. Second, SCID mice often present defects in DNA repair mechanisms which could limit cytotoxic testing. Third, and perhaps most significant, is the fact that these systems model cancer as if it was a disease of homogeneous rogue cells, as opposed to a complex organ system with distinct neoplastic and host components acting in concert to maintain the tumor. Thus, the reductionist nature of the xenograft model fails to involve the proper microenvironment. Finally, xenograft studies typically use only a few human tumor cell lines, which cannot mimic the whole oncogenomic profiles in disease [144]. Naturally, these models have their critics in opposition, and there is a low success rate of novel therapeutics in clinical study that showed promise in xenograft models.

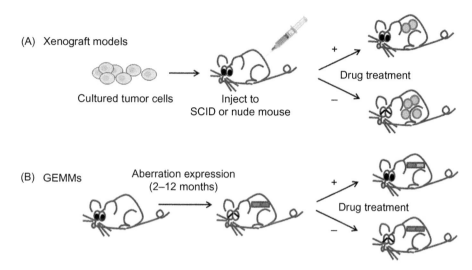

FIGURE 20.3 Xenograft versus GEMM testing for epigenetic inhibitors.

(A) In xenograft studies, cultured human tumor cells are generally injected subcutaneously into immunodeficient mice, which are then treated with the epigenetic inhibitors of interest for a time period during which subcutaneous xenograft tumors develop. The study is "positive" if the epigenetic inhibitors of interest reduce the rate of growth of new tumors. (B) In modern genetically engineered mouse models (GEMMs), oncogenes are activated and/or tumor-suppressor genes (TSGs) are inactivated somatically, generally through temporally controlled and tissue-specific expression of CRE recombinase. Animals then develop tumors in the tissue of interest. Tumor-bearing genetically engineered mice are then treated with the epigenetic inhibitors of interest and serially assessed for their responses.

The testing of novel compounds in these models will function best when the molecular genetics of the human tumor being targeted are well-understood and accurately engineered in these mouse models. In addition to showing reasonable penetrance and latency, the best murine models for drug testing will allow for ready determination of tumor progression or response. Therefore, murine models used for drug testing should be faithful to the neoplastic event being targeted, and the models should be as well-defined in molecular terms as possible. Of course, the ideal situation would be that the murine tumors being used for testing occur in the same compartment with the same molecular genetics as the human versions. Finally, optimal mouse models for new drug development will have to be simple to use [144].

The application of combinatorial genetic engineering strategies to generate tumor-prone genetic strains has been reviewed elsewhere [145]. In brief, it is possible to direct expression of a gene of interest throughout a tissue (e.g. through cell-specific transgene expression) or the entire organism (e.g. targeted germline mutations). New alleles can be produced through the use of transgenic technology, in which extra DNA that encodes the gene of interest is inserted into the mouse genome. In addition, researchers can target a portion of the mouse genome using knockin/knockout technology. Finally, using tet-regulated and CRE-inducible alleles, the timing, duration, and tissue compartment of gene expression can be further controlled. These technologies can be combined to yield faithful

GEMMs of specific cancers that overexpress or lack genes of interest in all cells or only in a specific tissue compartment and/or developmental stage of interest. Sophisticated models can be generated using these advances, such that they involve multiple alleles and conditional activation of gene expression, while still maintaining a simple breeding scheme. Thus, GEMMS have several important applications, allowing target validation, testing compounds by serial assessment of tumor response, pharmacodynamic markers of drug action, modelling resistance, and understanding toxicity.

20.3.1 USE OF ANIMAL MODELS TO STUDY DNMTs (DAC AND 5-AZA)

5-Azacytidine and 5-aza-2′-deoxycytidine (decitabine) were first synthesized in 1964 [146]. 5-Azacytidine was originally developed and tested as a nucleoside antimetabolite with clinical specificity for acute myelogenous leukemia [147]. The finding showed that as 5-azacytidine was incorporated into DNA, it inhibited DNA methylation which led to the widespread use of 5-azacytidine and decitabine to demonstrate the correlation between loss of methylation in specific gene regions and activation of the associated genes. More recently a renewed interest has emerged in the use of decitabine as a therapeutic agent for cancers in which epigenetic silencing of critical regulatory genes has occurred [148].

In 1968, researchers began studying the effect of decitabine against leukemic and hemopoietic tissues in AKR mice [149]. During that time, a number of studies used either xenograft mouse model or an AKR mouse strain to investigate the effect of decitabine as an antimetabolite [150−152]. (The AKR mouse strain is best known for its high incidence of lymphatic leukaemia, and for the Thy1a T-cell antigen, which is only present in this and a few other strains.) They found that decitabine administered daily at 1.5 mg/kg increased the life-span of P388 leukemia-beaming BALB/c × DBA/2 F1 mice by five times and that of second-generation lymphoma-beaning AKR mice by 2.5 times. Higher doses (total dose, 20 mg/kg) led to favorable results when administered in two portions on Days 4 and 5 after the s.c. inoculation of leukemic cell, while administering this dose consecutively for five days proved toxic. The drug was also effective in L1210 leukemia. Decitabine inhibited the phosphorylation of 2′-deoxycytidine in the acid soluble pool of cells from leukemic AKR mice as well as its incorporation into DNA [153].

The role of decitabine as a DNMT inhibitor was not established until after 1980. Meanwhile, the relationship between antineoplastic activity of decitabine in mice with L1210 leukemia and inhibition of DNA methylation was investigated. BALB/c X DBA/2 F1 mice with L1210 leukemia were given a 15-hr intravenous (IV) infusion of decitabine at a total dose ranging from 0.5 mg/kg (weak antineoplastic effect) to 22 mg/kg (very potent antineoplastic effect). The DNA of L1210 leukemia cells were isolated from decitabine treated mice and tested for its ability to accept methyl, which was found to be dependent on the dose of decitabine. At the start of the decitabine infusion, mice were given intraperitoneal (IP) injections of [6-3H] uridine, and the DNA of the L1210 leukemia cells were isolated at the end of therapy. Analysis of labeled pyrimidine bases showed that decitabine produced a dose-dependent reduction in the 5-methylcytosine content of the DNA. There appeared to be a correlation between the antileukemic activity of decitabine and its ability to inhibit DNA methylation [154]. Decitabine treatment is now applied in many disease models, especially MDS and leukemia.

Another example of the animal model providing insight to DNMTs involves studying the suppression of apoptosis by TP53 mutation which contributes to resistance of AML. Using differentiation to induce irreversible cell cycle exit in AML, cells could be a p53-independent treatment alternative. Therefore, *in vitro* and *in vivo* regimens of decitabine that deplete the chromatin modifying enzyme DNA methyltransferase 1 (DNMT1) without phosphorylating p53 or inducing early apoptosis were determined [155]. Subcutaneous decitabine and conventional cytotoxic cytarabine were compared in the xenotransplant model of p53-null human MLL-AF9 AML on a weekly basis. Nonirradiated NSG mice were transplanted with 3×10^6 THP1 cells by tail vein injection. Starting at day 5 after transplant, mice were treated with vehicle (PBS), cytarabine 75 mg/kg/day IP for five consecutive days (to model conventional chemotherapy [156]), or decitabine 0.2 mg/kg SC $3 \times$/week for 2 weeks then $2 \times$/week for 2 weeks then $1 \times$/week thereafter. Mice treated with decitabine had significantly longer median survival ($>20\%$ increase) than cytarabine and vehicle-treated mice. Furthermore, a xenotransplant model was established using fresh AML cells from a patient with relapsed/refractory AML. These AML cells contained multiple chromosome abnormalities including a t(8;18)(q22:q23) and t(11;13)(q21:q12). Nonirradiated NSG mice were transplanted with 1×10^6 patient cells by tail vein injection. Starting at day 5 after transplant, mice were treated with vehicle (PBS), cytarabine 75 mg/kg/day IP for five consecutive days [156], or decitabine 0.2 mg/kg SC $3 \times$/week for 2 weeks, then $2 \times$/week for 2 weeks, then $1 \times$/week thereafter. Mice treated with decitabine had significantly longer median survival ($>100\%$ increase) than cytarabine or vehicle-treated mice. Spleens of decitabine treated mice were significantly decreased in size compared to cytarabine or vehicle-treated mice. DNA-damaging cytarabine upregulated the key late differentiation factors CEBPε and p27/CDKN1B, induced cellular differentiation, and terminated AML cell-cycle, even in cytarabine-resistant p53- and p16/CDKN2A-null AML cells. Leukemia initiation by xenotransplanted AML cells were abrogated but normal HSC engraftment was preserved. *In vivo*, the low toxicity allowed frequent drug administration to increase exposure, an important consideration for S-phase specific-decitabine therapy. In xeno-transplant models of p53-null and relapsed/refractory AML, the noncytotoxic regimen significantly extended survival compared to conventional cytotoxic cytarabine. Modifying *in vivo* dose and schedule to emphasize this pathway of decitabine action can bypass a mechanism of resistance to standard therapy [155].

20.3.2 USE OF ANIMAL MODELS TO STUDY INHIBITORS OF HDATs

Huntington's disease (HD) is a progressive neurological disorder which currently has no available treatments on the market. One major molecular factor contributing to the progression of HD is transcriptional dysregulation, causing the development of HDAC inhibitors to be pursued for potential treatment. One such hydroxamic acid HDAC inhibitor is SAHA, which improved motor impairment in the R6/2 mouse model. To elucidate whether SAHA is a modifier of HDAC protein levels *in vivo*, Mielcarek et al. [157] performed two independent mouse trials. Both WT and R6/2 mice were chronically treated with SAHA and vehicle. They found that prolonged SAHA treatment causes the degradation of HDAC4 in the cortex and brain stem, but not hippocampus, without affecting its transcript levels *in vivo*. Similarly, SAHA also decreased HDAC2 levels while leaving no effect on mRNA expression. Consistent with previous data, SAHA treatment diminishes Hdac7 transcript levels in both wild type and R6/2 brains and unexpectedly was found to decrease Hdac11 in R6/2 but not wild type. The effects of SAHA administration was investigated on well-characterized molecular readouts of disease progression, finding that SAHA reduces SDS-insoluble aggregate load in the cortex and brain stem, while leaving the hippocampus of R6/s brains unaffected [157].

20.3.3 USE OF ANIMAL MODELS TO STUDY DOT1L INHIBITORS

Recent discoveries have demonstrated the importance of hypermethylation of histone lysines in myeloid leukemogenesis, and therefore these lysines can be targeted. As we mentioned previously, DOT1L-targeted therapy is an exciting prospect for the selective treatment of MLL-translocation leukemias.

Based on the chemical structures of the SAM substrate and S-adenosylhomocysteine (SAH) product, the reaction mechanism of DOT1L catalysis and the published crystal structure of the DOT1L active site, medicinal chemistry design tenets were established to facilitate mechanism-guided inhibitor discovery. Chemical analogs designed were synthesized and tested as inhibitors of DOT1L enzymatic activity. Based on these efforts, investigators from Epizyme, Inc. have identified EPZ004777 [94]. After a series of *in vitro* experiments, this inhibitor was tested in a MV4-11 subcutaneous xenograft model of MLL to determine if DOT1L activity would be inhibited. The compound was delivered via subcutaneously implanted mini-osmotic pumps capable of continuous infusion for a period of 7 days. Six female nude mice produced a wide range of tumors, from $300-400$ mm^3 and received 50 mg/mL solution of EPZ004777, while control mice did not receive any pumps. Six days later, tumors were harvested and immunoblot analysis of histones from MV4-11 tumors revealed decreased H3K79me2 in EPZ04777 mice in comparison to the control. Plasma samples indicated mice were exposed to 0.55 ± 0.12 nM EPZ04777. EPZ04777 efficacy was then tested in a more therapeutically relevant model in which MV4-11 cells are injected into the tail vein of immunodeficient mice resulting in disseminated leukemic disease. Female NSG mice were injected via tail vein with 1X106 MV4-11 cells stably expressing the firefly luciferase gene. Leukemia engraftment was confirmed by bioluminescence imaging five days after inoculation, at which point animals were divided into treatment and control groups (n = 8 per group). Optimization of EPZ004777 formulation enabled us to increase solubility to 150 mg/mL. Mice were then implanted with pumps loaded with vehicle alone, or containing EPZ004777 at concentrations of 50, 100, or 150 mg/mL. Pilot experiments indicated that mini-pumps containing 100 and 150 mg/mL EPZ004777 solutions could maintain average steady-state plasma concentrations of 0.64 ± 0.48 and 0.84 ± 0.45 mM, respectively. Osmotic pumps were replaced once to achieve a total of 14 days of *in vivo* compound exposure (from day 5 to day 19 postinoculum). Local irritation at the pump implant site in the 100 and 150 mg/mL groups precluded a second pump replacement. Despite the limited 14 day duration of EPZ004777 exposure achieved, they observed a dose-dependent and statistically significant increase in median survival ($p = 0.0002$, 0.0007, and 0.0285 for the 150, 100, and 50 mg/mL dose groups, respectively). Histopathological analysis of similarly xenografted mice confirmed that animals died of leukemic disease. EPZ004777 administration was well tolerated, and no significant weight loss was observed. Together, these results demonstrate that EPZ004777 has both pharmacodynamic and antitumor efficacy in a mouse xenograft model of MLL leukemia.

20.3.4 USE OF ANIMAL MODELS TO STUDY EZH2 INHIBITORS

More recently, hypertrimethylation of H3K27 has been identified in diffused large B cell lymphoma (DLBCL) cells heterozygous for point mutations recurrently targeting EZH2 tyrosine 641 (Y641) [158]. While wild-type (WT) EZH2 is most efficient at catalyzing monomethylation of non-methylated H3K27, the EZH2 Y641 mutant enzymes (Y641F, Y641N, Y641S, and Y641H) have the opposite substrate preference and are most efficient at catalyzing the conversion of di-methylated H3K27 (H3K27me2) into trimethylated H3K27 (H3K27me3) [159]. Thus, the cooperation between WT and Y641 mutant EZH2 drives the hypertrimethylation of H3K27 in these heterozygous DLBCL cells. While targeting EZH2 dysfunction has been pursued as a potential therapeutic strategy for the treatment of cancer, it is worth noting that EZH2 and PRC2 play an integral part in regulating stem cell pluripotency and differentiation [160]. Therefore, developing an EZH2 chemical probe that has suitable pharmacokinetic (PK) properties would be extremely useful for assessing therapeutic benefit(s) and potential toxicity of chronically inhibiting EZH2 in animal studies.

In 2013, the first orally bioavailable chemical probe of EZH2 and EZH1, UNC1999, was reported [106]. UNC1999 was highly potent and selective for EZH2 wild-type and Y641 mutant enzymes as well as EZH1 over a broad range of epigenetic and non-epigenetic targets. It was competitive with the cofactor and noncompetitive with the peptide substrate. In cell-based assays, UNC1999 successfully reduced the H3K27me3 mark and killed DB cells, a DLBCL cell line harboring the EZH2Y641N mutant. In mouse PK studies, UNC1999 was orally bioavailable, making it suitable for chronic animal studies. In this study, the *in vivo* PK properties of UNC1999 were evaluated [106]. A single IP injection of UNC1999 at 15, 50, or 150 mg/kg achieved high Cmax (9,700−11,800 nM) and exhibited linearity in male Swiss albino mice. Both the 150 and 50 mg/kg IP doses resulted in plasma concentrations of UNC1999 above its cellular IC50 over the entire 24 h period while the 15 mg/kg IP dose led to the plasma concentrations of UNC1999 above its cellular IC50 for approximately 12 h. Furthermore, they examined whether UNC1999 is orally bioavailable and found that a single 50 mg/kg oral dose of UNC1999 achieved high Cmax (4,700 nM) and good exposure levels in male Swiss albino mice. The plasma concentrations of UNC1999 were maintained above its cellular IC50 for approximately 20 h following this single oral dose. It is worth noting that all doses including the 150 mg/kg IP dose were well tolerated by all test mice, and no adverse effects were observed.

20.3.5 USE OF ANIMAL MODELS TO STUDY INHIBITORS OF KDM FAMILY AND LSD FAMILY

KDM1A (also known as AOF2, LSD1, KIAA0601, or BHC110) is a flavin adenine dinucleotide (FAD)-dependent lysine-specific demethylase with monomethyl- and dimethyl-histone H3 lysine-4 (H3K4) and lysine-9 (H3K9) substrate specificity. Poor prognosis in high-risk prostate cancer and breast cancer patients, as well as poor neuroblastoma differentiation, have correlated with KDM1A expression [161−163]. Researchers found KDM1A to be required for both MLL-AF9 AML cell's clonogenic and leukemia stem cell potential as well as sustained expression of the MLL-AF9 oncogenic program, making it a candidate for a therapeutic target, in theory. As KDM1A is inhibited *in vitro* at an IC50 of ~20 mM by the monoamine oxidase inhibitor TCP [164], it is likely to be of limited clinical use due to a lack of potency and selectivity.

Trans-*N*-((2-methoxypyridin-3-yl)methyl)-2-phenylcyclopropan-1-amine (Compound B) is synthesized as an analog of TCP to be used in mouse model study [21]. To assess whether KDM1A inhibition selectively targets MLL AML cells *in vivo*, CD45.2 + MLL-AF9 murine AML cells were transplanted into nearly lethally irradiated CD45.1 + congenic recipients and the mice were treated with DMSO vehicle, 10 mg/kg Compound B or 25 mg/kg Compound B twice daily by IP injection. Plasma levels of Compound B measured 1 h after a dose of 50 mg/kg IP were 2.0 ± 0.2 mM ($n = 3$), as determined by liquid chromatography and mass spectrometry. Following treatment, blood counts and samples revealed substantial leukemic leukocytosis with circulating blasts and leukemic myelomonocytic cells, modest anemia, and thrombocytopenia consistent with incipient AML. By contrast, mice treated with KDM1A inhibitor lacked evidence of circulating AML cells but were instead significantly more anemic and thrombocytopenic than vehicle-treated controls and exhibited circulating nucleated erythroid precursors and polychromasia consistent with the degree of anemia (mean $\pm$ SEM hemoglobin 5.0 ± 0.7 g/dL for mice receiving the 25 mg/kg dose vs. 9.0 ± 0.4 g/dL for vehicle-treated mice; $p < 0.001$; $n = 6$; mean $\pm$ SEM platelet count $89 \pm 9.3 \times 10^9$ L for mice receiving the 25 mg/kg dose vs. $592 \pm 55.3 \times 10^9$ L for vehicle-treated mice; $p < 0.001$; $n = 6$). Thus, treatment of mice with KDM1A inhibitor completely blocked progression of MLL-AF9 leukemia into circulation.

In another study, combined treatment of all-trans-retinoic acid (ATRA) with KDM1A inhibitor TCP was carried out for leukemia [165]. Acute promyelocytic leukemia (APL), a cytogenetically distinct subtype of AML, characterized by the t(15;17)-associated PML-RARA fusion, has been successfully treated with therapy utilizing ATRA to differentiate leukemic blasts. However, among patients with non-APL AML, ATRA-based treatment has not been effective. However, through epigenetic reprogramming, KDM1A inhibitors (including TCP) unlocked the ATRA-driven therapeutic response in non-APL AML. Treatment with ATRA and TCP diminished the engraftment of primary human AML cells *in vivo* in nonobese diabetic (NOD) — SCID mice, indicating that ATRA in combination with TCP may target leukemia-initiating cells. Furthermore, initiation of ATRA and TCP treatment 15 days after engraftment of human AML cells in NSG mice also revealed the ATRA plus TCP drug combination to have an antileukemic effect that was superior to treatment with either drug alone. This data identifies LSD1 as a therapeutic target and strongly suggests that it may contribute to AML pathogenesis by inhibiting the normal prodifferentiative function of ATRA, paving the way for new combination therapies for AML.

20.3.6 USE OF ANIMAL MODELS TO STUDY IDH1/2 INHIBITORS

As mentioned in part 2, 2-HG was demonstrated as a competitive inhibitor of multiple α-KG-dependent dioxygenases, including histone and DNA demethylases [135]. Several studies have shown that 2-HG producing IDH mutants are involved in global histone and DNA methylation alterations [134]. This implicates mutant IDH1 as an oncogene and a compelling drug target for new therapies for glioma and AML.

In 2012, a high-throughput screening for IDH1-P132H mutant homodimer inhibitors revealed a selective compound, often referred to as AGI-5198, which inhibits mutant IDH1 [R132H-IDH1], though not wildtype or examined isoforms [166]. AGI-5198, equipotent in both enzyme R132H and U87 cellular assays, was then selected for additional *in vivo* profiling in the U87 R132H tumor xenograft mouse model. *In vitro* and *in vivo* DMPK studies were conducted for AGI-5198, showing a rapid turnover in human and rat microsomal incubations with an estimated hepatic extraction ratio of 0.93 and 0.85. Plasma protein binding was 95.7% in mice using the equilibrium dialysis method. Reasonable plasma exposure was achieved via IP dosing at 50 mg/kg (AUC0–24 h = 20,800 h · ng/mL), enabling the use of AGI-5198 for further *in vivo* studies. Female nude mice bearing U87 R132H tumor xenografts were dosed via IP with 150 mg/kg of AGI-5198 formulated in 0.5% MC and 0.2% Tween, and compared to the vehicle control. Blood and tumor samples were taken throughout compound administration. The plasma and tumor concentrations of AGI-5198, as well as the corresponding tumor 2-HG concentrations, were determined using sensitive and specific LC/MS/MS methods. The unbound plasma concentration of AGI-5198 was calculated using the total plasma concentration of this inhibitor and free fraction of it in mouse plasma (4.3%) [166].

The effects of orally administered AGI-5198 on the growth of human glioma xenografts were examined in another study [167]. When given to mice with established R132H-IDH1 glioma xenografts daily, AGI-5198 (450 mg per kg of weight (mg/kg) per os) caused 50−60% growth inhibition. Treatment was tolerated with no signs of toxicity during 3 weeks of daily treatment. Tumors from AGI-5198 treated mice showed reduced staining with an antibody against the Ki-67 protein, a marker used for quantification of tumor cell proliferation in human brain tumors. In contrast, staining with an antibody against cleaved caspase-3 showed no differences between tumors from vehicle and AGI-5198 treated mice, suggesting that the growth-inhibitory effects of AGI-5198 were primarily due to impaired tumor cell proliferation rather than induction of apoptotic cell death. AGI-5198 did not affect the growth of IDH1 wild-type glioma xenografts.

20.3.7 USE OF ANIMAL MODELS TO STUDY EPIGENETIC INHIBITORS IN COMBINED THERAPY

Overexpression of P-glycoprotein causes renal cell carcinoma (RCC0) to be chemotherapy resistant, and hypermethylation of the RCC promoter region has shown to silence many tumor suppressor genes [168]. Cytotoxicity of vinblastine (VBL) was enhanced by pretreatment with de-demethylating agent, 5-aza-2′-deoxycytidine (Aza), in the RCC cell line, Caki-1 in a previous study, *in vivo* [169]. Therefore, the combined effect of Aza and VBL was investigated in a Caki-1 xenograft model in which tumor volume and mass, as well as P-glycoprotein, Bcl-2, and Cyclin B1, were significantly reduced in the co-treated group when compared with the control. Thus, this combined effect could be mediated by the accumulation of intracellular VBL and the enhancement of apoptosis and cell cycle arrest. Moreover, the cytotoxicity of VBL was enhanced *in vitro* in three RCC cell lines by Aza treatment. These findings suggest that the combination treatment with Aza and VBL is effective against RCC [168].

In 2011, Kalac et al. investigated the interactions between histone deacetylase inhibitors (HDACIs) and decitabine in diffuse large B-cell lymphoma (DLBCL) models [170]. DLBCL is the most common type of lymphoid neoplasm, representing 30−40% of all lymphomas [171]. It has been hypothesized that the combination of HDACI and hypomethylating agents might be synergistic by disrupting the previously mentioned transcription repressor complex consisting of MBDPs and HDACs. In their study, 5- to 7-week-old SCID beige mice were injected with Ly1 line (10^7 cells) in the posterior flank subcutaneously. When the tumors approached 50 mm^3, the mice were divided into five groups including: vehicle treated, decitabin treated, panobinostat treated, and two separate cohorts that were treated with differing drug schedules so toxicity could be compared. As a result, both of the combination groups had a statistically significant tumor growth inhibition compared to both control and cohorts treated with a single drug. This data strongly supports the potential therapeutic role of a combinatorial epigenetic platform for the treatment of B-cell lymphomas, in particular in patients with DLBCL [170].

20.4 PERSPECTIVE AND CONCLUSION

We have outlined the advantages and challenges associated with the use of animal models (especially mouse models) for discovery and development of epigenetic inhibitors. Although a number of epigenetic inhibitors have been evaluated *in vivo*, we encourage more in-depth studies of new inhibitors which have already shown promise in preclinical studies. There are broader explorations of *in vitro* or cell line sensitivity to epigenetic inhibitors, including many histotypes and oncogenotypes. Thus, we hope more labs will continue to take their *in vitro* leads through an evaluation process that will include determinations of efficacy in xenograft and GEMMs. However, the increasing use of these inhibitors, both in the laboratory and in clinical practice, also identified important limitations. These limitations are largely defined by a lack of specificity on several levels. This includes substrate specificity of the targeted enzymes as well as cancer specificity of the targeted epigenetic pathways. Careful characterization of these limitations will lead to the development of a diverse array of highly specific epigenetic inhibitors and specific delivery methods. More importantly, the "give the maximum tolerated dosage" paradigm needs to be reevaluated when it comes to epigenetic inhibitors. Although we may yet discover many silver bullets among epigenetic inhibitors, we believe it will be necessary to take a more rigorous look at dosing regimens and, more importantly, the timing and methods of administration for epigenetic inhibitors. If the goal is to reprogram the genome using epigenetic inhibitors, the effects may not be observed in short-term, high-dose evaluations, but may require longer term treatments at moderate dosing in animal models and later in patients.

Although the obstacles mentioned in this process still exist, many are improving. By contrast, the capacity to validate the efficacy of novel therapeutics in humans is becoming more difficult for various reasons. Given these changes in the landscape of human Phase I/II testing, the development of better preclinical methods to validate and prioritize novel epigenetic inhibitors has gained added significance. We believe that animal models (both GEMMs and and xenotransplant models) will become obligatory tools, along with cell culture-based systems, in the preclinical evaluation of epigenetic inhibitors and their associated compounds. Ongoing experiments of epigenetic analysis in GEMMs are beginning to appear with increasing frequency. We anticipate that these studies will collectively establish the superiority of GEMMs over xenograft models in bringing new epigenetic inhibitors and combinations to both basic and clinic practices with a cumulative focus to benefit the patients.

REFERENCES

[1] Robertson KD. DNA methylation and human disease. Nat Rev Genet 2005;6(8):597−610.
[2] Okano M, Bell DW, Haber DA, Li E. DNA methyltransferases Dnmt3a and Dnmt3b are essential for de novo methylation and mammalian development. Cell 1999;99(3):247−57.
[3] Kareta MS, Botello ZM, Ennis JJ, Chou C, Chedin F. Reconstitution and mechanism of the stimulation of de novo methylation by human DNMT3L. J Biol Chem 2006;281(36):25893−902.
[4] Reik W. Stability and flexibility of epigenetic gene regulation in mammalian development. Nature 2007;447(7143):425−32.
[5] Lan J, Hua S, He X, Zhang Y. DNA methyltransferases and methyl-binding proteins of mammals. Acta Biochim Biophys Sin 2010;42(4):243−52.
[6] Kouzarides T. Chromatin modifications and their function. Cell 2007;128(4):693−705.

[7]   Jenuwein T, Allis CD. Translating the histone code. Science 2001;293(5532):1074−80.

[8]   Ruthenburg AJ, Allis CD, Wysocka J. Methylation of lysine 4 on histone H3: intricacy of writing and reading a single epigenetic mark. Mol Cell 2007;25(1):15−30.

[9]   Sparmann A, van Lohuizen M. Polycomb silencers control cell fate, development and cancer. Nat Rev Cancer 2006;6(11):846−56.

[10]   Feinberg AP, Vogelstein B. Hypomethylation distinguishes genes of some human cancers from their normal counterparts. Nature 1983;301(5895):89−92.

[11]   Gnyszka A, Jastrzebski Z, Flis S. DNA methyltransferase inhibitors and their emerging role in epigenetic therapy of cancer. Anticancer Res 2013;33(8):2989−96.

[12]   Fouse SD, Costello JF. Epigenetics of neurological cancers. Future Oncol 2009;5(10):1615−29.

[13]   Villeneuve LM, Natarajan R. The role of epigenetics in the pathology of diabetic complications. Am J Physiol Renal Physiol 2010;299(1):F14−25.

[14]   Javierre BM, Fernandez AF, Richter J, Al-Shahrour F, Martin-Subero JI, Rodriguez-Ubreva J, et al. Changes in the pattern of DNA methylation associated with twin discordance in systemic lupus erythematosus. Genome Res 2010;20(2):170−9.

[15]   Adcock IM, Ito K, Barnes PJ. Histone deacetylation: an important mechanism in inflammatory lung diseases. COPD 2005;2(4):445−55.

[16]   Egger G, Liang G, Aparicio A, Jones PA. Epigenetics in human disease and prospects for epigenetic therapy. Nature 2004;429(6990):457−63.

[17]   Feng J, Fan G. The role of DNA methylation in the central nervous system and neuropsychiatric disorders. Int Rev Neurobiol 2009;89:67−84.

[18]   Yang X, Lay F, Han H, Jones PA. Targeting DNA methylation for epigenetic therapy. Trends Pharmacol Sci 2010;31(11):536−46.

[19]   Kautiainen TL, Jones PA. DNA methyltransferase levels in tumorigenic and nontumorigenic cells in culture. J Biol Chem 1986;261(4):1594−8.

[20]   Fathi AT, Abdel-Wahab O. Mutations in epigenetic modifiers in myeloid malignancies and the prospect of novel epigenetic-targeted therapy. Adv Hematol 2012;2012:469592.

[21]   Harris WJ, Huang X, Lynch JT, Spencer GJ, Hitchin JR, Li Y, et al. The histone demethylase KDM1A sustains the oncogenic potential of MLL-AF9 leukemia stem cells. Cancer Cell 2012;21 (4):473−87.

[22]   Figueroa ME, Lugthart S, Li Y, Erpelinck-Verschueren C, Deng X, Christos PJ, et al. DNA methylation signatures identify biologically distinct subtypes in acute myeloid leukemia. Cancer Cell 2010;17(1):13−27.

[23]   Ellinger J, Kahl P, von der Gathen J, Rogenhofer S, Heukamp LC, Gütgemann I, et al. Global levels of histone modifications predict prostate cancer recurrence. Prostate 2010;70(1):61−9.

[24]   Kelly TK, De Carvalho DD, Jones PA. Epigenetic modifications as therapeutic targets. Nat Biotechnol 2010;28(10):1069−78.

[25]   Steele VE, Lubet RA, Moon RC. Preclinical animal models for the development of cancer chemoprevention drugs. Cancer chemoprevention. Springer; 2005. p. 39−46.

[26]   Miyamoto K, Ushijima T. Diagnostic and therapeutic applications of epigenetics. Jpn J Clin Oncol 2005;35(6):293−301.

[27]   Zelent A, Waxman S, Carducci M, Wright J, Zweibel J, Gore SD. State of the translational science: summary of Baltimore workshop on gene re-expression as a therapeutic target in cancer January 2003. Clin Cancer Res 2004;10(14):4622−9.

[28]   Gilbert J, Gore SD, Herman JG, Carducci MA. The clinical application of targeting cancer through histone acetylation and hypomethylation. Clin Cancer Res 2004;10(14):4589−96.

[29]   Sato N, Maitra A, Fukushima N, van Heek NT, Matsubayashi H, Iacobuzio-Donahue CA, et al. Frequent hypomethylation of multiple genes overexpressed in pancreatic ductal adenocarcinoma. Cancer Res 2003;63(14):4158−66.

[30] Costello JF, Plass C. Methylation matters. J Med Genet 2001;38(5):285−303.

[31] Jones PA, Baylin SB. The fundamental role of epigenetic events in cancer. Nat Rev Genet 2002;3 (6):415−28.

[32] Kaminskas E, Farrell AT, Wang YC, Sridhara R, Pazdur R. FDA drug approval summary: azacitidine (5−azacytidine, Vidaza) for injectable suspension. Oncologist 2005;10(3):176−82.

[33] Gore SD, Jones C, Kirkpatrick P. Decitabine. Nat Rev Drug Discov 2006;5(11):891−2.

[34] Herman JG, Baylin SB. Gene silencing in cancer in association with promoter hypermethylation. N Engl J Med 2003;349(21):2042−54.

[35] Foubister V. Drug reactivates genes to inhibit cancer. Drug Discov Today 2003;8(10):430−1.

[36] Cheng JC, Yoo CB, Weisenberger DJ, Chuang J, Wozniak C, Liang G, et al. Preferential response of cancer cells to zebularine. Cancer Cell 2004;6(2):151−8.

[37] Guo D, Myrdal PB, Karlage KL, O'Connell SP, Wissinger TJ, Tabibi SE, et al. Stability of 5-fluoro-2′-deoxycytidine and tetrahydrouridine in combination. AAPS PharmSciTech 2010;11(1):247−52.

[38] Coronel J, Cetina L, Pacheco I, Trejo-Becerril C, González-Fierro A, de la Cruz-Hernandez E, et al. A double-blind, placebo-controlled, randomized phase III trial of chemotherapy plus epigenetic therapy with hydralazine valproate for advanced cervical cancer. Preliminary results. Med Oncol 2011;28(Suppl. 1):S540−6.

[39] Lee BH, Yegnasubramanian S, Lin X, Nelson WG. Procainamide is a specific inhibitor of DNA methyltransferase 1. J Biol Chem 2005;280(49):40749−56.

[40] Stresemann C, Brueckner B, Musch T, Stopper H, Lyko F. Functional diversity of DNA methyltransferase inhibitors in human cancer cell lines. Cancer Res 2006;66(5):2794−800.

[41] Datta J, Ghoshal K, Denny WA, Gamage SA, Brooke DG, Phiasivongsa P, et al. A new class of quinoline-based DNA hypomethylating agents reactivates tumor suppressor genes by blocking DNA methyltransferase 1 activity and inducing its degradation. Cancer Res 2009;69(10):4277−85.

[42] Nandakumar V, Vaid M, Katiyar SK. (-)-Epigallocatechin-3-gallate reactivates silenced tumor suppressor genes, Cip1/p21 and p16INK4a, by reducing DNA methylation and increasing histones acetylation in human skin cancer cells. Carcinogenesis 2011;32(4):537−44.

[43] Mittal A, Piyathilake C, Hara Y, Katiyar SK. Exceptionally high protection of photocarcinogenesis by topical application of (--)-epigallocatechin-3-gallate in hydrophilic cream in SKH-1 hairless mouse model: relationship to inhibition of UVB-induced global DNA hypomethylation. Neoplasia 2003;5(6):555−65.

[44] Zambrano P, Segura-Pacheco B, Perez-Cardenas E, Cetina L, Revilla-Vazquez A, Taja-Chayeb L, et al. A phase I study of hydralazine to demethylate and reactivate the expression of tumor suppressor genes. BMC Cancer 2005;5:44.

[45] Amato RJ, Stephenson J, Hotte S, Nemunaitis J, Bélanger K, Reid G, et al. MG98, a second-generation DNMT1 inhibitor, in the treatment of advanced renal cell carcinoma. Cancer Invest 2012;30(5):415−21.

[46] Garzon R, Liu S, Fabbri M, Liu Z, Heaphy CE, Callegari E, et al. MicroRNA-29b induces global DNA hypomethylation and tumor suppressor gene reexpression in acute myeloid leukemia by targeting directly DNMT3A and 3B and indirectly DNMT1. Blood 2009;113(25):6411−18.

[47] Villar-Garea A, Esteller M. Histone deacetylase inhibitors: understanding a new wave of anticancer agents. Int J Cancer 2004;112(2):171−8.

[48] Haggarty SJ, Koeller KM, Wong JC, Grozinger CM, Schreiber SL. Domain-selective small-molecule inhibitor of histone deacetylase 6 (HDAC6)-mediated tubulin deacetylation. Proc Natl Acad Sci USA 2003;100(8):4389−94.

[49] Yoshida M. [Potent and specific inhibition of mammalian histone deacetylase both in vivo and in vitro by trichostatin A]. Tanpakushitsu Kakusan Koso 2007;52(Suppl. 13):1788−9.

[50] Richon VM, Webb Y, Merger R, Sheppard T, Jursic B, Ngo L, et al. Second generation hybrid polar compounds are potent inducers of transformed cell differentiation. Proc Natl Acad Sci U S A 1996;93 (12):5705−8.

[51] Su GH, Sohn TA, Ryu B, Kern SE. A novel histone deacetylase inhibitor identified by high-throughput transcriptional screening of a compound library. Cancer Res 2000;60(12):3137−42.

[52] Remiszewski SW, Sambucetti LC, Bair KW, Bontempo J, Cesarz D, Chandramouli N, et al. N-hydroxy-3-phenyl-2-propenamides as novel inhibitors of human histone deacetylase with in vivo antitumor activity: discovery of (2E)-N-hydroxy-3-[4-[[(2-hydroxyethyl)[2-(1H-indol-3-yl)ethyl]amino]methyl]phenyl]-2-propenamide (NVP-LAQ824). J Med Chem 2003;46(21):4609−24.

[53] Atadja P, Gao L, Kwon P, Trogani N, Walker H, Hsu M, et al. Selective growth inhibition of tumor cells by a novel histone deacetylase inhibitor, NVP-LAQ824. Cancer Res 2004;64(2):689−95.

[54] Plumb JA, Finn PW, Williams RJ, Bandara MJ, Romero MR, Watkins CJ, et al. Pharmacodynamic response and inhibition of growth of human tumor xenografts by the novel histone deacetylase inhibitor PXD101. Mol Cancer Ther 2003;2(8):721−8.

[55] Vigushin DM, Coombes RC. Histone deacetylase inhibitors in cancer treatment. Anticancer Drugs 2002;13(1):1−13.

[56] Phiel CJ, Zhang F, Huang EY, Guenther MG, Lazar MA, Klein PS. Histone deacetylase is a direct target of valproic acid, a potent anticonvulsant, mood stabilizer, and teratogen. J Biol Chem 2001;276 (39):36734−41.

[57] Göttlicher M, Minucci S, Zhu P, Krämer OH, Schimpf A, Giavara S, et al. Valproic acid defines a novel class of HDAC inhibitors inducing differentiation of transformed cells. EMBO J 2001;20 (24):6969−78.

[58] Lea MA, Tulsyan N. Discordant effects of butyrate analogues on erythroleukemia cell proliferation, differentiation and histone deacetylase. Anticancer Res 1995;15(3):879−83.

[59] Sealy L, Chalkley R. The effect of sodium butyrate on histone modification. Cell 1978;14(1):115−21.

[60] Patnaik A, Rowinsky EK, Villalona MA, Hammond LA, Britten CD, Siu LL, et al. A phase I study of pivaloyloxymethyl butyrate, a prodrug of the differentiating agent butyric acid, in patients with advanced solid malignancies. Clin Cancer Res 2002;8(7):2142−8.

[61] Fournel M, Bonfils C, Hou Y, Yan PT, Trachy-Bourget MC, Kalita A, et al. MGCD0103, a novel isotype-selective histone deacetylase inhibitor, has broad spectrum antitumor activity in vitro and in vivo. Mol Cancer Ther 2008;7(4):759−68.

[62] Saito A, Yamashita T, Mariko Y, Nosaka Y, Tsuchiya K, Ando T, et al. A synthetic inhibitor of histone deacetylase, MS-27-275, with marked in vivo antitumor activity against human tumors. Proc Natl Acad Sci U S A 1999;96(8):4592−7.

[63] Younes A, Oki Y, Bociek RG, Kuruvilla J, Fanale M, Neelapu S, et al. Mocetinostat for relapsed classical Hodgkin's lymphoma: an open-label, single-arm, phase 2 trial. Lancet Oncol 2011;12 (13):1222−8.

[64] Hong J, Ishihara K, Yamaki K, Hiraizumi K, Ohno T, Ahn JW, et al. Apicidin, a histone deacetylase inhibitor, induces differentiation of HL-60 cells. Cancer Lett 2003;189(2):197−206.

[65] Cheong JW, Chong SY, Kim JY, Eom JI, Jeung HK, Maeng HY, et al. Induction of apoptosis by apicidin, a histone deacetylase inhibitor, via the activation of mitochondria-dependent caspase cascades in human Bcr-Abl-positive leukemia cells. Clin Cancer Res 2003;9(13):5018−27.

[66] Cole PA. Chemical probes for histone-modifying enzymes. Nat Chem Biol 2008;4(10):590−7.

[67] Lau OD, Kundu TK, Soccio RE, Ait-Si-Ali S, Khalil EM, Vassilev A, et al. HATs off: selective synthetic inhibitors of the histone acetyltransferases p300 and PCAF. Mol Cell 2000;5(3):589−95.

[68] Thompson PR, Kurooka H, Nakatani Y, Cole PA. Transcriptional coactivator protein p300. Kinetic characterization of its histone acetyltransferase activity. J Biol Chem 2001;276(36):33721−9.

[69] Zheng Y, Balasubramanyam K, Cebrat M, Buck D, Guidez F, Zelent A, et al. Synthesis and evaluation of a potent and selective cell-permeable p300 histone acetyltransferase inhibitor. J Am Chem Soc 2005;127(49):17182−3.

[70] Guidez F, Howell L, Isalan M, Cebrat M, Alani RM, Ivins S, et al. Histone acetyltransferase activity of p300 is required for transcriptional repression by the promyelocytic leukemia zinc finger protein. Mol Cell Biol 2005;25(13):5552−66.

[71] Stimson L, Rowlands MG, Newbatt YM, Smith NF, Raynaud FI, Rogers P, et al. Isothiazolones as inhibitors of PCAF and p300 histone acetyltransferase activity. Mol Cancer Ther 2005;4(10):1521−32.

[72] Balasubramanyam K, Swaminathan V, Ranganathan A, Kundu TK. Small molecule modulators of histone acetyltransferase p300. J Biol Chem 2003;278(21):19134−40.

[73] Balasubramanyam K, Varier RA, Altaf M, Swaminathan V, Siddappa NB, Ranga U, et al. Curcumin, a novel p300/CREB-binding protein-specific inhibitor of acetyltransferase, represses the acetylation of histone/nonhistone proteins and histone acetyltransferase-dependent chromatin transcription. J Biol Chem 2004;279(49):51163−71.

[74] Mantelingu K, Reddy BA, Swaminathan V, Kishore AH, Siddappa NB, Kumar GV, et al. Specific inhibition of p300-HAT alters global gene expression and represses HIV replication. Chem Biol 2007;14 (6):645−57.

[75] Arif M, Pradhan SK, Thanuja GR, Vedamurthy BM, Agrawal S, Dasgupta D, et al. Mechanism of p300 specific histone acetyltransferase inhibition by small molecules. J Med Chem 2009;52(2):267−77.

[76] Ravindra KC, Selvi BR, Arif M, Reddy BA, Thanuja GR, Agrawal S, et al. Inhibition of lysine acetyltransferase KAT3B/p300 activity by a naturally occurring hydroxynaphthoquinone, plumbagin. J Biol Chem 2009;284(36):24453−64.

[77] Bowers EM, Yan G, Mukherjee C, Orry A, Wang L, Holbert MA, et al. Virtual ligand screening of the p300/CBP histone acetyltransferase: identification of a selective small molecule inhibitor. Chem Biol 2010;17(5):471−82.

[78] Jenuwein T. The epigenetic magic of histone lysine methylation. FEBS J 2006;273(14):3121−35.

[79] Zagni C, Chiacchio U, Rescifina A. Histone methyltransferase inhibitors: novel epigenetic agents for cancer treatment. Curr Med Chem 2013;20(2):167−85.

[80] Shinkai Y, Tachibana M. H3K9 methyltransferase G9a and the related molecule GLP. Genes Dev 2011;25(8):781−8.

[81] Kubicek S, O'Sullivan RJ, August EM, Hickey ER, Zhang Q, Teodoro ML, et al. Reversal of H3K9me2 by a small-molecule inhibitor for the G9a histone methyltransferase. Mol Cell 2007;25(3):473−81.

[82] Liu F, Chen X, Allali-Hassani A, Quinn AM, Wasney GA, Dong A, et al. Discovery of a 2,4-diamino-7-aminoalkoxyquinazoline as a potent and selective inhibitor of histone lysine methyltransferase G9a. J Med Chem 2009;52(24):7950−3.

[83] Liu F, Chen X, Allali-Hassani A, Quinn AM, Wigle TJ, Wasney GA, et al. Protein lysine methyltransferase G9a inhibitors: design, synthesis, and structure activity relationships of 2,4-diamino-7-aminoalkoxy-quinazolines. J Med Chem 2010;53(15):5844−57.

[84] Chang Y, Ganesh T, Horton JR, Spannhoff A, Liu J, Sun A, et al. Adding a lysine mimic in the design of potent inhibitors of histone lysine methyltransferases. J Mol Biol 2010;400(1):1−7.

[85] Liu F, Barsyte-Lovejoy D, Allali-Hassani A, He Y, Herold JM, Chen X, et al. Optimization of cellular activity of G9a inhibitors 7-aminoalkoxy-quinazolines. J Med Chem 2011;54(17):6139−50.

[86] Kim JT, Li J, Jang ER, Gulhati P, Rychahou PG, Napier DL, et al. Deregulation of Wnt/beta-catenin signaling through genetic or epigenetic alterations in human neuroendocrine tumors. Carcinogenesis 2013;34(5):953−61.

[87] Konze KD, Pattenden SG, Liu F, Barsyte-Lovejoy D, Li F, Simon JM, et al. A chemical tool for in vitro and in vivo precipitation of lysine methyltransferase G9a. ChemMedChem 2014;9(3):549−53.

[88] Liu Y, Liu K, Qin S, Xu C, Min J. Epigenetic targets and drug discovery: part 1: histone methylation. Pharmacol Ther 2014;143(3):275−94.

[89] Sweis RF, Pliushchev M, Brown PJ, Guo J, Li F, Maag D, et al. Discovery and development of potent and selective inhibitors of histone methyltransferase g9a. ACS Med Chem Lett 2014;5(2):205−9.

[90] Rea S, Eisenhaber F, O'Carroll D, Strahl BD, Sun ZW, Schmid M, et al. Regulation of chromatin structure by site-specific histone H3 methyltransferases. Nature 2000;406(6796):593−9.

[91] Greiner D, Bonaldi T, Eskeland R, Roemer E, Imhof A. Identification of a specific inhibitor of the histone methyltransferase SU(VAR)3-9. Nat Chem Biol 2005;1(3):143−5.

[92] Frederiks F, Tzouros M, Oudgenoeg G, van Welsem T, Fornerod M, Krijgsveld J, et al. Nonprocessive methylation by Dot1 leads to functional redundancy of histone H3K79 methylation states. Nat Struct Mol Biol 2008;15(6):550−7.

[93] Meyer C, Kowarz E, Hofmann J, Renneville A, Zuna J, Trka J, et al. New insights to the MLL recombinome of acute leukemias. Leukemia 2009;23(8):1490−9.

[94] Daigle SR, Olhava EJ, Therkelsen CA, Majer CR, Sneeringer CJ, Song J, et al. Selective killing of mixed lineage leukemia cells by a potent small-molecule DOT1L inhibitor. Cancer Cell 2011;20 (1):53−65.

[95] Yao Y, Chen P, Diao J, Cheng G, Deng L, Anglin JL, et al. Selective inhibitors of histone methyltransferase DOT1L: design, synthesis, and crystallographic studies. J Am Chem Soc 2011;133 (42):16746−9.

[96] Anglin JL, Deng L, Yao Y, Cai G, Liu Z, Jiang H, et al. Synthesis and structure-activity relationship investigation of adenosine-containing inhibitors of histone methyltransferase DOT1L. J Med Chem 2012;55(18):8066−74.

[97] Basavapathruni A, Jin L, Daigle SR, Majer CR, Therkelsen CA, Wigle TJ, et al. Conformational adaptation drives potent, selective and durable inhibition of the human protein methyltransferase DOT1L. Chem Biol Drug Des 2012;80(6):971−80.

[98] Yu W, Smil D, Li F, Tempel W, Fedorov O, Nguyen KT, et al. Bromo-deaza-SAH: a potent and selective DOT1L inhibitor. Bioorg Med Chem 2013;21(7):1787−94.

[99] Deng L, Zhang L, Yao Y, Wang C, Redell MS, Dong S, et al. Synthesis, activity and metabolic stability of non-ribose containing inhibitors of histone methyltransferase DOT1L. MedChemComm 2013;4 (5):822−6.

[100] Daigle SR, Olhava EJ, Therkelsen CA, Basavapathruni A, Jin L, Boriack-Sjodin PA, et al. Potent inhibition of DOT1L as treatment of MLL-fusion leukemia. Blood 2013;122(6):1017−25.

[101] Mathews LA, Crea F, Farrar WL. Epigenetic gene regulation in stem cells and correlation to cancer. Differentiation 2009;78(1):1−17.

[102] Cao R, Zhang Y. The functions of E(Z)/EZH2-mediated methylation of lysine 27 in histone H3. Curr Opin Genet Dev 2004;14(2):155−64.

[103] Knutson SK, Wigle TJ, Warholic NM, Sneeringer CJ, Allain CJ, Klaus CR, et al. A selective inhibitor of EZH2 blocks H3K27 methylation and kills mutant lymphoma cells. Nat Chem Biol 2012;8 (11):890−6.

[104] Qi W, Chan H, Teng L, Li L, Chuai S, Zhang R, et al. Selective inhibition of Ezh2 by a small molecule inhibitor blocks tumor cells proliferation. Proc Natl Acad Sci U S A 2012;109(52):21360−5.

[105] McCabe MT, Ott HM, Ganji G, Korenchuk S, Thompson C, Van Aller GS, et al. EZH2 inhibition as a therapeutic strategy for lymphoma with EZH2-activating mutations. Nature 2012;492(7427):108−12.

[106] Konze KD, Ma A, Li F, Barsyte-Lovejoy D, Parton T, Macnevin CJ, et al. An orally bioavailable chemical probe of the Lysine Methyltransferases EZH2 and EZH1. ACS Chem Biol 2013;8 (6):1324−34.

[107] Kooistra SM, Helin K. Molecular mechanisms and potential functions of histone demethylases. Nat Rev Mol Cell Biol 2012;13(5):297−311.

[108] Mosammaparast N, Shi Y. Reversal of histone methylation: biochemical and molecular mechanisms of histone demethylases. Annu Rev Biochem 2010;79:155−79.

[109] Culhane JC, Wang D, Yen PM, Cole PA. Comparative analysis of small molecules and histone substrate analogues as LSD1 lysine demethylase inhibitors. J Am Chem Soc 2010;132(9):3164−76.

[110] Lee MG, Wynder C, Schmidt DM, McCafferty DG, Shiekhattar R. Histone H3 lysine 4 demethylation is a target of nonselective antidepressive medications. Chem Biol 2006;13(6):563−7.

[111] Maes T, Tirapu I, Mascaró C, Ortega A, Estiarte A, Valls N, et al. Preclinical characterization of a potent and selective inhibitor of the histone demethylase KDM1A for MLL leukemia. J Clin Oncol Abstr 2013;31:e13543.

[112] Chang KH, King ON, Tumber A, Woon EC, Heightman TD, McDonough MA, et al. Inhibition of histone demethylases by 4-carboxy-2,2′-bipyridyl compounds. ChemMedChem 2011;6(5):759−64.

[113] King ON, Li XS, Sakurai M, Kawamura A, Rose NR, Ng SS, et al. Quantitative high-throughput screening identifies 8-hydroxyquinolines as cell-active histone demethylase inhibitors. PLoS One 2010;5(11):e15535.

[114] Kruidenier L, Chung CW, Cheng Z, Liddle J, Che K, Joberty G, et al. A selective jumonji H3K27 demethylase inhibitor modulates the proinflammatory macrophage response. Nature 2012;488 (7411):404−8.

[115] Bedford MT, Richard S. Arginine methylation an emerging regulator of protein function. Mol Cell 2005;18(3):263−72.

[116] Yang Y, Bedford MT. Protein arginine methyltransferases and cancer. Nat Rev Cancer 2013;13 (1):37−50.

[117] Castellano S, Milite C, Ragno R, Simeoni S, Mai A, Limongelli V, et al. Design, synthesis and biological evaluation of carboxy analogues of arginine methyltransferase inhibitor 1 (AMI-1). ChemMedChem 2010;5(3):398−414.

[118] Dowden J, Hong W, Parry RV, Pike RA, Ward SG. Toward the development of potent and selective bisubstrate inhibitors of protein arginine methyltransferases. Bioorg Med Chem Lett 2010;20(7):2103−5.

[119] Sack JS, Thieffine S, Bandiera T, Fasolini M, Duke GJ, Jayaraman L, et al. Structural basis for CARM1 inhibition by indole and pyrazole inhibitors. Biochem J 2011;436(2):331−9.

[120] Hart P, Lakowski TM, Thomas D, Frankel A, Martin NI. Peptidic partial bisubstrates as inhibitors of the protein arginine N-methyltransferases. ChemBioChem 2011;12(9):1427−32.

[121] Wang J, Chen L, Sinha SH, Liang Z, Chai H, Muniyan S, et al. Pharmacophore-based virtual screening and biological evaluation of small molecule inhibitors for protein arginine methylation. J Med Chem 2012;55(18):7978−87.

[122] Liu F, Li F, Ma A, Dobrovetsky E, Dong A, Gao C, et al. Exploiting an allosteric binding site of PRMT3 yields potent and selective inhibitors. J Med Chem 2013;56(5):2110−24.

[123] Siarheyeva A, Senisterra G, Allali-Hassani A, Dong A, Dobrovetsky E, Wasney GA, et al. An allosteric inhibitor of protein arginine methyltransferase 3. Structure 2012;20(8):1425−35.

[124] Cao R, Tsukada Y, Zhang Y. Role of Bmi-1 and Ring1A in H2A ubiquitylation and Hox gene silencing. Mol Cell 2005;20(6):845−54.

[125] Gieni RS, Hendzel MJ. Polycomb group protein gene silencing, non-coding RNA, stem cells, and cancer. Biochem Cell Biol 2009;87(5):711−46.

[126] Alchanati I, Teicher C, Cohen G, Shemesh V, Barr HM, Nakache P, et al. The E3 ubiquitin-ligase Bmi1/Ring1A controls the proteasomal degradation of Top2alpha cleavage complex − a potentially new drug target. PLoS One 2009;4(12):e8104.

[127] Ismail IH, McDonald D, Strickfaden H, Xu Z, Hendzel MJ. A small molecule inhibitor of polycomb repressive complex 1 inhibits ubiquitin signaling at DNA double-strand breaks. J Biol Chem 2013;288 (37):26944−54.

[128] Plo I, Nakatake M, Malivert L, de Villartay JP, Giraudier S, Villeval JL, et al. JAK2 stimulates homologous recombination and genetic instability: potential implication in the heterogeneity of myeloproliferative disorders. Blood 2008;112(4):1402–12.

[129] Levine RL, Pardanani A, Tefferi A, Gilliland DG. Role of JAK2 in the pathogenesis and therapy of myeloproliferative disorders. Nat Rev Cancer 2007;7(9):673–83.

[130] Dawson MA, Bannister AJ, Gottgens B, Foster SD, Bartke T, Green AR, et al. JAK2 phosphorylates histone H3Y41 and excludes HP1alpha from chromatin. Nature 2009;461(7265):819–22.

[131] Pollack BP, Kotenko SV, He W, Izotova LS, Barnoski BL, Pestka S. The human homologue of the yeast proteins Skb1 and Hsl7p interacts with Jak kinases and contains protein methyltransferase activity. J Biol Chem 1999;274(44):31531–42.

[132] Liu F, Zhao X, Perna F, Wang L, Koppikar P, Abdel-Wahab O, et al. JAK2V617F-mediated phosphorylation of PRMT5 downregulates its methyltransferase activity and promotes myeloproliferation. Cancer Cell 2011;19(2):283–94.

[133] Amary MF, Bacsi K, Maggiani F, Damato S, Halai D, Berisha F, et al. IDH1 and IDH2 mutations are frequent events in central chondrosarcoma and central and periosteal chondromas but not in other mesenchymal tumours. J Pathol 2011;224(3):334–43.

[134] Lu C, Ward PS, Kapoor GS, Rohle D, Turcan S, Abdel-Wahab O, et al. IDH mutation impairs histone demethylation and results in a block to cell differentiation. Nature 2012;483(7390):474–8.

[135] Xu W, Yang H, Liu Y, Yang Y, Wang P, Kim SH, et al. Oncometabolite 2-hydroxyglutarate is a competitive inhibitor of alpha-ketoglutarate-dependent dioxygenases. Cancer Cell 2011;19(1):17–30.

[136] Helin K, Dhanak D. Chromatin proteins and modifications as drug targets. Nature 2013;502 (7472):480–8.

[137] Filippakopoulos P, Qi J, Picaud S, Shen Y, Smith WB, Fedorov O, et al. Selective inhibition of BET bromodomains. Nature 2010;468(7327):1067–73.

[138] van Alphen RJ, Wiemer EA, Burger H, Eskens FA. The spliceosome as target for anticancer treatment. Br J Cancer 2009;100(2):228–32.

[139] Wang Z, Burge CB. Splicing regulation: from a parts list of regulatory elements to an integrated splicing code. RNA 2008;14(5):802–13.

[140] Skotheim RI, Nees M. Alternative splicing in cancer: noise, functional, or systematic? Int J Biochem Cell Biol 2007;39(7–8):1432–49.

[141] Kotake Y, Sagane K, Owa T, Mimori-Kiyosue Y, Shimizu H, Uesugi M, et al. Splicing factor SF3b as a target of the antitumor natural product pladienolide. Nat Chem Biol 2007;3(9):570–5.

[142] Kaida D, Motoyoshi H, Tashiro E, Nojima T, Hagiwara M, Ishigami K, et al. Spliceostatin A targets SF3b and inhibits both splicing and nuclear retention of pre-mRNA. Nat Chem Biol 2007;3(9):576–83.

[143] Matsagas K, Lim DB, Horwitz M, Rizza CL, Mueller LD, Villeponteau B, et al. Long-term functional side-effects of stimulants and sedatives in *Drosophila melanogaster*. PLoS One 2009;4(8):e6578.

[144] Sharpless NE, DePinho RA. The mighty mouse: genetically engineered mouse models in cancer drug development. Nat Rev Drug Discov 2006;5(9):741–54.

[145] Van Dyke T, Jacks T. Cancer modeling in the modern era: progress and challenges. Cell 2002;108 (2):135–44.

[146] Šorm F, Piskala A, Čihák A, Veselý J. 5-Azacytidine, a new, highly effective cancerostatic. Cell Mol Life Sci 1964;20(4):202–3.

[147] Čihák A. Biological effects of 5-azacytidine in eukaryotes. Oncology 1974;30(5):405–22.

[148] Christman JK. 5-Azacytidine and 5-aza-2′-deoxycytidine as inhibitors of DNA methylation: mechanistic studies and their implications for cancer therapy. Oncogene 2002;21(35):5483–95.

[149] Sorm F, Vesely J. Effect of 5-aza-2′-deoxycytidine against leukemic and hemopoietic tissues in AKR mice. Neoplasma 1968;15(4):339–43.

[150] Veselý J, Čihák A, Šorm F. Association of decreased uridine and deoxycytidine kinase with enhanced RNA and DNA polymerase in mouse leukemic cells resistant to 5-azacytidine and 5-aza-2′-deoxycytidine. Cancer Res 1970;30(8):2180−6.

[151] Li L, Olin E, Buskirk H, Reineke L. Cytotoxicity and mode of action of 5-azacytidine on L1210 leukemia. Cancer Res 1970;30(11):2760−9.

[152] Momparler RL, Veselý J, Momparler LF, Rivard GE. Synergistic action of 5-aza-2′-deoxycytidine and 3-deazauridine on L1210 leukemic cells and EMT6 tumor cells. Cancer Res 1979;39(10):3822−7.

[153] Veselý J, Čihák A. Incorporation of a potent antileukemic agent, 5-aza-2′-deoxycytidine, into DNA of cells from leukemic mice. Cancer Res 1977;37(10):3684−9.

[154] Wilson VL, Jones PA, Momparler RL. Inhibition of DNA methylation in L1210 leukemic cells by 5-aza-2′-deoxycytidine as a possible mechanism of chemotherapeutic action. Cancer Res 1983;43(8):3493−6.

[155] Ng KP, Ebrahem Q, Negrotto S, Mahfouz RZ, Link KA, Hu Z, et al. p53 independent epigenetic-differentiation treatment in xenotransplant models of acute myeloid leukemia. Leukemia 2011;25(11):1739−50.

[156] Guo Y, Engelhardt M, Wider D, Abdelkarim M, Lubbert M. Effects of 5-aza-2′-deoxycytidine on proliferation, differentiation and p15/INK4b regulation of human hematopoietic progenitor cells. Leukemia 2006;20(1):115−21.

[157] Mielcarek M, Benn CL, Franklin SA, Smith DL, Woodman B, Marks PA, et al. SAHA decreases HDAC 2 and 4 levels in vivo and improves molecular phenotypes in the R6/2 mouse model of Huntington's disease. PLoS One 2011;6(11):e27746.

[158] Sneeringer CJ, Scott MP, Kuntz KW, Knutson SK, Pollock RM, Richon VM, et al. Coordinated activities of wild-type plus mutant EZH2 drive tumor-associated hypertrimethylation of lysine 27 on histone H3 (H3K27) in human B-cell lymphomas. Proc Natl Acad Sci 2010;107(49):20980−5.

[159] Yap DB, Chu J, Berg T, Schapira M, Cheng SW, Moradian A, et al. Somatic mutations at EZH2 Y641 act dominantly through a mechanism of selectively altered PRC2 catalytic activity, to increase H3K27 trimethylation. Blood 2011;117(8):2451−9.

[160] Chen Y-H, Hung M-C, Li L-Y. EZH2: a pivotal regulator in controlling cell differentiation. Am J Transl Res 2012;4(4):364.

[161] Lim S, Janzer A, Becker A, Zimmer A, Schüle R, Buettner R, et al. Lysine-specific demethylase 1 (LSD1) is highly expressed in ER-negative breast cancers and a biomarker predicting aggressive biology. Carcinogenesis 2010;31(3):512−20.

[162] Schulte JH, Lim S, Schramm A, Friedrichs N, Koster J, Versteeg R, et al. Lysine-specific demethylase 1 is strongly expressed in poorly differentiated neuroblastoma: implications for therapy. Cancer Res 2009;69(5):2065−71.

[163] Metzger E, Wissmann M, Yin N, Müller JM, Schneider R, Peters AH, et al. LSD1 demethylates repressive histone marks to promote androgen-receptor-dependent transcription. Nature 2005;437(7057):436−9.

[164] Hou H, Yu H. Structural insights into histone lysine demethylation. Curr Opin Struct Biol 2010;20(6):739−48.

[165] Schenk T, Chen WC, Göllner S, Howell L, Jin L, Hebestreit K, et al. Inhibition of the LSD1 (KDM1A) demethylase reactivates the all-trans-retinoic acid differentiation pathway in acute myeloid leukemia. Nat Med 2012;18(4):605−11.

[166] Popovici-Muller J, Saunders JO, Salituro FG, Travins JM, Yan S, Zhao F, et al. Discovery of the first potent inhibitors of mutant IDH1 that lower tumor 2-HG in vivo. ACS Med Chem Lett 2012;3(10):850−5.

[167] Rohle D, Popovici-Muller J, Palaskas N, Turcan S, Grommes C, Campos C, et al. An inhibitor of mutant IDH1 delays growth and promotes differentiation of glioma cells. Science 2013;340 (6132):626−30.

[168] Iwata H, Sato H, Suzuki R, Yamada R, Ichinomiya S, Yanagihara M, et al. A demethylating agent enhances chemosensitivity to vinblastine in a xenograft model of renal cell carcinoma. Int J Oncol 2011;38(6):1653−61.

[169] Takano Y, Iwata H, Yano Y, Miyazawa M, Virgona N, Sato H, et al. Up-regulation of connexin 32 gene by 5-aza-2′-deoxycytidine enhances vinblastine-induced cytotoxicity in human renal carcinoma cells via the activation of JNK signalling. Biochem Pharmacol 2010;80(4):463−70.

[170] Kalac M, Scotto L, Marchi E, Amengual J, Seshan VE, Bhagat G, et al. HDAC inhibitors and decitabine are highly synergistic and associated with unique gene-expression and epigenetic profiles in models of DLBCL. Blood 2011;118(20):5506−16.

[171] Pavan A, Spina M, Canzonieri V, Sansonno S, Toffoli G, De Re V. Recent prognostic factors in diffuse large B-cell lymphoma indicate NF-κB pathway as a target for new therapeutic strategies. Leuk Lymphoma 2008;49(11):2048−58.

Index

Note: Page numbers followed by "*f*" and "*t*" refer to figures and tables, respectively.

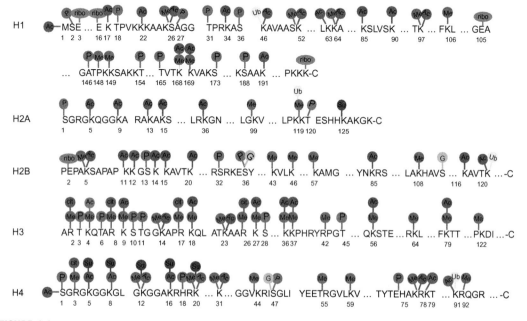

FIGURE 4.1

Histones are heavily decorated by PTMs. Acetylation is shown in blue, methylation is shown in red (mono, di, tri), phosphorylation in green, ubiquitination in yellow, citrullination in magenta, ADP-ribosylation in cyan, sumoylation in indigo, and GlcNAc in peach.

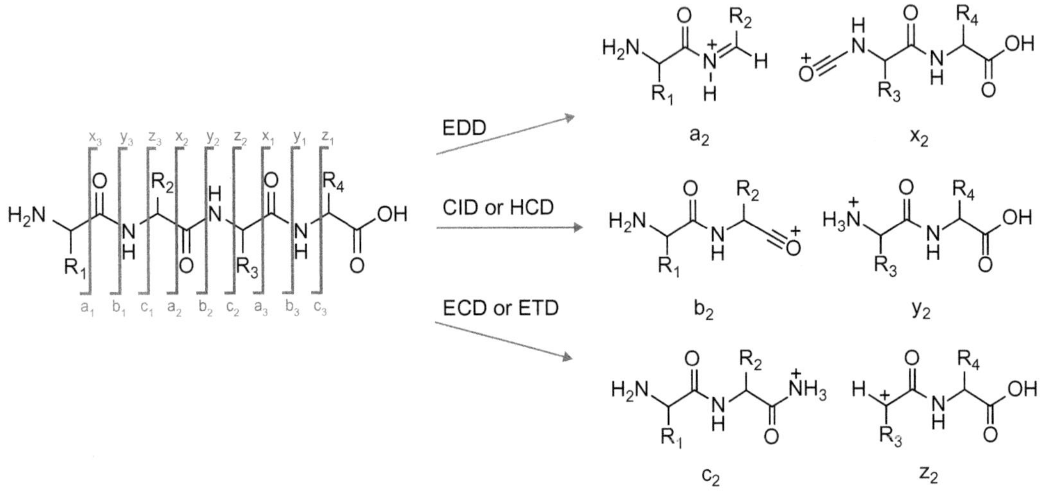

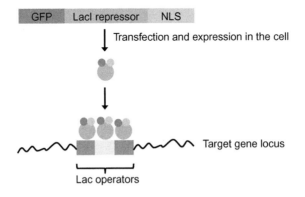

FIGURE 4.2

Generation of a, x, b, y, c, z type ions with the use of different fragmentation methods.

FIGURE 5.2

Lac repressor adapted into a powerful imaging technique. Lac repressor is engineered into a plasmid with GFP and nuclear localization signal at the same cassette. Then it is transfected into cells that are engineered with lac operator sequence repeats in the genome.

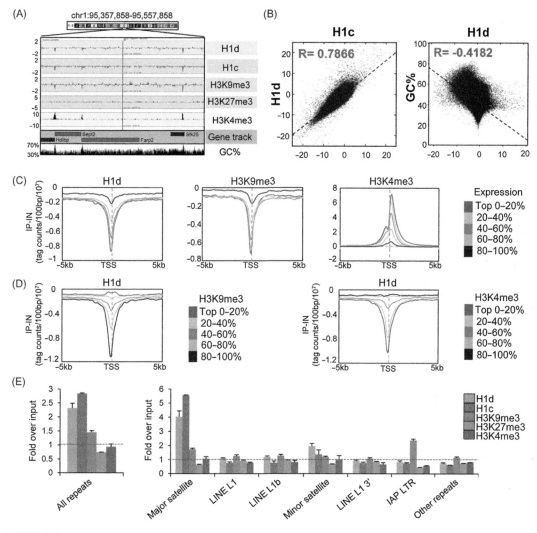

FIGURE 6.4

Genome-wide mapping of H1d and H1c in mouse ESCs. (A) An example of distribution of H1 variants and histone marks at a 200 kb region. The GC density track was obtained from the UCSC genome browser. Genes are color coded according to their transcription directions (Red: sense strand; Blue: anti-sense strand). (B) Genome-wide correlation scatter plots of H1d vs. H1c (left) and GC% vs. H1d (right). The correlation coefficient (R value) and the trend line are generated as described [64]. Pearson's correlation was used to perform the analysis. $P < 10^{-100}$ for all correlation coefficients. (C) Metagene analysis of H1d, H3K9me3, and H3K4me3 in relation to gene expression levels on a 10 kb window centered on TSSs. Genes were partitioned into five groups according to their expression levels. TSS: transcription start sites; Y axis: tag counts per 100 bp window per 10 million mappable reads; IP-IN: normalized signal values of ChIP-seq subtracted by that of input-seq. (D) Metagene analysis of H1d in relation to the levels of H3K9me3 (left) and H3K4me3 (right) on regions covering −5 kb to +5 kb of TSSs. Genes were evenly grouped into five categories according to the signals of the respective histone marks. (E) Enrichment of H1d and H1c at the major satellite sequences. Fold enrichment of percent mappable repeats (mapped to RepBase) from H1d, H1c, and histone marks of ChIP-seq libraries over that from the corresponding chromatin input-seq library on all repeats (left), six most abundant repetitive sequences and the remaining other repeats (right). The dashed lines indicate the level of normalized input signal. P values calculated with Fisher's exact test comparing ChIP-seq with input-seq libraries are less than 2.5×10^{-5} for all repeat classes shown. Error bars represent the differences between replicates. Data are presented as average ± SEM (standard error of the mean).

Adapted from Cao et al [64].

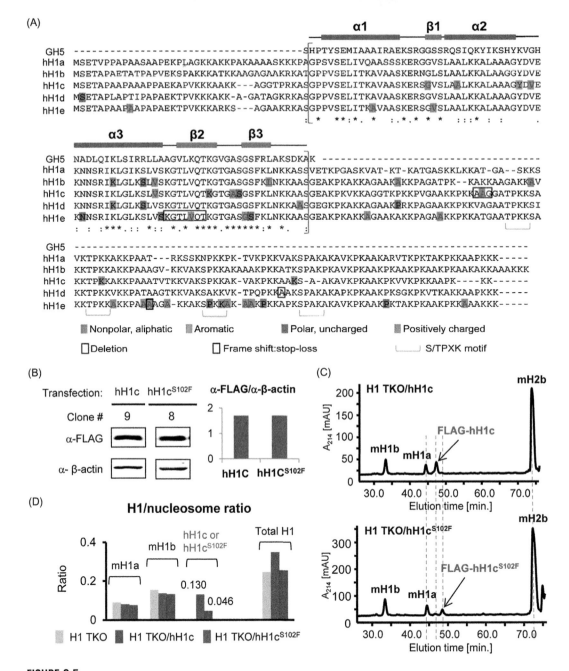

FIGURE 6.5

Functional analysis of H1 mutations identified in follicular lymphoma. (A) hH1a−hH1e sequence alignment and the distribution of H1 mutations. GH5, globular domain of chicken histone H5. H1 globular domain is marked with the red bracket. (B) Expression of FLAG tagged WT hH1c or hH1cS102F mutant in H1c/H1d/H1e triple null embryonic stem cells (H1 TKO ESCs). H1 TKO ESCs were transfected with vectors expressing FLAG-hH1c or FLAG-hH1c/S102F mutant. Stable ESC clones were picked for each transfection and screened using an anti-FLAG antibody. Immunoblotting with anti-β-ACTIN antibody were included as loading controls. Two clones with similar expression levels of hH1c and hH1c/S102F were selected for subsequent analysis. FLAG-hH1c and FLAG-hH1cS102F are expressed at the same level in selected clones shown. (C) Reverse-phase (RP)-HPLC profiles of histones extracted from chromatin isolated from histone H1 triple-knockout (H1 TKO) mouse ESCs

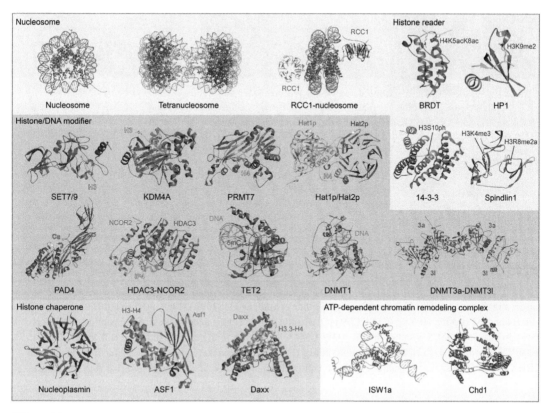

FIGURE 7.3

A structural gallery of nucleosomes and representative epigenetic regulators. Coordinates have PDB codes 1AOI (Nucleosome), 1ZBB (Tetranucleosome), 3MVD (RCC1−nucleosome complex, RCC1 is shown in purple), 1O9S (SET7/9), 2OQ6 (Catalytic domain of KDM4A), 4M38 (Catalytic domain of PRMT7), 4PSX (Hat1p/Hat2p, Hat1p is shown in green, Hat2p is shown in blue, Histone H4 peptide is shown in yellow), 1C3P (HDLP), 4A69 (HDAC3-NCOR2, HDAC3 is shown in blue, Deacetylase activation domain of NCOR2 [Nuclear receptor co-repressor 2] is shown in green), 4NM6 (TET2−DNA complex), 4DA4 (DNMT1−DNA complex), 2QRV (DNMT3a−DNMT3L complex, C-terminal domain of DNMT3L is shown in green, Catalytic domain of DNMT3a is shown in blue), 2WP2 (Bromodomain of Brdt), 1KNA (Chromodomain of HP1), 2C1J (14-3-3), 4MZG (Spindlin1), 1K5J (Nucleoplasmin), 2IO5 (ASF1-H3−H4 complex, ASF1 is shown in blue, H3−H4 is shown in green), 4H9N (DAXX H3−H4 complex, DAXX is shown in blue, H3−H4 is shown in green), 2Y9Z (ISW1a−DNA complex), and 3MWY (chromodomain-ATPase portion of Chd1).

◀ expressing wild-type or Ser102Phe human histone H1c. The Ser102Phe mutant demonstrated higher hydrophobicity than the wild-type protein. mH1a, mouse histone H1a; mH1b, mouse histone H1b; mH2b, mouse histone H2b; hH1c, human histone H1c. (D) Ratio of individual histone H1 variants (and total histone H1) to the nucleosome of the indicated ESCs. The ratio is calculated from the HPLC analysis in (C) and demonstrates that the total histone H1 levels in histone H1 triple-knockout ESCs expressing human histone H1c Ser102Phe were reduced compared to cells expressing wild-type human histone H1c, as a result of the weaker association of the mutant histone with chromatin (only 35% of wild-type association).

Adapted from Okosun et al. [94].

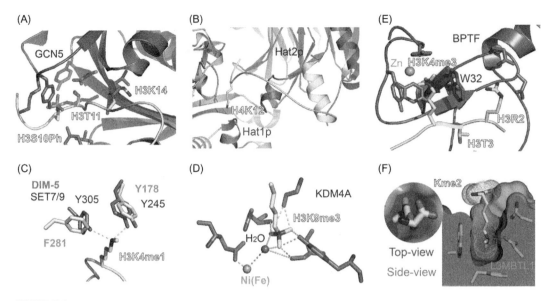

FIGURE 7.4

Site- and state-specific incorporation, elimination and readout of histone marks by representative histone writers, erasers and readers. (A) GCN5 in complex with peptide containing H3S10Ph (1PUA); (B) Hat1p/Hat2p in complex with H4 and H3 peptide (4PSX); (C) SET7/9 (1O9S) and DIM-5 (1PEG) in complex with H3 peptide. SET7/9 is shown in blue and DIM-5 is shown in gray; (D) KDM4A in complex with peptide containing H3K9me3 (2OQ6); (E) BPTF in complex with H3K4me3 peptide (2F6J); (F) L3MBTL1 pocket 2 with inserted dimethyllysine ligand (2RHX). In all panels, histone peptide is colored yellow; PDB entry code of each structure is listed in parentheses.

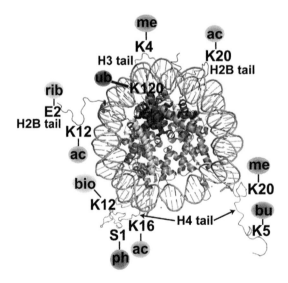

FIGURE 8.1

Histone tail modifications. X-ray crystal structure of a mononucleosome at 1.9 Å resolution (PDB code 1KX5) showing the histone tails and selected sites of posttranslational modification. *Xenopus laevis* core histones are shown in red (H2A), blue (H2B), amber (H3), and green (H4). Ac, acetyl; bio, biotinyl; bu, butyryl; me, methyl; ph, phosphoryl; rib, ADP-ribosyl groups.

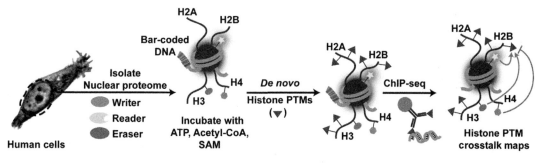

FIGURE 8.6

Systems-level studies of histone PTM crosstalk. Members of a DNA-barcoded library of differently modified nucleosomes (shown as star, ellipse, and hexagon) are incubated with the entire nuclear proteome of a human cell line in the presence of ATP, S-adenosyl methionine (SAM), and acetyl-CoA. The degree of *de novo* chromatin modification (shown as triangles) is analyzed by *in vitro* ChIP-seq in order to obtain histone crosstalk maps.

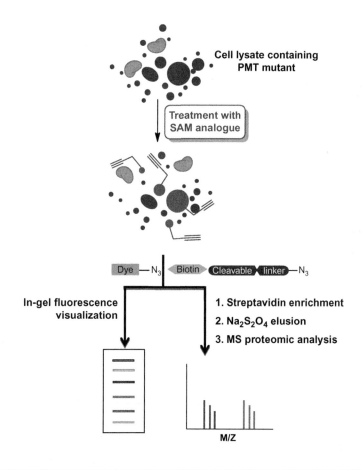

FIGURE 10.6

Descriptive workflow of methylome profiling by BPPM. Cells were transfected with a PMT mutant plasmid and then lysed, followed by treatment with SAM analogue cofactors. The labeled targets were then conjugated with the fluorescent dye for in-gel fluorescence or with cleavable azido-azo-biotin probe for target enrichment.

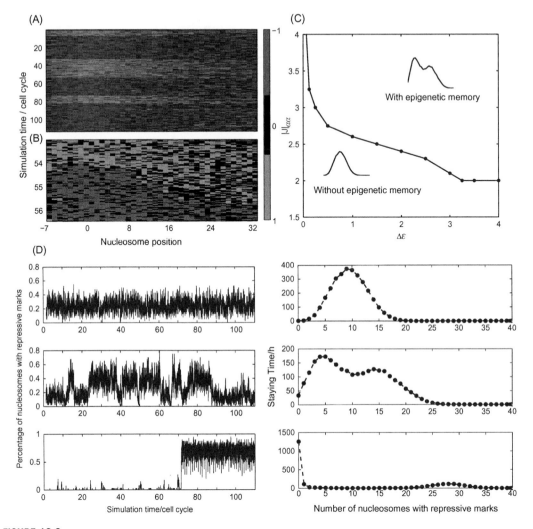

FIGURE 12.2

Simulation results using model parameters corresponding to Oct4. (A) Heat map representation of a typical simulation trajectory. (B) Zoom-in of the heat map in panel (A) showing epigenetic state transition. (C) Phase diagram on the $\Delta\varepsilon$ -J plane to illustrate bistability mechanism. (D) Typical trajectories of the fraction of nucleosomes with repressive marks (left) and the corresponding probability distribution of observing a given number of nucleosomes with repressive marks (right). All simulations are performed with $\Delta\varepsilon = 2$, but different $J_{\alpha\alpha}$ values, Upper panel, $J_{\alpha\alpha} = 0$; middle panel, $J_{\alpha\alpha} = 2{:}5$; lower panel, $J_{\alpha\alpha} = 3{:}5$. The dwelling time distribution is obtained by averaging over 100 trajectories, each started with a randomly selected initial histone modification configuration, simulated for 10^3 Gillespie steps, then followed by another 2×10^3 Gillespie steps for sampling.

Adapted from [27].

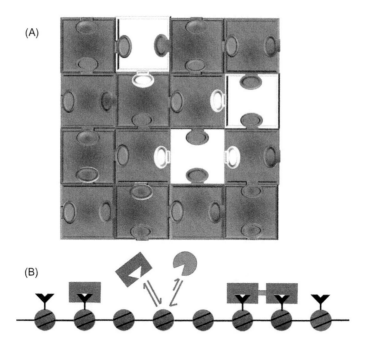

FIGURE 12.3

Schematic illustration of the reader/writer mechanism. (A) An analogous jigsaw puzzle reconstruction problem. (B) Reader-and-writer mechanism for epigenetic pattern reconstruction.

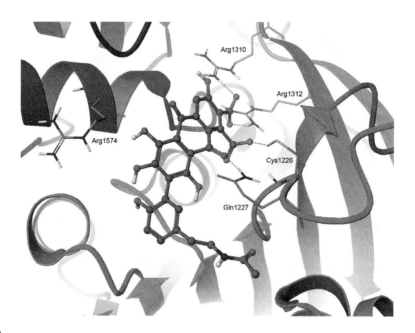

FIGURE 13.2

Induced-fit docking pose of laccaic acid A (Figure 13.1) within the substrate-binding site of human DNMT1. Selected amino acid residues of the binding pocket are shown. Hydrogen bonds between the ligand and side chains of DNMT1 are shown as dashed pink lines.

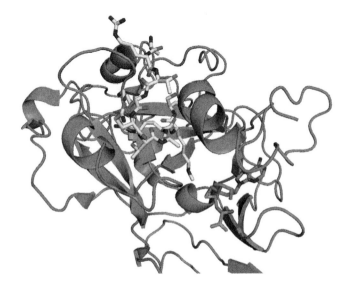

FIGURE 17.1

Crystal structure of GLP catalytic domain in complex with SAH and H3K9me1 (PDB 3HNA). The whole structure of GLP is colored in cyan. The H3K9me1 peptide is colored yellow and SAH is colored in cyan.

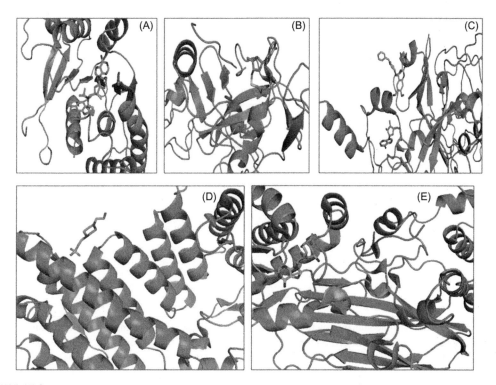

FIGURE 17.3

Co-crystal structures of inhibitors and the important cancer-related histone methyltransferases. (A) Crystal structure of DOT1L with EPZ-5676 [PDB 4HRA]. (B) Crystal structure of EZH2 SET domain [PDB 4MI5] aligned to domain of crystal structure of G9a with SAM [PDB 3HNA]; peptide residue from EED complex, molecule present is cofactor SAM. (C) Crystal structure of GLP [PDB 3FPD] in complex with SAM (green) and BIX-01294 (orange). (D) Crystal structure of menin-MLL complex [PDB 4GPQ] dissociated by small molecule MI-2. (E) Crystal structure of PRMT5:MEP50P complex [PDB 4GQB].

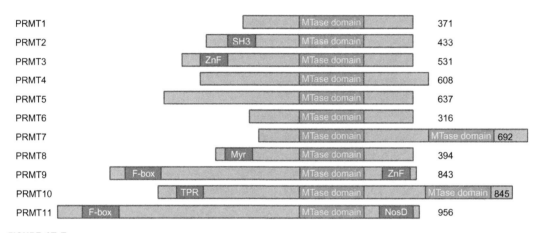

FIGURE 17.5

Schematic representation of the human protein arginine methyltransferase (PRMT) family. For each member, the length of the longest isoform is indicated on the right. All members of the family possess at least one conserved MTase domain with signature motifs I, post-I, II, and III and a THW loop. Additional domains are marked in maroon boxes: SH3, ZnF (zinc finger), Myr (myristoylation), F-box, TPR (tetratricopeptide), and NosD (nitrous oxidase accessory protein).

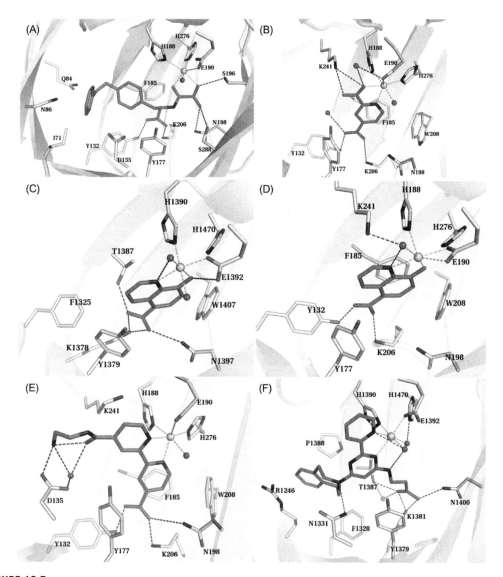

FIGURE 18.5

Crystal structures of LSD1-inhibitor complexes. (A) Five-membered ring adduct (cyan) of PCPA with FAD in LSD1 (white) (PDB ID 2Z5U (depicted) and 2UXX). (B) N(5)-adduct (cyan) of PCPA (PDB ID 2EJR (depicted), 2Z3Y, and 2XAJ). (C) The (+)-PCPA enantiomer forms a different adduct (cyan) with FAD and the phenyl group adopts a different orientation (PDB ID 2XAH). (D) LSD1 in complex with p-substituted PCPA derivatives. Compound **27**-FAD adduct is shown in cyan and the adduct of the branched derivative **28** in yellow (PDB ID 2XAS and 2XAQ). (E) LSD1 (white) in complex with ortho-benzyloxy-fluoro-PCPA − FAD adduct (cyan). (PDB ID 3ABU). In all complexes only side chains of LSD1 residues involved in the interaction are shown as white sticks for clarity.

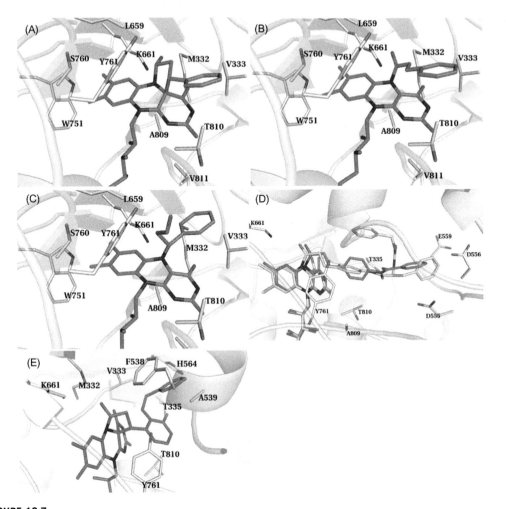

FIGURE 18.7

Crystal structures of LSD1-inhibitor complexes. (a) Five membered ring adduct (cyan) of PCPA with FAD in LSD1 (white) (PDB ID 2Z5U (depicted) and 2UXX). (b) N(5)-adduct (cyan) of PCPA (PDB ID 2EJR (depicted), 2Z3Y and 2XAJ). (c) The (+)-PCPA enantiomer forms a different adduct (cyan) with FAD and the phenyl group adopts a different orientation (PDB ID 2XAH). (d) LSD1 in complex with p-substituted PCPA derivatives. Compound **27**-FAD adduct is shown in cyan and the adduct of the branched derivative **28** in yellow (PDB ID 2XAS and 2XAQ). (e) LSD1 (white) in complex with ortho-benzyloxy-fluoro-PCPA − FAD adduct (cyan). (PDB ID 3ABU). In all complexes only side chains of LSD1 residues involved in the interaction are shown as white sticks for clarity.

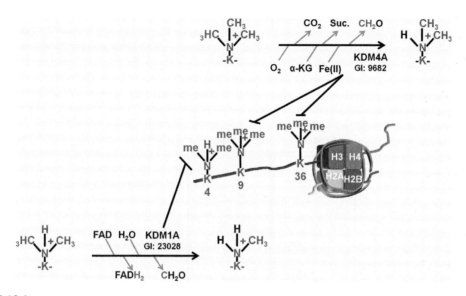

FIGURE 19.1

Schematic representation of the demethylase reaction. KDM4A demethylates di- and tri-methylated H3K9 and H3K36 by methyl group oxidation in the presence of cofactors α-ketoglutarate, molecular oxygen (O_2), and Fe (II). This reaction produces succinate, CO_2, and formaldehyde. KDM1A, with cofactors FAD and H_2O, demethylates mono- and di-methylated H3K4, producing $FADH_2$ and formaldehyde.

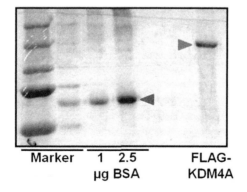

Marker	1	2.5	FLAG-
	μg BSA		KDM4A

FIGURE 19.2

Quantification of KDM4A purified from Sf9 cells. FLAG-tagged KDM4A was expressed in Sf9 cells, purified as described, run on an SDS-PAGE gel and Coomassie stained. The resulting enzyme prep (green arrow) is of high purity and can be quantified by comparison to a known concentration of BSA protein standard (red arrow).

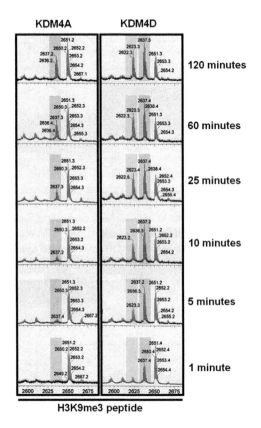

FIGURE 19.3

Comparison of KDM4A and KDM4D demethylation activity. Purified enzymes were incubated with H3K9me3 peptide for 120 minutes and analyzed by MALDI-TOF at six time points. KDM4D generates di-methylated H3K9 (red) within 1 minute, but the same amount is not observed with KDM4A until 25 minutes. KDM4D also produces H3K9me1 (green) at 1 minute, but this activity is undetectable for KDM4A.

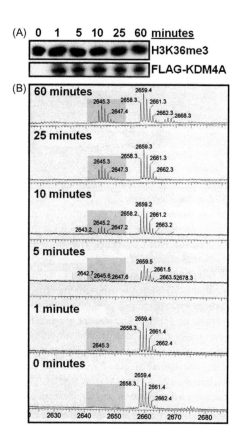

FIGURE 19.4

H3K36me3 demethylation by KDM4A. A. Western blot for H3K36me3 and the KDM4A peptide levels at each time point corresponding to the MALDI TOF spectra in panel B. B. Quantification of H3K36me3 demethylation reaction by MALDI-TOF spectrometry. Demethylation of the residue is observed beginning 1 minute after the addition of KDM4A to the reaction (red).

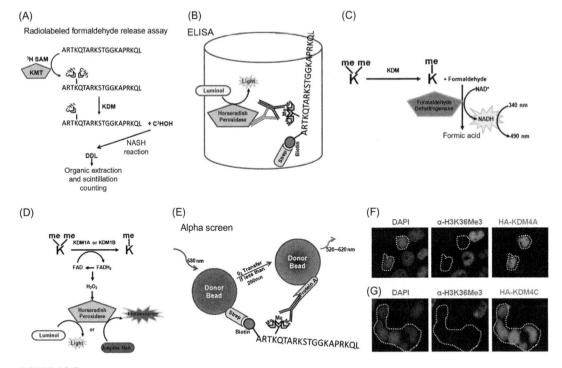

FIGURE 19.5

Methods to detect KDM activity. (A) Radiolabeled formaldehyde release assay. In the radiolabeled formaldehyde release assay a specific methyltransferase is used in conjunction with tritiated S-adenosyl-methionine to add tritiated methyl groups to a specific lysine residue on a histone tail peptide. Following demethylation the radioactivity is released as part of the formaldehyde by-product. The formaldehyde is converted to 3,5-diacethyl-1, 4-dihydrolutidine (DDL) through a NASH reaction. The DDL is organically extracted and the radioactivity is counted by a filter-binding assay followed by liquid scintillation. (B) ELISA. In the ELISA assay, biotinylated histone peptides with specific modifications (H3K4me3 is depicted) are affixed to streptavidin-coated microplate wells. Antibodies to specific histone modifications interact with the methylation and are subsequently bound by a secondary antibody coupled to horseradish peroxidase. The amount of methylated peptide is measured by the light produced from HRP conversion of luminol, with demethylation resulting in reduced signal. (C) Enzyme-coupled fluorescence assay. The JmjC-containing demethylases and FAD-dependent amine oxidases generate formaldehyde as a by-product of the demethylation reaction. Formaldehyde dehydrogenase uses NAD^+ to convert the formaldehyde to formic acid, which also reduces the NAD^+ to NADH. The amount of NADH can be assayed by absorbance at 490 nm. (D) Enzyme-coupled fluorescent and luminescent assays. The FAD-dependent amine oxidases generate hydrogen peroxide as a by-product of the demethylation reaction. The H_2O_2 allows horseradish peroxidase to convert luminescent substrates (Luminol) or fluorescent substrates (Amplex Red) into their light emitting compounds, which directly correlate with the amount of enzymatic activity. (E) AlphaScreen. In an AlphaScreen, donor beads coated with streptavidin are coupled to biotinylated histone peptides. The acceptor bead is coupled to antibodies directed against a specific histone modification, in this case H3K4me3. The binding of the antibody to the modification brings the acceptor bead within 200 nm of the donor bead, allowing an oxygen singlet to transfer to the acceptor bead causing emission of fluorescence. (F) Immunofluorescence. Cells are co-stained with an antibody specific to H3K36me3 and an antibody to HA for detection of tagged KDM4A (A) or KDM4C (B). Nuclei are visualized with DAPI. Cells expressing a high level of KDM (dashed line) have lower levels of H3K36me3 compared to cells not overexpressing the KDM.

Printed and bound by CPI Group (UK) Ltd, Croydon, CR0 4YY

08/05/2025

01864998-0001